The Colloidal Domain

The Colloidal Domain

Where Physics, Chemistry, Biology, and Technology Meet

Håkan Wennerström

Lund University
Lund, Sweden

Third Edition

WILEY

Published by John Wiley & Sons, Inc., Hoboken, New Jersey.

Library of Congress Cataloging-in-Publication Data:

Name: Wennerström, Håkan author | John Wiley & Sons publisher
Title: The colloidal domain : where physics, chemistry, biology, and
  technology meet / Håkan Wennerström, Lund University, Lund, Sweden.
Description: Third edition. | Hoboken, New Jersey : Wiley, [2026] |
  Includes bibliographical references and index.
Identifiers: LCCN 2025032171 | ISBN 9781394211098 hardback | ISBN
  9781394211104 adobe pdf | ISBN 9781394211111 epub
Subjects: LCSH: Colloids | Surface chemistry
Classification: LCC QD549 .W46 2026
LC record available at https://lccn.loc.gov/2025032171

Cover Design: Wiley
Cover Image: Courtesy of Kevin Roger, CNRS, Université de Toulouse, France

SKY10134670_121525

Contents

Preface to the First Edition

Colloidal interactions are omnipresent. They influence almost all technological processes, yet they are frequently invisible and often ignored. When scientists trained in the classical study of atoms and molecules, or engineers armed with correlations that focus on bulk properties, confront a problem, they may not recognize its colloidal source. Yet the colloidal domain forms a critical interface between micro and macroscopic regimes. Because colloids have a major impact on our ability to understand biological processes and to control technology, we must deal with colloidal phenomena to maintain a competitive technological base. This book offers a gateway to the colloidal domain.

For several decades, the study of colloids was an intellectual backwater, but significant advances in experiment and theory have combined to create a renaissance in colloidal science.

January 1994

D. Fennell Evans
Håkan Wennerström

Preface to the Second Edition

The second edition of *The Colloidal Domain* incorporates new topics that have emerged over the past 5 years and includes modifications that increase the clarity of the presentation and new additions that promote integration and coherence of the topics. Major changes include the addition and/or expansion of several chapters.

Epilogue (Chapter 12)

This new chapter seeks to provide an integrative perspective on *The Colloidal Domain*. By its very nature, the colloidal domain encompasses entities that contain many molecules. Thus, it occupies an intermediate position between the molecular and the macroscopic worlds. A major goal in this book has been to show how molecular interactions are manifested in the colloidal domain, as well as how colloidal phenomena are manifested in the macroscopic world that we perceive with our senses. In establishing these links, we were guided by the need to address four questions:

- How molecular, mesoscale, and macroscopic structures merge in the colloidal domain
- How molecular interaction forces determine colloidal behavior
- How the interplay between entropy and energy organizes colloidal entities
- How dynamic, kinetic, and transport properties influence colloidal processes

By using the second edition in manuscript form, we have encouraged our students to read this chapter both at the beginning of the course—knowing the answer before you start is a real help—and at the end of the course, where a perspective rooted in detailed analysis helps.

Solutes and Solvents, Self-assembly of Amphiphiles (Chapter 1)

The sections on regular solutions, chemical potential, and transport properties, which constitute the fundamental building blocks of the colloidal domain, have been expanded in order to provide a deeper understanding of the topic.

Electrostatic Interactions in Colloidal Systems (Chapter 3)

In response to feedback and evaluations from students, we have completely rewritten and reorganized this chapter. Chapter 3 provides the foundations for understanding long-range electrostatic interactions, which dominate in colloidal systems.

Forces in Colloidal Systems (Chapter 5)

The second half of this chapter has been extensively revised and now includes sections on electrostatic correlation forces, forces due to density variations, forces due to entropic effects, and hydrophobic forces.

Bilayer Systems (Chapter 6)

The focus in this chapter is on biological membranes with an emphasis on relationships between bilayer structure and function and between energetics and transmembrane transport processes.

Polymer Molecules in Colloidal Systems (Chapter 7)

The first part of this chapter has been reorganized, and a new section that analyzes polymer solubility in terms of the entropic contribution associated with polymer chain conformations has been added.

Micro- and Macroemulsions (Chapter 11)

A section on foams has been added and the presentation on emulsion stability has been reanalyzed in terms of new insights and data.

D. Fennell Evans
Håkan Wennerström

Preface to the Third Edition

Twenty-seven years have passed since the previous edition, and there has been a large expansion in the application of colloid science in technology, in medicine, and in life science in general. The emphasis in this book is and has been on the fundamental aspects of colloid science. It is, however, always a challenge to see the relation between basic phenomena and their bewildering manifestations in a complex colloidal system. The main aim of the present revision is to further elucidate this aspect of colloid science.

In this third edition, approximately 50% of the previous edition is kept unchanged except for clarifications and the adoption of a more consistent notation and an improvement of the figures. In addition, 25% of the previous text has been extensively reworked and made more up-to-date. Around 25% is new material with the ambition to discuss several topics that have been of growing interest during the last decades. The total length of the text is roughly unchanged, which implies that a fraction of the text in the previous edition has been eliminated.

The main new features are:

Chapter 1: The disposition of the chapter has been changed to highlight the sequence ideal solution, pairwise interactions as in the regular solution model, and the cooperative association exemplified by micelle formation. There is, in addition, an enlarged discussion of the osmotic pressure concept and its relation to other measures of solvent chemical potential.

Chapter 2: There is a new section on methods for studying surfaces and interfaces with the emphasis on the physical principles behind the methods. There is also a new section on evaporation at surfaces.

Chapter 3: A discussion of the interaction between oppositely charged surfaces has been added.

Chapter 4: The section on scattering has been generalized to also include X-ray scattering, neutron scattering, dynamic light scattering, and turbidity.

Chapter 5: The contact value theorem is derived and applied to the calculation of several types of surface forces. The disposition of the chapter is changed to give a more systematic view of the basis of surface forces. There is a new discussion of electrostatic interaction in concentrated salt solutions and ionic liquids. The discussion of the hydrophobic interaction is novel. In the last section, hydrodynamic interactions are discussed in greater detail compared to the previous edition.

Chapter 6: There is a new text on the energetics of vesicle fusion/fission.

Chapter 7: A discussion of coacervation has been added.

Chapter 8: A quantitative description of the size dependence of colloid stability has been added. There is also a discussion of colloidal stability in concentrated systems, including the living cell.

Chapter 9: This chapter has been extensively reworked, and it now contains descriptions of particle synthesis, including applications to latex particles, porous silica particles, lipid nanoparticles, and amyloid precipitates.

Chapter 10: There is an added description of liquid–liquid phase separation with applications to polymer and protein systems. The description of continuous phase transitions has been extended.

Chapter 11: The discussion of emulsion systems has been made broader. Foam stability is treated in more detail including some novel results.

This revision would not have been possible without the active support from my colleagues. Sven Engström, Emma Sparr, and Joakim Stenhammar have contributed throughout the revision. In addition, Lennart Piculell, Noemi Ferrante Carrante, Olle Söderman, Anna Stradner, Ulf Olsson, Karin Schillén, Magnus Ullner, and Björn Lindman have read parts of the manuscript, and they have suggested improvements to the text and illustrations. My coworkers have suggested many clarifications and corrected many errors, and I am responsible for the remaining ones. Dora Bartfalvy and Bengt Persson have contributed with technical support and many illustrations.

I dedicate this revision to my grandchildren Majken, Folke, Ingmar, Lo, Vanja, Nore, Ellen, and Ada.

October 2025 Håkan Wennerström

Physical Constants

Symbol	Name	Value
$\boldsymbol{R}$	Gas constant	$8.314\ J\ K^{-1}\ mol^{-1}$
k_B	Boltzmann's constant	$1.380\ 10^{-23}\ J\ K^{-1}$
e	Elementary charge	$1.602\ 10^{-19}\ C$
F	Faraday's constant	$96\ 500\ C\ mol^{-1}$
g	Gravity acceleration	$9.81\ m\ s^{-2}$
N_{Av}	Avogadro's number	$6.022\ 10^{23}\ mol^{-1}$
h	Planck's constant	$6.63\ 10^{-34}\ J\ s$
$\hbar$	Reduced Planck's constant $(h/2\pi)$	$1.05\ 10^{-34}\ J\ s$
ε_0	Permittivity of vacuum	$8.85\ 10^{-12}\ C\ V^{-1}\ m^{-1}$

Symbols

Symbol	Name	Unit	Defining Equation
A	Helmholtz free energy	J	
E	Energy	J	
G	Gibbs free energy	J	
H	Enthalpy	J	
S	Entropy	J/K	
U	Internal energy	J	
$\mathbf{U}^{\dagger}$	Effective internal energy	J	3.5.18
C	Number of components		10.1.3
F	Degrees of freedom		10.1.3
P	Number of phases		10.1.3
A	Area	m^2	
a_0	Surfactant head group area	m^2	1.5.1
B_2	Second virial coefficient	m^3	1.2.14
c	Concentration (molar)	M (mol/dm^3)	
c*	Concentration (per molecule)		
C	Capacitance	F	3.5.5
C*	Polymer overlap concentration	M	7.2.2
C_p	Heat capacity	J/K	
CCC	Critical coagulation concentration	M	8.2.11
CMC	Critical micelle concentration	M	4.1.11
D	Diffusion coefficient	m^2s^{-1}	1.7.5
E	Electrical field	Vm^{-1}	
E	Electromotive force	V	4.2.1
F	Force	N	
G_0	Static shear modulus	Nm^{-2}	7.2.16
G′	Storage shear modulus	Nm^{-2}	7.2.22
G″	Loss modulus	Nm^{-2}	7.2.22

(Continued)

(Continued)

Symbol	Name	Unit	Defining Equation
$g(r)$	Radial distribution function		4.3.9
g_i	Free energy density of system i	Jm^{-3}	5.5.8
H	Mean curvature	m^{-1}	1.5.8
H	Hamaker constant	J	5.2.5
H_0	Spontaneous curvature	m^{-1}	6.2.1
h	Separation distance	m	5.1.1
I	Current	A	
I_0	Intensity of incident radiation	Jm^{-2}	4.3.5
$I(q)$	Intensity of scattering	Jm^{-2}	4.3.5
J_A	Molecular flux	$m^{-2}s^{-1}$	1.7.6
K_A	Area expansion elastic modulus	J/m^2	11.1.1
K	Equilibrium constant		
K_0	Optical constant	m^{-2}	4.3.11
k_b	Bulk conductivity	$ohm^{-1}m^{-1}$	
k_d, k_{AB}	Diffusion-controlled rate constant	$M^{-1}s^{-1}$	1.7.11
k_n^+	Association rate constant	$M^{-1}s^{-1}$	4.2.6
k_n^-	Dissociation rate constant	s^{-1}	4.2.6
k_r	Binary reaction rate constant	$M^{-1}s^{-1}$; $m^{-3}s^{-1}$	8.3.2
L_r	Root mean square random walk length	m^2	7.1.11
l_p	Persistence length	m	7.1.15
l_r	Random walk step length	m	7.1.11
$\boldsymbol{l}$	Surfactant hydrocarbon chain length	m	1.3.3
M	Molecular weight		
$\langle M \rangle_{na}$	Number-averaged molecular weight		7.1.6
$\langle M \rangle_{wa}$	Weight-averaged molecular weight		7.1.5
$\mathbf{m}$	Dipole moment	D (Cm)	3.2.7
N	Micelle aggregation number		
$\langle N \rangle_{na}$	Number-averaged degree of polymerization		7.1.3
$\langle M \rangle_{wa}$	Weight-averaged degree of polymerization		7.1.4
N_s	Surfactant number		1.5.1
n	Refractive index		
n_i	Number of moles		
P	Membrane permeability	ms^{-1}	6.3.1
$P(q)$	Scattering form factor		4.3.6
p	Pressure	Nm^{-2}, P	

Symbol	Name	Unit	Defining Equation
$\boldsymbol{q}$	Scattering vector	m^{-1}	4.3.2
q	Charge	C	
q	Quenchers/Micelle		4.2.2
R	Electrical resistance	ohm (V/A)	
R	Radius	m	
R_g	Radius of gyration	m	
R_1, R_2	Radii of curvature	m	1.5.8
R_{1N}	Polymer end-to-end distance	m	7.1.8
R_θ	Rayleigh ratio	m^{-1}	4.3.14
$\boldsymbol{r}$	Distance	m	
S	Spreading coefficient		2.1.14
S_{CD}	NMR order parameter		6.2.1
S_n	Surfactant aggregate		
$S(q)$	Scattering structural factor		4.3.7
$[S]_{tot}$	Total surfactant		
s	Dimensionless charge parameter		3.9.14
T	Temperature	K	
u	Electrophoretic mobility	$m^2V^{-1}s^{-1}$	8.4.2
V_A	Molar volume	m^3	
$\boldsymbol{V(r)}$	Interaction potential	J	
$\boldsymbol{v}$	Surfactant hydrocarbon chain volume	m^3	1.5.2
$\boldsymbol{v}_i$	Molecular volume of compound i	m^3	
w	Interaction parameter	J/mol	1.1.16
W	Colloidal stability ratio		8.3.18
W_{ij}	Microscopic interaction energy i–j	J	1.1.12
w	Work	J	
X_i	Mole fraction of component i		
Z	Configuration integral		
z_i	Valency of ion i		
z_b	Number of nearest neighbors in bulk		1.1.12
z_s	Number of nearest neighbors at surface		2.1.2
Φ	Electrostatic potential	V	3.2.1
Φ_0	Surface potential	V	3.8.14
Θ	Fractional surface coverage		2.4.19
Π_{osm}	Osmotic pressure	Nm^{-2}	1.2.12
Π_s	Surface pressure	N/m	2.5.3

(Continued)

(Continued)

Symbol	Name	Unit	Defining Equation
Γ_i	Surface excess of molecule i	m^{-2}	2.4.11
Γ_0	Gouy–Chapman coefficient		3.8.14
Ω	Solid angle		3.3.1
Ω	Number of states accessible to a system		1.1.3
α	Polarizability	Fm^2	3.4.2
ε_r	Relative dielectric permittivity		3.5.12
γ_i	Surface tension of phase i	Jm^{-2}, Nm^{-1}	2.1.6
γ_{ij}	Interfacial tension of phases i and j	Jm^{-2}, Nm^{-1}	
γ	Shear strain		7.2.15
$\dot{\gamma}$	Shear rate	s^{-1}	7.2.17
δ	Distance of closest approach	m	5.2.11
λ	Wavelength	m	
λ_i	Decay length	m	5.5.2
κ	Inverse Debye length	m^{-1}	3.8.13
κ	Bending rigidity	J	6.2.1
$\hat{\kappa}$	Saddle splay constant	J	6.2.1
ϕ	Azimuthal angle		
ϕ	Volume fraction		
η	Viscosity	Nsm^{-2}	
$[\eta]$	Intrinsic viscosity		7.1.24
θ	Contact angle		2.1.13
θ	Polar angle		
$\rho(r)$	Charge density	Cm^{-3}	3.2.1
σ	Surface charge density	Cm^{-2}	3.5.1
σ	Shear stress	Jm^{-3}	7.2.18
μ	Friction coefficient	Jsm^{-2}	1.7.15
μ_i	Chemical potential of species i	J	1.2.5
μ_i^0	Standard state chemical potential of i	J	1.2.6
τ	Time of coagulation for colloids	s	8.3.9
τ_1	Fast micelle relaxation time	s	4.2.7
τ_2	Slow micelle relaxation time	s^{-1}	4.2.8
χ	Chi-parameter for polymer solutions		7.1.10
ξ	Zeta-potential	V	8.4.6

About the Author

Håkan Wennerström is an emeritus professor of Physical Chemistry at the Department of Chemistry, Lund University, Sweden. He is the author of more than 250 publications, mainly in Colloid Science in the areas of surfactant and lipid phase behavior, surface forces, electrostatic interactions, and nuclear magnetic resonance spectroscopy. He was a member of the Nobel Committee for Chemistry for 14 years and its chairman for 3 years.

Introduction

Why Colloidal Systems Are Important

The Colloidal Domain Encompasses Many Biological and Technological Systems

Evidence of humankind's use of colloids dates back to the earliest records of civilization. Stone Age paintings in the Lascaux caves of France and the written records of Egyptian pharaohs were produced with stabilized colloidal pigments. Many of our earliest technological processes, such as papermaking, pottery making, and the fabrication of soaps and cosmetics, involved the manipulation of colloidal systems. During the Middle Ages, artisans codified their knowledge and formed guilds that served as the prototypes of present-day professional organizations.

We can trace the establishment of colloid science as a scientific discipline to the mid-nineteenth century. In 1845, Francesco Selmi described the first examples of colloidal particles. By defining their common properties, he assigned these clear or slightly turbid "pseudosolutions" of silver chloride, sulfur, and Prussian blue in water to a class that also included alumina and starch.

In the 1850s, Michael Faraday made extensive studies of colloidal gold sols, which involve solid particles in water. He found that these sols are thermodynamically unstable. Once they are coagulated, the process cannot be reversed. For this reason, such insoluble dispersions are called *lyophobic* ("liquid-hating") colloids. Faraday's observations imply that such dispersions must be stabilized kinetically; if properly prepared, however, they can exist for many years. In fact, some of the colloidal systems Faraday prepared are still on display in the British Museum in London.

In 1861, Thomas Graham coined the term *colloid* (which means "glue" in Greek) to describe Selmi's "pseudosolutions." The term emphasized their low rate of diffusion and lack of crystallinity. Graham deduced that the low diffusion rates of colloidal particles implied that the particles were fairly large – at least 1 nm in diameter in modern terms. On the other hand, the failure of the particles to sediment under the influence of gravity implied that they had an upper size limit of approximately 1 μm. Graham's definition of the range of particle sizes that characterize the colloidal domain is still widely used today.

Colloidal phenomena played an important role in the genesis of physical chemistry by establishing a connection between descriptive chemistry and theoretical physics. For example, the discovery of Brownian motion resulted from the observation of colloidal-sized pollen particles by light microscopy. Einstein developed the relationship between Brownian motion and diffusion coefficients. In 1909, Jean Baptiste Perrin used this relationship to determine a definite value of Avogadro's number. Subsequently, Marian Smoluchowski derived an expression that related the kinetics of rapid coagulation of colloidal particles – where each Brownian encounter between two particles resulted in permanent contact and the formation of a larger dimer particle. His expression was extended to explain the role of diffusion in bimolecular reactions in general.

These fundamental developments paralleled increasingly sophisticated industrial use of colloidal systems. Formulations of paints and other coatings were improved; more sophisticated ceramics were developed; use of emulsions increased; and novel uses were found for newly discovered polymer molecules. However, the complexity of these practical systems far exceeded the ability of theorists to explain them, and to a large extent, colloids remained an empirically descriptive field.

Significant advances in explaining the stability of colloidal sols were made in 1945. Publication of the Derjaguin–Landau and Verwey–Overbeek (DLVO) theory provided a quantitative relationship between the attractive van der Waals forces, which lead to coagulation, and the repulsive electrostatic forces, which stabilize colloidal dispersions. The DLVO theory was studied extensively and verified; it now constitutes a cornerstone of colloid science.

Colloidal domains encompass much more than just the classical sols on which research was focused for almost a century. Polymers that have at least one molecular dimension greater than 1 nm exhibit many features of colloidal solutions. Unlike sols, polymer solutions form spontaneously and are thermodynamically stable. Such macromolecules (which include biopolymers like DNA and proteins) are called *lyophilic* ("liquid-loving") colloids. The great technical importance of polymers has led to the development of polymer science as a separate discipline.

Another major area of colloid science involves association colloids formed by amphiphilic (both "oil- and water-loving") molecules. It developed more slowly than the study of colloidal dispersions involving sols—partly because it depended on the synthesis and isolation of amphiphilic molecules in reasonably pure form. A key development in the study of association colloids was the observation by James McBain that the osmotic pressure of alkali metal fatty acid salts displays a pronounced break in the concentration curve, beyond which the osmotic coefficient remains almost constant. McBain attributed this fact to the self-association of these molecules into structures he called micelles. This theory did not meet with immediate success, however. When McBain first presented his observation at a symposium, the chairman is said to have remarked, "Nonsense, McBain."

Despite this unenthusiastic reception, McBain's observation stimulated considerable discussion concerning the structure of micelles. G.S. Hartley proposed that the structures are spherical, with a radius equal to that of the fully extended hydrocarbon chain. William Harkins considered the possibility of cylindrical micelles. McBain himself believed that micelles might be flat laminar sheets. In fact, when surfactants aggregate, they form structures of all these types. But in the case of the micelles McBain studied, Hartley was correct. Debate about the structure of amphiphilic assemblies continues to this day. In many ways, amphiphilic self-organizing systems are more complex than colloidal dispersions. One reason for this is that amphiphiles are associated physically, not chemically. Consequently, their microstructural size and shape can change in response to subtle variations in concentration or temperature. This facile response to changing environmental conditions contrasts strongly with the relatively immutable behavior of colloidal sols.

As the properties of association colloids were characterized more thoroughly, it became clear that the DLVO theory could not satisfactorily describe the interaction between such aggregates. In fact, our understanding of the complex interactions displayed by these systems has expanded significantly during the past two decades and in the process has brought new richness and sophistication to colloid science.

Large surface areas associated with the characteristic size of colloidal particles form an intrinsic property of all colloidal systems. For example, a typical micellar solution containing 0.1 M amphiphile has $\sim 4 \times 10^4$ m^2 of micellar–water interfacial area per liter of solution. It is often convenient to view such systems as being comprised of two phases: a dispersed phase (the colloidal particles) and the dispersing medium (the solvent), separated by a well-defined interface.

These interfaces are generally regions of high free energy, and many surface phenomena, such as adsorption and the formation of electrical double layers, occur in them. Because surface properties dominate the behavior of colloidal systems, surface phenomena play an integral role in colloid science. The organized study of surface phenomena can be traced back to Benjamin Franklin's observation of the effect of pouring oil on turbulent waters. In 1757, while watching a military fleet sail out of New York Harbor, Franklin observed that the wakes behind two of the ships were calmed after the ships' cooks dumped greasy refuse. Franklin later reproduced this phenomenon himself by pouring oil on Clapham Pond in England. Franklin's reports stimulated considerable scientific interest, which led to a new field of inquiry. Maritime patents for the process were issued, but the attempt to exploit the effect commercially proved impractical.

One of the more enigmatic characters in nineteenth-century scientific history is Agnes Pockels. A German woman with no formal scientific training, Pockels devised a series of experiments that remain a hallmark of surface science. Having developed a rudimentary surface balance in her kitchen sink, Pockels used this apparatus to determine surface pressure versus area curves. She sent her observations to Lord Rayleigh, who submitted them to *Nature* magazine. Publication of Pockels's work in 1891 set the stage for Langmuir's quantitative work on monolayers and laid the foundation for our ability to characterize them.

Understanding of Colloidal Phenomena Is Advancing Rapidly

Why should we be interested in colloidal systems? Everyday experience provides a variety of reasons.

The first reason is purely epistemological. As scientists, we are interested both in knowledge and in how we structure what we know. If we consider the three states of matter – gas, liquid, and solid – we can observe colloidal systems in all possible combinations. As Table I.1 illustrates, manifestations of the colloidal domain surround us every day. Our ability to understand, use, and control colloids depends on our mastery of their properties. Over the past few decades, what we know about matter at the molecular level has increased dramatically. Recent progress in our understanding of the forces that mediate colloidal interactions provides a new basis for further scientific advance.

The second reason is anthropological. Living systems are largely composed of proteins, polyelectrolytes, and amphiphilic molecules contained in an aqueous medium. By their very size, such molecules possess colloidal properties. More importantly, they self-assemble into a wide variety of organized structures that perform extremely sophisticated chemical transformations. Life processes involve the control and transformation of colloidal assemblies, and many diseases are associated with malfunctions at the colloidal level. The new level of knowledge made available by colloid science has considerably enhanced our understanding of interaction forces that control biological processes.

The third reason is technological. Nearly all industrial processes involve colloidal systems. The partial list given in Table I.1 makes their importance in a highly industrialized society seem almost self-evident. However, the challenge facing contemporary scientific research goes beyond these traditional applications. In many rapidly advancing areas of technology, such as the fabrication of new high-strength composite materials, the development of high-performance ceramics for use at

Table I.1 Types of colloidal system.

			Examples		
Medium	**Particle**	**Name**	**In nature**	**Biological**	**Technical**
Liquid	Solid	Colloidal sol	River water, glacial runoff, muddy water		Paint, ink Sol–gel processing
Liquid	Liquid	Emulsion (nonequilibrium) solution (equilibrium)		Fat digestion, biological membranes	Drug delivery, emulsion polymerization
Liquid	Gas	Foam	Polluted rivers	Vacuoles, insect excretions	Fire extinguishers, production of porous plastics
Gas	Solid	Aerosol	Volcanic smoke	Pollen	Inhalation of solid pharmaceuticals
Gas	Liquid	Aerosol	Clouds	Result of coughing	Hair spray, smog
Solid	Solid	Solid suspension	Wood	Bone	Composites
Solid	Liquid	Porous material	Oil reservoir rock, opals	Pearl	High-impact plastics
Solid	Gas	Solid foam	Pumice	Loofah	Styrofoam, zeolites

elevated temperatures, and the control of targeted drug delivery and medical diagnostic systems, progress depends on our ability to control colloidal interactions.

Association Colloids Display Key Concepts That Guided the Structure of This Book

Writing a textbook is an active process. An author must make difficult choices among subjects that vie for inclusion, like characters clamoring to take center stage during a play. Just as the play's action determines how long each character is allowed to remain in the spotlight or what staging is required, the author's creative philosophy determines the ideas in the text and the approach that is chosen. And just as the playwright becomes invisible once the actors begin to perform, publication usually draws a curtain over the decision-making process that shaped the content balance of a text. Because we believe our readers deserve to see backstage, so to speak, we want to explain the thought process that lies behind our work.

Emerging information about fundamental forces and interactions in the colloidal domain stimulated the writing of this book. We decided to focus on association colloids rather than on the more traditional lyophobic systems for two reasons. First, association colloids more readily illustrate the new concepts that we believe are the hallmark of modern colloid science. Second, concentrating on association colloids establishes a more immediate link between fundamental concepts, which are the organizing feature of this book, and their manifestations in the colloidal domain.

Most textbooks on colloids begin with lengthy discussions of the physical techniques, such as diffusion, light scattering, and viscosity, used to characterize colloidal systems. We find teaching from such books frustrating because they generally avoid addressing the real subject in a straightforward way. For this reason, we have chosen to include a succinct discussion of relevant experimental techniques in the appropriate chapters, but we focus our attention on describing fundamental physical phenomena, defining appropriate equations, and illustrating their applications through examples in the text and exercises at the end of each chapter. In addition, we cite relevant references at the end of each chapter, indicating the portions of the text to which each applies; additional literature citations appear as credits to figures and tables.

We have made a particular effort to articulate key concepts that form the basis of our understanding of colloidal phenomena. A major goal is to show how molecular interactions lead to colloidal behavior. We find that this connection between the molecular world and the colloidal domain emerges most clearly with association colloids. In some instances, we employ theoretical derivations that depart from those usually given. For example, instead of simply deriving expressions for free energy, we often obtain separate equations for the energy and entropy so that we can clarify the underlying fundamentals as we combine them. We also go to some lengths to point out where and how averaging our physical properties transforms energy expressions into expressions for free energy.

Our presentation presupposes an upper undergraduate level of mastery in physics, mathematics, and physical chemistry. While we have made every effort to provide linkages in the derivations, there is no substitute for writing them out as you read along. Learning is not a passive process. Forging enduring neurological connections in the brain requires a lot of groaning and sweating; it's just plain hard work.

Chapter 1 discusses the general characteristics of amphiphilic self-organizing systems. We describe their structural characteristics as well as the solvophobic interactions that drive aggregation. We also provide a simple geometric relationship between molecular and aggregate structures. This chapter includes a section on regular solutions, which summarizes many useful thermodynamic equations that we employ repeatedly throughout this book.

Chapter 2 addresses fundamental surface science, focusing on the aspects of the subject – particularly surface tension – that provide the background necessary to understand colloidal interactions. Amphiphilic monolayers, which hold considerable interest in their own right, are used in this chapter to illustrate many of the relevant principles of surface science.

Chapters 3 through 11 employ an alternating pattern in which a chapter that develops fundamental concepts is followed by one or more chapters, focusing on colloidal systems that illustrate these concepts.

For example, in Chapter 3, we derive the Gouy–Chapman theory and related equations that describe the properties of charged electrical double layers. Because the term *charged electrical double layer* describes both a charged surface and its adjacent solutions, we obtain relationships between the distribution of ions in a solution adjacent to a charged surface, then examine how the distribution varies with ionic valency, concentration, and surface potential or charge density. These relationships and the concepts they encompass form one of the bases of colloid science.

In Chapter 4, we discuss micellar systems in general but emphasize ionic micelles, because they illustrate electrical double-layer properties with particular clarity. In many ways, micelles are the simplest and best understood of all amphiphilic self-organizing systems. Equations describing micellar aggregation processes and micellar properties, dynamics, and kinetics provide a framework for dealing with more complex self-organizing structures, such as bilayers, vesicles, and emulsions, which we consider in subsequent chapters.

Chapter 5 develops equations that describe the repulsive and attractive forces that operate in colloidal systems. The picture that emerges reflects the myriad of interaction forces that come into play as colloidal particles approach one another. This is a challenging chapter, not only because of the inherent intricacy of the phenomena but also because it merges older classical concepts with newly emerging ones.

In Chapter 6, we discuss bilayer systems as an example of the interaction forces described in Chapter 5. The discussion also deals with association processes leading to the formation of liquid crystals, vesicles, and gels. Chapter 7 examines self-organizing properties of polymers, proteins, and polyelectrolytes.

Integrating the electrostatic and London dispersion forces gives an expression for the potential energy that we use in Chapter 8 to obtain the DLVO theory. We also consider the critical concentration of salt required for rapid coagulation and derive an expression for the kinetics of coagulation. This chapter develops the concept of the zeta potential and describes how to measure it.

Continuing our system of alternating theory with applications, Chapter 9 focuses on colloidal sols that exhibit typical properties and exemplify many of the concepts developed in previous chapters. Chapter 10 considers fundamental rules for understanding phase diagrams, and Chapter 11 concludes with a discussion of micro- and macroemulsions that illustrate these ideas.

As an aid to the reader, we have written all section headings as complete sentences. When gathered together in the contents, these sentences provide a succinct summary of each chapter. Key concept words appear in boldface the first time they are used. Symbols, units, and the equations that define them are listed following Chapter 11. In addition, all chapters begin with concept maps, which should be used to guide the detailed reading. We have found these maps to be of great help to students approaching the material for the first time. Spending 20–30 minutes internalizing the content of a concept map can significantly increase the information gained from reading the chapter.

Two final issues pertaining to our philosophy of authorship must be addressed. The first concerns the balance we have created between concepts and the systems that illustrate them. Our coverage of colloidal systems is clearly selective; the concepts under discussion dictated our choice of examples. However, references at the end of each chapter provide citations of alternative surveys of colloidal systems.

The second issue concerns our treatment of concepts that are still the subject of lively debate. We have tried to indicate which ideas are controversial, but we have not pursued the controversies in detail. All too often, a scholar who engages in such polemics is reminiscent of a player who "struts and frets his hour upon the stage ... and then is heard no more." Rather than attempting to present all the intellectual options currently available, we have identified the strategies we believe are most critical to the forward movement of the play and have allowed them to assume the leading roles in our text.

1

Solutes and Solvents, Self-assembly of Amphiphiles

Concept Map

Entropy of Mixing and Regular Solutions

Although the entropy of mixing and regular solutions are traditionally part of physical chemistry, they are reviewed here because they play an important role in understanding colloidal behavior.

- From elementary statistical mechanics, we obtain $\Delta \mathbf{S}_{\text{mix}} = -k_{\text{B}} \sum N_i \ln X_i$ (Eq. 1.1.8) for the ideal entropy of mixing.
- For regular binary solutions, $\Delta \mathbf{E}_{\text{mix}} = z_b N_{Av}\ n_{AB} w$ (Eq. 1.1.15), where the interaction parameter w depends on the microscopic interaction energies W_{AA}, W_{BB}, and W_{AB} (Eq. 1.1.16).
- From $\Delta \mathbf{G}_{\text{mix}} = \Delta \mathbf{H}_{\text{mix}} - \mathbf{T}\Delta \mathbf{S}_{\text{mix}}$ (Eq. 1.1.1b), we can predict the temperatures and compositions under which a binary solution will be homogeneous or form two phases.

Chemical Potentials

- The chemical potential μ_i (Eq. 1.2.5a) gives the free energy change in adding a molecule i to the system. It is the thermodynamic property that determines a component's ability to redistribute and is used throughout this text to analyze colloidal systems.
- For a solution, the chemical potential of the solvent is often expressed in terms of the osmotic pressure Π_{osm} (Eq. 1.2.12). In analogy with the nonideal gas, the second virial coefficient B_2 (Eq. 1.2.15) accounts for the interactions between the solute molecules.
- Chemical potentials (Eq. 1.2.24) can be generalized to show how an external gravitational field (Eq. 1.2.28) or an electrical field (Eq. 1.2.32) affects the distribution of components in a chemical system.

The Colloidal Domain: Where Physics, Chemistry, Biology, and Technology Meet, Third Edition. Håkan Wennerström.

Self-assembly Processes

- Amphiphilic molecules possess polar head groups and nonpolar chains (Table 1.1); to minimize unfavorable solvophobic (solvent-hating) interactions, they spontaneously aggregate to form a variety of microstructures. The sodium dodecyl sulfate (SDS) spherical micelle shown in Figure 1.6 is a specific example.
- Three general characteristics of the self-organizing capabilities of amphiphilic molecules—as illustrated by SDS spherical micelles—are the following:
  1. Aggregates form spontaneously—SDS micelles form at a well-defined critical micelle concentration (CMC).
  2. Aggregation is a start–stop process—adding more SDS results in the formation of more micelles of the same size.
  3. Aggregates have well-defined properties—the maximum radius of a spherical micelle is set by the length of the hydrocarbon chain.
- Amphiphilic molecules are associated physically, not chemically, and aggregate size and shape vary in response to changing solution conditions (temperature, pH, salt, etc.).

The Surfactant Number

- The surfactant number $N_s = \nu / l a_0$ (Eq. 1.5.1) relates amphiphilic molecular structure (Table 1.1) to aggregate architecture (Figure 1.10), where ν and l represent the volume and effective length of the hydrocarbon chains and a_0 is the area per head group.
- N_s relates properties of the amphiphile to the preferred curvature of the aggregate as given in Eqs. (1.5.4)–(1.5.7).

Solvophobicity

- Microphasic separation accompanying amphiphilic aggregation is driven by solvophobic interactions and requires (1) an amphiphile with well-defined polar and nonpolar groups and (2) either a very polar or nonpolar solvent.
- The Gordon parameter $\gamma / V_m^{1/3}$ N/m^2, where γ and V_m are the solvent's surface tension and molar volume, provides a useful measure of solvent cohesive energy, as illustrated in Figure 1.12.
- Micelle formation is observed in polar (water, hydrazine, and ethylene glycol, $\gamma / V_m^{1/3} > 1.3$) and nonpolar (oil and fluorocarbons, $\gamma / V_m^{1/3} < 0.3$) solvents, but not in intermediate solvents (alcohols, esters, and amides).

Brownian Motion, Intermolecular Interactions, and Diffusion

- In the Boltzmann distribution (Eq. 1.2.26), the interplay between Brownian motion as expressed by $k_B\mathbf{T}$ and an organizing force arising from intermolecular interactions is a key element in understanding interactions in the colloidal domain.
- Self-diffusion can be described by a random walk model that leads to the equation $\langle r^2 \rangle = 6Dt$ (Eq. 1.7.5), which relates the average distance traveled to the time t and the diffusion coefficient D.
- The flux of molecules in the presence of a concentration gradient is described by Fick's first law (Eq. 1.7.7) for the molecular flow and by Fick's second law (Eq. 1.7.12) for the concentration changes.

Most colloidal phenomena involve a liquid phase as at least one of the actors. In this liquid, we can typically identify a *solvent* and one or more *solutes*. Molecular interactions involving colloidal entities, solutes, and solvents determine the properties of the system.

Throughout this book, we try to maintain this molecular perspective on the colloidal phenomena. In this chapter, we first present the properties of solutions where molecules interact in a pairwise manner. We then use surfactant self-assembly to illustrate how surfactant molecules can associate in solution in a cooperative manner. There is a delicate interplay between solvent–solvent, solvent–solute, and solute–solute interactions. We will see that a characteristic feature of a surfactant molecule is that they have one part that likes the solvent and one part that the solvent dislikes. We say that such molecules are amphiphilic.

1.1 Understanding the Origin of Entropy and Enthalpy of Mixing Provides Useful Molecular Insight into Many Colloidal Phenomena

Throughout this book, we will be concerned with the thermodynamics of self-assembly of surfactant molecules into aggregates, the properties of polymer solutions, and the behavior of colloidal entities such as emulsions. To obtain a basis for these more complex systems, we start by considering the mixing of two pure liquids. Figure 1.1 shows three possible outcomes of such a mixing:

1. Formation of a homogeneous solution resulting from complete miscibility at all compositions. An example is *t*-BuOH and water at 25 °C.
2. Formation of two phases containing different concentrations of the two components. An example is *n*-BuOH and water at 25 °C.
3. Retention of two unmixed phases, illustrated by water and mercury.

In a thermodynamic description, the outcomes of these mixing processes differ because the free energies of mixing, $\Delta\mathbf{G}_{\text{mix}}$, are different. At constant temperature $\mathbf{T}$ and pressure $\mathbf{p}$, the Gibbs free energy is

$$\Delta\mathbf{G} = \Delta\mathbf{H} - \mathbf{T}\Delta\mathbf{S} \tag{1.1.1a}$$

and for a mixing process

$$\Delta\mathbf{G}_{\text{mix}} = \Delta\mathbf{H}_{\text{mix}} - \mathbf{T}\Delta\mathbf{S}_{\text{mix}} \tag{1.1.1b}$$

where $\Delta\mathbf{G}_{\text{mix}} = \mathbf{G}_{\text{final}} - \mathbf{G}_{\text{initial}}$. We can define $\Delta\mathbf{H}_{\text{mix}}$ and $\Delta\mathbf{S}_{\text{mix}}$ in an analogous manner. Our goal in this section is to develop expressions for $\Delta\mathbf{H}_{\text{mix}}$ and $\Delta\mathbf{S}_{\text{mix}}$, which reveal the reasons for the

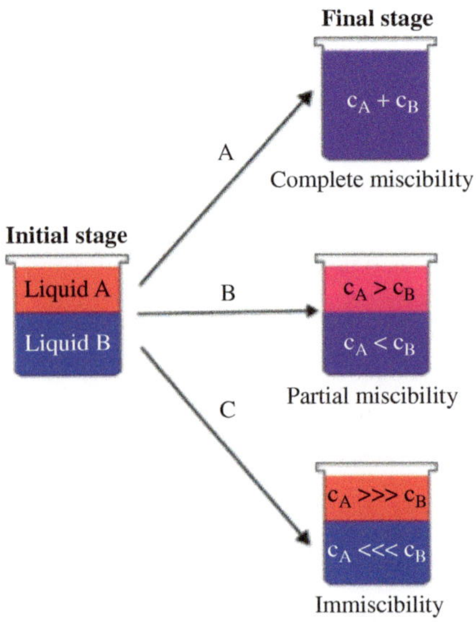

Figure 1.1 Three possible outcomes of mixing equal amounts of two liquids A and B: (a) formation of a homogeneous mixture; (b) formation of two phases, one solution of B in A and one solution of A in B; (c) retention of the two unmixed phases with the second component present only in small amounts.

behavior illustrated in Figure 1.1. This process leads us naturally to introduce a basic thermodynamic concept that we use repeatedly throughout the book, the chemical potential.

1.1.1 The Ideal Mixing Model Provides a Basis for Understanding the Formation of a Miscible Phase

We focus first on the $\mathbf{T\Delta S}_{\text{mix}}$ term in Eq. (1.1.1b). For the moment, we assume that $\Delta\mathbf{H}_{\text{mix}}$ is sufficiently small that it can be ignored so that

$$\Delta\mathbf{G}_{\text{mix}} = -\mathbf{T\Delta S}_{\text{mix}} \tag{1.1.2}$$

The relationship is called ideal mixing, and it applies only when we mix two molecular species that are virtually identical, such as CH_3OH and CD_3OH (where D denotes the heavier hydrogen isotope, deuterium). In the same way that the ideal gas model is very useful for the description of real gases, the corresponding ideal mixing model provides a basis for the understanding of real liquid mixtures.

When two miscible liquids, such as H_2O and *t*-BuOH, come into contact, the concentration gradient across the initial sharp interface causes *t*-BuOH molecules to diffuse into water and vice versa. We can gain insight into this molecular process by examining the Brownian motion and the resulting diffusion. In 1827, Robert Brown observed that pollen grains displayed a continuous random motion when viewed under a light microscope. All species in solution, whether colloidal or molecular, experience such Brownian motion. Given enough time, each molecule then moves throughout the entire liquid volume in a diffusion process. The mixing process results in an increased disorder, which means an increased entropy. If we assume that enthalpy changes could be neglected, $\Delta\mathbf{G}_{\text{mix}}$ is negative for the mixing process, as it should be for a spontaneous process.

Even if it is qualitatively clear that the ideal mixing process leads to an increase in entropy, later applications require a quantitative expression for the ideal entropy of mixing. To derive this expression, we start from one of the fundamental equations of statistical mechanics

$$\mathbf{S} = k_{\text{B}} \ln \Omega \tag{1.1.3}$$

which was first formulated by the Austrian physicist Ludwig Boltzmann in 1896. In this equation, k_{B} is the Boltzmann constant and Ω stands for the number of states accessible to the system under the given constraints.

The simplest approach to the task of calculating the entropy of mixing N_A molecules A with N_B molecules B in the liquid state is to adopt a so-called lattice model. In that case, we divide the total volume V into $N = N_A + N_B$ cells. Figure 1.2 shows, using a two-dimensional analog, how the A and B molecules initially are compartmentalized in the unmixed state and are arranged in a unique way.

When A and B are allowed to mix, the molecules interchange positions by diffusion. Since we assume that the interaction between all molecules is identical ($\Delta\mathbf{H}_{\text{mix}} = 0$), Brownian motion leads to a random distribution of molecules, which depends only on the fraction $X_A(X_B)$ of $A(B)$ molecules.

Given these rules, calculating Ω becomes an exercise in statistics. We start by distributing the N_A A molecules. There are

$$\frac{N!}{N_A!(N-N_A)!} \tag{1.1.4}$$

ways to distribute N_A identical objects on N identical positions.

All that remains is to place the N_B B molecules. However, since these molecules are equivalent and there are now N_B positions left, we have no choice except to put the B molecules on the empty positions. We can do this in only one way. The total number of ways of distributing the molecules is then

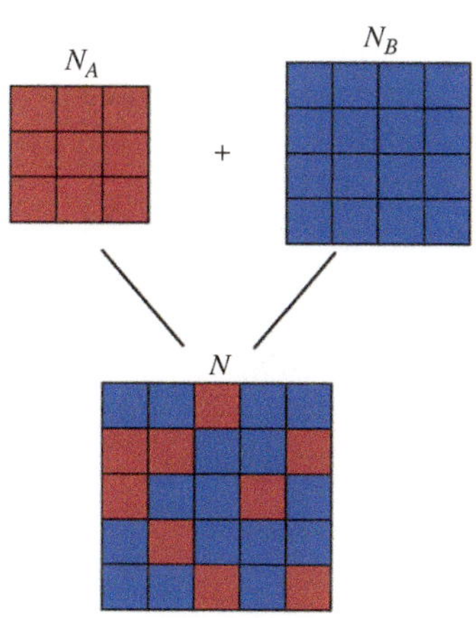

Figure 1.2 Lattice model for calculating the entropy of mixing. Nine red squares are mixed with 16 blue squares to form a mixture of 64% blue and 36% red squares. The mixed state shows one of the 25!/(16!9!) = 2 042 975 possible configurations.

$$\Omega = \frac{N!}{N_A!(N - N_A)!} = \frac{N!}{N_A!N_B!} \tag{1.1.5}$$

and using Eq. (1.1.3), the entropy is

$$\mathbf{S} = k_B(\ln N! - \ln N_A! - \ln N_B!) \tag{1.1.6}$$

For the specific mixing process shown in Figure 1.2, with $N = 25$, $N_A = 9$, and $N_B = 16$, $\Omega_{final} = 2\ 042\ 975$. Since $\Omega_{initial} = 1$, $\Delta\mathbf{S}_{mix} = k_B \ln \frac{\Omega_{final}}{\Omega_{initial}} = 15\ k_B$. With systems where N is very large, we can simplify further by using the Stirling approximation

$$\ln N! \approx N \ln N - N \tag{1.1.7}$$

Using this relation, Eq. (1.1.6) becomes

$$\Delta\mathbf{S}_{mix} = -k_B \left[N_A \ln \frac{N_A}{N} + N_B \ln \frac{N_B}{N}\right] = -k_B(N_A \ln X_A + N_B \ln X_B) \tag{1.1.8}$$

where X_A and X_B are the respective mole fractions. Equation (1.1.8) represents the desired expression for the ideal entropy of mixing of species A and B. Note that $\Delta\mathbf{S}_{mix}$ has a maximum at $X_B = 0.5$ because the largest number of ways of arranging the molecules in the solution occurs when there are equal numbers of the two species.

In deriving Eq. (1.1.8), we made a number of assumptions that do not seem to apply, even approximately, to a real solution. However, the situation is not as bad as one might suspect. In fact, Eq. (1.1.8) has a more general validity than the restricted lattice model shown in Figure 1.2 would suggest. For example, it correctly describes the mixing of two ideal gases, where the notion of a lattice is totally irrelevant.

The minimal assumption needed to obtain Eq. (1.1.8) as an exact result is that when we distribute the molecules in space, the same distribution rule must apply to both the A and B species. In our derivation, we used the specific constraint of placing one and only one molecule into each cell, with no correlations. This restriction is applied to both A and B. For an ideal gas mixture, we use the rule that we can place any molecule anywhere within the volume V irrespective of the location of the other molecules. Again, this rule applies equally to the A and B species.

We can also obtain Eq. (1.1.8) as a limiting case at high dilution for inherently nonideal systems. Nonideality formally enters such systems in two ways. First, selective interactions between the molecules develop correlations in the solution so that A molecules prefer to be surrounded by A molecules and B molecules by other B molecules. However, when $N_A \gg N_B$, dilution ensures that solvent A molecules surround solute B molecules anyway.

When the two species have different sizes, the geometric factor creates a second source of nonideality. In the derivation of Eq. (1.1.8), we assumed that A and B occupied equal volumes. However, at high dilution, unequal volumes $V_A \leq V_B$ change Eq. (1.1.8) by a linear term $N_B \ln(V_B/V_A)$, which is concentration-independent and can be included in the standard free energy.

1.1.2 The Regular Solution Model Provides a Simple Description of Nonideal Mixing, Which Ultimately Leads to the Formation of a Two-phase System

If we repeat the mixing experiment using water and *n*-BuOH instead of *t*-BuOH, the following sequence of events occurs:

1. Mutual diffusion results in the smearing of the initially sharp interface, just as with *t*-BuOH.
2. Instead of forming a homogeneous solution, a sharp interface separates the water-rich and *n*-BuOH-rich phases. This constitutes the equilibrium state.

In this case, the driving force to randomize that arises from the entropy of mixing is balanced by a second driving force that organizes the system. Direct molecular interactions contribute to making $\Delta\mathbf{H}_{\text{mix}}$ positive, which favors segregation.

The regular solution theory is the simplest model that describes liquid mixtures, including interactions that lead to a nonzero $\Delta\mathbf{H}_{\text{mix}}$ and a deviation from ideal behavior. In its simplest form, this theory deals with an incompressible mixture of two molecular species of equal size. These molecules interact only with their nearest neighbors whose number is fixed to z_b (the bulk coordination number). As a result, the total interaction energy, which is assumed to be pairwise additive, depends exclusively on the numbers n_{AA}, n_{BB}, and n_{AB} of the three different types of molecular contacts. These, in turn, can be related to the numbers N_A and N_B of A and B molecules in the system.

Due to the (near) incompressibility of liquids, the energy and enthalpy only differ by a constant term $\mathbf{pV}$. Therefore, if we compare two states at the same pressure, the enthalpy change equals the change in internal energy

$$\Delta\mathbf{H} = \Delta\mathbf{U} \tag{1.1.9}$$

Because we usually consider incompressible systems, Eq. (1.1.9) applies in most of this book. It also follows that changes in Helmholtz, **A**, and Gibbs, **G**, free energies are equal under these circumstances. Furthermore, for a molecular system, the thermodynamic internal energy, **U**, equals the total intermolecular interaction energy, E, plus a kinetic energy term that is unaffected by mixing. For pairwise additive interactions, such as in the regular solution theory, E is the sum over all pair interactions W_{ij}

$$E = \sum_{i>j} W_{ij} \tag{1.1.10}$$

Figure 1.2 illustrates the mixing of two pure liquids into a randomized mixture. The energy of mixing is

$$\Delta\mathbf{E}_{\text{mix}} = E_{\text{final}} - E_{\text{initial}} = E_{\text{mix}} - E_0 \tag{1.1.11}$$

where E_0 is the energy of the unmixed liquids:

$$E_0 = \frac{1}{2} z_a N_A W_{AA} + \frac{1}{2} z_b N_B W_{BB} \tag{1.1.12}$$

Since $z_A N_A/2$ and $z_B N_B/2$ give, respectively, the number of AA or BB pairs in the neat liquids, and W_{AA} and W_{BB} are the microscopic pair interaction energies. We can estimate these energies from the enthalpy of vaporization, $\Delta\mathbf{H}_{\text{vap}}$, since in this process molecules are separated from each other and become isolated molecules in the vapor phase.

Problem: Given that the enthalpy of vaporization, $\Delta \mathbf{H}_{vap}$, for carbon tetrachloride is 29.7 kJ/mol, calculate W_{AA}, assuming that each molecule has six nearest neighbors, $z_b = 6$.

Solution: $-z_b W_{AA}/2 = \Delta \mathbf{E}_{vap} = \Delta \mathbf{H}_{vap} - \boldsymbol{RT} = 27.2$ kJ/mol, where the factor of 2 avoids double counting the interactions. Counted per molecule, $z_b W_{AA} = -2 \times 27.2 \times 10^3/6 \times 10^{23} = -9.1 \times 10^{-20}$ J, which corresponds to $-22k_B\mathbf{T}$ at room temperature. With $z_b = 6$, W_{AA} is $-3.7k_B\mathbf{T}$ per pair interaction.

We obtain the value of ΔE_{mix} by summing over all interactions on a pairwise basis

$$\Delta \mathbf{E}_{mix} = n_{AA} W_{AA} + n_{BB} W_{BB} + n_{AB} W_{AB} \tag{1.1.13}$$

The values of n_{AA}, n_{BB}, and n_{AB} are related to the total number of molecules by

$$2n_{AA} + n_{AB} = z_b N_A \tag{1.1.14a}$$

$$2n_{BB} + n_{AB} = z_b N_B \tag{1.1.14b}$$

The first equality follows, for example, by realizing that an A molecule is engaged in z_b pairs, either of AA or AB molecules. When we sum over all N_A A molecules, we double count the AA pairs but not the AB ones. Solving Eq. (1.4.14 a,b) for n_{AA} and n_{BB} and substituting back into Eq. (1.1.13) yields after some algebraic manipulation

$$\Delta \mathbf{E}_{mix} = E_{mix} - E_0 = n_{AB}\left(W_{AB} - \frac{1}{2}W_{AA} - \frac{1}{2}W_{BB}\right) \equiv n_{AB}\, w/(z_b N_{Av}) \tag{1.1.15}$$

where a single effective interaction parameter

$$w = z_b N_{Av}\left(W_{AB} - \frac{1}{2}W_{AA} - \frac{1}{2}W_{BB}\right) \tag{1.1.16}$$

emerges naturally. Here N_{Av} denotes Avogadro's number so that w is the interaction per mole. Even though all the interaction energies are negative, w can easily be positive. In fact, this is normally the case. This case requires only that the heterogeneous interaction, W_{AB}, is weaker than the average of the interactions in the respective pure phases.

The thermodynamic properties of the solution are determined by the free energy rather than the energy. Thus, we need an estimate of the entropy of mixing. In a real system, this mixing involves changes in internal conformations, in the density, and in orientational correlations, as well as in the mixing of the two species. The regular solution theory makes the simplifying assumption that mixing affects only the positional order at constant density. Furthermore, it assumes that the molecules mix in a fully random way.

These assumptions produce two consequences. First, the entropy of mixing is that of an ideal mixture, as in Eq. (1.1.8). Second, we can calculate the number n_{AB} of AB pairs in Eq. (1.1.15) using the assumed random distribution. Of the z_b nearest neighbors to an A molecule, $z_b X_B$ are B molecules. Since there are $N_A = X_A(N_A + N_B)$ A molecules, the number of AB pairs is

$$n_{AB} = z_b X_A X_B (N_A + N_B) \tag{1.1.17}$$

Combining Eqs. (1.1.8), (1.1.15), and (1.1.17) permits us to calculate the Gibbs free energy of mixing as $\Delta \mathbf{G}_{mix} = \Delta \mathbf{A}_{mix} = \Delta \mathbf{E}_{mix} - \mathbf{TS}_{mix}$, so that

$$\Delta\mathbf{G}_{\text{mix}} = (n_A + n_B)[X_A X_B w + \boldsymbol{RT}(X_A \ln X_A + X_B \ln X_B)] \tag{1.1.18}$$

for n_A and n_B moles of A and B. This free energy expression forms the basis of the regular solution theory. It contains only a single unknown parameter, w. Figure 1.3 shows plots of $\Delta\mathbf{G}_{\text{mix}}$, $\Delta\mathbf{H}_{\text{mix}}$, and $\mathbf{T}\Delta\mathbf{S}_{\text{mix}}$, versus composition obtained within the regular solution model that demonstrates the three types of behavior shown in Figure 1.1.

A totally miscible system implies that not only is $\Delta\mathbf{G}_{\text{mix}}$ negative for all mixing ratios, it is also minimal for the equimolar mixture. This qualitative behavior is observed for $w < 2\boldsymbol{RT}$ (Figure 1.3a, b). When w is slightly larger than $2\boldsymbol{RT}$, a local maximum appears in $\Delta\mathbf{G}_{\text{mix}}$ at the equimolar point and one has a partially miscible system (Figure 1.3c). For mixing ratios between $X_B \approx 0.1$ and $X_B \approx 0.9$, the total free energy for the system is minimized when two separate phases form with a composition corresponding to the two minima in the free energy form. Here complete mixing does not occur in spite of the fact that $\Delta\mathbf{G}_{\text{mix}}$ is negative at all compositions. With a completely immiscible system (Figure 1.3d), at least one component has strong internal interactions, W_{AA} and/or W_{BB}, while W_{AB} is relatively weak. As a result, w is large and positive and $\Delta\mathbf{G}_{\text{mix}}$ positive except for very dilute solutions, not discernible in the diagram.

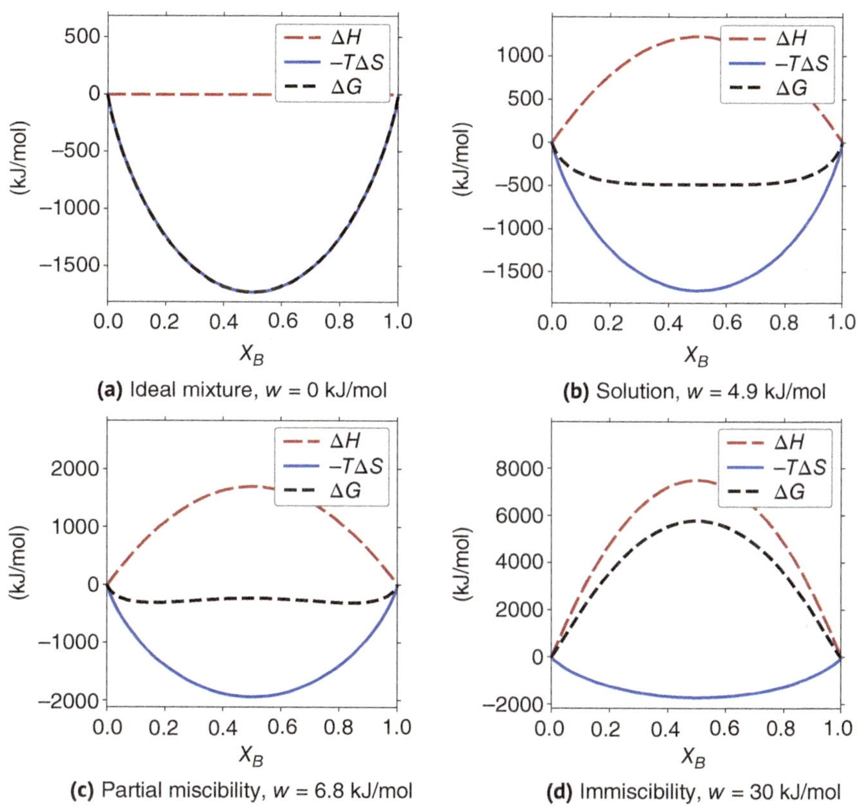

Figure 1.3 Predictions of the regular solution theory for $\Delta\mathbf{G}_{\text{mix}}$, $\Delta\mathbf{H}_{\text{mix}}$, and $\Delta\mathbf{S}_{\text{mix}}$ as a function of the composition for four values of the interaction constant w and at $\mathbf{T} = 298$ K. (a) $w = 0$ ideal mixing and complete miscibility; (b) $w = 4.9$ kJ/mol $= 1.98\boldsymbol{RT}$, miscibility; (c) $w = 6.8$ kJ/mol $= 2.75\boldsymbol{RT}$ phase coexistence in the interval $0.10 < X_B < 0.90$; (d) $w = 30$ kJ/mol $= 12\boldsymbol{RT}$ coexistence of two nearly pure liquids.

1.2 The Chemical Potential Is a Central Thermodynamic Concept in the Description of Multicomponent Systems

At constant temperature and pressure, the equilibrium state is the one that minimizes the Gibbs free energy. We used this fact (along with the free energy functions given in Figure 1.3) to find that for $w \geq 2\boldsymbol{RT}$, two phases separate at room temperature. In systems with two or more components, the individual constituents always have the option to rearrange into a separate phase or just spatially respond to an external potential. To arrive at the chemical potential, the thermodynamic property that determines a component's ability to redistribute, let us analyze the n-BuOH–H_2O system in detail.

At equilibrium, molecules continue to move across the interface, but since the system's composition does not change with time, the net fluxes of each component must be zero. Now let us determine the change in free energy associated with transferring one BuOH molecule from the upper to the lower phase at constant **T** and **p**. The free energy before transfer is

$$\mathbf{G}_\mathrm{i} = \mathbf{G}\left(N^l_{\mathrm{H_2O}}, N^l_{\mathrm{BuOH}}\right) + \mathbf{G}\left(N^u_{\mathrm{H_2O}}, N^u_{\mathrm{BuOH}}\right) \tag{1.2.1}$$

while after transfer, it becomes

$$\mathbf{G}_\mathrm{f} = \mathbf{G}\left(N^l_{\mathrm{H_2O}}, N^l_{\mathrm{BuOH}} + 1\right) + \mathbf{G}\left(N^u_{\mathrm{H_2O}}, N^u_{\mathrm{BuOH}} - 1\right) \tag{1.2.2}$$

where the superscripts l and u denote the lower and upper phases, respectively. The change in free energy is $\Delta\mathbf{G} = \mathbf{G}_\mathrm{f} - \mathbf{G}_\mathrm{i}$

$$\begin{aligned}\mathbf{G}_\mathrm{f} &= \mathbf{G}\left(N^l_{\mathrm{H_2O}}, N^l_{\mathrm{BuOH}} + 1\right) + \mathbf{G}\left(N^u_{\mathrm{H_2O}}, N^u_{\mathrm{BuOH}} - 1\right) - \left[\mathbf{G}\left(N^l_{\mathrm{H_2O}}, N^l_{\mathrm{BuOH}}\right) + \mathbf{G}\left(N^u_{\mathrm{H_2O}}, N^u_{\mathrm{BuOH}}\right)\right] \\ &= \mathbf{G}\left(N^l_{\mathrm{H_2O}}, N^l_{\mathrm{BuOH}} + 1\right) - \mathbf{G}\left(N^l_{\mathrm{H_2O}}, N^l_{\mathrm{BuOH}}\right) + \mathbf{G}\left(N^u_{\mathrm{H_2O}}, N^u_{\mathrm{BuOH}} - 1\right) - \mathbf{G}\left(N^u_{\mathrm{H_2O}}, N^u_{\mathrm{BuOH}}\right)\end{aligned} \tag{1.2.3}$$

Now, N_{BuOH} is a large number in the thermodynamic limit. Adding or subtracting one molecule, therefore, represents an infinitesimal change in the concentration of the respective phases. This enables us to interpret the two differences given in Eq. (1.2.3) as two derivatives that define the chemical potential μ_{BuOH} so that per molecule

$$\Delta\mathbf{G} = \left(\frac{\partial \mathbf{G}}{\partial N^l_{\mathrm{BuOH}}}\right)_{\mathbf{T},\,\mathbf{p},\,N_{\mathrm{H_2O}}} - \left(\frac{\partial \mathbf{G}}{\partial N^u_{\mathrm{BuOH}}}\right)_{\mathbf{T},\,\mathbf{p},\,N_{\mathrm{H_2O}}} = \mu^l_{\mathrm{BuOH}} - \mu^u_{\mathrm{BuOH}} \tag{1.2.4}$$

where μ_i^α is the chemical potential of component i in phase α with the general definition

$$\mu_i^\alpha = \left(\frac{\partial \mathbf{G}^\alpha}{\partial N_\mathrm{i}}\right)_{\mathbf{T},\,\mathbf{p},\,N_{j\neq i}} \tag{1.2.5a}$$

per molecule or

$$\mu_i^\alpha = \left(\frac{\partial \mathbf{G}^\alpha}{\partial n_\mathrm{i}}\right)_{\mathbf{T},\,\mathbf{p},\,n_{j\neq i}} \tag{1.2.5b}$$

per mole. We will use the definitions 1.2.5a or 1.2.5b, which differ by a factor of N_{Av}, alternatively throughout this text.

Equation (1.2.4) shows that the free energy change in transferring a butanol molecule from the upper to the lower phase is simply the difference in chemical potential of the butanol in the two phases. At equilibrium, $\Delta\mathbf{G}$ should be zero, and we arrive at the important conclusion that the chemical potentials are equal in the two phases.

1.2.1 Having an Expression for the Free Energy, We Can Determine the Chemical Potential by Differentiation

It is usually convenient to write the chemical potential as a sum of one term that is independent of concentration plus one or more dependent on concentration. For example, for an ideal solution

$$\mu_i = \mu_i^\theta + \boldsymbol{R}\mathbf{T}\ln X_i \tag{1.2.6}$$

The concentration-independent standard chemical potential μ_i^θ is an essential quantity that enters, for example, into the expression for chemical equilibrium constants and partition coefficients. The concentration-dependent term $\boldsymbol{R}\mathbf{T}\ln X_i$ arises from the changes in the entropy that occur on mixing.

Problem: Derive Eq. (1.2.6) for a binary system $A + B$ with ideal mixing.

Solution: The free energy of the mixture is

$$\mathbf{G} = n_A \mathbf{G}_A + n_B \mathbf{G}_B + n_A \boldsymbol{RT} \ln X_A + n_B \boldsymbol{RT} \ln X_B$$

where $\mathbf{G}_A$ and $\mathbf{G}_B$ are the free energy per mole of pure A and B, respectively. Taking the derivative with respect to n_A yields

$$\begin{aligned}
\mu_A &= \left(\frac{\partial \mathbf{G}}{\partial n_A}\right)_{n_B} \\
&= \mathbf{G}_A + \boldsymbol{RT}\ln X_A + n_A \boldsymbol{RT}\frac{1}{X_A}\left(\frac{\partial X_A}{\partial n_A}\right)_{n_B} + n_B \boldsymbol{RT}\frac{1}{X_B}\left(\frac{\partial X_B}{\partial n_A}\right)_{n_B} \\
&= \mathbf{G}_A + \boldsymbol{RT}\ln X_A + (n_A + n_B)\boldsymbol{RT}\left[\left(\frac{\partial X_A}{\partial n_A}\right)_{n_B} + \left(\frac{\partial (1 - X_A)}{\partial n_A}\right)_{n_B}\right] \\
&= \mathbf{G}_A + \boldsymbol{RT}\ln X_A
\end{aligned}$$

which reduces to Eq. (1.2.6) with $\mathbf{G}_A$ identified with μ_A^θ.

In the regular solution model, an additional interaction term, $wX_A X_B$, occurs in the expression for the free energy of mixing in Eq. (1.1.18). This expression introduces an extra term into the chemical potential and

$$\mu_A = \mu_A^\theta + \boldsymbol{R}\mathbf{T}\ln X_A + wX_B^2 \tag{1.2.7}$$

We obtain the corresponding expression for μ_B by exchanging the subscripts A and B.

Problem: For the regular solution model, derive the contribution in Eq. (1.2.7) to μ_A from the term $w(n_A + n_B) X_A X_B$ in Eq. (1.1.18).

Solution:

$$\begin{aligned}
w\left(\frac{\partial[(n_A + n_B)X_A X_B]}{\partial n_A}\right)_{n_B} &= wn_B\left(\frac{\partial[n_A/(n_A + n_B)]}{\partial n_A}\right)_{n_B} \\
&= wn_B\left[\frac{1}{(n_A + n_B)} - \frac{n_A}{(n_A + n_B)^2}\right] = wn_B\left[\frac{n_B}{(n_A + n_B)^2}\right] = wX_B^2
\end{aligned}$$

1.2.2 Mixtures and Solutions Differ Through the Choice of Standard States

Intermolecular interactions fundamentally determine the value of the concentration-independent term μ^θ of the chemical potential. However, μ^θ is also affected by two choices in the description. One needs to choose a standard state, which for liquids is normally the pure liquid of the component or the component totally surrounded by solvent molecules; this is termed the infinite dilution standard state. The choice of the concentration variable also affects the value of the standard chemical potential, mole fraction units, and molar or possibly molal concentrations. This point is a classical source of confusion!

We can use the regular solution model to illustrate the factors that determine the value of the standard chemical potential. In this model, the natural standard state is the neat liquid, and we use mole fraction units to characterize the concentration. In this case, the standard chemical potential is simply the free energy of the liquid divided by the number of molecules, so for example

$$\mu_A^\theta = \frac{\mathbf{G}_\mathrm{A}(\text{neat liquid})}{N} \tag{1.2.8}$$

In Eq. (1.1.12), we obtained an expression for the energy E_0 in the neat liquid. The strict lattice model exhibits no entropy for a pure system, but it is clear that a real liquid has substantial disorder, and we assign an entropy $\mathbf{S}_\mathrm{liq}^\mathrm{A}$ to this effect. The regular solution model implicitly assumes that this contribution is constant under mixing. Combining the energy and entropy contributions, we get

$$\mu_A^\theta = \frac{z_b}{2} N_\mathrm{Av} W_{AA} - \mathbf{T}\mathbf{S}_\mathrm{liq}^\mathrm{A} + \mathbf{p}V_A + \mathbf{U}_\mathrm{A}(\text{int}) \tag{1.2.9}$$

where V_A is the molar volume of A, $\mathbf{p}$ is the pressure, and $\mathbf{U}_\mathrm{A}(\text{int})$ is the intramolecular energy of A.

In the regular solution model, the standard chemical potential of B is as in Eq. (1.2.9) with subscript and superscript A substituted with B. However, in some cases, we prefer not to treat A and B on an equal footing but to consider A as the solvent and B as the solute. Then the infinitely dilute limit becomes the standard state for B and

$$\mu_B^\theta(X_B \to 0) = z_b N_\mathrm{Av}\left(W_{AB} - \frac{1}{2}W_{AA}\right) - \mathbf{T}\mathbf{S}_\mathrm{liq}^\mathrm{B} + \mathbf{p}V_B + \mathbf{U}_\mathrm{B}(\text{int}) \tag{1.2.10a}$$

still measuring concentrations in mole fractions. If we instead measure the concentrations in moles per liter, an additional term $\boldsymbol{RT}\ln V_A$ enters $\mu_B^\theta(c_B \to 0)$.

$$\mu_B^\theta(c_B \to 0) = z_b N_\mathrm{Av}\left(W_{AB} - \frac{1}{2}W_{AA}\right) - \mathbf{T}\mathbf{S}_\mathrm{liq}^\mathrm{B} + \mathbf{p}V_B + \mathbf{U}_\mathrm{B}(\text{int}) + \boldsymbol{RT}\ln V_A \tag{1.2.10b}$$

1.2.3 The Chemical Potential of the Solvent Can Be Expressed in Several Ways

We will encounter many cases where it is essential to characterize the chemical potential of the solvent. In most applications of practical significance, water is the solvent and we therefore use water to illustrate different concepts used to characterize the solvent chemical potential. The most straightforward is the activity a_w defined as

$$\mu_w = \mu_w^\theta + \boldsymbol{RT}\ln a_w \tag{1.2.11}$$

For an ideal solution, $a_w = X_w$, as seen from Eq. (1.2.6). For nonideal systems, the activity is particularly useful when water is not in excess relative to other components, and can, for example, be used to characterize moisture content in food. A much-used alternative to the activity is the osmotic pressure, Π_{osm}. Even though the osmotic pressure is typically discussed in terms of a pressure difference across a semipermeable membrane, it is more useful to consider it as a measure of the solvent chemical potential through the relation

$$\Pi_{\text{osm}} = \frac{\left(\mu_w^\theta - \mu_w\right)}{V_w} \tag{1.2.12}$$

where V_w is the molar volume of water. For an ideal solution with solute concentration c_s,

$$\Pi_{\text{osm}} = \boldsymbol{RT}c_s \tag{1.2.13}$$

which is called the van't Hoff equation and is formally identical to the ideal gas law, $\mathbf{pV} = n\boldsymbol{RT}$. The analogy with a gas also suggests a way of accounting for deviations from the ideal solution case by a so-called *virial expansion*:

$$\Pi_{\text{osm}} = \boldsymbol{RT}c_s(1 + B_2 c_s + \ldots) \tag{1.2.14}$$

Here, B_2 is called the *second virial coefficient* and is determined by the (effective) pair interaction $V(r)$ between two solute molecules. If we assume an orientation-independent potential $V(r)$ that only depends on the radial separation r between two molecules, B_2 can be expressed as

$$B_2 = -\frac{1}{2}\int_0^\infty \left[\exp\left(-\frac{V(r)}{k_\text{B}\mathbf{T}}\right) - 1\right] 4\pi r^2 \mathrm{d}r \tag{1.2.15}$$

This shows that, for attractive solute–solute interactions $V < 0$, the second virial coefficient B_2 is negative, while positive values of B_2 correspond to net repulsive interactions between solute molecules.

A useful way to characterize nonideal solutions is to introduce an effective concentration called the *osmolarity*, c_{osm}, formally defined by inverting the van't Hoff equation so that

$$c_{\text{osm}} = \frac{\left(\mu_w^\theta - \mu_w\right)}{\boldsymbol{RT}V_w} \tag{1.2.16}$$

Physiological saline corresponds to a 0.150 M solution of NaCl, whose osmolarity can be experimentally determined to be 0.285 M. If sodium and chloride ions would mix ideally, the osmolarity would be 0.300 M (0.150 M of each species), but due to ion–ion interactions, it is somewhat reduced compared to the ideal value.

We will encounter situations where an aqueous solution is in contact with another phase containing water. At a gas–liquid interface, there is an exchange of water molecules between the solution and the gas phase, usually air. The water chemical potential in the gas can be conveniently expressed through the *relative humidity*, *RH*, which is the ratio between the partial vapor pressure of water, $\mathbf{p}_\text{w}$, and its saturation pressure, $\mathbf{p}_\text{w}^{\text{eq}}$:

$$\mu_w = \mu_w^\theta = \boldsymbol{RT}\ln\frac{\mathbf{p}_\text{w}}{\mathbf{p}_\text{w}^{\text{eq}}} = \mu_w^\theta + \boldsymbol{RT}\ln RH \tag{1.2.17}$$

A comparison with Eq. (1.2.11) shows that the relative humidity equals the water activity in the gas phase.

To summarize, there are several different ways of expressing the chemical potential of water, as shown in Table 1.1. Figure 1.4 illustrates the values of osmolarity in different environments of practical relevance.

Table 1.1 Summary of relations between different measures of the chemical potential of water.

$\mu_{H_2O} \equiv \mu^0_{H_2O} + k_B T \ln(a_w)$ (per molecule)
$\mu_{H_2O} \equiv \mu^0_{H_2O} + \boldsymbol{R}\mathbf{T} \ln(a_w)$ (per mole)
$RH \equiv (p_w/p_{sat}) = a_w$
$\mu_{H_2O} - \mu^0_{H_2O} = -N_{Av} V_w \cdot 10^3 \boldsymbol{R}\mathbf{T} c_{osm}$ (per mole)
$a_w = \exp(-N_{Av} V_w \cdot 10^3 c_{osm})$
$\Pi_{osm} = 10^3 \boldsymbol{R}\mathbf{T} c_{osm}$
$\mu_{H_2O}(T) \equiv \mu^0_{ice}(T)$, $(T < 273\ K)$
$c_{osm}(T) \approx 0.55 \times (273 - T) - 6 \times 10^{-4}(273 - T)^2$, $(250\ K < T < 273\ K$ (from values of water vapor pressure)

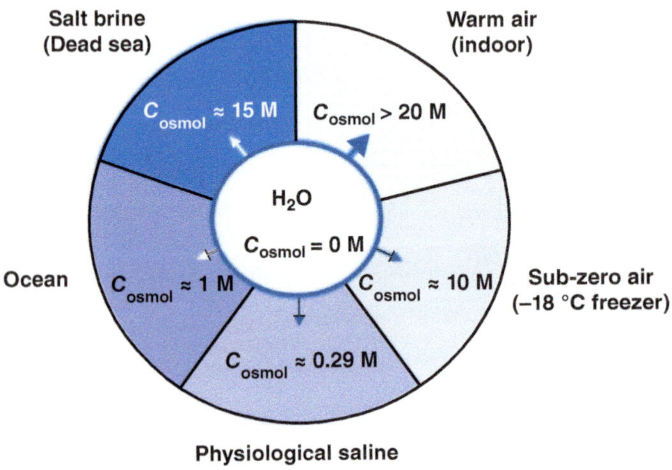

Figure 1.4 The water chemical potential in different environments expressed in terms of the osmolarity. *Source*: Adapted from Wennerström and Oliveberg, 2022.

1.2.4 The Chemical Potential Enters into Many Thermodynamic Equalities

For further reference, we now state how the chemical potential enters into some central thermodynamic relations. Textbooks in physical chemistry contain more complete discussions. Temperature $\mathbf{T}$, pressure $\mathbf{p}$, and the amount, n_i, of the components, i, are the natural independent variables for the Gibbs free energy, and by formal differentiation

$$dG = \left(\frac{dG}{d\mathbf{p}}\right)_{\mathbf{T},n_i} d\mathbf{p} + \left(\frac{dG}{d\mathbf{T}}\right)_{\mathbf{p},n_i} d\mathbf{T} + \sum_i \left(\frac{dG}{dn_i}\right)_{\mathbf{T},\mathbf{p},n_i} dn_i \tag{1.2.18}$$

where the thermodynamic relations provide the expression for the partial derivatives and

$$dG = V d\mathbf{p} - \mathbf{S} d\mathbf{T} + \sum_i \mu_i dn_i \tag{1.2.19}$$

Since the chemical potentials are intensive, i.e., independent of system size, the quantities defined in Eq. (1.2.19) can be integrated at a constant **T** and **p** to yield

$$\mathbf{G} = \sum n_i \mu_i \tag{1.2.20}$$

Problem: Show explicitly that the chemical potentials of Eq. (1.2.7) for the regular solution model obey Eq. (1.2.20).

Solution: $\mu_i = \mu_i^\theta + \boldsymbol{RT}\ln X_i + (1 - X_i)^2 w$ so that

$$\begin{aligned} n_A\mu_A + n_B\mu_B &= n_A\mu_A^\theta + n_A\mu_A^\theta + n_A\boldsymbol{RT}\ln\ X_A + n_B\boldsymbol{RT}\ln\ X_B + \left(n_A X_B^2 + n_B X_A^2\right)w \\ &= \mathbf{G}^\theta + \boldsymbol{RT}(n_A\ln\ X_A + n_B\ln\ X_B) + (n_A + n_B)X_A X_B w \\ &= \mathbf{G}^\theta + \Delta\mathbf{G}_{\text{mix}}\ (\text{regular solution}) \end{aligned}$$

since

$$n_A X_B^2 + n_B X_A^2 = \frac{n_A n_B^2 + n_B n_A^2}{(n_A + n_B)^2} = \frac{n_A n_B(n_A + n_B)}{(n_A + n_B)^2} = (n_A + n_B) X_A X_B$$

Using the defining relation $\mathbf{G} = \mathbf{U} + \mathbf{pV} - \mathbf{TS}$ between the internal and Gibbs energies, we find

$$\mathbf{U} = \mathbf{TS} - \mathbf{pV} + \sum_i n_i \mu_i \tag{1.2.21}$$

We can also formally differentiate Eq. (1.2.20) and arrive at the Gibbs–Duhem equation at constant **T** and **p**

$$\sum_i n_i\, d\mu_i = 0 \tag{1.2.22}$$

This equation shows that the chemical potentials for a two-component system cannot be varied independently. A prescribed change $d\mu_1$ of the chemical potential of component 1 necessarily changes the chemical potential of component 2 by $-(n_2/n_1)d\mu_2$.

1.2.5 Chemical Potentials Can Be Generalized to Include the Effect of External Fields

So far, we have considered the chemical potential in terms of a system's intermolecular interactions expressed, for example, in terms of w and the entropy of mixing. Clearly, external fields can also exert forces on molecules that affect their distribution in a system. Two such external forces that are important in colloidal systems are gravitational and electric fields.

Suppose we prepare an aqueous suspension containing Fe_2O_3 particles and examine it first on a space shuttle and again when it returns to Earth. In space, the distribution of particles is uniform and random, but on Earth gravitational forces generate a vertical concentration gradient of particles; that is, they create a more ordered system. We can use the concept of chemical potential to derive the concentration profile of the suspension. An external potential acts to give an additional energy, $\mathbf{E}_{\text{ext}}^i(\mathbf{r})$, to a molecule i at position $\mathbf{r}$. Consider a volume element containing N_i' molecules of component i. If it is small enough that the variation in the external potential over the volume element is negligible, then its free energy, $\mathbf{G}$, is

$$\mathbf{G}(\mathbf{r}) = \mathbf{G}_n'(\{X_i'\}) + \sum_i N_i' \mathbf{E}_{\text{ext}}^i(\mathbf{r}) \tag{1.2.23}$$

where $\mathbf{G}_n'$ is the free energy of the corresponding system in the absence of the external field and at a composition described by the mole fractions X_i'. We obtain the chemical potential of component j by differentiating with respect to N_j'

$$\mu_j(\mathbf{r}) = \mu_j^n(\{X_i'\}) + \mathbf{E}_{\text{ext}}^j(\mathbf{r}) \tag{1.2.24}$$

as a sum of one term derived from the bulk liquid and one due to the external field.

Through Brownian motion, molecules seek out the concentration profile that minimizes the free energy. This occurs when the chemical potential is uniform throughout the system, i.e., when $\mu_j^n(\{X_i(\mathbf{r})\}) + \mathbf{E}_{\text{ext}}^j(\mathbf{r})=$ constant. In the case of an ideal bulk solution,

$$\mu_j^n(\{X_i\}) + E_{\text{ext}}^j(\boldsymbol{r}) = \mu_j^\theta + k_\text{B}\mathbf{T}\ln X_j(\mathbf{r}) + \mathbf{E}_{\text{ext}}^j(\mathbf{r}) = \text{constant} \tag{1.2.25}$$

and the condition of constant chemical potential results in a concentration profile

$$X_j(\boldsymbol{r}) = \text{constant} \times \exp\left(-\frac{\mathbf{E}_{\text{ext}}^j(\mathbf{r})}{k_\text{B}\mathbf{T}}\right) \tag{1.2.26}$$

which varies exponentially in the external potential. This form of the Boltzmann distribution describes how the interplay between entropy of mixing (as expressed by $k_\text{B}\mathbf{T}$) and the external field generates a concentration variation in the system at equilibrium. In a constant gravitational field in the x direction, we have

$$\mathbf{E}_{\text{ext}}^i = m^i g x \tag{1.2.27}$$

where m^i is the mass of molecule i and g is the standard acceleration of gravity. For a gas where X^i is proportional to the partial pressure p_i

$$p_i(x) = p_i(x=0) \times \exp\left(-\frac{m^i g x}{k_\text{B} T}\right) \tag{1.2.28}$$

This equation is often called the barometric formula. When we apply the same argument to a liquid, a complication arises. Due to the (near) incompressibility of liquids, we cannot move a liquid molecule from one position to another without moving another molecule in the opposite direction. In the simple case of perfect incompressibility with the total volume given as a sum of concentration-independent molecular volumes, we can take care of the interdependence by writing

$$\mathbf{E}_{\text{ext}}^i = m_{\text{eff}}^i g x \tag{1.2.29}$$

where m_{eff} is the effective mass corrected for the buoyancy effect of the medium.

Problem: Calculate the effective mass for both components in an equimolar mixture of CH_3OH and CD_3OH.

Solution: In effect, this is an ideal solution in which both components have the same molecular volumes. The masses differ by 3.0 unit masses $u = 1.66 \times 10^{-27}$ kg. CH_3OH will float so that, relative to the average composition, it has $m_{\text{eff}} = -0.5 \times 3.0 \times 1.66 \times 10^{-27}$ kg = -2.5×10^{-27} kg while for CD_3OH, $m_{\text{eff}} = 2.5 \times 10^{-27}$ kg.

The ultracentrifuge was probably the first advanced instrument specifically designed to study colloidal systems. In this case, the gravitational potential is

$$\mathbf{E}^i_{\text{ext}} = m^i_{\text{eff}}\,\omega^2 r^2 \qquad (1.2.30)$$

where ω is the rotational velocity and r is the distance from the rotational axis. To conduct accurate studies, one must include effects due to concentration-dependent partial molal volumes in calculating the effective mass.

Electrical fields play an important role in colloidal systems. They influence all charged species, and for an ion j of charge $z_j e$ at position $\mathbf{r}$, the external interaction energy is

$$E^j_{\text{ext}}(\mathbf{r}) = -z_j e\,\Phi(\mathbf{r}) \qquad (1.2.31)$$

where $e = 1.602 \times 10^{-19}$ C is the elementary charge, z_j the ion valency ($z_j = +1$ for Na^+ and $z_j = -1$ for Cl^-), and Φ is the electrostatic potential at position $\mathbf{r}$.

Assuming ideal mixing in the bulk, the chemical potential is

$$\mu_j(\mathbf{r}) = \mu_j^\theta + k_B T \ln c_j(\mathbf{r}) + z_j e\Phi(\mathbf{r}) \qquad (1.2.32)$$

In many textbooks, μ_j as given in Eq. (1.2.32) is called the electrochemical potential. Such a distinction between the electrochemical and the chemical potential indicates that electrostatic interactions provide a very special contribution to the free energy. This point of view is not compatible with the spirit of this book, and we will therefore avoid using the concept of an electrochemical potential.

Sections 1.2.1 through 1.2.5 have introduced and developed the abstract thermodynamic concept of the chemical potential. We will use it extensively in this book and hopefully demonstrate that this concept is a gateway for understanding the thermodynamics of multicomponent systems. Using the chemical potential as a key concept, we can unify the description of a range of phenomena occurring in the colloidal domain.

See Atkins et al. (2022, focus 5), Doi (2013, Chapter 2), and Israelachvili (2011, Chapter 2) for additional discussion over Solutions.

1.3 Amphiphilic Self-assembly Processes Are Spontaneous, Are Characterized by Start–Stop Features, and Produce Aggregates with Well-defined Properties

In the regular solution model, one considers the interaction between two molecules and how this dictates the thermodynamic properties of the macroscopic mixture. The rich behavior of colloidal systems typically has its source in interactions involving many molecules that act in a cooperative manner. Surfactant molecules provide a clear illustration of such effects. Surfactants have countless practical applications as key components in, for example, soaps, detergents, and shampoos. Their basic role in these products is to increase the effective solubility of apolar compounds like fat. This process is accomplished since the surfactant molecules self-assemble into finite aggregates called *micelles*. The aggregation process is driven by the *amphiphilic* character of the surfactant molecules, i.e., they contain both polar (water-soluble) and apolar (oil-soluble) parts. Amphiphilic molecules, in fact, spontaneously self-organize into a variety of structures, where the simplest and most well-understood is the spherical micelle. To characterize the amphiphilic aggregation process, we begin by considering how adding a surfactant to water leads to the formation of micelles.

One example of a dual-character molecule possessing the well-defined polar head and nonpolar tail needed to produce amphiphilic behavior is SDS, whose structure is given in Table 1.2. In aqueous solutions up to 8×10^{-3} M, most of the properties of SDS are similar to those we can observe for a typical electrolyte such as NaCl. Figure 1.5 illustrates an exception: decreasing surface tension, denoted as γ.

The insolubility of SDS's hydrocarbon chain in water causes the molecule to concentrate at the air–water interface, with its hydrocarbon chain oriented toward the vapor phase. Since the surface tension of water is higher than that of hydrocarbons, adding surfactant results in a decrease in γ. We will analyze the concentration dependence of the surface tension in more detail in Chapter 2. Here, we note that above 8×10^{-3} M, the surface tension becomes almost constant. As we add more surfactant to the solution, the surface tension, and thus the surface, remains virtually unchanged. So what is happening?

One possibility is analogous to the phase separation that occurs when water is saturated with nonpolar or apolar compounds such as benzene or dodecanol. The same force works to diminish unfavorable hydrocarbon–water interactions in SDS, but in this case, the formation of two bulk phases would also produce an energetically unfavorable result because it would remove the polar ionic head groups from the water. What actually happens, therefore, is a compromise between the extremes of a complete phase separation and a molecularly dispersed solution: the surfactant molecules self-assemble to create a "microphase" in which the hydrocarbon chains sequester themselves inside the aggregate and the polar head groups orient themselves toward the aqueous phase, as illustrated in Figure 1.6. In the case of SDS, groups of approximately 60 dodecyl sulfate molecules self-assemble in this way to form spherical micelles. We denote the minimal surfactant concentration at which we can discern this micellization process, the CMC.

The occurrence of a CMC is the result of two competing factors. Transfer of the hydrocarbon chain from water into the oil-like interior of the micelle drives micellization, while repulsion between head groups as they come close together opposes it. Under these conditions, the growth process has specific limits, and aggregates cease to grow when they have reached a certain size. Above the CMC, adding more surfactant simply produces more micelles rather than the growth of existing micelles.

The way SDS behaves when added to water provides a specific example of the self-organizing capabilities of amphiphilic molecules, but we can generalize the characteristics of this process in the following three points:

1. Aggregates form spontaneously.
2. Aggregation is a start–stop process.
3. Aggregates have well-defined properties.

Self-assembly must be understood as a physicochemical process: amphiphilic molecules are associated physically, not chemically, and for this reason, they can change the size or shape of their microstructure in response to small changes in concentration, salt content, temperature, pH, and pressure. However, they still retain their start–stop characteristics, and the size of the molecule limits the aggregate's growth in at least one dimension.

Table 1.2 Structures of typical amphiphilic compounds.

Types and class	Structure	Amphiphile[a]	Symbol	CMC M[b]
Ionic	$C_8H_{17}OSO_3Na$	Sodium octylsulfate	C_8SO_4Na	1.33×10^{-1}
(sodium alkylsulfate)	$C_{10}H_{21}OSO_3Na$	Sodium decylsulfate	$C_{10}SO_4Na$	3.3×10^{-2}
	$C_{12}H_{25}OSO_3Na$	Sodium dodecylsulfate	$C_{12}SO_4Na$ (SDS)	8.1×10^{-3}
(sodium alkylcarboxylate)	$C_{12}H_{25}CO_2Na$	Sodium dodecanoate	$C_{11}CO_2Na$	2.5×10^{-3}
Sodium 1,4-dialkyl-3-sulfonate-succinate	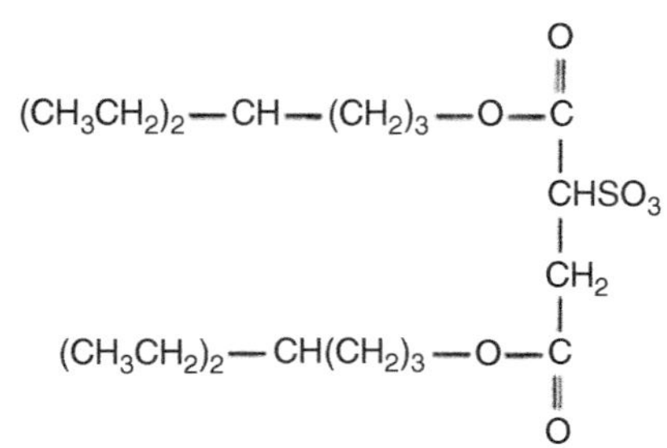	Aerosol OT	AOT	
Alkyltrimethyl-ammonium bromide	$C_{12}H_{25}N(Me)_3Br$	Dodecyl trimethyl-ammonium bromide	$C_{12}N(Me)_3Br$	1.5×10^{-2}
	$C_{16}H_{32}N(Me)_3Br$	Hexadecyl trimethyl-ammonium bromide	$C_{16}N(Me)_3Br$ (CTAB)	9.2×10^{-4}
Dialkyldimethyl-ammonium bromide	$(C_{12}H_{25})_2N(Me)_2Br$	Didodecyldimethyl-ammonium bromide	$(C_{12})_2N(Me)_2Br$ (DDAB)	
Dialkyldimethyl-ammonium acetate	$(C_{16}H_{33})_2N(Me)_2CH_3CO_2$	Dihexadecyldimethy1-ammonium acetate	$(C_{16})_2N(Me)_2OAc$ (DHDAA)	

(Continued)

Table 1.2 (Continued)

Types and class	Structure	Amphiphile[a]	Symbol	CMC M[b]
Zwitterionic				
Dialkylphosphatidylcholine	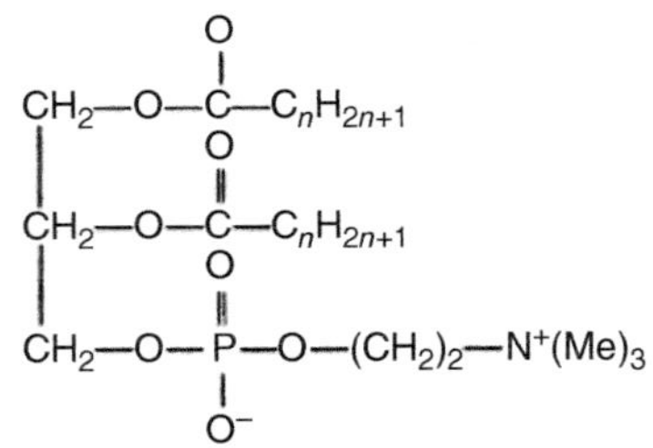	Dimyristoyl phosphatidylcholine Dipalmitoyl phosphatidylcholine Distearoyl phosphatidylcholine	DMPC ($n = 14$) DPPC ($n = 16$) DSPC ($n = 18$)	$<10^{-9}$
Alkyldimethylpropane sultaine	$C_{12}H_{25}N(CH_3)_2(CH_2)_3SO_3$	Dodecanedimethyl-propanesultaine	$C_{12}N(Me)_2(CH_2)_3SO_3$	3.6×10^{-3}(30° C)
Nonionic				
Alkylethylene glycol	$C_{10}H_{21}(OCH_2CH_2)_8OH$		$C_{10}E_8$	1.0×10^{-3}
	$C_{12}H_{25}(OCH_2CH_2)_6OH$		$C_{12}E_6$	8.7×10^{-5}
Alkyldimethylamine oxides	$C_{12}H_{23}(CH_2)NO$	*Dodecyldimethylamine* oxide	$C_{12}(Me)_2NO$	1.6×10^{-3}
Alkylglucoside	$C_{10}H_{21}C_6O_6H_{11}$	β-d-decylglucoside		2.2×10^{-3}

a. The naming of organic compounds is often a source of considerable frustration to the less informed. Two or three names are used interchangeably for many common alkyl chains. For example, the formal IUPAC (International Union of Pure and Applied Chemistry) name for the carboxylic acid, $C_5H_{11}COOH$, is hexanoic acid, while the common name is caporic acid because its rancid odor suggested the smell of goats to early chemists. For the $C_{16}H_{32}$ alkyl chain, the IUPAC name is hexadecyl. However, surface chemists often use "cetyl," while lipid chemists use "palmitoyl." Throughout this book, we will use the symbols given in column 4. Table 6.1 and Figure 6.2 in Chapter 6 give additional information on the names and structures of lipids.

b. Temperature 25°C unless otherwise noted.

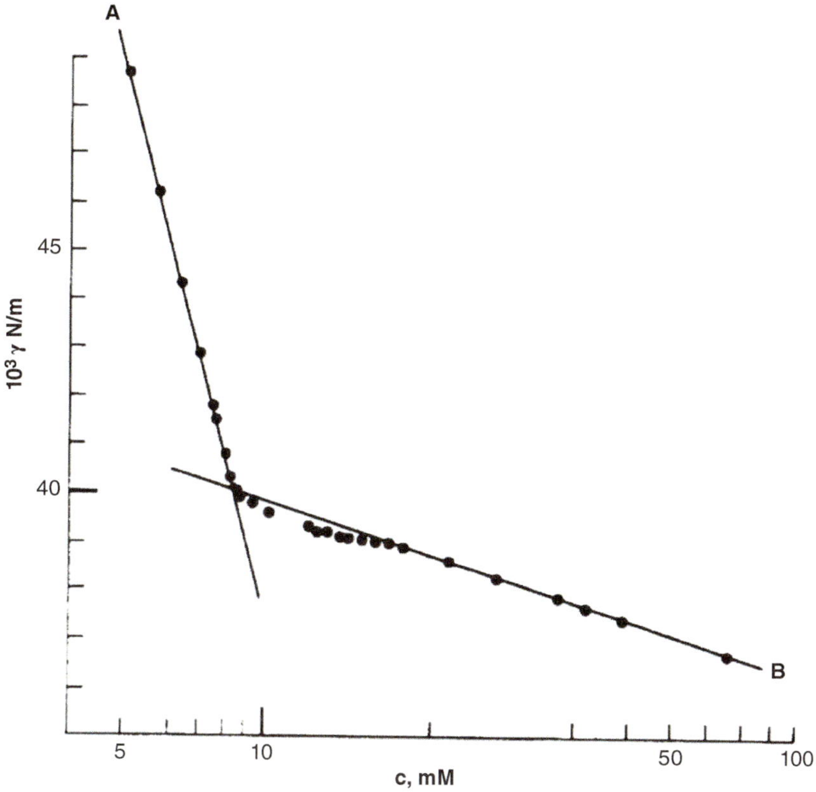

Figure 1.5 Solution surface tension versus SDS concentration. As SDS concentration increases, the surface tension γ of the solution initially decreases and then becomes almost constant. (Note the logarithmic scale.) The CMC is determined as the intersection of two linearly extrapolated lines, **A** and **B**.

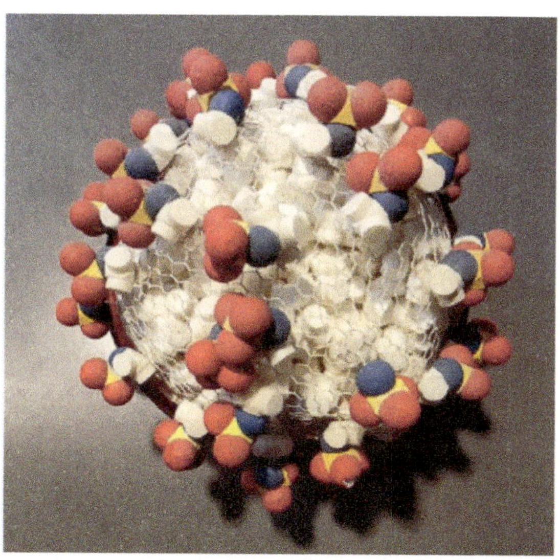

Figure 1.6 Space-filling model of an ionic spherical micelle. Photo by Christopher Ward.

1.4 Amphiphilic Molecules Are Liquid-like in Self-assembled Aggregates

As we have seen, amphiphilic behavior requires molecules that possess separate, distinctly polar and nonpolar regions. They should not be readily soluble as individual molecules in solvents that are either strongly polar or strongly apolar. To achieve this property, the nonpolar tail of the typical molecule is a hydrocarbon or a fluorocarbon with at least eight carbons. The polar head group can be ionic, zwitterionic, or dipolar. Table 1.2 lists typical amphiphiles along with the abbreviations we will use to designate them in the text.

Organic molecules with functional groups that are not polar enough to serve as effective amphiphilic head groups undergo phase separation instead of forming micelles or other self-associated structures in water. Examples include alcohols, ketones, aldehydes, and simple ethers.

Temperature plays an important role in the behavior of amphiphilic molecules. For example, the results are quite different when we measure the surface tension of SDS below or above 23 °C. At lower temperatures, the surfactant precipitates out of solution as a hydrated crystal instead of forming micelles. In other words, at 18 °C, its solubility limit lies below the CMC, as illustrated in Figure 1.7.

The point at which solubility equals the CMC is called the **Krafft temperature, $\mathbf{T}_{Kr}$**. Just above this point the solubility increases rapidly, and a solution of almost any composition becomes a single homogeneous phase. Surfactant solutions cooled below their $\mathbf{T}_{Kr}$ can remain in a metastable homogeneous state for days.

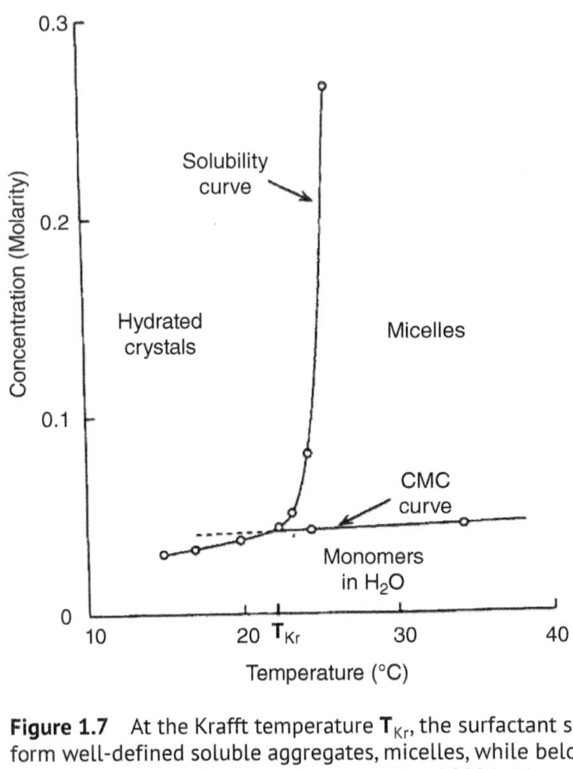

Figure 1.7 At the Krafft temperature $\mathbf{T}_{Kr}$, the surfactant solubility equals the CMC. Above $\mathbf{T}_{Kr}$, surfactants form well-defined soluble aggregates, micelles, while below $\mathbf{T}_{Kr}$, they crystallize out of solution as hydrated crystals. *Source*: Lindman and Wennerström, 1980/with permission of Springer Nature.

The Krafft phenomenon reflects an equilibrium between surfactants in solution and in hydrated crystals. A familiar complication is the "ring around the tub" caused by precipitation of Ca^{2+} and Mg^{2+} soaps in hard water. The Krafft temperature is much higher for divalent than for monovalent counterions, and it is also sensitive to small changes in the chemical structure of the surfactant because of the rather stringent packing requirements of the molecules in the solid (crystalline) phase. Saturated crystalline hydrocarbons display an all-trans configuration shown in Figure 1.8.

Table 1.3 summarizes the thermodynamics for the melting of crystalline hydrocarbons. These values illustrate packing differences between molecules with odd and even numbers of carbons in the solid state. Disturbance of the packing caused by a *cis* (see Figure 1.8), and to a lesser extent by a *trans*, double bond weakens the stability of the ordered phase, thus decreasing both the chain melting temperature and $\mathbf{T}_{\mathrm{Kr}}$. Strong head group interactions, on the other hand, contribute substantially to a crystal's stability. Thus, divalent counterions stabilize the crystals of carboxylates, resulting in high $\mathbf{T}_{\mathrm{Kr}}$ and precipitation at ambient temperatures. For sulfate and sulfonate head groups with their more delocalized charge, this difference between monovalent and divalent counterions is less pronounced.

As hydrocarbons melt, they may take many conformations. Rotation around the individual C—C bonds results in either retention of the *trans* configuration or a change into one of the two possible *gauche* states that are 3 to 4 × 10^{-21} J higher in energy than the *trans* state. This energy is very close to $k_B\mathbf{T} = 4.1 \times 10^{-21}$ J at room temperature, and a balance between energetic and entropic factors determines the overall conformation of the chain. This raises questions about the state of the alkyl chains in micelles. Are they solid-like or liquid-like? Are they coiled or straight?

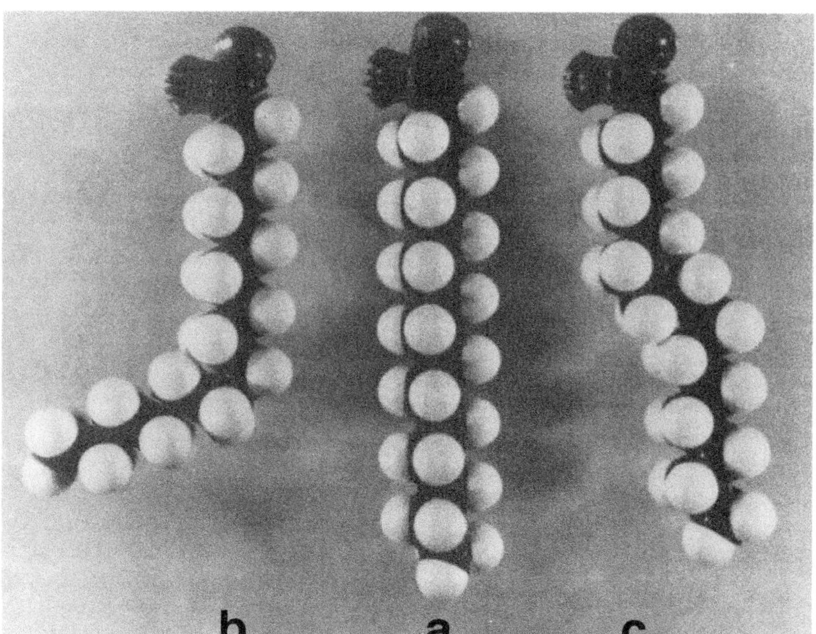

Figure 1.8 Configurations of hydrocarbon chains. Space-filling models of hydrocarbon chains displaying (a) an all-trans configuration as it would exist in the crystalline state, (b) one gauche orientation in an otherwise all-trans chain, and (c) two gauche orientations resulting in a kink in the chain. *Source*: Tanford, 1980/with permission of John Wiley & Sons.

Table 1.3 Thermodynamics of hydrocarbon chain melting.[a]

Structure	Melting point °C	ΔH^{0} kJ/mol	ΔS^{0} J/(K · mol)	ΔG^{0} at 25 °C kJ/mol
$C_{13}H_{28}$	−5.4	28.5	106	−3.2
$C_{14}H_{30}$	5.9	45	162	−3.1
$C_{15}H_{32}$	10.0	35	122	−1.8
$C_{16}H_{34}$	18.2	54	183	−1.3
$C_{17}H_{36}$	22.0	41	137	−0.4
$C_{18}H_{38}$	28.2	62	206	+0.7
$C_{19}H_{40}$	32.0	50	151	+1.0
$C_{20}H_{42}$	36.7	70	227	+2.6

a. Data are for saturated *n*-paraffins based on compilations by the American Petroleum Institute (1953). Values of ΔH^{0} and ΔS^{0} refer to the melting point. Values of ΔG^{0} at 25°C are calculated assuming constant ΔH^{0} and ΔS^{0} between 25°C and the melting point. Crystals formed by unsaturated hydrocarbons melt at much lower temperatures.

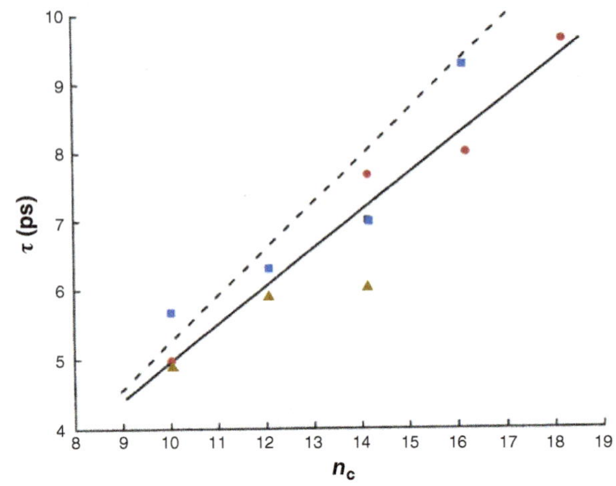

Figure 1.9 Micelle interiors show liquid phase properties. Nuclear magnetic resonance data showing characteristic time τ (ps) for rotation as a function of hydrocarbon chain length n_c for the micellar surfactants: ●, $C_nH_{2n+1}N(CH_3)_3Cl$; ■, $C_nH_{2n+1}N(CH_3)_3Br$, ▲, $C_nH_{2n+1}SO_4Na$; and hydrocarbons, $C_nH_{2n+1}C_6H_5$. The micellar times (solid line) are on the same scale as those in anhydrous alkylbenzene solution (dashed line), demonstrating the liquid-like properties of the micelle's hydrocarbon interior. If the micellar hydrocarbons were in the solid phase, the characteristic time would be many orders of magnitude longer. *Source*: Stilbs et al., 1983/with permission of American Chemical Society.

One way to determine the state of hydrocarbon chains in the pure phase and in the interior of surfactant aggregates is to track the rate of motion of the hydrocarbon molecules themselves or of dissolved molecules. In a medium of low viscosity, diffusional rotation of small molecules occurs with a characteristic time on the order of 10^{-11} s. This time defines a relevant time scale for a spectroscopic measurement. Several spectroscopic techniques for accessing dynamics on these timescales are available. Some, like fluorescence depolarization and electron spin resonance

(ESR), require the addition of a suitable probe molecule. Others like nuclear magnetic resonance (NMR) can measure hydrocarbon chain motion directly.

As Figure 1.9 illustrates, relaxation measurements using ^{13}C and ^{2}H NMR (with specifically deuterated chains) find that, to a good approximation, the local chain motion in a micelle is as rapid as in a corresponding hydrocarbon. These direct measurements definitively confirm an inference already made by G.S. Hartley in the 1930s that micelles have a liquid-like interior—that is, the molecular motion in the interior is as rapid as in a bulk liquid.

1.5 Surfactant Numbers Provide Useful Guides for Predicting Aggregate Structures

So far we have limited our discussion of amphiphilic self-assembly to spherical micelles. However, a variety of aggregates can actually form. Possible structures are restricted by an amphiphile's need to keep its nonpolar and polar parts surrounded in a favorable way. To make *a priori* predictions about the size and shape of amphiphilic aggregates, we need a conceptual framework that relates molecular parameters (such as hydrophobic volume, chain length, and head group area) and intensive variables (such as temperature and ionic strength) to aggregate structure. Such theories seek to determine the optimal aggregate or aggregates for a given set of molecular parameters and intensive variables—sometimes called *field variables*—through free energy minimization arguments.

As in the example of SDS micellization discussed in Section 1.3, we can assume that the free energy of surfactant self-assembly in dilute solution is made up of three terms:

- A favorable **hydrophobic contribution**, due to the hydrocarbon chains sequestering themselves within the interior of the aggregates.
- A **surface term** that reflects the opposing tendencies of the surfactant head groups to crowd close together to minimize hydrocarbon–water contacts and to spread apart, as a result of electrostatic repulsion, hydration, and steric hindrance.
- A **packing term** which, at its simplest level, requires that the hydrophobic interior of the aggregate excludes water and head groups, thus limiting the geometrically accessible forms available to the aggregate.

Surface and packing terms take on different functional forms for each specific aggregate geometry. The form giving the minimum free energy for a given set of conditions determines the optimal aggregate.

For dilute solutions in which interactions between aggregates are not important, we can conveniently subsume these ideas into the **surfactant number**

$$N_s = \frac{v}{la_0} \tag{1.5.1}$$

where v stands for the volume of the hydrophobic portion of the surfactant molecule, l is the length of the hydrocarbon chain, and a_0 is the effective area per head group.

The volume of the hydrocarbon core of a saturated hydrocarbon chain, expressed in cubic nanometers, is

$$v = 0.027(n_c + n_{Me}) \tag{1.5.2}$$

where n_c is the total number of carbon atoms per chain and n_{Me} is the number of methyl groups, which are twice the size of a CH_2 group. For amphiphilic structures with closed hydrocarbon cores, such as micelles and vesicles, the total hydrocarbon volume is given by Nv (for single-chain surfactants) and $2Nv$ (for double-chain surfactants), where N is the number of amphiphiles contained in the aggregate.

We can estimate the length (in nanometers) of a fully extended hydrocarbon chain from

$$l = 0.15 + 0.127n_c \tag{1.5.3}$$

where the first term (0.15 nm) comes from the van der Waals radius of the terminal methyl group (0.21 nm) minus half the bond length of the first atom not contained in the hydrocarbon core (0.06 nm). The second term (0.127 nm) is the carbon–carbon bond length (0.154 nm) projected onto the direction of the chain in the all-*trans* configuration.

Micelles can reach $r_{\text{micelle}} \approx l$ without a major free energy cost because only a few of the hydrocarbon chains need to extend the length of the radius. For bilayers, the average length of a chain is half the bilayer thickness, which seldom exceeds 1.8 l for chains in the liquid state.

The most problematic quantity in the definition of the surfactant number, N_s, is the area per polar group, a_0. For ionic surfactants, a_0 depends on both electrolyte and surfactant concentration, and the surfactant number has only limited usefulness for quantitative descriptions. For nonionic or zwitterionic surfactants, a_0 is less sensitive to external conditions, but it is still difficult to make an *a priori* calculation of its value as we did for v and l. However, in all cases, we can predict trends in a_0 for a specific change in the system variables, and this ability can be extremely useful for qualitative interpretation of experimental data.

Surfactant numbers relate the properties of the molecule to the preferred curvature properties of the aggregates. Small values of N_s imply highly curved aggregates, but when N_s is close to unity, planar bilayers usually form. For spheres, cylinders, and bilayers, a simple relation exists between volume, area, and radius (thickness), and a comparison of the surfactant numbers shows that optimal stability of the different aggregates occurs as follows:

$$\text{Spherical micelles } N_s = 0.33 \tag{1.5.4}$$

$$\text{Infinite cylinders } N_s = 0.5 \tag{1.5.5}$$

$$\text{Planar bilayers } N_s = 1 \tag{1.5.6}$$

$$\text{Inverted cylinders and micelles } N_s > 1 \tag{1.5.7}$$

Various amphiphilic structures are illustrated in Figure 1.10 and they will be further discussed in later chapters.

Several things can happen when the value of N_s lies between 0.33 and 0.5 or between 0.5 and 1. The molecules may assemble into a highly symmetrical aggregates that are slightly off the optimal conditions or into aggregates of lower symmetry, or they may undergo phase separation. All these variations exist in experimental systems, and one of the challenges in research on surfactant systems is to determine when and why each of these outcomes occurs.

Obviously, surfactant numbers can represent only an approximate model of the nature of surfactant self-assembly. In spite of this, the model provides valuable insight into the way that changes in solution conditions and molecular structure affect an aggregate's size and shape. For example, surfactant numbers enable us to predict the effect of adding a second hydrophobic chain to a polar head group. Because amphiphiles with single hydrophobic chains have surfactant numbers of less than 0.5, they are constrained—at least in dilute solution—to form micellar aggregates. Adding a second hydrocarbon chain doubles v, while a_0 and l remain essentially the same. Thus, the addition doubles $v/(a_0 l)$ into the range 0.5–1. Hence, double-chain surfactants tend to form bilayer structures such as vesicles and lamellar liquid crystals, whose curvatures are inherently lower than those formed by their single-chain counterparts.

A comparable, although quantitatively different, approach to analyzing aggregate structure is to explicitly use the concept of curvature. In this approach, the crucial parameter is not the surfactant number N_s but the preferred mean curvature of the surfactant film. We can define the *mean curvature*, H, at a point on a surface as

$$H = \frac{1}{2}\left(\frac{1}{R_1} + \frac{1}{R_2}\right) \tag{1.5.8}$$

where R_1 and R_2 are the radii of curvature in two perpendicular directions, as illustrated in Figure 1.11.

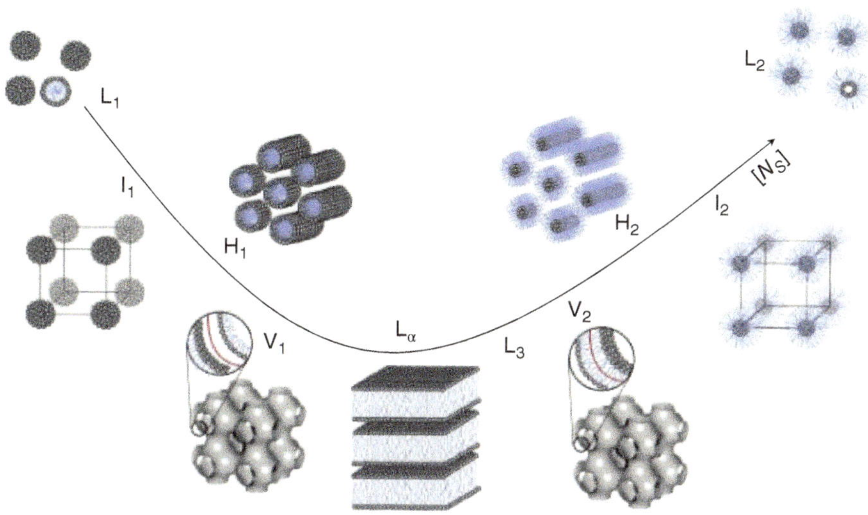

Figure 1.10 Illustration of how amphiphile self-assembly structures change as the surfactant number is increased. From left: spherical micelles (L1) $N_s \approx 0.33$, in which the micelle's radius $R_{mic} \approx l_{max}$; cylindrical micelles (H1) $N_s \approx 0.5$, in which $R_{mic} \approx l_{max}$; planar bilayers (Lα) $N_s \approx 1.0$, in which the bilayer thickness is $\approx 1.6\ l_{max}$ in the liquid state; cubic structures with N_s slightly larger (V2) or smaller (V1) than 1; inverted hexagonal phase (H2) $N_s \geq 1.3$ and reversed micelles (L2) $N_s \geq 1.5$. Courtesy of J Barauskas.

Figure 1.11 Radius of curvature. For a surface in three dimensions, two mutually perpendicular radii of curvature, R_1 and R_2, can be specified at each point. On a saddle-shaped surface, the two radii of curvature have opposite signs. Here R_1 and R_2 are shown at two different points on the surface.

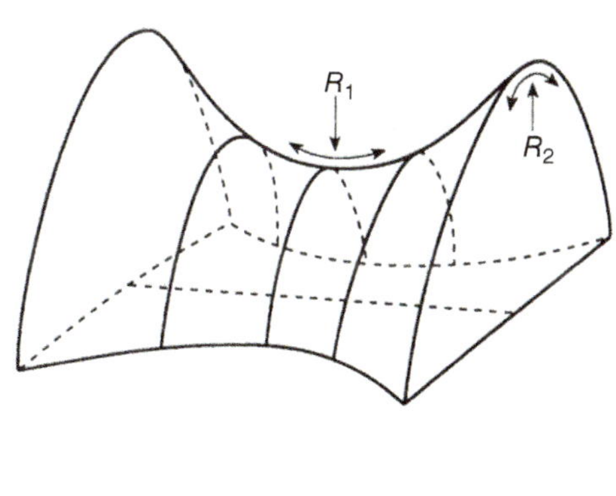

For a sphere, $R_1 = R_2 = R$ and $H = 1/R$; for a cylinder, $R_1 = R$, $R_2 = \infty$ and $H = 1/(2R)$; while for a planar bilayer, $H = 0$. A value of $H = 0$ can also occur on a saddle-shaped surface on which $R_1 = -R_2$. To assign a sign to the radii of curvature, we must define a normal direction, $\boldsymbol{n}$, for the surface. By convention, $\boldsymbol{n}$ is usually defined pointing toward the polar region, and, therefore, the curvature of inverted aggregates is negative.

See Berg (2010, Chapter 3), Cosgrove (2010, Chapter 4), Israelachvili (2011, Chapter 19), and Kronberg et al. (2014, Chapter 4) for additional discussion over Amphiphile Self-assembly.

1.6 Solvophobicity Drives Amphiphilic Aggregation

Amphiphilic aggregation results from the low solubility of either the amphiphile's hydrocarbon chain or its polar head group. This means that to promote the self-organization of amphiphilic molecules, the solvent must be either very nonpolar (like oil) or very polar (like water). Because water is widely used in industrial applications and biological processes, we will consider its solvent properties first.

Low solubility of the amphiphilic hydrocarbon chains drives aggregation in polar solvents like water. It arises because we try to introduce a weakly interacting species into a liquid of high cohesive energy. The Gordon parameter, which is defined as $\gamma / V_m^{1/3}$ (J/m^3), where γ is the surface tension and V_m the molar volume, is very useful for characterizing the cohesive energy of solvents and even allows us to consider ionic liquids such as ethylammonium nitrate (EAN). Figure 1.12 summarizes the free energy of transfer of various nonpolar gas molecules from the vapor phase to selected solvents. A roughly linear relationship between $\Delta \mathbf{G}^\theta$ and $\gamma / V_m^{1/3}$ appears in all of them

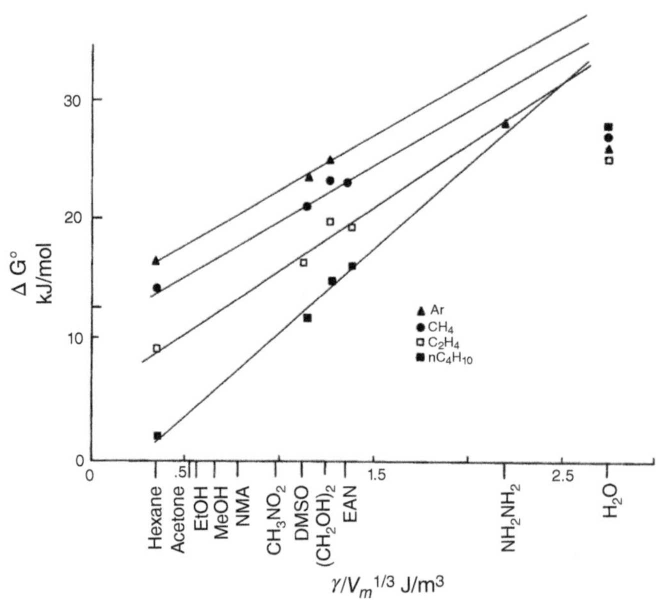

Figure 1.12 The free energy $\Delta \mathbf{G}^\theta$ of transferring nonpolar gases into solvents increases almost linearly with the Gordon parameter $\gamma / V_m^{1/3}$, which is a measure of the solvent cohesive energy. The $\Delta \mathbf{G}^\theta$ values for water, to the right in the figure, fall distinctly below the fitted lines and reflect water's unusual response to nonpolar groups. *Source*: Evans, Langmuir, 1988/with permission of American Chemical Society.

except water, whose solute order shows a different sequence than in other solvents. We will return to this point in Chapter 5.

Table 1.4 summarizes the physical properties of water, hydrazine, ethylammonium nitrate, formamide, and ethylene glycol, which are polar solvents in which micelle formation has been documented. For this series, the Gordon parameter decreases from 2.7 to 1.3 J/m^3. As a solvent's cohesive energy density decreases, the driving force for aggregation also decreases, causing the CMC to increase and become less well-defined. No evidence of amphiphilic aggregation exists for solvents with a Gordon parameter below 1.3 J/m^3.

We will finally briefly consider oil solvents. Here, insolubility drives the polar head groups to be oriented toward the micelle interior, while the nonpolar chains point out into the oily solvent. For the most part, these inverted structures form gradually, and we do not observe sharply defined CMCs. In the case of ionic surfactants, the concentration of a third polar component, such as water, governs the formation and size of inverted micelles.

The behavior of the partially fluorinated hydrocarbon $F(CF_2)_8(CH_2)_{12}H$ in the solvent perfluorotributylamine $[CF_3(CF_2)]_3N$ clearly illustrates the role of solvophobicity in driving amphiphilic aggregation. We can detect a Krafft temperature at 24 °C, and well-defined changes in NMR chemical shifts, light scattering, and fluorescence at 6 wt% $F(CF_2)_8(CH_2)_{12}H$ define a CMC consistent with the formation of small spherical micelles. Fluorocarbons and hydrocarbons are mutually insoluble. In a fluorinated solvent, a partially fluorinated hydrocarbon behaves as a dual-character molecule and aggregates to form a micelle with a hydrocarbon core containing fluorocarbon head groups.

These observations clearly establish that the insolubility of one part of the amphiphilic molecule drives the physicochemical process of amphiphilic aggregation. Such an aggregation can appear for many combinations of solutes and solvents. However, the most common case is that of apolar molecules or groups in water. In this case, the general term solvophobic is changed to *hydrophobic*, which is a central concept for understanding many physicochemical properties of aqueous systems. Unfortunately, as used in the scientific literature, hydrophobicity has meant different things to different people at different times. Therefore, a few distinctions between various aspects of this loose concept may be useful.

Hydrophobic hydration refers to the water structure around a single apolar solute in water. Many experimental and theoretical studies manifest the existence of such a hydration structure.

Table 1.4 Properties of polar solvents.

Polar solvents	ε_r[a]	γ (kN/m)	m(D)	$\gamma/V^{1/3}$ (J/m³)
Water	78.4	72.5	1.85	2.75
Hydrazine	51.7	66.4	1.86	2.10
Ethylammonium nitrate	∞	46.0	—	1.40
Formamide	109.0	57.2	3.73	1.50
Ethylene glycol	37.7	47.0	2.28	1.20
3-Methylsydnone	144.0	57.0	7.30	1.40

Source: From Beesley et al., 1988/with permission of American Chemical Society.

a. Here, ε_r is the relative dielectric permittivity, γ is the surface tension, m is the molecular dipole moment in Debye units, and $\gamma/V_m^{1/3}$ is the Gordon parameter for the cohesive energy density.

Hydrophobic interactions refer to the attractive forces between two or more apolar solutes in water. They simply manifest the insolubility of oil in water, whatever the molecular mechanism. The third term, *hydrophobic effect*, has more than one meaning. Sometimes it refers to the combined phenomena observed for aqueous solutions of apolar substances and sometimes it describes specific anomalous thermodynamic properties, particularly partial molar enthalpies, heat capacities, and volumes of apolar solutes in water.

Hydrophobic interactions occur as a general phenomenon when apolar substances are dissolved in water. However, hydrocarbon chains have particular importance in applications, especially in any discussion of surfactants. For this reason, this section concludes by summarizing some quantitative data on the thermodynamics of the dissolution of hydrocarbon compounds in water.

Table 1.5 summarizes data for the transfer of alkanes, aromatics, and alkanols from water to the neat liquid. The first column gives the difference in standard chemical potential, which we can view as a manifestation of the hydrophobic interaction. The remaining four columns show manifestations of the hydrophobic effect. The change in enthalpy is small. In fact, at 25 °C, it is so small that the chemical variations between the hydrocarbons result in a change of the sign. Having independently measured $\mu^\theta(l) - \mu^\theta(aq)$ and $\mathbf{H}^\theta(l) - \mathbf{H}^\theta(aq)$, the values of the third column follow from the relation

$$\mathbf{S}^\theta(l) - \mathbf{S}^\theta(aq) = \{\mathbf{H}^\theta(l) - \mathbf{H}^\theta(aq) - \mu^\theta(l) + \mu^\theta(aq)\}/\mathbf{T} \tag{1.6.1}$$

A comparison between the two last columns shows that the partial molal heat capacity is anomalously high in the aqueous solution. It can be more than a factor of three larger than in the neat liquid! At this stage, we accept the data given in Table 1.5 as established experimental thermodynamic facts and postpone our analysis of the underlying molecular causes until Chapter 5.

Table 1.5 Thermodynamic data for the transfer of small apolar compounds from a dilute aqueous solution (*aq*) to the neat liquid (*l*) at 25 °C.

Compound	$\mu^\theta(l)-\mu^\theta(aq)$ kJ/mol	$H^\theta(l)-H^\theta(aq)$ kJ/mol	$S^\theta(l)-S^\theta(aq)$ J/(mol · K)	$C_p^\theta(aq)$ J/(mol · K)	$C_p^\theta(l)$ J/(mol · K)
C_4H_{10}	−25.0	3.3	96	414	142
C_5H_{12}	−28.7	2.1	104	573	171
C_6H_{14}	−32.4	0.0	108	635	196
C_6H_6	−19.2	−2.1	59	359	134
$C_6H_5CH_3$	−22.6	−1.7	71	418	155
$C_6H_5C_2H_5$	−25.9	−2.0	79	501	184
$C_6H_5C_3H_8$	−28.8	−2.3	88	602	209
C_3H_7OH	−6.6	10.1	56	384	148
C_4H_9OH	−10.0	9.4	65	478	178
$C_5H_{11}OH$	−13.5	7.8	71	558	209

Source: Tanford, 1980/with permission of John Wiley & Sons.

Table 1.6 The standard free energy of transfer $\mu^{\theta}(l) - \mu^{\theta}(aq)$ of compounds $R(CH_2)_{n-1}CH_3$ from a dilute aqueous solution to the neat liquid.[a]

R	*A* (kJ/mol)	*B* (kJ/mol)
$-CH_3$	−10.2	3.70
$H_2C{=}CH-$	−6.3	3.70
$H(-CH{=}CH-)_2$	−3.1	3.60
−OH	3.5	3.43
COOH	17.8	3.45

Source: Tanford, 1980/with permission of John Wiley & Sons.

a. Values of the coefficients *A* and *B* in Eq. (1.6.2); concentrations in mol fraction units, and n_c is the total number of carbon atoms.

An important observation in Table 1.5 is that, for a homologous series of compounds like the alkanes, aromatics, and alkanols, the thermodynamic values vary linearly with the number of carbon atoms. This relationship indicates that a simple additivity rule applies so that we can extrapolate to molecules with longer chains, where direct measurements are difficult to achieve. Most importantly, the difference in standard chemical potential can be written as

$$\mu^{\theta}(l) - \mu^{\theta}(aq) = A - Bn_c \tag{1.6.2}$$

where n_c represents the number of carbon atoms. Table 1.6 summarizes the values of *A* and *B* for a series of apolar compounds. The constant *B* depends only weakly on the nature of the end group, *R*, showing that a substitution at the end of the chain does not significantly influence the chain-length dependence. In contrast, substitution strongly influences the constant term, *A*, and this term contains a large contribution from the free energy of solvation of the group, *R*.

See Israelachvili (2011, Chapter 8) and Kronberg et al. (2014, Chapter 3) for additional discussion on Solvophobicity.

1.7 Brownian Motion Gives Rise to Molecular Diffusion

1.7.1 The Diffusional Motion of Individual Molecules Can Be Analyzed in Terms of a Random Walk

The nineteenth-century British botanist Robert Brown observed in his microscope that pollen particles moved in a random manner in spite of the fact that the liquid medium was at rest. It was later realized that this "Brownian motion" not only occurred for particles of colloidal size but also for molecules in liquids and gases. The basic source of the motion is the thermal translational energy of a molecule, which is given by

$$E_{kin} = \frac{m\langle v^2\rangle}{2} = \frac{3k_B\mathbf{T}}{2} \tag{1.7.1}$$

Here m is the molecular mass and $\langle v^2\rangle$ the mean-square molecular speed. Molecules are in constant motion at a considerable velocity. For a water molecule at room temperature (298 K), the average speed is 642 m/s. Importantly, this value applies equally to a water molecule in a gas and in liquid water, since it depends only on the molecular mass and the temperature. This random molecular translational motion has two important related consequences. Consider a single tagged molecule in a gas or in a liquid, which at time $t = 0$ is at a position $\mathbf{r}(t = 0)$. Due to Brownian motion, it will move around and at some time $t = t_1$ later it will be at a position $\mathbf{r}(t_1)$. Since Brownian motion is a random process, we can take the average of a large number of such trajectories for various propagation times t and construct its probability distribution $f(r,t)$, where $r = |\mathbf{r}|$ is the distance traveled.

To find how the distribution function $f(r,t)$ varies in time, we make use of the mathematical concept of a *random walk*. In the same spirit that we used the lattice model to determine the ideal entropy of mixing, we can, for simplicity, use a two-dimensional lattice to describe the random walk. Each position is characterized by integer coordinates x and y. We start the walk at $x = y = 0$, and at each step either x or y changes by one unit. After N_p steps, the position is $x = i, y = j$, so that $r = \left(i^2 + j^2\right)^{1/2}$. For N_p steps on the lattice, the distribution $f(r, N_p)$ is found to be

$$f(r,N_p) = \frac{2r}{N_p}\exp\left(-\frac{r^2}{N_p}\right) \tag{1.7.2}$$

which encodes a Rayleigh distribution that becomes wider with N_p. The average distance traveled after N_p steps is given by

$$\langle r\rangle = \int_0^{\infty} rf(r,N_p)\,\mathrm{d}r = \frac{\sqrt{\pi}}{2}N_p^{1/2} \tag{1.7.3}$$

This shows that the distance traveled in a random walk grows as the square root of the number of steps. This result holds qualitatively both for one- and three-dimensional random walks, and for off-lattice situations where the step length can vary between steps.

The molecular translational motion in a gas has a random walk character. Molecules are on average far apart and thus travel a large distance between collisions that result in a change of direction and speed. Now assume that the collisional cross section is the effective molecular area $4\pi R^2$, where R is the effective molecular radius. It is then possible to calculate the distance traveled between collisions, called the mean free path length λ, as

$$\lambda = 0.05\frac{k_B\mathbf{T}}{R^2\mathbf{p}} \tag{1.7.4}$$

The mean free path is thus inversely proportional to the pressure $\mathbf{p}$, since, for an ideal gas, $\mathbf{p}$ is directly proportional to the number density N/V. For water molecules at room temperature, λ varies from 1 m at $\mathbf{p} = 10^{-7}$ atm to 10^{-7} m at $\mathbf{p} = 1$ atm. The path of a gas molecule differs from that of an ideal random walk since the step size varies and the direction of the velocity is not perfectly randomized. Still the two processes share the basic stochastic nature and share the same scaling $\langle r\rangle \propto N_p^{1/2}$. For times t large relative to the time between collisions, we have that $N_p \propto t$, and thus $\langle r^2\rangle \propto t$, and the expression for the *mean-square displacment* is

$$\langle r^2\rangle = 2dDt \tag{1.7.5}$$

Here, d is the system dimensionality, and the proportionality constant D is called the *diffusion constant* and has units of (area/time). For water vapor at room temperature, D varies from 3.4×10^2 $\mathrm{m^2/s}$ at $\mathbf{p} = 10^{-7}$ atm to $3.4 \times 10^{-5}\,\mathrm{m^2/s}$ at $\mathbf{p} = 1$ atm. D is inversely proportional to $\mathbf{p}$ due to the scaling $\lambda^2 \propto \mathbf{p}^2$ and a collision frequency proportional to $\mathbf{p}$.

In a liquid, molecules are constantly in contact with neighboring molecules and, as a consequence, the notion of individual collisions loses significance. Nevertheless, the diffusion in a liquid can be treated as a random walk with very small steps and the linear proportionality between $\langle r^2 \rangle$ and N_p is present also in the case of liquids and Eq. (1.7.5) remains valid. The molecular translation is, however, significantly more obstructed in the liquid so that the diffusion coefficient becomes much smaller. In the neat liquid at room temperature, D(water) = 2.8×10^{-9} m^2/s, D(benzene) = 2.2×10^{-9} m^2/s, and D(tetradecane) = 6×10^{-10} m^2/s. For water, this value is around 10^4 times smaller than in the gas phase at $\mathbf{p} = 1$ atm! This shows that the diffusion of molecules in a liquid is a much slower process than in a gas, not because the velocity of the individual molecules is slower but because their trajectories are much more tortuous.

In a crystalline solid, where molecules are confined to distinct positions in a periodic array, the mechanism leading to diffusion is somewhat different. In a perfect crystal, diffusion can involve an exchange of positions between two nearest neighbors or a more complex concerted positional exchange. This is normally a rare event and diffusion by this mechanism is very slow. In practice, diffusion usually takes place at defects such as unoccupied lattice sites or stacking faults, where successive motion of molecules into neighboring empty lattice sites leads to a diffusional transport of molecules. Typical values at room temperature are D(Cu) = 6×10^{-40} m^2/s and D(Na) = 6×10^{-13} m^2/s. This massive difference shows that for solids, the diffusion coefficient is much more system dependent than for liquids and gases. For example, the large difference between copper and sodium reflects mainly the fact that sodium has a much lower melting temperature.

1.7.2 Self-diffusion and Mutual Diffusion Are Related but Not Identical Processes

The fundamental relation in Eq. (1.7.5) describes how an individual molecule moves in a medium in thermal equilibrium. This is called a self-diffusion process and D is therefore called the self-diffusion coefficient. It can be measured by labeling a small fraction of the molecules and monitoring how they spread in time due to their diffusional motion. Such labeling can be made without perturbing the system by introducing (radioactive) isotopes or by selective spin excitation in a nuclear magnetic resonance experiment. Labeling can also be made by selectively exciting some molecules using light, which, however, introduces a small perturbation of the system that can affect its dynamics.

For a system out of equilibrium, Brownian motion provides a route for relaxing toward equilibrium. As a simple example, consider a solution with an uneven concentration of its molecular components. The random molecular motions will have the effect of mixing the system until a uniform concentration is established. Neglecting specific molecular interactions between the species, the net flow J_i of species i is according to Fick's first law:

$$J_i = -D_i \frac{\mathrm{d}c_i}{\mathrm{d}x} \tag{1.7.6}$$

where we assume that the concentration gradient is in the x direction. Thus, if we can measure the net flow J_i in a known concentration gradient, we can determine the diffusion coefficient D_i. Fick's first law in the form of Eq. (1.7.6) implicitly assumes ideal solution behavior. However, the fundamental driving force for the flux is a gradient in the chemical potential μ_i, yielding a more general form of Fick's law:

$$J_i = -\frac{D_i}{k_\mathrm{B}\mathbf{T}} c_i \frac{\mathrm{d}\mu_i}{\mathrm{d}x} \tag{1.7.7}$$

In the presence of an external field, as discussed earlier, we can thus maintain a concentration gradient even at equilibrium, while the chemical potential of each species is always homogeneous.

Problem: In a gravitational field, the concentration of species i varies with position x as

$$c_i = c_{io} \exp\left(-m_{\text{eff}}^i gx/k_B\mathbf{T}\right)$$

Using $\mu_i = \mu_i^\theta + k_B\mathbf{T}\, ln\, c_i + m_{\text{eff}}^i gx$, show by explicit calculation that J_i in Eq. (1.7.7) is zero.

Solution:

$$\begin{aligned}\frac{d\mu_i}{dx} &= \frac{d\mu_i^\theta}{dx} + k_B\mathbf{T}\frac{d\, ln\, c_i}{dx} + \frac{d}{dx}\left(m_{\text{eff}}^i gx\right)\\ &= 0 + k_B\mathbf{T}\left[\frac{d\, ln\, c_{io}}{dx} - \frac{d}{dx}\left(\frac{m_{\text{eff}}^i gx}{k_B\mathbf{T}}\right)\right] + m_{\text{eff}}^i g\\ &= -k_B\mathbf{T}\frac{m_{\text{eff}}^i g}{k_B\mathbf{T}} + m_{\text{eff}}^i g = 0\end{aligned}$$

Thus, the flux is zero.

In an incompressible liquid, the diffusion of one species i in a concentration gradient necessarily involves the diffusion of another species in the opposite direction. In this case, the D_i values entering Eqs. (1.7.6) and (1.7.7) are referred to as *mutual* diffusion coefficients. For example, in a solution, the net diffusion of the solute in one direction must be coupled to a net diffusion of the solvent in the other direction. Although self-diffusion and mutual diffusion both originate from the thermal Brownian motion, they have somewhat different characteristic features. In the dilute regime, the mutual and self-diffusion coefficients of the solute are virtually the same. At concentrations where the solute molecules are abundant enough so that they interact with each other, the two types of diffusion coefficients start to differ. As an example, we assume the solutes to have a net-repulsive interaction. In a concentration gradient, the solutes then tend to push each other apart so that, in regions of high solute concentration, the molecules experience a net force pushing them toward regions of lower concentration. The diffusional process is therefore faster than in the absence of solute–solute interactions, implying an increase in the mutual diffusion coefficient compared to the self-diffusion coefficient. In contrast, the self-diffusion of the solute, i.e., in the absence of a concentration gradient, is obstructed by the presence of the other solute molecules leading to a lowering of D due to interactions.

A special mutual diffusion effect occurs when the solute is a charged species, such as a charged colloidal particle. The charged particle is accompanied by neutralizing counterions, which are typically much smaller than the particle. When placed in a concentration gradient, both the particle and its counterions start to diffuse. The smaller counterions diffuse individually much faster than the particle; however, they cannot leave the particle completely behind since this would break local electroneutrality. The net effect is therefore that the counterion cloud will "pull" the larger particle somewhat like dogs pulling a large sled. The resulting mutual diffusion coefficient is intermediate between the self-diffusion coefficients of the ions and that of the particle.

1.7.3 The Association Rate in Colloidal Systems Can Be Diffusion Controlled

When two molecular species A and B associate, we can typically write the reaction rate v_{AB} as

$$v_{AB} = k_{AB}[A][B] \tag{1.7.8}$$

where k_{AB} is the rate constant. For chemical reactions involving small molecules, the association typically involves the passing of a reaction barrier, while if A is a colloidal particle and B a molecular species, it is more common that there is no such barrier and the rate is determined by the rates of the diffusive motion of A and B. Assume that A is a colloidal particle and that association between A and B occurs whenever the center-to-center separation is smaller than R_{AB}. If we place an A particle at the origin, B molecules will diffuse toward it with a rate given by Fick's first law 1.7.6. Since the particle is isotropically surrounded by B molecules in all directions, we can rewrite Eq. (1.7.6) in spherical symmetry, yielding the flux J_B at radial position r as

$$J_B \times \text{area} = -4\pi r^2 D_{AB} \frac{\mathrm{d}[B]}{\mathrm{d}r} \tag{1.7.9}$$

Here $D_{AB} = D_A + D_B$ is the effective diffusion constant for the relative motion of A and B. In three dimensions, a steady state will be established at times $t > \frac{r^2}{D_{AB}}$, whereby the flow across a spherical area is independent of r so that J_B times area is constant. From Eq. (1.7.9), this implies that $[B] \propto \frac{1}{r}$. At large r, $[B] = [B]_0$, where $[B]_0$ is the bulk concentration, while at $r = R_{AB}$ the reaction fully eliminates B so that $[B] = 0$. Inserting these boundary conditions yields the steady-state reaction rate

$$v_{AB} = [A](J_B \times \text{area})_{\text{steady-state}} = 4\pi D_{AB} R_{AB} [A][B] \tag{1.7.10}$$

Comparison with Eq. (1.7.8) yields the rate constant for the diffusion-controlled reaction as

$$k_{AB} = 4\pi R_{AB} D_{AB} \tag{1.7.11}$$

1.7.4 Fick's Second Law States How Concentrations Change in Both Time and Space

Fick's first law describes the diffusional flow in an external concentration gradient and can be applied directly for steady-state conditions where the gradient remains constant. In general, the resulting flux will, however, result in a reduction of the gradient, and it is then necessary to explicitly consider the time dependence of the concentrations. Since the amount of each component is conserved, one can combine Fick's first law 1.7.6 with the corresponding equation of continuity to arrive at Fick's second law:

$$\frac{\mathrm{d}c_i(x,t)}{\mathrm{d}t} = D_i \frac{\mathrm{d}^2 c_i(x,t)}{\mathrm{d}x^2} \tag{1.7.12}$$

This is a partial differential equation, which is difficult to solve in general. However, it has a reasonably simple solution for the case of a point source at $t = 0$ with diffusion in one, two, or three spatial dimensions. As an example, consider the case where one initially puts a (negligibly) thin film of a solute on the surface of a solvent. The solute will diffuse into the solvent in the positive x direction, while the concentration will be uniform in the yz-plane parallel to the surface. The solution of 1.7.12 satisfying the boundary condition is

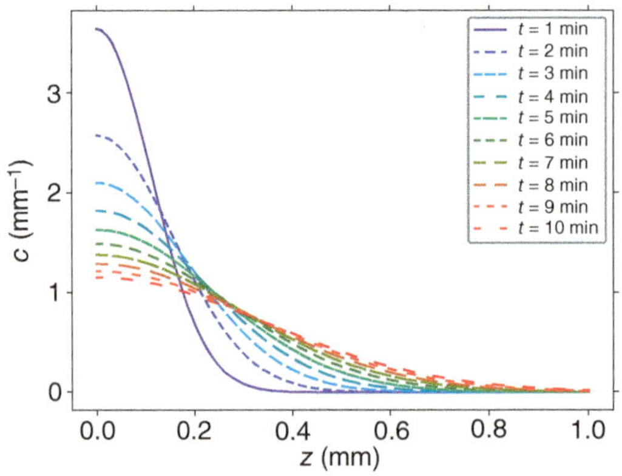

Figure 1.13 Illustration of how the concentration profile evolves in time into a solution when applied at the surface $z = 0$ at $t = 0$, from an initial state at the surface.

$$c_i(x,t) = \alpha t^{-1/2} \exp\left(-\frac{x^2}{4D_i t}\right) \tag{1.7.13}$$

The normalization constant α is determined by the conservation of mass and is given by

$$\alpha = \frac{\sigma_i}{\sqrt{\pi D}} \tag{1.7.14}$$

where σ_i is the initial surface concentration. Figure 1.13 shows how the concentration profile develops in time by diffusion when a substance is added at a surface.

1.7.5 In a Liquid, Molecules Are Transported by Diffusion, Convection, and Flow

In the macroscopic world, molecules are typically transported by a liquid flow driven by pressure gradients. Take as example whiskey left to age in an oak barrel: when the tap is opened, the liquid flows out driven by the gravitational force, but for most of the time, the liquid is at rest inside the barrel. Brownian motion generates diffusional transport in the whiskey and enables molecular exchange between the liquid and the soaked oak wood. It is straightforward to estimate the time it takes for a molecule to diffuse from the wall to the center of the barrel. Assume that the barrel diameter is 1 m and that the diffusion coefficient is approximately $D = 5 \times 10^{-10}$ m^2/s at the storage temperature. The average time to diffuse a distance of 0.5 m using Eq. (1.7.5) is $0.5^2/(2 \times 5 \times 10^{-10}) \approx 10^8$ s $= 2.8 \times 10^4$ h = 32 years!

Thus, if diffusion were the only transport process, even a whiskey stored 18 years would not be significantly matured by the barrel. However, even in liquids that from the outside appear to be at rest, there can still be convective flows, typically caused by temperature gradients, which in turn generate density gradients. The gravitational force then gives rise to a liquid flow, which is circular since the liquid as a whole is at rest. Such a process provides an effective mixing on a larger length scale and is an efficient complement to diffusion, which is efficient in mixing on smaller length scales.

Even if one makes an effort to have a homogeneous temperature, it is in practice very difficult to avoid convection in vessels of macroscopic size – and even in an ordinary test tube! A simple demonstration of the efficiency of convection relative to diffusion can be made using two identical beer bottles. Place one of them carefully on the table, but shake one vigorously for a few seconds before placing it on the table, and open both. There was no visible difference between the bottles before opening, but one is foaming while the other only produces a limited number of gas bubbles. The bubbles form at glass walls and interfaces, and with only diffusional transport, only a limited number of bubbles form before the boundary layers are depleted of dissolved carbon dioxide. However, the relaxation times for the convective flows caused by shaking are tens of seconds to a minute, so that, if the shaken bottle is opened within this time frame, the surface-depleted layers are replenished with carbon dioxide, leading to the formation of many more bubbles than for the unshaken bottle.

In a living cell, the aqueous phase is confined by the cell wall and diffusion is the main source of molecular transport supporting metabolic processes. As was seen in the whiskey barrel example, diffusional transport becomes less efficient the longer the distance that needs to be covered. For small bacterial cells with diameters of order 10^{-6} m, diffusion is very efficient, but for more complex cells with dimensions in at least one direction of order 10^{-4} m, diffusional transport starts to limit the metabolic efficiency. One possible solution to this problem is to use metabolic energy to mechanically excite convective modes, for example, using the cytoskeleton. Another possibility, particularly relevant for elongated cells such as nerve cells, is to have specific transport machineries for motion along the cell.

1.7.6 Hydrodynamic Interactions Determine the Diffusion Coefficient of Colloidal Particles

A colloidal particle in an aqueous medium will constantly collide with water molecules, giving rise to the motion that Robert Brown first observed in his microscope. Intuitively, it is tempting to conclude that the small water molecules will push equally in all directions so that, for a large enough particle, the net force will be negligible and no Brownian motion will occur. This scientific dilemma was resolved by two major contributions. The first step was taken by Sir George Stokes, who considered the problem of what force is needed to pull a spherical particle through a liquid medium. The motion of the particle gives rise to a fluid flow and is associated with an energy penalty from the viscous forces in the fluid. This leads to an effective friction force $F = \mu v$ proportional to the velocity v of the particle. Stokes derived that the friction coefficient μ is

$$\mu = 6\pi\eta R \tag{1.7.15}$$

where η is the viscosity of the liquid and R the particle radius.

The second step is due to Einstein, who realized that there is a general relation between dissipation (*via* friction) and fluctuations (*via* diffusion), which for the case of a spherical particle is given by

$$D = \frac{k_B\mathbf{T}}{\mu} \tag{1.7.16}$$

The resulting Stokes-Einstein expression for the diffusion coefficient of a sphere in a pure liquid is

$$D = \frac{k_B\mathbf{T}}{6\pi\eta R} \tag{1.7.17}$$

For a particle with $R = 50$ nm diffusing in water, the calculated value of the diffusion coefficient is $D \approx 4 \times 10^{-12}$ m^2/s, which agrees well with experimental values. Even when the formula is applied to a water molecule in water, one obtains a reasonable estimate of $D = 1 \times 10^{-9}$ m^2/s, which is only a factor of two smaller than the experimental value. A notable property of Eq. (1.7.17) is that the diffusion coefficient decreases only moderately with increasing particle size.

See Atkins et al. (2022, Focus 16), Doi (2013, Chapter 4), and Ohshima (2024, Chapter 19) for additional discussion on Diffusion.

Exercises

1.1 Calculate the aggregation numbers of micelles formed by single hydrocarbon chain surfactants with 10, 12, and 16 carbon atoms by assuming that the hydrocarbon core forms a sphere with a radius equal to the extended length of the hydrocarbon chain.

1.2 Calculate the average area per polar group for micelles in Exercise 1.1 by assuming that the centers of the polar groups are located 0.25 nm outside the hydrocarbon core.

1.3 The table shows the chemical potential of a fluorocarbon, C_4F_{10}, in a mixture with the corresponding hydrocarbon, C_4H_{10}. The neat liquid is used as the standard state, and the chemical potential μ is expressed in terms of the activity, $a = \exp\{-\Delta\mu/(\mathbf{RT})\}$

Mole fraction	Activity	Mole fraction	Activity
0.90	0.91	0.30	0.68
0.80	0.86	0.20	0.62
0.70	0.81	0.10	0.54
0.60	0.77	0.075	0.47
0.50	0.75	0.050	0.40
0.40	0.72	0.25	0.25

Obtain a fit of the data to the regular solution theory. What is the value of the interaction parameter? Estimate the change in standard chemical potential if the reference state is chosen as infinite dilution in C_4H_{10}.

1.4 Calculate the second virial coefficient for a spherical solute with radius R so that the interaction potential is $V(r) = \infty$ for $r < 2R$ and $V(r) = 0$ otherwise.

1.5 Water is transported from the roots to the leaves of trees. Assume that the osmotic pressure in the roots corresponds to that of physiological saline. What is the relative humidity in the air at the canopy 20 m above ground level when there is osmotic equilibrium between the roots and the leaves in the canopy?

1.6 The table shows the osmotic pressure of a series of aqueous glucose solutions prepared from a crystalline hydrate, $(C_6H_{12}O_6)(H_2O)_n$. What is the number n? ($\mathbf{T} = 298$ K).

Crystal % (w/w)	Osmotic pressure ($\times 10^{-5}$ Pa)
1.1	1.46
2.2	2.97
3.3	4.53
4.4	6.14
6.6	9.02

Crystal % (w/w)	Osmotic pressure ($\times 10^{-5}$ Pa)
8.8	12.9
11.0	16.9

1.7 Calculate the aqueous solubilities of decane, 1-decene, decanol, and decanoic acid (at low pH) using Table 1.5 (**T** = 298 K).

1.8 Determine the temperature at which benzene and hexane have their lowest aqueous solubilities.

1.9 Calculate the activity of water and the corresponding osmotic pressure for physiological saline c(NaCl) = 0.150 M. Assume ideal solutions.

1.10 Consider a molecule in a solution where it exchanges between being free in solution with a diffusion constant D_f and being bound to an aggregate with a diffusion constant D_a. Determine the effective diffusion constant for times that are much larger than the exchange time.

1.11 For an aqueous solution of a solute, concentration c, the osmotic pressure is $\prod_{osm} = k_B Tc + B_2 c^2$. Determine the corresponding expression for the chemical potential of the solute.

1.12 Consider a dilute (1% V/V) aqueous dispersion of colloidal particles of radii $R = 500$ nm and density $\rho = 900$ kg/m^3 in a tube of height 0.1 m in a gravitational field. What value of the gravitational constant g is required to assemble 90% of the particles in the upper 1 cm section of the tube (**T** = 298 K).

1.13 Estimate the difference in the bilayer thickness between the phospholipids DMPC and DSPC.

1.14 Consider a cylindrically symmetric surface which between $-x_0 < x < x_0$ have the radius $r = r_0 \cosh(x)$. (a) Determine the mean curvature at $x = 0$. (b) Determine the mean curvature at an arbitrary point on the surface.

References

Atkins, P., de Paula, J., and Keeler, J. (2022). *Atkins Physical Chemistry*, 12th ed. Oxford, UK: Oxford University Press.

Beesley, A.H., Evans, D.F., and Laughlin, R.G. (1988). *J. Phys. Chem.* 92, 791.

Berg, J.C. (2010). *An Introduction to Interfaces & Colloids. The Bridge to Nanoscience*. World Scientific.

Cosgrove, T. (2010). *Colloid Science. Principles, Methods and Applications*. Chichester, UK: Wiley.

Doi, M. (2013). *Soft Matter Physics*. Oxford, UK: Oxford University Press.

Evans, D.F. (1988). *Langmuir*. 4, 6.

Israelachvili, J.N. (2011). *Intermolecular and Surface Forces*, 3rd ed. London, UK: Academic Press.

Kronberg, B., Holmberg, K., and Lindman, B. (2014). *Surface Chemistry of Surfactants and Polymers*, 3rd ed. Chichester, UK: Wiley.

Lindman, B. and Wennerström, H. (1980). *Topics in Current Chemistry*, Vol. 87. Berlin: Springer-Verlag.

Ohshima, H. (2024). *Fundamentals of Soft Interfaces in Colloid and Surface Chemistry*. London, UK: Academic Press.

Stilbs, P., Walderhaug, H., and Lindman, B. (1983). *J. Phys. Chem.* 87, 4762

Tanford, C. (1980). *The Hydrophobic Effect, Formation of Micelles and Biological Membranes*, 2nd ed. New York: Wiley.

Wennerström, H. and Oliveberg Quart., M. (2022). *Rev. Biophys. Discovery*, 3: e12, 1–10.

2

Surface Chemistry and Monolayers

Concept Map

Surface Chemistry and Monolayers
• By analogy with the thermodynamic equation for bulk material, we can write $d\mathbf{G} = \gamma dA + \Sigma \mu_i dn_i$ for surfaces and interfaces at constant temperature and pressure. A system can minimize its free energy by: **1.** Decreasing its surface area—this branch of surface science leads to capillarity. **2.** Changing its composition—this phenomenon leads to the Gibbs adsorption isotherm. • We pursue each topic sequentially; in multicomponent systems, both area and composition can change at an interface to minimize $\Delta\mathbf{G}$.

Surface Tension
Surface tension reflects the unfavorable change in interaction energy, $\mathbf{E}_s \propto W_{AA}(z_s - z_b)/(2a_0)$ (Eq. 2.1.3), which occurs upon moving a molecule from the bulk to the surface, where the number of nearest neighbors decreases from z_b to z_s. Surface tension can be written either as the change in free energy per unit surface area, $\gamma = (\partial\mathbf{G}/\partial A)_{\mathbf{T,p}}$ (Eq. 2.1.10) or as the force per unit length, $\gamma = F/(2l)$ (Eq. 2.1.7), as defined in Figure 2.1. These two points of view provide useful perspectives in interpreting γ. Interfacial tension γ_{AB} between two immisible phases A and B can be expressed (Eq. 2.1.5) in terms of the interaction parameter w (Eq. 1.1.16) and the number of $A - B$ interactions.

Work of Cohesion and Adhesion
The free energy of cohesion, $\Delta\mathbf{G}_{\alpha\alpha}$/area $= 2\gamma_{\alpha v}$ (Eq. 2.1.11), measures the free energy associated with creating unit areas of surface by cleaving a homogeneous phase, as illustrated in Figure 2.2a.

(Continued)

Work of Cohesion and Adhesion (Continued)

The free energy of adhesion, $\Delta \mathbf{G}_{\alpha\beta}$/area $= \gamma_A + \gamma_B - \gamma_{AB}$, measures the free energy required to separate the unit area of the interface into two liquid–air interfaces, as illustrated in Figure 2.2b.

The spreading coefficient, $S = \gamma_{sv} - (\gamma_{lv} + \gamma_{sl})$ (Eq. 2.1.14a), permits us to predict whether a drop of one liquid will spontaneously spread on the surface of a solid, or a second liquid (Eq. 2.1.14b).

Young–Laplace Equation and Interfacial Free Energies

The Young–Laplace equation, $\Delta \mathbf{p} = \gamma(1/R_1 + 1/R_2) = \gamma 2H$ (Eq. 2.1.24), relates pressure differences across a curved interface to radii of curvature and surface tension. Useful applications of this equation include explaining how surface tension is measured using capillaries, sessile and pendant drops, and a Wilhelmy plate.

Calculations of changes in surface free energies explain the following:

- Why and under what conditions vapors will condense into capillaries, into surface pores and cracks, or onto rough surfaces at vapor pressures p below the equilibrium value p_0.
- Why small particles in heterogeneous colloidal dispersions decrease in size and disappear, while larger ones grow at their expense, a process called Ostwald ripening.
- Why, in terms of the Kelvin equation (Eq. 2.3.12), the nucleation of a liquid phase from a pure vapor phase require vapor pressures considerably greater than the equilibrium value $\mathbf{p}_0$.
- Why a liquid film can rise on the walls of a capillary tube.

Gibbs Adsorption Isotherm

- It is difficult to assign a volume to an interfacial region because physical properties change continuously across the interface, as illustrated in Figure 2.16.
- In the Gibbs model, we treat the interface as a mathematical plane of area A, containing n_i^σ molecules, and we define a surface concentration by $\Gamma_i = n_i^\sigma/A$ (Eq. 2.4.4).
- We can relate the change in surface tension with $\ln c_i$ of an added component to a surface excess concentration, Γ_2 (Eq. 2.4.14).
- In a binary solution, if we choose $\Gamma_1 = 0$, we focus on the behavior of the solute.
- Γ_i can have either sign reflecting concentration (+) or depletion (−) at the interface.

Langmuir Isotherm

The Langmuir isotherm is the simplest model for the adsorption of molecules onto solid surfaces. Assuming that solvent and solute molecules occupy the same adsorption area, that the surface solution is ideal, and that the solution is dilute, we can relate bulk solute concentration to fractional surface coverage by Eq. (2.4.19).

Surface Monolayers

Monolayers are two-dimensional, self-organizing systems.

- When surface-active molecules are in equilibrium with those in the adjacent solution, we can use surface tension measurements and the Gibbs adsorption isotherm to calculate surface concentrations and surface areas per molecule.
- With insoluble amphiphiles, we can use the Langmuir balance to measure the surface pressure, Π_s. Plots of Π_s versus surface area per molecule, a_0, yield two-dimensional isotherms.
- Insoluble monolayers are two-dimensional analogs of bulk systems that display dilute and concentrated liquid phases as well as solid states.

Experimental Studies of Surfaces and Interfaces

- The Langmuir–Blodgett method can be used to create monolayers on surfaces.
- Atomic force microscopy can be used to investigate the topology of surfaces.
- Ellipsometry, Brewster angle microscopy, and quartz crystal microbalances are used to study surface adsorption.
- Second harmonic generation and sum frequency spectroscopy monitor molecular orientation at interfaces.

Drying Is a Nonequilibrium Process

- Drying at a water surface is limited by both heat transport and diffusion.
- Drying of solutions gives rise to concentration gradients in the liquid.
- New phases can appear at a drying surface.

A typical colloidal system shows a large surface-to-volume ratio. These surfaces may consist of proper interfaces between different phases, as in an emulsion, or they may constitute boundaries between regions that possess distinctly different molecular properties, as in a micellar solution. Liquid–liquid, liquid–solid, liquid–gas, and solid–gas interfaces all appear in different applications of colloid science.

Although textbooks often treat surface and colloid science together, this text focuses specifically on colloid science. However, surfaces and interfaces play such an important role in colloidal science that we present some of the more fundamental aspects of surface chemistry with an emphasis on those of particular importance for understanding colloids. As a basis for the understanding of surface activity, we first discuss the energy associated with an interface between single-component phases. Then follows a description of how selective surface adsorption in a multicomponent liquid occurs to reduce the free energy of interfaces.

2.1 We Can Comprehend Surface Tension in Terms of Surface Free Energy

2.1.1 Molecular Origins of Surface Tension Can Be Understood in Terms of the Difference in Interaction Between Molecules in the Bulk and at the Interface

Differences between the energies of molecules located at the surface and in the bulk phase of a material manifest themselves as surface tension. Historically, this concept has been a key factor in understanding molecular behavior in liquids. In Chapter 1, we examined the interaction of molecules in the bulk phase in terms of a pair interaction energy W_{AA}. If we now take as an example a pure liquid or solid in equilibrium with its vapor and assume its interaction energies to be pairwise additive, the energy $E_{A,\text{bulk}}$ per molecule in the bulk phase is (cf. Eq. 1.1.12)

$$E_{A,\text{bulk}} = \frac{z_b}{2} W_{AA} \tag{2.1.1}$$

where z_b stands for the number of nearest neighbors in the bulk phase.

The corresponding expression for molecules at a surface is

$$E_{A,\text{s}} = \frac{z_s}{2} W_{AA} \tag{2.1.2}$$

where z_s is the number of near neighbors for a surface molecule.

Because W_{AA} is negative and $z_b > z_s$, moving a molecule from the bulk to the surface increases the internal energy. In other words, work must be done to create a new surface. The energy E_s required to create an area A is then

$$\mathbf{E}_s/A = \frac{E_{A,\text{s}} - E_{A,\text{bulk}}}{a_0} = \frac{W_{AA}(z_s - z_b)}{2a_0} \tag{2.1.3}$$

where a_0 equals the area per molecule at the surface. If we neglect changes in entropy, E_s/A provides an estimate of the surface tension of the liquid phase. We will in the following, use the term surface tension for a condensed phase exposed to a gas, while interfacial tension refers to the corresponding quantity for an interface separating two condensed phases, solid or liquid.

Problem: Calculate the surface tension for carbon tetrachloride using the value of $z_b W_{AA} = -9.1 \times 10^{-20}$ J obtained from the heat of vaporization. Assume that CCl_4 can be treated as a cube with $z_b = 6$.

Solution: In going from the bulk to the surface, one face of the cube is exposed, so $z_b = 6$ and $z_s = 5$. Using the density of CCl_4, $\rho = 1.6 \times 10^3$ kg/m^3, we can calculate the area for each side of the cube to be 3×10^{-19} m^2. Thus

$$\frac{\mathbf{E}_s}{A} = \frac{(5-6) \times (-9.1 \times 10^{-20})}{6 \times 2 \times 3 \times 10^{-19}} = 25 \text{ mN/m}$$

a value that is fortuitously close to the experimental value of 26 mN/m for the surface tension of CCl_4. This calculation illustrates the major source of surface tension in a liquid.

If we consider the interface between two phases containing molecules A and B, we can see that when an A molecule moves to the interface, it loses $z_b - z_s$ interactions with A but gains about an equal number of interactions with B. Similarly, B molecules lose interactions with B and gain interactions with A. The net energy change for N_s surface molecules of each kind is

$$\mathbf{E}_s = N_s(z_s - z_b)\left\{\frac{W_{AA} + W_{BB}}{2} - W_{AB}\right\} = \frac{N_s}{N_{\mathrm{Av}}}\frac{z_b - z_s}{z_b}w \tag{2.1.4}$$

with w defined as in Eq. (1.1.16). We have assumed that the coordination number z_b is the same in the two bulk phases and at the interface, a reasonable assumption for molecules of similar size. The energy per unit area is then

$$\frac{\mathbf{E}_s}{A} = \frac{1}{a_0}\frac{z_b - z_s}{z_b}\frac{w}{N_{\mathrm{Av}}} \tag{2.1.5}$$

Problem: For two strongly immiscible liquids with $w = 6\boldsymbol{RT}$, estimate the interfacial tension and compare it to the value of the surface tension calculated for CCl_4. Assume that $a_0 = 3 \times 10^{-19}$ m^2 and that $z_s = z_b - 1(\mathbf{T} = 298$ K).

Solution: Similar to the problem with carbon tetrachloride

$$\mathbf{E}_s/A = 6 \times 8.31 \times 298/\left(6 \times 3 \times 10^{-19} \times 6 \times 10^{23}\right) = 14 \text{ mN/m}^2$$

This example demonstrates the general rule that interfacial energies are smaller than surface energies.

In deriving Eq. (2.1.4), we assumed that A and B are pure liquids. Then E_s is the energy of the instantaneous formation of the interface. At equilibrium, some B molecules will mix into the A liquid and vice versa. This exchange decreases the interfacial energy, and for miscible liquids, $w < 2\boldsymbol{RT}$, the interface disappears altogether (see Section 1.1.2).

Previously, we have estimated the energy contribution to the formation of a surface or an interface. To calculate the surface free energy or surface tension, we must also consider an entropy contribution that arises, for example, from the orientation of surface molecules and capillary waves at the surface. However, the energy contribution is usually the largest.

2.1.2 Two Complementary Concepts Define Surface Tension: Line of Force and Energy Required to Create New Surface Area

As the analysis of Eq. (2.1.3) indicates, creating a new surface requires free energy $\Delta\mathbf{G}$, which is proportional to the number of molecules transported to the surface and thus to the area of the new surface. We can express the linear relation between $\Delta\mathbf{G}$ and area A as defining the surface tension γ

$$\gamma = \Delta\mathbf{G}/A \tag{2.1.6}$$

We can view γ in two ways: either as the free energy per area or as a surface tension corresponding to force per length. Historically, the existence of these different perspectives has posed a dilemma, but they represent two ways of looking at the same thing, and both have their uses. Understanding them allows us to determine when one point of view or the other provides a more effective cognitive interpretation of the data.

Figure 2.1 shows a simple device consisting of a wire loop with a movable slide that can clarify the two ways of viewing surface tension.

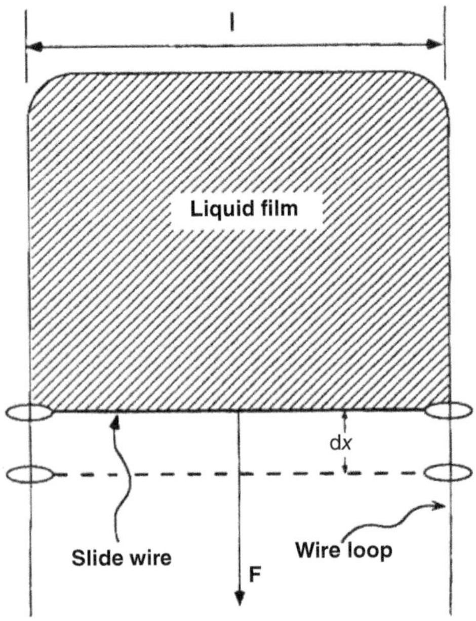

Figure 2.1 Wire loop with a slide wire on which a liquid film can be formed and stretched by an applied force *F*.

We can assume that the device behaves like an idealized frictionless apparatus. Dipping the wire loop into a liquid forms a liquid film. Surface tension causes the slide wire of the apparatus to move in the direction of decreasing film area unless we apply an opposing force. This force, F, operates along the entire edge of the film, varies linearly with the length l of the slide wire, and has a characteristic value for each liquid. If γ represents an intrinsic property of the liquid surface, the apparatus measures the surface tension as

$$\gamma = F/(2l) \tag{2.1.7}$$

The factor of two in the denominator appears since both sides of the film are exposed to air.

The loop-and-slide wire device is a two-dimensional analog of the cylinder–piston apparatus used to evaluate pressure · volume products in gas cycles in thermodynamics. This analogy suggests that γ represents a two-dimensional pressure, a point we will develop in Section 2.5. However, one major difference separates the two situations: if we remove the restraining force in the three-dimensional system, the piston expands, whereas in the two-dimensional system, the slide wire contracts.

We can write the work associated with expanding the interfacial area as

$$\mathrm{d}w = F\mathrm{d}x = \gamma 2l\mathrm{d}x = \gamma \mathrm{d}A \tag{2.1.8}$$

At constant temperature and pressure, the work of increasing the area contributes to the differential Gibbs free energy. Now it is possible to relate Eq. (2.1.8) to thermodynamic quantities and write

$$\mathrm{dG} = \gamma \mathrm{d}A \tag{2.1.9}$$

where

$$\gamma = \left(\frac{\partial \mathrm{G}}{\partial A}\right)_{\mathrm{T,p}} \tag{2.1.10}$$

This expression identifies surface tension as the increase in Gibbs free energy per unit increment in area. Equation (2.1.10) is equally valid for liquids and solids and it refers to an equilibrium situation. However, it is difficult to stretch a solid maintaining equilibrium, so for that case γ is best seen as a surface free energy, while the tension aspect is less relevant. In a strict analysis, the "surface tension" of a solid or liquid depends on the component forming the gaseous medium above the surface. It is usually assumed to be air at a pressure of 1 atm, and at such pressures, the molecular density of the gas is so low that the surface tension depends only marginally on the precise molecular composition of the gaseous phase.

2.1.3 The Work of Adhesion and Cohesion Is Related to Surface Tension

The surface free energy γ provides a direct measure of the cohesion in a liquid or solid. Figure 2.2 illustrates the case where substance α is cleaved into two parts (see Figure 2.2a).

The free energy change in this process is

$$\Delta \mathbf{G}_{\alpha\alpha}/\text{area} = 2\gamma_{\alpha v} \tag{2.1.11}$$

For the case where there is a phase boundary between substances α and β with an interfacial free energy $\gamma_{\alpha\beta}$ (see Figure 2.2b), the cleavage energy is

$$\Delta \mathbf{G}_{\alpha\beta}/\text{area} = \gamma_{\alpha v} + \gamma_{\beta v} - \gamma_{\alpha\beta} \tag{2.1.12}$$

This is called the Dupree equation, and it is widely used to describe the free energy of adhesion between two materials.

2.1.4 The Wetting Behavior of Liquids Is Determined by Surface Free Energies

How a liquid behaves on a solid surface has many practical applications. Paint is spread on surfaces and we normally want the film to cover the surfaces completely. In contrast, when it is raining, we prefer the case when water does not wet, spread on, the surface of our jackets. Consider the case when we place a drop of a pure liquid on a clean solid surface. There are two possible outcomes: (1) It remains as a finite lens, as shown in Figure 2.3.

In this case, the liquid does not wet the surface. (2) The liquid spreads uniformly over the surface with a liquid film covering the solid surface. The liquid wets the surface. The outcome of the process is determined by a minimization of the surface free energies γ_{sl}, γ_{sv}, and γ_{lv}. Consider first the case when the solid has such a high surface free energy that $\gamma_{sl} + \gamma_{lv} < \gamma_{sv}$. In this case, it is favorable for the liquid to spread on the solid surface and form a uniform liquid film. In the

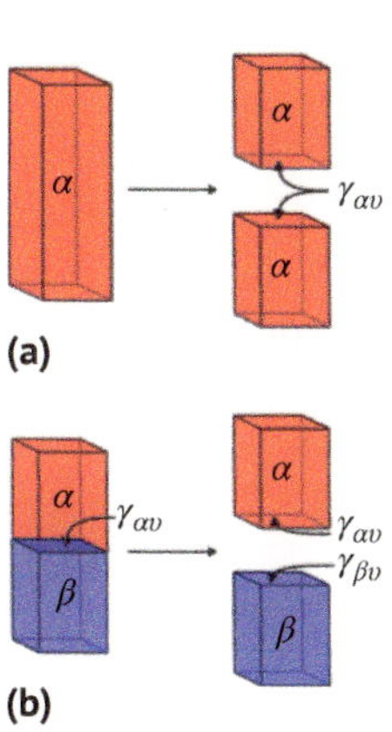

Figure 2.2 The free energy changes required to cleave a homogeneous phase (a) and at a phase boundary (b) are determined by the surface free energy and the interfacial energy, respectively.

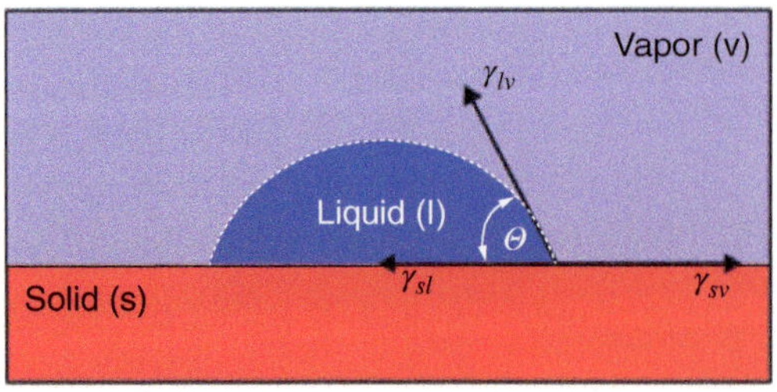

Figure 2.3 Contact angle on a solid surface. γ_{sl}, γ_{sv}, and γ_{lv} are the solid–liquid, solid–vapor, and liquid–vapor interfacial energies, respectively.

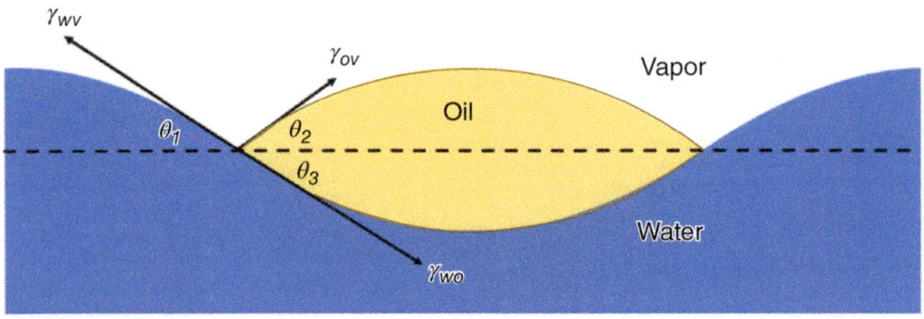

Figure 2.4 Shape of an oil lens on a water surface. The shape may change with time if the oil is able to spread or if the two liquids are mutually soluble to some extent.

opposite case, when $\gamma_{sl} + \gamma_{lv} > \gamma_{sv}$, the liquid will adopt the shape of a spherical cap, neglecting gravity effects. The equilibrium position of the contact line is most easily obtained from a force balance, as illustrated in Figure 2.3, leading to the important relation:

$$\gamma_{sv} = \gamma_{sl} + \gamma_{lv} \cos\theta \tag{2.1.13}$$

This is the Young equation relating the contact angle, θ, to the surface free energies. It is valid under the condition that equilibrium has been established. In practice, this can be difficult to achieve. A practical way to test this is to measure the contact angle both when the volume of the drop is increased so that the contact line advances, and when it is decreased, making the contact line recede. If the measured angle is equal for both the advancing and the receding case, it is a strong indication of equilibrium conditions. It is the rule rather than the exception that the advancing angle is larger than the receding one. This can be due to slow adsorption–desorption processes, but it is more often caused by inhomogeneities on the surface. If the surface free energies of the solid and the liquid are known, the Young equation provides a useful method to measure the interfacial free energy between a solid material and a liquid.

When a drop of a liquid is placed on the surface of another liquid, the situation is more complex than for the case of a solid surface as illustrated in Figure 2.4. If the liquids are miscible, the drop is simply dissolved. If the density of the liquid in the drop is higher than that in the bulk, the drop will usually fall to the bottom of the container, as for a mercury drop on water. When the density of the drop is lower than that of the bulk liquid, it can remain as a drop on the surface. However, if the surface and interfacial free energies obey the relation $\gamma_{\alpha\beta} + \gamma_{\beta v} < \gamma_{\alpha v}$, the liquid drop will spread on the surface forming a thin film.

To distinguish spreading from nonspreading on a solid or liquid surface, it is convenient to introduce the spreading coefficient S:

$$S = \gamma_{sv} - (\gamma_{lv} + \gamma_{sl}), \text{ (solid–liquid)} \tag{2.1.14a}$$

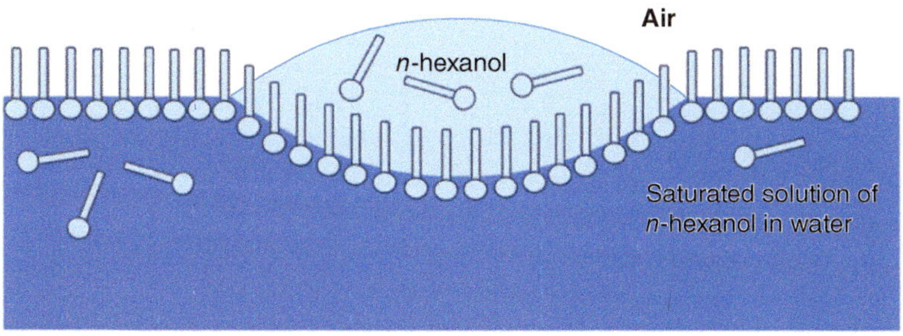

Figure 2.5 When a drop of hexanol is placed on water, the drop initially spreads; that is, the spreading coefficient (Eq. 2.1.14 a,b) is positive. However, when equilibrium is achieved, the water surface contains a monolayer of hexanol, and the spreading coefficient becomes negative as a consequence of the lowering of γ_{wv}. Then the hexanol forms a lens as shown. *Source*: Shaw, 1992/with permission of Elsevier.

$$S = \gamma_{\alpha v} - (\gamma_{\beta v} + \gamma_{\alpha\beta}), \text{ (liquid–liquid)} \tag{2.1.14b}$$

If $S > 0$ spreading will occur, while for $S < 0$ the drop will remain on the surface. Figure 2.4 illustrates, for the specific case of oil on water, how a lens is formed when S in Eq. (2.1.14b) is negative. Due to the combined effect of surface free energies and gravity, the liquid in the drop forms two combined spherical caps, and at the contact position between oil and water, there are now three different angles to be characterized as shown in the figure. The force balance in the lateral direction yields in this case:

$$\gamma_{wv} \cos\theta_1 = \gamma_{wo} \cos\theta_3 + \gamma_{ov} \cos\theta_2 \tag{2.1.15}$$

This can be seen as an extension of the Young equation.

With a liquid drop on a liquid medium, there is a possibility of diffusive transport across the interfacial boundary. Even if the liquids are immiscible, there is some mutual solubility. As the compositions of the liquid phases change, there are also changes in the surface energies. It can take considerable time before equilibrium is achieved. This process can have a dramatic effect that can lead to a change from initial wetting to nonwetting at equilibrium, as illustrated in Figure 2.5.

Problem: Given that $\gamma_{H_2O/vapor} = 72.8$, $\gamma_{water/benzene} = 35.0$ and $\gamma_{benzene/vapor} = 28.9$, calculate the initial spreading coefficient. After equilibrium is reached, $\gamma_{H_2O/vapor}$ decreases to a value of 62.4 because the small amount of benzene that dissolves in water is surface active. Will the benzene remain spread after equilibrium is achieved?

Solution: $S(\text{init}) = 72.8 - (28.9 + 35.0) = 8.9$ mJ/m^2, which corresponds to initial positive spreading coefficient. At equilibrium

$$S(\text{eq}) = 62.4 - (28.9 + 35.0) = -1.5 \text{ mJ/m}^2$$

which means that the initial continuous benzene film will roll back to form a flat lens.

The spreading of hexanol on water provides another example of the dynamic effect. Initially, the spreading coefficient is strongly positive

$$S(\text{init}) = 72.8 - (24.8 + 6.8) = 41.2 \text{ mJ/m}^2$$

and the hexanol initially forms a film on the surface, but at equilibrium it becomes negative

$$S(\text{eq}) = 28.5 - (24.8 + 6.8) = -3.1 \text{ mJ/m}^2$$

as a result of the pronounced surface activity of hexanol in the aqueous phase. In the final states, hexanol forms very flat lenses, which go over into a near monolayer at the aqueous–vapor interface.

Trace amounts of impurity can cause dramatic changes in spreading behavior. For oil on water, impurities in the oil phase can reduce γ_{ow} enough to make S positive. Impurities in the aqueous phase generally reduce S, since the impurity lowers γ_{wv} more than γ_{ow}, especially if γ_{ow} is low already. Thus, n-octane will spread on a clean water surface but often not on a contaminated one.

One can trace the first scientific paper on surface chemistry back to a publication by Benjamin Franklin, one of the founding fathers of the United States, who in 1774 discussed the spreading of oil on water on a pond in Clapham Common in London. Franklin noted that just a "teaspoon" of oil spread out on an area of half an acre, with the visible effects that leaves were pushed away quickly and that the surface was less agitated by waves in the wind. He speculated about the forces acting between the "particles" in the oil and concluded that they must have both a repulsive and an attractive character! We know nowadays that the rapid expansion of the film is due to the Marangoni effect, where a film of lower surface tension quickly spreads on a surface of higher surface tension. The apparent repulsive interaction originates from differences in attractive interactions as discussed in Chapter 1 in connection with the regular solution model.

2.1.5 The Young–Laplace Equation Relates Pressure Differences Across a Surface to Its Curvature

Surface tension operates in a liquid film, and to obtain mechanical equilibrium, this tension must be balanced by some equal and opposite force. For example, to blow a bubble in a liquid medium, an excess pressure is applied. In an isolated particle, such as a droplet, the balancing force comes from stresses within the particle. How is the excess pressure related to surface tension and surface shape?

Figure 2.6 illustrates a circular cross section of a bubble with radius R. When the bubble is expanded infinitesimally to a radius $R + \text{d}R$, the concomitant area change is

$$\text{d}A = 4\pi\left\{(R + \text{d}R)^2 - R^2\right\} = 8\pi R\text{d}R \tag{2.1.16}$$

The corresponding change in free energy is

$$\text{d}\mathbf{G} = \gamma\text{d}A = \gamma 8\pi R\text{d}R \tag{2.1.17}$$

At equilibrium, this free energy is balanced by a pressure–volume work, $\text{d}w = \Delta\mathbf{p}\text{d}\mathbf{V}$ because of the pressure difference across the film. The volume change is

$$\text{d}\mathbf{V} = \frac{4\pi}{3}\left\{(R + \text{d}R)^3 - R^3\right\} \approx 4\pi R^2\text{d}R \tag{2.1.18}$$

Balancing the work of increasing the surface with the pressure–volume work

$$4\pi R^2\Delta\mathbf{p}\text{d}R = 8\pi R\gamma\text{d}R \tag{2.1.19}$$

leads to a pressure difference

$$\Delta\mathbf{p} = \gamma\frac{2}{R} \tag{2.1.20}$$

per interface. The pressure difference in a soap bubble is twice this value, due to the two surfaces.

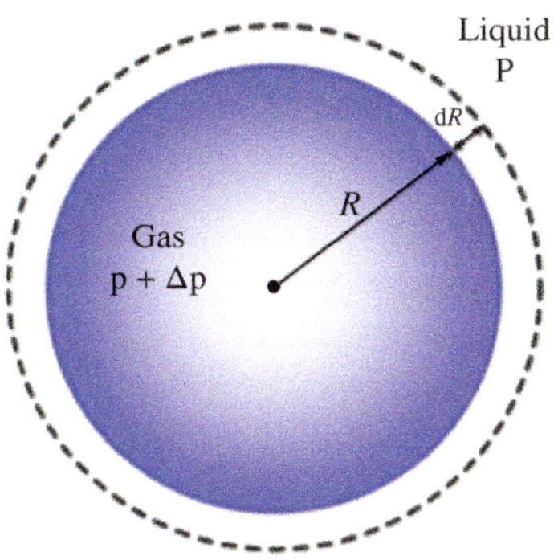

Figure 2.6 Cross section of a gas bubble of radius R suspended in a liquid used to calculate the increase in free energy as the bubble expands from R to $R + dR$.

Problem: What is the excess pressure in a soap bubble of radius 1 cm blown from an SDS solution above the CMC?

Solution: From Figure 1.5, the surface tension is $\gamma \approx 38$ mM/m and

$$\Delta \mathbf{p} = 2 \times \gamma \times \frac{2}{R} = 4 \times 38 \times 10^{-3}/10^{-2} \approx 15 \text{ Pa} = 1.5 \times 10^{-4} \text{ atm}$$

For a bubble with a large radius, there is only a small Laplace pressure, while it can be substantial for radii of colloidal dimensions.

If we consider Figure 2.6 as the cross section of a uniform cylinder of length L, the area change is

$$dA = 2\pi L(R + dR - R) = 2\pi L dR \tag{2.1.21}$$

and the volume change is

$$d\mathbf{V} = \pi L\left\{(R + dR)^2 - R^2\right\} = 2\pi L R dR \tag{2.1.22}$$

A balance between the surface and the pressure–volume work now results in

$$\Delta \mathbf{p} = \frac{\gamma}{R} \tag{2.1.23}$$

per interface for the cylindrical geometry.

We can generalize the equations to interfaces of arbitrary shape, and the result is

$$\Delta \mathbf{p} = \gamma\left(\frac{1}{R_1} + \frac{1}{R_2}\right) = \gamma 2H \tag{2.1.24}$$

where R_1 and R_2 are the radii of curvature of the interface, and H represents the mean curvature as defined in Eq. (1.5.8). Equation 2.1.24 is called the Young–Laplace equation. For a sphere $R_1 = R_2 = R$, while for a cylinder $R_1 = R$, and $R_2 = \infty$, and Eq. (2.1.24) reduces to Eqs. (2.1.20) and (2.1.23), for spheres and cylinders, respectively.

A consequence of Eq. (2.1.24) is that if no other forces are affecting a liquid, its surface has a constant curvature. For example, a drop on a surface will take the shape of a spherical cap. If we form a tubelike soap film between two circular wires, the pressure difference across the film is zero, since the ends of the tube are open. According to Eq. (2.1.24), the mean curvature of the film then becomes zero everywhere. Locally, the surface is saddle-like with the curvatures along and perpendicular to the main axis of equal magnitude but opposite sign, as illustrated in Figure 1.11. We will encounter such surfaces in later chapters.

2.2 Several Techniques Measure Surface Tension

2.2.1 Surface Tension Governs the Rise of a Liquid in a Capillary Tube

We can use the phenomenon of capillary rise illustrated in Figure 2.7 to measure the surface tension of a liquid. A simple relation exists between surface tension γ, capillary rise h, capillary radius r, and contact angle θ.

In the capillary tube, the liquid surface curves as a spherical cap. As Figure 2.7a shows, the radius of curvature R is related to the radius r of the tube and the contact angle:

$$R \cdot \cos\theta = r \tag{2.2.1}$$

Across the interface is a Laplace pressure

$$\Delta\mathbf{p} = \frac{-2\gamma}{R} = \frac{-2\gamma\cos\theta}{r} \tag{2.2.2}$$

which is balanced by a hydrostatic pressure

$$\Delta\mathbf{p} = \Delta\rho g h \tag{2.2.3}$$

where $\Delta\rho$ represents the density difference, $\rho_{liq} - \rho_{vap} \approx \rho_{liq}$, and g is the acceleration due to gravity. Combining the two equations, we find

$$\gamma = \frac{\Delta\rho g h r}{2\cos\theta} \tag{2.2.4}$$

This equation is particularly simple to apply when the liquid wets the solid surface so that $\cos\theta = 1$ as in Figure 2.7b. In this case, it suffices to know $\Delta\rho$ and r and measure h to be able to calculate the liquid–vapor surface tension, γ.

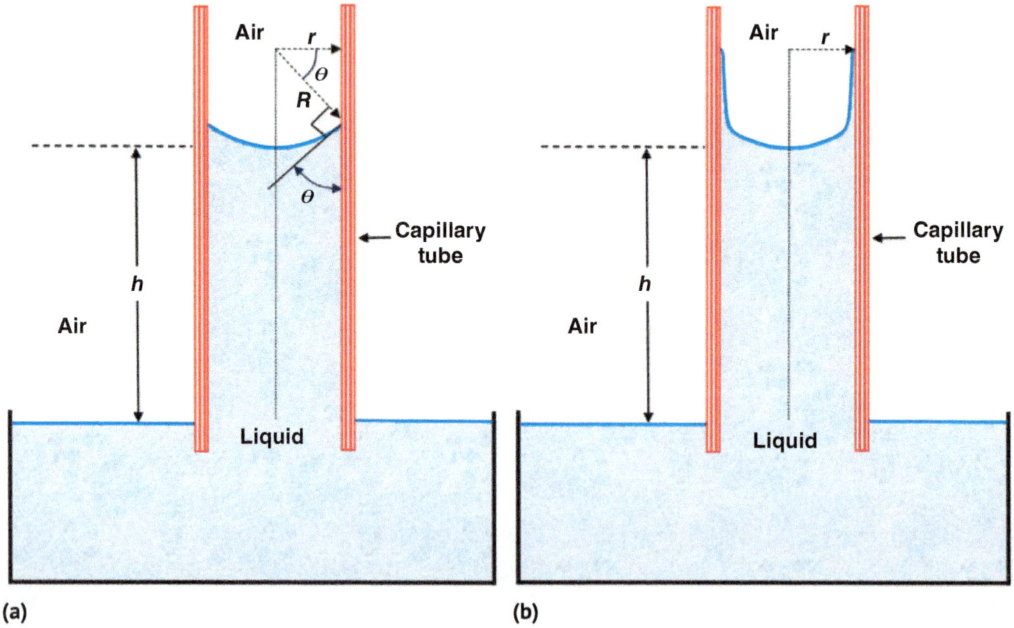

Figure 2.7 Schematic illustration of capillary rise in a cylindrical tube seen in the cross section of radius r. Nonwetting (a) and wetting (b) conditions.

In using the Young–Laplace equation to obtain the pressure forcing the liquid up the capillary tube, we have sidestepped the question of the molecular cause of the capillary rise. Even though the surface tension of the liquid is finally measured, the phenomenon is basically the result of the interfacial properties of both the solid and the liquid.

Consider an infinitesimal rise dh of the liquid in the tube at a given contact angle. The liquid–vapor contact area remains constant, while the solid–liquid interface increases as

$$\mathrm{d}A = 2\pi r\mathrm{d}h \tag{2.2.5}$$

and the solid–vapor interface displays a corresponding decrease. The change in surface free energy is

$$\mathrm{d}\mathbf{G}_s = (\gamma_{sl} - \gamma_{sv})2\pi r\mathrm{d}h \tag{2.2.6}$$

where γ_{sl} and γ_{sl} are the respective interfacial tensions. A change in the gravitational energy

$$\mathrm{d}\mathbf{G}_{grav} = \Delta\rho h d\mathbf{V} = \pi r^2\Delta\rho g h \mathrm{d}h \tag{2.2.7}$$

also occurs. At equilibrium, these two free energy contributions add to zero so that

$$\gamma_{sv} - \gamma_{sl} = \frac{\Delta\rho g h r}{2} \tag{2.2.8}$$

To arrive at Eq. (2.2.4), we recall that according to the Young equation

$$\gamma_{sv} - \gamma_{sl} = \gamma_{lv}\cos\theta \tag{2.1.13}$$

This derivation explains Eq. (2.2.4) in terms of surface free energies for nonwetting conditions, so that $\theta > 0$. When the liquid wets the surface, the argument is slightly different. In this case, a wetting film of thickness δ will rise higher than h in the tube. The exact height to which the wetting film rises is determined by the difference $\gamma_{sv} - \gamma_{sl}$, but for the measurement of the liquid–vapor surface tension, it suffices to know that the rise is higher than that for the liquid in the center of the capillary.

As bulk liquid rises in the tube a distance dh, the change in interfacial free energy is

$$\mathrm{d}\mathbf{G}_s = \gamma\mathrm{d}A = \gamma 2\pi(r - \delta)\,\mathrm{d}h \tag{2.2.9}$$

because the solid surface is already covered. At equilibrium, this energy balances with the gravitational energy and

$$\gamma = \Delta\rho g h(r - \delta)/2 \tag{2.2.10}$$

Provided the film thickness δ is negligible relative to radius r, we regain Eq. (2.2.4) for the case $\cos\theta = 1$.

By deriving Eq. (2.2.4) in two different ways, we have illustrated the complementarity between seeing γ mechanically as a surface tension and seeing it as a surface free energy. The former point of view is usually advantageous when determining the mechanical equilibrium, while the latter reveals more about the molecular mechanisms behind the observed phenomena.

2.2.2 The Surface Tension of Liquids Can Be Measured from a Mechanical Force Balance

In Section 2.1.5, we showed that the Young equation could be used to measure the interfacial tension between solid and liquid. The extended Young equation (2.1.15) shows how the same principle can be utilized to determine the interfacial tension between two liquids. A prerequisite

Table 2.1 Methods of measuring surface tension.

Method	Principle	Performance
Capillary height	Capillary rise	Best when the contact angle $\theta = 0$.
Wilhelmy plate	Capillary force on a plate	Convenient, but susceptible to atmospheric contamination. Best when $\theta = 0$.
Drop profile	Analysis of the geometric drop shape	Works also when a long time is needed to reach equilibrium.
DeNouy ring	Capillary force on a ring	Convenient. Best when $\theta = 0$. Often used for surfactant systems.
Spinning drop tensiometer	Analysis of the geometric shape of a centrifugally distorted drop	Liquid–liquid systems. Appropriate only for very low surface tensions (not measurable with the other methods).

for using these equations is a knowledge of surface tensions. Table 2.1 lists some methods used to determine the surface tensions of liquids. The principles behind the measurements are illustrated in Figures 2.8–2.10.

See Atkins et al. (2022, Focus 5), Berg (2010, Chapter 3), Cosgrove (2010, Chapter 10), Doi (2013, Chapter 4), and Lyklema (2000, Vol. 3, Chapters 1 and 2) for additional discussion on Surface Tension.

2.3 Capillary Condensation, Ostwald Ripening, Nucleation, and Particle Adsorption on Interfaces Are Practical Manifestations of Surface Phenomena

The surface tension and surface free energy concepts discussed in Sections 2.1 and 2.2 provide a basis for understanding a number of practically important phenomena. We illustrate this by four examples: capillary condensation of a liquid on a rough surface, growth of colloidal precipitates through Ostwald ripening, the nucleation of a new phase under supersaturated conditions, and the adsorption of particles at interfaces.

2.3.1 Surface Energy Effects Can Cause a Liquid to Condense on a Surface Prior to Saturation in the Bulk Phase

Consider a vapor in contact with a (solid) surface. If the corresponding liquid wets the surface, the vapor will have an extra tendency to condense on the surface to form a wetting film. At a given temperature, this film will be formed at a vapor pressure $\mathbf{p}$ lower than the bulk saturation pressure $\mathbf{p}_0$. The chemical potential in the vapor is

$$\mu_g = \mu_g^\theta + \boldsymbol{RT}\ln \mathbf{p} \tag{2.3.1}$$

while the chemical potential of the condensed bulk liquid is

$$\mu_l^\theta = \mu_g^\theta + \boldsymbol{RT}\ln \mathbf{p}_0 \tag{2.3.2}$$

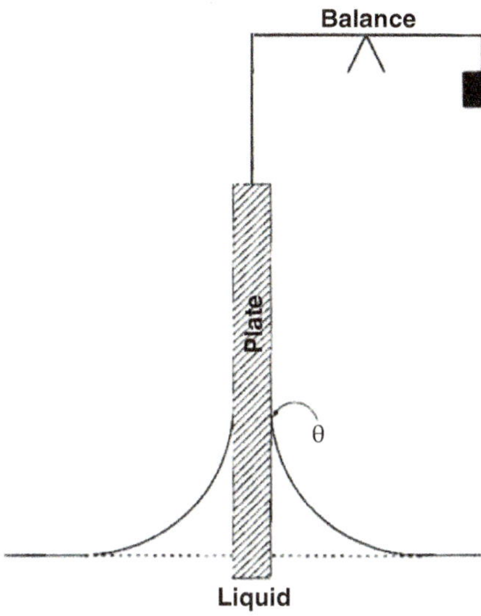

Figure 2.8 The Wilhelmy plate used for measuring γ. A thin plate, seen from the side, is dipped into the liquid, and the balance gives the force needed to lift the plate. A vertical force balance shows that the weight of the meniscus must equal the upward force, f, provided by the surface. This force, in turn, equals the vertical component of the surface tension $\gamma \cos \theta$ times the perimeter P of the plate and $f = \gamma P \cos \theta$.

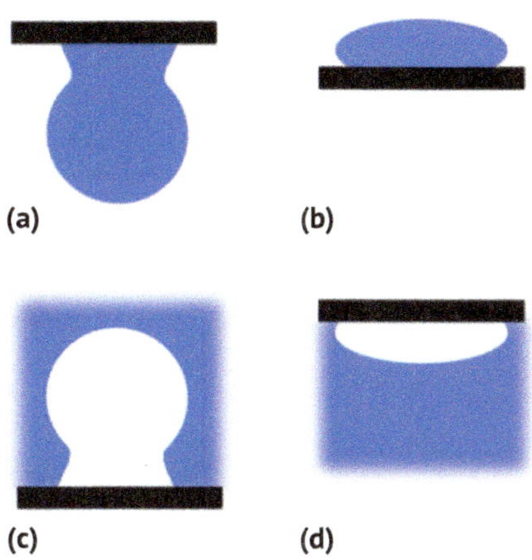

Figure 2.9 Shapes of sessile and hanging drops and of bubbles: (a) pendent drop, (b) sessile drop, (c) hanging bubble, and (d) sessile bubble. Detailed analysis of the shapes of the drops or bubbles makes it possible to determine the surface tension.

The free energy change in forming a liquid film of thickness δ of an area A involves a transfer from vapor to liquid of $\delta A/V_L$ moles, where V_L represents the molar volume of the liquid. A change in surface free energy $(\gamma_{lv} + \gamma_{sl} - \gamma_{sv})A$ also occurs. The total free energy change is

$$\Delta \mathbf{G} = \frac{\delta A}{V_L} \boldsymbol{RT} \ln\left(\frac{\mathbf{p}_0}{\mathbf{p}}\right) + (\gamma_{lv} + \gamma_{sl} - \gamma_{sv})A \tag{2.3.3}$$

This free energy change is zero when the vapor pressure $\mathbf{p}$ is given by

$$\boldsymbol{RT} \ln\left(\frac{\mathbf{p}_0}{\mathbf{p}}\right) = \frac{V_L}{\delta}(\gamma_{sv} - \gamma_{sl} - \gamma_{lv})A \tag{2.3.4}$$

where we recognize the spreading coefficient S of Eq. (2.1.14) in the right-hand side. Clearly, surface condensation occurs only when the liquid wets the surface, because then the right-hand side of Eq. (2.3.4) is positive. In the derivation, we have treated the thickness of the film δ as given. In fact, to determine δ, we must go beyond the macroscopic thermodynamic level and discuss intermolecular and surface forces in detail. The background for such a description is presented in Chapter 5.

Figure 2.10 In the DeNouy ring method, the force F needed to lift the film is balanced by surface tension and $\gamma = \frac{F}{4\pi R}$, where R is the radius of the ring. For precision measurements, correction factors must be applied. Photo by Justin Bergmann.

Even for a liquid that does not wet the surface, condensation at $\mathbf{p} < \mathbf{p}_0$ can occur if the surface arrangement is more favorable than that of a single planar surface. Consider, for example, two planar parallel surfaces with a gap of thickness h between them and in equilibrium with a bulk vapor phase. Under what conditions does the liquid condense in the gap? In the condensed state, the gap contains hA/V_L moles and the free energy change on condensation is

$$\Delta \mathbf{G} = \frac{hA}{V_L} \boldsymbol{RT} \ln\left(\frac{\mathbf{p}_0}{\mathbf{p}}\right) + 2(\gamma_{sl} - \gamma_{sv})A \tag{2.3.5}$$

in analogy with Eq. (2.3.3). The factor of 2 arises owing to the presence of two interfaces. Condensation occurs at the threshold pressure where $\Delta \mathbf{G} = 0$ and

$$\boldsymbol{RT} \ln\left(\frac{\mathbf{p}_0}{\mathbf{p}}\right) = \frac{2V_L}{h}(\gamma_{sv} - \gamma_{sl}) = \frac{2V_L}{h}\gamma_{lv} \cos\theta \tag{2.3.6}$$

We have used the Young equation (2.1.13) to obtain the last equality.

Problem: Calculate the relative humidity at 25 °C at which water will condense in a gap of thickness 100 Å on a surface for which (a) the contact angle is small (<10°), and (b) for a typical hydrophobic surface with $\theta = 115°$.

Solution:

a. When $\theta < 10°$ we have $\cos\theta \approx 1$

$$\ln(\mathbf{p}_0/\mathbf{p}) = \frac{2(18 \times 10^{-6}\ \mathrm{m}^3)(72 \times 10^{-3}\ \mathrm{J/m}^2)}{(1 \times 10^{-8}\ \mathrm{m})(2.5 \times 10^3\ \mathrm{J})} = 0.1$$

$\mathbf{p} = 0.91\mathbf{p}_0$ or 91% relative humidity.

b. When $\theta = 115°$, $\cos\theta = -0.42$, and $\mathbf{p} = 1.04\mathbf{p}_0$ and water will only condense in the gap when the air is 4% supersaturated with water vapor.

An interesting aspect of Eq. (2.3.6) is that for $\theta > 90°$ (so that $\cos\theta < 0$), the vapor will be less likely to condense in the gap than in the bulk. This tendency implies that if we consider the reverse process of a liquid filling the system that approaches the boiling point, the vapor is preferentially formed in the gap. Because the vapor has a much lower density, the contribution from the surface free energy is much larger, counted per molecule in the gap. Vaporization for $\cos\theta < 0$ typically occurs at a larger critical separation than for condensation for $\cos\theta > 0$.

An actual solid surface is normally rough. It may contain cracks, and pores can lead to the interior of the solid. For such a surface in equilibrium with a vapor, the more macroscopic phenomenon of a condensation of a nonwetting liquid may occur in addition to physical and chemical adsorption. The final result of the condensation events depends on general thermodynamic properties, such as vapor pressure, contact angle, and surface tension, as well as on the local geometry of the surface. Clearly, the finer the cracks or pores, or the rougher the surface, the more the liquid is likely to condense. Ordinary stones contain small amounts of condensed water at grain boundaries.

When snow falls on a roof, it forms a fluffy layer of ice crystals that are usually branched in a repeatedly hexagonal pattern. If now the weather changes to temperatures just above 0 °C, under humid conditions, water vapor can start to condense on the snow crystals. The effective weight of the snow gradually increases, resulting in an increasing mechanical load on the roof construction. It happens that roofs collapse by this mechanism.

2.3.2 Surface Free Energies Govern the Growth of Colloidal Particles

Consider a dispersion of colloidal particles that have been formed by precipitating a new phase from solution. The "particles" may be either solid or liquid; they may even take the form of gas bubbles. The equilibrium state is to form a separate bulk phase, which can be accomplished by particle associating. However, this can be inhibited by kinetic barriers. In this case, there is another mechanism, called Ostwald ripening, that leads toward the equilibrium state. The constituent molecules are present as a solute in the liquid at a slightly supersaturated concentration. Molecules exchange between the particles and the solution so that each particle is in local equilibrium with the solution close to it.

The free energy of a particle i with n_i molecules and a surface area A_i consists of a bulk term, $n_i\mu_p^\theta$, plus a surface term $A_i\gamma_{ps}$, where γ_{ps} stands for the particle–solution surface free energy. The free energy change $\Delta\mathbf{G}_i$ caused by taking a molecule from the solution at concentration c_i and adding it to particle i is

$$\Delta\mathbf{G}_i = \left(\mu_p^\theta - \mu_i^\theta\right) - k_\mathrm{B}\mathbf{T}\ln c_i + \gamma_{ps}\Delta A_i \tag{2.3.7}$$

This free energy change equals zero at local equilibrium, making $c_i = c_i^{\mathrm{eq}} + \Delta c_i$.

The area change, ΔA_i, upon adding a molecule to the particle depends on the size of the particle. For spherical particles of radius R_i, the volume is

$$\frac{4\pi}{3}R_i^3 = n_i V_m \tag{2.3.8}$$

where V_m represents the volume per molecule. Thus, increasing n_i by one leads to a change in the radius by

$$\Delta R_i = \frac{V_m}{A_i} \tag{2.3.9}$$

and the corresponding area change is

$$\Delta A_i = \frac{2V_m}{R_i} \tag{2.3.10}$$

This equation shows that the larger the particle, the smaller will be the area change and the smaller the surface energy term in Eq. (2.3.7). At local equilibrium, Δc_i becomes smaller as the particle becomes larger.

In a dispersion containing both large and small particles, the solute concentration is higher close to the small ones, and a diffusional flow of solute moves from the small particles to the large ones. As a result, the small particles decrease in size and dissolve and the large ones grow (see Figure 2.11).

The net movement of monomers from smaller particles to larger ones leads to a drift in R_m, and it can be shown that R_m is proportional to $t^{-1/3}$. The Ostwald ripening process provides one important mechanism for destabilizing colloidal dispersions, as will be discussed in later chapters.

2.3.3 Surface Free Energies Oppose the Nucleation of a New Phase

It is a common observation that liquids can be heated somewhat above the equilibrium boiling point before vapor bubbles actually form. Similarly, a vapor can be brought to a temperature/pressure below the boiling point before liquid drops actually appear in the bulk phase, and a liquid can remain intact even when cooled to below the freezing point. The time stability of metastable states depends on the kinetics of the different processes leading to the formation of the stable new phase. We say that the new phase must nucleate and that the surface free energy of the new interface makes nucleation a sluggish process.

There are three distinctly different mechanisms for the formation of a new phase. One possibility is homogeneous nucleation, where the new phase appears in the bulk phase from an initially metastable state. Another possibility is that a new phase starts to form on a surface, either at the walls of the container or on some particle present in the system. This is called heterogeneous nucleation. A third, less common, possibility is that the system has been brought out of equilibrium in

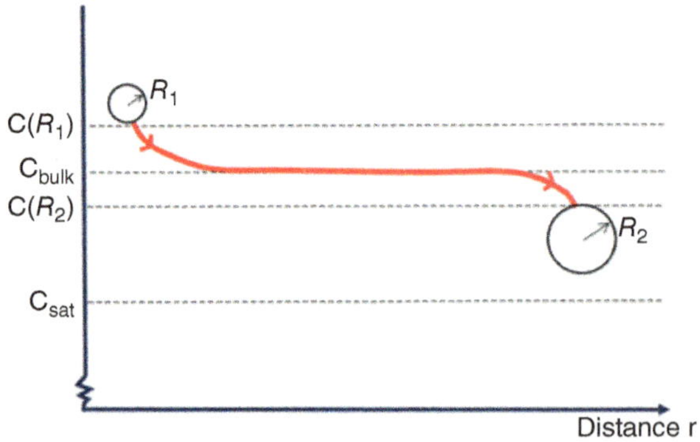

Figure 2.11 Illustration of the Ostwald ripening process. At a given moment, there is a bulk concentration c_{bulk} and particles with a radius R_m are in equilibrium with the bulk. Smaller particles with $R_1 < R_m$ decrease in size by releasing molecules to the bulk, while larger particles with $R_2 > R_m$ grow in size by a net addition of dissolved molecules from the bulk.

such a way that the initial state is locally unstable. In such a case, there is an instantaneous spinodal decomposition leading to phase separation throughout the system. Spinodal decomposition will be further discussed in Chapter 10.

As an example of homogeneous nucleation, consider the formation of a liquid phase from a supersaturated gas of pressure $\mathbf{p}$. The free energy change of forming a liquid drop of radius R is

$$\Delta\mathbf{G} = -nk_B\mathbf{T}\ln\left(\frac{\mathbf{p}}{\mathbf{p}_0}\right) + 4\pi R^2\gamma = -\frac{4}{3}\pi\frac{R^3}{V_M}k_B\mathbf{T}\ln\left(\frac{\mathbf{p}}{\mathbf{p}_0}\right) + 4\pi R^2\gamma \tag{2.3.11}$$

in analogy with Eq. (2.3.5). When $\mathbf{p} > \mathbf{p}_0$, the first term on the right-hand side will make $\Delta\mathbf{G} < 0$ for sufficiently large R, but for sufficiently small R, the positive second term will dominate. Consequently, a maximum in $\Delta\mathbf{G}$ occurs at some intermediate value of R. Figure 2.12 illustrates the variation of $\Delta\mathbf{G}$ with R. At the maximum, $\mathbf{p}$ is related to the radius by

$$k_B\mathbf{T}\ln\left(\frac{\mathbf{p}}{\mathbf{p}_0}\right) = \frac{2\gamma V_M}{R_c} \tag{2.3.12}$$

which is called the Kelvin equation where R_c is the Kelvin radius.

The pressure $\mathbf{p}$ given by this equation is often interpreted as the equilibrium vapor pressure over a surface of curvature $1/R$ and surface tension γ. A better description is that $\mathbf{p}$ represents the maximum possible vapor pressure in the presence of drops of radius R_c before the bulk liquid phase is necessarily formed.

If we introduce a drop of radius R_1 into a vapor of a given vapor pressure $\mathbf{p}$, what is the fate of the drop? If R_i is smaller than the radius R_c, given by the Kelvin equation, we are on the left-hand

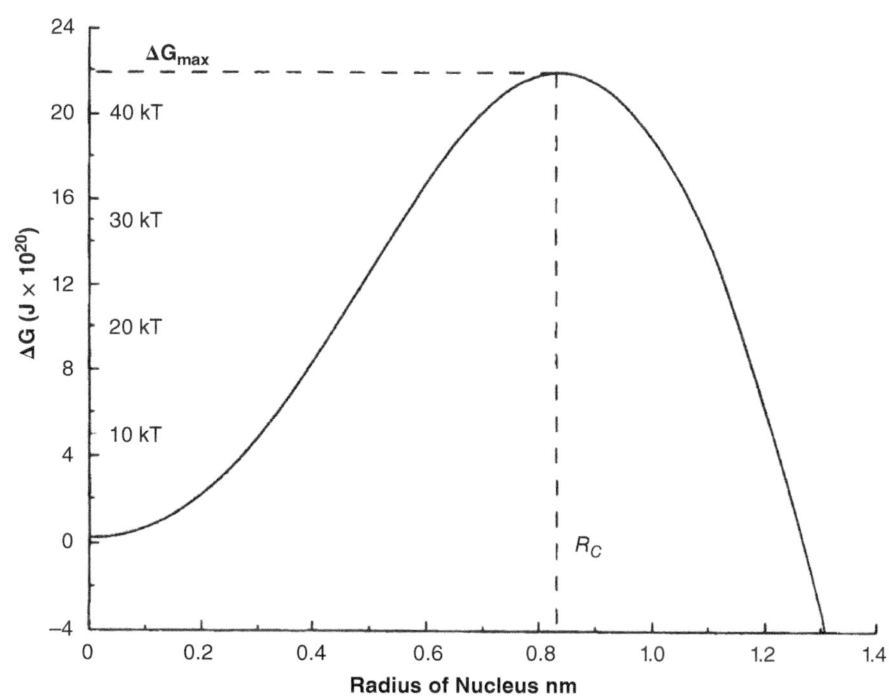

Figure 2.12 The free energy of a water droplet in steam as a function of the radius calculated from Eq. (2.3.11). The temperature is 0 °C and the supersaturation $\mathbf{p}/\mathbf{p}_o = 4$. Note that the free energy maximum corresponds to several tens of $k_B\mathbf{T}$.

side of the maximum in Figure 2.12, and vaporizing the liquid decreases the free energy. On the other hand, if $R_1 > R_c$, condensing more vapor on the drop decreases the free energy and leads to the formation of a bulk liquid phase.

For a solute in a liquid slightly above its saturation concentration c_s, a similar nucleation problem is at hand. If we were to repeat the derivation of the preceding paragraph, giving the solute concentration c the role of the vapor pressure $\mathbf{p}$, and making c_s correspond to $\mathbf{p}_0$, the resulting Kelvin equation is

$$k_B \mathbf{T} \ln\left(\frac{c}{c_s}\right) = \frac{2\gamma V_s}{R_c} \tag{2.3.13}$$

where γ equals the surface free energy solution-precipitate, and V_s is the solute volume. R_c is the particle radius that by thermodynamic necessity nucleates a bulk solid.

Similar nucleation problems arise when a pure liquid crystallizes or when a liquid is heated above its boiling point. In each case, the thermodynamic driving force for forming the new bulk phase must balance the surface free energy of the small nuclei of the new phase. In the formation of vapor bubbles, pressure equilibrium provides an extra complication and the corresponding Kelvin equation contains more terms.

When in a supersaturated system, there already exists some interface, and the formation of the nucleus can be facilitated. Figure 2.13 illustrates the case of a nucleus of a liquid, l, forming on a solid, s, from a vapor medium.

Assume, for simplicity, that the contact angle is 90°. Then the liquid drop takes the form of a hemisphere. The change in free energy forming a liquid drop of radius R_h is

$$\Delta\mathbf{G}_s = 2\pi R_h^2 \gamma_{lv} + \pi R_h^2(\gamma_{sl} - \gamma_{sv}) = 2\pi R_h^2 \gamma_{lv} \tag{2.3.14}$$

where the second equality follows from the Young equation (2.1.13) since $\cos\theta = 0$. A spherical drop with the same volume has $R = 2^{-1/3} R_h$ and with this relation, the surface free energy term in Eq. (2.3.11) is replaced by the value in Eq. (2.1.14) to yield

$$\Delta\mathbf{G} = -\frac{4}{3}\pi \frac{R^3}{V_M} k_B\, \mathbf{T} \ln \frac{\mathbf{p}}{\mathbf{p}_0}\, 2^{5/3}\pi\gamma$$

Since $2^{5/3} \approx 3.17 < 4$, this implies that in this example, there is a smaller cost in surface energy to form a drop of given volume on the surface than in the bulk. The barrier in Figure 2.12 is lowered and the critical radius is moved to lower values. The smaller the contact angle, the more favorable is the heterogeneous nucleation. In the limit of zero contact angle, the liquid wets the surface and can condense below saturation as discussed in Section 2.3.1.

Carbonated drinks like soda, beer, and champagne provide ample opportunities to study heterogeneous nucleation. After opening a bottle of soda pop, bubbles tend to form at the surface and on the container walls. There is also a visible difference in heterogeneous bubble nucleation when one compares drinks in a plastic cup or in proper glasses. There is a considerable difference in the surface free energies on these two types of materials. In champagne, one can often observe the phenomenon that bubbles seem to rise slowly but regularly from a few spots at the bottom of the glass. This is

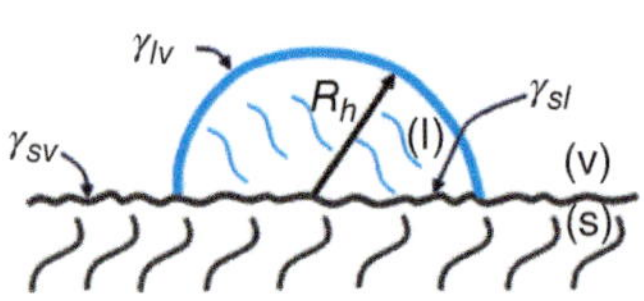

Figure 2.13 Heterogeneous nucleation occurs when a liquid drop emerges on a solid surface from a supersaturated gas phase.

Figure 2.14 Part of a spider web on an early summer morning. Photo by Olof Wennerström.

caused by bubble nucleation on a few small particle contaminants, often cellulose, rather than on the glass surface.

A direct test of the qualitative difference between heterogeneous and homogeneous nucleation rates can be performed on a cold day. While still indoors, you can put your eyeglasses, or some other glass material, close to the mouth and exhale. Since the air you exhale is practically saturated with water vapors at 37 °C, the air becomes supersaturated when it is cooled by the ambient air. Cooling is a faster process than convection/diffusion, so heterogeneous nucleation can occur on the glass surface since the air hitting the surface is supersaturated. Now you go outside where the temperature is below 0 °C. When you exhale, there is such a high degree of supersaturation that water droplets can nucleate homogeneously, and a little cloud is temporarily formed each time you exhale. This does not happen indoors since the homogeneous nucleation rate is too low.

In late summer, the temperature often decreases during the night below the dewpoint, and water condenses on surfaces. Figure 2.14 shows a detail of a spider net, where water droplets have condensed on the circumferential threads, while there are few drops on the threads in the radial direction, signifying a difference in the surface properties of the different types of threads.

2.3.4 Colloidal Particles Adsorb at Liquid–Liquid Interfaces

When we add colloidal particles to a two-phase liquid system, they will partition between the bulk phases α and β. For a particle of typical colloidal size, $R > 10$ nm, the solvation free energy in the two phases can be quantified by the particle surface free energy $4\pi R^2\gamma_{p\alpha(p\beta)}$. The partition coefficient $K_{\alpha\beta}$ for the particle is then $K_{\alpha\beta} = \exp\{-4\pi R^2\Delta\gamma/k_B\mathbf{T}\}$, where $\Delta\gamma = \gamma_{p\alpha} - \gamma_{p\beta}$. For typical values of the radius R, $K_{\alpha\beta}$ has a value that almost exclusively favors the particles being in one of the phases. This argument is based on bulk thermodynamics, but in practice, there is often an important

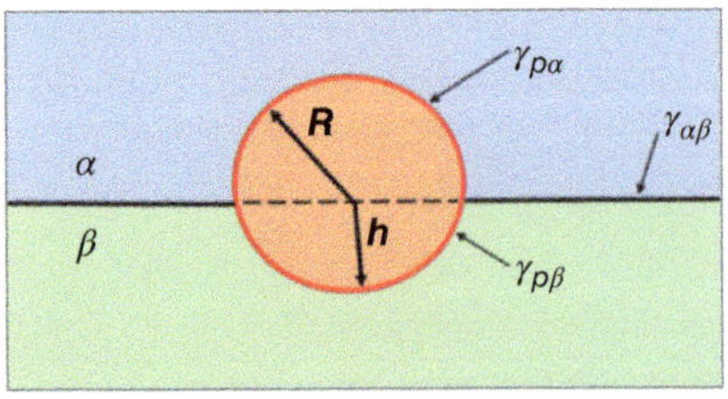

Figure 2.15 A particle can adsorb on a liquid–liquid interface to reduce the surface free energy.

surface effect. The most favorable position for a particle is often at the interface. Figure 2.15 shows a spherical colloidal particle adsorbing at the interface.

Assume that $\gamma_{p\alpha}$ is smaller than $\gamma_{p\beta}$ so that the particle favors the α-phase. Now, at the interface, it can penetrate at a distance h into the β-phase. The area of a spherical cap is $2\pi hR$ so that the particle solvation free energy increases by $2\pi hR\Delta\gamma$. In the process, an interfacial area of $\pi(2Rh - h^2)$ is eliminated. The net change in free energy is

$$\Delta\mathbf{G}(h) = 2\pi hR(\gamma_{p\beta} - \gamma_{p\alpha}) - \pi(2Rh - h^2)\gamma_{\alpha\beta} \tag{2.3.15}$$

The optimal value of h is then $h = R(1 - \Delta\gamma/\gamma_{\alpha\beta})$, and the free energy change is

$$\Delta\mathbf{G} = -\pi R^2\gamma_{\alpha\beta}\left(1 - \frac{\Delta\gamma}{\gamma_{\alpha\beta}}\right)^2 \tag{2.3.16}$$

Particles will be attracted to the interface when the interfacial free energy is larger than the difference between the interfacial free energies of the particle in the two liquids. This is, as a rule, indeed the case and particles often adhere to liquid–liquid interfaces. In Eq. (2.3.16), $\Delta\mathbf{G}$ is proportional to R^2 so that the larger the particle, the stronger is the tendency to go to the interface. One consequence is that the particles decrease the net interfacial free energy of the system, leading to emulsion stabilization, which can have large practical consequences in, for example, oil drilling.

The aforementioned discussion referred to a situation with an interface between two liquids. It also applies to the case when one of the phases is air (gas), with the small difference that the particle solubility in the gas is negligible. In the mining industry, flotation is widely used to separate valuable ore from the embedding mineral material. It is, for example, possible to separate out small gold particles present in the ppm range in the raw material by choosing an appropriate flotation system. This is done by, to the slurry of milled ore, adding a surfactant that selectively adsorbs on gold surfaces. Gold particles become hydrophobic and $\gamma_{\alpha\beta}$ is then reduced to make $\Delta\mathbf{G}$ of Eq. (2.3.15) sufficiently negative to ensure surface adsorption. When the slurry is aerated, gold particles float in the foam, which can be separated from the rest of the slurry.

See Berg (2010, Chapter 5), Cosgrove (2010, Chapter 6), Israelachvili (2011, Chapter 19), and Kronberg et al. (2014, Chapter 19) for additional discussion on Nucleation and Ripening.

2.4 Thermodynamics Can Be Extended to Include Surface Contributions

Conventional thermodynamics deals with, in principle, infinite systems, and surface effects become negligible. In Section 2.3, we combined thermodynamic bulk descriptions with effects due to surfaces and/or interfaces. One of the pioneers of thermodynamic theory, J.W. Gibbs, managed to extend the description to also include surface effects in a rigorous way.

2.4.1 The Gibbs Model Provides a Powerful Basis for Analyzing Surface Phenomena by Partitioning a System into Two Bulk Phases and a Dividing Surface

The properties of a surface depend on all the thermodynamic parameters of the bulk system. Gibbs developed a model for analyzing surface effects by dividing the system into three parts: the volumes of the bulk phases V' and V'' plus the surface that separates them. At equilibrium, the intensive variables take a uniform value. Extensive variables have a well-defined value for the total system, and we can define the surface component by seeing the total value as a sum of two bulk terms plus an interfacial one. For the internal energy, we define the surface part $\mathbf{U}_\sigma$ as

$$\mathbf{U}_\sigma = \mathbf{U} - u'V' - u''V'' \tag{2.4.1}$$

where $\mathbf{U}$ represents the total energy of the system, and u' and u'' are the energies per unit volume in the two phases. Other extensive surface quantities are defined using the procedure of Eq. (2.4.1).

By dividing the total volume of the bulk phases into precise quantities V' and V'', we construct an imaginary system that has the same thermodynamic properties as the real system, yet separates the two phases by an infinitesimally thin dividing surface. Although in a real system physical and thermodynamic properties change rapidly (but not discontinuously) at the interface, in this imaginary system they remain constant up to the surface. Replacing the real system with the model makes it easier to think of such extensive quantities as energy and volume in discrete segments rather than continuously varying quantities.

Molecular composition also changes across the interface. In a two-phase multicomponent system with volumes V' and V'', we can write c_i' and c_i'' as the concentration of component i in each phase. Then the number of moles of component i in each phase becomes

$$n_i' = c_i' V' \text{ and } n_i'' = c_i'' V'' \tag{2.4.2}$$

The amount of component i at the interface is

$$n_i^\sigma = n_i - c_i' V' - c_i'' V'' \tag{2.4.3}$$

where n_i is the total number of moles in the whole system. We can define a surface concentration or surface excess for component i as

$$\Gamma_i = \frac{n_i^\sigma}{A} \tag{2.4.4}$$

Equation (2.4.3) makes it clear that n_i^σ (and thus Γ) can be either positive or negative. Varying the location of the dividing surface by only a few angstroms can change the magnitude and sign of the value of Γ_i dramatically. For the surface of an aqueous solution, the convention is to fix the location of the dividing surface so that the surface excess of the solvent, H_2O is zero; $\Gamma_{H_2O} = 0$. For solutes that are attracted to the surface like surfactants, the surface excess becomes strongly positive, while for substances that are repelled from the surface, like most electrolytes, the surface excess is negative.

2.4.2 The Gibbs Adsorption Equation Relates Surface Excess to Surface Tension and the Chemical Potential of the Solute

We can develop thermodynamic relations for the surface variables in a way that is analogous to those used for bulk phases. For example, the differential of the surface internal energy is

$$\mathrm{d}\mathbf{U}^\sigma = \mathbf{T}\mathrm{d}\mathbf{S}^\sigma + \gamma \mathrm{d}A + \sum_i \mu_i(\mathbf{T})\mathrm{d}n_i^\sigma \tag{2.4.5}$$

where the bulk work term $\mathbf{p}dV$ has been replaced with the surface work term γdA. The same manipulations that lead to the relation $\mathbf{U} = \mathbf{TS} + \mathbf{pV} + \sum_i \mu_i n_i$ for the bulk yield (see Eq. 1.2.21)

$$\mathbf{U}^\sigma = \mathbf{TS}^\sigma + \gamma A + \sum_i \mu_i n_i^\sigma \tag{2.4.6}$$

for the surface quantities. If we then differentiate Eq. (2.4.6) and compare it with Eq. (2.4.5), we obtain the Gibbs adsorption equation:

$$\mathbf{S}^\sigma \mathrm{d}\mathbf{T} + A\mathrm{d}\gamma + \sum_i n_i^\sigma \mathrm{d}\mu_i = 0 \tag{2.4.7}$$

At a constant temperature, this gives the Gibbs adsorption isotherm

$$-\mathrm{d}\gamma = \sum_i \frac{n_i^\sigma}{A}\mathrm{d}\mu_i = \sum_i \Gamma_i \mathrm{d}\mu_i \tag{2.4.8}$$

which relates the change in surface tension, $\mathrm{d}\gamma$, to changes in the chemical potential, $\mathrm{d}\mu_i$, through the surface excess. This thermodynamic equation is generally valid, but in practical applications it is useful to judiciously specify the location of the dividing surface in order to arrive at a conceptual and algebraic simplification of Eq. (2.4.8). Rather than keeping the general thermodynamic description, we consider an interface between a vapor and a binary solution so that Eq. (2.4.8) reduces to

$$-\mathrm{d}\gamma = (\Gamma_1 \mathrm{d}\mu_1 + \Gamma_2 \mathrm{d}\mu_2) \tag{2.4.9}$$

In many solid systems, interfaces between phases are sharply defined and concentration changes abruptly on an angstrom scale in slight contrast to liquid systems, where concentrations vary across an interfacial region that is typically a few molecular diameters thick, that is, on the order of a nanometer. Figure 2.16 illustrates this variation by showing how the water concentration changes from 55 M in bulk liquid to 1×10^{-3} M in the vapor phase (20 °C). Note that although the concentration of water changes dramatically, its chemical potential remains equal in the two bulk phases and also remains constant across the interface. This situation is analogous to the relation between concentration and chemical potential in external fields discussed in Section 1.2.5.

To determine the surface excess of water, Γ_{H_2O}, in moles per unit area, we first locate in the interface a dividing plane situated at a specific distance (Figure 2.16a), z_0, and define

$$\begin{aligned} \Delta c(z) &= c(z) - c_{\mathrm{liq}} \quad \text{for} \quad z < z_0 \\ \Delta c(z) &= c(z) - c_{\mathrm{vap}} \quad \text{for} \quad z > z_0 \end{aligned} \tag{2.4.10}$$

Then we obtain the surface excess, Γ_{H_2O}, as an integral

$$\Gamma = \int_{-\infty}^{+\infty} \Delta c(z)\,\mathrm{d}z \tag{2.4.11}$$

or

$$\Gamma = \int_{z_0}^{+\infty} (c(z) - c_{\mathrm{vap}})\,\mathrm{d}z + \int_{-\infty}^{z_0} (c(z) - c_{\mathrm{liq}})\,\mathrm{d}z \tag{2.4.12}$$

across the interfacial region. The integral is finite since $\Delta c = 0$ well into the respective bulk phases. By projecting an interfacial volume onto an interfacial area, we have formulated the concept of a surface concentration or surface excess.

If we choose to locate z_0 as shown in Figure 2.16a, then $\Gamma_{H_2O} = 0$. The area below z_0 is defined by the second integral of Eq. (2.4.12). This area is negative because $c(z) < c_{\mathrm{liq}}$, while the area for

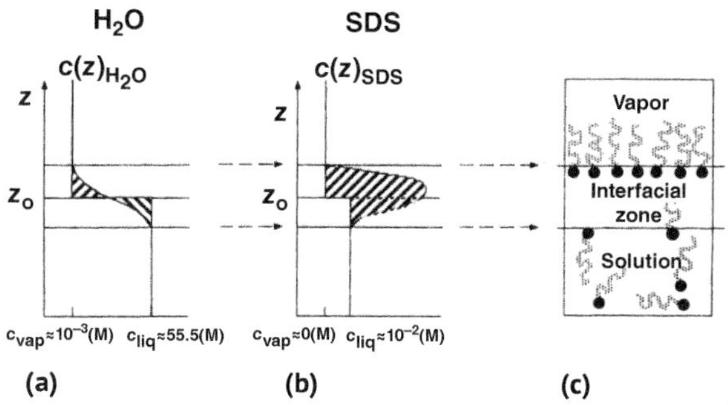

Figure 2.16 Illustration of how the concentrations of water, $c(z)_{H_2O}$ (a), and SDS, $c(z)_{SDS}$ (b), vary across the interphase (vertical axis, height relative to surface, horizontal axis, concentration). The position z_0 of the Gibbs dividing surface is chosen so that the surface excess $\Gamma_{H_2O} = 0$. For SDS, this choice of z_0 has the result that there is a large surface excess $\Gamma_{SDS} \gg 0$. Panel (c) shows a molecular picture of the interfacial zone.

$z > z_0$, defined by the second integral, is positive because $c(z) > c_{vap}$. Since the two areas are equal but opposite in sign, Γ_{H_2O} is zero. This is the conventional choice of the position of the Gibbs dividing surface for aqueous systems. For SDS, this choice of z_0 leads to $\Gamma_{SDS} > 0$ since both areas make positive contributions.

The usefulness of this specific choice $\Gamma_{H_2O} = 0$ for the dividing surface is that when we apply Eq. (2.4.9) to a binary system, this reduces to

$$d\gamma/d\mu_2 = -\Gamma_2 \tag{2.4.13}$$

We can relate Eq. (2.4.13) to accessible experimental quantities. Assume that one has dilute conditions so that the chemical potential of the solute is $\mu_2 = \mu_2^\theta + \boldsymbol{RT} \ln c_2$. Differentiation at constant temperature yields $d\mu_2 = \boldsymbol{RT}$ dln c_2. Combining this with Eq. (2.4.13) yields

$$\Gamma_2 = -\frac{1}{\boldsymbol{R}T}\frac{d\gamma}{d\ln c_2} \tag{2.4.14}$$

For an ionic system like a SDS solution, there are both anions and cations, so the chemical potential of a salt AC is at ideal mixing

$$\mu_{AC} = \mu_{AC}^\theta + \boldsymbol{RT}(\ln c_A + \ln c_C) = \mu_{AC}^\theta + 2\boldsymbol{RT}\ln c_{AC} \tag{2.4.15}$$

For such a case, like for SDS, there is an extra factor ½ in the expression for Γ in Eq. (2.4.14).

Problem: According to Figure 1.5, the surface tension of SDS solutions decreases from 49 mN/m^2 at c = 5 mM to 40 mN/m^2 at c = 8 mM. Calculate the surface excess in this concentration regime.

Solution: The surface excess Γ is then, according to Eqs. (2.4.14) and (2.4.15), $\Gamma = -(49 - 40)\times 10^{-3}/\{2\boldsymbol{RT}\ \ln(5/8)\} = 3.9 \times 10^{-6}$ mol/m^2 = 2.2×10^{18} molecules/m^2, which corresponds to 46×10^{-20} m^2/molecule. This value is somewhat smaller than the value relevant for SDS micelles.

We note that the reason why the surface tension decreases with increasing concentration for SDS in the mM concentration regime is not that the surface becomes more covered by surfactant molecules. The constant slope in the region c = 5 mM to 8 mM SDS seen in Figure 1.5 implies, according to Eq. (2.4.14), that the surface coverage is constant in this region. The molecular interpretation is then that creating a new surface involves recruiting surfactant molecules, including counterions, from the bulk. The lower the bulk concentration, the larger is the free energy cost in bringing molecules from bulk to surface.

Figure 1.5 also shows that the surface tension varies only slightly with increasing concentration above the CMC. To see why, we go back to the more general Eq. (2.4.13). Above the CMC, molecules enter micelles and the chemical potential of the surfactant remains virtually unaffected by a change in concentration, and the chemical potential is practically constant. According to Eq. (2.4.13), there is no change in surface tension if there is no change in chemical potential of the solute. Seen in molecular terms, the free energy cost of recruiting a molecule from the bulk to the surface is independent of concentration since the product of surfactant and counterion concentrations is virtually unchanged.

In NaCl solutions, the surface tension increases from 72 mJ/m^2 in water to 74.4 at 1 M, making Γ_{NaCl} negative. This increase corresponds to a surface concentration that is less than that of the bulk solutions. Ions are repelled from the liquid–vapor interface by the lower dielectric constant of the vapor.

Problem: What is the value of Γ_{H_2O} when the reference place z_0 is chosen so that $\Gamma_{SDS} = 0$?

Solution: To equalize the areas defined by Eq. (2.4.12), we must locate z_0 far into the vapor phase. Since we can choose to locate it wherever we want, this location is formally acceptable, but in practice, such a choice could easily lead to conceptual confusion. In order to find the value of Γ_{H_2O}, we extrapolate the bulk concentration of water up to the reference plane and observe that $\Gamma_{H_2O} \ll 0$.

2.4.3 The Langmuir Equation Describes Adsorption at Solid Interfaces Where We Cannot Measure Surface Tension Directly

Solubilization, detergency, and flotation are among the many applications that involve adsorption of surfactant molecules onto solid surfaces. We cannot characterize the adsorption process in these situations by the Gibbs isotherm because changes that occur in surface tension as a result of adsorption at solid interfaces are not experimentally accessible. Instead, we describe the behavior of such systems in terms of adsorption isotherms.

The Langmuir adsorption isotherm is conceptually simple and widely used to represent experimental data. It is most easily derived using a two-dimensional version of the lattice model introduced in Section 1.1. Assume that there are N independent binding sites on a surface in equilibrium with a dilute solution (or low-pressure gas) of the adsorbent. Then we can follow Eq. (1.1.18) and write the free energy of the surface as

$$\boldsymbol{G}_s = \boldsymbol{G}_s^\theta + N_2\mu_s^\theta + \boldsymbol{RT}\{N_2 \ln X_2 + (N - N_2)\ln(1 - X_2)\} \tag{2.4.16}$$

where N_2 stands for the number of adsorbed solute molecules, N is the total number of sites, and $X_2 = N_2/N$. The chemical potential at the surface is thus

$$\left(\frac{\partial \boldsymbol{G}}{\partial N_2}\right) = \mu_2(s) = \mu_2^\theta(s) + \boldsymbol{RT}\ln\left\{\frac{X_2}{1 - X_2}\right\} \tag{2.4.17}$$

In the bulk solution (b),

$$\mu_2(b) = \mu_2^\theta(b) + \boldsymbol{RT} \ln c_2(b) \tag{2.4.18}$$

where $c_2(b)$ is the concentration of solute in the solution.

At equilibrium, the two chemical potentials are equal, and Θ, the fraction of the surface site occupied by the solute molecules, becomes

$$\Theta = X_2 = \frac{Kc_2(b)}{Kc_2(b)+1} \tag{2.4.19}$$

where K stands for the equilibrium constant,

$$\boldsymbol{RT} \ln K = \{\mu_2^\theta(b) - \mu_2^\theta(s)\} \tag{2.4.20}$$

Equation (2.4.19) is called the Langmuir adsorption isotherm.

At high solute concentrations or high surface affinities, K large, $Kc_2(b) \gg 1$ and $\Theta \cong 1$, which means that the surface is saturated. At the other limit, $Kc_2(b) \ll 1$ and Θ is proportional to $c_2(b)$.

The assumptions used to derive the Langmuir equation are rather stringent. For example, ideal behavior on the surface implies a homogeneous surface that lacks dislocations or any other structural nonidealities that might induce preferred adsorption. Interactions between adsorbed molecules are ignored, and only monolayer formation can be included. Although we almost never realize these conditions in practice, the Langmuir isotherm provides, in analogy with the ideal mixing model for solutions, a conceptual basis for thinking about surface adsorption as well as a basis for modeling the adsorption process.

The Bragg–Williams model, which is an application of the regular solution description, Section 1.1, to adsorption, provides a simple extension of the Langmuir picture. The interaction between adsorbed molecules on neighboring sites is assigned a value W. One identifies component A in the regular solution case with an adsorbed molecule on the surface, while component B represents an empty site. The number of coordinated neighbors is lower on the surface, but the basic equations are the same in the two models. For attractive neighbor interactions $W > 2\boldsymbol{RT}$, the model predicts a two-dimensional phase separation on the surface, in analogy with the bulk phase separation occurring for $w > 2\boldsymbol{RT}$.

For many applications to colloids, the notion of surface binding sites is less relevant, and the Langmuir model is quantitatively not relevant. In the typical case, one has a uniform surface/interface to which colloidal particles can adsorb. For a low surface coverage, the Langmuir equation applies, since the two-dimensional system behaves ideally as long as one can neglect direct particle–particle interactions. For higher coverages, the behavior depends on the character of the lateral interaction between the adsorbed species, but in the same way as in the Langmuir model, there is an upper limit to the coverage.

For the case of adsorption to a uniform surface in the xy-plane, we can relate the equilibrium constant in Eq. (2.4.20) to the particle surface interaction potential $V(z)$, and in analogy with the discussion in Chapter 8, one can show that the equilibrium constant is related to the interaction potential $V(z)$ as

$$K = \int_{z_b}^{z_a} \exp\left\{-\frac{V(z)}{k_B\mathbf{T}}\right\} dz \tag{2.4.21}$$

Here, $z_b < z < z_a$ represents the interval at the surface where the particle is considered as bound; $V(z) < 0$.

See Atkins et al. (2022, Focus 18), Berg (2010, Chapter 3), and Lyklema (2000, Vol. 3, Chapter 4) for additional discussion on Surface Thermodynamics.

2.5 Monolayers of Insoluble Amphiphiles Form Independent Two-dimensional Systems

In the discussion of surfaces in the previous sections, we have assumed that one can, in practice, achieve equilibrium between surface and bulk phase. For amphiphiles with sufficiently long hydrocarbon chains or those with two chains, the aqueous solubility can be extremely low, making the exchange between bulk and surface so slow that equilibrium is not reached over experimentally realizable times. Once formed, the monolayer on the surface can be treated as an independent system.

2.5.1 Monolayers Formed by Insoluble Amphiphiles Behave as a Separate System and Can Be Characterized by a Langmuir Balance

Amphiphiles orient at the water–air interface and form, as we found for SDS, a rather dense monolayer. There is an equilibrium between SDS molecules in the bulk and on the surface. Considering the surface film as a separate phase, one can assign a standard free energy μ_s^θ to this state. The equilibrium with the bulk is then obtained as

$$\mu_s^\theta = \mu_w^\theta + \boldsymbol{RT} \ln (c) \tag{2.5.1}$$

In 1917, Langmuir analyzed data for several homologous families of amphiphiles. He assumed that adding surfactant to lower γ enables us to measure the relative values of μ_s^θ directly from the values of the concentration that produce an equal lowering of the surface tension in each solution. Langmuir showed that all the results for a series of fatty acids, esters, and alcohols with a varying number of carbon atoms μ_s^θ could be expressed by

$$\mu_s^\theta - \mu_w^\theta = \text{constant} - 2.6\ n_C \left(\frac{\text{kJ}}{\text{mol}}\right) \tag{2.5.2}$$

where the constant depends only on the head group. An analysis for the transfer of similar compounds from water to the neat phase shows that $\mu_s^\theta - \mu_w^\theta$ decreases by 3.4 to 3.7 kJ/mol per methylene group (see Table 1.6). The solubility decreases nearly an order of magnitude per two CH_2-groups, and for large n_C, the bulk concentration becomes very low. It then takes a considerable time to achieve equilibrium between the surface film and the bulk due to the slow diffusional transport between surface and bulk. Thus, for amphiphiles with more than 22 carbons in the chain(s), the monolayer can be treated as a separate system on a typical experimental timescale. The surface film can then be characterized thermodynamically by the variables temperature, area, amount of amphiphile, and a two-dimensional pressure Π_s, which has the same dimension as the surface tension (force/area).

Using the Langmuir balance, we can directly measure Π_s as a function of vapor–liquid interfacial area. As Figure 2.17 shows, the balance employs floats, which act as barriers and separate the film from the clean solvent surface. The floats can move in response to an applied pressure, Π_s, while the area of the film is monitored from the position of the floats.

Before the film is deposited, the entire surface can be cleaned by sweeping the movable barriers back and forth across the trough. A known amount of the amphiphile, dissolved in a volatile solvent, is then carefully pipetted onto the aqueous surface. After the organic solvent evaporates, Π_s is measured as a function of area.

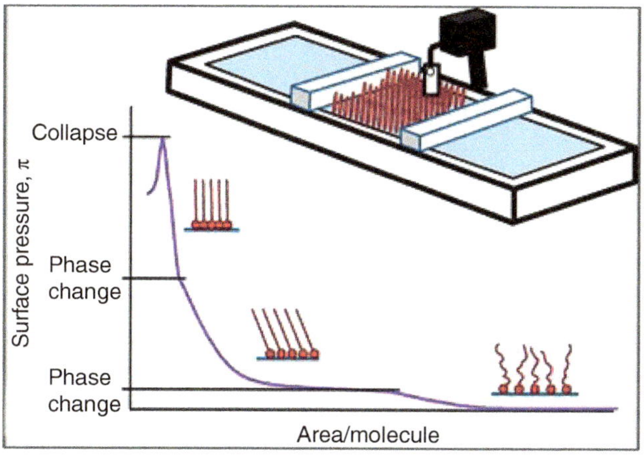

Figure 2.17 A Langmuir film balance with a trough, two floats, a mirror, and a main torsion wire. The deflection of the float is measured optically using the mirror. The inset shows a schematic pressure–area relation.

2.5.2 Insoluble Monolayers Are Characterized by Π_s Versus Area Isotherms

Figure 2.18 displays a schematic of features of pressure–area isotherms observed for monolayers containing a single amphiphile. At very low pressures, molecules are far apart, and as Π_s approaches zero, we can apply a two-dimensional equivalent of the van't Hoff for osmotic pressure (Eq. 1.2.13):

$$\Pi_s A = n\mathbf{RT} \tag{2.5.3}$$

or in a version more useful for our purposes

$$\Pi_s a_0 = k_B \mathbf{T} \tag{2.5.4}$$

where $a_0 = A/(nN_{Av})$ signifies the area that each molecule occupies. In this so-called gaseous region of the isotherm, the hydrocarbon chains lie flat on the surface. These equations are valid when the dominant contribution to the free energy comes from the entropy of the individual molecule confined to an area A in the same way as for an ideal gas whose only free energy contribution is due to the entropy of a gas molecule confined to a volume V.

In analogy with the condensation of real gases to liquids, the gaseous surface film, with an area per molecule of order 400 nm^2, condenses to a liquid film with an area per molecule in the range 0.3–0.5 nm^2 for single-chain molecules. This liquid-expanded, l_e, state gradually transforms to a liquid-condensed, l_c, state under further increase of the pressure. Finally, we observe at the end of the isotherm high Π_s, low a_0, a solid-like behavior. In this case, the hydrocarbon chains are oriented vertically and packed together. The film is relatively incompressible. For alcohols and esters, $a_0 \approx 0.19$ nm^2, which corresponds to the close-packed cross section we obtain for bulk, solid hydrocarbons by X-ray scattering. Compressing the solid region further leads to the collapse of the film as shown in Figure 2.18. In the two-dimensional system, packing constraints are less severe than in three dimensions. A more careful study of the region with liquid-expanded and liquid-condensed phases often reveals a rich variation in packing behavior, resulting in complex phase properties, which will be further discussed in Chapter 10.

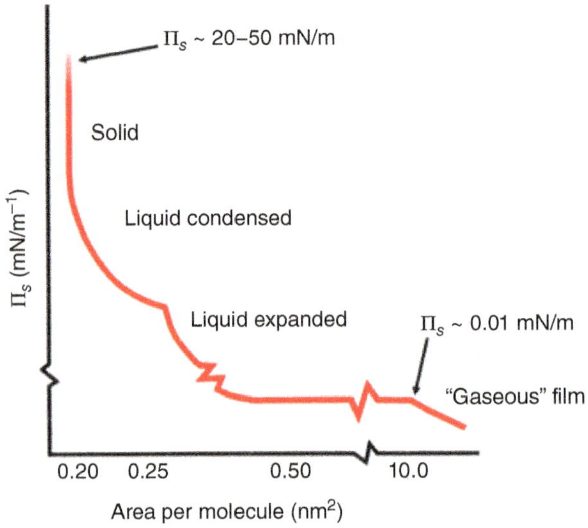

Figure 2.18 Composite two-dimensional pressure π_s–area a_0 isotherm, which illustrates a wide variety of liquid-like and solid-like monolayer phenomena. At π_s of order 0.01 mN/(nm)^2, there is a transition from a low coverage to a liquid surface film. At $\pi_s > \frac{20\text{ mN}}{(\text{nm})^2}$, there is a collapse of the film.

2.6 A Range of Experimental Methods Can Be Used to Study Surfaces and Interfaces

Surfaces and interfaces have a limited extension in one direction so that they contain relatively few molecules. It is an experimental challenge to selectively study the properties of the molecules in the interfacial region. In this section, we describe some methods that can be used to study surface and interfaces of particular relevance for colloidal systems. Conditions differ at the air–liquid surface from those at a liquid–liquid interface, which in turn are different from those at a liquid–solid boundary. Some methods are useful for all three situations, while others work best only for one of the cases.

2.6.1 The Langmuir–Blodgett Technique Provides a Way to Deposit Monolayers or Multilayers onto Solid Surfaces

By modifying the Langmuir balance, we can deposit monolayers on a solid substrate to create mono- or multilayer films. Modifications include a dipping device to lower or raise the substrate through the monolayer, an automated movable barrier, which moves during the deposition process so as to maintain a control value of Π_s and a surface pressure sensor, which controls the movable barrier.

The Langmuir–Blodgett technique provides a way to construct organized systems of molecules—one monolayer at a time, as illustrated in Figure 2.19. Potential applications of the technique include electron beam resists for micro- and macrolithography, insulating films in semiconductor devices, nonlinear optical elements, coatings for communication devices, biosensors, and highly selective membranes for biotechnology. These application opportunities have stimulated considerable research interest in the basic physics and chemistry of the assembly process, synthesis of appropriate amphiphilic molecules, and assessment of the character and

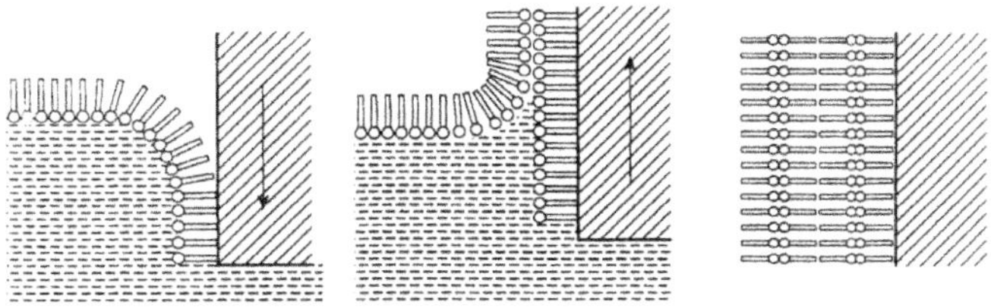

Figure 2.19 Deposition of multiple monolayers by the Langmuir–Blodgett technique. *Source*: Weissberger and Rossiter, 1971/with permission of John Wiley & Sons.

durability of the resulting films. A major practical problem in these applications is the long-term stability of the film.

2.6.2 Fluorescence Microscopy Can Visualize the Aggregation State of Monolayers Directly

The ability to visualize amphiphilic microstructures deposited on the surface of the Langmuir balance is a tremendous aid in understanding monolayer systems. Fluorescence microscopy can detect molecules at exceedingly low concentrations and permits us to visualize phase transformations in monolayer films. Figure 2.20 shows how a liquid-condensed phase, LC, appears as domains in a liquid-expanded state.

The added fluorescent dye preferentially dissolves in the LE phase, and the dark regions are domains of the LC phase. Since the molecules are oriented in the film, the monolayers possess a dipole moment, which causes an electrostatic repulsion between the domains (see Chapter 3). The DPPC molecules are chiral, and the LC domains acquire an in-plane chirality. The "kidney-shaped" domain geometry is caused by an interplay between the dipole–dipole interaction and the chirality, although the detailed mechanism is not fully understood.

2.6.3 Atomic Force Microscopy Permits Imaging of Structures at Interfaces and Surfaces

In the atomic force microscope (AFM), a molecularly sharp tip is interacting with the molecules on a preferably solid surface. The tip is mounted on a cantilever spring, whose mechanical deviation is monitored and controlled by a piezoelectric device. When the tip is allowed to scan the surface at constant load, the molecular topography of the surface is reflected by the deflection of the tip. Figure 2.21 shows an AFM image of a hydrophobic surface covered with elongated SDS micelles. These are packed in large, ordered domains interrupted at domain boundaries. The consequence of such lateral surface inhomogeneities will be further discussed in Chapter 5.

2.6.4 X-ray and Neutron Scattering Can Be Used to Study Molecular Organization at Surfaces and Interfaces

X-rays and effectively also neutrons have a wavelength that is short on the molecular scale and can be used to monitor structure on a molecular scale. The general discussion of scattering methods is postponed to Chapter 4 and here we discuss only the specific challenges met in studying surfaces/interfaces. There are much fewer molecules on a surface than in a bulk sample, and there is a general

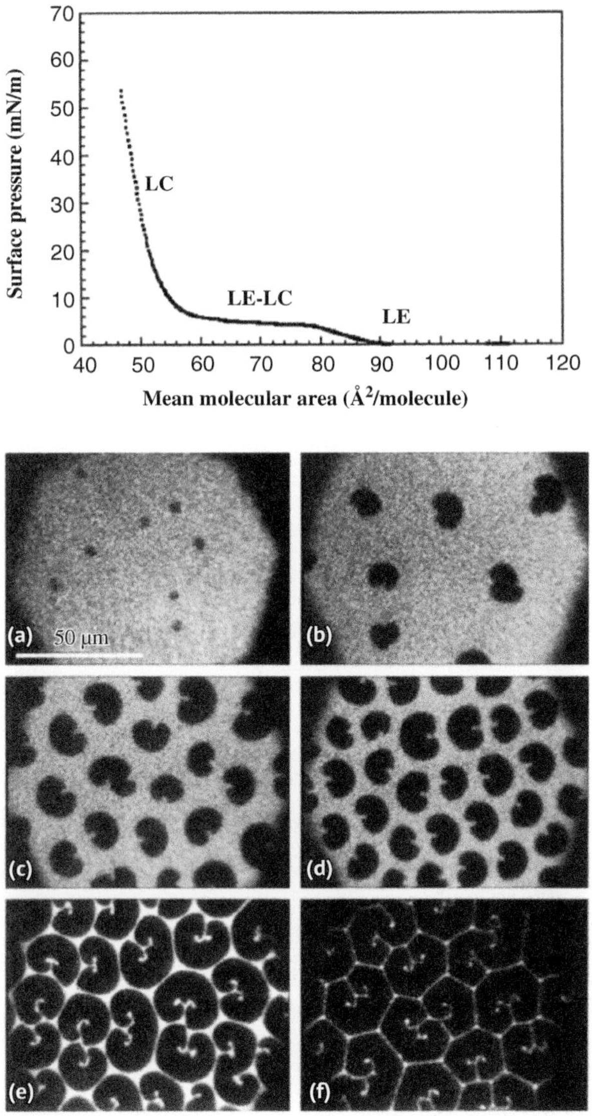

Figure 2.20 Upper panel: surface pressure versus molecular area of a dipalmitoylphosphatidylcholine (DPPC) film at the air–water interface. LE denotes the liquid-expanded and LC the liquid-condensed phases (see Figure 2.18). Lower panel: fluorescence microscope images of the film obtained using a fluorescent dye at low concentration. The images (a)–(f) are for compositions with a coexistence of the LE and LC phases, with the largest molecular area for (a) and the lowest for (f). *Source*: C W McConlogue et al., 1997/with permission of American Chemical Society.

intensity problem in the study of surfaces. In X-ray scattering and neutron reflectivity studies, this problem can be partly solved by maximizing the path length in the interfacial region through having the incoming beam in a direction that is nearly parallel to the interface. This provides a good resolution of the molecular organization perpendicular to the interface. An X-ray study can, for example, determine the order of alkyl chains in an amphiphilic layer. In neutron reflectivity studies, one has the extra possibility of contrast variation by isotope labeling, which can yield even more detailed information.

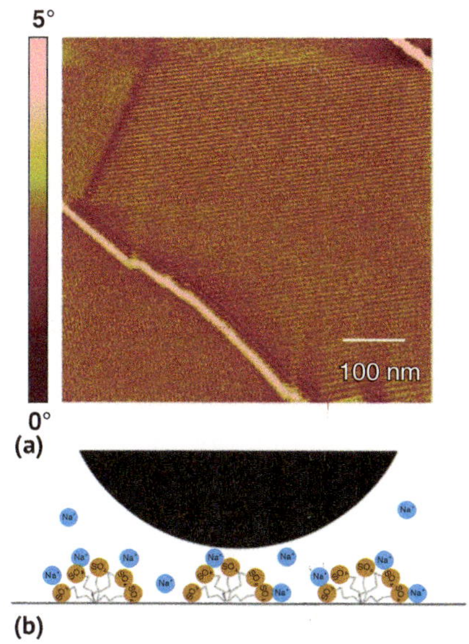

Figure 2.21 (a) AFM imaging of a hydrophobic surface on which SDS micelles have adsorbed. There are large domains where the micelles are ordered in stripes. (b) Schematic picture of how the tip, solid black, interacts with the micelles. *Source:* Jaschke et al., 1997/with permission of American Chemical Society.

2.6.5 Visible Light Shows a Range of Surface/Interface Specific Phenomena

Visible light is sensitive to the abrupt variation in the refractive index, n_r, across a phase boundary. This variation in n_r generates scattering effects that depend on the polarization of the light, which has a wavelength that is much larger than molecular sizes. Adsorption of molecules or aggregates at the surface/interface gives rise to local variations in the refractive index. There exist several techniques that can be used to monitor adsorption in a noninvasive and rapid way. Generally, when light falls on an interface, one part of the beam is refracted into the new medium and another part is reflected. The intensity ratio between the two parts depends on both the angle of incidence and the direction of the polarization. At the Brewster angle, one polarization component is not reflected, resulting in a perfect polarization of the reflected beam. This effect can be used in Brewster angle spectroscopy, where the adsorption of molecules causes a slight change in the optical properties and consequently in the value of the Brewster angle.

When light falls on an interface from a low to a higher value of the refractive index n_r, there is an incidence angle below which the light is totally reflected, and the interface appears as a mirror. Even if the light is totally reflected, this occurs over a range in the perpendicular direction. This generates an evanescent light wave, which decays exponentially in amplitude into the reflecting medium. The evanescent wave selectively probes the interfacial region, and adsorption of particles/molecules gives rise to observable effects, particularly using fluorescent markers. This method is called total internal reflection fluorescence (TIRF).

Elliptically polarized light represents an intermediate stage between completely polarized and unpolarized light. Since the reflection amplitude at an interface depends on the direction of polarization, the reflected light is, in general, elliptically polarized. The degree of ellipticity depends on the refractive index at the interface, which in turn is affected by molecular adsorption. This is the basis for using ellipsometry to observe adsorption processes in real time (see Figure 2.22 in Section 2.6.7).

2.6.6 Sum-frequency Spectroscopy Provides a Surface-selective Spectroscopic Method

Visible, UV, and IR spectroscopy are widely used to study molecular properties in bulk phases. It is very demanding in a bulk system to reveal properties of an interface since there is a close-to-negligible contribution to the total signal from those molecules. However, at the interface, the inversion symmetry is broken. This provides a contrast to the property of an isotropic medium like a liquid. There are, consequently, spectroscopic transitions that can occur for interfacial molecules, but not for molecules in the solution. Focusing on such transitions, one can selectively study the molecules in the immediate vicinity of a surface/interface. In practice, there are two methods used to study interfaces using such nonlinear optical effects.

In surface second harmonic generation (surface SHG), two incoming photons combine to double the frequency. Such transitions are symmetrically forbidden in an isotropic medium but allowed at the interface. By combining this effect with normal spectroscopical transitions, one can obtain information about the amount and partial orientation of molecules at the interface.

In a development of this method, two laser beams are focused on the interface, and photons from the two beams can combine to form a photon with a frequency that is the sum. This is called sum frequency generation (SFG) spectroscopy, and it is often a more versatile method than surface SHG. The two methodologies are similar in the type of molecular information that can be obtained.

2.6.7 The Adsorption on a Solid Surface Can Be Measured by a Quartz Crystal Microbalance and Surface Plasmon Resonance

A quartz crystal is macroscopically dipolar, which generates the piezoelectric effect. When exposed to an alternating voltage, it resonates at a very precise frequency, which depends on the crystal mass. When a substance is adsorbed on the surface, the resulting change in mass gives rise to a frequency shift that is readily monitored. It is possible to follow an adsorption process in real time after a proper calibration of the device. When a crystal resonates in a liquid medium, there is a viscous damping broadening the resonance frequency. Measuring both frequency shift and broadening in a QCMB device provides information on both the adsorbed amount and the dynamics at the interface. By combining with an ellipsometry measurement on the same system, one can get detailed information on the amount and the character of the adsorbed layer as illustrated in Figure 2.22.

The surface plasmon spectroscopy is a useful alternative to the QCMB method. The surface, whose properties one wants to monitor, is deposited on a thin metal film, usually gold or silver. This film has, in turn, been deposited on one side of a glass prism. When laser light, visible or infrared, passes through the prism, it is reflected by the metal film. However, at a specific angle of incidence, the light couples strongly to collective modes of the conducting electrons called surface plasmons. At such resonance conditions, the photons are transiently trapped in the metal film, leading to multiple reflections at the two interfaces: metal–glass and metal–sample. As described in Section 2.6.5, there is an exponentially decaying evanescent wave into the adjacent medium at such reflections. The net result is that a single photon probes the properties of the sample multiple times, leading to a substantial enhancement of the sensitivity. Adsorption at the surface leads to a change in refractive index of the interfacial region, resulting in a measurable shift of the resonance condition.

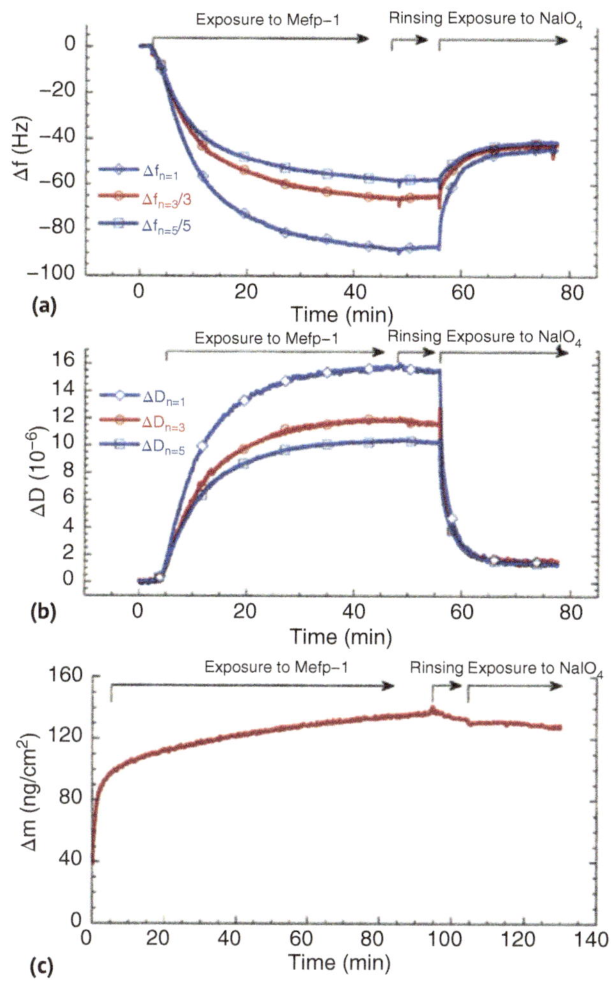

Figure 2.22 Adsorption of the protein Mefp-1 on a silica surface. (a) Frequency shift in a quartz crystal microbalance (QCMB) measurement. (b) Damping of the QCMB signal. (c) Adsorbed amount measured by ellipsometry. The drastic drop in the damping (panel (b)) caused by the addition of $NaIO_4$ is due to a transition from an extended protein layer to a compact one. *Source*: Höök et al., 1999/with permission of American Chemical Society.

See Atkins et al. (2022, Focus 18), Berg (2010, Chapter 3), Israelachvili (2011, Chapter 19), and Lyklema (2000, Vol. 3, Chapter 3) for additional discussion on Monolayers and Adsorption.

2.7 Evaporation at a Surface Leads to Intriguing Nonequilibrium Phenomena

It is the rule rather than the exception that a liquid sample is not in equilibrium with the ambient gas, which typically is air. As pointed out in Chapter 1, indoor air has a low relative humidity, corresponding to a very low chemical potential of water. This is often also the case for outdoor air. As a result, water molecules tend to evaporate from aqueous phases and other moist materials. The equilibration is slow, and it gives rise to other transport processes coupled to the evaporation.

2.7.1 Evaporation at a Water Surface Yields a Local Cooling Effect

A common situation is that essentially pure water is exposed to an atmosphere of moderate to low humidity. This happens not only in lakes and water reservoirs but also in a glass of water. The low water vapor pressure in the air gives rise to continuous water evaporation. Yet it is found that it takes days for a glass of water to dry out. The process is rather slow, although there is no obvious kinetic barrier to pass. There are essentially two possible rate-limiting processes. One is the transport of water molecules in the gas phase. As they leave the liquid phase, there is a diffusive transport away from the surface. As discussed in Section 1.7, diffusion coefficients are much higher in a gas than in liquid, but this is compensated by the fact that the volume concentration of water molecules is much lower in the gas. Diffusion is only the dominant transport process in a thin layer closest to the surface, while in the bulk gas convective flow ensures efficient mixing. At steady state, a gradient of the water vapor pressure is established across the surface gas layer. At the upper end of the diffusion layer, the vapor pressure equals that of the bulk, while the boundary condition at the interface is determined by the other rate-limiting process. Water has a substantial heat of vaporization, and the surface is constantly cooled.

Heat transport is more efficient in the condensed phase. Below the surface, convective flow ensures temperature equilibration with the bulk liquid. In analogy with the situation in the gas phase, there is an unstirred layer below the surface, where heat conduction is the main mechanism for heat transport. At steady state, a dip in temperature is established at the interfacial region, as illustrated in Figure 2.23.

An illustration of the importance of heat conduction in water vaporization is found in the so-called "coffee stain effect." It is an everyday experience that if you spill a small quantity of coffee on a table and let it dry out, it leaves a dark ring at the edges of the original liquid drop with a lighter coloring toward the center. This shows that the evaporation mainly occurs where the drop edge meets the table surface. Heat conduction is efficient in the solid, and the best thermal contact between the table material and the air–liquid surface occurs at the rim of the drop. When the water dries out, the "coffee particles" are sedimented on the surface. The liquid–solid contact line is pinned by the precipitate so that the evaporation at the rim gives rise to a liquid flow from the center to the periphery of the drop. As the process continues, the thickness of the liquid film becomes more homogeneous, and the evaporation becomes equally efficient across the remaining liquid.

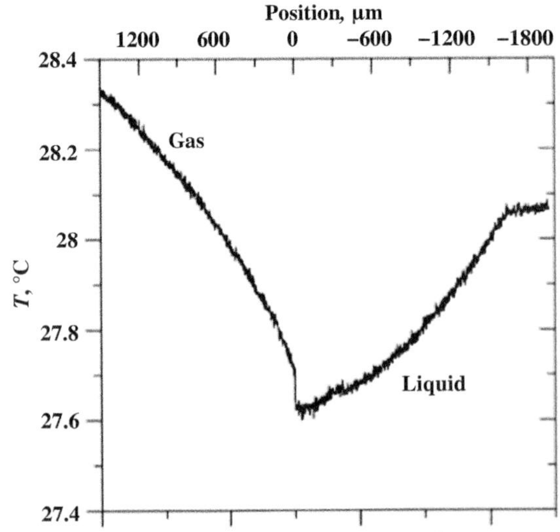

Figure 2.23 Temperature profile across an air–water interface at steady state. The liquid water is heated from below to compensate for the heat of vaporization. *Source*: Gatapova et al., 2017/with permission of Elsevier.

2.7.2 Tears of Wine Are Caused by Selective Evaporation

Soon after pouring wine or even better brandy into a glass, one can observe drops of liquid flowing down the glass surface above the bulk liquid surface. This phenomenon is called tears of wine, and observation can reputedly be traced as far back as to the Bible (Proverb 3.22). The dynamic process is driven by the preferential evaporation of ethanol from the water–ethanol mixture in the wine or brandy. The liquid coming from the bottle wets the surface and climbs the wall of the glass in the manner illustrated in Figure 2.7.

Evaporation is efficient in such a thin liquid film, and the evaporated liquid is compensated by an upward flow of liquid within the film. Since ethanol evaporates more easily than water, the composition in the film gradually changes and water is accumulated. If the evaporation is rapid, back-diffusion of water is too slow to compensate for the upward flow of liquid and one encounters an instability as illustrated in Figure 2.24. The upper part of the film contracts laterally to form drops, which in turn flow downward due to gravity. This process continues as long as the bulk liquid holds a sufficiently high alcohol content. In the steady state, there is an upward flow of liquid in the surface film covering most of the circumference of the glass, ethanol evaporation in the film, and a downward flow of water-rich drops.

The existence of tears of wine is often given as an example of the Marangoni effect. This states that the presence of a gradient in surface tension drives a liquid flow to eliminate the gradient. The explanation provided earlier would be valid even if the ethanol evaporation has no effect on the surface free energy. It is the accumulation of the non-evaporating component (water) in the surface film that drives the instability.

A similar spectacular phenomenon is observed when liquid helium is kept in a container. The low surface tension of liquid helium ensures that it wets the surface of the container, forming a thin film. Now, liquid helium has a very low density, with the consequence that the film climbs high up on the container wall. Furthermore, the viscosity is also low, so if care is not taken, the liquid simply flows out of the container via the wetting film!

(a)

(b)

Figure 2.24 (a) Tears of wine in a glass of brandy. (b) The lower panel shows a close-up of the liquid rim and the tears floating down. *Source*: Courtesy of Olof Wennerström.

2.7.3 Evaporation from an Aqueous Solution Generates Gradients of Chemical Potential Below the Surface

Drying is an important technical process, but it is also an everyday phenomenon. The air around us is often of low relative humidity (see Figure 1.4). Living cells cannot function properly at low chemical potentials of water, and many strategies have been invented by evolution to cope with life on land. Mammals and birds have a skin that substantially decreases the evaporation rate, and they have, in addition, feathers and fur as protection. Simpler organisms seek humid places and wait for humid conditions to explore surroundings that are usually dry. Bacteria shut down their metabolism in a dry environment and start to reproduce only when the relative humidity is high. Although it is very difficult to prevent water from vaporizing from a liquid medium, the rate can be much reduced from that typically of a pure water–air system discussed in Section 2.7.1.

In an aqueous solution that is exposed to dry air, evaporation occurs. Assume now that the solute is nonvolatile. Then the solution close to the interface will become more concentrated since water is disappearing. Rapidly, a situation is established where the solute concentration is high enough so that the water chemical potential at the surface is only marginally higher than in the air. The gradient in solute concentration will cause the solute molecules to diffuse away from the surface.

Figure 2.25 shows the situation for a case with a tube of length, d, in a liquid reservoir with bulk concentration c_b of a solute. Assume ideal mixing and that there is no convection in the liquid in the tube. The reservoir is large, so the concentration at the entrance of the tube is fixed. There is an evaporation of water at the surface of the tube, which leads to a net liquid flow, V_l, in the tube and at steady state, the solute has a constant concentration profile. This is accomplished when the transport toward the surface of the solute through liquid flow is compensated by diffusional transport in the opposite direction so that

$$D\frac{dc_s}{dx} = -c_s V_l^{SS} \tag{2.7.1}$$

Using the boundary conditions at the surface, $x = 0$, and at the bottom of the tube, $x = d$, the steady-state concentration profile is exponential:

$$c_s(x) = c_s(0)\exp\left\{\ln\left[\frac{c_s(d)}{c_s(0)}\right]\frac{x}{d}\right\} \tag{2.7.2}$$

The resulting steady-state evaporation rate per unit area is

$$V_l^{SS} = \frac{D}{d}\ln\left[\frac{c_s(d)}{c_s(0)}\right] \tag{2.7.3}$$

Problem: Show that Eqs. (2.7.2) and (2.7.3) satisfy Eq. (2.7.1).

Solution: Equations (2.7.1) and (2.7.3) combine to $\frac{dc_s}{dx} = -\left\{\frac{1}{d}\ln\left[\frac{c_s(d)}{c_s(0)}\right]\right\}c_s(x)$, which is integrated to give $c_s(x) = c_s(0)\exp\{\ln[c_s(d)/c_s(0)(x/d)\}$.
This satisfies the boundary conditions at $x = 0$ and $x = d$.

A crucial assumption in this calculation is that there is a position below the surface where the concentration is kept constant. The system in Figure 2.25 has the somewhat artificial feature of a

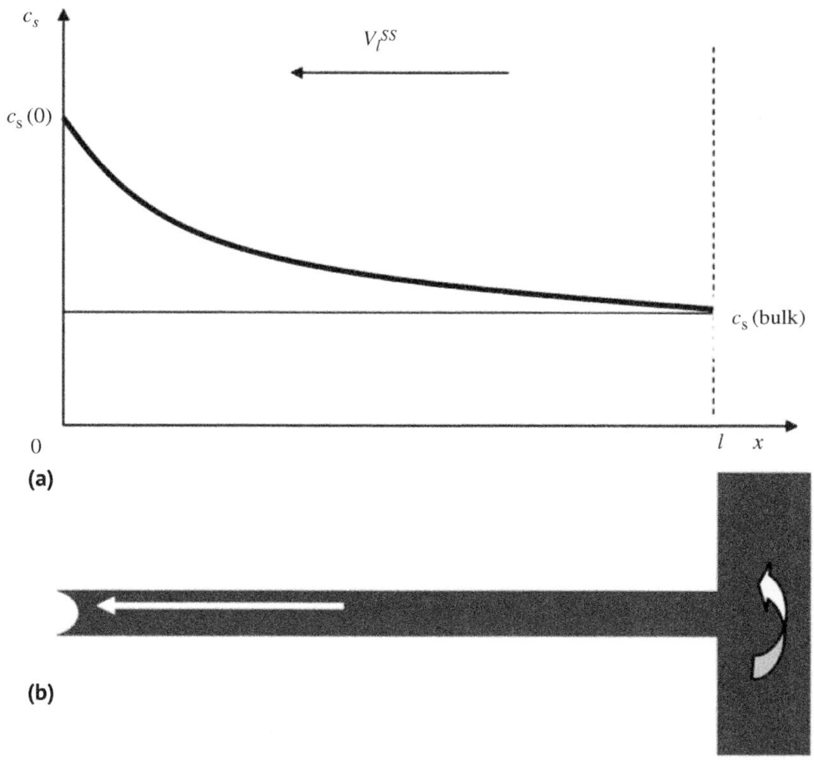

Figure 2.25 Illustration of the solute concentration profile (a) at steady state, where the evaporation rate is given by V_l^{SS}, and (b) liquid flow in capillary tube. *Source*: Kabalnov and Wennerström, 2009/with permission of Royal Society of Chemistry.

tube extending into the liquid. For an open water surface, there is always some convection, and one has introduced the concept of an "unstirred layer" to semi-quantitatively account for this effect. The tube can be seen as a device to illustrate the effect of an unstirred layer.

In the case of normal dry indoor air, it can easily happen that solute concentrations near the surface reach such high values that the solubility limit is exceeded. Precipitation or the formation of other types of aggregates strongly affects the diffusion properties. This surface "skin" reduces the evaporation rate considerably. The phenomenon can be beneficial in water reservoirs, but it can also cause problems in industrial drying processes and, for example, in inkjet printers.

In Eq. (2.7.2), we solved for the steady-state condition, without considering the time it takes to reach this state. By rewriting the diffusion equation in dimensionless quantities, the characteristic time for the dynamics is measured by a dimensionless quantity $\tau = \mathrm{t}\, D/d^2$. For large values of the tube length d, this equilibration time is very long. For $D \approx 10^{-9}$ m²/s, $d \approx 1$ cm, the equilibration time τ equals one at $t \approx 10^5$ s. For such a case, a steady state is not reached in practice, and one finds a dynamic concentration profile like the one shown in Figure 1.13. Under such conditions, there is a constant shape of the concentration profile but with a width that grows in time. For systems that have a phase change in the range μ_w(air) to μ_w(bulk), a phase will appear near the surface. Figure 2.26 illustrates how the surface of a micellar solution of dodecyl maltoside in water

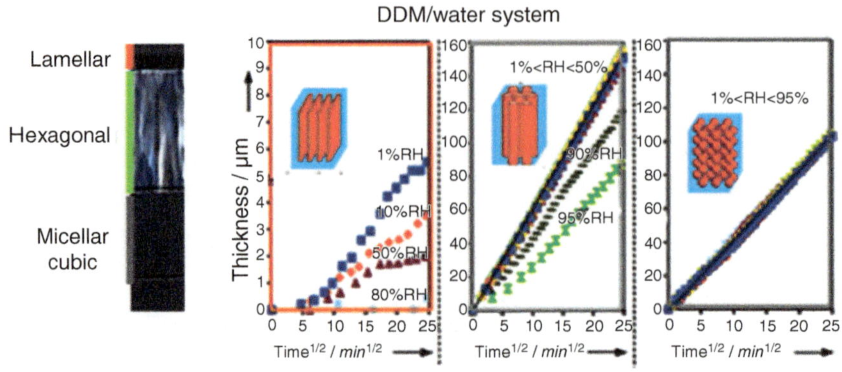

Figure 2.26 The left panel shows a recording in polarized light of a capillary (gravity along the direction of the arrow). The right panels show how the thickness of different phases increases with time. *Source*: Adapted from Roger et al., 2018.

develops with time at different external relative humidities, RH. At 95% RH, the phase change from micelle solution to micellar cubic phase occurs rapidly, and at the beginning of the recording after ≈1 min, a transition to a hexagonal phase has already occurred, and the system develops under quasi steady-state conditions. At a water chemical potential corresponding to ≈80% RH, there is an appearance of a lamellar phase in a bulk system, and when the relative humidity is lower than 80%, an oriented lamellar phase, which is seen as black between crossed polarizers, appears at the surface. The figure shows that the thicknesses of the respective phases grow proportional to the square root of time. Furthermore, for the phases not in contact in direct contact with the surface, the growth rate is independent of the relative humidity since the chemical potentials at the interphases remain constant in time.

Exercises

2.1 Two liquid phases, A and B, separate when mixed into one phase consisting of 87.5% A and 12.5% B and one phase of 12.5% A and 87.5% B. Use the regular solution model to estimate the interfacial free energy. $\mathbf{T} = 298$ K, the interaction parameter w is 6 kJ/mol, and the area per molecule at the interface is 3×10^{-19} m^2 and $z_b = 6$.

2.2 In Philosophical transactions of the Royal Society 1774, Benjamin Franklin reports on a simple experiment he performed on a pond in Clapham Common, London. He took "one teaspoon" of oil ($\approx 2 \cdot 10^{-6}$ m^3) and observed a film that spread "half an acre" ≈ 2000 m^2. What thickness of the film does this correspond to? Do you think Franklin did this calculation?

2.3 Two soap bubbles blown from the same solution have radii R_1 and R_2. The bubbles are cautiously joined by a thin tube. What is the outcome if one can avoid bursting the bubbles?

2.4 An air bubble of radius 1×10^{-6} m is generated in water with a syringe. The bubble is gently transferred to a solid hydrophobic surface. Calculate the contact angle. Is the volume of the bubble significantly changed upon adsorbing on the surface? Use $\gamma_{H_2O,v} = 72$ mN/m, $\gamma_{s,H_2O} = 50$ mN/m, and $\gamma_{s,v} = 20$ mN/m.

2.5 What is the critical drop size, the Rayleigh radius, for nucleating liquid water at $\mathbf{p} = 1$ atm and (a) $\mathbf{T} = 99.9$ °C, (b) $\mathbf{T} = 99.0$ °C, and (c) $\mathbf{T} = 90$ °C. Between 90°C and 100°C, the surface tension of water varies linearly from 61 to 59×10^{-3} J/m^2.

2.6 The Langmuir model for adsorption isotherms can be extended to allow for interactions between adsorbed molecules. Using the same arguments as in the regular solution theory, an extra term, $N_2 X_2 W_{22}$, contributes to the free energy. Derive the adsorption isotherm: that is, X_2 as a function of $c_2(b)$.

2.7 The table shows the measured surface tension of aqueous solutions of dodecyldimethylammonium chloride at 293 K for (a) no added salt and (b) 0.20 M NaCl solution.

No salt		0.20 M NaCl	
c (M)	γ(mJ/m^2)	c (M)	γ(mJ/m^2)
1×10^{-5}	72	1×10^{-5}	72
8×10^{-5}	71.5	1.5×10^{-5}	70
3×10^{-4}	71	5×10^{-5}	66
1.0×10^{-3}	68	1.6×10^{-4}	58
2.0×10^{-3}	64	4.0×10^{-4}	52
3.5×10^{-3}	58	9.0×10^{-4}	46
5.6×10^{-3}	52	1.5×10^{-3}	40
9.5×10^{-3}	45	2.5×10^{-3}	35
1.1×10^{-2}	41	3.9×10^{-3}	34
1.4×10^{-2}	37.5	5.0×10^{-3}	34
2.0×10^{-2}	37.5	4.0×10^{-2}	34
4.0×10^{-2}	37.5	–	–

Why does the surface tension go to a constant value at high concentration? Give an answer both with respect to molecular events and in relation to the Gibbs adsorption equation.

2.8 Use the data of Exercise 2.7 to estimate the surface excess and area per molecule at $c = 1 \times 10^{-3}$ M and $c = 1.4 \times 10^{-2}$ M for the salt-free case.

2.9 Use the data of Exercise 2.7 to determine the surface excess and the area per molecule at $c = 1 \times 10^{-3}$ M and $c = 2 \times 10^{-2}$ M for the case with $c_{NaCl} = 0.2$ M.

2.10 The table gives the measured surface pressure of a dipalmitoylphosphatidylcholine film at the n-heptane-water (0.1 M NaCl) interface for a range of surface concentrations. Verify that $\Pi_s = k_B T$ at low surface coverage. Determine also the second surface virial coefficient defined analogously to B_2 in Eq. (1.2.14) (**T** = 20 °C).

$10^{-17} \times \Gamma(m^{-2})$	$10^3 \Pi_s (J/m^2)$
2	1.02
4	2.97
6	5.86
8	10.0
10	14.8
12	21.6

2.11 The surface tensions of aqueous solutions of potassium iodide, KI, have the following values at 20 °C.

KI (mol/L)	0.1	0.5	1.0	2.0
Surface tension, γ(mN/m)	72.59	72.50	71.35	69.95

Using the Gibbs adsorption equation, calculate the surface excess concentration Γ_{KI} in molecules per 1000 Å^2 for a 0.8 molar concentration of KI. Surface tension of pure water at 20 °C is $\gamma_0 = 72.83$ mN/m.

2.12 Given the following surface and interfacial tensions, at 20 °C, work out answers to (a)–(c).

Interface	γ (mN/m)	Interface	γ (mN/m)
Air–water	72	Mercury–water	375
Air–octanol	28	Mercury–octanol	348
Air–hexane	18	Mercury–hexane	378
Air–mercury	476	Water–octanol	9
		Water–hexane	50

a. Will octanol spread at a mercury–water interface?
b. If it were to spread, how would the polar group in the alcohol be oriented at the water and at the mercury interface? Explain.
c. Will hexane spread on the water–mercury interface?

2.13 An air bubble of radius 0.1 mm is produced 10 m below the surface of a lake. What is the radius when it arrives at the surface if
a. The water is pure, so that the surface tension remains at 72 mN/m?
b. The bubble picks up surface active contaminants in the water so the surface tension is reduced to 40 mN/m when it arrives at the surface. Assume that there is no gas exchange between bubble and water.

2.14 In a two-phase system, a particle of radius R adsorbs at the interface. Assume that the particle has the same surface free energy with respect to both phases and that this is also the value of the interfacial free energy. Calculate the equilibrium constant relating the surface and bulk concentrations of the particles.

References

Atkins, P., de Paula, J., and Keeler, J. (2022). *Atkins Physical Chemistry*, 12th ed. Oxford, UK: Oxford University Press.

Berg, J.C. (2010). *An Introduction to Interfaces & Colloids. The Bridge to Nanoscience.* World Scientific.

Cosgrove, T. (2010). *Colloid Science. Principles, Methods and Applications.* Chichester, UK: Wiley.

Doi, M. (2013). *Soft Matter Physics.* Oxford, UK: Oxford University Press.

Gatapova, E., Graur, I., Kabov, O., Aniskin, V., Filipenko, M., Sharipov, F., and Tadrist, L. (2017). *Int. J. Heat Mass Transfer.* 104, 800–812.

Höök, F., Kasemo, B., Nylander, T., Fant, C., Sott, K., and Elwing, H. (1999). *Anal. Chem.* 73, 5796–5804.

Israelachvili, J.N. (2011). *Intermolecular and Surface Forces*, 3rd ed. London, UK: Academic Press.

Janschke, M., Butt, H.J., Gaub, H.E., and Mannae, S. (1997). *Langmuir.* 13, 1381–1384.

Kabalnov, A. and Wennerström, H. (2009). *Soft Matter.* 5, 4712.

Kronberg, B., Holmberg, K., and Lindman, B. (2014). *Surface Chemistry of Surfactants and Polymers*, 3rd ed. Chichester, UK: Wiley.

Lyklema, H. ed. (2000). *Fundamentals of Interface and Colloid Science. Liquid-Fluid Interfaces, Vol. 3.* Amsterdam, The Netherlands: Elsevier.

McConlogue, C.W. and Venderlick, T.K. (1997). *Langmuir.* 13 (26), 7158–7164.

Roger, K., Sparr, E., and Wennerström, H. (2018). *Phys. Chem. Chem. Phys.* 20, 10430–10438.

Shaw, D.J. (1992). *Introduction to Colloid and Surface Chemistry*, 4th ed. London: Butterworths.

Weissberger, A. and Rossiter, B.W. eds. (1971). *Physical Methods of Chemistry, Vol. 1, Part IIIB.* New York: Wiley-Interscience.

3

Electrostatic Interactions in Colloidal Systems

Concept Map

Intermolecular Interactions

We can obtain the interaction energies V_{ij} between two molecules by solving the Schrödinger equation. It is, however, more transparent to write V_{ij} as the sum of five interactions: (1) overlap repulsion, (2) charge transfer, (3) electrical multipole–electrical multipole, (4) electrical multipole–induced multipole, and (5) induced dipole–induced dipole, because we can express some of these interactions in classical rather than quantum mechanical terms.

Electrostatic Intermolecular Interactions in the Vapor Phase

Interactions between isolated molecules in the vapor phase provide the simplest way to understand the basic concepts. Table 3.1 summarizes the essential equations.

Electrostatic interactions arise when the electrical field emanating from one molecule couples to the existing charge distribution of other molecules. They can be attractive or repulsive, depending on the charge and orientation of the molecules.

The ion–ion interaction between charges q_1 and q_2 $V_{\text{ion–ion}} \propto q_1 q_2 / R$ is long-ranged.

The ion–dipole interaction $V_{\text{ion–dipole}} \propto q_1 m_2 / R^2$ is orientation-dependent and proportional to the magnitude of the dipole moment m (Figure 3.2).

The dipole–dipole interaction $V_{\text{dipole–dipole}} \propto m_1 m_2 / R^3$ depends on the angles θ and ϕ, as depicted in Figure 3.3.

The hydrogen bond is electrostatic in origin also involving interactions of higher multipole moments.

Induced interactions arise when the electrical field emanating from one molecule induces charge polarization in other molecules, and they are always attractive:

$$V_{\text{ion–ind dipole}} \propto q_1 \alpha_2 / R^4$$

where α is the polarizability (Eq. 3.4.5).

$V_{\text{dipole–ind dipole}} \propto m_1^2 \alpha_2 / R^6$ (Eq. 3.4.8),

If $V_{\text{coulombic}} \propto 1/R^n$, then $V_{\text{ind coulombic}} \propto 1/R^{2n}$ because one $1/R^n$ term induces the response while the second produces the interaction.

The Colloidal Domain: Where Physics, Chemistry, Biology, and Technology Meet, Third Edition. Håkan Wennerström.

The Dielectric Approximation

1. The Poisson equation (Eq. 3.2.1) is difficult to apply in a condensed phase where multiple interactions occur. However, averaging over solute–solvent and solvent–solvent interactions transforms Poisson's equation into a useful form (Eq. 3.5.13) by introducing the relative permittivity ε_r.

   In this way, one can determine electrostatic solvation energies and solute–solute interaction free energies.
2. The Born equation (Eq. 3.6.2) gives the free energy associated with transferring an ion from one phase to another.

Angle-averaged Potentials

Interactions involving dipoles are often less than $k_B\mathbf{T}$, and we must average over all orientations using the Boltzmann distribution to determine their interaction energy. Such averaging yields free energies half as large as the corresponding energies.

Angle-averaged interactions are always attractive and for dipoles:

$\left\langle V_{\text{dipole-dipole}} \right\rangle \propto (m_1 m_2)^2/\left(k_B\mathbf{T}R^6\right)$, which is called the Keesom interaction.

Chemical Origins of Surface Charges

1. Virtually all interfaces acquire charge by dissociation or adsorption of ionic constituents.
2. In addition to electrostatic interactions, short-range chemical interactions between ions and surface can lead to pronounced effects. These can be described by a conventional chemical equilibrium when electrostatic contributions are included.
3. Many ionic crystals in contact with a polar solvent acquire charge because two ionic species possess different equilibrium adsorption affinities for the surface.
4. The Hofmeister series describes specific ion binding in many systems.
5. Because electrostatic forces are long-ranged, charged interfaces play an important role in many interfacial processes.

Charged Interfaces: The Poisson–Boltzmann Equation

We can describe entropic and energetic interactions between charges fixed at a surface and those free to move in a solution by combining the Poisson and Boltzmann equations to yield the Poisson–Boltzmann equation (Eq. 3.8.4).

1. The **Gouy–Chapman theory** describes the properties of a charged planar interface by relating the surface charge density σ and the surface potential Φ_0 (Eq. 3.8.15). It shows how the potential at long range decays exponentially with distance (Eq. 3.8.19; Figure 3.7) with a decay constant given by the Debye length $1/\kappa$ (Eq. 3.8.13).

2. The **Debye length** is a key parameter that measures how the combination of valency, concentration, and dielectric constant contributes to the screening of interactions between charges in solutions.
3. Near a charged surface, counterions concentrate, while co-ions are repelled, forming the **electrical double layer.** The Gouy–Chapman theory permits us to calculate how ionic concentrations vary in the solution (Figure 3.8).

The **Debye–Hückel theory** describes how the potential decreases exponentially with distance around a spherical ion (Eq. 3.8.24).

The Graham equation (Eq. 3.8.26), which follows from the Poisson–Boltzmann equation, relates how the concentration of ions close to a charged surface varies with charge density and electrolyte concentration.

Electrostatic Free Energy of a Charged Surface

By separately calculating the entropic and energetic contributions to the electrostatic free energy of a solution using the Poisson–Boltzmann equation, we gain considerable insight into the properties of charged surfaces (Figure 3.11).

By evaluating the electrostatic lateral pressure associated with a charged interface, we obtain a very useful description of how the free energy and the energy depend on the surface charge density (Eq. 3.9.16).

Intensive physical properties such as concentration, energy density, and dielectric polarizability vary over a short distance at the interface between two phases. For example, at the air–water interface, the dielectric constant changes over a few molecular diameters from 78 to 1. The orientation and polarization of interfacial molecules that result from this change constitute a microscopic separation of charge in the localized area. Preferential adsorption of electrolytes or ionic surfactants produces a charged interface, and polar interactions involving head groups and solvent play a critical role in self-assembled structures, such as monolayers, micelles, bilayers, vesicles, and emulsions. They also dominate self-assembly and stability in other colloidal systems such as charged latex particles, silica particles, and clays. For this reason, the study of electrostatic interactions between molecules and particles is a central theme of colloidal science.

Electrostatic interactions play an essential role in many industrial and biological processes. Electrochemical reactions, including the splitting of water into oxygen and hydrogen, the reduction of aluminum, the purification of copper, and the production of Cl_2 gas, all occur at charged interfaces. Operation of most microelectronic solid-state devices depends on the charge separation that occurs at the interface between materials. In biology, selective adsorption and transport are governed by electrical properties associated with biological membranes or other surfaces. In fact, understanding the behavior of charged interfaces is a major step toward understanding many phenomena that affect our daily lives.

Most textbooks on colloids begin their discussion of electrostatics with a form of the Poisson equation (see Section 3.5) containing the solvent dielectric constant to describe the ion–ion interaction. Then other important electrostatic interactions, such as ion–dipole or dipole–dipole interactions, are considered on an ad hoc basis. Although, in one sense, this approach is appealing in its directness, it glosses over many fundamental issues that are prerequisites for understanding the subject. We therefore instead begin our discussion with the equations in their most basic form and show how electrostatic interactions emerge in a unified and consistent way. Although this approach is more abstract and requires deeper contemplation on the part of the reader, we are convinced that the insight that emerges makes the effort worthwhile. It provides a concrete link between molecular and colloidal interactions, one of the main themes of this book.

3.1 Intermolecular Interactions Can Be Expressed as the Sum of Five Terms

The notion of intermolecular interactions is based on the idea that in a condensed phase we can identify molecular units that remain largely unaffected by their immediate environment. By solving the Schrödinger equation for the isolated molecular unit, we can obtain the intramolecular energies and charge distributions. We can also determine V_{ij}, the interaction between two molecules, by this procedure. In practice, it is convenient to interpret the total interaction potential energy as the sum of five parts:

$$\begin{aligned} V_{ij} = &\ (1)\ \text{overlap repulsion} + (2)\ \text{charge transfer} \\ &+ (3)\ \text{electrical multipole} - \text{electrical multipole} \\ &+ (4)\ \text{electrical multipole} - \text{induced electrical multipole} + (5)\ \text{dispersion} \end{aligned}$$

Each of these five terms depends on the molecular properties derived from the Schrödinger equation. However, the advantage of organizing V_{ij} in this way is that some of the terms can be expressed more simply in classical rather than quantum mechanical terms.

1. Overlap repulsion occurs when the electron clouds of two closed-shell molecules overlap. In accord with the Pauli exclusion principle, guest electrons are forced into excited state orbitals of the host molecule and produce a strong increase in energy. The resulting repulsion is approximately proportional to the square of the overlap, which increases strongly as the separation between molecules decreases. This effect determines a molecule's size in terms of an effective diameter σ. We can express the overlap repulsion for atoms or molecules as a power law

$$V(r) \propto (\sigma/r)^n \tag{3.1.1}$$

   where r is the center-to-center separation between the two entities. For the case $n = 12$, we obtain the repulsive term in the Lennard–Jones potential. When $n = \infty$, we obtain an expression for hard spheres, since for $r > \sigma$, the value of $V(r)$ is zero, whereas for $r < \sigma$, it is infinite.
2. Charge transfer interactions occur when one molecule donates excess electrons to an acceptor molecule with an electron deficiency. This type of interaction occurs when the molecules are in close proximity, but because its effect is relatively unimportant in colloidal systems, we will not consider it further.
3. Electrical multipole–electrical multipole interactions occur between molecules that possess net charges and/or an asymmetrical distribution of electrons or nuclei. We can describe this term by classical electrostatics.
4. Electrical multipole–induced multipole interactions occur between one molecule with a permanent electrical multipole and a polarizable molecule. This term is semiclassical in the sense described in Section 3.4.
5. The dispersion term can be described only in quantum mechanical terms. It corresponds to an instantaneous dipole–induced dipole interaction. We will postpone discussing it until Chapter 5, where we develop equations for London dispersion attractive interactions.

Overlap and dispersion terms describe properties common to all molecules (both polar and non-polar). Multipole and induced multipole interactions occur more selectively and are thus potentially more effective in promoting an organized system. They are the focus of this chapter.

Many textbooks treat the hydrogen bond as a special type of interaction and include an additional term in the expression for V_{ij}. For reasons that will emerge later in this chapter, we will instead consider the hydrogen bond to be a special case of an electrostatic interaction.

3.2 Multipole Expansion of the Charge Distribution Provides a Convenient Way to Express Electrostatic Interactions Between Molecules

In an electrostatic description, we can represent a molecule by a fixed charge distribution, $\rho(\boldsymbol{r})$, set up by its electrons and nuclei. This charge distribution generates an electrostatic potential Φ according to the Poisson equation

$$\varepsilon_0 \nabla^2 \Phi(\boldsymbol{r}) = -\rho(\boldsymbol{r}) \tag{3.2.1}$$

where $\nabla^2 = \frac{\partial^2}{\partial x^2} + \frac{\partial^2}{\partial y^2} + \frac{\partial^2}{\partial z^2}$ is the Laplace operator and ε_0 represents the dielectric permittivity of vacuum. When discussing electrostatic effects, we will consistently use SI units, while, in older literature, cgs units are customary. In the Poisson equation (Eq. 3.2.1), a difference by a factor $4\pi\varepsilon_0$ exists between the two systems of units, which should be considered when comparing equations in different texts.

We can formally solve Eq. (3.2.1) for a fixed charge distribution $\rho(\boldsymbol{r})$ by first noting that it is a linear equation so that the principle of superposition applies. That is, the potential Φ generated by two charge distributions, $\rho_1(\boldsymbol{r}) + \rho_2(\boldsymbol{r})$, is the sum $\Phi = \Phi_1 + \Phi_2$ of the potentials generated by $\rho_1(\boldsymbol{r})$ and $\rho_2(\boldsymbol{r})$ separately. Second, we know that the potential from a point charge q_0 at $\boldsymbol{r}_0$ is $\Phi(\boldsymbol{r}) = q_0/(4\pi\varepsilon_0|\boldsymbol{r} - \boldsymbol{r}_0|)$, which consequently is the solution of Eq. (3.2.1) when $\rho(\boldsymbol{r})$ represents a point charge. We can look at the actual charge distribution, $\rho(\boldsymbol{r})$, as a collection of point charges, and, using the principle of superposition, integrate over $\rho(\boldsymbol{r})$ to get the potential

$$\Phi(\boldsymbol{r}) = \frac{1}{4\pi\varepsilon_0} \int \frac{\rho(\boldsymbol{r}')}{|\boldsymbol{r} - \boldsymbol{r}'|} \, \mathrm{d}V' \tag{3.2.2}$$

A charge q_2 at $\boldsymbol{r}$ will interact with an external potential Φ_1 to yield the energy $V_{12}^{el} = q_2 \Phi_1(\boldsymbol{r})$. Thus, the charge distribution $\rho_1(\boldsymbol{r})$ of one molecule will sense the potential $\Phi_2(\boldsymbol{r})$ from another molecule to yield the interaction energy

$$\begin{aligned} V_{12}^{el} &= \int \rho_1(\boldsymbol{r})\Phi_2(\boldsymbol{r})\mathrm{d}V = \frac{1}{4\pi\varepsilon_0} \iint \frac{\rho_1(\boldsymbol{r})\rho_2(\boldsymbol{r}')}{|\boldsymbol{r} - \boldsymbol{r}'|} \mathrm{d}V\mathrm{d}V' \\ &= \int \rho_2(\boldsymbol{r}')\Phi_1(\boldsymbol{r}')\mathrm{d}V' \end{aligned} \tag{3.2.3}$$

whose coordinates are defined in Figure 3.1. The equality shows that we could equivalently consider the potential generated by molecule 1 at the position of molecule 2. This deceptively simple formula contains a multitude of effects. If, by solving the Schrödinger equation for the molecules, we determine the charge distributions $\rho_1(\boldsymbol{r})$ and $\rho_2(\boldsymbol{r})$, we could numerically integrate Eq. (3.2.3) to obtain the interaction energy. Even though this gives accurate numbers, it does not provide a conceptual basis for understanding the qualitative properties of electrostatic intermolecular interactions. For molecules, we are accustomed to characterizing their electrical properties through the

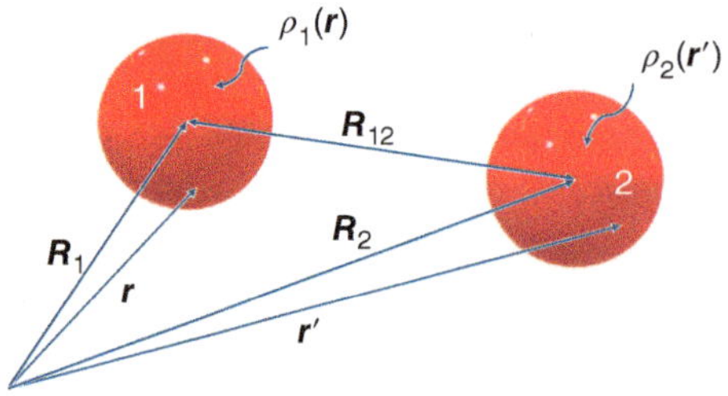

Figure 3.1 Coordinates used to calculate the electrostatic interaction between two molecules, 1 and 2, with charge distributions $\rho_1(\boldsymbol{r})$ and $\rho_2(\boldsymbol{r})$, respectively.

net charge, the dipole moment, and possibly the quadrupole moment. This point of view is based on a mathematical device called the multipole expansion.

Choose a central point $\boldsymbol{R}_1$, within $\rho_1(\boldsymbol{r})$, and a central point, $\boldsymbol{R}_2$, within $\rho_2(\boldsymbol{r})$. The denominator in Eq. (3.2.3) can be rewritten as

$$|\boldsymbol{r} - \boldsymbol{r}'| = |\boldsymbol{R}_2 - \boldsymbol{R}_1 + (\boldsymbol{r}' - \boldsymbol{R}_2) - (\boldsymbol{r} - \boldsymbol{R}_1)| \tag{3.2.4}$$

When the terms $(\boldsymbol{r}' - \boldsymbol{R}_2)$ and $(\boldsymbol{r} - \boldsymbol{R}_1)$ are small relative to the intermolecular separation $\boldsymbol{R}_{12} = \boldsymbol{R}_2 - \boldsymbol{R}_1$, we can eliminate $(\boldsymbol{r}' - \boldsymbol{R}_2)$ and $(\boldsymbol{r} - \boldsymbol{R}_1)$ from the denominator in Eq. (3.2.3) by a series expansion. This expansion is straightforward but somewhat involved mathematically. Here we simply state the expressions for the most important terms. The general result can be written as a sum over different multipole terms having a progressively stronger distance and angular dependence:

$$V_{12}^{el} = \sum_{m,n=1}^{\infty} V_{12}^{m,n}(R_{12}) = \sum_{m,n=1}^{\infty} C_{m,n} R_{12}^{-m-n+1} \tag{3.2.5}$$

Here, the summation indices denote the order of the multipoles. The coefficient $C_{m,n}$ is a product of molecular multipole moments and an angle-dependent term. We will consider the first few terms of this series and give the appropriate expressions for $C_{m,n}$ rather than deriving them. In Sections 3.2–3.4, we focus on interactions between molecules that occur in the gas phase; the most important interaction formulas are summarized in Table 3.1. In Section 3.5, we generalize these results to condensed phases.

The lowest order term is obtained for $m = n = 1$, and it simply represents the Coulomb ion–ion interaction:

$$V_{\text{ion-ion}} = \frac{q_1 q_2}{4\pi\varepsilon_0 R_{12}} = \frac{z_1 z_2 e^2}{4\pi\varepsilon_0 R_{12}} \tag{3.2.6}$$

where the total charge $q_i = \int \rho_i \, \mathrm{d}V$ is the **monopole** moment in the general expansion. We can use Eq. (3.2.6) to describe the interaction between two ions, where the magnitude and sign of each ionic charge $q_i = z_i$ are given by a product of the valency z_i and the elementary charge $e = 1.602 \times 10^{-19}$ C.

Problem: Calculate the Coulomb interaction between a sodium ion and a chloride ion at contact, $R_{12} = 0.276$ nm.

Solution:

$$V_{\text{ion-ion}} = \frac{(1)(-1)\left(1.602 \times 10^{-19}\text{C}\right)^2}{4\pi\left(8.854 \times 10^{-12}\ \text{C/Vm}\right)\left(0.276 \times 10^{-9}\ \text{m}\right)} = -8.4 \times 10^{-19}\ \text{J}$$

Table 3.1 Electrostatic interaction energies between molecules in the gas phase.

Value of *m*, *n* in Eq. (3.2.5)	Name	Interaction energy	Comment
1, 1	Coulomb ion–ion	$\frac{q_1 q_2}{4\pi\varepsilon_0 R_{12}}$	See Figure 3.1
1, 2	Ion–dipole	$\frac{q_1 m_2 \cos\theta}{4\pi\varepsilon_0 R_{12}^2}$	See Figure 3.2
2, 2	Dipole–dipole	$\frac{-m_1 m_2 f(\theta,\phi)}{4\pi\varepsilon_0 R_{12}^3}$	$f(\theta,\phi) = 2\cos\theta_1\cos\theta_2 - \sin\theta_1\sin\theta_2\cos(\phi_2-\phi_1)$ See Figure 3.3
–	Ion–induced dipole	$\frac{-\alpha q_1^2}{2(4\pi\varepsilon_0)^2}\frac{1}{R_{12}^4}$	Attractive α polarizability
–	Dipole–induced dipole (Debye)	$\frac{-m_1^2\alpha(1+3\cos^2\theta)}{2(4\pi\varepsilon_0)^2 R_{12}^6}$	Attractive
2, 2	Thermally averaged dipole–dipole (Keesom)	$\frac{-2m_1^2 m_2^2}{3(4\pi\varepsilon_0)^2 k_B \mathbf{T} R_{12}^6}$	Attractive $k_B\mathbf{T} > \frac{m_1 m_2}{4\pi\varepsilon_0 R_{12}^3}$

In the aforementioned problem, we found that the Coulombic interaction energy between two typical ions at contact was -8.4×10^{-19} J. A comparison with the thermal energy $k_B\mathbf{T} = \left(1.38 \times 10^{-23}\ \text{J/K}\right)(300\ \text{K}) = 4.1 \times 10^{-21}$ J at 300 K shows that the interaction per ion pair in a vacuum is around $-200\ k_B\mathbf{T}$. Thus, Coulomb interactions are very strong. In fact, only at separations greater than 56 nm will the Coulomb interaction be less than $k_B\mathbf{T}$. In the gas phase, even a small net charge on a dust particle strongly influences its interactions with other particles.

A noncharged polar molecule has a dipole moment, $\boldsymbol{m}_i$, defined by

$$\boldsymbol{m}_i = \int \rho_i(\mathbf{r})\mathbf{r}\mathrm{d}V \tag{3.2.7}$$

In the simple case of ρ equal to two opposite charges, $\pm q$, separated by a distance l, the dipole moment $m = ql$. For two elementary charges, $q = \pm e$ separated by $l = 0.1$ nm, the dipole moment is $m = \left(1.60 \times 10^{-19}\ \text{C}\right)\left(1 \times 10^{-10}\ \text{m}\right) = 1.6 \times 10^{-29}$ Cm = 4.8 D. The unit of Debye, D, is often used for dipole moments and equals 3.336×10^{-30} Cm. Permanent dipole moments occur in asymmetric molecules and thus not in single atoms.

As an example, consider the case illustrated in Figure 3.2 where an ion (1) of charge ze ($m = 0$) interacts with a dipole (2) with two charges, $\pm q$, separated by a distance l. We obtain the electrostatic interaction energy by summing the two Coulomb terms for the charge ze interacting with the charges $\pm q$ so that $m = lq$.

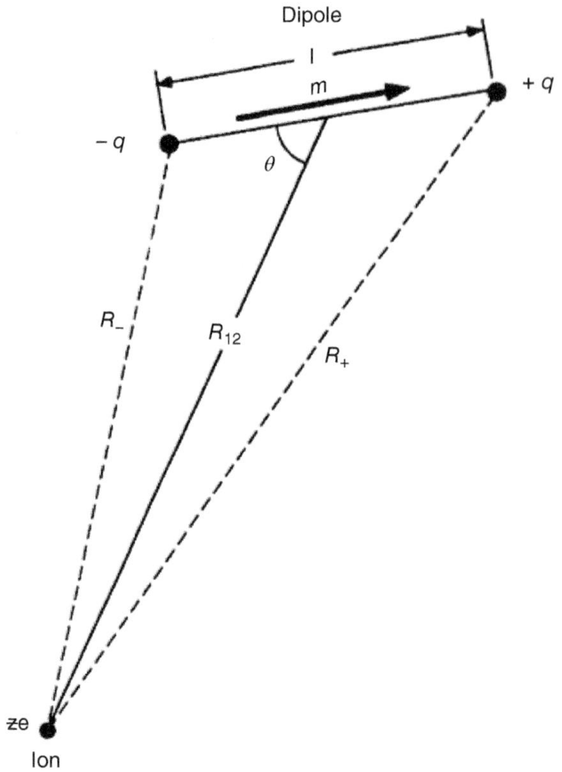

Figure 3.2 Ion–dipole interactions. The ion–dipole distance $\boldsymbol{R}_{12}$ is the distance between the ion and the center of the dipole with charges ±q a distance l apart. The interaction angle θ is defined as the acute angle between $\boldsymbol{R}_{12}$ and m.

The interaction potential is

$$\begin{aligned} V^{el} &= \frac{zeq}{4\pi\varepsilon_0}\left(\frac{1}{\boldsymbol{R}_+} + \frac{1}{\boldsymbol{R}_-}\right) \\ &\approx \frac{zeq}{4\pi\varepsilon_0}\frac{1}{\boldsymbol{R}_{12}}\left(\frac{1}{1+\frac{l}{2}\cos\theta} + \frac{1}{1-\frac{l}{2}\cos\theta}\right) \\ &\approx \frac{zeq}{4\pi\varepsilon_0}\frac{1}{\boldsymbol{R}_{12}^2} l\cos\theta = \frac{zem\cos\theta}{4\pi\boldsymbol{R}_{12}^2} \end{aligned} \tag{3.2.8}$$

If a cation is near the dipole, the maximum attraction occurs when the dipole vector points away from the ion ($\theta = 0$), while the maximum repulsion occurs when the dipole points toward the ion ($\theta = 0$). The general expression emerging from Eq. (3.2.5) with $m = 1, n = 2$ and $m = 2, n = 1$ is

$$V_{\text{ion-dipole}} = \frac{1}{4\pi\varepsilon_0 R_{12}^3}(-q_1\boldsymbol{m}_2 \cdot \boldsymbol{R}_{12} + q_2\boldsymbol{m}_1 \cdot \boldsymbol{R}_{12}) \tag{3.2.9}$$

Problem: Estimate the Coulombic interaction between a water molecule and a sodium ion using Eq. (3.2.8). The dipole moment of a water molecule equals 1.85 *D*, and assume a separation of 0.23 nm for a water molecule in contact with a sodium ion.

Solution: The maximum interaction occurs for $\theta = 0$ in Eq. (3.2.8) and

$$V_{\text{ion-dipole}}(\boldsymbol{R}_{12}, \theta = 0) = \frac{-\left(1.602 \times 10^{-19}\ \text{C}\right)\left(1.85 \times 3.336 \times 10^{-30}\ \text{C}\cdot\text{m}\right)}{4\pi\left(8.854 \times 10^{-12}\ \text{CV}\cdot\text{m}\right)\left(0.23 \times 10^{-9}\ \text{m}\right)^2}$$
$$= -1.7 \times 10^{-19}\ \text{J} = -40k_B\mathbf{T} \quad (\mathbf{T} = 300\ \text{K})$$

Dipole–dipole interactions dominate the electrostatic interactions in polar liquids. They correspond to $m = 2$, $n = 2$ in Eq.(3.2.5) and are given by

$$V_{\text{dipole-dipole}}(\boldsymbol{R}_{12}, \theta_1, \theta_2, \phi_2 - \phi_1) = -\frac{m_1 m_2}{4\pi\varepsilon_0 \boldsymbol{R}_{12}^3}\left[2\cos\theta_1 \cos\theta_2 - \sin\theta_1 \sin\theta_2 \cos(\phi_2 - \phi_1)\right] \quad (3.2.10)$$

The two dipoles are separated by $\boldsymbol{R}_{12}$ and are oriented with polar angles θ_1, ϕ_1 and θ_2, ϕ_2 with respect to $\boldsymbol{R}_{12}$, as illustrated in Figure 3.3. This interaction is strongly orientation-dependent. For example, with $\boldsymbol{m}_1$ fixed but $\boldsymbol{m}_2$ rotated isotropically over all orientations, the net effect is zero. In a liquid with only weak orientational correlations, this effect substantially reduces the net averaged interaction. Maximum attraction occurs when two dipoles are aligned parallel to $\boldsymbol{R}_{12}$, $\theta_1 = \theta_2 = 0$. This attraction is given by

$$V(\boldsymbol{R}_{12}, 0, 0, \phi) = -\frac{m_1 m_2}{2\pi\varepsilon_0 \boldsymbol{R}_{12}^3} \quad (3.2.11)$$

For two equal parallel dipoles with moments of 1 D, the interaction energy in vacuum equals $k_B\mathbf{T}$ at either $\boldsymbol{R}_{12} = 0.36$ nm (when the dipoles are in line with $\boldsymbol{R}_{12}$) or at $\boldsymbol{R}_{12} = 0.29$ nm (when they are

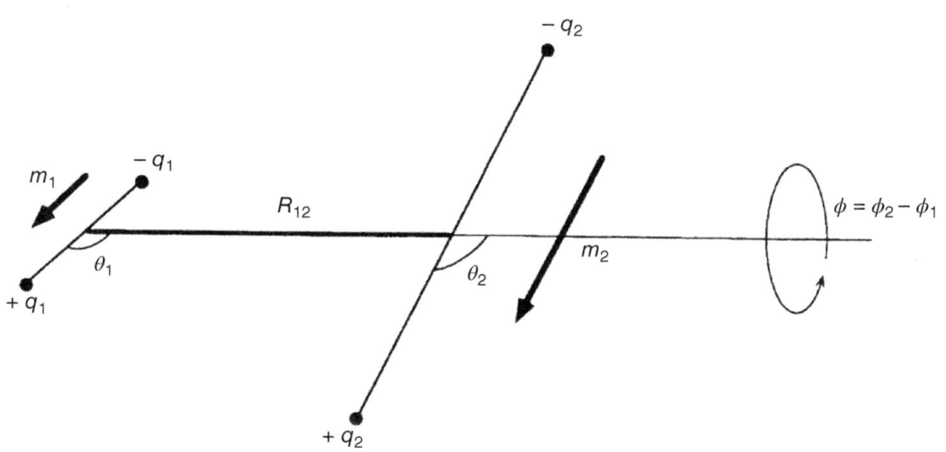

Figure 3.3 Dipole–dipole interactions. The dipole–dipole distance $\boldsymbol{R}_{12}$ represents the distance between the centers of dipoles. Relative orientations of the dipoles are specified by the polar (θ_1, θ_2) and azimuthal (ϕ_1, ϕ_2) angles.

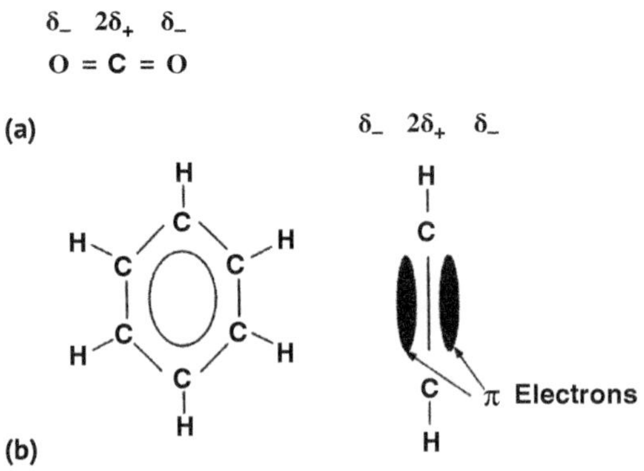

Figure 3.4 Illustration of the charge distributions in carbon dioxide and benzene. For (a) CO_2, the oxygens have a partial negative charge, making the central carbon strongly positive. For (b) benzene, the negative π-electrons are located on both sides of the ring plane, with the compensating positive nuclear charge residing in the plane. Note that despite the structural differences between the molecules, the charge distributions along the symmetry axes are similar.

perpendicular to $\boldsymbol{R}_{12}$). Since these distances are on the order of molecular separations in liquids and solids, dipole–dipole interactions generally are not strong enough to lead to substantial mutual alignment of polar molecules in the liquid state.

The quadrupole–quadrupole interaction, $m = n = 3$, forms the dominant electrostatic term in liquids formed by molecules having zero dipole moment. Although the quadrupolar term is shorter-ranged and more orientation-dependent than the dipole–dipole term, as the general properties of the series in Eq. (3.2.5) indicate, it is of utmost importance for understanding the properties of the condensed phase for molecules such as CO_2, N_2, and C_6H_6. Figure 3.4 illustrates the charge distributions of the quadrupolar molecules carbon dioxide (CO_2) and benzene (C_6H_6). For example, one major reason for the noticeably greater water solubility of benzene compared to other hydrocarbons of similar size lies in its slight polar character caused by its high quadrupolar moment.

Problem: Calculate the quadrupolar moment of CO_2. Assume that the charge distribution can be represented by partial charges on each atom. Take the oxygen charge to be $-0.32e$ and use a bond length of 0.12 nm. The quadrupolar moment Q is given by (for an axially symmetric molecule)

$$Q = \sum_i q_i x_i^2$$

where x_i denotes the coordinate along the symmetry axis and the sum is over all charges.

Solution: Choose $x_i = 0$ on the carbon atom (the final result doesn't depend on this choice)

$$\begin{aligned} Q &= -0.32e \times (-0.12)^2 \times 10^{-18} - 0.32e \times (0.12)^2 \times 10^{-18} \\ &= -4.6 \times 10^{-21} \times 1.602 \times 10^{-19} = -7.4 \times 10^{-40}\ \text{Cm}^2 \end{aligned}$$

This is a typical magnitude of a molecular quadrupole moment.

The hydrogen bond often affects molecular association in colloidal systems. In most textbooks, it is presented as an intermolecular interaction different from the five parts discussed in Section 3.1. In fact, the hydrogen bond is dominated by electrostatics but with the special property that it does not fit into the multipole expansion scheme given in Eq. (3.2.5). To be useful, the multipole expansion for the interaction energy should converge rapidly. Usually, we thus want to include only the first nonzero term and perhaps the second one. If convergence is not reached within two terms, there are good reasons to look for an alternative formulation.

The electron density around a hydrogen atom is low, particularly when it is attached to an electronegative atom such as N, O, or F. Then the overlap repulsion with other atoms is relatively weak, and it allows for a close approach of $\approx 0.15 - 0.20$ nm to atoms in other molecules. A strong electrostatic interaction results if these atoms have a high electron density, as do N, O, and F. With such small values of R_{12}, the series expansion in Eq. (3.2.5) does not converge rapidly. When higher-order terms in the expansion provide substantial contributions to the interaction energy V_{12}, it follows that the interaction is of shorter range and is more orientation-dependent. The relative weight of the higher-order terms changes from one type of hydrogen bond to the other so one cannot give a general expression for either the precise form of the distance dependence or of the angular dependence. At large separations, it is the dipole–dipole term that dominates, provided the molecules are neutral, but around the most favorable separation, this is not true. The term *hydrogen bond* is thus a convenient way to denote a strong, short-range, and highly directional electrostatic interaction. The wide application of this term, not least in biological systems, has led to its electrostatic origin being partly forgotten.

In a hydrogen bond, there is a donor, which is a hydrogen atom covalently bonded to an electronegative atom, and an acceptor, which is another electronegative atom. In liquids like water, hydrazine, and ethylene glycol, the molecule can act as both a donor and an acceptor, and hydrogen bonds act to give a strong association and cohesion in such liquids. We discussed some of the consequences of this cohesion in Section 1.6.

We can use the water dimer to illustrate some properties of hydrogen bonds. Figure 3.5a shows the energy of a water dimer, where the hydrogen atoms, for simplicity, have been confined to the same plane as the oxygens. Note that the hydrogen atom of the hydrogen bond falls nearly but not perfectly on the oxygen–oxygen line at the energy minimum. Figure 3.5b shows the strong orientation dependence of the interaction at the equilibrium separation. This is perfectly described, except for a constant, by only considering the charge–charge interaction as in Eq. (3.2.3), while the dipolar component by itself fails to give the qualitatively correct behavior. The minimum in the intermolecular interaction (Figure 3.5a) occurs for an oxygen–oxygen distance of 0.286 nm where $V_{12} = 4.8 \times 10^{-20}$ J or $V_{12}/k_B\mathbf{T} \approx 12$ at $\mathbf{T} = 298$ K. In the gas phase or in an apolar medium, the strength of the hydrogen bond is such that it promotes association with a strong orientation component. In a polar medium, like water, there is a competition between the solvation (free) energy and the specific hydrogen bond energy. When an internal hydrogen bond is formed during the folding of a protein or the formation of a DNA double helix in an aqueous environment, there is only a small contribution to the net binding energy. However, it has a strong influence on the specificity of the folding, since there is a large energy penalty for introducing hydrogen bond donors or acceptors that are not matched in the folded structure.

3.3 When Electrostatic Interactions Are Smaller than the Thermal Energy, We Can Use Angle-averaged Potentials to Evaluate Them and Obtain the Free Energy

Molecules in a liquid or gas tumble through the thermal motion. Interactions involving dipoles and higher multipoles are angle-dependent, and the rotation results in an averaging of the interaction. If a molecule adopts all orientations with equal probabilities, the average multipole

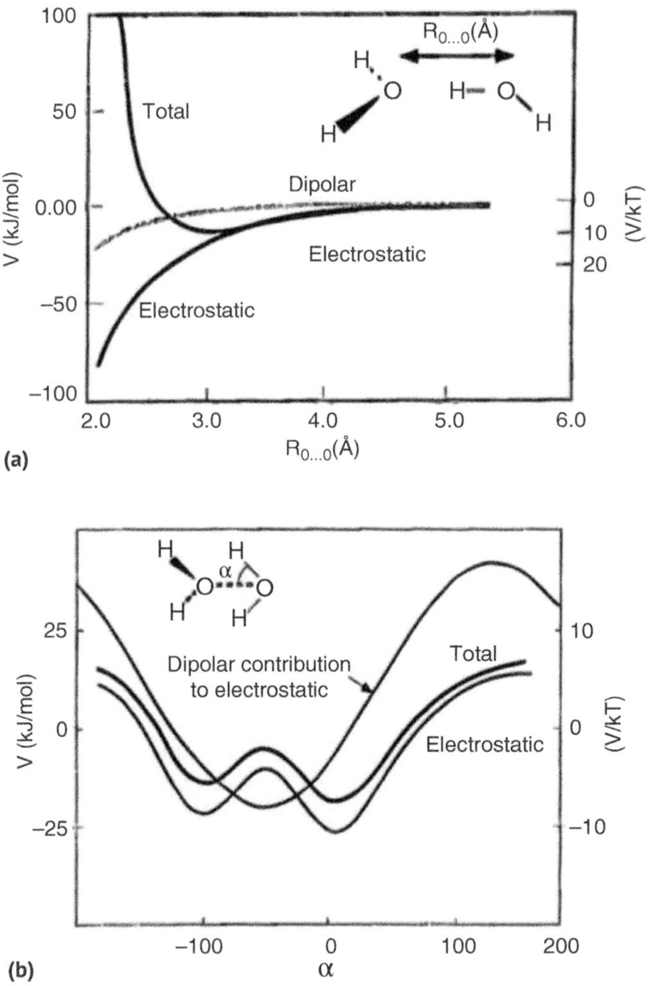

Figure 3.5 (a) The interaction potential between two water molecules as a function of the oxygen–oxygen distance $R_{(O-O)}$ calculated using a quantum chemical formalism. The molecules are kept at a fixed orientation corresponding to that of the lowest energy configuration. The three curves show the total interaction, the electrostatic interaction, and its dipolar part. At long range, the interaction is dominantly electrostatic with other components coming in close to the optimal separation of $R_{O-O} = 0.286$ nm. (b) The water–water interaction energy at a fixed separation $R_{O-O} = 0.286$ nm as a function of the orientation of the hydrogen bond donor molecule illustrates the electrostatic nature of the hydrogen bond. The curves show that the angular dependence of the total interaction is completely determined by the electrostatic interaction. The two minima correspond to the two hydrogen bond configurations, while the dipolar part has its minimum value at the maximum between the two hydrogen bond minima. Note that the barrier separating the two minima is a few $k_B\mathbf{T}$ so that the interaction is strong enough to cause a net orientation in a thermally averaged system. Courtesy of Gunnar Karlström.

interaction is zero. However, in a thermal average, the low-energy configurations occur preferentially as determined by a Boltzmann weighting factor.

We can calculate the energy contribution to first order by multiplying the interaction energy for each configuration by the corresponding Boltzmann factor and then averaging over all orientations and dividing by the same average in the absence of an interaction. The angle-averaged energy component $U(R)$ of an orientation-dependent interaction energy can be calculated from the relationship

$$U(R) = \int V(R,\Omega)\exp\left(-\frac{V(R,\Omega)}{k_B\mathbf{T}}\right)\mathrm{d}\Omega \Big/ \int \exp\left(-\frac{V(R,\Omega)}{k_B\mathbf{T}}\right)\mathrm{d}\Omega \tag{3.3.1}$$

where the integration is performed over polar, θ, and azimuthal, ϕ, angles; $\mathrm{d}\Omega = \sin\theta\mathrm{d}\theta\mathrm{d}\phi$. When

$$V(R,\Omega) \ll k_B\mathbf{T} \tag{3.3.2}$$

we can expand the exponential functions in Eq. (3.3.1) and

$$e^{-V(R,\Omega)/k_B\mathbf{T}} = 1 - \frac{V(R,\Omega)}{k_B\mathbf{T}} \tag{3.3.3}$$

Eq. (3.3.1) then simplifies to

$$U(R) = -\int V(R,\Omega)^2 d\Omega/(4\pi k_B\mathbf{T}) \tag{3.3.4}$$

provided that

$$\int V(R,\Omega)d\Omega = 0 \tag{3.3.5}$$

By using Eq. (3.2.8) for $V_{\text{ion-dipole}}(\mathbf{R}_{12}, \theta)$, we can obtain the angle-averaged energy for the charge–dipole interaction

$$U(R) = \int\left(\frac{(-ze)m\cos\theta}{4\pi\varepsilon_0 R^2} - \left(\frac{(ze)m}{4\pi\varepsilon_0 R^2}\right)^2\frac{\cos^2\theta}{k_B\mathbf{T}} + \dots\right)\frac{\mathrm{d}\Omega}{4\pi} \tag{3.3.6}$$

Substituting the spatially averaged values of $\langle\cos\theta\rangle = 0$ and $\langle\cos^2\theta\rangle = 1/3$ gives

$$U(R) \approx -\frac{(ze)^2 m^2}{3(4\pi\varepsilon_0)^2 k_B\mathbf{T}R^4} \quad \text{when} \quad k_B\mathbf{T} > \frac{(ze)m}{4\pi\varepsilon_0 R^2} \tag{3.3.7}$$

This is always attractive, a general property of second-order terms. The angle-averaged energy for the dipole–dipole interaction we obtained from Eq. (3.2.10) leads to

$$U(R) = -\frac{2m_1^2 m_2^2}{3(4\pi\varepsilon_0)^2 k_B\mathbf{T}R^6} \quad \text{for} \quad k_B\mathbf{T} > \frac{m_1 m_2}{4\pi\varepsilon_0 R^3} \tag{3.3.8}$$

Angle-averaged energies in Eqs. (3.3.7) and (3.3.8) result from a compromise between energy, which favors perfect alignment, and entropy, which strives toward a total randomization of the angular variables. $U(R)$ contains only the energy, and to obtain the *free* energy representing the effective potential, we must also include the entropy contribution. We can use a general result from linear response theory, which states that for a perturbation λV so small that the energy change ΔU = constant times λ, the corresponding change in free energy $\Delta\mathbf{A} = \Delta\boldsymbol{U}/2$. Thus, half of the energy gain is compensated by a decrease in $\mathbf{T}\Delta\mathbf{S}$.

The effective potential is then for the ion–rotating dipole interaction

$$V_{\text{eff}}(R) = \frac{(ze)^2 m^2}{6(4\pi\epsilon_0)^2 k_B\mathbf{T}R^4} \tag{3.3.9}$$

For the angle-averaged dipole–dipole interaction, we obtain:

$$V_{\text{eff}}(R) = \frac{m_1^2 m_2^2}{3(4\pi\epsilon_0)^2 k_B \mathbf{T} R^6} \tag{3.3.10}$$

This is sometimes called the Keesom interaction, and it constitutes one of three contributions to the van der Waals interaction.

In the subsequent sections, we will encounter more examples of the fact that electrostatic interactions also contain an entropic contribution as soon as one averages over molecular degrees of freedom that respond thermally to the electrical fields.

3.4 Induced Dipoles Contribute to Electrostatic Interactions

When describing the interaction between two charge distributions using multipole expansions as in Eq. (3.2.5), we assumed that the interaction did not change the charge distributions ρ_1 and ρ_2. However, in a real system, one molecule responds internally to the presence of another. The separation between strong intramolecular and weak intermolecular interactions suggests that we can treat this response in terms of a perturbation so that

$$\rho_i(R_{12}) = \rho_i(R_{12} \to \infty) + \delta\rho_i \tag{3.4.1}$$

where $\delta\rho_i \ll \rho_i$. Induced charges $\delta\rho_i$ lower the interaction energy with the second charge distribution by an amount

$$\int \Phi_1 \delta\rho_2 \, \mathrm{d}V + \int \Phi_2 \delta\rho_1 \, \mathrm{d}V$$

However, setting up $\delta\rho$ involves an intramolecular energy penalty. For small perturbations, in which the response is linear, this penalty is simply half of the energy gain.

The two most important induced electrostatic interactions are ion–induced dipole and dipole–induced dipole interactions. Both involve polarization of one molecule by the electrical field emanating from nearby molecules (ions). We define the polarizability α of an atom or molecule by the magnitude m_{ind} of the induced dipole moment $\boldsymbol{m}_{\text{ind}}$ acquired in an electrical field $\boldsymbol{E}$ of magnitude E as

$$m_{\text{ind}} = \alpha E \tag{3.4.2}$$

The polarizability is nonzero for all atoms and molecules and plays an important role in determining colloidal forces. It has the dimensionality of ε_0 times volume, and its magnitude is typically somewhat smaller than the molecular volume times ε_0. For example, in water, $\frac{\alpha}{\varepsilon_0} = 1.86 \times 10^{-29}$ m^3, while the molecular volume is 3.0×10^{-29} m^3.

With multi-atom molecules, the induced dipole moment depends on the orientation of the polarizing field relative to the molecule, and we should express α as a direction-dependent tensor quantity. To simplify our calculations, we can assume that α is isotropic, which is an acceptable approximation for many molecules.

The energy of a dipole $\boldsymbol{m}$ in a field $\boldsymbol{E}$ is given by the scalar product $\boldsymbol{m} \cdot \boldsymbol{E}$. When the dipole is induced by the field, half of this energy is used to achieve the polarization so that the net energy change is

$$V_{\text{field-ind dipole}} = -\alpha E^2/2 \tag{3.4.3}$$

This factor of ½ is a general result of linear response theory, and we already encountered the effect for the Keesom interaction in the relation between the energy in Eq. (3.3.8) and the free energy in Eq. (3.3.10).

We are now in the position to calculate $V_{\text{ion-ind dipole}}$ by combining Eq. (3.4.3) with the expression for the electric field associated with an ion of charge ze at $\boldsymbol{R} = 0$:

$$\boldsymbol{E}(\boldsymbol{R}) = \frac{ze}{4\pi\varepsilon_0}\frac{\boldsymbol{R}}{R^3} \tag{3.4.4}$$

to give

$$V_{\text{ion-ind dipole}}(R_{12}) = -\frac{\alpha(ze)^2}{2(4\pi\varepsilon_0)^2 R_{12}^4} \tag{3.4.5}$$

The range is R^{-4}, rather than R^{-2} as in Eq. (3.2.8), because one R^{-2} term induces the response, and a second R^{-2} term produces the interaction.

The field from a dipole depends on the angle θ between the dipole and the vector $\boldsymbol{R}$ joining the two molecules. The magnitude of the field is

$$E(\cos\theta) = \frac{m(1+3\cos^2\theta)^{1/2}}{4\pi\varepsilon_0 R^3} \tag{3.4.6}$$

The dipole–induced dipole interaction thus becomes

$$V_{\text{dipole-ind dipole}}(R_{12},\theta) = -\frac{1}{2}\alpha E^2 = -\frac{m^2\alpha(1+3\cos^2\theta)}{2(4\pi\varepsilon_0)^2 R_{12}^6} \tag{3.4.7}$$

by combining Eqs. (3.4.3) and (3.4.6). The angle-averaged energy follows directly since $\langle\cos^2\theta\rangle = 1/3$ and

$$V_{\text{dipole-ind dipole}}(R_{12}) = -\frac{m^2\alpha}{(4\pi\varepsilon_0)^2 R_{12}^6} \tag{3.4.8}$$

This result represents the Debye attractive interaction and constitutes the second of the three inverse sixth-power contributions to the total van der Waals interaction energy between molecules.

Problem: Calculate the dipole–induced dipole interaction between two water molecules at a separation of 0.29 nm, dipole moment $m = 1.86$ D, and polarizability $\alpha/\varepsilon_0 = 1.86$ m^3.

Solution:

$$V = -\frac{m^2\alpha}{(4\pi\varepsilon_0)^2 R^6} = -\frac{(1.86)^2 \times (3.3\times 10^{-30})^2 \times 1.86\times 10^{-29}}{(4\pi)^2 8.85\times 10^{-12} \times (0.29\times 10^{-9})^6} = -8.4\times 10^{-22}\ \text{J} \leftrightarrow -0.2k_B\mathbf{T}$$

or $V \sim -0.2k_B\mathbf{T}$, which is only a small contribution to the total interaction in Figure 3.5.

Induced interactions are not pairwise additive, and this fact adds an important complication to our calculations. For permanent moments, the interaction V_{123} between three molecules at fixed orientations is simply

$$V_{123} = V_{12} + V_{13} + V_{23} \tag{3.4.9}$$

or the sum over the individual pairs. But summing pairs does not hold true for induced interactions. For example, a molecule midway between two similar ions does not experience any electrical field, so the contribution from the ion–induced dipole term is zero for this configuration. If we assume pairwise additivity as in Eq. (3.4.9) and use Eq. (3.4.8), we will get a quite different result—one that amounts to an artificial stabilization of this particular molecular configuration.

See Atkins et al (2022, Focus 14) and Israelachvili (2011, Chapters 4–8) for additional discussion on Intermolecular Interactions.

3.5 Separating Ion–Ion Interactions from Contributions of Dipoles and Higher Multipoles in the Poisson Equation Simplifies Dealing with Condensed Phases

The electrostatic interaction between two molecules in an isolated pair shows substantial but manageable complexity, as demonstrated earlier. In a condensed phase, the situation becomes much more intricate. Attempts to use the Poisson equation given in Eq. (3.2.1) would require evaluating all multipole interactions simultaneously and would result in an intractable problem. This section develops a procedure to circumvent this difficulty, which permits us to rewrite the Poisson equation in a more convenient form.

Figure 3.6 shows a plate capacitor where the gap in one case is filled with (a) low-density gas or (b) a liquid. The plates are conducting, and when we apply an electric potential difference $\Delta\Phi$, a surface charge $\pm\sigma$ is induced on the plates. From the one-dimensional version of the Poisson equation (Eq. 3.2.1),

$$\varepsilon_0 \frac{\mathrm{d}^2\Phi}{\mathrm{d}x^2} = -\rho(x) \tag{3.5.1}$$

It follows that when there is no charge in the gap between the plates, $\rho = 0$, the field in the x direction is

$$E_x \stackrel{\text{def}}{=} -\frac{\mathrm{d}\Phi}{\mathrm{d}x} = \text{constant} \tag{3.5.2}$$

The value of the constant is given by the boundary conditions at $x = \pm h/2$. To find this, we integrate Eq. (3.5.1) between $x = h/2 - \Delta x$ and $x = h/2 + \Delta x$ so that

$$\frac{\mathrm{d}\Phi}{\mathrm{d}x}\left(\frac{h}{2} + \Delta x\right) - \frac{\mathrm{d}\Phi}{\mathrm{d}x}\left(\frac{h}{2} - \Delta x\right) = -\frac{1}{\epsilon_0}\int_{\frac{h}{2}-\Delta x}^{\frac{h}{2}+\Delta x} \rho \mathrm{d}x = -\frac{\sigma}{\varepsilon_0} \tag{3.5.3}$$

The field inside a conductor is zero, so Eq. (3.5.2) yields the boundary conditions

$$\frac{\mathrm{d}\Phi}{\mathrm{d}x}\left(\pm\frac{h}{2}\right) = -E_x\left(\pm\frac{h}{2}\right) = -\frac{\sigma}{\varepsilon_0} \tag{3.5.4}$$

The second equality follows since, at the two boundaries, the field goes from zero to the finite value in opposite directions (see Eq. 3.5.3). We will use these boundary conditions repeatedly in the coming sections. The field is constant in the gap, so the potential difference between the plates is

$$\Delta\Phi = h\sigma/\varepsilon_0 \tag{3.5.5}$$

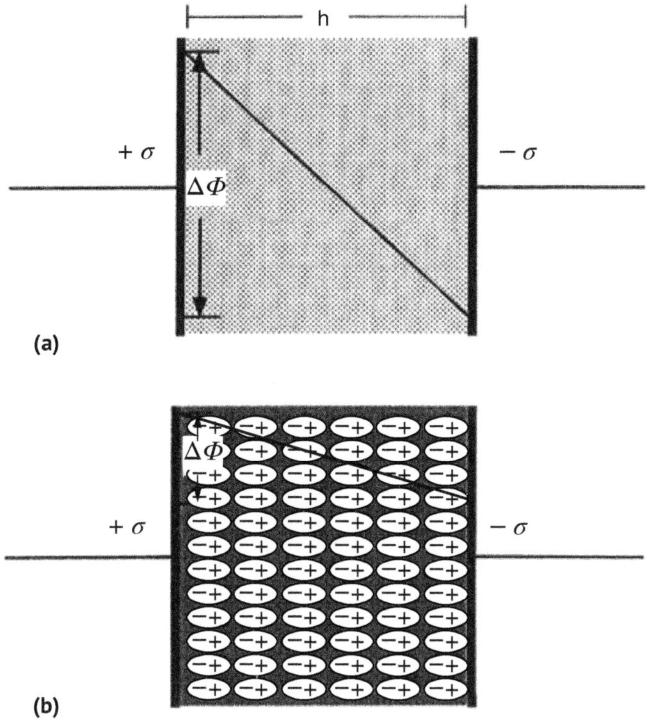

Figure 3.6 Illustration of the drop in electrostatic potential across a parallel plate capacitor with a surface charge density $\pm\sigma$ at the two plates. (a) With air in the gap, the potential varies linearly with a slope $\frac{d\Phi}{dx} = -\frac{\sigma}{\varepsilon_0}$ determined solely by the surface charge density. (b) With a liquid medium in the gap, the molecules are oriented and polarized by the field. As illustrated, this occurs uniformly throughout the gap with the result that the potential still varies linearly, but with a smaller slope $\frac{d\Phi}{dx} = -\frac{\sigma}{\varepsilon_r\varepsilon_0}$ so that the potential drop $\Delta\Phi$ becomes smaller (see Eq. 3.5.5).

For a capacitor, the capacitance, C, per unit area equals the ratio between the charge density and the applied potential, so from Eq. (3.5.5) we get $\frac{C}{\text{area}} = \varepsilon_0/h$. In the capacitor, it is the applied potential difference that sets the charge density. However, in colloidal systems, it is generally the charge density that sets the potential at the surface.

Equation (3.5.3) represents a specific example of Gauss' law, which provides a very useful relation between electrical charge and the associated electrical field. In its general form, Gauss' law states that the electric field perpendicular to the surface A of a closed volume V is determined by the enclosed charge; in mathematical terms

$$\varepsilon_0 \int \nabla\Phi \cdot \boldsymbol{n} dA = -\int \rho dV \tag{3.5.6}$$

where $\boldsymbol{n}$ signifies a surface normal vector pointing outward. We can often use this relation to obtain a qualitative understanding (and sometimes also a quantitative solution) to an electrostatic problem. We will return to illustrations of the use of Gauss' law.

We can calculate the energy required to charge the capacitor, i.e., the energy stored, U, by considering a charging process in which an infinitesimal amount of charge dq is transferred from one plate to the other. As the degree of charging increases, the potential difference between the plates increases; the differential energy increase is the product of charge transferred, dq, and the potential difference $\Delta\Phi$, which depends on the charge density σ'

$$dU = dq\Delta\Phi(\sigma') = d\sigma' A\frac{h\sigma'}{\varepsilon_0}$$

By integrating from zero charge density to the final value σ, we find

$$\begin{aligned} U &= \frac{Ah}{\varepsilon_0}\int_0^{\sigma}\sigma' d\sigma' = \frac{Ah}{\varepsilon_0}\frac{\sigma^2}{2} = \frac{A\sigma\Delta\Phi}{2} \\ &= A\frac{\varepsilon_0}{2}\frac{(\Delta\Phi)^2}{h} \end{aligned} \tag{3.5.7}$$

In a more general form one has, for an arbitrary globally electroneutral charge distribution,

$$U = \frac{1}{2}\int \rho\Phi dV = \frac{\varepsilon_0}{2}\int(\nabla\Phi)^2 dV \tag{3.5.8}$$

Since the potential is generated by the charge distribution itself, there is a factor of one-half entering the expression for the energy to compensate for double counting.

Problem: Calculate the energy of the capacitor by considering a separation of the two plates at constant charge density from zero distance to h using Eq. (3.2.3):

$$U = \frac{1}{4\pi\varepsilon_0}\int\frac{\rho_1(r)\rho_2(r')}{|\mathbf{r}-\mathbf{r}'|}dVdV'$$

Solution: For a capacitor, ρ_1 and ρ_2 are charge distributions σ and $-\sigma$ confined to two parallel planes. Let a surface element dA on surface 1 initially be located at the origin

$$\rho_1(0)dV = \sigma dA$$

The distance from dA to dA' on plate 2 is $(x^2+y^2+z^2)^{1/2}$. The energy of bringing the surface element dA from $z = 0$ to $z = h$ is

$$\begin{aligned} d(U(h) - U(0)) &= \frac{1}{4\pi\varepsilon_0}\sigma dA(-\sigma)\iint_0^{\infty}\left\{\frac{1}{\left(x^2+y^2+h^2\right)^{1/2}} - \frac{1}{(x^2+y^2)^{1/2}}\right\}dxdy \\ &= \frac{-\sigma^2}{2\varepsilon_0}dz\left\{\int_0^{\infty}\frac{rdr}{\left(r^2+h^2\right)^{1/2}} - \int_0^{\infty}dr\right\} \\ &= \frac{-\sigma^2}{2\varepsilon_0}dA\left[\left(r^2+h^2\right)^{1/2} - r\right]_0^{\infty} = \frac{\sigma^2}{2\varepsilon_0}dA\cdot h \end{aligned}$$

Integrating over plate 1

$$U(h) - U(0) = \frac{\sigma^2}{2\varepsilon_0}A\cdot h$$

as in Eq. (3.5.7).

In this example we express the energy relative to that of zero separation ($x = 0$) between the plates. In Eq. (3.2.3), we have, as the reference state, the charges at infinite separation. The latter choice is only practical for finitely charged systems.

In a commercial capacitor, a nonconducting (dielectric) medium fills the gap between the plates, as shown in Figure 3.6b. This arrangement increases the capacitance because the electrical field polarizes the medium so that the induced charge distribution partly compensates for the surface charges on the conducting plates. In other words, the response of the molecules results in a smaller potential drop $\Delta\Phi$ for a given σ and thus to a higher capacitance.

To formally describe this effect, we have to go back and rewrite the Poisson equation by dividing the total charge distribution ρ into two parts as follows:

$$\rho = \rho_{\text{free}} - \nabla \cdot \boldsymbol{P} \tag{3.5.9}$$

where $\boldsymbol{P}$ is called the polarization vector and is a dipole moment density. The point of separating ρ into two parts is that ρ_{free} describes local net charges, while $\nabla \cdot \boldsymbol{P}$ describes the effects of dipoles and higher moments. Combining Gauss' law (Eq. 3.5.6) and the Poisson equation (Eq. 3.2.1), we have

$$\int_V \nabla \cdot \boldsymbol{P}\, \mathrm{d}V = \int_A \boldsymbol{P} \cdot \boldsymbol{n}\, \mathrm{d}A \tag{3.5.10}$$

Thus, the polarization carries an effective net charge only at the surface of the volume. Inserting Eq. (3.5.9) into the fundamental Poisson equation yields

$$\varepsilon_0 \nabla \cdot \nabla\Phi = \nabla \cdot \boldsymbol{P} - \rho_{\text{free}} \tag{3.5.11}$$

Further progress requires introducing some approximation that relates polarization to the electrical field. By analogy with the concept of the polarizability of an isolated molecule, we can assume that $\boldsymbol{P}$ relates in a linear fashion to the field $\boldsymbol{E} = -\nabla\Phi$ to yield

$$\boldsymbol{P} = \varepsilon_0(\varepsilon_r - 1)\boldsymbol{E} \tag{3.5.12}$$

The proportionality factor in this linear relation is chosen so that $\varepsilon_r = 1$ corresponds to no polarization or the vacuum case, and ε_r represents the relative permittivity, or simply the dielectric constant. For a time-independent field $\boldsymbol{E}$, ε_r is always larger than unity.

Combining Eq. (3.5.12) with Eq. (3.5.11) yields a Poisson equation:

$$\varepsilon_0\varepsilon_r \nabla^2\Phi = -\rho_{\text{free}} \tag{3.5.13}$$

Because $\nabla\boldsymbol{E} = -\nabla \cdot \nabla\Phi$, this equation differs from the fundamental Poisson equation only by a factor ε_r. Consequently, the calculation of the capacitance in the dielectric case applies as long as we replace ε_0 by $\varepsilon_0\varepsilon_r$. Thus, the capacitance for the capacitor in Figure 3.6b increases by a factor ε_r and $\dfrac{C}{\text{area}} = \dfrac{\varepsilon_r\varepsilon_0}{h}$.

To see the molecular origin of the increased capacitance, we note that the field is $E = -\dfrac{\sigma}{\varepsilon_r\varepsilon_0}$ and the magnitude P of the polarization is

$$P = -\varepsilon_0(\varepsilon_r - 1)\frac{\sigma}{\varepsilon_0\varepsilon_r} = -\left(1 - \frac{1}{\varepsilon_r}\right)\sigma \tag{3.5.14}$$

using Eq. (3.5.12). Note that charge per area is dimensionally equivalent to dipole moment per volume. In the capacitor, there is thus a constant polarization of magnitude $(1 - 1/\varepsilon_r)\sigma$ oriented to compensate for the charges on the capacitor plates as illustrated in Figure 3.6b. For example, liquid water has a dielectric constant of approximately 80, showing that it sets up a compensating polarization amounting to $(1 - 1/80) \approx 0.9875$ of the charge on the plate.

Problem: Calculate the maximum dipolar polarization in liquid water, given that a water molecule has a dipolar moment of 1.85D.

Solution: A density of 10^3 kg/m³ implies a volume of 3.0×10^{-29} m³ per molecule. In SI units, 1.85D corresponds to $1.85 \times 3.336 \times 10^{-30}$ C · m. For perfectly oriented water molecules, the dipolar density and thus the polarization is $P_{max} = 1.85 \times 3.336 \times 10^{-30}$ C · m/3.0×10^{-29} m³ = 0.21 C/m². To this comes the contribution from the electronic polarization.

Three additional points must be noted to complete the argument in this section. The first is that Eq. (3.5.12) treats the dielectric permittivity $\varepsilon_0\varepsilon_r$ as a constant, irrespective of the properties of the electric field. In fact, our discussion of the capacitor implicitly assumes a DC field. If we apply a slowly alternating field, the dipoles turn as the polarity of the electrodes varies. When the AC frequency increases to a point at which the dipoles can no longer follow the field, the dielectric constant decreases. In the limit of a frequency that is high relative to molecular reorientation but lower than typical electronic excitation frequencies, the dielectric constant equals the square of refractive index of the material.

A second point relates to the gap size h in our discussion of the capacitor. Although the gap is small, it has a macroscopic magnitude. Describing the solvent as a dielectric continuum may be invalid on a microscopic scale; nevertheless, experience has shown that it gives a useful approximation. Therefore, we will assume that the Poisson equation holds true on a molecular scale and that ρ_{free} contains the charges that we consider explicitly. The factor ε_r includes the effects of the medium. Under these circumstances, all the equations we derived for the electrostatic interactions between molecules in isolated pairs are still valid in a medium, provided we replace ε_0 by $\varepsilon_0\varepsilon_r$.

For example, the Coulomb interaction between two charges q_1 and q_2 in a solvent with a relative dielectric permittivity ε_r is

$$V_{\text{coulomb}}(R_{12}) = \frac{q_1 q_2}{4\pi\varepsilon_0\varepsilon_r}\frac{1}{R_{12}} \tag{3.5.15}$$

in analogy with Eq. (3.2.6). From a molecular perspective, this simple equation describes a very complicated reality. We can identify four different molecular contributions to V_{coulomb} in Eq. (3.5.15). There is a direct ion–ion interaction

$$V_{\text{ion–ion}} = \frac{q_1 q_2}{4\pi\varepsilon_0 R_{12}} = \frac{z_1 z_2 e^2}{4\pi\varepsilon_0 R_{12}} \tag{3.2.6}$$

In addition, there is an ion–solvent interaction of both the ion-dipole and the ion-induced dipole type. How the ions orient the solvent molecules depends on R_{12}, which gives rise to a separation-dependent solvent–solvent interaction. Finally, for a dipolar solvent like water, the orientation of the solvent molecules involves a decreased entropy relative to that in a homogeneous fluid. To represent a consistent description, the three solvent-mediated contributions have to add up to the difference between Eqs. (3.5.15) and (3.2.6)

$$-\frac{q_1 q_2}{4\pi\varepsilon_0}\frac{1}{R_{12}}\left(1 - \frac{1}{\varepsilon_r}\right).$$

The interaction energy of Eq. (3.5.15) represents the total free energy change in bringing the charges from infinite separation up to R_{12}. During the process, the solvent molecules adjust to equilibrium. This illustrates the third point; with ε_r present, the interaction energies represent free energies because the solvent's response involves both energy and entropy contributions. With ε_r, we describe an average over some molecular degrees of freedom in much the same way as we do for the Keesom interaction of Eq. (3.3.6). The basic relation between the free energy and the internal energy is

$$\mathbf{A} = \mathbf{U} - \mathbf{TS} = \mathbf{U} + \mathbf{T}\left(\frac{\partial \mathbf{A}}{\partial \mathbf{T}}\right) \tag{3.5.16}$$

In Eq. (3.5.15), V_{coulomb} is an effective interaction and has the character of a free energy, which we can separate into an energy and entropy part using the relation in Eq. (3.5.16) as a basis. In Eq. (3.5.15), the relative permittivity is temperature-dependent and we can decompose it into an energy and a entropy part

$$V_{\text{coulomb}} = V^E_{\text{coulomb}} + V^S_{\text{coulomb}} = \frac{q_1 q_2}{4\pi\varepsilon_r\varepsilon_0}\left\{1 + \frac{\mathbf{T}}{\varepsilon_r}\frac{d\varepsilon_r}{d\mathbf{T}}\right\} - \frac{q_1 q_2}{4\pi\varepsilon_r\varepsilon_0}\frac{\mathbf{T}}{\varepsilon}\frac{d\varepsilon_r}{d\mathbf{T}} \tag{3.5.17}$$

Thus, if the relative permittivity, ε_r, changes with temperature, as it does for all polar solvents including water but not appreciably for nonpolar solvents like hydrocarbons, the interaction potential has an entropic component even if it does not have an explicit temperature dependence. It follows from Eq. (3.5.17) that we can obtain the energy part V^E_{coulomb} by multiplying Eq. (3.5.15) with the factor $\left\{1 + \frac{\mathbf{T}}{\varepsilon_r}\frac{d\varepsilon_r}{d\mathbf{T}}\right\}$. For the case that $\varepsilon_r \propto 1/\mathbf{T}$, we have that the energy part is zero since in that case $\frac{d\varepsilon_r}{d\mathbf{T}} = -\frac{\varepsilon_r}{\mathbf{T}}$. For water at room temperature, the permittivity has an even stronger temperature dependence and $\frac{\mathbf{T} d\varepsilon_r}{\varepsilon_r d\mathbf{T}} \approx -1.35$ so that $V^E_{\text{coulomb}} \approx -0.35 V_{\text{coulomb}}$ and the energy and free energy parts have opposite sign.

Problem: Calculate the changes $\Delta\mathbf{U}$ and $\mathbf{T}\Delta\mathbf{S}$ in internal energy and entropy when a sodium ion and a chloride ion are brought from a large separation into contact ($R_{\text{NaCl}} = 0.276$ nm) in water at 20 °C ($\varepsilon_r = 80$).

Solution: The free energy change is

$$\Delta A = \frac{-e^2}{4\pi\varepsilon_0\varepsilon_r}\left[\frac{1}{R_{\text{NaCl}}} - \frac{1}{R_\infty}\right]$$

$$= -\frac{\left(1.6 \times 10^{-19}\right)^2}{4\pi \times 80 \times 8.85 \times 10^{-12}} \times \frac{1}{0.279 \times 10^{-9}} = -1.05 \times 10^{-20} (\sim -2.6\text{kT})$$

or − 6.6 kJ/mol. Then

$$\Delta U = -0.35\ \Delta A = 3.7 \times 10^{-21}\ \text{J or } 2.3\ \text{kJ/mol}$$

Despite the fact that the chloride ion attracts the sodium ion, the internal energy increases and heat is taken up from the surrounding thermal bath when the ions approach one another.

The entropic part of the free energy is $-\mathbf{T}\Delta\mathbf{S} = \Delta\mathbf{A} - \Delta\mathbf{U} = 1.35\Delta\mathbf{A} = -1.42 \times 10^{-20}$ or −8.9 kJ/mol. This large increase in entropy comes from water dipolar orientation around the ions as they come closer to each other.

For a general charge distribution ρ in a medium, the charge–charge interaction energy remains as in Eq. (3.5.8):

$$U_{el}^{\dagger} = \frac{1}{2}\int \rho\Phi \, dV = \frac{1}{2}\varepsilon_0\varepsilon_r \int (\nabla\Phi)^2 dV \tag{3.5.18}$$

with the difference that ε_r enters into the second equality. We have added a dagger superscript on $U_{el}^{\dagger}$ to show that the charge–charge interaction is a free energy for a medium where ε_r is temperature-dependent. The entropic contribution comes from the orientation of solvent molecules, which is implicitly included in ε_r.

3.6 The Poisson Equation Containing Solvent-averaged Properties Describes the Free Energy of Ion Solvation

The Born model of ion solvation illustrates the use of the continuum dielectric model of a solvent down to molecular length scales. For a spherical ion, we can calculate the free energy difference of an ion in the solvent relative to the gas phase by using either of the two expressions in Eq. (3.5.18). Although the second expression requires somewhat more computation, it is conceptually clearer.

Using Gauss' law (Eq. 3.5.6), we can determine that the field in the radial direction outside a spherically symmetrical charge distribution such as in an atomic ion with a net charge q is

$$E_r = \frac{q}{4\pi\varepsilon_0}\frac{1}{r^2} \tag{3.6.1}$$

in a vacuum and a factor $1/\varepsilon_r$ smaller in the solvent. Inside the ion, the electrical field is unaffected by the presence of a medium in spherical symmetry. We can see this by using the principle of superposition and consider only the potential due to the medium. In the space occupied by the ion the field across a spherical shell is still, according to Gauss' law, determined by the enclosed charge. Since the medium is absent in the central space, the enclosed medium charge is zero and that part of the field is zero. Thus, we need to consider the electrical field only for $r \geq R_{\text{ion}}$, where R_{ion} represents the radius of the ion

$$U_{\text{vac}} = \frac{1}{2}\varepsilon_0 \int E^2 dV = \frac{\varepsilon_0}{2}\int_{R_{\text{ion}}}^{\infty} \left(\frac{q}{4\pi\varepsilon_0}\right)^2 \frac{1}{r^4} 4\pi r^2 dr = \frac{1}{8\pi\varepsilon_0} q^2 \frac{1}{R_{\text{ion}}} \tag{3.6.2}$$

We can obtain the free energy in a medium simply by replacing ε_0 with $\varepsilon_0\varepsilon_r$ in Eq. (3.6.2). The Born solvation free energy, which is the difference between these expressions, is

$$\begin{aligned}\mathbf{A}_{\text{solv}} &= \mathbf{U}_{\text{med}}^{\dagger} - \mathbf{U}_{\text{vac}} = \frac{1}{8\pi\varepsilon_0\varepsilon_r} q^2 \frac{1}{R_{\text{ion}}} - \frac{1}{8\pi\varepsilon_0} q^2 \frac{1}{R_{\text{ion}}} \\ &= \frac{-q^2}{8\pi\varepsilon_0}\frac{1}{R_{\text{ion}}}\left(\frac{\varepsilon_r - 1}{\varepsilon_r}\right)\end{aligned} \tag{3.6.3}$$

Another way to derive Eq. (3.6.3) is by writing $\Phi(\boldsymbol{r}) = \Phi_{\text{ion}}(\boldsymbol{r}) + \Phi_{\text{medium}}(\boldsymbol{r})$. Using Gauss' law once more, we can determine that for $r < R_{\text{ion}}$, the potential from the medium is constant. Therefore, the charge distribution of the ion does not change when it passes from vacuum to the medium. As a result, Φ_{ion} is the same in the two cases, and $\mathbf{A}_{\text{solv}} = \left(\frac{1}{2}\right) q\Phi_{\text{medium}}(r \leq R_{\text{ion}})$. For $r \geq R_{\text{ion}}$, $\Phi_{\text{medium}} = \Phi - \Phi_{\text{ion}} = -\left(\frac{q}{4\pi\varepsilon_0}\right)\left(1 - \frac{1}{\varepsilon_r}\right)\left(\frac{1}{r}\right)$, and by inserting $r = R_{\text{ion}}$, we regain Eq. (3.6.3).

For a monovalent ion, $q = \pm 1.6 \times 10^{-19}$ C, with a radius $R_{ion} = 0.2$ nm in water at 20°C, where $\varepsilon_r = 80$ the calculated solvation free energy is $\mathbf{A}_{solv} = -5.7 \times 10^{19}\text{J} \Leftrightarrow 140\ k_B\mathbf{T}$. This calculation demonstrates why we need a polar solvent with a high ε_r to dissolve electrolytes. Changing ε_r from 20 to 80 changes the solvation energy by more than 6 $k_B\mathbf{T}$ per ion. For a 1:1 electrolyte, this would imply a change in solubility of a factor $\{\exp(12)\}^2 \approx 1.6 \times 10^5$!

A biological lipid membrane sets up a sheath of low dielectric permittivity, $\varepsilon_r \approx 3$, between two aqueous regions. Passive diffusion of ions across the membrane is slow because the concentration of ions in an oil-like layer is very small. For example, we can use the Born model to estimate the difference in solvation energy for a sodium ion between oil and water, $e^2/(8\pi\varepsilon_0 R_{ion}) \times (79/80 - 2/3) \approx 44 k_B\mathbf{T}$, which corresponds to a distribution coefficient $K = \exp(44) \approx 10^{19}$. This result overestimates the distribution coefficient, but it provides a qualitative explanation of the phenomenon.

Even though the Born model of ion solvation relies on a simplistic model of a solvent, it conveys a great deal of important information. It gives numbers of the right order of magnitude, although the quantitative agreement is on the order of ±50%, depending on the ion and on the choice of R_{ion}. Equation (3.6.3) shows that solvation energy depends on the charge squared, and that the polarization effects in the medium are long-ranged. For example, the solvation energy of an ion in a solvent droplet would depend on $R_{droplet}^{-1}$.

See Israelachvili (2011, Chapter 3) for additional discussion on Dielectric Description.

3.7 Self-assembly, Ion Adsorption, and Surface Titration Play an Important Role in Determining Properties of Charged Interfaces

In practice, colloidal interfaces are often charged, and one of the themes of this book is to demonstrate how these charges are essential in determining the physical and chemical properties of the colloidal systems in question. There are three basic mechanisms that generate charged interfaces. The first, and most direct, is when the colloid is formed by intrinsically charged units. Examples include surfactant systems where the self-assembly process generates a charged colloidal micelle by collecting the charged amphiphilic molecules into a unit; polyelectrolytes, where the charge is due to covalently linked ionic groups distributed along the polymer chain; latex particles, where charged groups are covalently attached by sulfonation or carboxylation; and clay surfaces, where trivalent aluminum atoms replace tetravalent silicon atoms in the oxygen–silicon network generating negatively charged colloidal particles.

A second mechanism for generating charged surfaces arises when titratable groups are attached to the surface. Proteins contain carboxylic and amino groups that make their net charge strongly pH-dependent. Weak polyelectrolytes like polyacrylic acid and polyamines also show this behavior. Silica surfaces contain silanol groups that titrate and a similar situation is found at most metal oxide surfaces.

The third mechanism occurs when an intrinsically uncharged colloidal particle acquires a charge from species adsorbing from solution. A minute contamination of a strongly adsorbing species can generate interfaces containing appreciable charge. Fatty acids, or the corresponding ions, as well as surfactants, adsorb strongly on apolar surfaces and insert into lipid bilayers. Large inorganic anions like I^-, ClO_4^-, or SCN^- adsorb to most interfaces. All ionic crystals tend to acquire a surface charge because the ionic species forming the crystal have slightly different affinities to the surface layer.

For most charged surfaces, the actual charge is determined by an equilibrium process involving the exchange of ionic species between the bulk solution and the interface. The final equilibrium state is determined by an interplay between three free energy contributions; the short-range chemical interactions at the surface that promote adsorption; the long-range electrostatic interactions that act to prevent the formation of highly charged surfaces; and the entropy of the adsorbing ion, which typically favors desorption and the formation of the charged surface.

We can describe the effect of short-range chemical interactions in terms of a conventional chemical equilibrium. For example, an ionic surface group, S^-, can bind a counterion, M^+, (or a proton) to form a neutral species, SM, and establish an equilibrium,

$$S^- + M^+ \rightleftharpoons SM \tag{3.7.1}$$

The scheme works equally well for S neutral and SM charged, giving us

$$S + M^+ \rightleftharpoons SM^+ \tag{3.7.2}$$

To write down the proper equilibrium equation for the process described in Eq. (3.7.1), we must take into account two additional complications that normally do not arise in bulk solution, where the equilibrium can be written as $K_B = X_{SM}/X_S X_M$. First, association with the ionic species changes the surface charge density and as a result changes the electrostatic ion–ion interaction. Second, since we assume that both S^- and SM (or S and SM^+) are confined to the surface, we must consider the entropy of mixing contribution to the free energy with some care.

Bearing these factors in mind, we can write the chemical potential of a surface group as

$$\mu_S = \mu_S^\theta(\text{surface}) + \mu_S(\text{entropy of mixing}) + \mu_S^{el} \tag{3.7.3}$$

If we assume a random (ideal) mixing of S and SM groups, the entropic term becomes

$$\mu_S(\text{entropy of mixing}) = k_B\mathbf{T}\ln X_S \tag{3.7.4}$$

where $X_S = n_S/(n_S + n_{SM})$ represents the fraction of charged groups, S^-, relative to the total number of *S* groups, that is, the surface mole fraction. Similarly, for the neutral *SM* group with mole fraction X_{SM}, we have

$$\mu_{SM} = \mu_{SM}^\theta(\text{surface}) + k_B\mathbf{T}\ln X_{SM} \tag{3.7.5}$$

The electrostatic contribution, μ_S^{el}, to the chemical potential of the charged surface group represents the free energy required to bring a charge $z_S e$ from the bulk to the surface. This amounts to

$$\mu_S^{el} = z_S e\Phi_0 \tag{3.7.6}$$

where Φ_0 is the surface potential, which we define as the difference in electrical potential between surface and bulk, where the potential Φ is set to zero by convention.

The chemical potential of the ion M^+ is uniform and assuming ideal solution behavior

$$\mu_M = \mu_M^\theta(\text{solution}) + k_B\mathbf{T}\ln c_M^* \tag{3.7.7}$$

By requiring that chemical potentials are equal for the left- and right-hand sides of the scheme in Eq. (3.7.1), i.e, $\mu_S + \mu_M = \mu_{SM}$, we obtain the equilibrium equation

$$\frac{X_{SM}}{X_S c_M} = K_S \exp\left(\frac{z_S e\Phi_0}{k_B\mathbf{T}}\right) \tag{3.7.8}$$

which explicitly brings out the electrostatic contribution to the equilibrium. The surface equilibrium constant is related to the standard chemical potentials through

$$k_B\mathbf{T}\ln\, K_S = \mu^\theta_M(\text{solution}) + \mu^\theta_S(\text{surface}) - \mu^\theta_{SM}(\text{surface}) \tag{3.7.9}$$

In this expression, K_S differs from the corresponding equilibrium constant K_B for the association in the bulk solvent, because of two factors. First, K_B includes a substantial contribution from the long-range electrostatic interaction between S^- and M^+. The surface potential Φ_0 includes this contribution, however, so it is absent in K_S. Second, confining $S^-$and SM to a standard state of a surface environment leads to additional differences between K_B and K_S. The solvation energy at the surface, particularly for the charged species, S^-, can be substantially smaller than it is in the bulk if we assume an aqueous or polar solvent. Nevertheless, K_S and K_B usually do not differ by more than an order of magnitude, since the two effects tend to cancel in part.

Surface group titration provides the clearest example of a substantial nonelectrostatic contribution to the binding between an ion and a surface. In surface titration, a proton binds to a carboxylate ion, an amine, or a silanol group with a covalent bond. A titration curve shows that the pH dependence of the degree of ionization for surface charges does not have the same sigmoid shape as observed in the bulk. Because of electrostatic interactions, increasing the number of charged groups becomes progressively more difficult, as the surface charge density increases.

Equation (3.7.8) expresses the effect of these interactions in the factor $\exp(-z_i e\Phi_0/k_B\mathbf{T})$. We can calculate the expected titration behavior at a surface if we combine Eq. (3.7.8) with an independent estimate that gives the relation between surface charge density and surface potential. We show in the following section of this chapter how this is accomplished.

It is also possible to estimate the surface potential of a colloidal particle through the electrophoretic mobility as we discuss in detail in Chapter 8. As an example, consider the protein enzyme ribonuclease. Like all proteins, it has both carboxylic and amino groups, whose degree of protonation changes with pH. Figure 3.7 shows the surface potential in the range of pH from 4.0 to 11.0 measured at 25 °C and in a 50 mM KNO_3 solution. Ribonuclease is a small protein (molecular weight 12.600 Daltons) and the net charge varies from approximately +7e to −6e over this pH range. The surface potential goes through zero at a pH just above 9. This is called the isoelectric point or the point of zero charge. Ribonuclease has an unusually high fraction of amino groups so that the isoelectric point is at a higher pH than for most proteins. From a functional point of view, this is certainly correlated with the fact that ribonuclease has as its substrate the negatively charged ribonucleic acid, RNA.

It is a general observation that colloidal interfaces are more often negatively than positively charged. For example, it is striking that in the living cell, all membranes and organelles carry a negative charge. There are several reasons why interfaces more easily acquire a negative than a positive charge. One of them is that anions tend to bind nonspecifically to interfaces. A general sequence called the Hofmeister series describes the ion-binding effects in a large number of systems. In order of decreasing surface affinity, the sequence for more common anions is

$$I^- > ClO_4^-, NO_3^- > Br^- > Cl^- > OH^-, F^-, SO_4^{2-}$$

Although the effect of the ions is sometimes ascribed to their influence on water structure, their relative affinity to interfaces seems to offer a more appropriate explanation.

As a consequence of specific ion binding effects, the surface charge density, σ, and surface potential, Φ_0, are difficult to control in any particular system. These properties vary with conditions,

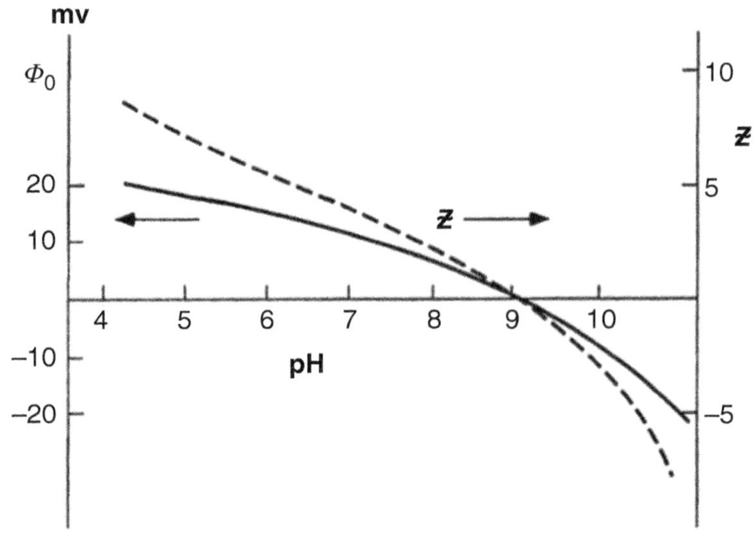

Figure 3.7 The measured surface potential Φ_0 and charge *ze*, from electrophoresis, of the protein ribonuclease as a function of pH in 50 mM KNO_3. At the isoelectric point, just above pH 9, the protein is net neutral, that is, it carries on average the same number of positive and negative charges. The pH dependence of the surface potential is due to the titration of COO^- and $-NH_2$ groups. *Source:* Adapted from Norde, 1976.

which we have to remember when we consider the electrostatic contribution to the interaction between surfaces.

See Berg (2010, Chapter 6), Cosgrove (2010, Chapter 2), Doi (2013, Chapter 10), Israelachvili (2011, Chapter 14), Kronberg et al (2014, Chapter 3), and Ohshima (2024, Chapter 3) for additional discussion on Charged Interfaces.

3.8 The Poisson–Boltzmann Equation Can Be Used to Calculate the Ion Distribution in Solution

3.8.1 The Gouy–Chapman Theory Relates Surface Charge Density to Surface Potential and Ion Distribution Outside a Planar Surface

Interactions that occur between charges fixed at the surface and those that are free in solution play an important role in colloidal systems. Charged surfaces are characterized by a surface charge density σ and a potential Φ_0. The adjacent solution contains electrolytes and is characterized by the bulk concentration c_{io} and valency z_i of each ion, and the solvent dielectric constant, ε_r. We want to determine the relationship between σ and Φ_0 and also how the potential and distribution of ions in the solutions vary with distance from the charged interface.

To do this, we solve the Poisson equation. In the region of the solution, this equation takes the form

$$\varepsilon_0 \varepsilon_r \nabla^2 \Phi = -\rho_{(\text{free ions})} \tag{3.5.13}$$

where the solution's charge distribution is expressed as

$$\rho_{(\text{free ions})} = e\sum_i z_i c_i^*(\boldsymbol{r}) \tag{3.8.1}$$

The asterisk on c indicates that the concentration is measured in molecules per cubic meter rather than in moles per liter, and $c_i^*(\boldsymbol{r})$ represents the local concentration of ions. Because the ions in the solution are free to respond to the electrical fields, the solution's charge distribution, ρ, is not independently known. In addition to the electrostatic interaction energy, we must also consider the entropy associated with the solution's ion distribution. The electrostatic interaction energy favors an ordered and very localized ion arrangement, while entropic factors, in contrast, strive to generate a random uniform distribution of ions.

In an external potential, this compromise between energy and entropy results in a Boltzmann distribution as discussed in Section 1.2.5:

$$c_i^*(\boldsymbol{r}) = c_{i0}^* \exp\left(\frac{-z_i e\Phi}{k_B \mathbf{T}}\right) \tag{3.8.2}$$

where z_i stands for the ion valency and c_{i0}^* is the concentration of ion species i at the reference point where $\Phi = 0$, which we usually choose to be the bulk. We can now combine Eqs. (3.8.1), (3.8.2), and (3.5.13) to give

$$\rho_{(\text{free ions})} \approx e\sum_i z_i c_{i0}^* \exp\left(\frac{-z_i e\Phi}{k_B \mathbf{T}}\right) \tag{3.8.3}$$

and

$$\varepsilon_r \varepsilon_0 \, \Delta\Phi = e\sum_i z_i c_{i0}^* \exp\left(\frac{-z_i e\Phi}{k_B \mathbf{T}}\right) \tag{3.8.4}$$

where the latter is the **Poisson–Boltzmann equation** describing the ion distribution in an electrolyte solution outside a charged interface or particle. There is, in fact, a fundamental approximation involved in the step connecting Eqs. (3.8.2) to (3.8.4): The former equation is strictly valid only when the potential is generated by an external charge, while in the Poisson–Boltzmann equation, the potential is also generated by the charge of the ions whose distribution we want to determine. In a real electrolyte solution outside a charged surface, the ions move and constantly change position. In each moment, there are certain positions of all ions and we could, in principle, solve the Poisson equation for that particular charge configuration. In the next moment, the ions have changed place slightly and we can solve for the new spatial variation of the potential. If we follow the system in this way over a molecularly long time, we get a distribution of values for the potential at each position in space. The average value of this potential is called the *mean* potential, which is the relevant potential that enters the Poisson–Boltzmann equation. Thus, in deriving the Poisson–Boltzmann equation, we have made a so-called *mean field approximation*, replacing true ion distributions by their distribution in the mean, and the electrostatic potential by its average value. This is a very useful and convenient approximation, but there are circumstances of practical importance where it gives quantitatively incorrect results and it can even fail qualitatively as will be discussed in Chapter 5.

We can illustrate the properties of the ion distribution around a single colloidal particle by choosing as our model an infinite, uniformly charged planar surface exposed on one side to an electrolyte solution, as shown in Figure 3.8. Because the surface extends uniformly in the y and z

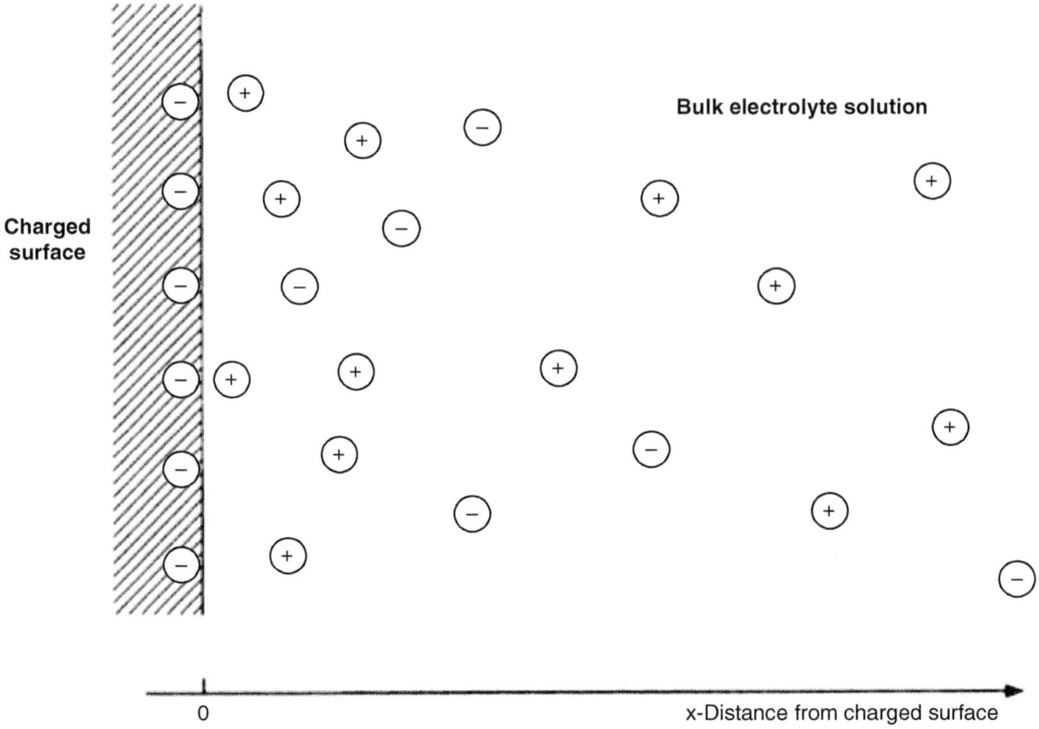

Figure 3.8 Model used to derive the Gouy–Chapman theory from the Poisson–Boltzmann equation. A negatively charged wall that is infinite in the y and z directions has associated with it a charge density σ and a potential Φ. The valency and concentration of electrolyte and the dielectric permittivity of the solvent determine the decrease in potential in the x direction away from the wall ($\Phi = \Phi_0$ at $x = 0$,) ($\Phi = 0$ for large x).

directions, only variations in the x direction are important. As a consequence of this lateral symmetry, the Poisson–Boltzmann equation reduces to an ordinary differential equation

$$\varepsilon_r \varepsilon_0 \frac{d^2\Phi}{dx^2} = -e \sum_i z_i c_{i0}^* \exp\left(\frac{-z_i e \Phi}{k_B \mathbf{T}}\right) \tag{3.8.5}$$

To obtain a unique solution, we need two boundary conditions plus a reference point for the potential. Electroneutrality requires that the surface charges are fully neutralized by ions in the solution, and at sufficiently large distances from the surface, we have

$$\left.\frac{d\Phi}{dx}\right|_{x\to\infty} = 0 \tag{3.8.6}$$

which follows from Gauss' law. Letting $\Phi = 0$ at large z is a natural choice because then c_{io}^* simply represents the bulk electrolyte concentration. Now we can apply the second boundary condition at the surface, where the situation is similar to the capacitor discussed in Section 3.5, and the boundary condition is in fact the same

$$\left.\frac{d\Phi}{dx}\right|_{x\to 0} = -\frac{\sigma}{\varepsilon_r \varepsilon_0} \tag{3.8.7}$$

This equation follows from Gauss' law if the field is zero for $z < 0$. No charged species are allowed for $z < 0$, and because the system is electroneutral, $\mathrm{d}\Phi/\mathrm{d}z = 0$ for $z < 0$, provided no external field is applied. A more elaborate description would be necessary if the surface is porous so that ions could penetrate into the surface material.

Solving a nonlinear second-order differential equation generally causes difficulties, but in this case, a little trick can help us. By using the identity

$$\frac{\mathrm{d}}{\mathrm{d}x}\left(\frac{\mathrm{d}f}{\mathrm{d}x}\right)^2 = 2\frac{\mathrm{d}^2f}{\mathrm{d}x^2}\frac{\mathrm{d}f}{\mathrm{d}x} \tag{3.8.8}$$

we can transform Eq. (3.8.5) into the integrable form

$$\frac{\mathrm{d}}{\mathrm{d}x}\left(\frac{\mathrm{d}\Phi}{\mathrm{d}x}\right)^2 = -\frac{2e}{\varepsilon_r\varepsilon_0}\frac{\mathrm{d}\Phi}{\mathrm{d}x}\sum_i z_i c_{i0}^* \exp\left(\frac{-z_ie\Phi}{k_B\mathbf{T}}\right) = \frac{2k_B\mathbf{T}}{\varepsilon_r\varepsilon_0}\sum_i c_{i0}^*\frac{\mathrm{d}}{\mathrm{d}x}\exp\left(\frac{-z_ie\Phi}{k_B\mathbf{T}}\right) \tag{3.8.9}$$

When integrating Eq. (3.8.9), we choose the limits of integration to be x and ∞ and

$$\left(\frac{\mathrm{d}\Phi}{\mathrm{d}x}\right)^2 = \frac{2k_B\mathbf{T}}{\varepsilon_r\varepsilon_0}\sum_i c_{i0}^*\left\{\exp\left(\frac{-z_ie\Phi}{k_B\mathbf{T}}\right) - 1\right\} \tag{3.8.10}$$

If we now specialize to a symmetric $z : z$ electrolyte such as NaCl ($z = 1$) or $MgSO_4$ ($z = 2$) at a concentration c_{i0}^*, then the right-hand side of Eq. (3.8.10) contains an even square and

$$\frac{\mathrm{d}\Phi}{\mathrm{d}x} = -\left(\frac{8k_B\mathbf{T}c_0^*}{\varepsilon_r\varepsilon_0}\right)^{1/2}\sinh\left(\frac{ze\Phi}{2k_B\mathbf{T}}\right) \tag{3.8.11}$$

Here $\sinh(x) = \{\exp(x) - \exp(-x)\}/2$ stands for the hyperbolic sine function. The minus sign emerges from the boundary condition at $x = 0$. The nonlinear first-order differential equation in Eq. (3.8.11) can be integrated to yield the Gouy–Chapman expression for the electrostatic potential. In spite of its name, the equation was first derived by the Nobel Laureate O. Stern

$$\Phi(x) = \frac{2k_B\mathbf{T}}{ze}\ln\left\{\frac{1 + \Gamma_0\exp(-\kappa x)}{1 - \Gamma_0\exp(-\kappa x)}\right\} \tag{3.8.12}$$

Here, the **Debye screening length**, $1/\kappa$, given by

$$\frac{1}{\kappa} = \left(\frac{\varepsilon_r\varepsilon_0 k_B\mathbf{T}}{\sum_i(z_ie)^2c_{i0}^*}\right)^{1/2} \tag{3.8.13a}$$

emerges naturally. It is a central concept for the description of electrostatic interactions in electrolyte solutions and we will encounter it repeatedly throughout this book. For many applications, one considers an aqueous medium at 25 °C where for future reference

$$\frac{1}{\kappa} = 3.04 \times 10^{10}/\sqrt{c_0}\ (\mathrm{m}) \tag{3.8.13b}$$

for a 1:1 electrolyte at a molar concentration c_0. For a 0.1 M 1:1 electrolyte solution, $1/\kappa$ is 0.96 nm, and for a 1 mM solution, $1/\kappa = 9.6$ nm. These lengths are comparable to colloidal sizes and they set the length scale for the range of electrostatic interactions.

The second parameter Γ_0 in Eq. (3.8.12) is related to the potential at the charged surface, that is, the surface potential, Φ_0, through

$$\Gamma_0 = \frac{\exp(ze\Phi_0/2k_B\mathbf{T}) - 1}{\exp(ze\Phi_0/2k_B\mathbf{T}) + 1} = \tanh(ze\Phi_0/4k_B\mathbf{T}) \tag{3.8.14}$$

In Chapters 5 and 8, it will turn out that this parameter is important in determining the interaction between surfaces. However, for now, we merely note that as $\Phi_0 \to 0, \Gamma_0$ approaches zero, while as $\Phi_0 \to \infty, \Gamma_0$ approaches a limiting value of unity.

We can obtain a relation between the surface charge density, σ, and the surface potential by combining Eqs. (3.8.7) and (3.8.11) to yield

$$\sigma = (8k_B\mathbf{T}c_0^*\varepsilon_r\varepsilon_0)^{1/2}\sinh\left(\frac{ze\Phi_0}{2k_B\mathbf{T}}\right) \tag{3.8.15}$$

The quantity $ze\Phi/k_B\mathbf{T}$ is often encountered in describing charged colloidal systems. At 25 °C, $e\Phi/k_B\mathbf{T} = 1$ when $\Phi = 25.7$ mV. Keeping this number in mind provides an easy way to estimate the ratio $V/k_B\mathbf{T}$ in charged systems and also to evaluate the exponentials. As $|\sigma|$ increases, $|\Phi_0|$ increases, and vice versa. Thus, we have obtained the desired relationships between σ and Φ_0 and can see how Φ varies with electrolyte concentration and distance from the charged surface.

Problem: Determine the surface charge density for a charged interface with a surface potential of −75 mV in an aqueous NaCl solution of 0.15 M at 25 °C.

Solution: According to Eq. (3.8.15)

$$\sigma = (8k_B\mathbf{T}c_0^*\varepsilon_r\varepsilon_0)^{1/2}\sinh\left(\frac{ze\Phi_0}{2k_B\mathbf{T}}\right)$$

$$c_0 = 0.15\ \mathrm{M} \to c_0^* = 0.15 \times 6.023 \times 10^{26}\ \mathrm{m}^{-3}$$

$$\varepsilon_r = 78.5,\ \varepsilon_0 = 8.85 \times 10^{-12}\ \mathrm{C/Vm}$$

$$z = 1,\ k_B\mathbf{T} = 4.1 \times 10^{-21} \to (8k_B\mathbf{T}\varepsilon_r\varepsilon_0)^{\frac{1}{2}} = 4.77 \times 10^{-15}$$

$$\to \sigma = \left(0.15 \times 6.023 \times 10^{26}\right)^{1/2} \cdot 4.77 \times 10^{-15}$$
$$\times \frac{1}{2}\left\{\exp\left(\frac{1.6 \times 10^{-19} \times -75 \times 10^{-3}}{2 \times 4.1 \times 10^{-21}}\right) - \exp\left(\frac{-1.6 \times 10^{-19} \times -75 \times 10^{-3}}{2 \times 4.1 \times 10^{-21}}\right)\right\}$$
$$= 4.5 \times 10^{-2} \times (-2.04) = -0.09\ \mathrm{C/m^2}$$

which corresponds to one negative charge per 180 Å^2.

The charged surface and the neutralizing diffuse layer of counterions are said to form an **electrical double layer**. The thickness of the diffuse layer is of the order $1/\kappa$, the Debye screening length. Viewed this way, the arrangement of charges resembles that in a parallel plate capacitor discussed in Section 3.5, with a gap size of $1/\kappa$.

3.8.2 Linearizing the Poisson–Boltzmann Equation Leads to Exponentially Decaying Potentials and the Debye–Hückel Theory

We can gain considerable insight into the relationships between σ, Φ, and κ by first examining the limiting forms of Eq. (3.8.12), for the potential profile, and Eq. (3.8.15), for the surface potential, in weakly charged or highly screened (large κ) systems. In the limit of low potential, we can linearize the Poisson–Boltzmann equation by writing the right-hand side of Eq. (3.8.4) as

$$e\sum_i z_i c_{i0}^* \exp\left(\frac{-z_i e\Phi}{k_B\mathbf{T}}\right) \approx -e\sum_i z_i c_{i0}^*\left\{1-\left(\frac{-z_i e\Phi}{k_B\mathbf{T}}\right)\dots\right\} = -\varepsilon_r\varepsilon_0\kappa^2\Phi \tag{3.8.17}$$

where the first term vanishes because of electroneutrality in the bulk case. Then Eq. (3.8.5) simplifies to

$$\frac{d^2\Phi}{dx^2} = \kappa^2\Phi \tag{3.8.18}$$

which upon integration using the boundary conditions given in Eq. (3.8.6) gives

$$\Phi(x) = \Phi_0 \exp(-\kappa x) \tag{3.8.19}$$

From the boundary condition Eq. (3.8.7), the surface charge density is proportional to the surface potential

$$\sigma = \varepsilon_0\varepsilon_r\kappa\Phi_0 \tag{3.8.20}$$

We can picture the counterions that neutralize the charge on the surface to behave on average as though they were located at a distance of $1/\kappa$ away from it. As the concentration of ions increases and consequently κ decreases, the potential decays more rapidly and the diffuse layer moves closer to the charged surface.

To obtain Eqs. (3.8.19) and (3.8.20), we assumed that the linearization in Eq. (3.8.17) holds for all x values. This condition is rather restrictive. However, the exponential decay of the potential in Eq. (3.8.19) always applies sufficiently far from the surface, where the potential approaches zero. We can see this by making a series expansion of the solution in Eq. (3.8.12). When $\Gamma_0\exp(-\kappa x) \ll 1$, which occurs for a sufficiently large x

$$\Phi(x) = \frac{4k_B\mathbf{T}}{ze}\Gamma_0 \exp(-\kappa x) \tag{3.8.21}$$

Problem: Expand Γ_0 assuming $\frac{ze\Phi_0}{k_B\mathbf{T}} < 1$ to show that Eqs. (3.8.21) and (3.8.19) are consistent.

Solution:

$$\Gamma_0 = \frac{\exp\left(\frac{ze\Phi_0}{2k_B\mathbf{T}}\right)-1}{\exp\left(\frac{ze\Phi_0}{2k_B\mathbf{T}}\right)+1} \approx \frac{1+\frac{ze\Phi_0}{2k_B\mathbf{T}}-1}{1+\frac{ze\Phi_0}{2k_B\mathbf{T}}+1} \approx \frac{ze\Phi_0}{4k_B\mathbf{T}}$$

$$\Phi(x) = \frac{4k_B\mathbf{T}}{ze}\cdot\frac{ze\Phi_0}{4k_B\mathbf{T}}\exp(-\kappa x) = \Phi_0\exp(-\kappa x)$$

It is only for a few special cases that the nonlinear Poisson–Boltzmann equation has a closed-form solution. However, when it can be linearized, the mathematical problems reduce substantially and analytical solutions are obtained for a range of conditions and geometries. Of special practical importance is the solution for a spherically symmetrical case. For example, it forms the basis of the Debye–Hückel theory of electrolyte solutions. This theory focuses on a central ion j and calculates the distribution of the other ions around it. We assume that the central ion with radius R_{ion} is of the same nature as the free ions and that we can describe the interactions by the linearized Poisson–Boltzmann equation. Because the distribution of other ions around the central ion is spherically symmetrical, we transform the Laplace operator of Eq. (3.8.4) to spherical coordinates and combine with Eq. (3.8.17) to

$$\frac{1}{r^2}\left(\frac{\mathrm{d}}{\mathrm{d}r}\left(r^2\frac{\mathrm{d}\Phi}{\mathrm{d}r}\right)\right) = \kappa^2\Phi \tag{3.8.22}$$

Here Φ is the potential in a coordinate system fixed at the central ion as it moves through the solution.

Integration of Eq. (3.8.22) yields

$$\Phi = \frac{A\exp(-\kappa r)}{r} + \frac{B\exp(\kappa r)}{r} \tag{3.8.23}$$

Combined with the boundary conditions $\Phi \to 0$ as $r \to \infty$ and $\mathrm{d}\Phi/\mathrm{d}r = -z_j e/(4\pi\varepsilon_r\varepsilon_0 r^2)$ when $r = R_{\text{ion}}$ (Eq. 3.6.1), this gives

$$\Phi = \frac{z_j e}{4\pi\varepsilon_0\varepsilon_r}\frac{\exp\left[-\kappa(r - R_{\text{ion}})\right]}{r(1 + \kappa R_{\text{ion}})} \tag{3.8.24}$$

which is the Debye–Hückel expression for the potential. Thus, in the electrolyte solution, the potential decreases exponentially with distance from the central ion rather than the slow $1/r$ decay for the bare ion. The ion cloud screens the potential and effectively makes the electrostatic interactions more short-ranged.

As a result of these calculations, a picture emerges of a central ion surrounded by its ionic atmosphere located at a distance of $1/\kappa$ from the ion's surface. We can extend this idea to the description of weakly charged particles in general and we will return to it in later chapters.

3.8.3 The Gouy–Chapman Theory Provides Insight into the Ion Distribution Near Charged Surfaces

Using the physical insight gained from the linearized equations, we now delve into a more thorough discussion of the complete Gouy–Chapman Eqs. (3.8.12) and (3.8.15). Figure 3.9 displays how the potential varies as a function of distance at high and low electrolyte concentrations for the two limiting cases of constant potential and constant charge density. Recalling Eq. (3.8.7), which established the limiting slope of Φ as proportional to the negative of the charge density, we can see from the figures that, with a constant charge density, the initial slopes are identical, but Φ_0 decreases upon the addition of more electrolyte. At constant potential, the slope becomes larger, that is, σ increases with added electrolyte.

We can use the data shown in Figure 3.9, along with the Boltzmann expression given in Eq. (3.8.2), to calculate the local concentration of counter- and coions as a function of bulk concentration and distance (see Figure 3.10). The charge density is the difference between the concentrations of these

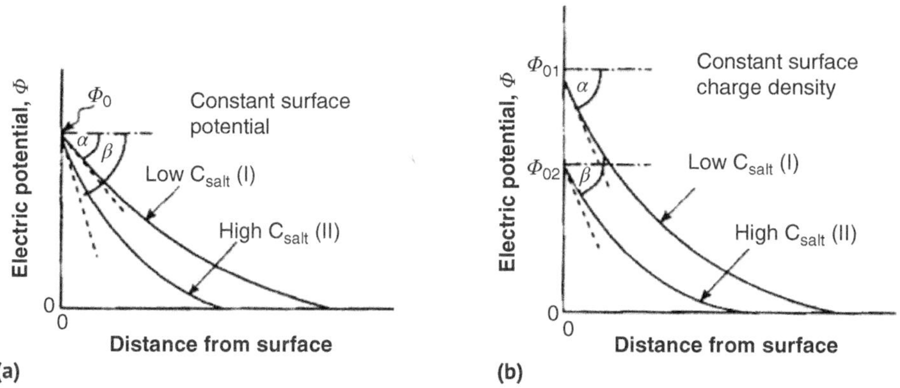

Figure 3.9 The change in potential as a function of distance and electrolyte concentration at (a) constant surface potential Φ_0 and (b) constant surface charge density σ calculated using Eqs. (3.8.12), (3.8.14), and (3.8.15). At constant σ, the initial slopes α and β of $\frac{d\Phi}{dx}\left(\frac{d\Phi}{dx}\right)_{x=0} = -\frac{\sigma}{\varepsilon_r\varepsilon_0}$ (Eq. 3.8.7) are identical, but Φ_0 decreases upon the addition of electrolyte. At constant Φ_0, adding electrolyte causes the surface charge density to increase. *Source:* van Olphen/with permission of John Wiley & Sons.

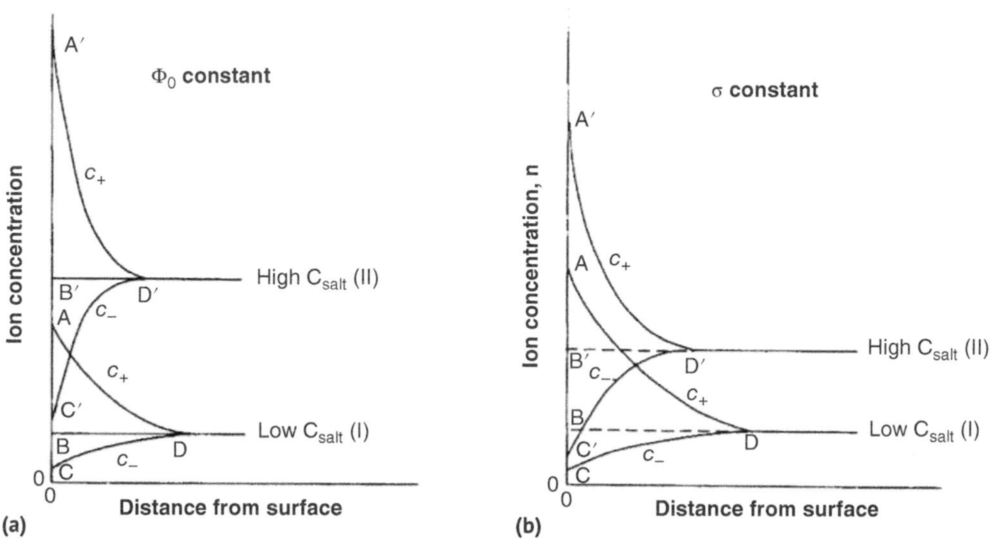

Figure 3.10 (a) Ion concentration in the Gouy–Chapman diffuse double layer at constant surface potential calculated using Eq. (3.8.12) and the Boltzmann distribution. The distance D and D′ where the local concentrations c_+ and c_- begin to depart from the bulk value c_0 decrease with increasing electrolyte concentration. The areas of A′D′C′ and ADC are proportional to the net charge in the solution and thus to the charge density on the surface so that A′D′C′ > ADC. (b) Ion distribution at constant surface charge density. The general features are similar to those in (a), except that the areas A′D′C′ and ADC are equal and the difference A′–A equals twice the difference in bulk ion concentration according to the Graham equation. *Source:* van Olphen/with permission of John Wiley & Sons.

two species, which yields σ when integrated out from the surface. This shows that the system is electroneutral on a length scale on the order of $1/\kappa$. Figure 3.10 also indicates how the charge distribution ρ concentrates more toward the surface when the electrolyte concentration (or ion valency) increases.

In practical applications, we sometimes want to know the ion concentration close to the surface. This concentration is relevant when a chemical process, such as micellar catalysis, ion adsorption, or surface group titration, occurs at the surface. We can obtain a simple expression for the total ion surface concentration by combining Eq. (3.8.10) evaluated at $z = 0$ with the boundary condition of Eq. (3.8.7) so that

$$\begin{aligned}\left(\frac{-\sigma}{\varepsilon_o\varepsilon_r}\right)^2 &= \frac{2k_B\mathbf{T}}{\varepsilon_o\varepsilon_r}\sum_i c_{io}^*\{\exp(-z_ie\Phi_o/k_B\mathbf{T})-1\}\\ &= \frac{2k_B\mathbf{T}}{\varepsilon_o\varepsilon_k}\sum_i\{c_i^*(0)-c_{io}^*\}\end{aligned} \tag{3.8.25}$$

Or, rearranged to yield the sum of all surface concentrations $c_i(0)$ including both counter- and coions

$$\sum_i c_i^*(0) = \frac{\sigma^2}{2k_B\mathbf{T}\varepsilon_r\varepsilon_0} + \sum_i c_{i0}^* = \frac{\sigma^2}{2k_B\mathbf{T}\varepsilon_r\varepsilon_0} + \frac{\Pi_{\text{osm}}}{k_B\mathbf{T}} \tag{3.8.26}$$

where the second equality comes from the expression $\sum_i c_{i0}^* = \Pi_{\text{osm}}/(k_B\mathbf{T})$, which follows from the value of the osmotic pressure in an ideal solution (cf. Eq. 1.2.13). When derived from the Gouy–Chapman equation, it is called the **Graham equation**, but it also follows from an exact statistical mechanical description and is then called the contact value theorem, which will be discussed in Section 5.1.

In the limit of high charge densities, the term $\sigma^2/(2k_B\mathbf{T}\varepsilon_r\varepsilon_0)$ dominates the right-hand side of Eq. (3.8.26). The surface potential is also high, so few coions reach the surface, and in the sum $\sum c_i^*(0)$, the counterion terms dominate. For example, with $\sigma = 0.1$ C/m^2, corresponding to one unit charge per 1.6 nm^2, $\sigma^2/(2k_B\mathbf{T}\varepsilon_r\varepsilon_0) = 2.9$ M in water at room temperature. In this regime, neither the valency of the ion nor the bulk ionic concentrations < 0.5 M appreciably affect the surface concentration for fixed σ. For aqueous systems, where $\mathbf{T}\varepsilon_r$ is approximately constant, the surface concentration is also insensitive to temperature changes. These predictions are counterintuitive but well-verified by experiments.

In fact, for highly charged surfaces, a large fraction of the neutralizing counterions reside in close vicinity of the charged surface by purely electrostatic forces. Earlier, we found that the counterion concentration was 2.9 M in the solution in direct contact with the surface. Consider a slice of the solution of thickness $\Delta \approx 0.3$ nm, i.e., roughly the size of a solvent molecule. If we neglect the decrease in counterion concentration over this slice, the total surface charge density in the layer is, for monovalent ions,

$$\begin{aligned}\sigma_{\text{counterions}}(\Delta) &= 2.9\times10^3 N_{Av}\times\Delta\times e\\ &= 2.9\times6.023\times10^{26}\times3\times10^{-10}\times1.6\times10^{-19} = 0.084\ \text{C/m}^2\end{aligned}$$

which amounts to 84% of the surface charge. Although this represents an overestimation, the calculation shows that, for highly charged surfaces, $|\sigma| > 0.05$ C/m^2, a majority of the counterions reside in close molecular proximity to the surface.

Problem: Recalculate $\sigma_{\text{counterions}}(\Delta)$ by approximately taking the decay in counterion concentration into account, using the known field at the surface.

Solution: Assume the surface to be negatively charged:

$$\begin{aligned}\sigma_{\text{counterions}}(\Delta) &= e\int_0^{\Delta} c_{\text{counterions}}(z)\mathrm{d}z \\ &= ec_o\int_0^{\Delta}\exp(-e\Phi/k\mathbf{T})\mathrm{d}z = ec_o\exp(-e\Phi_0/k\mathbf{T})\int_0^{\Delta}\exp\left[\frac{e}{k\mathbf{T}}(\Phi_0-\Phi)\right]\mathrm{d}z \\ &\simeq ec(0)\int_0^{\Delta}\exp\left[-\frac{e}{k\mathbf{T}}z\left.\frac{d\Phi}{dz}\right|_0\right]\mathrm{d}z \\ &= ec(0)\int_0^{\Delta}\exp\left(\frac{e\sigma}{k\mathbf{T}\varepsilon_o\varepsilon_r}z\right)\mathrm{d}z \simeq ec(0)\int_0^{\Delta}\left(1+\frac{e\sigma z}{k\mathbf{T}\varepsilon_o\varepsilon_r}\right)\mathrm{d}z \\ &= ec(0)\Delta\left(1+\frac{e\sigma\Delta}{2k\mathbf{T}\varepsilon_o\varepsilon_r}\right) = 0.084(1-0.76) \simeq 0.02\ \mathrm{C/m^2}\end{aligned}$$

Since we use the field at the surface, we overestimate the drop in potential and thus underestimate $\sigma_{\text{counterion}}$. An exact integration of the Gouy–Chapman solution yields

$$\sigma_{\text{counterion}} = 0.048\ \mathrm{C/m^2}$$

for the particular choice of parameters.

The high concentration of counterions close to the surface raises some questions on the validity of the approximation, leading to the Poisson–Boltzmann equation. At this close contact between counterions and surface ions and between the counterions themselves, short-range "chemical" interactions are likely to play a role. In practice, these show up as changes in the dependence of the properties of the colloidal system on the specific nature of the counterion. These ion-specific effects are larger for positively than for negatively charged surfaces; in other words, anionic counterions show more ion specificity than cationic ones do. Furthermore, it becomes difficult to determine a surface charge density experimentally since the counterions are so closely associated with the surfaces. That these surfaces in practice are also molecularly rough makes the problem even more complex.

One way of dealing with these difficulties is to say that the solution layer closest to a charged surface has properties so different from the bulk that it should be treated as a separate entity. This view was introduced around 1920 by the German physicist Otto Stern (Nobel Prize in Physics in 1943), and the surface layer is commonly referred to as the Stern layer, and its properties are specified by a number of empirical parameters. It is the opinion of the authors of this book that the Stern layer concept represents an intellectual cul-de-sac in the description of electrostatics in colloidal systems. One reason for this point of view is that experimental results show that molecular properties are not dramatically changed for a liquid close to a charged surface. As long as the surface charge is generated by molecular charges, and not through electrons of a metal, counterion concentrations predicted from the Poisson–Boltzmann equation may be high but remain physically reasonable.

An important aspect of charged surfaces is an interesting competition between counterions of different valencies due to purely electrostatic factors. By a direct application of the Boltzmann distribution, the surface concentration of an ion i with charge $z_i e$ is

$$c_i^*(0) = c_{i0}^* \exp\left(\frac{-z_i e\Phi_0}{k_B\mathbf{T}}\right) \tag{3.8.2}$$

As an application, we can see how negatively charged clays or latex resins can act as ion exchangers to soften water by removing divalent cations such as calcium. Highly charged ions in solution preferentially concentrate in the electrical double layer adjacent to the surface. If the clay or resin is preloaded with sodium ions, exchange of Ca^{2+} for Na^+ occurs. This ionic selectivity is enhanced when $e\Phi_0 > k_B\mathbf{T}$ and it becomes much less pronounced as Φ_0 becomes small.

Problem: Calculate the ratio between Ca^{2+} and Na^+ surface concentrations relative to the same quantity in the bulk given a surface potential of −100 mV at **T** = 298 K.

Solution:

$$\frac{c_{\mathrm{Ca}}(0)}{c_{\mathrm{Na}}(0)} = \frac{c_{\mathrm{Ca}}(\mathrm{bulk})}{c_{\mathrm{Na}}(\mathrm{bulk})}\exp(-e\Phi_0/k\mathbf{T}) = \frac{c_{\mathrm{Ca}}(\mathrm{bulk})}{c_{\mathrm{Na}}(\mathrm{bulk})}\exp\left(\frac{+1.6\times10^{-19}\times10^{-1}}{4.1\times10^{-21}}\right)$$

$$= \frac{c_{\mathrm{Ca}}(\mathrm{bulk})}{c_{\mathrm{Na}}(\mathrm{bulk})}\times 50$$

There is thus an enhancement of the relative Ca^{2+} concentration by a factor of 50. Such a charged surface truly has the ability to "fish" Ca^{2+} ions from, for example, tap water where the electrolyte concentration is low so that the surface potential is high.

3.9 The Electrostatic Free Energy Is Composed of One Contribution from the Direct Charge–Charge Interaction and One Due to the Entropy of the Nonuniform Distribution of Ions in Solution

3.9.1 There Are Several Equivalent Expressions for the Electrostatic Free Energy

We have, for the Gouy–Chapman and the Debye–Hückel cases, calculated the electrostatic potential and the distribution of ions outside a charged unit. With this information, we can also determine the electrostatic contributions to thermodynamic properties. The fundamental quantity of interest is the free energy, which, as we emphasized in Chapter 1, can be analyzed most effectively in terms of its energetic and entropic contributions.

$$\mathbf{A}_{\mathrm{el}} = \mathbf{U}_{\mathrm{el}} - \mathbf{T}\mathbf{S}_{\mathrm{el}} \tag{3.9.1}$$

The energy part (which, however, also includes the entropy of the solvent) derives from the direct interaction between charges, modified by the presence of the solvent. We considered this effect in Section 3.5 and derived

$$\mathbf{U}_{\mathrm{el}}^{\dagger} = \frac{1}{2}\int \rho\Phi \mathrm{d}V = \frac{1}{2}\varepsilon_r\varepsilon_o\int \left(\vec{\nabla}\Phi\right)^2 \mathrm{d}V \tag{3.5.18}$$

As discussed in connection with the Boltzmann distribution of Eq. (3.8.2), an entropy term is also associated with the inhomogeneous ion distribution in the presence of an electric field. Apart from the effect of the field, ions mix ideally with the solvent; therefore, the entropy of mixing term has the same value locally as for an ideal solution (see Section 1.1). For a small volume element $\mathrm{d}V$ containing $\mathrm{d}n_s^*$ solvent molecules and $\mathrm{d}n_i^*$ molecules of ion species i (see Eq. 1.1.8)

$$\mathrm{d}\mathbf{S} = -k_B\left(\mathrm{d}n_s^* \ln X_s + \sum_i \mathrm{d}n_i^* \ln X_i\right) \approx -k_B\left\{\sum_i(-c_i^* + c_i^* \ln c_i^* + c_i^* \ln v_s)\right\}\mathrm{d}V \tag{3.9.2}$$

In the second equality, we have assumed that the solvent is in excess so that the series expansion becomes possible

$$\ln X_s = \ln\left(1 - \sum_i X_i\right) \approx -\sum_i X_i \tag{3.9.3}$$

We also use $\mathrm{d}n_s \approx (\mathrm{d}V)/v_s$, $\mathrm{d}n_i^* \approx c_i^*\,\mathrm{d}V$, and $X_i \approx v_s c_i^*$, where v_s is the solvent molecular volume. The subscript s denotes the solvent. The last term in Eq. (3.9.2) is constant and can be included in the standard chemical potential. Integrating Eq. (3.9.2) leads to an expression for the entropy of mixing of the total system. To explicitly obtain the entropy due to the inhomogeneous ion distribution close to a charged surface, we must subtract the entropy of the corresponding homogeneous ideal system (see Eq. 1.1.8). Then the electrostatic contribution to the entropy becomes

$$\mathbf{S}_{\mathrm{el}} = \mathbf{S} - \mathbf{S}_{\mathrm{ideal}} = -k\sum_i\left\{\int c_i^*(\ln c_i^* - 1)\mathrm{d}V - n_i^*\left(\ln\left(\frac{n_i^*}{V}\right) - 1\right)\right\} \tag{3.9.4}$$

This formula simply reflects the excess entropy of mixing due to the inhomogeneous ion distribution, assuming local ideal mixing. The second term on the right-hand side represents the ideal entropy of mixing in the absence of surface charges.

One advantage of using Eq. (3.9.1) to calculate $\boldsymbol{A}_{\mathrm{el}}$ is that one needs to know the potential and ion distribution in the actual system only. In the alternative ways of determining $\boldsymbol{A}_{\mathrm{el}}$, one performs an integration through a range of states. The basis for these methods is a statistical mechanics relation

$$\mathbf{A}(\alpha) = \mathbf{A}(\mathrm{ref}) + \int_{\alpha(\mathrm{ref})}^{\alpha} \frac{\boldsymbol{U}(\alpha')}{\alpha'}\mathrm{d}\alpha' \tag{3.9.5}$$

where α is a linear scaling parameter entering the expression for the energy.

For charged systems, the natural choice for α is a scaled charge $\alpha = q' = \lambda q (0 \le \lambda \le 1)$. We say that the free energy is determined by the work of a charging process. In any colloidal system, there are several kinds of charged species, and when applying Eq. (3.9.5), we have the choice of charging them simultaneously or sequentially. In the former case,

$$\boldsymbol{U}^{\dagger}(q') = \frac{1}{2}\int \rho(q')\Phi(q')\mathrm{d}V \tag{3.9.6}$$

in analogy with Eq. (3.5.17). If, on the other hand, we consider a single charged species i

$$\boldsymbol{U}^{\dagger}(q') = q_i\Phi(\boldsymbol{r}_i, q') \tag{3.9.7}$$

Here, $\Phi(\boldsymbol{r}_i, q)$ is the potential from the other charges at the position of charge i. There is no double counting of the ion–ion interaction and hence the factor 1/2 is not present.

3.9.2 In the Debye–Hückel Theory, the Electrostatic Contribution to the Chemical Potential of an Ion Is Obtained by a Charging Process

In Section 3.8.2, we solved the linearized Poisson–Boltzmann equation for a spherically symmetrical case. We can use this result to calculate the electrostatic contribution to the chemical potentials of simple ions in solution. To accomplish this, we fix the attention on a single ion of charge $z_i e$. The potential outside the ion is

$$\Phi(r) = \frac{z_i e}{4\pi\varepsilon_0\varepsilon_r} \frac{\exp\left[-\kappa(r - R_{\text{ion}})\right]}{r(1 + \kappa R_{\text{ion}})} \tag{3.8.24}$$

which is composed of one contribution

$$\Phi_{\text{ion}}(r) = \frac{z_i e}{4\pi\varepsilon_0\varepsilon_r} \frac{1}{r} \tag{3.9.8}$$

from the ion itself plus one generated by the surrounding ion cloud

$$\Phi_{\text{ion cloud}} = \Phi - \Phi_{\text{ion}} \tag{3.9.9}$$

In an application of Eq. (3.9.5), we now consider the charging of the single central ion from zero to $z_i e$ while keeping the other charged ions intact. This process amounts to adding a single new ion to the solution. The free energy change then, by definition, corresponds to the electrostatic contribution to the chemical potential of this charged species:

$$\begin{aligned} \mu_i^{el} &= \int_0^1 z_i e \Phi_{\text{ion cloud}}(R_{\text{ion}}) \mathrm{d}\lambda \\ &= \frac{(z_i e)^2}{4\pi\varepsilon_r\varepsilon_0 R_{\text{ion}}} \left\{ \frac{1}{1 + \kappa R_{\text{ion}}} - 1 \right\} \int_0^1 \lambda \mathrm{d}\lambda \\ &= -\frac{(z_i e)^2 \kappa}{8\pi\varepsilon_r\varepsilon_0} \frac{1}{1 + \kappa R_{\text{ion}}} \end{aligned} \tag{3.9.10}$$

The second equality follows from the fact that $\Phi_{\text{ion cloud}}$ is proportional to λ. In most applications, the Debye screening length κ^{-1} is large relative to ion sizes, so we can neglect the term κR_{ion} in the denominator, leading to

$$\mu_i^{el}(\text{Debye} - \text{Hückel}) = -\frac{(z_i e)^2 \kappa}{8\pi\varepsilon_r\varepsilon_0} \tag{3.9.11}$$

which is the renowned Debye–Hückel ion activity correction term.

3.9.3 The Electrostatic Free Energy of a Planar Charged Surface Can Be Calculated in Closed Form

For a planar uniform charged surface, the Gouy–Chapman theory provides expressions for the electrostatic potential and the ion distribution. However, the expressions for these are fairly complex and it is not evident that one can analytically integrate these further by either Eq. (3.9.1) or

(3.9.5) to obtain the free energy. Let us thus first look at the simpler case when the linearization approximation can be used. Then there is a linear relation between surface charge density and surface potential:

$$\sigma = \varepsilon_0 \varepsilon_r \kappa \Phi_0 \tag{3.8.20}$$

which makes the integration in a charging process straightforward. Let the surface charge density grow from zero to σ with the surface exposed to an inert electrolyte solution. When calculating the free energy needed to charge the surface, we can use the first equality of Eq. (3.5.8) and consider the surface charge σ and the neutralizing charge distribution ρ_{bulk}. The latter will generate a surface potential Φ_0 interacting with the surface charge so that

$$\frac{\mathbf{A}_{\text{el}}}{\text{area}} = \frac{1}{2}\sigma\,\Phi_0 = \frac{\sigma^2}{2\varepsilon_r \varepsilon_0 \kappa} \tag{3.9.12}$$

The factor ½ eliminates double counting of interactions in a globally neutral system.

In the general nonlinear case, we have the more complex relation between σ and Φ_0 given in Eq. (3.8.15). With some difficulty, one can perform the integration as in Eq. (3.9.12). However, let us use another route to the final result and start from Eq. (3.9.1). For a single planar surface, we can use the Gouy–Chapman solution for the potential profile. For a 1:1 electrolyte, using Eq. (3.5.18)

$$\begin{aligned}\frac{\boldsymbol{U}^{\dagger}_{\text{el}}}{\text{area}} &= \frac{\varepsilon_0\varepsilon_r}{2}\int_0^{\infty}\left(\frac{\mathrm{d}\Phi}{\mathrm{d}z}\right)^2 \mathrm{d}z = \frac{\varepsilon_0\varepsilon_r}{2}\int_{\Phi_0}^{0}\frac{\mathrm{d}\Phi}{\mathrm{d}z}\mathrm{d}\Phi \\ &= -(2k_B\mathbf{T}c_0^*\varepsilon_0\varepsilon_r)^{1/2}\int_0^{\Phi_0}\sinh\left(\frac{e\Phi}{2k_B\mathbf{T}}\right)\mathrm{d}\Phi \\ &= \frac{k_B\mathbf{T}\sigma}{e}\frac{1}{s}\left[\left(s^2+1\right)^{1/2}-1\right]\end{aligned} \tag{3.9.13}$$

where we introduce the dimensionless parameter

$$s \equiv \sigma(8k_B\mathbf{T}c_0\varepsilon_0\varepsilon_r)^{-1/2} = \sinh\left(\frac{e\Phi_0}{2k_B\mathbf{T}}\right) \tag{3.9.14}$$

In Eq. (3.9.13), the first equality follows from Eq. (3.5.17); the second results from a change of variable, $x \to \Phi$; the third is a consequence of Eq. (3.8.11), while the fourth involves an integration of the $\sinh(x)$ function combined with the definition in Eq. (3.9.14).

For $s \gg 1$ in Eq. (3.9.13), which occurs for high charge densities, the electrostatic energy is $\mathbf{U}^{\dagger}_{\text{el}}/\text{area} = k_B\mathbf{T}\sigma/e$. If N unit charges generate the surface charge, the energy $\mathbf{U}^{\dagger}_{\text{el}}$ per charge is simply $k_B\mathbf{T}$. We might intuitively expect the energy per charge to increase with increasing charge density, but this does not happen for high charge densities.

In a similar but slightly more involved computation, we can using Eq. (3.9.4) determine the entropic contribution to be

$$-\frac{\mathbf{TS}_{\text{el}}}{\text{area}} = \frac{2k_B\mathbf{T}\sigma}{e}\left\{\ln\left[s+\left(s^2+1\right)^{1/2}\right]-\frac{3}{2s}\left[\left(s^2+1\right)^{1/2}-1\right]\right\} \tag{3.9.15}$$

Thus, the total electrostatic free energy is

$$\frac{\mathbf{A}_{\text{el}}}{\text{area}} = \frac{2k_B\mathbf{T}\sigma}{e}\left\{\ln\left[s+\left(s^2+1\right)^{1/2}\right]+\frac{1}{s}-\frac{(s^2+1)^{1/2}}{s}\right\} \tag{3.9.16}$$

Expressions for the different free energy terms are surprisingly simple to calculate algebraically, considering the involved expressions needed to obtain the potential and concentration profiles. Figure 3.11 plots these different terms as a function of s. In the linear regime of the Gouy–Chapman solution, where s is small, the energy term $\mathbf{U}_{\text{el}}^{\dagger}$ equals the entropy term $-\mathbf{T}\boldsymbol{S}_{\text{el}}$. For large s, however, $\mathbf{U}_{\text{el}}^{\dagger}$ tends toward a constant value of one $k_{\text{B}}\mathbf{T}$ per charge while $-\mathbf{T}\boldsymbol{S}_{\text{ell}}$ increases. In this nonlinear regime, a change in s through a change in salt concentration c_{o}^{*} leaves the ion–ion interaction $\mathbf{U}_{\text{el}}^{\dagger}$ unchanged, despite a substantial change in the entropy of the ion distribution. What does this mean?

According to the Graham equation, Eq. (3.8.26), the ion concentration close to a highly charged surface changes only marginally when the bulk concentration changes. The entropy, however, increases upon setting up the inhomogeneous distribution, because the ratio between the surface and the bulk concentrations changes. Usually, the decrease in the free energy of the electrical double layer when salt is added is attributed to a screening effect on $\mathbf{U}_{\text{el}}^{\dagger}$. This explanation is

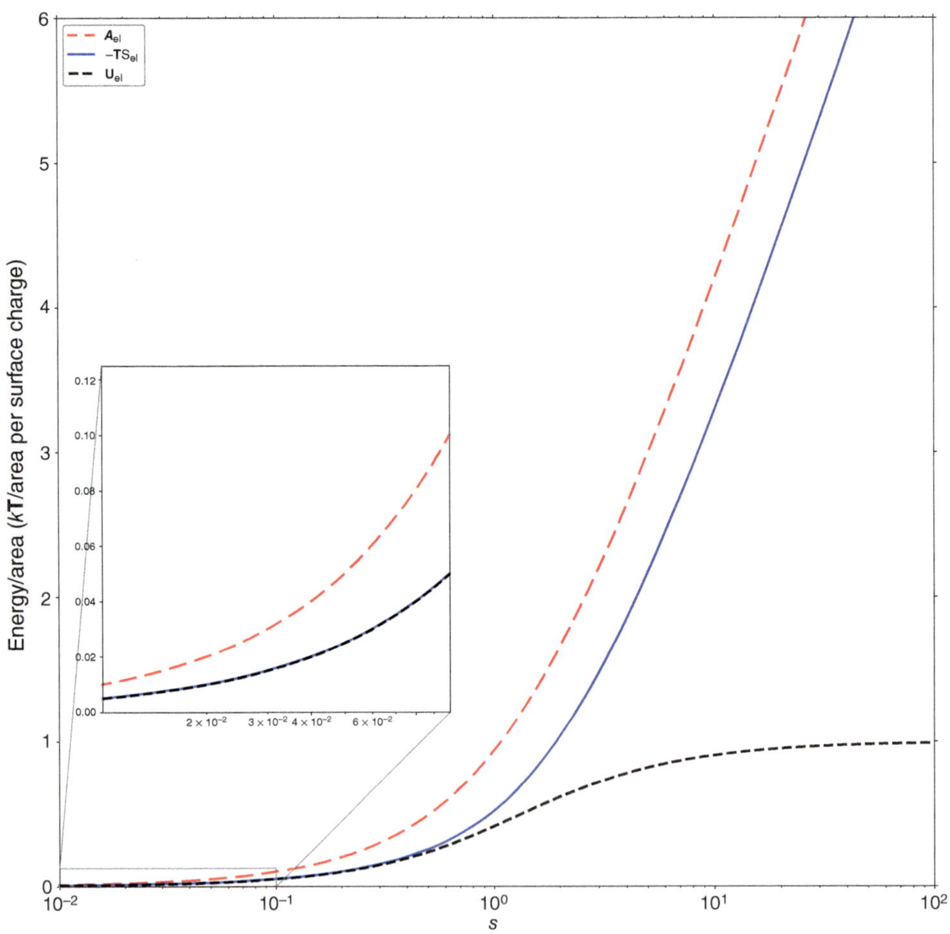

Figure 3.11 The free energy (Eq. 3.9.16, red), energy (Eq. 3.9.13, black dashed), and entropy times **−T** (Eq. 3.9.15, blue) associated with forming a charged surface as a function of the parameter s (Eq. 3.9.14). The inset shows the low s linear regime.

reasonable in the linear regime with small s; for high s values, however, the effect is an entropic one, resulting from a change in the reference concentration in the bulk. Equation (3.9.5) represents the true thermodynamic internal energy only when ε_r is independent of temperature. As discussed in Section 3.5, a factor of

$$\left\{1 + \frac{\mathbf{T}}{\varepsilon_r}\frac{d\varepsilon_r}{d\mathbf{T}}\right\}$$

is used to multiply $\mathbf{U}_{\text{el}}^{\dagger}$ of Eq. (3.9.13) to obtain the true thermodynamic internal energy.

Clearly, electrostatic forces alone would not cause ionic surfactants to self-assemble. At the aggregate surface, an attractive interaction, due mainly to hydrophobic forces, balances the repulsion from the electrostatic interactions. The free energy of the electrical double layer also contributes to the surface pressure for charged monolayers and bilayers. We can calculate this electrostatic lateral pressure $\mathbf{p}_{\text{el}}(\text{lat})$ from the derivative $\partial\mathbf{A}_{\text{el}}/\partial(\text{area})$ at constant charge. Because σ/e times area represents the number of charges, the area derivative of Eq. (3.9.16) is

$$\begin{aligned}
\mathbf{p}_{\text{el}}(\text{lat}) &= -\frac{\partial\mathbf{A}_{\text{el}}}{\partial(\text{area})} \\
&= -\frac{2k_B\mathbf{T}\sigma}{e}\cdot\text{area}\,\frac{\partial s}{\partial(\text{area})}\frac{d}{ds}\left\{\ln\left[s+\left(s^2+1\right)^{\frac{1}{2}}\right]+\frac{1-(s^2+1)^{\frac{1}{2}}}{s}\right\} \\
&= 2k_B\mathbf{T}\frac{\sigma}{es}\left[\left(s^2+1\right)^{\frac{1}{2}}-1\right] = 2\,\frac{\mathbf{U}_{\text{el}}^{\dagger}}{\text{area}}
\end{aligned} \tag{3.9.17}$$

The second-to-last equality requires some algebra, noting that $\dfrac{\partial s}{\partial(\text{area})} = -s/\text{area}$ and the last equality follows from Eq. (3.9.13).

In the nonlinear regime of the Gouy–Chapman solution, where $s \gg 1$, $\mathbf{U}_{\text{el}}^{\dagger} = k_B\mathbf{T}$ per charge so that the electrostatic component of the surface pressure is constant and equals $2k_B\mathbf{T}/\text{area}$. Equation (3.9.17) represents one in a series of simple relations involving electrostatic free energies, and its validity goes beyond the Gouy–Chapman case. By permitting us to obtain analytical expressions like these for the different physical quantities, the Gouy–Chapman theory contributes a great deal to our conceptual understanding of the electrical double layer.

See Berg (2010, Chapter 6), Cosgrove (2010, Chapter 2), Doi (2013, Chapter 10), Israelachvili (2011, Chapter 14), Lyklema (1995, Vol. 2, Chapter 3), and Ohshima (2024, Chapters 1 and 2) for additional discussion on Poisson–Boltzmann Description.

Exercises

3.1 In a cube of side a, equal charges q are located at all eight corners. Calculate the total electrostatic interaction energy. Compare this with the electrostatic energy of a uniformly charged spherical shell of charge $8q$ and a radius of $\left(\frac{\sqrt{3}}{2}\right)a$ such that it touches the corners of the cube.

3.2 A simple model to represent the charge distribution in a water molecule is to consider charges q_O, q_H at the positions of the atoms. Since the molecule is net neutral, $q_{H1} =$

$q_{H2} = -0.5q_O$. What is the value of q_O consistent with a dipole moment 1.87D? The bond length is 0.096 nm and the H—O—H bond angle is 104°.

3.3 Calculate the interaction between a water molecule and a spherical anion A^- for both the hydrogen-bonded, A^-...HO on a straight line, and the dipole-bonded, A—O along the dipole configurations. Make the calculation for the two $A^-\cdots O$ distances of 0.25 and 0.30 nm. Use the electrostatic model of Exercise 3.2 for the water molecule.

3.4 Calculate the electrostatic interaction energy between two water molecules in a hydrogen bond conformation with O—H···O on a straight line and an oxygen–oxygen distance of 0.28 nm. Assume a planar complex. Use the model of Exercise 3.2. Compare the interaction with the optimal dipole–dipole interaction of the same oxygen–oxygen separation.

3.5 The Keesom interaction energy for rotating dipoles is obtained from a series expansion, which is not useful when the dipole–dipole interaction can reach a value of $k_B\mathbf{T}$. At what magnitude of the dipole does this happen at room temperature and at a separation between the two equivalent dipoles of 0.30 nm?

3.6 Calculate the hydrogen-bond-acceptor dipole–donor-induced dipole interaction for two water molecules in the hydrogen bond configuration of Exercise 3.4. Assume that the polarizabilities and dipole moments are located on the oxygen atoms.

3.7 The polarizability of a free Cl^- ion can be estimated to 5.0×10^{-39} C · m²/V. What is the polarization interaction in an Na^+Cl^-pair and in a linear $Na^+Cl^-Na^+$complex? Use a distance of 0.28 nm and neglect the polarization of Na^+.

3.8 Show that the electrostatic potential outside an infinite uniformly charged cylinder varies logarithmically in the radial direction.

3.9 The table gives the hydration free energies and the ionic radii determined from crystal data for a number of cations. Explain the trends and comment on the quantitative accuracy of the Born model ($\mathbf{T}$ = 298 K).

Ion	R_{ion} (nm)	ΔG(hydr) (kJ/mol)
Li^+	0.068	-505
Na^+	0.098	-406
K^+	0.133	-332
Mg^{2+}	0.065	-1895
Cr^{3+}	0.069	-4209

3.10 Calculate the Debye screening length κ^{-1} at 298 K for the aqueous solutions: (a) 0.1 M NaCl, (b) 10 mM NaCl, (c) 1 mM NaCl, (d) 0.1 M $MgSO_4$, and (e) pure water, assuming $K_w = 10^{-14}$. Discuss what the relevant κ value is in a 28 mM solution of sodium dodecyl sulfate.

3.11 A bilayer of amphiphilic molecules separates two aqueous electrolyte solutions. When a charge is brought from one of the solutions to the other, the bilayer will act as a capacitor separating two conducting media. What is the capacitance per unit area if the bilayer is 4 nm thick and has a relative permittivity of $\varepsilon_r = 3$.

3.12 Consider an ion of charge $-e$ located at the center of a spherical particle of radius 4 nm and relative permittivity of 3 and surrounded by water. What is the contribution of the water medium to the solvation free energy?

3.13 A spherical ion of charge $-e$ is located with its center on an interface between two media, 1 and 2, of relative permittivities $\varepsilon_r(1)$ and $\varepsilon_r(2)$. (a) Show that the electric field from the ion is in each half-space simply the corresponding field found for the respective homogeneous media. (b) Calculate the solvation free energy for the ion.

3.14 An amphiphilic bilayer in an aqueous solution can close upon itself to form a spherical shell, called a vesicle. Calculate the resulting potential difference between inside and outside when 10 potassium ions have been transported across the bilayer. Assume excess electrolyte in the water and a vesicle radius of 20 nm. Use the bilayer characteristics of Exercise 3.11.

3.15 In Exercises 2.7–2.9 of Chapter 2, we found that the area per molecule at the air–water interface of an aqueous solution of dodecyldimethylammonium chloride was 0.43 nm^3 for the pure surfactant and 0.415 nm^2 when 0.20 M NaCl had been added. These values refer to conditions at the CMC. What is the concentration of Cl^- at the charged surface in the two cases, assuming that the amphiphile is fully protonated?

3.16 Dodecyldimethylammonium chloride has a proton that can titrate to form the uncharged primary amine, and the pK_a is around 10.6. At a bulk concentration of 15 mM, the pH is approximately 6.2. Estimate the surface charge density at the air–water interface. The area per molecule is 0.43 nm^2.

3.17 Calculate the electrostatic free energy involved in creating the charged interfaces of Exercise 3.15. Ignore possible deprotonation of the surfactant ($\mathbf{T}$ = 298 K).

References

Atkins, P., de Paula, J., and Keeler, J. (2022). *Atkins Physical Chemistry*, 12th ed. Oxford, UK: Oxford University Press.

Berg, J.C. (2010). *An Introduction to Interfaces & Colloids. The Bridge to Nanoscience*. World Scientific.

Cosgrove, T. (2010). *Colloid Science. Principles, Methods and Applications*. Chichester, UK: Wiley.

Doi, M. (2013). *Soft Matter Physics*. Oxford, UK: Oxford University Press.

Israelachvili, J.N. (2011). *Intermolecular and Surface Forces*, 3rd ed. London, UK: Academic Press.

Kronberg, B., Holmberg, K., and Lindman, B. (2014). *Surface Chemistry of Surfactants and Polymers*, 3rd ed. Chichester, UK: Wiley.

Lyklema, H. ed. (1995). *Fundamentals of Interface and Colloid Science. Solid-Liquid Interfaces*, Vol. 2. Amsterdam, The Netherlands: Elsevier.

Norde, W. (1976). *Proteins at Interfaces Wageningen*. The Netherlands: The Agricultural University.

Ohshima, H. (2024). *Fundamentals of Soft Interfaces in Colloid and Surface Chemistry*. London, UK: Academic Press.

van Olphen, H. (1963). *Introduction to Clay Colloid Chemistry*. New York: Wiley Interscience, p. 34.

4

Structure and Properties of Micelles

Concept Map

Micellar Cooperative Association Models
• Micelles are the simplest and most thoroughly characterized self-organizing structures. A stepwise association process, $S + S_{n-1} \rightarrow S_n$ (Eq. 4.1.3), describes their aggregation in general. However, because it is almost impossible to specify all the K_n equilibrium steps, approximate models are used. The **isodesmic model** assumes that K_n is independent of n and describes the association of some dyes in water. However, it does not capture the cooperativity associated with amphiphilic aggregation. The **phase separation model** approximates aggregation as a phase separation process in which the activity of the monomer remains constant above the CMC. It captures the start mechanism of aggregation but not the stop mechanism. The **closed-association model** assumes that one aggregation number N dominates and yields $\Delta \mathbf{G} = \boldsymbol{RT}$ ln CMC (Eq. 4.1.13), thus relating the free energy of micellization to the measured CMC. This model captures both the stop and start cooperative features associated with aggregation processes. • Examination of Eq. (4.1.13) shows that as N becomes larger, the CMC becomes more sharply defined. As $N \rightarrow \infty$ the phase separation model becomes a true representation of the system. • Examination of the thermodynamics of micellization shows that above the CMC, the monomer's concentration increases very slowly (Eq. 4.1.21) and it also provides a way to evaluate, $\Delta \mathbf{H}_{\text{mic}}$, $\Delta \mathbf{S}_{\text{mic}}$ and the change in CMC with temperature (Eq. 4.1.24). • The closed association model for ionic micelles (Eq. 4.1.16) involves an empirical counterion binding parameter, and expressions involving the PB equation are more useful.

The Colloidal Domain: Where Physics, Chemistry, Biology, and Technology Meet, Third Edition. Håkan Wennerström.

Characterization of Micelles

Micelles are characterized by:

- **CMC**—many physical properties undergo abrupt change at the CMC, including (1) surface tension, which also provides a sensitive measure of surfactant purity, (2) conductance, and (3) surfactant ion activity as measured by specific ion electrodes.
- Aggregation number, *N*, which can be measured by either light-scattering or fluorescence quenching techniques.
- Small-angle X-ray and neutron scattering, SAXS or SANS, are useful to characterize micelle size and internal structure as well as micelle–micelle interactions.
- Time scales for dynamic processes, obtained by relaxation measurements, which yield two characteristic relaxation times: $\tau_1\left(10^{-6} \text{ to } 10^{-3}\text{ s}\right)$ measures the rate of exchange of monomer between micelles, while $\tau_2\left(10^{-3} \text{ to } 1\text{ s}\right)$ measures the rate of micelle formation and disintegration. Because micelles must pass through a kinetic bottleneck as they disintegrate, τ_2 is much longer than τ_1.
- Diffusion of micellar components, which involves an interplay between monomers, counterions, and micelles.

Quantitative Analysis of Micellar Solutions

We can interpret variations in CMCs by writing $\Delta\mathbf{G}_{\text{mic}} = \Delta\mathbf{G}(HP) + \Delta\mathbf{G}(\text{contact}) + \Delta\mathbf{G}(HG)$ where

- $\Delta\mathbf{G}(HP)$ represents the free energy associated with transferring the hydrocarbon chain out of water and into the oil-like interior of the micelle.
- $\Delta\mathbf{G}(\text{contact})$ represents the surface free energy attributed to solvent–hydrocarbon contact in the intact micelle. It is proportional to area, A_{mic} and takes the form $\Delta G(\text{contact}) = \gamma' A_{\text{mic}}/N$ (Eq. 4.1.2).
- $\Delta\mathbf{G}(HG)$ represents the positive contribution associated with head group interactions, including electrostatic as well as head group conformation effects.

Ionic Micelles

Increasing the hydrocarbon chain length lowers the CMC and increases *N*.

For ionic surfactants, the solution of the PB equation provides quantitative expressions for $\Delta\mathbf{G}(HG) = \Delta\mathbf{G}_{\text{el}}$ that account for these experimental observations.

Figure 4.4 compares the calculated and observed changes in CMC with chain length and added salt.

Table 4.2 compares the calculated and observed changes in CMC with chain length, salt, and temperature.

Nonionic Micelles

Nonionic micelles with zwitterionic or nonionic head groups have substantially lower CMCs than corresponding ionic surfactants (Table 1.1).

Micellar solutions of surfactants with oligoethylene oxide head groups often phase separate when the temperature is increased. The temperature where this happens is called the cloud point.

Added salt has relatively little effect on either the CMC or the aggregation number N. Nonionic micelles are more difficult to model than ionic micelles because head group interactions are dominated by more local interactions.

Nonspherical Micelles

Surfactants with surfactant number $N_S \approx 0.3$ form spherical micelles and show the start (well-defined CMCs), stop (adding more surfactant leads to more micelles not larger ones), and well-defined properties ($R_{\text{mic}} \propto l_c$) as described in Section 1.5.

Many surfactants with $N_S > 0.3$ form rods, polymer-like threads, or branched infinite aggregates near the CMC or at high concentrations. Aggregate growth can be promoted:

- In nonionic systems, by increasing the temperature, which diminishes favorable head group–solvent interactions.
- In ionic systems by adding a cosurfactant with a small head group, adding an electrolyte that reduces head group repulsion, or adding an amphiphilic counterion.

Applications

Many biological and industrial applications involve mixed micelles containing:

- Similar surfactants with different chain length, where the relative concentration of surfactants in the micelles changes with changes in the concentration and can be understood in terms of a mixing model (Figure 4.16).
- Adding an ionic surfactant to a nonionic micellar system raises the cloud point, while adding a nonionic surfactant to an ionic system decreases the CMC.
- Hydrocarbons and fluorocarbons that form two distinct micelles because of the mutual insolubility of these two chains.

Examples involving surfactants include digestion of fats by bile salt micellar solubilization, dissolution mechanisms of fats and oils by aqueous surfactants, and micellar catalysis leading to increased reaction rates.

Micelles are the simplest of all amphiphilic self-organizing structures. They are also the best understood, because many experimental techniques have been used to characterize them. Critical micelle concentrations (CMC) occur at experimentally accessible concentrations, and we can study micelles in dilute solution, in which interaggregate interactions can be ignored. Because they display rather narrow size distributions, we can often characterize micelles by a single aggregation number. All these factors make them amenable to theoretical analysis, and quantitative relationships are available. The insights gained from studying micellar systems provide the basis for understanding the more complex bilayers and emulsions, as well as self-organization in general.

4.1 Micelle Formation Is a Cooperative Association Process

4.1.1 The Attraction Between Apolar Chains Promotes a Cooperative Association While the Head Group Repulsion Limits Aggregate Growth

A typical micelle-forming amphiphile has an apolar hydrocarbon chain with 10–18 CH_2 units. The head group can be either charged or strongly polar. We use the ionic SDS molecule with 12 carbons in the chain to illustrate micellar behavior. The dodecyl chains have a low solubility in the aqueous environment and tend to aggregate. A stepwise association leads to larger and larger aggregates. This hydrophobic, or more generally, solvophobic association would continue indefinitely in the absence of the polar groups. They repel each other, however, and tend to limit the aggregate growth. For SDS, the balance between the chain attraction and the head group repulsion effectively stops the growth at an aggregation number, N, of around 60. This value of N corresponds to the situation where the aggregate has to deform from a spherical shape when a monomer is added in order to keep the charged head group at the aggregate surface, as illustrated in Figure 4.1.

We can approximately describe the energetics of aggregate formation in terms of three contributions.

$$\Delta\mathbf{G}^{\theta}_{\mathrm{mic}} = \Delta\mathbf{G}(HP) + \Delta\mathbf{G}(\mathrm{contact}) + \Delta\mathbf{G}(HG) \tag{4.1.1}$$

The hydrophobic free energy change, $\Delta\mathbf{G}(HP)$, of bringing a hydrocarbon chain from water to a completely apolar surrounding was discussed in Chapters 1 and 2, and it depends linearly on the chain length. In a micelle, there remains a contact between the aqueous medium and the apolar

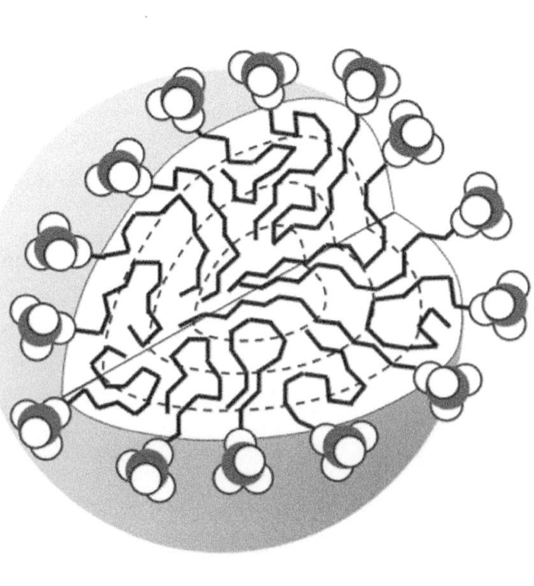

Figure 4.1 Illustration of the alkyl chain packing in a spherical SDS micelle. *Source:* Kronberg et al., 2014/with permission of John Wiley & Sons

part of the aggregate and the associated free energy is, to first order, proportional, with a proportionality constant γ', to the exposed area

$$\Delta\mathbf{G}(\text{contact}) = \gamma'\{A_{\text{mic}} - na_0\} \tag{4.1.2}$$

Here a_0 represents the sterically determined head group area. $\Delta\mathbf{G}(\text{contact})$ is a positive contribution to the free energy and counted per molecule added; it decreases as the aggregate grows since the area-to-volume ratio decreases. The positive contribution $\Delta\mathbf{G}(HG)$ from the head group repulsion can for ionic amphiphiles be calculated from the Poisson–Boltzmann equation as described in Section 3.9, and the issue is further elaborated in Section 4.1.4.

4.1.2 Several Models Usefully Describe Micellar Aggregation

When surfactants associate into micelles, they form a liquid-like aggregate. Because no obvious mechanism leads to a specific aggregation number, it is natural to describe the association in terms of a stepwise addition of a monomer, S, to the aggregate, S_{n-1} as in

$$S + S_{n-1} \rightleftarrows S_n \tag{4.1.3}$$

If we neglect additional interactions between aggregates and between monomers, we can write the equilibrium given in Eq. (4.1.3) as

$$K_n = \frac{[S_n]}{[S][S_{n-1}]} \tag{4.1.4}$$

Equation (4.1.4) provides a general description of any stepwise association process in dilute solution, and it can be applied to many other colloidal systems. In situations that involve aggregation numbers n of order 100, we must specify a rather intractable number of equilibrium constants K_n. For this reason, we can obtain a more useful qualitative sense of the characteristic features of different association patterns by considering some simplified models.

We will first consider the isodesmic model, where the assumption is that K_n is independent of n. This occurs when the free energy change of adding a monomer is independent of aggregate size. This property is typical for systems that grow in one dimension only. If $[S]K$ is larger than one, Eq. (4.1.4) predicts that the larger the aggregate size, the higher its concentration. Since the total amount of S is finite, this would be inconsistent and one necessarily has $[S]K < 1$. From Eq. (4.1.4), it follows that the aggregate distribution function

$$f(n) = \frac{[S_n]}{\sum_{n=1}^{\infty}[S_n]} \tag{4.1.5}$$

decays monotonically with $[S_1] > [S_2] > [S_n]$.

Problem: Show that for $f(n) = C\exp(-\alpha n)$ the equilibrium in Eq. (4.1.4) is satisfied for all n.

Solution: Inserted $f(n)$ into Eq. (4.1.4) yields $K[S_1] = \exp(-\alpha n)/\exp\{-\alpha(n-1)\} = \exp(-\alpha)$ and the equality is satisfied for all n when $\alpha = -\ln\,(K[S_1])$. This shows that $f(n)$ decays exponentially.

In the isodesmic model, aggregation is a continuous process that does not show the abrupt onset in a narrow concentration range that typifies micelle formation.

If we instead analyze what happens when an apolar liquid reaches its solubility limit in water, we can appreciate the particular features of micelle aggregation. The free energy of a small droplet can be written as the sum of a bulk term $n\mu^{\theta}(\infty)$ plus a surface term γA_n. When a monomer in solution, with volume v_s and chemical potential μ (solution), is added to a droplet of n monomers of radius R_n to form one with $n + 1$ monomers, the change in free energy is

$$\mathrm{G}(n+1) - \mathrm{G}(n) = (n+1)\mu^{\theta}(\infty) + \gamma A_{n+1} - n\mu^{\theta}(\infty) - \mu(\text{solution}) - \gamma A_n$$
$$= \mu^{\theta}(\infty) - \mu(\text{solution}) + \gamma\frac{\partial A_n}{\partial n} = \mu^{\theta}(\infty) - \mu(\text{solution}) + \frac{2\gamma v_s}{R_n} \rightarrow \tag{4.1.6a}$$

$$K_n = K_{\infty}\exp\left\{-2v_s^{2/3}\gamma(4\pi/3)^{1/3}/(k_B\mathrm{T}n^{1/3})\right\} \tag{4.1.6b}$$

Problem: Show that $\partial A/\partial n = 2v_s/R_n$ for a sphere!

Solution: Adding dn molecules of volume v_s increases the sphere volume with $dV = v_s\,dn$. For a sphere the area $A = 4\pi R^2$ and the volume $V = (4\pi/3)R^3$ so that

$$\partial A/\partial n = v_s \partial A/\partial V = v_s (dA/dR)(dV/dR)^{-1} = v_s\, 8\pi R\left(12\pi R^2/3\right) = 2v_s/R.$$

The important feature of Eq. (4.1.6a) is that the surface energy term makes the free energy change in adding a monomer smaller, as the droplet becomes larger. Compared to the isodesmic model, the equilibrium constant in Eq. (4.1.6b) has a n-dependent term, making the equilibrium constant larger the larger the n-value. This implies that adding a monomer becomes more and more favorable as the aggregate enlarges. Cooperativity of this kind is precisely what drives a drop to grow indefinitely and form a separate phase.

In micelle formation, the aggregation pattern is qualitatively the same for low values of n as in the example with proper phase separation. When the micelle has reached a spherical size with a radius corresponding to the length of a monomer, the cooperativity ceases, and the aggregation pattern is more similar to that of the isodesmic model. The net result is that there is an optimal aggregate size with a rather narrow size distribution. It is demanding to construct an accurate model for how the equilibrium constant K_n varies with n for a substance that forms micelles. Figure 4.2 shows a calculated aggregate distribution function for the surfactant SDS for a number of concentrations in the range close to the CMC.

Although it is feasible, albeit demanding, to model micelle formation in detail, it is conceptually very useful to have simplified descriptions that catch the main features of the association process. In the closed association model, one makes the simplifying assumption that the micellar aggregates can be represented by a mean aggregation number and that one has only two species, monomers and micelles, with aggregation number N. Then

$$NS \leftrightarrows S_N;\; K_N = \frac{[S_N]}{[S]^N} \tag{4.1.7}$$

The total surfactant concentration is

$$[S]_T = N[S_N] + [S] = NK_N[S]^N + [S] \tag{4.1.8}$$

Experiments typically identify the CMC as the concentration at which added surfactant preferentially starts to enter into aggregates. A good measure of this concentration point is where an

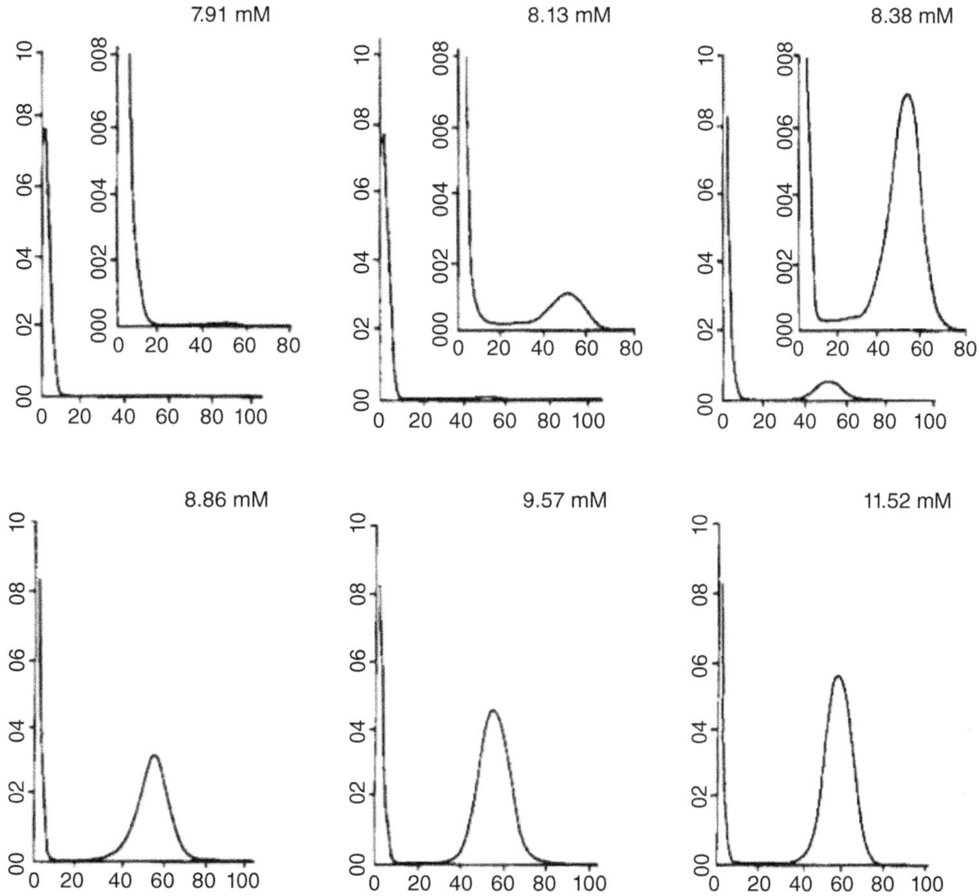

Figure 4.2 The size distribution function f(*n*) for SDS obtained in a model calculation. At a concentration of 7.91 mM, there is a small fraction of micelles appearing, which grows when the concentration increases. The calculated aggregation number has a distribution in the micellar region with a standard deviation of approximately 7. *Source*: Gunnarsson, 1979/with permission of Lund University.

added monomer is as likely to enter a micelle as to remain in solution so that it falls within the closed association model

$$\left.\frac{\partial\{N[S_N]\}}{\partial[S]_T}\right|_{\mathrm{CMC}} = \left.\frac{\partial[S]}{\partial[S]_T}\right|_{\mathrm{CMC}} = \frac{1}{2} \tag{4.1.9}$$

Solving for the inverse of Eq. (4.1.8), $\partial[S]_T/\partial[S] = 2$ in Eq. (4.1.9), we find

$$[S]_{\mathrm{CMC}}^{N-1} = \left(N^2K_N\right)^{-1} \tag{4.1.10}$$

where $[S]_{\mathrm{CMC}}$ is the monomer concentration at the CMC. The critical micelle concentration refers to the total surfactant concentration, and if we combine Eqs. (4.1.8) and (4.1.10)

$$\mathrm{CMC} = [S]_{\mathrm{CMC}}\left(1 + N^{-1}\right) = \left(N^2K_N\right)^{-1/(N-1)}\left(1 + N^{-1}\right) \tag{4.1.11}$$

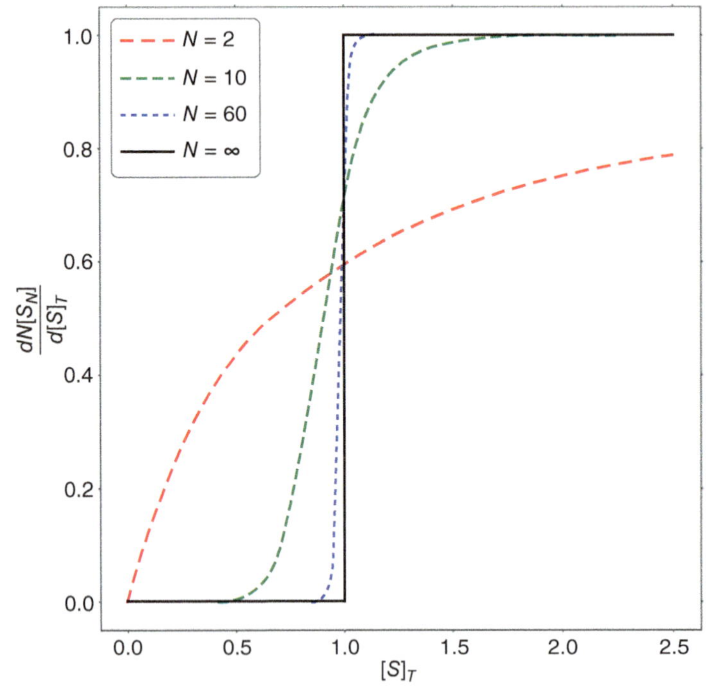

Figure 4.3 Critical micelle concentration. As the aggregation number N increases, the fraction of added surfactant that goes to the micelle, $dN[S_N]/d[S]_T$, varies more and more steeply with total concentration $[S]_T$. In the limiting case in which the aggregation number becomes infinite, the transition becomes a step function that unambiguously defines the critical micelle concentration CMC, while small micellar aggregation numbers lead to less well-defined values of CMC.

From the first equality, we see that the amount of micellized surfactant at the CMC is $[S]_{\mathrm{CMC}}/N$ which becomes more negligible as the value of N increases. Figure 4.3 illustrates how the fraction of added surfactant approaches unity more steeply as the aggregation number N increases. The limit $N \to \infty$ represents the behavior at a phase transition. For $N = 60$, as for an SDS micelle, the deviation from the phase transition limit is clearly visible but not substantial. For $N = 10$, on the other hand, the transition is more gradual and does not represent a typical micellization behavior.

We can use the Eq. (4.1.11) to obtain an approximate relation between the CMC and the standard free energy of micelle formation $\Delta\mathbf{G}^{\theta}_{\mathrm{mic}}$. Noting that $\Delta\mathbf{G}^{\theta} = -\boldsymbol{RT}\ln K_N$ and K_N refers to the association of N monomers to a micelle,

$$-\boldsymbol{RT}\ln K_N = \Delta\mathbf{G}^{\theta} = N(\mu^{\theta}(\text{micelle}) - \mu^{\theta}(\text{solvent})) = N\Delta\mathbf{G}^{\theta}_{\mathrm{mic}} \tag{4.1.12}$$

Combining Eq. (4.1.8) with Eq. (4.1.11) and neglecting the term N^{-1} in the second equation gives

$$\begin{aligned}\boldsymbol{RT}\ln \mathrm{CMC} &\approx \boldsymbol{RT}\ln [S]_{\mathrm{CMC}}\\ &= -\frac{\boldsymbol{RT}}{(N-1)}\ln\left(K_N N^2\right)\\ &= \frac{1}{N-1}\left(N\,\Delta\mathbf{G}^{\theta}_{\mathrm{mic}} - 2\boldsymbol{RT}\ln N\right) \approx \Delta\mathbf{G}^{\theta}_{\mathrm{mic}}\end{aligned} \tag{4.1.13}$$

In the last equality, we have introduced two approximations whose validity depends on a large value of N. First, we replaced $N - 1$ by N in the denominator, and second, we neglected the logarithmic term relative to the one linear in N.

Micelle formation has several features in common with the formation of a separate liquid phase. This fact provides a basis for a model in which micelles formally constitute a separate phase. Figure 4.3 shows that as the aggregation number N grows the association becomes more and more like a phase transition. Seeing the micelle as a pseudo-phase, the surfactant possesses a certain chemical potential μ^θ(micelle) in the aggregate. When

$$\mu^\theta(\text{micelle}) = \mu^\theta(\text{solvent}) + \mathbf{RT}\ \ln\ [S] \tag{4.1.14}$$

monomers and aggregates coexist in equilibrium and $[S]$ is the CMC, neglecting the existence of dimers and oligomers. The standard free energy of micelle formation $\Delta\mathbf{G}^\theta_{\text{mic}}$ represents the free energy difference between a monomer in the micelle and the standard chemical potential in dilute solution and

$$\Delta\mathbf{G}^\theta_{\text{mic}} = \mu^\theta(\text{micelle}) - \mu^\theta(\text{solvent}) = \mathbf{RT}\ \ln\ \text{CMC} \tag{4.1.15}$$

as in Eq. (4.1.15). This equation provides a useful approximation for obtaining $\Delta\mathbf{G}^\theta_{\text{mic}}$ and the phase separation model captures several, but not all, essential features of micelle formation. Although it describes the start mechanism of the self-assembly process as discussed in Chapter 1, it does not describe the stop mechanism.

So far, we have limited our discussion to uncharged surfactants, and for charged surfactants, one has the complication that counterions are attracted to the micelle. One can take this into account by stating that a fraction of the counterions are so closely associated that they can be seen as bound to the micelle. We can then generalize Eq. (4.1.7) to include ionic surfactants by writing an equilibrium between surfactant monomers, S^- counterions, C^+ and micelles, S_N with P bound counterions

$$(N-P)C^+ + NS^- \rightleftarrows S_N^{-P} \tag{4.1.16}$$

for which

$$K_N = \frac{\left[S_N^{-P}\right]}{\left[S^-\right]^N \left[C^+\right]^{N-P}} \tag{4.1.17}$$

or

$$\Delta\mathbf{G}^\theta_{\text{mic}} = \frac{\Delta\mathbf{G}^\theta}{N} = -\frac{\mathbf{RT}}{N} \ln\left[S_N^{-P}\right] + \mathbf{RT}\ \ln\ \left[S^-\right] + \mathbf{RT}\left(1 - \frac{P}{N}\right)\ \ln\ \left[C^+\right] \tag{4.1.18}$$

When N lies in the range 50–100, the $\ln\left[S_N^{-P}\right]/N$ term becomes small and consequently negligible by an argument analogous to the one made in Eq. (4.1.13). If no salt is added, we can replace both $\left[S^-\right]$ and $\left[C^+\right]$ with C_{CMC} in the second and third terms to give

$$\Delta\mathbf{G}^\theta_{\text{mic}} = (2-\beta)\mathbf{RT}\ \ln\ C_{\text{CMC}} \tag{4.1.19}$$

Here $\beta = P/N$ is the degree of dissociation of the micelle or $1-\beta$ is the degree of counterion binding. For a completely ionized micelle, $\beta = 1$ and for a neutral micelle, $\beta = 0$. Thus, this equilibrium model pictures the micelle as a charged entity consisting of N surfactant molecules and $(N-P)$ counterions with a net charge of $\pm Pe$.

The notion that counterions bind to a charged surfactant aggregate has played a major historical role in the description of micellar systems and other charged colloids. Counterions drawn into the regions of the charged head groups reduce the repulsive electrostatic interactions between them, and this is the heuristic physical basis for the model of counterion binding. The idea provides a good qualitative picture of the phenomenon, but it is not amenable to quantitative analysis. It is important to realize that counterions are not "bound" in the conventional chemical sense. They reside outside the micelle surface attached by electrostatic forces. A quantitative alternative to Eq. (4.1.19) emerges naturally from the Poisson–Boltzmann equation as described in Section 4.1.4.

Earlier, we analyzed surfactant association into micelles in some detail. Apart from being an important phenomenon in itself, we will see in later chapters that model descriptions can be applied to many other self-association processes occurring in colloidal systems. A key feature of the micellization is the cooperative association, similar to the separation at a phase boundary, occurring for aggregate sizes up to an aggregate with a radius of approximately the length of the surfactant molecule. Once this size has been reached, the association is no longer cooperative and further growth in size can be approximately described by the isodesmic model. For typical surfactants, this results in a steep decay of the size distribution function, but under some circumstances the decay in size is not pronounced and micelles grow to elongated cylinders as will be discussed in Section 4.4. The longer the apolar part of a surfactant, the larger the size of the spherical micelle and the larger the aggregation number, which is approximately proportional to the surfactant length l squared.

Problem: Calculate how the aggregation number N varies with the surfactant of length l assuming a constant area per polar group a_0.

Solution: Micelle area $A_{\text{mic}} = 4\pi l^2 = Na_0$ so that $N = 4\pi l^2/a_0$.

From Figure 4.3, it is apparent that the larger the aggregation number, the more well-defined the CMC is, and one expects that the size distribution function becomes more narrow, on a relative scale. Table 4.1 shows measured aggregation numbers together with the standard deviation of the size distribution function for a series of sodium alkyl sulfates. The mean aggregation number is approximately proportional to l^2 and the ratio between the standard deviation σ and the mean aggregation number N decreases with increasing alkyl chain length.

4.1.3 Thermodynamics Provide Useful Relationships Between Free Energies, Chemical Potentials, Enthalpies, and Entropies for Micellar Solutions

4.1.3.1 Chemical Potential

In the preceding section, we analyzed the relationship between stepwise aggregation of amphiphiles and related the corresponding equilibrium constants to standard free energies. We saw

Table 4.1 The CMC, the mean aggregation number N, and the standard deviation σ of the micelle size distribution for a series of sodium alkyl sulfates, determined from kinetic measurements (see Section 4.2.3).

Surfactant	Temperature (°C)	CMC (M)	N	σ
NaC_6SO_4	25	0.42	17	6
NaC_7SO_4	25	0.22	22	10
$NaC_{12}SO_4$	25	8.2×10^{-3}	64	13
$NaC_{14}SO_4$	40	2.05×10^{-3}	80	16.5

Source: Aniansson et al., 1976/with permission of American Chemical Society.

that as a consequence of the cooperativity of the micelle formation process, aggregates start to form at a relatively well-defined concentration, the CMC. Cooperative association also affects the concentration dependence of the surfactant chemical potential μ_S.

At equilibrium, μ_s is uniform throughout the system, so we can obtain a value for it by focusing on the free monomer in solution. Neglecting activity coefficient corrections, which are too small to be of consequence for the basic conclusions, we write

$$\mu_s = \mu_s^\theta(\text{solvent}) + \boldsymbol{RT} \ln [S] \tag{4.1.20}$$

Below the CMC, $[S] = [S]_\text{T}$ and the surfactant chemical potential increases logarithmically with concentration as it normally does for any dilute solution. Far above the CMC, where $[S] \ll [S]_\text{T}$ we find from Eq. (4.1.8) that

$$[S] = [S]_\text{T}{}^{1/N}(NK_N)^{-1/N} \tag{4.1.21}$$

Substituted into Eq. (4.1.20), this gives

$$\mu_s = \mu_s^\theta(\text{solvent}) + \frac{\boldsymbol{RT}}{N} \ln [S]_\text{T} - \frac{\boldsymbol{RT}}{N} \ln (NK_N) \tag{4.1.22}$$

This equation shows that the surfactant chemical potential increases only marginally with increasing concentration above the CMC. To obtain the same change in μ_S above the CMC that doubling the concentration produces below the CMC, we need to increase $[S]_\text{T}$ by 2^N. For typical aggregation numbers, this is simply impossible. Thus, the surfactant chemical potential remains nearly constant above the CMC up to very high values of the total concentration.

The weak concentration dependence of μ_S above the CMC has important consequences in applications of surfactants. A surface or a bulk phase in equilibrium with a surfactant solution registers the chemical potential, and at the CMC, the surfactant already reaches its maximum influence. For example, in cleaning processes, a detergent realizes its maximum effect at the CMC. Adding more detergent only creates a reservoir that can compensate for adsorption losses.

An even more subtle mechanism operates in ionic surfactants, and we can illustrate this mechanism qualitatively by using the association model of Eq. (4.1.16). When N-charged monomers, S^- and $N - P$ counterions, C^+ associate with the micelles, the surfactant chemical potential carries contributions from both ions and

$$\begin{aligned} \mu_{SC} &= \mu_S^- + \mu_C^+ \\ &= \mu_S^\theta(\text{solvent}) + \boldsymbol{RT}\ln[S^-] + \mu_{C^+}^\theta(\text{solvent}) + \boldsymbol{RT}\ln\left[C^+\right] \\ &= \mu_{SC}^\theta(\text{solvent}) + \boldsymbol{RT}\ln[S^-]\left[C^+\right] \end{aligned} \tag{4.1.23}$$

As with nonionic surfactants, μ_{SC} is virtually independent of the total concentration above the CMC, which means that the product $[S^-]\left[C^+\right]$ remains constant. Because a fraction of the counterions does not associate with the micelle, $\left[C^+\right]$ increases with total surfactant concentration. Paradoxically, for the product $[S^-]\left[C^+\right]$ to remain constant, the free charged monomer concentration $[S^-]$ must decrease with increasing concentration, as does the chemical potential μ_S of the surfactant ion.

One consequence of the decreasing μ_S is that adsorption on positively charged surfaces actually decreases with increasing surfactant concentration above the CMC. As a result, maximum adsorption should occur close to the CMC.

4.1.3.2 Enthalpy and Entropy

The relative contributions of enthalpy determine the temperature dependence of the CMC. We can see this effect most easily if we use the approximate equation derived in Section 4.1.2

$$\Delta\mathbf{G}^{\theta}_{\mathrm{mic}} = \boldsymbol{RT}\ln \mathrm{CMC} \tag{4.1.15}$$

which combined with the Gibbs–Helmholtz equation provides an expression for the enthalpy of micelle formation,

$$\Delta\mathbf{H}^{\theta}_{\mathrm{mic}} = -\boldsymbol{RT}^2\frac{\partial}{\partial\mathbf{T}}\ln \mathrm{CMC} \tag{4.1.24}$$

For an ionic surfactant, there is an additional factor $(2-\beta)$ as in Eq. (4.1.19). For SDS, $\Delta\mathbf{H}^{\theta}_{\mathrm{mic}}$ is close to zero at 25 °C, and the CMC has a minimum value at this temperature. The small value of $\Delta\mathbf{H}^{\theta}_{\mathrm{mic}}$ at room temperature is due to two effects. In Section 3.6, we concluded that $\Delta\mathbf{H}_{\mathrm{el}}$ was close to zero in aqueous systems due to the near $1/\mathbf{T}$ dependence of the permittivity, and in Section 5.7, we will see that it is a notable feature of the hydrophobic interaction that $\Delta\mathbf{H} \approx 0$ around 25 °C.

4.1.4 The Electrostatic Free Energy of Micelle Formation Can Be Reliably Calculated Using the Poisson–Boltzmann Equation

When an ionic surfactant like SDS associates to form a micelle with approximately 60 monomers, a highly charged interface is formed. There is a considerable electrostatic free energy cost in forming such an interface. In Section 3.9, we derived the electrostatic free energy for a planar interface and showed that it could be divided into a negative, direct charge–charge part E^{*}_{el} and positive entropy part $-\mathbf{TS}_{\mathrm{ion}}$. This result is equally valid in spherical geometry and the two terms can be evaluated provided we find a solution to the Poisson–Boltzmann equation, which for a 1:1 electrolyte and in spherical geometry reduces to an ordinary differential equation.

$$\Delta\Phi = \frac{1}{r^2}\left\{\frac{\mathrm{d}}{\mathrm{d}r}\left(r^2\frac{\mathrm{d}\Phi}{\mathrm{d}r}\right)\right\} = \frac{2ec^{*}_{0}}{\varepsilon_r\varepsilon_0}\sinh\left(\frac{e\Phi}{k_{\mathrm{B}}\mathbf{T}}\right) \tag{4.1.25}$$

This equation has no analytical solution, but after tedious algebra, one can obtain a series expansion in terms of the product of the ratio between the Debye screening length κ^{-1} and micelle radius R_{mic} valid for $(\kappa R_{\mathrm{mic}})^{-1} < 1$ and evaluate the electrostatic free energy per charged amphiphile

$$\mathbf{G}_{\mathrm{el}} \approx \mathbf{G}_{\mathrm{el}}(\mathrm{plane}) - \frac{4k_{\mathrm{B}}\mathbf{T}}{s\kappa R_{\mathrm{mic}}}\ln\left\{\frac{1+(1+s^2)^{1/2}}{2}\right\} \tag{4.1.26}$$

Here $\mathbf{G}_{\mathrm{el}}(\mathrm{plane})$ is given by Eq. (3.9.16) and s by Eq. (3.9.14). The first term on the right-hand side of Eq. (4.1.26) is the electrostatic free energy of a planar double layer, and the second term is an explicit correction due to the curvature of the surface. The curvature term is negative and shows that the electrical free energy is smaller on the (convex) curved surface because in addition to the screening effects, the electrical field decays out into the solution on the curved surface for geometrical reasons. (For a concave surface, the correction term is positive.)

If we need to calculate the electrostatic free energy for a range of electrolyte concentrations and micellar radii, including cases when $\kappa R < 1$, we ultimately must solve the PB equation numerically. Today, this is straightforward. However, in a sense, numerical results are less transparent because the dependence on the different parameters appears through explicit numbers rather than in analytical formulae.

The addition of salt reduces the electrostatic penalty of forming the highly charged surface with the result that the CMC increases. The monomer surfactant contributes to the electrostatic screening so that below the CMC, the total electrolyte concentration is $c_{salt} + c_{surf}$. One can assume that the addition of salt only affects the electrostatic contribution to micelle formation and from the general relation between ΔG^{θ} and CMC in Eq. (4.1.15), it follows that the salt dependence of the CMC is

$$\begin{aligned}\Delta\Delta \mathbf{G}^{\theta}_{mic} &= \Delta \mathbf{G}^{\theta}_{mic}(c_{salt}) - \Delta \mathbf{G}^{\theta}_{mic}(c_{salt} = 0) \\ &= \mathbf{G}_{el}(c_{salt}) - \mathbf{G}_{el}(c_{salt} = 0) \\ &= RT\{\ln\left[\mathrm{CMC}(c_{salt})\right] - \ln\left[\mathrm{CMC}(c_{salt} = 0)\right]\}\end{aligned} \quad (4.1.27)$$

This electrostatic free energy can be calculated with reasonable accuracy using Eq. (4.1.26), but a more reliable result is obtained from a numerical solution of Eq. (4.1.25). Figure 4.4 shows that the PB equation accurately accounts for the experimentally observed salt dependence of the CMC for a series of alkyl sulfates. This calculation provides a rather stringent test of the approximations, leading to the Poisson–Boltzmann equation, and we can with some confidence apply it to more complex systems in the coming chapters.

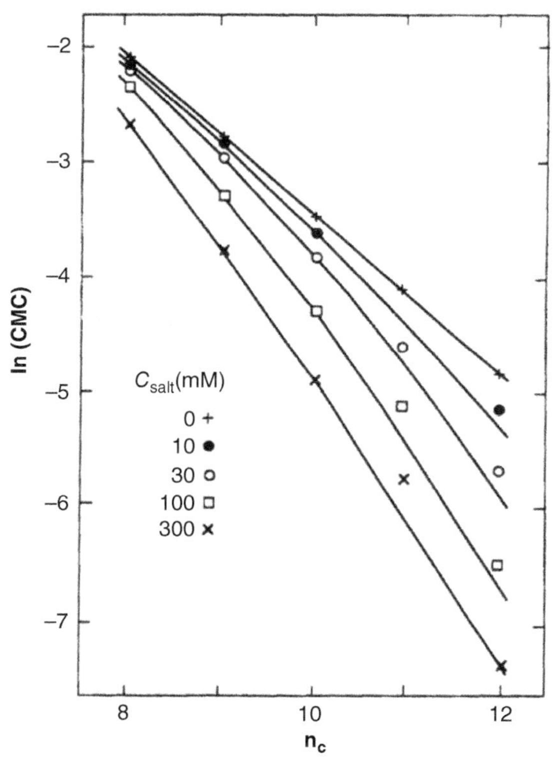

Figure 4.4 Effect of hydrocarbon chain length and added salt on the CMC for sodium alkyl sulfates. As the chain length increases for a given head group, $\Delta \mathbf{G}(HP)$, see Eq. (4.1.1), becomes increasingly negative, leading to a lower CMC. Adding salt diminishes the head group repulsions, $\Delta \mathbf{G}(HG)$, which also lowers the CMC. This effect is more pronounced for the longer chain surfactants because the total electrolyte concentration involves contributions from free surfactant and from salt. The solid lines calculated using Eq. (4.1.25) show that the theory quantitatively accounts for the variation of CMC. *Source*: Gunnarsson et al., 1980/with permission of American Chemical Society.

See Berg (2010, Chapter 3), Cosgrove (2010, Chapter 4), Doi (2013, Chapter 4), Israelachvili (2011, Chapter 19), Kronberg et al. (2014, Chapter 4), and Lyklema (2005, Vol. 5, Chapter 4) for additional discussion on Micellization.

4.2 We Can Measure Critical Micelle Concentrations, Aggregation Number, Micelle Structure, and Characteristic Lifetimes by a Number of Methods

A large variety of experimental techniques has been used to study micellar properties. Rather than surveying the literature, this section cites key experimental methods that provide insightful information on the properties of micellar solutions. We will focus first on spherical micelles.

4.2.1 Measurements of the CMC Are Based on the Difference in Molecular Properties Between Surfactants in Solution and in a Micellar Aggregate

Many physical properties of surfactant solutions undergo an abrupt change in the concentration dependence at the CMC, as shown in Figure 4.5. There is thus a wide choice of methods when one wants to determine the CMC. It is instructive to briefly analyze the physico-chemical basis of the concentration dependence of the different properties. At the CMC, the solubility of an apolar compound starts to increase approximately linearly with micelle concentration. The micelles provide an apolar environment and this property has large practical use in, for example, detergents. The turbidity, a measure of the total amount of scattered light, is negligible below the CMC, but the micellar objects scatter light. This effect will be further discussed in Section 4.3. The micellization process can readily be followed by ^{13}C and ^{1}H nuclear magnetic resonance. Both the chemical shift and the relaxation rates are affected by

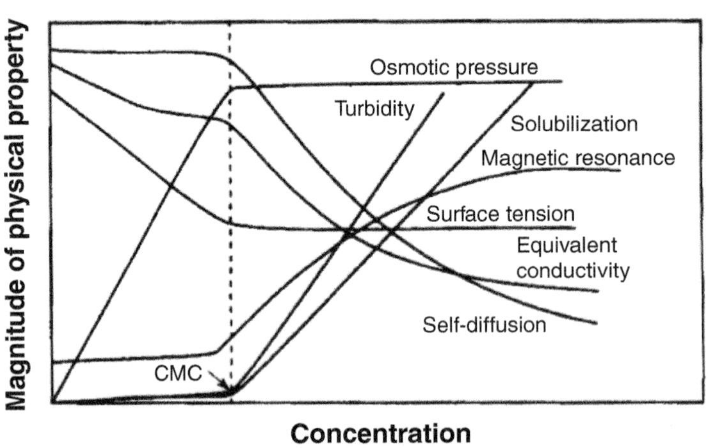

Figure 4.5 Effect of surfactant concentration on various physical properties. Over a fairly narrow surfactant concentration range, virtually all physical solution properties display a pronounced change in slope that can be used to identify a CMC. Variations in CMC remind us that micellization is a cooperative process that occurs over a finite concentration range, in contrast to an abrupt phase transition. *Source:* Lindman et al., 1980/with permission of Springer Nature.

micellization. A closer analysis of such data provides detailed insights into the change in molecular behavior when a surfactant enters a micelle. Measurements of surface tension were historically an important method for establishing micellar behavior. It will be further discussed in the following text. For a solution of a charged surfactant, the electrical conductivity increases less rapidly with increasing concentration beyond the CMC. This implies that the equivalent conductivity decreases since added surfactant molecules form micelles, which also attract counterions. The number of conducting species thus decreases relative to what is found in a solution of monomers. When a surfactant molecule enters a micelle, it stays there for a considerable time and the self-diffusion rate is decreased. It is a characteristic feature of micellization that the chemical potential of the surfactant increases only slowly beyond the CMC. It follows from the Gibbs–Duhem relation (1.2.22) that the chemical potential of the solvent, as measured by the osmotic pressure, remains virtually constant in this concentration range. The osmotic pressure curve in Figure 4.5 refers to a nonionic surfactant, while for an ionic one, the contribution from the unbound counterions complicates the situation. We now discuss in more detail a few methods that have been particularly useful in revealing different aspects of the molecular properties of micellar solutions.

Surface Tension: Surface tension γ decreases from 72 mN/m for pure water to an almost constant value beyond the CMC. It might be tempting to conclude that the surface is saturated by the surfactant at the CMC. However, the proper explanation of the constant γ value above the CMC follows by combining the basic form of the Gibbs adsorption isotherm (Eq. 2.4.13) and Eq. (4.1.22) for the surfactant chemical potential. Above the CMC, the chemical potential of the surfactant scarcely changes, and as a result, conditions do not change at the surface. The limiting value of the surface tension ranges from 40 to 30 mN/m depending on the surfactant. The intersection resulting from linear extrapolation of the two straight lines shown in Figure 4.5 provides a measure of CMC value.

Measurements of surface tension constitute an extremely sensitive test for surfactant purity. For example, a 0.01% hydrolysis impurity of 1-dodecanol, $CH_3(CH_2)_{10}CH_2OH$ in SDS results in curve B shown in Figure 4.6. Although both compounds adsorb at the air–water interface, alcohol concentrates preferentially at the interface because it is more surface active. Since γ(alcohol) < γ(SDS), the surface tension curve falls below that observed for pure SDS. In the bulk phase, alcohol initiates micelle formation at a value below the CMC for the pure surfactant. The aggregation process also extends over a much broader concentration range. Above the CMC, alcohol desorbs from the interface as more surfactant is added and solubilizes back into the micelles as the micellar concentration increases. The result is a gradual increase in γ. A small amount of surface impurity, therefore, leads to a pronounced dip in the surface tension curve, such as we find for curve B in Figure 4.6.

The slope of the γ versus $\ln C_T$ for concentrations below the CMC provides a way to determine the interfacial area per surfactant at the air–water interface, as discussed in Chapter 2.

Surfactant Ion Electrodes: Ion-selective electrodes provide a direct measure of surfactant monomer and counterion chemical potentials. Thus, we can use them to determine patterns of surfactant aggregation, including premicellar dimerization. We can measure the change in electromotive forces (emf) as a function of concentration using the Nernst equation

$$E = E^0 - \frac{(\mu - \mu^0)}{F} \approx E^0 - \left(\frac{\mathbf{RT}}{F}\right) \ln c \qquad (4.2.1)$$

where E^0 represents the standard state emf, and F the Faraday constant.

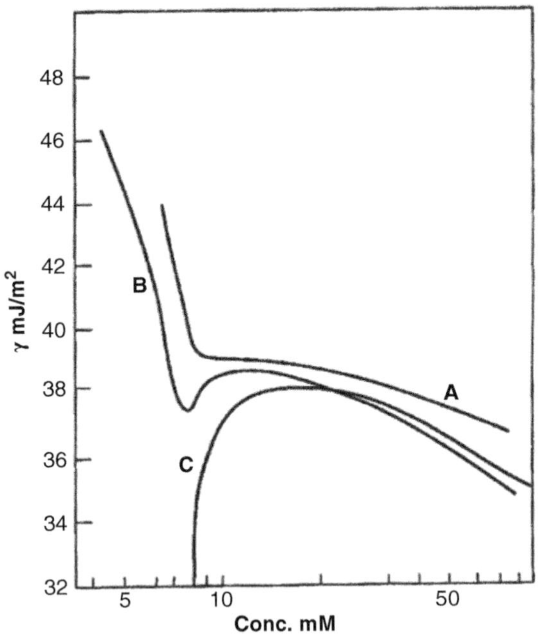

Figure 4.6 Effect of dodecanol on the surface tension of solutions of SDS: curve A, pure SDS; curve B, unpurified SDS; curve C, pure SDS +0.05% dodecanol of the amount of SDS. *Source*: Elworthy and Mysels, 1966/with permission of Elsevier.

Electrodes can be arranged in two ways. If we connect the surfactant and counterion electrodes in series, the resulting emf cell directly determines the combined chemical potentials. Alternatively, if we operate the ion electrodes individually against a common reference electrode, we can measure μ_+ and μ_- separately.

The emf data for decyltrimethylammonium bromide in Figure 4.7 illustrates measurements using ion-specific electrodes. When we plot the combined $(\mu - \mu^\theta)_{C_{10}N(Me)_3Br}/\boldsymbol{RT}$ against $\log[C_{10}\,N(Me)_3\,Br]$ below the CMC, the results show a straight line with a slope of unity. Above the CMC, the data produces a straight line with a minute positive slope.

When the ion-selective electrodes operate separately against a reference electrode, the measured chemical potentials in Figure 4.7 show a slope of 0.98 below the CMC. After the sharp change in slope at the CMC, the chemical potential of the DTA^+ ion decreases and that of the bromide increases. This provides an experimental demonstration of the conclusions of the concentration dependence of chemical potentials presented in Section 4.1.3.

Fluorescence Measurements: Fluorescent molecules can be detected on a single-molecule basis and this can be done with a high spatial resolution. Fluorescence methods are widely used today in biophysics and also in colloidal science. The application of the method to micellar systems can serve as an introduction to its usefulness. Typical fluorescent molecules are apolar and are nearly exclusively solubilized in the micelles when added to a micellar solution.

A useful method of determining micelle aggregation numbers is obtained by first solubilizing a known amount of fluorescent molecules in a micellar solution in such a way that only a fraction of the micelles contains a fluorescent molecule. If we know the total surfactant concentration and the CMC, we can determine the average aggregation number after determining the fraction of tagged micelles. One can determine the fraction of micelles containing a fluorescent molecule by adding a fluorescence quencher to the solution. This will also be distributed among the micelles. In the presence of

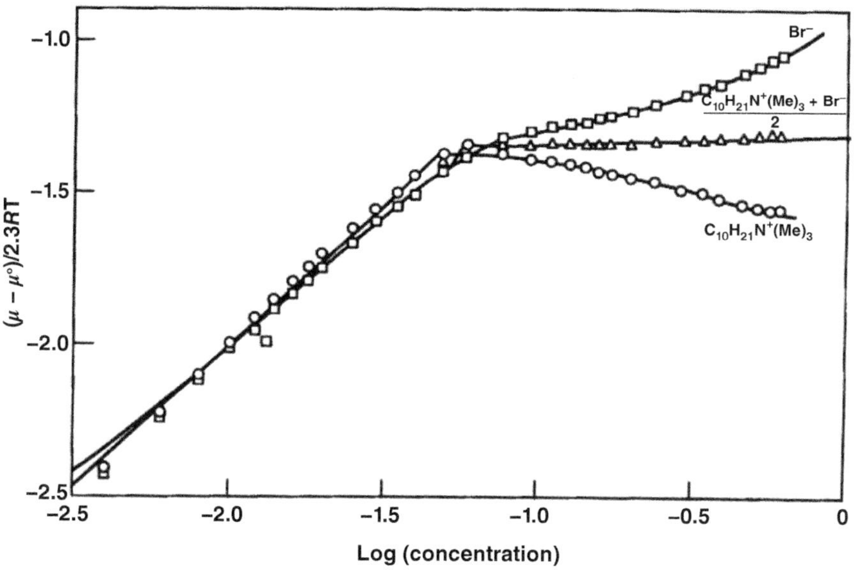

Figure 4.7 Chemical potential of decyltrimethylammonium (DTA^+) and bromide ions measured by ion-selective electrodes. When the surfactant and bromide electrodes are connected in series, we determine the chemical potential of the neutral surfactant, which increases minutely above the CMC. With a common reference electrode, we can determine the single-ion chemical potentials of surfactant and counterion, which above the CMC decrease and increase, respectively. *Source*: Kale et al., 1980/with permission of American Chemical Society.

a quencher, the fluorescence lifetime will be drastically reduced. The fluorescence is then dominated by the contribution from micelles containing a fluorescent molecule but no quencher.

Figure 4.8 shows how pyrene fluorescence intensity varies with the addition of Cu^{2+} quencher ions for an SDS solution. The initial rapid decay in the intensity is due to the effect of the quencher, electrostatically bound to the micelle, while after 200 ns fluorescence comes only from micelles lacking a quencher. Extrapolating the straight lines back to $t = 0$ gives a direct measure of the probability that a micelle is free of quenchers. If we assume that the number of m quenchers (objects) is placed randomly in n micelles (boxes), the result is a Poisson distribution with a probability P of having p objects in a box

$$P(p) = q^p \frac{\exp(-q)}{p!} \tag{4.2.2}$$.

Here q is equal to the ratio between quencher and micelle concentrations. The probability of a micelle free of quencher is $(0! = 1)$

$$P(0) = \exp(-q) \tag{4.2.3}$$

The curves in Figure 4.8 provide three independent measures of q from which the micelle concentration $c(\text{micelles}) = c(\text{quencher})/q$. The aggregation number follows from $N = c\,(\text{micelles})/\{c(\text{surfactant}) - \text{CMC}\}$.

Nuclear magnetic resonance: Surfactant molecules contain several nuclei that can be used in nuclear magnetic resonance investigations. These include ^1H, ^2H, ^{13}C, ^{15}N, ^{17}O, ^{31}P and in addition nuclei in counterions like ^7Li, ^{23}Na, ^{35}Cl, ^{79}Br and ^{123}I. To obtain sufficient intensity it is sometimes necessary

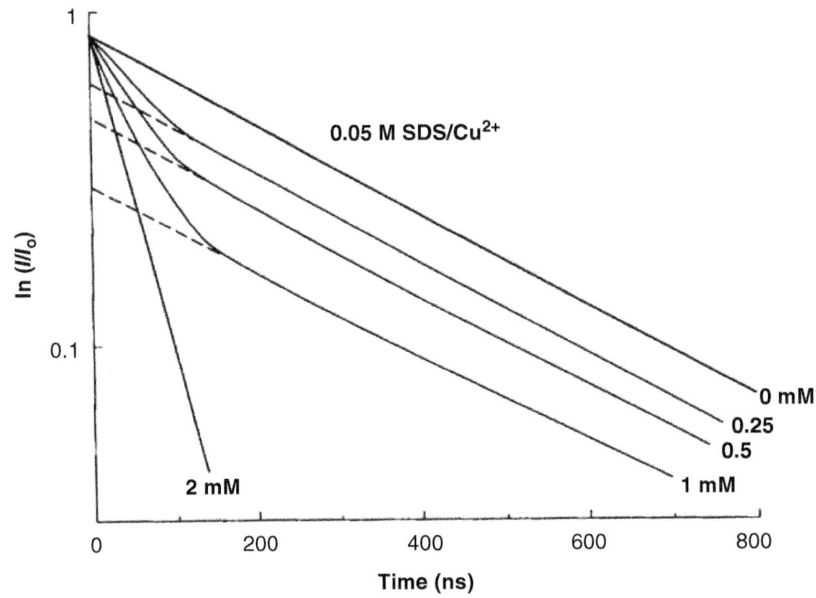

Figure 4.8 Plot of relative fluorescence intensity as $\ln(I/I_0)$ for pyrene in 0.05 M SDS versus time at different amounts of added quencher, cupric ion, Cu^{2+}. In the absence of quencher, the slope of the line measures the lifetime of the fluorescent state of pyrene in the micelles. As quencher is added, we observe a double decay curve. Rapid decay is a consequence of quenching the excited pyrene by Cu^{2+}. Analysis of the data using Eq. (4.2.3) provides the average aggregation number $N = 56$. *Source*: Grieser and Tauschtreml, 1980/with permission of American Chemical Society.

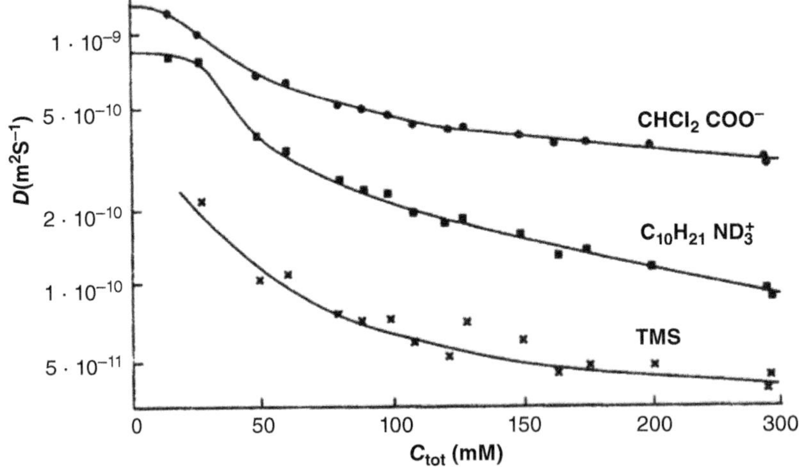

Figure 4.9 Self-diffusion coefficients of all the molecular species in a micellar solution of decylammonium dichloroacetate determined by NMR. The self-diffusion of the solvent HDO, counterion $CHCl_2COO^-$, surfactant ion $C_{10}H_{21}NH_3^+$, and a hydrophobic solubilizate tetramethyl silane (TMS) varies with total surfactant concentration. The micelle determines the diffusion of the TMS. *Source*: Lindman and Stilbs, 1985/with permission of Italian Physical Society.

to invoke isotope labeling. The NMR method is very versatile and can be used to study many aspects of molecular behavior. For an isotropic system like a micellar solution, a main observable is the chemical shift, which is sensitive to molecular conformation and to solvation environment. More specific information is obtained by measuring spin relaxation rates, which report on the molecular motions. The relaxation rates for molecules in micelles are similar to those of a pure liquid showing unequivocally that the micelle interior is liquid-like. However, the polar heads of the surfactant molecules are fixed to the micelle surface, which results in a local restriction of the molecular reorientation.

A third use of the NMR method is to measure self-diffusion. When molecules diffuse in a magnetic field gradient, there is a shift in the resonance frequency, which can be monitored by a proper pulse sequence. A simultaneous measurement of the diffusion of several molecular species provides a detailed picture of the molecular motions in a system that is locally inhomogeneous.

Figure 4.9 shows the measured self-diffusion coefficient of all components in a micellar solution of decylammonium dichloroacetate with an added trace amount of tetramethylsilane (TMS). A noticeable break in the curves around 25 mM signals a CMC at this concentration. We can use the diffusion data to separate the contribution to the diffusion coefficient from free and micellized species with the formula:

$$D_{\text{obs}} = \frac{c^{\text{mic}}}{c} D^{\text{mic}} + \frac{c^{\text{free}}}{c} D^{\text{free}} \tag{4.2.4}$$

where c stands for the total concentration. Figure 4.10 shows the deduced values for c^{mic} and c^{free} for both surfactant and counterion. The free monomer concentration decreases above the CMC, yet the degree of counterion binding remains constant at 0.80 up to concentration 10 times the CMC.

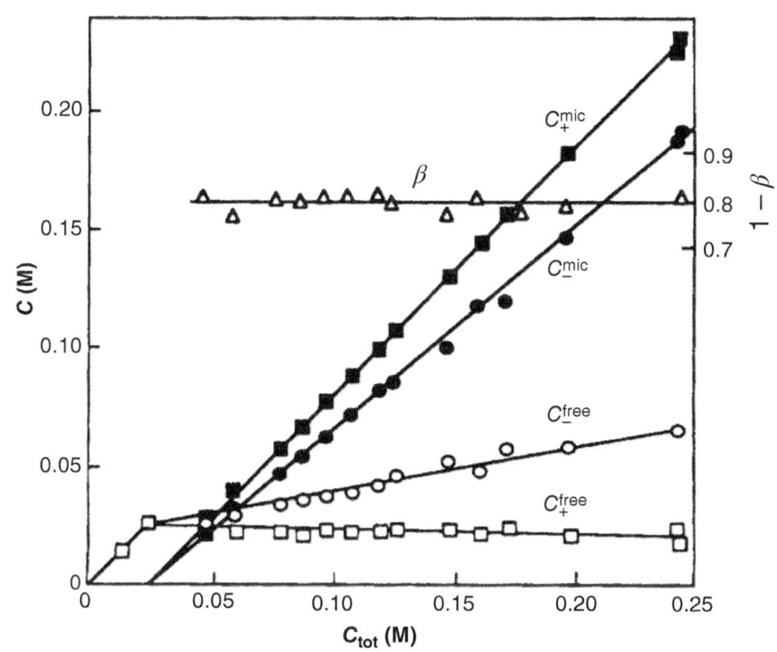

Figure 4.10 Concentration of various micellar species in a decylammonium dichloroacetate solution. Using the data given in Figure 4.10, we can calculate the concentration of micellized amphiphile ions, c_+^{mic} ■, free amphiphilic ions c_+^{free} □; bound counterions c_-^{mic} ●; and free counterions c_-^{free}, ○, plotted as a function of c_{tot}. β is the degree of dissociation (cf. Eq. 4.1.19). *Source*: Stilbs and Lindman, 1981/with permission of American Chemical Society.

The micelle self-diffusion coefficient, which is measured through the self-diffusion of the completely solubilized TMS (tetramethyl silane) in Figure 4.9 decreases strongly with increasing concentration. Strong intermicellar repulsion, mainly of electrostatic origin, causes this decrease.

4.2.3 Kinetic Experiments Provide Valuable Insights into the Time Scales of Dynamic Processes in Micellar Solutions

Micellar formation, dissolution, and size redistribution can be described by a series of elementary kinetic steps as in Eq. (4.1.3) in analogy to the nucleation processes discussed in Section 2.3.3:

$$S + S_{n-1} \rightleftarrows S_n \tag{4.2.5}$$

The ratio between the two rate constants is fixed by the equilibrium constant of Eq. (4.1.4), which ensures that the forward and backward rates are equal at equilibrium

$$\frac{k_n^+}{k_n^-} = K_n \tag{4.2.6}$$

We cannot use classical kinetic experiments to determine the rates of micelle formation because the process occurs so rapidly. Instead, we must use fast relaxation procedures, such as temperature jump, pressure jump, or ultrasonic absorption. The idea is to take a system that is initially at equilibrium, subject it to a small but rapid change in an intensive parameter (e.g., **T** or **p**), and then follow the relaxation to the new equilibrium state.

To conduct a **T**-jump experiment, we place the solution in a spectrometric cell containing two large electrodes. A huge capacitor charged up to 30 kV is shorted across the electrodes; discharged electrical energy causes the temperature of the solution to increase by $\approx 10\,°\text{C}$ in a few seconds. Any process with a sizable $\Delta\mathbf{H}$ will respond to the temperature change at its characteristic rate, which we can follow using changes in the spectrum of appropriately chosen dye additives.

Different fast relaxation methods sample different time scales. Therefore, to characterize a process with multiple equilibria, each with its characteristic relaxation time, we must use several measuring techniques. As shown in Table 4.2, measurements on a variety of surfactant solutions yield a biexponential relaxation process with two well-defined times: τ_1 which typically falls in the range $10^{-6} - 10^{-3}$ s and τ_2 in the range $10^{-3} - 1$ s. The cooperative nature of the micelle formation process is the cause of the existence of separate relaxation times.

Table 4.2 Relaxation times τ_1 and τ_2 for some sodium alkyl sulfates.

Surfactant	Temperature (°C)	Concentration (M)	τ_1 (μs)	τ_2 (ms)
$NaC_{16}SO_4$	30	1×10^{-3}	760	350
$NaC_{14}SO_4$	25	2.1×10^{-3}	320	41
	30	2.1×10^{-3}	245	19
	35	2.1×10^{-3}	155	7
	25	3×10^{-3}	125	34
$NaC_{12}SO_4$	20	1×10^{-2}	15	1.8
	20	5×10^{-2}	—	50

Source: Aniansson et al., 1976/with permission of American Chemical Society.

It has been shown that experimental observations were fully consistent with the kinetic scheme in Eq. (4.2.5), in which micelles change size by the addition or removal of one monomer at a time. A micellar system responds to a **T**-or a ***p***–jump perturbation in two steps. First, it relaxes rapidly to a conditional equilibrium state under the constraint that the number of micelles remains constant. The perturbation causes a small shift in the optimal size distribution. Monomers leave some micelles and associate with others, restoring an optimal size distribution. However, such processes leave the number of micelles unchanged.

Detailed analysis of the fast relaxation process results in the expression

$$\tau_1^{-1} = \frac{k^-}{\sigma^2} + k^- \frac{(c - \mathrm{CMC})}{(\mathrm{CMC} \cdot N)} \tag{4.2.7}$$

Here, N is the most probable aggregation number and σ stands for the standard deviation of the micelle size distribution function (cf. Figure 4.2). Experiments confirm the predicted inverse linear dependence between the relaxation time τ_1 and the surfactant concentration c. We can determine k^- and σ in a τ_1 versus CMC plot if we independently know the aggregation number N. After the rapid relaxation caused by monomer exchange between existing micelles, a slower process takes place that involves a change in the number of micelles. For a particular micelle to dissolve when it starts from an aggregation number that is close to optimal, it must pass through a stage involving very unfavorable aggregation numbers, analogously with the nucleation process discussed in Section 2.3.3. The result is a kinetic bottleneck consisting in the formation of the least-probable aggregation numbers between N and unity (cf. Figure 4.2). A detailed analysis yields

$$\tau_2^{-1} = \frac{N^2}{\mathrm{CMC}} \frac{1}{R_1} \left(1 + \sigma^2 \frac{c - \mathrm{CMC}}{\mathrm{CMC}} \frac{1}{N}\right)^{-1} \tag{4.2.8}$$

where the equilibrium and kinetic properties of micelles with the least-probable aggregation numbers determine the "resistance" R_1. Through the concentration dependence of τ_2, we can get an independent measure of σ^2/N.

Table 4.3 summarizes the parameters obtained from an analysis of kinetic data for a series of sodium alkyl sulfates. Note that the dissociation constant k^- changes by a factor of 1400 in going from the C_6 to the C_{14} compound, while the association constant k^+ changes only by a factor of 7. In fact, k^+ is

Table 4.3 Kinetic parameters of association and dissociation of alkyl sulfates from their micelles.

Surfactant	N	CMC (M)	k^- (s^{-1})	k^+ ($M^{-1}s^{-1}$)
NaC_6SO_4	17	0.42	1.32×10^9	3.2×10^9
NaC_7SO_4	22	0.22	7.3×10^8	3.3×10^9
NaC_8SO_4	27	0.13	1.0×10^8	7.7×10^8
NaC_9SO_4	33	6×10^{-2}	1.4×10^8	2.3×10^9
$NaC_{11}SO_4$	52	1.6×10^{-2}	4×10^7	2.6×10^9
$NaC_{12}SO_4$	64	8.2×10^{-3}	1×10^7	1.2×10^9
$NaC_{14}SO_4$	80	2.05×10^{-3}	9.6×10^5	4.7×10^8

Source: Aniansson et al., 1976/with permission of American Chemical Society.

close to the diffusion-limited rate (Eq. 1.7.11). Because the micelle is essentially a liquid-like drop, the monomer association reaction involves no slow structural reorganization or bond formation. The sole barrier felt by a monomer entering the micelle arises from long-range electrostatic repulsions due to the micellar charge. With longer alkyl chains, the CMC is lower, and the electrolyte concentration, consisting of monomer surfactants and counterions, is lower. The charge of micelle surface is less screened and slows down the association rate for long-chain surfactants relative to short-chain ones.

For the most probable aggregation numbers, $N-1$ and N, we have at equilibrium $k^+ + [S_1][S_{N-1}] = k^-[S_N]$. Since the monomer concentration equals the CMC, this reduces to CMC $k^+ = k^-$. Thus, k^- decreases strongly with increasing hydrocarbon chain length, due to the lowering of the CMC and a weakly decreasing k^+.

We can learn several things from the analysis of the micellar kinetics. Micelles actually grow in a stepwise manner, and two aggregates rarely associate into one big aggregate. (For micelles that grow to rods, this conclusion might not be valid.) No kinetic barriers to micellar growth exist on the molecular scale, and most processes can be seen as diffusion-controlled if we correct for the effects of long-range electrostatic interactions.

An additional useful insight is that the cooperativity of micelle formation causes the slow relaxation time, τ_2. This process exists only because intermediate aggregation numbers are unlikely. The existence of the fast relaxation time demonstrates the existence of a distribution of aggregation numbers, but the experimental values reflect a moderately narrow distribution of the aggregation number. For typical ionic micelles, the standard deviation, σ, in the distribution is of order 10.

See Berg (2010, Chapter 3), Cosgrove (2010, Chapter 4), Doi (2013, Israelachvili (2011, Chapter 19), and Kronberg et al. (2014, Chapter 6) for additional discussion on Micelle Properties.

4.3 Scattering Provides Very Useful Techniques for Studying Micellar Structure and Colloid Systems in General

Scattering of light, X-rays, and neutrons provide ways to determine structure and order in molecular and colloidal systems. In this section, we present the basic principles of the methods and illustrate how they are used to study micellar systems. Other examples will appear in later chapters on more complex systems with more sophisticated experimental designs.

4.3.1 Scattering of Light, X-rays, and Neutrons Share Many Features but Differ in the Source of the Contrast

In the typical experimental scattering setup, there is a source, a sample, and a detector as shown in Figure 4.11. The source, whether it is visible light, X-ray, or neutrons, should produce photons or neutrons, which have as well-defined momentum $\boldsymbol{p}_0$ as possible with respect to both magnitude and direction. It is convenient to characterize energy and direction of the source beam by its wave vector $\boldsymbol{q}_0 = (2\pi/h)\boldsymbol{p}_0$. There must always be some spread in the energy and direction of the beam, but we ignore this complication in the present discussion. The wave vector is related to the momentum $\boldsymbol{p}$ of the photon or neutron and the corresponding wavelength λ through

$$\boldsymbol{q} = 2\pi\boldsymbol{p}/h;\ q = 2\pi/\lambda;\ \lambda = h/p;\ E(q) = hcq/2\pi \tag{4.3.1}$$

When the incoming beam hits the target, the photon or neutron can interact with the electrons/nuclei of the sample and get scattered into a new direction, or it can pass through unaffected. In

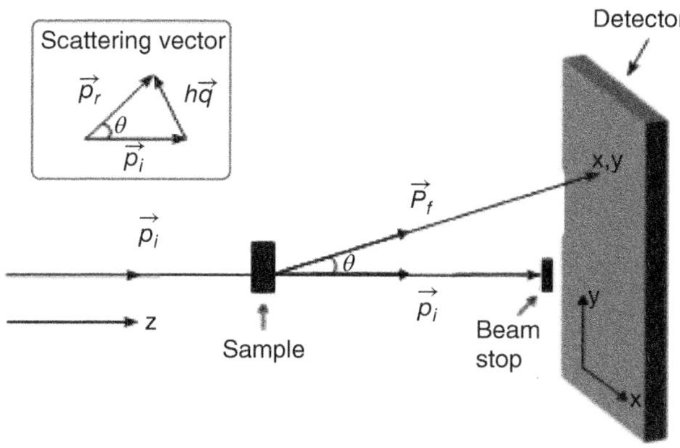

Figure 4.11 An incoming particle with momentum $\boldsymbol{p}_i$ is either passing through the sample with unchanged momentum or scattered with a new momentum $\boldsymbol{p}_f$. The scattering vector $\boldsymbol{q} = 2\pi(\boldsymbol{p}_f - \boldsymbol{p}_i)/h$ with an orientation given by the polar angle θ and the azimuthal angle ϕ. *Source*: Adapted from Wennerström, 2014.

general, scattering is an improbable event and in a typical experimental design the sample thickness is adjusted so that photons and neutrons are only scattered once or not at all as they pass the sample. In most scattering setups, one has a beam stop that collects the photons/neutrons that pass through on a straight path. The focus is then on the photons/neutrons that are scattered away from the primary beam and observed on the detector screen where a position is characterized by the angles θ and ϕ (see Figure 4.11). It is only the momentum transfer in the perpendicular direction that affects the final position on the detector. For a sample that has a long-range directional order as in a crystal, both angles θ and ϕ are relevant, but in isotropic colloidal systems, there is no ϕ dependence on the scattering. Define a scattering vector as in Figure 4.11:

$$h\boldsymbol{q} = \mathbf{p}_f - \mathbf{p}_i \tag{4.3.2}$$

Most scattering events are elastic, so no energy is exchanged between sample and probe. In this case, the scattering vector has an amplitude

$$q = 2q_i \sin(\theta/2) \tag{4.3.3}$$

as illustrated in Figure 4.11. Scattered photons or neutrons are registered one by one on the detector, and as the experiment progresses, a scattering pattern emerges, which for isotropic samples is described by a scattering intensity $I(\boldsymbol{q})$. The challenge in the analysis of $I(\boldsymbol{q})$ is to extract useful information about the properties of the sample.

A fundamental aspect of quantum mechanics is the particle-wave duality. The description of scattering experiments provides a clear demonstration of this duality. Conceptually, it appears as most transparent to describe the scattering event in terms of a particle picture. The incoming particle has a given momentum when it meets the sample. In a quantum description, collision between incoming particle and electrons or nuclei in the sample is a discrete event that occurs at most once during the passage through the sample with adjusted thickness. Elastic collisions have a higher probability than inelastic ones so that the collision between probing particle and sample results in a momentum transfer and it is only the component in the direction perpendicular to that of the incoming beam that affects the scattering angle. Conservation of total momentum implies that there is a corresponding change in the momentum of the sample.

There is a fundamental quantum mechanical relation between position and momentum involving a Fourier transform and it applies to the state of the sample. Thus, a nonuniform distribution of electrons or nuclei in the sample necessarily implies a nonuniform distribution of their momenta. This in turn results in a distribution of the probability of a momentum transfer between particles and the sample. For a sample with periodic order as in a crystal, there are only discrete values of the momenta with the consequence that there are only discrete momentum transfers possible. This results in discrete peaks in the scattered intensity. For samples with order on a shorter length scale, as in an isotropic colloidal system, there is a more gradual variation of the momenta of nuclei and electrons in the sample, and the intensity $I(\boldsymbol{q})$ varies more smoothly. A prerequisite for scattering to occur is that the sample is inhomogeneous in the direction perpendicular to the incoming beam.

In this conceptual explanation of the appearance of a scattering pattern, the quantum nature of the sample plays an essential role. In the more conventional wave picture, on the other hand, the Fourier transform relation between position and momentum transfer emerges as an interference effect in the particle wave. It is in practice easier to conceptually think in terms of such an interference when extracting structural information from the scattering pattern.

The essential difference between the scattering of light, X-rays, and neutrons is found in the source of the contrast. Although light and X-ray photons represent electromagnetic radiation, they interact with the electrons in the sample in a qualitatively different way. The high-energy X-ray photons interact directly with individual electrons and the scattering amplitude is proportional to the variation in electron density squared. Photons with a lower energy, typical for the visible spectrum, see the electrons more collectively and the scattering intensity depends on variations in the local refractive index n. Neutrons interact at short range with the nuclei in an isotope-specific way. The information about the sample structure is similar to the one obtained from X-ray scattering since nuclei and electrons colocalize in the sample. However, the isotope specificity of the neutron–nucleus interaction can be used to great advantage in neutron scattering since the contrast can be varied by isotope substitutions.

The theoretical description of the scattering of light, X-rays, and neutrons can be put into a common formalism by introducing the concept of a scattering length b_j of a particle j as

$$b_j(\boldsymbol{q}) = \int \{\Delta C(\boldsymbol{r})\} \exp\left(-i\boldsymbol{q} \cdot \boldsymbol{r}_j\right) dr^3 \tag{4.3.4}$$

Here ΔC denotes contrast, a cross-section specific for each type of scattering method. The total scattering has contributions from both correlations within a colloidal particle j at position $\boldsymbol{r}_j$ and correlations between particles j and k. The resulting scattering intensity $I(\boldsymbol{q})$ observed a distance R from the sample containing N particles in the scattering volume and for an incoming beam of intensity I_0 is

$$\frac{I(q)R^2}{I_0} \propto \left\langle \sum_{j=1}^{N} \sum_{k=1}^{N} b_j(\boldsymbol{q}) b_k^*(\boldsymbol{q}) \exp\left[-i\boldsymbol{q} \cdot \left(\boldsymbol{r}_j - \boldsymbol{r}_k\right)\right] \right\rangle \tag{4.3.5}$$

Here the angular bracket denotes an ensemble, in practice time, average.

In a dilute system, particles are spatially separated and uncorrelated, so that when averaging the exponential factors in Eq. (4.3.5), they cancel when $j \neq k$. The sum over j gives N identical terms for identical particles, and the observed scattering intensity can be written

$$\frac{I(q)R^2}{I_0} \propto N\left\langle |b(0)|^2 \right\rangle P(q);\ P(q) \equiv \frac{\left\langle |b(q)|^2 \right\rangle}{\left\langle |b(0)|^2 \right\rangle} \tag{4.3.6}$$

$P(q)$ is called the form factor of the particle, and it follows from the definition that it takes the value of unity at $q = 0$.

For systems of higher concentrations, particle positions are correlated, and there is a contribution to the scattering from pairs j, k. For uniform particles, one can separate this contribution from that of the form factor by introducing a structure factor $S(q)$.

$$\frac{I(q)R^2}{I_0} \propto P(q)S(q) \tag{4.3.7}$$

The limit q goes to zero, representing large separations, and in this limit $S(q \approx 0)$ is due to concentration fluctuations in the thermodynamic limit and one can show that

$$S(q = 0) \propto \left\{ \frac{\mathrm{d}\Pi_{\mathrm{osm}}}{\mathrm{d}c_{\mathrm{p}}} \right\}^{-1} \tag{4.3.8}$$

Here $\mathrm{d}\Pi_{\mathrm{osm}}/\mathrm{d}c_p$ is called the inverse osmotic compressibility. The structure factor is related to the pair distribution function $g(r)$ by a Fourier transform and for isotropic systems

$$S(q) = 1 + c\int [g(r) - 1]\, \frac{\sin(qr)}{qr}\, r^2\mathrm{d}r \tag{4.3.9}$$

For a random distribution of particles $g(r) = 1$ and the structure factor reduces to one, while if particles interact the nonuniform distribution makes $S(q)$ deviate from unity.

We have now introduced several key concepts that are common to the analysis of scattering studies of colloidal systems. These include the scattering vector $\boldsymbol{q}$, characterizing the momentum transfer in a collision, the scattering length b characterizing the interaction between the probing particle and scattering particles in the sample, the form factor $P(q)$ determining the scattering property of an isolated colloidal particle, the structure factor $S(q)$ accounting for the scattering consequences of the correlation between colloidal particles. In the limit $q \to 0$, the structure factor is proportional to osmotic compressibility, which is a thermodynamic quantity. In the following text, we illustrate how scattering methods can be used to study micellar systems. Other applications of scattering methods will be discussed in later chapters.

4.3.2 Micellar Sizes and Interactions Can Be Determined by Light Scattering

The wavelength λ of visible light is two orders of magnitude larger than the radii of typical spherical micelles. A light beam hitting a micellar solution will see the micelles as small objects of refractive index n_{mic} more or less randomly distributed in a solvent of refractive index n_{solv}. Provided $n_{\mathrm{mic}} \neq n_{\mathrm{solv}}$ the refractive index averaged over a volume of size λ^3 will vary in accordance with the local concentrations of micelles. Variations in n will cause the light beam to scatter.

The intensity of the scattered light I_s depends on four factors, so that

$$\begin{aligned} I_s(\text{sample}) - I_s(\text{solvent}) &= \text{instrumental constant} \times \text{angular factor} \\ &\quad \times \text{optical constant} \times \text{concentration fluctuation factor} \end{aligned} \tag{4.3.10}$$

The instrumental constant is proportional to the intensity I_0 of the incoming unpolarized light beam and to the scattering volume (width of the beam times sample thickness). It is furthermore inversely proportional to the square of the distance $\mathrm{R_s}$ between the sample and the detector.

The optical constant K_0 depends on the wavelength λ, the solution's index of refraction, n, and its variation with concentration dn/dc, which can be measured independently in a differential refractometer. The expression for K_0 is

$$K_0 = 2\pi^2 \frac{n^2}{\lambda^4}\left(\frac{dn}{dc}\right)^2 / N_{Av} \tag{4.3.11}$$

Note that the scattering intensity depends strongly on the wavelength. This wavelength dependence is the source of the perceived blue color of the sky as well as that of the color of the iris of blue-eyed persons.

We can most easily understand the dependence of scattering intensity on θ, the angle between the primary beam and the scattered beam, by considering the unpolarized incoming light as composed of two perpendicularly polarized components. For small scattering angles, θ, the polarized components contribute equally to the scattering. However, for light scattered perpendicular to the primary beam ($\theta = 90°$), only the component polarized perpendicular to the direction of the scattered beam can contribute, because no polarized component occurs parallel to the direction of propagation. For micelles with a radius $R << \lambda$, this effect is the only contribution to the θ dependence of the scattering. The explicit form of the angular factor in Eq. (4.3.10) is then $(1 + \cos^2\theta)$.

From a chemical perspective, the concentration fluctuation term is the most interesting factor in Eq. (4.3.10) because it contains the desired information on the properties of the sample. The free energy cost in setting up an inhomogeneity in the micelle concentration is proportional to the derivative of the surfactant chemical potential with respect to concentration, $d\mu_s/dc$. Using the Gibbs–Duhem relation, $n_{solvent}\, dn_{solvent} = -n_s\, d\mu_s$ (Eq. 1.2.22), giving

$$\frac{d\mu_s}{dc} = -\frac{1}{V_{solv}c}\frac{d\mu_{solvent}}{dc} = \frac{1}{c}\frac{d\Pi_{osm}}{dc} \tag{4.3.12}$$

Here Π_{osm} represents the osmotic pressure and $(d\Pi_{osm}/dc)^{-1}$ is called the osmotic compressibility.

The smaller the $d\mu_s/dc$, the smaller the free energy cost of generating a concentration fluctuation and the larger the scattering intensity. An explicit derivation based on Einstein's fluctuation theory shows that the concentration fluctuation factor in Eq. (4.3.10) is

$$\mathbf{RT}c\left(\frac{d\Pi_{osm}}{dc}\right)^{-1} \tag{4.3.13}$$

We have now specified all the factors in Eq. (4.3.10). Separating the factors that depend on the experimental setup makes reporting data more convenient. We can do this by defining a Rayleigh ratio

$$R_\theta \equiv \frac{1}{1 + \cos^2\theta}\frac{R_s^2 I_s}{I_0 V_{sc}} \tag{4.3.14}$$

Here I_s is the scattered intensity per area at the detector at a distance R_s from the sample. I_0 is the intensity of the incoming light and V_{sc} is the scattering volume. Thus R_θ has the dimension of inverse length. In terms of the Rayleigh ratio, the scattering at $\theta = \pi/2$ caused by the presence of the micelles becomes

$$\Delta R_\theta = R_\theta(\text{solution}) - R_\theta(\text{solvent}) = \frac{2\pi^2 n^2}{\lambda^4}\left(\frac{dn}{dc}\right)^2 \mathbf{RT}c\left(\frac{d\Pi_{osm}}{dc}\right)^{-1} \tag{4.3.15}$$

If we know λ, n, and dn/dc, the light-scattering experiment measures ΔR_θ to give the thermodynamic quantity $(d\Pi_{osm}/dc)$. The Rayleigh ratio depends, in general, on the angle θ, but for small micelles, it is a constant.

To obtain the molecular weight and thus the micelle aggregation number, we need a model for the concentration dependence of Π_{osm}. With well-defined particles, we can use a virial expansion of the osmotic pressure as in Eq. (1.2.14). The molar concentration c is related to the mass concentration, c_m, through the molecular weight M and $c = 10^3\, c_m/M$. Thus,

$$\Pi_{\text{osm}} = 10^3 \boldsymbol{RT} c_m \left(\frac{1}{M} + \frac{10^3 B_2 c_m}{M^2} + \dots \right) \tag{4.3.16}$$

and

$$\frac{d\Pi_{\text{osm}}}{dc_m} = \boldsymbol{RT} \left(\frac{10^3}{M} + 2B_2' c_m + \dots \right) \tag{4.3.17}$$

where $B_2' = B_2/M^2$. By extrapolating $d\Pi_{\text{osm}}/dc$ toward $c_m = 0$, we can determine the molecular weight M from the intercept and the virial coefficient B_2 from the slope.

For micellar systems, we cannot use Eq. (4.3.17) directly because below the CMC there are only monomers in the solution, while above the CMC the scattering is dominated by micelles. A reasonable approximate procedure is to formally include the monomers in the solvent and replace c_m by $c_m - \text{CMC}_m$ (counted as mass concentration). Then we can modify Eq. (4.3.17) to

$$\frac{d\Pi_{\text{osm}}}{dc_m} = 10^3 \frac{\boldsymbol{RT}}{M} + 2\boldsymbol{RT} B_2'(c_m - \text{CMC}_m) + \dots \tag{4.3.18}$$

By substituting Eqs. (4.3.11) and (4.3.18) into Eq. (4.3.15) we find

$$\Delta R_{\pi/2} = \frac{K_0(c_m - \text{CMC}_m)}{10^3/M + 2B_2'(c_m - \text{CMC}_m)} \tag{4.3.19}$$

which upon rearrangement gives

$$10^{-3} \frac{K_0(c_m - \text{CMC}_m)}{\Delta R_{\pi/2}} = \frac{1}{M} + 2 \times 10^{-3} B_2' (c_m - \text{CMC}_m) \tag{4.3.20}$$

In Figures 4.12 and 4.13, we illustrate the procedure for determining the aggregation number of tetradecyltrimethylammonium bromide micelles. The concentration dependence of the scattering intensity is shown in Figure 4.12. Around $c_m \approx 1$ kg/m^3 a sharp increase in I_s signals the onset of micelle formation, but the scattering intensity is perceptible even below the CMC. Such a scattering could be caused by small amounts of high molecular weight impurities in water. In fact, removing impurities from the system constitutes a major experimental difficulty. In Figure 4.13, the data have been replotted using Eq. (4.3.20) to obtain a straight line, where the intercept is the inverse molecular weight of the micelle and the slope yields the second virial coefficient.

4.3.3 Small-angle X-ray Scattering Can Be Used to Characterize the Structure of Individual Micelles and Correlations Between Micelles

The wavelength of X-rays is of the order of a few tenths of a nanometer. This is clearly smaller than a typical micellar radius. For micellar systems and isotropic colloids in general, the significant structural information appears at small scattering angles so that the corresponding $\mathbf{q}$ vector is of order an inverse micellar radius. Measuring at small scattering angles requires a dedicated experimental setup, and the technique is called small-angle X-ray scattering (SAXS).

By measuring the scattering from a dilute system, one can measure the form factor $P(q)$ of a micelle. In the simplest case, a micelle can be seen as a homogeneous sphere where the contrast

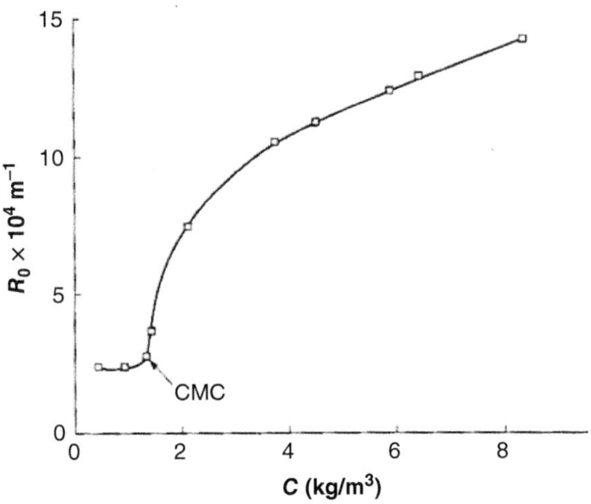

Figure 4.12 Light-scattering data for tetradecyltrimethylammonium bromide $C_{14}N(Me)_3Br$ as a function of surfactant concentration at 25 °C. Below the CMC, the Rayleigh ratio R_θ is almost constant. Above the CMC, R_θ reflects the increase in concentration fluctuations when *N* molecules associate into a single micelle. *Source*: Beesley, 1990/with permission of University of Minnesota.

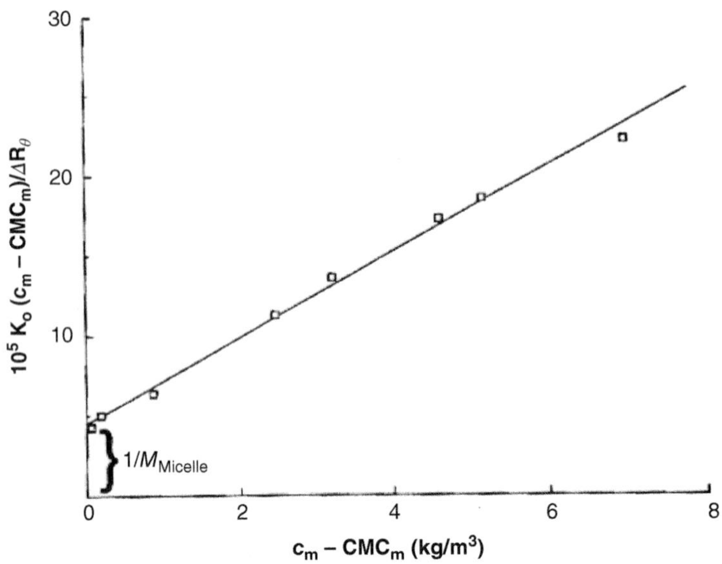

Figure 4.13 Plot of the light-scattering data of the previous figure that yields, using Eq. (4.3.20), the molecular weight of the micelle $M_{mic} = 2.3 \times 10^4$ from the inverse of the intercept and the second virial coefficient $B_2 = 5.4$ m^3 from the slope. The micelles have a radius of 1.9 nm and occupy a volume of 0.018 m^3 per mole. The value of B_2 shows that micelles repel, $B_2 > 0$, and excludes a volume 300 times their own size by electrostatic interactions. *Source*: Beesley, 1990/ University of Minnesota.

factor $\Delta C(R)$ is uniform inside the sphere and zero otherwise. The scattering length in Eq. (4.3.4) then reduces to

$$b(q) = \Delta C \int_0^R \exp(-i\mathbf{q}.\mathbf{r})\, \mathrm{d}^3 r = \Delta C \int_0^R \frac{\sin(qr)}{qr} r^2 \mathrm{d}r \tag{4.3.21}$$

The integral is solved by converting to spherical coordinates, and the resulting form factor is

$$P(q) \equiv \frac{\langle |b(q)|^2 \rangle}{\langle |b(0)|^2 \rangle} = \left\{ \frac{3}{(qR)^3} [\sin(qR) - qR\cos(qR)] \right\}^2 \tag{4.3.22}$$

Problem: Solve Eq. (4.3.21) to obtain Eq. (4.3.22) for the form factor of a sphere.

Solution: In the limit $q = 0, b = \Delta C \int_0^R r^2 dr = \frac{\Delta C R^3}{3}$. For nonzero q

$$
\begin{aligned}
b(q) &= \frac{\Delta C}{q} \int_0^R \sin(qr) r dr \\
&= \frac{\Delta C}{q} \left\{ \frac{1}{q} \left[-R\cos(qR) + \int_0^R \cos(qr) dr \right] \right\} \\
&= \frac{\Delta C}{q^3} \{\sin(qR) - Rq\cos(qr)\}
\end{aligned}
$$

When inserted into the expression for the form factor, one obtains Eq. (4.3.22).

$P(q)$ in Eq. (4.3.22) has zeros at $\tan(qR) = qR$ so that the predicted scattering intensity has a sequence of zeros.

Figure 4.14 shows a comparison between the ideal scattering pattern of Eq. (4.3.22) and an experimental curve. Such a measurement provides a rather accurate value for the radius R of the micelle, but there is a discrepancy between the curve for a perfect sphere, which shows that, on average, the micelle is spherical, but not at each moment. In addition, there is a size distribution resulting in a small variation in the radius R.

At higher micelle concentrations, the structure factor deviates significantly from unity since micelle positions correlate. In the previous section, we saw that for ionic micelles, the second virial coefficient was large, showing that micelles repel one another. A consequence is that there is a preference for micellar distributions that maximize the separation between micelles so that the most probable separation between micelles varies as the inverse cube of the concentration. As a consequence, the structure factor has a maximum at q-value proportional to $c^{-1/3}$ at higher micellar concentrations.

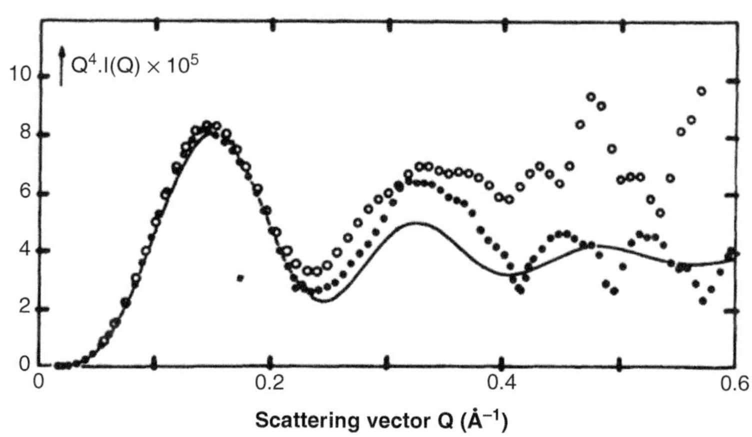

Figure 4.14 SANS scattering from a solution diluted in SDS micelles illustrating the oscillatory behavior predicted by Eq. (4.3.22). The scattering amplitude is multiplied by q^4 to amplify amplitude at large q values. (Note capital Q is used instead of q.) *Source*: Cabane et al., 1985/with permission of EDP Sciences.

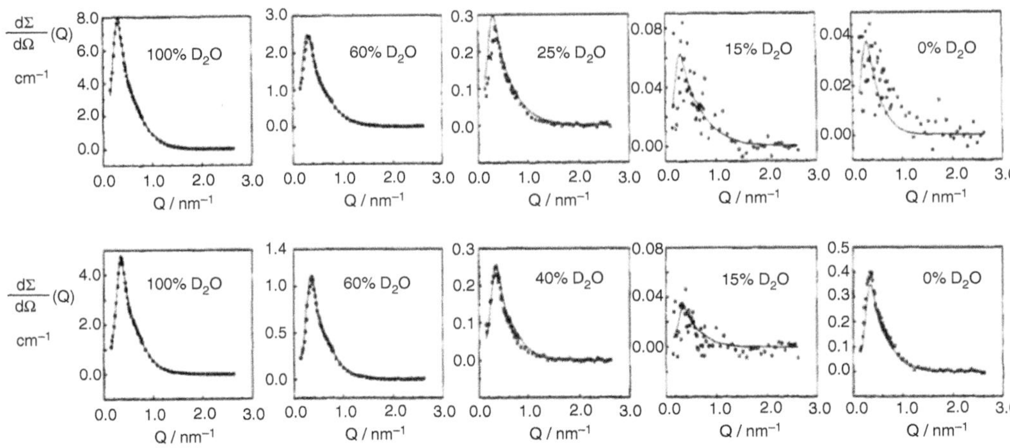

Figure 4.15 SANS differential cross sections measured for a mixed micelle with an 8:2 ratio between tetradecyldimethylamine oxide and SDS. Upper panels protonated; lower panels deuterated SDS. Fully deuterated water to the left and ordinary water to the right in each panel row. The scattering curves are consistent with the assumption that in this case the two surfactants form one type of micelle with a homogeneous distribution of the two components. *Source*: Pils et al., 1993/with permission of American Chemical Society.

4.3.4 Using Contrast Matching Neutron Scattering Can Provide Specific Information on Micellar Structure

Neutrons interact with the nuclei of the sample. Colloidal systems are typically rich in hydrogen atoms, and they have a strong interaction with the probing neutrons. Neutrons as well as hydrogen isotopes have a spin, and this makes the interaction between them spin-dependent. For a colloidal sample, the spin states of individual ^{1}H and ^{2}H nuclei are random, and for the spin-dependent part of the neutron–hydrogen interaction, there is no correlation between different nuclei. In the expression for the intensity in Eq. (4.3.5), it is only the $j = k$ self-term that survives. There is no structural information from this contribution to the scattering except for the average density. This contribution is called incoherent scattering. It turns out that this part is particularly large for the ^{1}H isotope with a cross section, b^2, of 80×10^{-28} m^2, while it is 40 times smaller for ^{2}H. Replacing ordinary water with deuterated water as a solvent gives a substantial reduction in the incoherent scattering, which facilitates the measurement of the coherent part, which contains the structural information.

It is the coherent part of the scattering that is of interest. The scattering length b for the coherent scattering is determined by the spin-independent interaction between neutron and nucleus. This interaction can be both attractive and repulsive, and b can have either sign. For ^{1}H $b = -0.38$, while for ^{2}H, it is +0.66 in units of 10^{-14} m. It is possible to systematically vary the contrast of a sample by isotopic labeling of surfactant or polymer molecules and also by varying the isotope composition of the solvent. This strategy provides opportunities to obtain structural information that is not amenable using SAXS. Figure 4.15 shows a study of micellar structure using the contrast variation method.

4.3.5 Turbidity Measurements Provide a Simple Tool to Measure the Presence of Aggregates

Earlier, we had focused on the case of thin samples to avoid the occurrence of multiple scattering. For a thick sample, more and more photons are scattered away from the primary beam. Seen from the perspective of the primary beam, this is analogous to absorption, and its intensity decreases exponentially with increasing distance. The intensity $I(z_s)$ passing through a sample depends on the thickness z_s as

$$I(z) = I_0 \exp\left(-\tau z_s\right) \tag{4.3.23}$$

in analogy with Beer's law for absorption of light. Here τ is called the turbidity, and its magnitude is determined by the scattering, expressed in terms of the Rayleigh ratio, integrated over all angles

$$\tau = \int_0^{\pi}\int_0^{2\pi} \frac{\left\{I_s(\theta)R_s^2\right\}}{I_0} \sin(\theta)\,\mathrm{d}\varphi\mathrm{d}\theta \tag{4.3.24}$$

In the general case, $I_s(\theta)$ can depend strongly on the scattering angle θ. Here, we consider the simplest case with a dilute solution of spheres with a uniform radius R, which is much smaller than the wavelength, λ, of the light. With those simplifying assumptions, we have from Eq. (4.3.14) that

$$\frac{I_s(\theta)R_s^2}{I_0} = R_{\pi/2}(1 + \cos^2\theta) \tag{4.3.25}.$$

For this simple angular dependence, the integral in Eq. (4.3.24) is readily performed and

$$\tau = \frac{16\pi}{3} R_{\pi/2} \tag{4.3.26}$$

Assuming an ideal solution, the Rayleigh ratio caused by scattering from the particles is according to Eq. (4.3.15)

$$R_{\pi/2} = \frac{2\pi^2 n_s^2}{\lambda^4}\left(\frac{\mathrm{d}n}{\mathrm{d}\phi}\right)^2 c_{\text{sphere}} V_{\text{sphere}}^2 \tag{4.3.27}$$

Here $\phi = c_{\text{sphere}} V_{\text{sphere}}$ is the volume fraction of particles and c_{sphere} is the concentration of particles per volume. If the particles don't affect the refractive index of the solution, there is no scattering. Our final expression for the turbidity

$$\tau = \frac{16\pi}{3}\frac{2\pi^2 n_s^2}{\lambda^4}\left(\frac{\mathrm{d}n}{\mathrm{d}\phi}\right)^2 c_{\text{sphere}} V_{\text{sphere}}^2 \tag{4.3.28}$$

shows that the larger the spheres (particles), the larger is the contribution to turbidity (and scattering). Although Eq. (4.3.28) is quantitatively valid only for spheres much smaller than the wavelength of the light, this quantitative result remains qualitatively valid also for larger particles in the system. It is in fact a considerable practical problem to eliminate small amounts of large contaminants when performing light-scattering experiments. One of the curves of Figure 4.5 shows how the turbidity of a solution starts to increase at the CMC. This would be an easy method of determining the CMC if it were not for the problem with contaminants. Turbidity measurements play an important role in assessing the quality of drinking water. In this case, it is precisely the contamination by particular matter that is the issue.

4.3.6 Dynamic Light Scattering Reveals the Diffusional Motion of Micelles

Time-resolved measurements of the scattering from colloidal systems show that the intensity fluctuates in time around an average value. To see the origin of this effect, consider a solution of colloidal particles at a given moment in time. The incoming light is focused on a fixed scattering volume and the number of particles within that volume fluctuates, resulting in a time dependence of the scattering. In addition, each particle is surrounded by other particles in an irregular way, and the scattering from this configuration differs from that of the mean particle distribution. In solution, the particles diffuse to new positions, resulting in a slightly different scattering pattern. When averaged over a sufficiently long time, the relative positions adopt the equilibrium distribution. The discussion of scattering in the previous sections has been based on this average distribution.

In contrast to the situation in a disordered solid, the scattering from a liquid system depends on the motion of the particles. This implies a coupling between particle translation and photons involving a possible energy exchange. We have based the discussion of scattering in the previous sections on the assumption that the collisions are elastic, but this is an approximation. There is in fact an exchange of particle kinetic energy with the photon energy. However, the kinetic energy of a colloidal particle is around $k_B\mathbf{T}$, which is much smaller than the energy of a probing photon, so the possible energy exchange is small as seen from the perspective of the photon. One describes in this case the collisions as quasi-elastic. The method to study time-dependent phenomena using time-resolved light scattering goes under the name dynamic light scattering (DLS). Sometimes one uses the term quasi-elastic light scattering (QELS), but this is a questionable terminology since the quasi-elastic nature of the scattering is typically not explicitly explored.

The time dependence of the scattering is characterized by the correlation function $f_I(t)$

$$f_I(\tau, q) = \langle\, [I(t, q) - I_{av}(q)]\, [(I(t + \tau, q) - I_{av}(q)]\,\rangle \tag{4.3.29}$$

The scattering at a given value of q probes the inhomogeneities in the sample on a length scale $1/q$. For a sample containing colloidal particles, the particle diffusion generates the time dependence of the scattering. In Chapter 1, Eq. (1.7.5), it was found that the characteristic length scale probed during a time τ is proportional to the square root of the diffusion coefficient $D^{1/2}$. A detailed derivation shows that for a diffusive process

$$f_I(\tau, q) = \left\{\left\langle I(t, q)^2\right\rangle - I_{av}(q)^2\right\} \exp(-q^2 D\tau) \tag{4.3.30}$$

The diffusive processes giving rise to the time-dependent scattering involve a relaxation from an instantaneous nonequilibrium state to a more uniform equilibrium distribution. Thus, D in Eq. (4.3.30) is the mutual diffusion coefficient, discussed in Section 1.7. The diffusion coefficient of SDS micelles measured by DLS is substantially larger than the self-diffusion coefficient. This discrepancy is an illustration of the effect mentioned in Section 1.7.2 that counterions tend to pull on charged colloidal particles with the result that the latter shows an anomalously rapid mutual diffusion. As salt is added to the solution, the coupling between the micellar and the counterion diffusion is weakened and D(mutual) approaches D(self).

See Berg (2010, Chapter 5) and Cosgrove (2010, Chapter 13) for additional discussion on Scattering Methods.

4.4 Micelles Is Formed by Surfactants with a Variety of Head Groups and Can Adopt Several Shapes

4.4.1 Surfactants Can Have Ionic, Zwitterionic, or Polar Head Groups

Earlier, we used SDS (sodium dodecyl sulfate) as the typical micelle-forming substance. SDS belongs to a general group of anionic surfactants, to which deprotonated fatty acids, soaps, also belong. Cationic surfactants often have positively charged ammonium head groups as the protonated decyl amine of Figure 4.12. Typical of ionic micellar systems, the CMC is higher than that for nonionic micelle-forming substances. The CMC is furthermore strongly dependent on the electrolyte concentration in the solution as shown in Figure 4.4. Micelle formation occurs as a compromise between the hydrophobicity of the alkyl chains and the hydrophilicity of the head group. The separation to a bulk phase is always an alternative to micelle formation. For ionic surfactants, such a separate phase must contain the counterions and this is best achieved in an ordered solid phase. The longer the alkyl chains, the more easily they crystallize. The Krafft temperature denotes the temperature where the solubility relative to the crystalline phase matches the CMC. For SDS, it is around 23°C, preventing studies of micellization of this surfactant below room temperature. In a crystal, a divalent ion like Ca^{2+} interacts more strongly with the surfactant head group than Na^{+}. The consequence is that in the presence of divalent ions, ionic surfactants tend to precipitate rather than form micelles. One has, for example, to use special soaps when washing in sea water which contains rather high concentrations of divalent cations.

There is a large variety of micelle-forming nonionic surfactants. If the head group is not sufficiently polar, the molecule can separate as a liquid phase. This happens, for example, for long-chain alcohols with only an OH polar group. Table 4.4 shows the more common types of head groups for nonionic micelle formers. A zwitterion ensures strong short-range ion–solvent interactions. For strong dipolar groups, the dipole–dipole interaction with the solvent prevents phase separation. For alkyl sugars, the number of OH groups per head group add up to give micelle formation as the preferred structure. For ethylene oxide surfactants, C_nE_m, each ethylene oxide group is only weakly polar, but the number of units combined with the conformational freedom ensures micellar properties when there is a sufficient number m of ethylene oxide units in the molecule.

For nonionic surfactants the CMC varies in a systematic way with the number, n_C of CH_2-units in the alkyl chain. We can write the free energy of micelle formation as (cf. Eq. 4.1.1)

$$\Delta \mathbf{G}^0_{\text{mic}} = \text{constant} + n_C\,\Delta \mathbf{G}^0_{\text{CH2}} \tag{4.4.1}$$

Table 4.4 CMC values for some noncharged surfactants.

Surfactant	T (°C)	CMC
$C_{10}E_8$	25	1.0×10^{-3}
$C_{11}E_6$	25	3.0×10^{-4}
$C_{12}E_8$	15	9.7×10^{-5}
	25	7.1×10^{-5}
	40	5.8×10^{-5}
$C_{14}E_8$	25	9.0×10^{-6}
$C_{12}E_6$	25	6.8×10^{-5}
α-*d*-Octylglucoside	25	1.0×10^{-2}
β-*d*-Octylglucoside	25	2.5×10^{-2}
β-*d*-Decylglucoside	25	2.2×10^{-3}
β-*d*-Dodecylglucoside	25	1.9×10^{-4}
Sucrose monolaurate	20	1.85×10^{-4}
Dodecyl betaine	25	1.4×10^{-3}

Source: Mukerjee and Mysels, 1971/National Institute of Standards and Technology/Public Domain.

By determining the CMC for a range of surfactants with the same head group but differing chain length, one finds

$$\Delta \mathbf{G}^0_{\mathrm{CH2}} \approx 2.8\ \mathrm{kJ/mol} \tag{4.4.2}$$

This implies that the CMC decreases by a factor of 3.1 for each added CH_2-group. Table 4.4 shows the CMC of some nonionic surfactants and illustrates, for example, the Eq. (4.4.2).

4.4.2 Mixed Micellar Systems Show a Complex Concentration Behavior Around the CMC of the Respective Components

Commercial detergents, like a washing-up liquid or a soap bar, contain a mixture of surfactant molecules and additives. The main function of the detergent is to desorb apolar compounds from surfaces and solubilize them into micelles as well as stabilize particles in the liquid phase. A mixture of surfactants can show a broader spectrum of properties, which is beneficial for the general cleaning performance. There is also a cost aspect in not having to purify the ingredients.

In the detergent solution, the components form mixed micelles. To see the qualitative association properties of a mixed system, consider two surfactants that differ in the length of their alkyl chains but have the same head group. Because the molecules are very similar, we can assume that they mix ideally in the micellar aggregate. However, in the monomer state, the surfactant with the longer chain has a higher chemical potential at equal concentrations. As we increase the concentration of the surfactants, the one with the longer chain will have the highest tendency to form micelle. We can calculate the resulting CMC using the phase separation model. For the two surfactants, index 1 and index 2, the chemical potential in the micellar pseudophase is

$$\mu_i(\mathrm{mic}) = \mu_i^\theta(\mathrm{mic}) + \boldsymbol{RT}\ln X_i \tag{4.4.3}$$

where X_i is the mole fraction of surfactant i in the micelle so that $X_1 + X_2 = 1$. In solution, we have

$$\mu_i(\mathrm{aq}) = \mu_i^\theta(\mathrm{aq}) + \boldsymbol{RT}\ln c_i \tag{4.4.4}$$

Before any micelles have formed,

$$c_i = \alpha_i C_{\mathrm{tot}} \tag{4.4.5}$$

where α_i is the fraction of surfactant i in the original mixture and $\alpha_1 + \alpha_2 = 1$. From the pure systems, we know that an equilibrium between the micellar pseudophase and the solution exists at the CMC so that

$$\mu_i(\mathrm{mic}) = \mu_i^\theta(\mathrm{aq}) + \boldsymbol{RT}\ln \mathrm{CMC}_i \tag{4.4.6}$$

where CMC_i is the CMC of component i. Equating the chemical potentials in the micellar and aqueous phases

$$\mu_i(\mathrm{mic}) = \mu_i(\mathrm{aq}) \quad i = 1, 2 \tag{4.4.7}$$

we can solve for $\mathrm{CMC}_{\mathrm{tot}}$ where micelles first appear and the composition in the micellar aggregates. Combining Eqs. (4.4.3) to (4.4.7), we find

$$\mathrm{CMC}_{\mathrm{tot}} = \left\{ \frac{\alpha}{\mathrm{CMC}_1} + \frac{(1-\alpha)}{\mathrm{CMC}_2} \right\}^{-1} \tag{4.4.8}$$

and

$$X_i = \alpha\{\alpha + (1-\alpha)\mathrm{CMC}_1/\mathrm{CMC}_2\}^{-1} \tag{4.4.9}$$

For example, if $\alpha = 0.5$ and $\text{CMC}_1 = 0.1 \cdot \text{CMC}_2$, then $\text{CMC}_{\text{tot}} = \text{CMC}_1 \cdot 20/11$ and $X_i = 10/11$. In this case, the CMC is reached just below the point where component 1 reaches its CMC. The micellar aggregate contains surfactant 1 as the dominant component. As further surfactant is added, component 1 continues to enter the micelles preferentially so that $c_2(aq)$ increases more with c_{tot} than $c_1(aq)$. For $c_{\text{tot}} \gg \text{CMC}_1, \text{CMC}_2$, the composition of the micellar aggregates must reflect the composition of the mixture so that $X = \alpha$. In this concentration range, $c_i(aq) = \alpha\text{CMC}_i$.

Figure 4.16 illustrates how the monomer concentration varies with the total concentration for a particular case. Clearly, the micelle formation process is much more drawn out in this example than for a pure surfactant, and the properties of the solution change less dramatically at the onset of micelle formation. Furthermore, the monomer concentration of the longer chain surfactant peaks around the CMC and then slowly approaches the asymptotic value αCMC_i. Since we are using the phase separation model, we cannot directly predict aggregate sizes. However, the micelles have a high content of component 1 just above the CMC, and having a longer chain should produce larger micelles. Thus, we also see that the micelles that are formed first are larger than those formed at higher concentrations. This effect is rather subtle, and in cases where component 1 forms large aggregates, such as bilayers, by itself, we can observe large aggregates at low concentrations, which dissolve into small micelles as the concentration increases. The peak in monomer concentration explains the dip in the surface tension of curve B of Figure 4.6 for an SDS solution contaminated with dodecanol.

Fluorocarbon and hydrocarbon surfactants both form micelles. What happens when we mix such surfactants with the same polar head? Hydrocarbons and fluorocarbons do not mix readily, so we definitely have a nonideal mixing in the micelles. We can apply the regular solution model to describe this nonideality. If we use the phase separation model again, we can derive how the CMC is changed in the mixture. Depending on the value of the interaction parameter, w, the regular solution model predicts mixing $w < 2\boldsymbol{RT}$ or phase separation $w > 2\boldsymbol{RT}$.

In the latter case, we should expect two populations of micelles, one rich in hydrocarbon and one rich in fluorocarbon tails. Although the phase separation model gives qualitatively reasonable

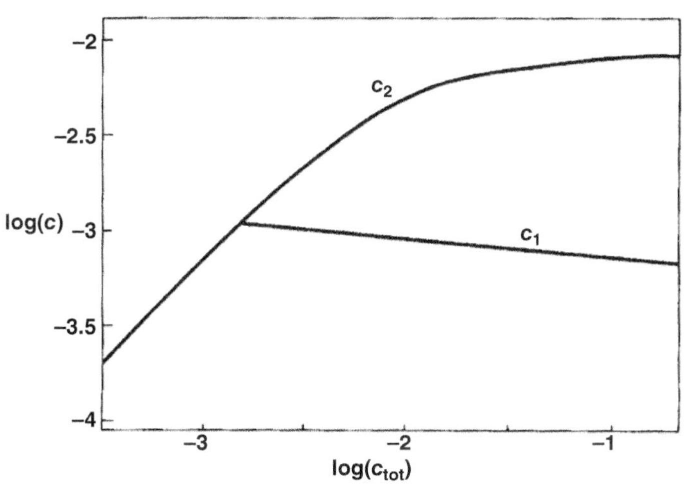

Figure 4.16 The calculated monomer concentrations as a function of the total concentration of an equimolar mixture of two surfactants 1 and 2. For the pure components, we have $\text{CMC}_1 = 1 \times 10^{-3}$ M and $\text{CMC}_2 = 1 \times 10^{-2}$ M. For $c_{\text{tot}} < 1.8 \times 10^{-3}$ M, there are only monomers in the solution and $c_1 = c_2$. At $c_{\text{tot}} = 1.8 \times 10^{-3}$ M, micelles start to form that preferentially contain surfactant 1, whose monomer concentration actually decreases with increasing total concentration for $c_{\text{tot}} > 1.8 \times 10^{-3}$ M.

predictions, it has a serious shortcoming in this application. Typical micellar aggregation numbers are on the order of 50–100, and the micelles cannot be considered as macroscopic systems. The Stirling approximation (see Section 1.1) does not apply, and there will be a distribution of micellar compositions. If the interaction parameter, w, is clearly larger than $2\boldsymbol{RT}$, this distribution function has two peaks, while if $w < 2\boldsymbol{RT}$, the average composition is also the one most likely to occur in the micelles.

Mixing surfactants with different head groups can induce a multitude of effects. If we add a noncharged surfactant to an ionic surfactant system, we can reduce the electrostatic repulsion, and the CMC will decrease below the value we derived for ideal mixing. We can estimate the change in electrostatic free energy on inserting one noncharged monomer into an ionic micelle. The monomer reduces the average charge density and thus lowers the free energy. For typical micellar charges, the dimensionless parameter $s \approx 10$ (see Eq. (3.9.14)) and it follows from Figure 3.11 that the electrostatic free energy per surface charge is around $4k_B\mathbf{T}$. Thus, replacing a charged surfactant with an uncharged surfactant results in a substantial additional stabilization of the micelles.

We can obtain an even more dramatic effect by mixing anionic and cationic surfactants to create so-called catanionic systems. An oppositely charged surfactant is twice as effective in reducing the charge density as a noncharged one. In these mixtures, we can easily produce aggregate growth, bilayer formation, and crystals tend to form in the vicinity of equimolar mixtures.

We get another perspective of electrostatic interactions by considering the reverse situation: adding an ionic surfactant to a noncharged system. Whether this addition increases or decreases, head group repulsion will depend on the details of the short-range interaction between the two types of head groups. However, it is clear that even slightly charging the micelles increases long-range repulsion.

4.4.3 Polymers Can Facilitate Micelle Formation

When an amphiphile is added to a dilute polymer solution, we often observe a cooperative self-assembly process below the CMC of the pure amphiphile solution. A micelle-like aggregate is formed, which interacts with the polymer coil. Formation of these polymer-bound micelles occurs at a critical association concentration (CAC), which varies with the nature of the amphiphile and the polymer.

The micelles associated with the polymer can form a coiled string of beads. However, micelle–micelle repulsion will occur for charged amphiphiles in particular, so each polymer coil can accommodate only a limited number of micelles. When the coils are saturated with micelles, new amphiphiles will go into the solution again until the CMC is reached. Figure 4.17 illustrates this point by a schematic plot of the amphiphile chemical potential versus concentration.

What is the nature of the combined amphiphile polymer aggregate? Table 4.5 lists CACs for a series of amphiphile–polymer pairs. Typically, the CAC is smaller than the CMC by a factor of 10–1000, and in a homologous series, it seems to decrease somewhat faster than the CMC with increasing chain length, particularly when the polymer is a polyelectrolyte of opposite charge.

Combined observations show that the amphiphile aggregate is a micelle, and we can understand that the CAC is lower than the CMC due to a stabilization of the aggregate by interaction with the polymer chain. The nature of this interaction varies from system to system. With SDS and polyethylene oxide, the polymer adsorbs on the micelle surface primarily to reduce hydrocarbon–water contact. Polyelectrolytes will bind micelles of opposite charge electrostatically. For a polyelectrolyte like polystyrene sulfonate, an additional hydrophobic interaction takes place between the styrene part and the interior of the micelle.

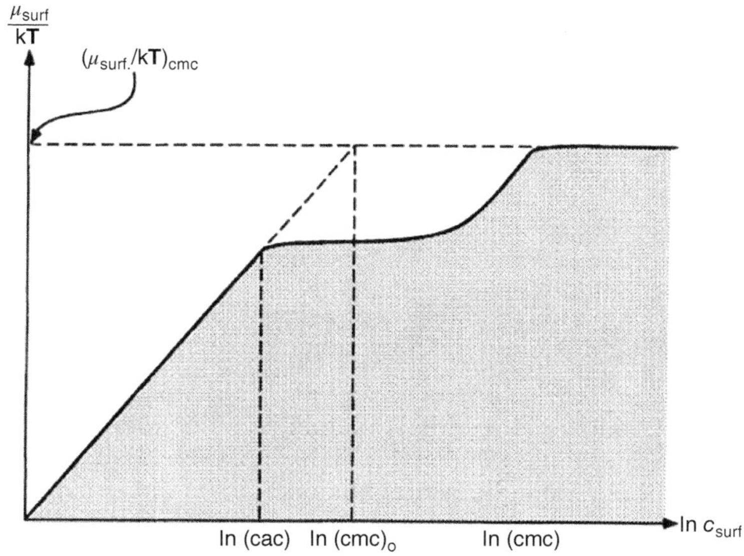

Figure 4.17 Schematic picture of the variation of the surfactant chemical potential with increasing surfactant concentration in a solution containing a constant amount of polymer. At the CAC, surfactants self-assemble on the polymer. At the critical micelle concentration for the pure surfactant $(CMC)_0$, aggregates continue to form on the polymer. The surfactant chemical potential first reaches the value that allows for the formation of free micelles at higher concentrations. This CMC depends on the polymer concentration. *Source*: Kronberg et al., 2014/with permission of John Wiley & Sons.

Table 4.5 Critical association concentration (CAC) for some surfactant polymer systems.

Surfactant	Polymer	CAC (mM)	CMC (mM)
$C_{12}H_{25}OSO_3Na$	Polyethylene oxide	5.7	8.3
	Polyvinylpyrrolidone	2.5	8.3
$C_{10}H_{21}OSO_3Na$	Polyvinylpyrrolidone	10.0	32.0
$C_{12}H_{25}N(CH_3)_3Br$	Sodium hyaluronate	7.0	16.0
	Sodium alginate	0.4	16.0
	Sodium polyacrylate	0.03	16.0
$C_{14}H_{29}N(CH_3)_3Br$	Sodium hyaluronate	0.4	3.8
	Sodium alginate	0.03	3.8
	Sodium polyacrylate	0.0025	3.8

A more specific micelle–polymer interaction can be obtained by using a chemical modification process to add a few long alkyl side chains to the polymer main chain. These side chains will strengthen the polymer–polymer interaction and also serve as a strong binding site for micelles. Dramatic changes in the viscosity signal indicate that the micelles are able to link two polymer chains by incorporating side chains from the different main chains. As a result, a transient polymer network is formed. Such polymers are used as thickeners in paint and in other colloid dispersions for which a high viscosity is desirable.

4.4.4 Micelles Can Grow into Cylinders and Discs

We have stressed that the formation of a spherical micelle is a cooperative process. When and why does the association stop at the spherical shape? The condition that the head group must be at the surface of the aggregate valid for a sphere also applies to a capped cylinder and a disc. To see the conditions when growth might occur, we recall the concept of the surfactant number N_s (Section 1.5). For a spherical micelle to grow as a cylinder, surfactants have to enter the cylindrical part, leaving the (hemispherical) cap unchanged as illustrated in Figure 4.18.

This has two consequences: first, the added monomer n and monomer $n + 1$ involves the same change in the aggregate, which implies non-cooperative isodesmic growth (Section 4.1). Second, the monomer is added to a structure where $N_s = 1/2$ is the optimal value of the surfactant number. Thus, as N_s increases from 1/3 to 1/2, the tendency to grow should increase, but growth should result in an exponentially decreasing aggregate size distribution as in the isodesmic model. We can once more use the SDS system to illustrate this effect. For the ionic surfactants, the electrostatic repulsion between head groups primarily determines the effective area per polar group. When an electrolyte is added, this repulsion is partially screened, and the head group area decreases, leading to an increase in the surfactant number. There is a growth in micellar size on the addition of SDS and at ambient temperatures, micelles contain around 80 monomers at NaCl concentration of 0.2 M.

For a disk-like geometry, the situation is different. In the central part of the disc, the situation is like in a lamellar structure, and the locally preferred surfactant number corresponds to $N_s = 1$. On the rim, the packing is different, corresponding more to $N_s \approx 1/2$. Now, as a monomer is introduced, the length of the rim increases slightly and in such a way that the increase is smaller the larger the aggregation number. There is then a cooperativity in the aggregation process, and there is a large tendency for the system to grow to an infinite size, forming extended bilayers and, in practice, a lamellar liquid crystal. It is difficult to find conditions where, for a one-component system, the surfactant number is large enough to promote the formation of disk-shaped micelles and still small enough to prohibit the cooperative association into a macroscopic aggregate size. However, in a mixed system with one component with $N_s \approx 1$ and one component with $N_s < 1/2$, the association can result in disk-like micelles sometimes called "bicelles," which, for example, can be used to solubilize membrane proteins in a local bilayer-like environment.

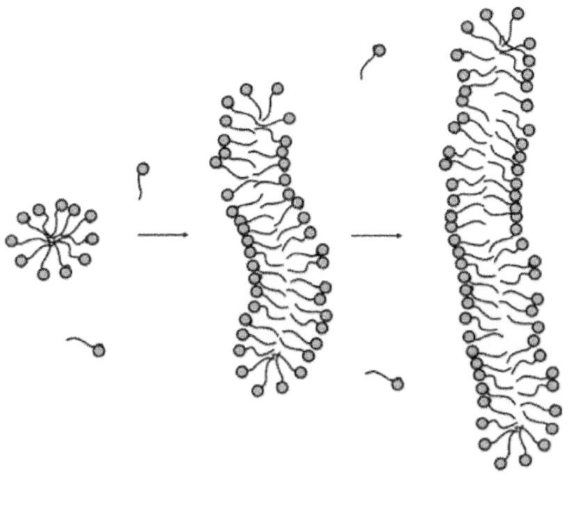

Figure 4.18 When a spherical micelle grows into a cylinder, the semispherical endcaps remain, and monomers enter the cylindrical part with a free energy change that is independent of cylinder length.

4.5 Micelles Are Used to Solubilize Apolar Substances

Solubilization refers to the phenomenon that the presence of micelles leads to a greatly enhanced solubility of an apolar compound. In the typical case, one or a few apolar molecules enter the micelle, but there is no sharp demarcation between solubilization, formation of mixed micelles, and the formation of microemulsions discussed in Chapter 11. Solubilization can be used to an advantage in a wide variety of technological and scientific contexts. We have already observed that it is a central phenomenon in cleaning processes. In water-based inks, solubilization provides one method of keeping dyes in solution. In the formulation of pharmaceuticals, the active (apolar) compound can be dissolved through micelle solubilization. Micelles can be used to solubilize membrane proteins in biochemical studies. The solubilization of oxygen in fluorocarbon-based surfactant micelles can be used as a temporary substitute for blood in emergency situations at catastrophes. Such micellar solutions work equally well as hemoglobin molecules in carrying oxygen, but such a treatment cannot be sustained for a longer period.

4.5.1 Using the Phase Separation Model, We Can Describe How a Solubilizate Partitions Between Micelles and the Bulk Solution

The distribution of an apolar substance between bulk and micelles can be seen as a partition equilibrium similar to the one obtained for the distribution of a molecule between two phases. The ratio between the concentrations in two phases is given by the partition constant K_D. For a concentration of micelles, c_M, the total available solubilization volume is $c_M V_M$, where V_M is the micellar volume. The substrate concentration, c_s, in this volume is then

$$c_s(\text{micelle}) = K_D c_s(\text{bulk}) \tag{4.5.1}$$

and the average number of solubilized molecules, n_s, per micelle is then

$$n_s = c_s(\text{micelle})\, V_M = K_D c_s(\text{bulk})\, V_M \tag{4.5.2}$$

In this description, it is tacitly assumed that n_s is of order unity. When n_s is a substantial fraction of the total aggregation number, one has the case of mixed micelles.

The solubilization association rate is typically diffusion-controlled. In Chapter 1, we derived a general expression for the rate constant of such a reaction and applied it to the case of a small molecule of radius R_{mol} associating with a micelle of radius R_M; the combination of Eqs. (1.7.11) and (1.7.17) yields the rate constant

$$k_+(\text{mol, micelle}) = (2k_B\mathbf{T}/3\eta)R_M/R_{\text{mol}} \tag{4.5.3}$$

Using this condition, the residence time, τ, of a solubilizate in a micelle is

$$\tau = \frac{3\eta K_D V_M R_{\text{mol}}}{2k_B\mathbf{T}R_M} \tag{4.5.4}$$

Here, the reaction radius is assumed to be the micellar radius R_M. It follows that a spectroscopic measurement of the residence time yields an estimate of the distribution coefficient K_D (see Table 4.6).

The larger the aromatic compound, the larger the distribution coefficient K_D reflecting the fact that the hydrophobic interaction increases with increasing molecular size. The CTAB micelles consistently have a K_D-value two to four times that of the SDS micelles.

Table 4.6 Measured residence times, τ, of aromatic fluorescent molecules in SDS and CTAB $\left(C_{16}H_{33}N(CH_3)^{+}Br^{-}\right)$ micelles and calculated distribution coefficients assuming a diffusion-controlled association reaction.

Probe	τ (SDS)	τ (CTAB)	K_D (SDS)	K_D (CTAB)
Benzene	0.23	1.3	120	430
Naphthalene	4	13	1600	3400
Biphenyl	10	62	4100	16×10^3
Anthracene	59	303	21×10^3	69×10^3
Pyrene	243	588	80×10^3	12×10^4

Source: Adapted from Almgren, 1979.

4.5.2 Micellar Catalysis Exploits the Large Surface Areas Associated with Micelles and Also Illustrates the Graham Equation

In micellar catalysis, we use the special properties of micellar solutions to facilitate chemical reactions. Typically, one of the reactants (e.g., an ion) is highly polar, while the other is apolar and sparingly soluble in water. Under such circumstances, finding a suitable solvent for performing the reaction can be difficult.

In a micellar solution, however, the apolar reactant can be solubilized in the micelles while the polar one is dissolved in the aqueous region. The two reactants can meet at the micellar surface, where the chemical transformation occurs. An ester hydrolysis by acid can serve to illustrate the principle:

$$\mathrm{R\overset{\overset{\displaystyle O}{\|}}{C}{-}O{-}R' + H_2O \xrightarrow[k^2]{H^+} R\overset{\overset{\displaystyle O}{\|}}{C}{-}OH + R'OH} \tag{4.5.5}$$

What type of surfactant should be used to catalyze this reaction? To maximize the local concentration of both ester and H^+ in the same region, we can use an anionic surfactant because the H^+ ions will be attracted to the micellar surface electrostatically. The ester is solubilized with the relatively polar ester group localized at the micellar surface.

Figure 4.19 shows how the rate constant varies with surfactant concentration. A dramatic increase in the rate constant occurs at the CMC, with an enhancement factor of approximately 50. This increase illustrates one of the principles behind enzymatic catalysis. Simply by positioning the two reactants in close proximity, we can substantially enhance the rate of a bimolecular reaction.

A feature of Figure 4.19 is that we observe the maximum rate of ester hydrolysis around the CMC, and adding more surfactant leads to a decrease in the rate. At first sight, this rate decrease seems counterintuitive, but it can be readily explained by using the Graham equation (3.8.26). Surface charge density determines the counterion concentration at the highly charged surface. Assuming no chemical preference, with a CMC of 8 mM and a pH of 6, implying a bulk concentration of H^+ of 1×10^{-6} M, the ratio of surface concentration of Na^+ to H^+ is 8×10^3 M. The micelle has approximately one charge per 0.75 nm^2. According to the Graham equation, this yields a total counterion surface concentration of 9.3 M. We can obtain the concentration of H^+ at the surface as

$$[H^+]_{\text{surface}} = \frac{1}{8 \times 10^3} \times 9.3 = 1.2\ \text{mM} \tag{4.5.6}$$

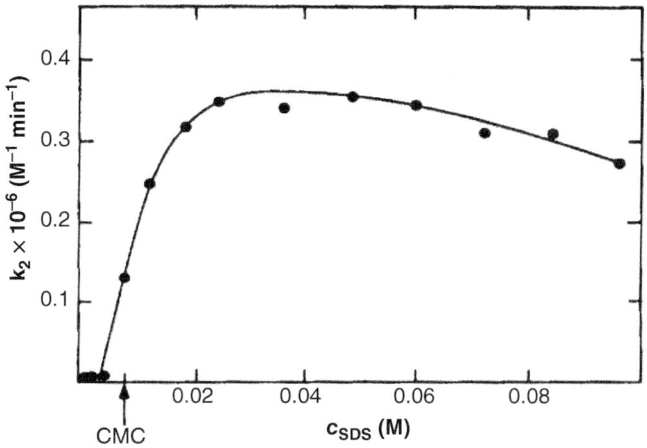

Figure 4.19 Measured second-order rate constant k_2 for the acidic hydrolysis of methyl orthobenzoate in an aqueous solution at 25 °C as a function of the concentration of SDS (CMC = 8 mM). *Source*: Dunlap and Cordes, 1968/with permission of American Chemical Society.

which is 300 times higher than the bulk concentration.

As more surfactant is added, the concentration of the Na^+ counterion increases. It becomes more and more difficult for the H^+ to compete, and its surface concentration gradually decreases, leading to a decrease in the overall rate constant. The changes in the rate constant shown in Figure 4.19 are not caused by some change in local chemistry; they are due simply to changes in the concentrations of H^+ at the location of the ester.

4.5.3 Digestion of Fats Requires Solubilization by Bile Salt Micelles

Digestion of fats provides a major source of energy for basic metabolic processes. Figure 4.20 shows how enterohepatic circulation transports bile salts, which act as the body's detergent, through the digestive tract. This process illustrates several interesting features of solubilization and the dynamic behavior of micellar solutions.

The liver synthesizes bile salts from cholesterol and excretes them into the common bile duct as mixed micelles, also containing lecithin and cholesterol. During periods of fasting, bile is concentrated by a factor of 10 to become a 30 wt% solution stored in the gallbladder. Figure 4.20 illustrates seven steps of the fat digestion process:

Step 1: Fats from the stomach—in the form of triglycerides with hydrocarbon chains ($n = 16$–20)—are squirted into the duodenum, the upper part of the small intestine, and emulsified by mechanical aggregation.

Step 2: The presence of fats in the duodenum stimulates the release of pancreatic enzymes that hydrolyze the triglycerides to fatty acids and 2-monoglycerides. Simultaneously, the gall bladder expels bile into the intestine.

Step 3: Because fatty acids are insoluble in water at low pH, they combine with the bile salts to form mixed micelles containing fatty acids and 2-monoglyceride lipids.

Step 4: Diffusion and convection transport the mixed micelles to the walls of the intestine.

Step 5: The lipids adsorb into the bilayer membranes of the epithelial cells and diffuse across the intestinal wall, where enzymes catalytically reesterify the fatty acids and 2-monoglycerides to triglycerides. Bile salts, which are more polar, are not adsorbed in this process.

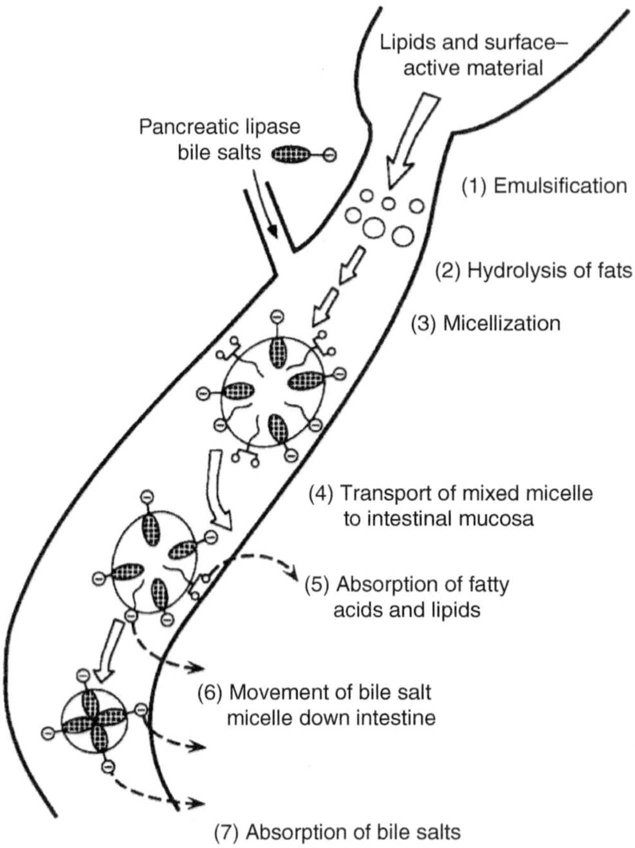

Figure 4.20 Schematic illustration of the sequential steps in the solubilization and digestion of fats in the stomach and upper intestine.

Step 6: Bile salts continue to shuttle hydrolyzed lipids to the intestinal wall as food moves down the intestine.

Step 7: In the ileum, the last segment of the small intestine, bile salts are adsorbed by an active transport and flow to the liver via the portal vein. In the liver, they are separated from the blood and returned to the common bile duct. If food remains to be digested, bile salts flow back into the intestine. Otherwise, they are stored in the gallbladder.

Concentrated detergents disrupt membranes. So how does the gallbladder tolerate concentrated bile salt solutions? The body's trick is to lower the activity of the bile salts by forming mixed micelles containing lecithin and cholesterol. Often, bile becomes supersaturated with cholesterol, which then precipitates out to form cholesterol gallstones, a disease that affects a substantial fraction of the adult population. Since gallstones can be dissolved under some conditions by feeding the patient selected bile salts, the micellar dissolution mechanism is interesting also from a medical standpoint.

See Kronberg et al. (2014, Chapters 6 and 13) for additional discussion on solubilization.

Exercises

4.1 Calculate the equilibrium constant K_N in the closed association model for a system with (a) CMC = 1.0 M and $N = 10$ and (b) CMC = 8.0 mM and $N = 60$. Estimate the error in the approximation used in Eq. (4.1.13) to relate the CMC and $\Delta G^{\theta}_{\text{mic}}$. Use Eq. (4.1.11) to identify the CMC.

4.2 In the isodesmic model of stepwise aggregation

$$[S_n] = [S][S_{n-1}]K$$

a. Show that the total surfactant concentration is

$$[S]_{\text{tot}} = \sum_{1}^{\infty} n[S_n] = \frac{[S]}{(1 - [S]K)^2}$$

Hint: Use the formulas for a geometric series repeatedly.

b. Show that in the limit of $K[S]_{\text{tot}} \gg 1$, it follows that

$$K[S] \approx 1 - \{K[S]_{\text{tot}}\}^{-1/2}$$

4.3 We can combine the closed-association model and the isodesmic model by stipulating that for $n \leq N$, the equilibrium can be described by $NS \leftrightarrow S_N$ with an equilibrium constant K_N, while for $n > N$, one has $S + S_n \Leftrightarrow S_{n+1}$ with equilibrium constant K for $n > N$. Show that

$$[S]_{\text{tot}} = [S] + \frac{NK_N[S]^N(1 - [S]K + [S]K/N)}{(1 - [S]K)^2}$$

4.4 In Exercise 2.7 of Chapter 2, we gave the concentration dependence of the surface tension for dodecyldimethylammonium chloride solutions with no salt and with 0.20 M NaCl. Determine the CMC for the two cases.

4.5 Use the Langmuir adsorption isotherm model to calculate the adsorption of SDS to a solid surface from a solution with excess electrolyte where the CMC is 1.0 mM. Assume the closed association model for SDS in the bulk. The adsorption equilibrium constant is $K = 1.2 \times 10^2\ \text{M}^{-1}$.

4.6 In pressure-jump experiments on micellar solutions of sodium hexadecyl sulfate (SHS) at 45 °C, one observes a fast relaxation process with a time constant that varies with concentration as follows:

Concentration SHS (mM)	τ_1 (μs)	Concentration SHS (mM)	τ_1 (μs)
0.58	807	1.0	365
0.66	665	1.1	325
0.70	603	2.0	114
0.90	430	5.0	38

The CMC is 0.52 mM and the aggregation number $N = 100$. Determine the rate constant k_N^- for a monomer leaving the micelle. Estimate also the standard deviation σ of the micelle size distribution. What is the association rate constant k_N^+? Calculate the average lifetime of a monomer in the micelle.

4.7 The aggregate size of potassium dodecanoate micelles in an aqueous solution containing 27% (w/w) dodecanoate was determined using time-resolved fluorescence. The molar ratio of dodecanoate to fluorescent pyrene was $93 \times 10^3 : 1$. The fluorescence decayed exponentially with a time constant $\tau_0 = 200$ ns. In a second sample, a quencher (benzophenone) was added at a 65:1 molar ratio of dodecanoate to quencher. The fluorescence at long times also decayed exponentially in this sample with the same time constant $\tau_0 = 200$ ns, while the decay was faster at short times. Extrapolating the long-time behavior back to $t = 0$ resulted in an apparent zero time intensity of 0.33 times the value in the absence of the quencher. Calculate the mean aggregate size.

4.8 The table gives the measured CMC of an SDS aqueous solution as a function of the concentration of sodium chloride in the solution. The table also gives the measured mean aggregation numbers.

C_{NaC1} (M)	CMC (mM)	N
0.00	8.1	58
0.01	5.7	64
0.03	3.1	71
0.10	1.5	93
0.30	0.71	123

Calculate for all five systems:

a. The volume of the hydrocarbon core.
b. The effective radius of the hydrocarbon core, assuming spherical geometry.
c. The cross-sectional area per chain at the aggregate surface (spherical geometry).
d. The area per charge, assuming it is located 0.3 mM outside the hydrocarbon core.

4.9 Light-scattering measurements on tetradecyltrimethylammonium bromide solutions at 25 °C give the following Rayleigh ratios, R_θ:

H_2O $K_0 = 3.7 \times 10^{-5}$ m^2/kg^2		0.05MKBr $K_0 = 3.7 \times 10^{-5}$ m^2/kg^2	
10^3C (g/mL)	10^4R_θ m^{-1}	10^3C (g/mL)	10^4R_θ m^{-1}
0.406	2.318	0.0275	2.292
0.910	2.347	0.0556	2.332
1.298	2.788	0.0857	2.194
1.426	3.702	0.173	2.332

H_2O $K_0 = 3.7 \times 10^{-5}$ m^2/kg^2		0.05MKBr $K_0 = 3.7 \times 10^{-5}$ m^2/kg^2	
10^3C (g/mL)	10^4R_θ m^{-1}	10^3C (g/mL)	10^4R_θ m^{-1}
2.116	7.428	0.291	3.634
3.764	10.665	0.436	5.599
4.534	11.270	0.829	11.195
5.962	12.500	1.406	18.527
6.526	12.982	1.917	25.270
8.408	14.331	2.498	31.671
		3.010	37.410

Beesley, 1990/University of Minnesota.

a. Using these data, verify that in water the CMC is 1.2 kg/m^3 and the Rayleigh ratio at the CMC is $R_\theta = 2.3 \times 10^{-4}$ m^{-1} and in 0.05M KBr the CMC is 0.20 kg/m^3 and $R_\theta = 2.3 \times 10^{-4}$ m^{-1}.

b. Calculate the mean aggregation number N and second virial coefficient B_2. How does the value of B_2 compare to the estimated volume of a micelle?

c. Explain qualitatively why N and B_2 change in the way they do upon the addition of salt.

4.10 Use the Stokes–Einstein relation Eq. (1.7.17) to estimate the self-diffusion coefficient of decylammonium micelles in water at **T** = 298 K. Compare this value with the measured D for the hydrophobic substance TMS shown in Figure 4.9.

4.11 Determine the coefficient R_g in the series expansion $P(q) = 1 - 3R_g^2q^2 +$ of the expression for $P(q)$ in Eq. (4.3.22).

4.12 Use the data of Table 4.4 to estimate the CMC of the surfactant $C_{12}E_5$ at 25°C.

References

Almgren, M., Grieser, F., and Thomas, J.K. (1979). J. *Am. Chem.* Soc. 101, 279.

Aniansson, E.A.G., Wall, S.N., and Almgren, M., Hoffmann, H., Kielmann, I., Ulbricht, W., Zana, R., Lang, J., and Tondre, C. (1976). *J. Phys. Chem.* 80, 905.

Beesley, A.H. (1990). Ph.D. Thesis, University of Minnesota.

Berg, J.C. (2010). *An Introduction to Interfaces & Colloids. The Bridge to Nanoscience.* World Scientific.

Cabane B., Duplessix, R., Zemb, T. (1985). *J Physique* 46, 2161.

Cosgrove, T. (2010). *Colloid Science. Principles, Methods and Applications.* Chichester, UK: Wiley.

Doi, M. (2013). *Soft Matter Physics.* Oxford, UK: Oxford University Press.

Dunlap, R.B. and Cordes, E.H. (1968). *J. Am. Chem. Soc.* 90, 4395.

Elworthy, P. and Mysels, K. (1966). *J. Colloid Interface Sci.* 21, 339.

Grieser, F. and Tauschtreml, R. (1980). *J. Am. Chem. Soc.* 102, 7258.
Gunnarsson, G. (1979). Thesis, Lund University.
Gunnarsson, G., Jonsson, B., and Wennerström, H. (1980). *J. Phys. Chem.* 84, 3114.
Israelachvili, J.N. (2011). *Intermolecular and Surface Forces*, 3rd ed. London, UK: Academic Press.
Kale, K.M., Cussler, E.L., and Evans, D.F. (1980). *J. Phys. Chem.* 84, 59.
Kronberg, B., Holmberg, K., and Lindman, B. (2014). *Surface Chemistry of Surfactants and Polymers*, 3rd ed. Chichester, UK: Wiley.
Lindman, B. and Stilbs, P. (1985). *Proceedings of Enrico Fermi School of Physics*, C.V. Degiorgio and M. Corti, eds., North-Holland Physics Publishing, Amsterdam.
Lindman, B. and Wennerström, H. (1980). *Current Topics in Chemistry*, Springer-Verlag, Berlin.
Lyklema, H. ed. (2005). *Fundamentals of Interface and Colloid Science*. Soft Colloids, *Vol. 5*. Amsterdam, The Netherlands: Elsevier.
Mukerjee, P. and Mysels, K.J. (1971). *CMC of Aqueous Surfactant Systems*, National Standards Reference Data Service, U.S. National Bureau of Standards.
Pilsl, H., Hoffmann, H., Hoffmann, S., Kalus, J., Kencono, A.W., Lindner, P., and Ulbricht, W. (1993). *J. Phys. Chem.* 97, 2745.
Stilbs, P. and Lindman, B. (1981). *J. Phys. Chem.* 85, 2587.
Wennerström, H. (2014). *Adv. Colloid and Interface Sci.* 205, 105–112, 2014.

5

Forces in Colloidal Systems

Concept Map

Forces in Colloidal Systems
Describing colloidal interaction forces at a molecular level remains an unsolved challenge. Because the distances involved in colloidal systems are often large compared to molecular sizes, we can often average over molecular structures. In this chapter, we primarily consider the force between parallel surfaces in a liquid, and then we can relate the force per area, A, to the osmotic pressure of the solvent in the space between the surfaces $F/A = \prod_{\text{osm}}$ (Eq. 5.1.2).

Electrostatic Double-layer Forces

- The force per area on a hard wall is determined by the contact density and $\frac{F}{A} = k_B \mathbf{T} \sum_i c_i(h)$ Eq. (5.1.14). When applied to the interaction between a charged and a neutral surface (Figure 5.2) the force is determined by the ion density at the neutral wall (Eq. 5.1.17), which reflects the fact that at the neutral surface with zero field, it is only the entropy of mixing that determines the free energy. When the two surfaces move together, the ions are forced into a smaller space, decreasing their entropy and giving rise to a repulsive interaction force.
- Two surfaces with identical charges (Figure 5.3) repel each other when their double layers overlap. By symmetry, the field is zero at the midplane and the force is as for the case with one neutral wall, provided one interprets h as the position of the midplane (Eq. 5.1.29).
- To evaluate the force, we need to solve the Poisson–Boltzmann equation, and for the counterion-only case, there exists an analytical solution giving $\frac{F}{A} = \frac{2(k_B \mathbf{T})^2 \varepsilon_r \varepsilon_0}{(ze)^2} (s/h)^2$ (Eq. 5.1.24).
- For a medium containing electrolyte ions, a useful approximate expression for the force is $\frac{F}{A} = 64 k_B \mathbf{T} c_0^* \Gamma_0^2 \exp(-\kappa h)$ (Eq. 5.1.36), which shows that (1) the force decays

(Continued)

The Colloidal Domain: Where Physics, Chemistry, Biology, and Technology Meet, Third Edition. Håkan Wennerström.

Electrostatic Double-layer Forces (Continued)

exponentially with separation h; (2) the spatial decay is determined by Debye length κ^{-1}; and (3) the surface potential enters as $\Gamma_0 = \tanh\left(\frac{ze\Phi_0}{4k_B\mathbf{T}}\right)$ so that when $\Phi_0 < \frac{k_B\mathbf{T}}{ze}$, $\Gamma_0 \approx ze\Phi/(4k_B\mathbf{T})$ and when $\Phi_0 \gg \frac{k_B\mathbf{T}}{ze}$, $\Gamma_0 \approx 1$ and the force becomes independent of Φ_0.

- Two surfaces with equal signs but different magnitudes always repel each other. When two surfaces bearing opposite signs move together, a long-range attraction can change into a repulsion at shorter separations.

London Dispersion Forces: Induced Dipole–Induced Dipole Interactions

The motion of electrons in atoms and molecules creates instantaneous dipoles that generate a field at a second atom or molecule, which in turn induces an instantaneous dipole. Quantum mechanics shows that this interaction leads to an attractive force $V_{12}^{\text{disp}} = -C_{12}/R_{12}^6$ (Eq. 5.2.1).

- The coefficient C_{12} depends on polarizabilities and ionization potentials (Eq. 5.2.3).
- London dispersion, Keesom, and Debye forces combine to constitute the attractive van der Waals force.

London Dispersion Forces Between Colloidal Particles

Two macroscopic plates separated by a slit of width h (Figure 5.6) serve as a model.

- Assuming pairwise additivity, we can add all interactions between plates (separated by a vacuum) to obtain $\mathbf{U}_{12}/A \approx -H_{12}/(12\pi h^2)$ (Eq. 5.2.7), where $H_{12} = \pi^2 C_{12}\rho_1\rho_2$ is the Hamaker constant and ρ represents the density of material. This summation transforms a short-range interaction between atoms decaying as $1/R^6$ into a long-range interaction decaying as $1/h^2$.
- Bringing two identical surfaces into contact measures the work of cohesion and thus relates dispersion energies to surface tension, as illustrated by Eq. (5.2.12).
- When the solvent fills the slit, we can obtain an effective Hamaker constant by considering the interaction energy between plates of media 1 and 3 separated by a liquid 2 (Figure 5.11):
- For identical plates, $H_{12} \cong (\sqrt{H_{11}} - \sqrt{H_{22}})^2$, where H_{11} and H_{22} are the values for individual materials. This interaction is always attractive, but greatly diminished compared to the interactions in a vacuum.
- For dissimilar plates, $H_{123} \cong \left\{\sqrt{H_{22}} - \sqrt{H_{11}}\right\}\left\{\sqrt{H_{22}} - \sqrt{H_{33}}\right\}$ resulting in attractive or repulsive interactions, depending on the magnitudes of the individual Hamaker constants.
- Lifshitz theory, which is based on a continuum electrodynamics point of view, provides an alternative and more general way to calculate attractive interactions.

Attractive Forces Generated by Correlated Electrostatic Interactions

Examining interactions between ionic or dipolar groups at a level beyond the mean field approximation shows that correlations can generate attractive forces.

- Lateral inhomogeneities generated in the distribution of ions located between two similarly charged surfaces induce the formation of instantaneous fields. With divalent ions at highly charged surfaces, these fields generate an attractive component to the force, which can overwhelm repulsive double-layer forces (Figure 5.14).
- When two surfaces containing dipolar groups approach, correlations lead to attractive forces. The thermally averaged force varies with a distance dependence of h^{-5} (Eq. 5.3.1); while for a rigid lattice, the force decays exponentially (Eq. 5.3.2).
- When a surface containing a nonuniform charge distribution is formed, for example, by adsorbed micelles or polyelectrolytes or surface phase domains, an electric field is generated that extends into the adjacent solution. This field can induce a correlated response from a second surface, giving rise to a long-range attractive force.

The Derjaguin Approximation

The Derjaguin approximation relates the force $F_{12}(h)$ between curved surfaces to the interaction free energy $\mathbf{G}_{12}(h)/A$ between flat surfaces, provided the radius of curvature R is larger than the range of the interactions.

- The relations between $F_{12}(h)$ and $\mathbf{G}_{12}(h)/A$ for a sphere and a plate (Eq. 5.4.5), two crossed cylinders (Eq. 5.4.6a), and two spheres (Eq. 5.4.6b) involve products of the radii of curvature, R_1 and R_2.

Solid Surfaces—Forces Arising from Density Variations

When two surfaces approach, their molecular components play an increasingly important role in determining interaction forces.

- Molecularly flat solid surfaces show a strong adhesion due to dispersion. However, most solids have rough surfaces, strongly reducing the net interaction.
- The surface force apparatus can accurately measure forces between solid surfaces based on mica. The atomic force microscope (AFM) provides a versatile, but less accurate way of measuring forces between colloidal objects.
- Surface-induced molecular variations in solvent density lead to an oscillatory force when two smooth surfaces approach (Eq. 5.5.2 and Figure 5.21).
- When a solute does not adsorb on a surface, it becomes excluded between two closely separated surfaces. The resulting difference in osmotic pressure generates an attractive force, typically called the depletion force (Eq. 5.5.3).

(Continued)

Solid Surfaces—Forces Arising from Density Variations (Continued)

- Solute adsorption leads to a net repulsive contribution to the interactions as surfaces approach contact. The long-range character of the force depends on the properties of the adsorbed molecules.
- Capillary-induced phase separation (CIPS) occurs when two surfaces approach and induce a transition to a phase with lower surface energy, as illustrated in Figure 5.23. Such transitions generate attractive forces (Eqs. 5.5.10 and 5.5.16).

Fluid-like Surfaces—Forces Arising from Entropic Interactions

With liquid interfaces, entropic effects strongly influence the interaction forces.

- When two surfaces covered with polymer approach, polymer tails reaching out into the solution impede the configurational freedom of the polymers attached to the other surface. This reduction in configurational entropy generates a repulsive overlap force which can overwhelm both van der Waals and double-layer forces.
- Flexible bilayers undergo undulations, which are damped out when bilayers approach. This creates a repulsive force of entropic origin, which varies as h^{-3} (Eq. 5.6.1).
- When phospholipid bilayers approach, a short-range repulsive force (Figure 5.27) is observed at 2–3 nm that increases exponentially with decreasing distance of approach.

The Hydrophobic Interaction

- Because of its technological and biological importance, water is the most well-characterized solvent with a strong internal cohesion.
- The hydrophobic interaction provides an example of a solvophobic interaction, which is also present in other solvents with strong cohesion like ionic liquids.
- To accommodate an apolar solute, water molecules engage in more hydrogen bonding around the solute. This effect, which is more pronounced at lower temperatures, makes water a less poor solvent for apolar molecules.
- Apolar molecules attract in water with an effective potential that has both a primary and a secondary minimum (Figure 5.28).
- The secondary minimum gives rise to the formation of clathrates, in which water molecules surround an apolar molecule in a crystalline state.
- The strength of the hydrophobic interaction decreases with decreasing temperature.
- Macroscopic, purely hydrophobic surfaces represent a high free energy state, and the interaction between them gives rise to a capillary-induced vaporization at very long range. The interaction between surfaces with hydrophobic patches has a range that corresponds to molecular sizes.
- Careful attention should be exercised to distinguish between the terms *hydrophobic hydration, hydrophobic interaction,* and *hydrophobic effect.*

Hydrodynamic Interactions

- When particles move past one another, hydrodynamic interactions create flow in the surrounding liquid (Eq. 5.8.6), which in turn affects the motion of other particles.
- When two particles move toward one another in a medium, fluid flow generates a repulsive force (Eq. 5.8.8). This is called the lubrication force.

Many practical applications of colloid chemistry revolve around the problem of controlling the forces between colloidal particles, between particles and surfaces, or between two surfaces. Consequently, colloid scientists have devoted considerable experimental and theoretical effort to resolving this problem. Forces between mesoscopic objects result from a complex interplay between many interacting molecules or atoms, both in the particles and in the media separating them. Thus, the total force depends on intermolecular forces of all the types discussed in the introduction to Chapter 3.

In a colloidal system, the interplay between forces is so intricate that describing it in full molecular detail constitutes a major challenge that has not yet been resolved. However, most distances involved in the colloidal domain are large compared to molecular sizes, and this fact allows us to partially average over the local molecular structure. If we execute such a scheme in a suitable way, forces of different types between colloidal particles emerge naturally. In this chapter, we will discuss these forces separately and then return to the problem of the effects of their combined action in subsequent chapters.

The force F between two surfaces or particles at a separation h and a constant external pressure is determined by the change in Gibbs free energy $\mathbf{G}$ with separation

$$F = -\left(\frac{\partial \mathbf{G}}{\partial h}\right)_{\mathrm{T}} = -\left(\frac{\partial \mathbf{H}}{\partial h}\right)_{\mathrm{T}} + \mathbf{T}\left(\frac{\partial \mathbf{S}}{\partial h}\right)_{\mathrm{T}} \tag{5.1.1}$$

The sign convention is such that the repulsive force is positive. As discussed already in Chapter 3, we usually consider condensed phases to be incompressible, in which case the force in Eq. (5.1.1) could equally well be related to the derivative of Helmholtz free energy, and we can then replace the enthalpy $\mathbf{H}$ by the energy $\mathbf{U}$ in the right-hand side of the equation.

The conceptually most transparent illustration of colloidal forces is obtained by considering the case of two parallel planar surfaces. In this case, our main interest is the force per unit area, which can be obtained by dividing all terms in Eq. (5.1.1) by the area. For this case, F/area has the dimension of a pressure and is sometimes termed the *disjoining pressure*, but we will not use this terminology further.

If the two surfaces are separated by vacuum, or more realistically by a low-density gas, we can interpret Eq. (5.1.1) as giving the experimentally relevant force directly. However, in most applications, the surfaces are separated by a liquid medium as shown in Figure 5.2. As the two surfaces approach each other, the liquid must leave the region between the surfaces. For most systems, we can assume that the liquid is practically incompressible. Then, we can rewrite Eq. (5.1.1) for planar parallel surfaces as an equation for the force per unit area, A

$$\frac{F}{A} = -\frac{1}{A}\left(\frac{\partial \mathbf{G}}{\partial h}\right)_{\mathrm{T}} = -\left(\frac{\partial \mathbf{G}}{\partial V}\right)_{\mathrm{T}} = -\frac{1}{V_s}\left(\frac{\partial \mathbf{G}}{\partial n_s}\right)_{\mathrm{T}} = -\frac{1}{V_S}(\mu_S - \mu_S^{\theta}) = \Pi_{\mathrm{osm}} \tag{5.1.2}$$

where using Eq. (1.2.12), we see that the force per unit area is simply the osmotic pressure of the solvent between the surfaces.

The force in Eq. (5.1.2) represents the case when the liquid is expelled into a reservoir of pure liquid at the same temperature. If it ends up in a different state, transferring the pure solvent to a solution, vapor, or whatever the final state is makes an extra contribution to the change in free energy. However, the key quantity to calculate to obtain the net force is

$$\left(\frac{\partial \mathbf{G}}{\partial h}\right)_{\mathrm{T}}$$

in Eq. (5.1.2) or equivalently, since we are dealing with an incompressible system,

$$\left(\frac{\partial \mathbf{A}}{\partial h}\right)_{\mathrm{T}}$$

In calculating this derivative, we will make an effort to always directly connect the intermolecular interactions that combine into this thermodynamic quantity.

5.1 Electrostatic Double-layer Forces Are Long-ranged

5.1.1 The Density at a Hard Wall Reports on the Pressure on the Wall

To arrive at a convenient expression for the electrostatic interaction between two surfaces, we first develop a general expression for the force exerted by an incompressible liquid on two parallel confining walls. A subtle consequence of the incompressibility assumption is that there is a mapping between a two-component liquid and a single-component gas. In this case, the volume of the gas corresponds to the volume of the solvent, and the concentration of the solute corresponds to the density of the gas. The point of this mapping is that one can use results from the statistical mechanics of classical gases, which have been thoroughly worked out theoretically. In Chapter 1, we pointed out that the average translational kinetic energy per molecule is $3k_B\mathbf{T}/2$ for both gases and liquids. The fact that the kinetic energy is completely determined by the temperature and is thus independent of the structure of the material reflects the general circumstance that, for classical systems which have thermally excited degrees of freedom, the dynamic and positional degrees of freedom are mutually independent. A central result from classical statistical mechanics is that we can write the Helmholtz free energy $\mathbf{A}$ associated with the positional coordinates as

$$\mathbf{A} = -k_{\mathrm{B}}\mathbf{T}\ln Z \tag{5.1.3}$$

where Z is the configurational integral, defined as

$$Z = \int \exp\left(-\frac{U(\boldsymbol{r}_1,\ldots\boldsymbol{r}_N)}{k_{\mathrm{B}}\mathbf{T}}\right)\mathrm{d}\boldsymbol{r}_1\ldots\mathrm{d}\boldsymbol{r}_N \tag{5.1.4}$$

Here $U(\boldsymbol{r}_1,\boldsymbol{r}_2,\ldots\boldsymbol{r}_N)$ encodes the interaction energy between all N particles in the system. The average energy can then be obtained as a statistical–mechanical average according to (c.f. Eq. 3.3.1)

$$\langle U\rangle = Z^{-1}\int U\exp\left(-\frac{U(\boldsymbol{r}_1,\ldots\boldsymbol{r}_N)}{k_B\mathbf{T}}\right)\mathrm{d}\boldsymbol{r}_1\ldots\mathrm{d}\boldsymbol{r}_N \tag{5.1.5}$$

The concentration $c_i^*(\boldsymbol{r})$ of particle i in the point $\boldsymbol{r}$ is similarly given by the average

$$c_i^*(\boldsymbol{r}) = Z^{-1}\int \delta(\boldsymbol{r}-\boldsymbol{r}_i)\,\exp\left(-\frac{U(\boldsymbol{r}_1,\ldots\boldsymbol{r}_N)}{k_B\mathbf{T}}\right)\mathrm{d}\boldsymbol{r}_1\ldots\mathrm{d}\boldsymbol{r}_N \tag{5.1.6}$$

where $\delta(\boldsymbol{r})$ is the three-dimensional Dirac delta function. We can now perform the integral over $\boldsymbol{r}_i$ to yield

$$c_i^*(\boldsymbol{r}) = Z^{-1} \int \exp\left(-\frac{U(\boldsymbol{r}_i = \boldsymbol{r})}{k_B \mathbf{T}} \right) \mathrm{d}\boldsymbol{r}_1 \ldots \mathrm{d}\boldsymbol{r}_{i-1}\, \mathrm{d}\boldsymbol{r}_{i+1} \ldots \mathrm{d}\boldsymbol{r}_N \tag{5.1.6b}$$

Since all particles are identical, the average concentration at a given position is the same for each particle, and the total concentration $c^*(\boldsymbol{r})$ is simply

$$c^*(\boldsymbol{r}) = N c_i^*(\boldsymbol{r}) \tag{5.1.7}$$

In evaluating Z from Eq. (5.1.4), we use that the system is isotropic, and we can thus choose the position $\boldsymbol{r}_i$ of particle i arbitrarily and integrate over it, yielding a prefactor V:

$$Z = V \int \exp\left(-\frac{U(\boldsymbol{r}_i = \boldsymbol{r})}{k_B \mathbf{T}} \right) \mathrm{d}\boldsymbol{r}_1 \ldots \mathrm{d}\boldsymbol{r}_{i-1}\, \mathrm{d}\boldsymbol{r}_{i+1} \ldots \mathrm{d}\boldsymbol{r}_N \tag{5.1.4b}$$

Here, the positions of the remaining $N - 1$ particles are given relative to the position of particle i. The integral in Eq. (5.1.4b) exactly matches the integral in (5.1.6b), and the net result is thus the expected homogeneous density

$$c^*(\boldsymbol{r}) = \frac{N}{V} \tag{5.1.8}$$

In this chapter, we will be interested in inhomogeneous systems, where the liquid phase is confined by surfaces or interfaces. A simple, but very useful, case is when the confinement is due to two parallel walls at $x = 0$ and $x = h$. In the case when the walls do not interact with the liquid except for a hard wall confinement, the configuration integral is now most conveniently written in Cartesian coordinates $\boldsymbol{r}_i = (x_i, y_i, z_i)$:

$$Z(h) = \iint_{-\infty}^{\infty} \int_0^h \exp\left(-\frac{U(x_1, y_1, z_1 \ldots x_N, y_N, z_N)}{k_B \mathbf{T}} \right) \mathrm{d}x_1\, \mathrm{d}y_1\, \mathrm{d}z_1 \ldots \mathrm{d}x_N\, \mathrm{d}y_N\, \mathrm{d}z_N \tag{5.1.9}$$

where the three integration limits correspond to the respective sets of all x-, y-, and z-coordinates. For a hard-wall confinement, U does not depend on h but only on the internal coordinates (x_i, y_i, z_i), and the h dependence of the configuration integral lies purely in the upper integration limits of Eq. (5.1.9). The force F between the walls mediated by the intervening liquid can now be obtained as the variation of the free energy with respect to separation:

$$F = \frac{\mathrm{d}\mathbf{A}}{\mathrm{d}h} = -k_B \mathbf{T} \frac{\mathrm{d} \ln Z}{\mathrm{d}h} = -k_B \mathbf{T} \frac{1}{Z} \frac{\mathrm{d}Z}{\mathrm{d}h} \tag{5.1.10}$$

where we have used Eq. (5.1.3). Similar to the homogeneous case, after using the Leibniz integral rule, the derivative of Z with respect to h can be shown to consist of N identical integrals:

$$\frac{\mathrm{d}Z}{\mathrm{d}h} = N \times \text{area} \times \iint_{-\infty}^{\infty} \int_0^h \exp\left(-\frac{U(x_i = h)}{k_B \mathbf{T}} \right) \mathrm{d}x_1 \ldots \mathrm{d}z_{i-1}\, \mathrm{d}x_{i+1} \ldots \mathrm{d}z_N \tag{5.1.11}$$

Now we can identify the integral in Eq. (5.1.11) with the expression for the particle density $c^*(h)$ at $x = h$ obtained from Eq. (5.1.6b):

$$c^*(h) = Z^{-1} N \iint_{-\infty}^{\infty} \int_0^h \exp\left(-\frac{U(x_i = h)}{k_B \mathbf{T}} \right) \mathrm{d}x_1 \ldots \mathrm{d}z_{i-1}\, \mathrm{d}x_{i+1} \ldots \mathrm{d}z_N \tag{5.1.12}$$

obtained from the general expression for the density in Eq. (5.1.6b). Combining Eqs. (5.1.10) and (5.1.12) we obtain the result that the force per unit area between the hard walls is simply related to the concentration at the walls through

$$\frac{F}{A} = k_B \mathbf{T} c^*(h) \tag{5.1.13}$$

For the case of several components, α, in the liquid, the derivation is easily generalized, provided that the hard-wall criterion is the same (Figure 5.1). The force per area is

$$\frac{F}{A} = k_B \mathbf{T} \sum_{\alpha} c_{\alpha}^*(h) \tag{5.1.14}$$

This is a very useful and exact equation called the contact value theorem, and it is generally valid for incompressible liquids/solutions confined by hard walls. In this section, we are concentrating on charged surfaces, and we will use Eq. (5.1.14) for the case when the density at the wall is calculated from the (approximate) Poisson–Boltzmann equation. However, the expression in (5.1.14) is general and can be used for any type of interactions, not just those due to charges.

5.1.2 A Repulsive Electrostatic Force Exists Between a Charged and a Neutral Surface

When two charged surfaces approach each other in an electrolyte medium, an effective interaction becomes appreciable on a length scale usually determined by the Debye screening length, $1/\kappa$ (see Eq. 3.8.13). In effect, the counterion distributions, or as it is usually expressed, the electrical double layers, start to overlap. To better understand the physical mechanisms behind such interactions, we will first consider the somewhat simpler case of one charged and one neutral surface, aligned in parallel at a separation h, as illustrated in Figure 5.2. This case has practical applications, but it also provides important general insight into the mechanisms of interaction between charged objects in solution.

When a neutral surface approaches a charged surface, the dissolved counterions associated with the latter have gradually less available space in which to move. This spatial constraint implies a

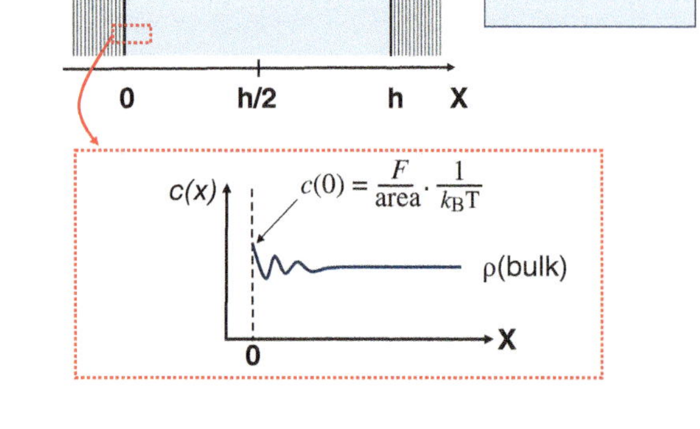

Figure 5.1 The contact density at a hard wall determines the force per area relative to a bulk reservoir (see Eq. 5.1.13).

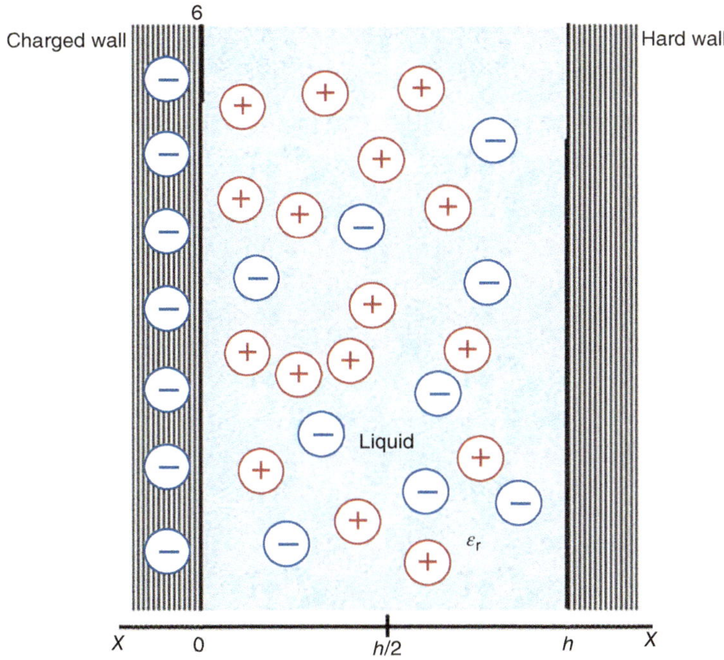

Figure 5.2 A negatively charged and a neutral surface separated by a distance h in an electrolyte solution. The surface charge density is σ, the solvent dielectric constant ε_r The mean concentrations, $\langle c_{\pm} \rangle$ of cations (counterions) and anions (coions) are related by $\langle c_+ \rangle = \langle c_- \rangle + \sigma/h$. Two surfaces that moved together encounter a repulsive force of a magnitude given by Eq. (5.1.17).

decrease in the entropy of the ion distribution and, as we will see, results in a repulsive force. To determine the force, we must consider the free energy **A** as a function of distance. In Section 3.9, we showed how to calculate the electrostatic part $\mathbf{A}_{\text{el}}$ for a single charged surface exposed to an electrolyte solution using the Gouy–Chapman theory. Clearly, the same method could be applied for the case of a charged and a neutral wall. However, the solution to the Poisson–Boltzmann (PB) equation is not quite the same. Instead of Eq. (3.8.6) stating that the electrical field is zero far from the charged surface, we now have a setup in which, due to electroneutrality, the field must be zero at the neutral wall, i.e.,

$$\left.\frac{\mathrm{d}\Phi}{\mathrm{d}x}\right|_{x=h} = 0 \tag{5.1.15}$$

For convenience, we also chose the potential to be zero at the same point

$$\Phi(h) = 0 \tag{5.1.16}$$

In principle, we could now proceed as in Sections 3.8 and 3.9 by first solving the PB equation and then calculating the electrostatic free energy using Eqs. (3.9.1), (3.5.18), and (3.9.4). However, we have in the previous section derived a general result for the force on a hard wall placed at $x = h$. Since we treat the aqueous phase as a continuous medium, the only species in solution are the ions at a concentration c_i^* and it follows from Eq. (5.1.14) that

$$\frac{F}{A} = k_{\text{B}}\mathbf{T}\sum_i c_i^*(h) \tag{5.1.17}$$

The term $\sum_i c_i^*(h)$ can be interpreted as the osmotic pressure resulting from the ion concentration at the neutral surface. Because concentrations are always positive, the force is also positive, which implies a repulsive interaction. Equation (5.1.17) is an exact result, but we will apply it later by calculating the density within the PB approximation.

We can understand the source of the repulsive force as follows. The entropy in the counterion distribution explains why a diffuse double layer initially forms at a charged surface. A configuration where all counterions are bound to the surface would minimize the energy and correspond to the ground state at $\mathbf{T} = 0$. On the other hand, releasing the ions lowers the free energy because it results in an increase in entropy. However, the counterions cannot leave the surface entirely because of the high energies involved in this charge separation. Therefore, the nonuniform ion distribution outside the charged surface represents an optimal compromise between these two competing effects.

When a confining neutral surface approaches a charged one, the ion distribution between them must rearrange in the smaller space that is available, so that the entropy of ions decreases. A decrease in energy partially compensates for this change. For the ions at the neutral wall, the interaction with the charged wall is completely screened by the intermediary ions, shown by the fact that $d\Phi/dx = 0$ at $x = h$; thus, only the entropy term comes into play here.

We could also use the ion concentration at the *charged* wall to calculate the force with equal success, provided we correct for the effects resulting from the nonzero electrical field. Equation (3.8.10), which remains valid for the two-wall case, demonstrates this use. We can apply the general relation

$$\left(\frac{\mathrm{d}\Phi}{\mathrm{d}x}\right)^2 = \frac{2k_B\mathbf{T}}{\varepsilon_r\varepsilon_0}\sum_i c_i^*(h)\left\{\exp\left(\frac{-z_i e\Phi}{k_B\mathbf{T}}\right) - 1\right\} \tag{3.8.10}$$

to arrive at

$$\left(\frac{\sigma}{\varepsilon_r\varepsilon_0}\right)^2 = \frac{2k_B\mathbf{T}}{\varepsilon_r\varepsilon_0}\{c_i^*(0) - c_i^*(h)\} \tag{5.1.18}$$

Here we have used the field $-\mathrm{d}\Phi/\mathrm{d}x$ at the charged surface ($x = 0$) relates to the surface charge density through the boundary condition 3.8.7. Now we use Eq. (5.1.14), which relates $c_i^*(h)$ to the force and find

$$\frac{F}{A} = \Pi_{\mathrm{osm}} = k_B\mathbf{T}\sum_i c_i^*(0) - \frac{\sigma^2}{2\varepsilon_r\varepsilon_0} \tag{5.1.19}$$

which is equivalent to Eq. (5.1.17). Equation (5.1.19) remains valid irrespective of the charges on the two surfaces, a fact that proves useful in the following.

5.1.3 We Can Solve the Poisson–Boltzmann Equation When Only Counterions Are Present Outside the Charged Surface

To actually determine the force between two surfaces, we still must solve the PB equation to obtain the concentrations $c_i^*(h)$ at the neutral surface. In general, this solution is quite complex, but if only counterions are involved, we can employ a simple analytical solution of

$$\nabla^2\Phi(x) = \frac{ze}{\varepsilon_r\varepsilon_0}c^*(h)\exp\left(\frac{-ze\Phi(x)}{k_B\mathbf{T}}\right) \tag{5.1.20}$$

Here $c^*(h)$ signifies the concentration of the ionic species at the neutral wall. Electroneutrality requires that

$$\int_0^h c^*(x)\mathrm{d}x = -\frac{\sigma}{ze} \tag{5.1.21}$$

where σ is the surface charge density. We can solve the PB equation for this special case in much the same way that we solved the Gouy–Chapman case in Section 3.8. The counterion distribution comes out as

$$c^*(x) = 2k_\mathrm{B}\mathbf{T}\varepsilon_r\varepsilon_0 \frac{s^2}{(zeh)^2} \frac{1}{\cos^2\left[s\left(1-\frac{x}{h}\right)\right]} \tag{5.1.22}$$

In this equation, s represents a dimensionless parameter analogous to s in Eq. (3.9.14), but here the boundary conditions imply

$$s\ \tan(s) \ = -\frac{ze\sigma h}{2k_\mathrm{B}\mathbf{T}\varepsilon_r\varepsilon_0} \tag{5.1.23}$$

Because $\tan(s)$ diverges at $\pm\pi/2$, s can take only values within the range $-\pi/2 < s < \pi/2$. We can then obtain the force from $c^*(h)$ according to Eqs. (5.1.17) and (5.1.22)

$$\frac{F}{A} = \frac{2(k_\mathrm{B}\mathbf{T})^2\varepsilon_r\varepsilon_0}{(ze)^2}(s/h)^2 \tag{5.1.24}$$

For high charge densities or large h, s approaches an asymptotic value of $\pm\pi/2$ in Eq. (5.1.16), and in this nonlinear regime of the PB equation, we find the following simple expression for the force

$$\frac{F}{A} = \frac{\pi^2(k_\mathrm{B}\mathbf{T})^2\varepsilon_r\varepsilon_0}{2(ze)^2}\frac{1}{h^2} \tag{5.1.25}$$

It is a somewhat surprising feature that the force is independent of the surface charge density. We will see that this feature is typical of highly charged surfaces. The force is of long range, decaying as the inverse square of the separation h. This is a consequence of the fact that there are only counterions present in the medium and, in contrast, an exponentially decaying force is obtained when additional electrolyte is present, as we will show in the following sections. In the limit of small s, corresponding to weakly charged surfaces, $\tan(s) \approx s$, and the concentration entering Eq. (5.1.17) simply represents the average concentration, $\bar{c} = -\sigma/zeh$, so that the force shifts to an h^{-1} dependence.

As a practical example, let us calculate the thickness of the aqueous film on a charged surface exposed to humid air. This surface might be clean glass or one with an adsorbed ionic surfactant layer. With a relative humidity of p/p_0 in the air, where p_0 is the saturation pressure, some water will condense on the surface layer and free the (monovalent) counterions. At equilibrium, the water chemical potential in the film, which is related to Π_osm (i.e., F/A), is the same as in air, and

$$V_{H_2O}\,\Pi_\mathrm{osm} = -k_\mathrm{B}\mathbf{T}\,\ln\frac{p}{p_0} \tag{5.1.26}$$

where V_{H_2O} is the molecular volume of liquid water. Unless h is very small, the highly charged surface allows us to use Eq. (5.1.25), noting that the force per unit area equals the osmotic pressure.

Equating the chemical potentials in the gas and liquid yields the following expression for the equilibrium film thickness h_{eq}

$$h_{eq} = \pi \left\{ \frac{V_{H_2O} k_B \mathbf{T} \varepsilon_r \varepsilon_0}{2z^2 e^2 \ln\left(\frac{p_0}{p}\right)} \right\}^{\left(\frac{1}{2}\right)} \tag{5.1.27}$$

At 99% relative humidity, $h_{eq} = 1.3$ nm, but the thickness increases strongly as p approaches saturation.

5.1.4 The Ion Concentration at the Midplane Determines the Force Between Two Identically Charged Surfaces

When two equally charged surfaces approach each other in a liquid medium, their diffuse double layers overlap and a repulsive force results, in a mechanistically similar fashion as between one charged and one neutral surface. However, the repulsion between two equally charged surfaces sets in at longer separations, as we will show in the following. Consider two identical, parallel charged surfaces, as shown in Figure 5.3. The boundary conditions of the PB equation are the same ones that we used for the Gouy–Chapman solution in Eq. (3.8.7), but because of the symmetry about the midplane, the field at the midplane must be zero. This fact allows us to replace the boundary condition at one of the surfaces by a condition at the midplane ($x = h/2$ in Figure 5.3) to give

$$\left.\frac{d\Phi}{dx}\right|_{x=h/2} = 0; \quad \Phi(h/2) = \Phi_d \tag{5.1.28}$$

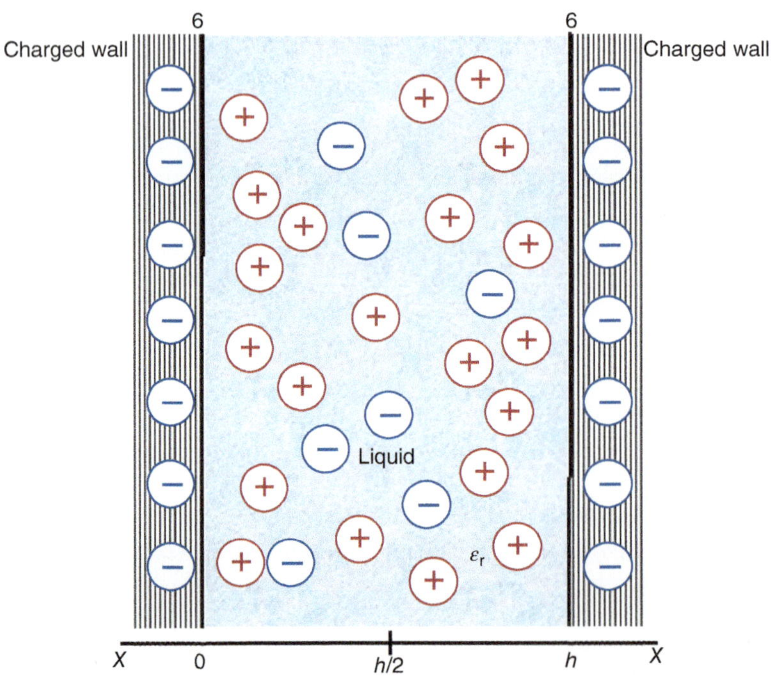

Figure 5.3 Two interacting parallel negatively charged surfaces separated by a distance h in an electrolyte solution experience a repulsive force (Eq. 5.1.29) as they are pushed together.

The reason for fixing the potential at the midplane to a nonzero value will become apparent in the following paragraphs.

The PB equation for one half of the gap shown in Figure 5.3 is the same as for the system in Figure 5.2, which we discussed in Section 5.1.2. In that situation, electrical neutrality between the charged wall and the neutral wall occurred of necessity; if two identical double layers overlap, however, symmetry brings about neutrality relative to the midplane. Whatever the physical cause of the boundary conditions, the solution of the PB equation is the same: The ion distribution is identical and so is the free energy. Thus, the derivation of Eq. (5.1.17) remains valid and

$$\Pi_{\text{osm}} = k_{\text{B}}\mathbf{T}\sum_i c_i^*(h/2) \tag{5.1.29}$$

For two charged surfaces, this result is derived by applying the PB approximation, but it is an exact equation. We can explain the repulsion between the two surfaces in Eq. (5.1.29) with the same mechanistic reasoning we used to explain the charged and neutral surfaces in Section 5.1.2: when the two double layers start to overlap, ion concentrations increase, lowering the entropy. A decrease in energy offsets some of the resulting increase in free energy, but at the midplane—where no fields act on the ions—only the entropy effect is present.

Equation 5.1.29 applies directly to the swelling of both lamellar liquid crystals and some clays with a basic lamellar structure. As long as the double-layer forces dominate, the charged surfaces in a lamellar liquid crystal repel each other, and the liquid crystal swells indefinitely in pure water. In practice, although we usually do observe a strong swelling, other factors also play a role in the process.

As an illustration, we can consider a clay with a plate-like structure that has swollen in pure water. How does the thickness of the water layer vary as we squeeze the clay mechanically? For a pure water medium, the only ions present are the counterions, and we can use Eq. (5.1.22) to calculate their distribution. If we assume that the surface is highly charged, we can use Eq. (5.1.25) to describe the osmotic pressure. Because this pressure balances the applied pressure p_{ext}, we can obtain the water layer thickness h_{eq} at mechanical equilibrium from

$$h_{\text{eq}} = \frac{\pi k_{\text{B}}\mathbf{T}}{ze}\left(\frac{2\varepsilon_r\varepsilon_0}{p_{\text{ext}}}\right)^{1/2} \tag{5.1.30}$$

noting that h in Figure 5.2 corresponds to $h/2$ in Figure 5.3. When the clay is under mechanical pressure from a tractor wheel or by the feet of playing children, the clay can release water, creating a slippery slurry. Conversely, in a dry atmosphere, clays retain water, which is a beneficial property in an agricultural soil.

5.1.5 The Bulk Solution Often Provides a Suitable Reference for the Potential

The solvent and the ions between the surfaces of colloidal particles in a medium are in direct contact with the bulk solution. As the surfaces are pushed toward each other, solvent is expelled from the gap. Equation (5.1.29) gives the local osmotic pressure between the surfaces. In the bulk, the osmotic pressure is $\Pi_{\text{osm}}(\text{bulk}) = k_{\text{B}}\mathbf{T}\sum_i c_{i0}^*$ where c_{io}^* equals the bulk concentration of ion i. The net force acting between the surfaces is then

$$\frac{F}{A} = \Pi_{\text{osm}}(h/2) - \Pi_{\text{osm}}(\text{bulk}) = k_{\text{B}}\mathbf{T}\left\{\sum_i c_i^*(h/2) - \sum_i c_{i0}^*\right\} \tag{5.1.31}$$

The bulk sets the chemical potential of the ionic species so that it serves as a natural reference point for the electrostatic potential, i.e., $\Phi(\text{bulk}) = 0$. The potential Φ_d at the midplane is then different from zero. Here, the concentration of the counterion species is a bit higher than in the bulk, while that of the coions is lower. The chemical potential of both types of ions is equal in the bulk and at the midplane when

$$c_i^*(h/2) = c_{i0}^* \exp\left(-\frac{z_i e\Phi_d}{k_B \mathbf{T}}\right) \tag{5.1.32}$$

Thus, we find from Eqs. (5.1.31) and (5.1.32)

$$\frac{F}{A} = k_B \mathbf{T} \sum_i c_{i0}^* \left[\exp\left(-\frac{z_i e\Phi_d}{k_B \mathbf{T}}\right) - 1\right] \tag{5.1.33}$$

Equation (5.1.33) constitutes a central equation for the force between two equally charged surfaces in equilibrium with a bulk electrolyte solution. This equation has been derived in many different ways. The derivation chosen here has the advantage of explicitly describing the force as a derivative of the free energy and demonstrating the relation to the ion concentration. In this way, we gain insight into the molecular origin of the force.

One virtue of having such physical insight is that it brings out some shortcomings in the theory. In Sections 5.1.2 and 5.1.4, we considered the close analogy between the cases of charged and neutral walls. The discussion of identically charged walls in Section 5.1.4 shows that in this case, the two halves do not interact directly. The identity between the system shown in Figure 5.2 and the halves of Figure 5.3 is a consequence of using the mean field approximation of the PB equation. In reality, instantaneous electrical fields go from one half to the other in Figure 5.3, since the lateral charge distribution is inhomogeneous at each particular moment in time. Only the average distribution is laterally homogeneous. In systems with strong interactions between individual ions (such as those with multivalent ions or for low ε_r), these instantaneous fields are large enough to permit the development of substantial correlations between the ions in the two halves. We postpone a detailed discussion of this effect until Section 5.3.

To use Eq. (5.1.33) to obtain an expression for the force in terms of known quantities, we still need to solve the PB equation. In the general case with several ionic species present, it is best to use numerical methods. However, limiting cases that are tractable to analytical approximations are still of considerable interest. For large separations, when the two double layers just start to overlap, the potential Φ_d deviates only slightly from zero, since $\Phi_d \to 0$ as $h \to \infty$. Thus, for weakly overlapping double layers, we can expand the exponent in Eq. (5.1.33) in a power series. The bulk term cancels the first term, and the second term cancels as a result of electroneutrality. The first nonzero contribution, therefore, comes from the third term in the expansion, and

$$\frac{F}{A} \approx \frac{1}{2} \sum_i \frac{c_{i0}^* (ze)^2 \Phi_d^2}{k_B \mathbf{T}} = \frac{\varepsilon_r \varepsilon_0 \kappa^2 \Phi_d^2}{2} \tag{5.1.34}$$

The second equality follows from the definition (Eq. 3.8.13a) of the Debye screening length κ^{-1}. In this weak overlap limit, we can estimate Φ_d by taking advantage of the fact that the potential profile differs only slightly from the Gouy–Chapman solution for a single surface, as illustrated in Fig. 5.4. We can then assume that the potential Φ_d at the midplane simply equals the sum of the Gouy–Chapman potentials from the two surfaces, which follows from the superposition principle applied to the Poisson equation as illustrated in Figure 5.4.

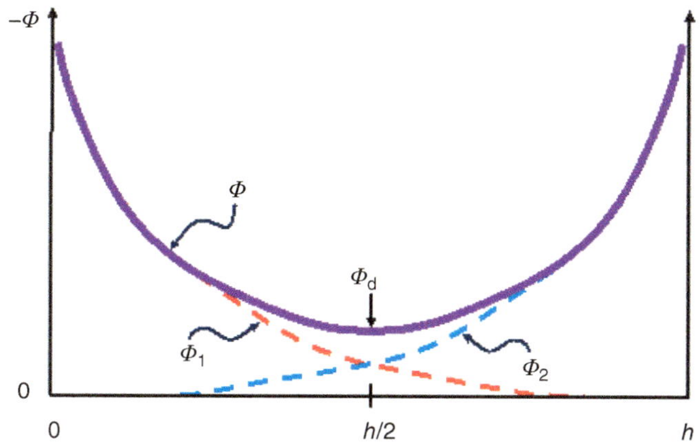

Figure 5.4 Superposition approximation for two similarly charged surfaces. If the charge distributions ρ_1 and ρ_2 outside the respective surfaces are assumed to be unaffected by the interaction between the surfaces, the resulting charge distribution is $\rho = \rho_1 + \rho_2$. By the linearity of the Poisson equation (see Chapter 3), the resulting potential Φ represents the sum $\Phi = \Phi_1 + \Phi_2$ of the potentials Φ_1 and Φ_2 outside the unperturbed surfaces. In this approximation, $\Phi_d = \Phi_1(h/2) + \Phi_2(h/2)$.

Using the linearized form of the Gouy–Chapman solution as given in Eq. (3.8.21), we find

$$\Phi_d = 2\Phi\left(\frac{h}{2}\right) = \frac{8k_B\mathbf{T}}{ze}\Gamma_0 \exp\left(-\frac{\kappa h}{2}\right) \tag{5.1.35}$$

valid for a symmetrical $z : z$ electrolyte.

As the two double layers start to overlap, they experience a repulsive force

$$\frac{F}{A} \approx 32(k_B\mathbf{T})^2 \frac{\varepsilon_r\varepsilon_0\kappa^2}{z^2e^2}\Gamma_0^2 \exp(-\kappa h) = 64k_B\mathbf{T}c_0^*\Gamma_0^2 \exp(-\kappa h) \tag{5.1.36}$$

where c_o^* represents the bulk electrolyte concentration and $\Gamma_0 = \tanh(ze\Phi_0/4k_B\mathbf{T})$ as in Eq. (3.8.14).

Equation (5.1.36) reveals several important facts. At long range, the double-layer force between two equally charged surfaces in an electrolyte solution decays exponentially with a decay constant equal to the Debye screening length. The force is sensitive to the electrolyte concentration through the factor c_o^* and more significantly through κ. For a $z : z$ electrolyte, κ^{-1} is proportional to z, so divalent ions weaken the force substantially compared to monovalent ions.

The nature of the charged surface enters the calculation through the quantity Γ_0. For highly charged surfaces Γ_0 approaches unity, and the force at large separations is independent of the surface charge density. In the opposite limit of low charge densities, where $ze\Phi_0/k_B\mathbf{T} \ll 1$, we can instead linearize the tanh function in Eq. (3.8.14) and combine with Eq. (3.8.20) to obtain

$$\Gamma_0^2 = \frac{1}{32k_B\mathbf{T}}\frac{\sigma^2}{\varepsilon_r\varepsilon_0}\frac{1}{c_0^*} \tag{5.1.37}$$

In this limit, the force is proportional to the charge density squared and given by

$$\frac{F}{A} = \frac{2\sigma^2}{\varepsilon_r\varepsilon_0}\exp(-\kappa h) \tag{5.1.38}$$

Equation (5.1.36) can only describe the double-layer force when it is so weak that the linearization approximation applies. Within this limit, the two double layers are perturbed only slightly relative to infinite separation, and possible chemical rearrangements at the surface are negligible. The situation becomes more complex when the surfaces interact strongly. Then we must solve the PB equation numerically, except in the limit of short separations of highly charged surfaces, so that the coions are completely expelled from the region between the surfaces. In that limit, we go back to the case of counterions only, and we can use Eqs. (5.1.24) or (5.1.25) for the force.

However, we face a problem with boundary conditions. Normally, we do not know whether the ionization of surface groups or the adsorption of ions generates the charge at the surfaces. As two surfaces approach each other, these surface groups can titrate, or the ions can desorb or adsorb and thus affect the force. Measurements of the force in the strong interaction regime can therefore provide an indirect method for studying the chemical processes at the surface. Even though these surface effects are intrinsically complex, limiting cases exist where they can be modeled. For example, if we assume that the charge is constant, we totally neglect the surface charge rearrangement effects and overestimate the force at shorter range. In the other extreme, if we assume an infinitely adjustable electrical surface so that the electrostatic potential remains constant, we obtain a lower limit to the force. The reality is usually somewhere between constant charge and constant potential. We presented a more detailed model for this possibility in Section 3.7, where the titration of surface groups was explicitly included. When we include this effect when treating the force problem, we call it the *charge regulation model.*

5.1.6 Two Surfaces with Equal Signs but Different Magnitudes of Charge Always Repel Each Other

So far, our focus on the standard case of two planar and equally charged parallel walls has allowed us to develop a reasonably simple quantitative theoretical description of the double-layer interaction. As a rule, reality is, however, more complex than our idealized examples. For instance, particle surfaces are not always planar, and in Section 5.4.1, we discuss the complications that arise from different surface geometries. Another complication relates to so-called heterocoagulation, in which two dissimilar surfaces or particles can come together. For the sake of clarity, we will maintain the parallel planar configuration in our discussion of this complication.

For an asymmetric system with two different surface charge densities σ_1 and $\sigma_2 \neq \sigma_1$ (see Figure 5.5), the ion concentration at the midplane is no longer the relevant parameter. The boundary conditions at the two charged surfaces $x = 0$ and $x = h$ are

$$\left.\frac{\mathrm{d}\Phi}{\mathrm{d}x}\right|_0 = -\frac{\sigma_1}{\varepsilon_r\varepsilon_0} \tag{5.1.39}$$

and

$$\left.\frac{\mathrm{d}\Phi}{\mathrm{d}x}\right|_h = \frac{\sigma_2}{\varepsilon_r\varepsilon_0} \tag{5.1.40}$$

We will now treat separately the two cases in which σ_1 and σ_2 bear the same or the opposite signs, respectively.

For charged surfaces of the same sign of their charge densities, the double layers still overlap, just as for the fully symmetrical case. At some position x_0 between the surfaces, different from the geometrical midplane, the field is zero, and this plane now represents the dividing surface between the two double layers (see Figure 5.5). In this plane, only the entropic effect operates and

$$\frac{F}{A} = k_\mathrm{B}\mathbf{T}\sum_i \{c_i^*(x_0) - c_{i0}^*\} \tag{5.1.41}$$

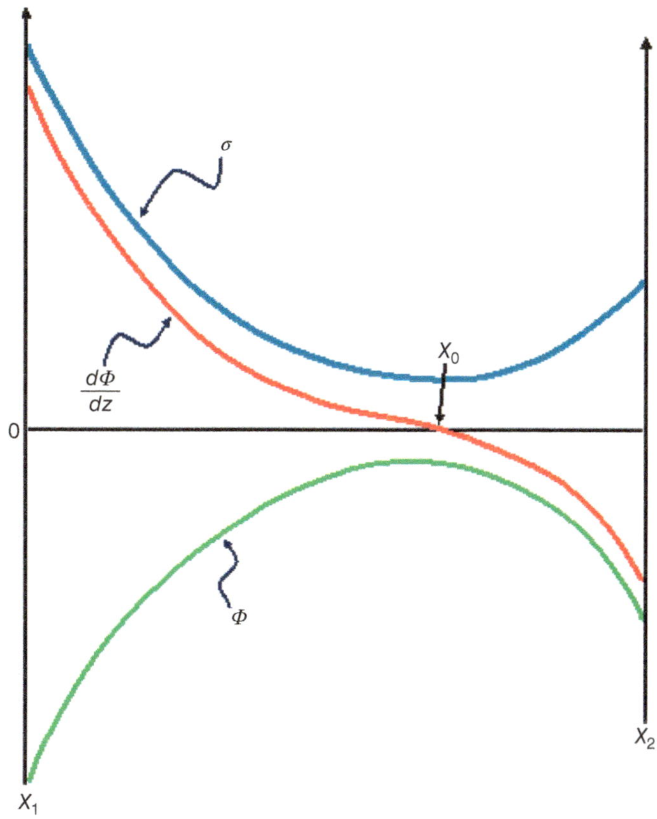

Figure 5.5 Schematic illustration of the charge distribution, σ (blue), the electrostatic potential Φ (green), and its derivative $d\Phi/dx$ (red) between two negatively charged surfaces with $|\sigma_1| > |\sigma_2|$. At $x = x_0$, the derivative $d\Phi/dx$ is zero, while $|\Phi|$ and ρ have their maximal/minimal values. The force is obtained from Eq. (5.1.29) with $\Phi_d = \Phi(x_0)$.

in analogy with Eq. (5.1.29). In general, to find x_0 and $c_i^*(x_0)$ we must, however, solve the PB equation for the given surface charge densities.

For weakly overlapping double layers, Eq. (5.1.31) still applies if we interpret Φ_d as the potential at x_0. At x_0 we can view the electric field, which is zero by definition, as being composed of two compensating parts. In the region of weak overlap, these two fields approximately equal those of the unperturbed fields from each surface separately. Using Eq. (3.8.21), we can obtain x_0 from

$$\Gamma_{01}\exp\left[-\kappa(x_0 - x_1)\right] = \Gamma_{02}\exp\left[-\kappa(x_2 - x_0)\right] \tag{5.1.42}$$

which leads to

$$x_0 = \frac{1}{2\kappa}\ln\left(\frac{\Gamma_{01}}{\Gamma_{02}}\right) + (x_1 + x_2)/2 \tag{5.1.43}$$

For these formulas to apply, the charge density on the less charged wall must be large enough that $|\sigma_1| > \varepsilon_r\varepsilon_0|\Phi_{01}|\exp\left\{-\kappa(x_2 - x_1)\right\}$. Otherwise we locate x_0 outside the gap, which shows that the weak overlap approximation is not applicable.

Using the same approximations that led to Eq. (5.1.34) and Eq. (5.1.35) for the symmetrical case, Eq. (5.1.43) for x_0 results for the unsymmetrical case in

$$\frac{F}{\text{area}} = 64k_B\mathbf{T}c_0^*\Gamma_{01}\Gamma_{02}\exp(-\kappa h) \tag{5.1.44}$$

which simply replaces Γ_0^2 in Eq. (5.1.36) by the square of the geometrical mean, $\Gamma_{01}\Gamma_{02}$, in Eq. (5.1.44).

From Eq. (5.1.44), we can see that the force between the two surfaces is always repulsive. For constant potential boundary conditions, the corresponding calculation, however, results in attraction for some parameter values and separations. A closer inspection shows that this attraction is due to the charge density on one of the surfaces changing sign, due to the charge regulation intrinsic in the constant-potential approximation.

5.1.7 As Surfaces Bearing Opposite Signs Move Closer Together, Long-range Electrostatic Attraction Changes to Repulsion

An attractive force at long separations arises when σ_1 and σ_2 have opposite signs. One surface can then act as a giant counterion for the other, and the small counterions are released into the bulk solution with a concomitant increase in entropy. In this case, the solution of the PB equation follows from the same boundary conditions we used with surfaces of the same sign of σ in Eqs. (5.1.39) and (5.1.40). However, because no point with zero field exists in the intervening solution for this setup, we must determine the force in a different way. Equation (5.1.19) remains valid, so if we know the ion concentration at the wall and the surface charge density, the force follows from

$$\frac{F}{\text{area}} = -\frac{\sigma^2}{2\varepsilon_r\varepsilon_0} + k_B\mathbf{T}\sum_i\left\{c_i^*(\text{wall}) - c_{i0}^*\right\} \tag{5.1.45}$$

Because the two surfaces partially neutralize each other, counterions leave the gap between the surfaces, and initially the counterion concentration at the wall will decrease, leading to an attraction of entropic origin. If $\sigma_2 = -\sigma_1$, this attraction will be sustained up to close contact, where all small ions are released into the bulk solution. At short separations, $h < \kappa^{-1}$, the bulk term cancels the wall concentration in Eq. (5.1.45), and the force becomes $\frac{F}{\text{area}} = -\frac{\sigma^2}{2\varepsilon_r\varepsilon_0}$ which simply represents the force between two planar capacitor plates.

When $|\sigma_1| \neq |\sigma_2|$, some counterions instead must remain in the gap. These counterions concentrate when the two surfaces come close and ultimately produce a repulsive force. This effect shows that, if electrostatic forces dominate, we should expect heterocoagulation of oppositely charged particles of constant charge to result in a complex where the particles do not reach molecular contact.

Figure 5.6 shows an example of how the force between two oppositely charged surfaces varies with the magnitude of surface charge density and screening length. When surface charge densities don't match, there is an optimal separation, leaving the liquid medium between the surfaces. For a particle system, this would leave the possibility of a sliding motion. It can then have the effect that such a mixed system forms a concentrated liquid phase rather than a precipitate with direct molecular contact between the particles. Such systems are sometimes referred to as coacervates, as discussed further in Chapter 7.

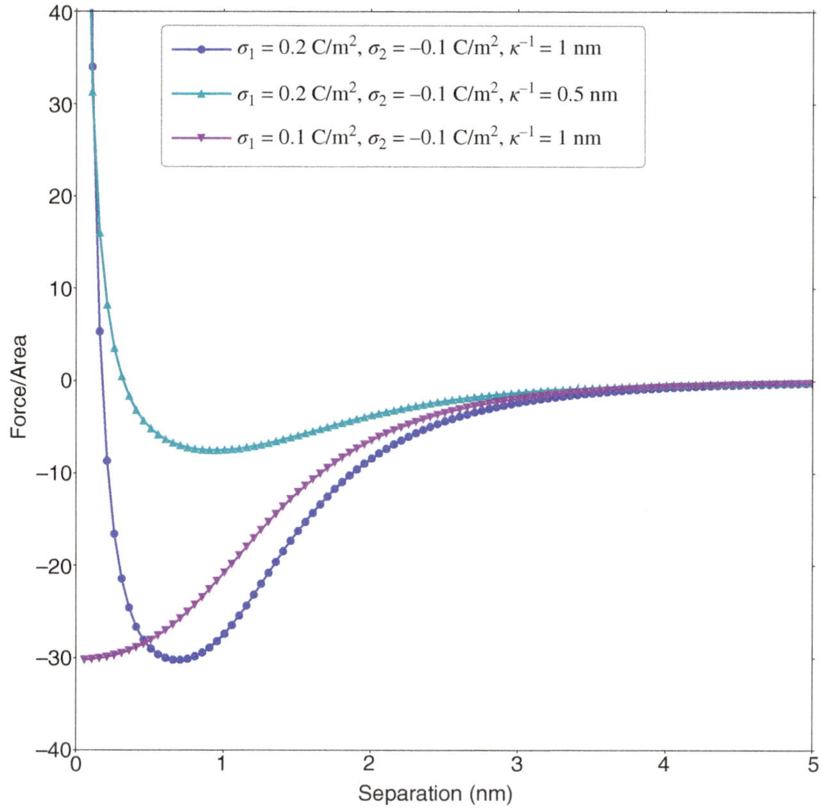

Figure 5.6 Force per unit area versus separation between two parallel surfaces with opposite signs of their respective charge densities. The force is given in dimensionless units, normalized by $k_B * T * c_0^*$. For exactly matched charge magnitudes, the attractive force persists until contact. For unmatched charge magnitudes, the force changes from attractive to repulsive, roughly at the separation where the density of the remaining counterions corresponds to the difference between the two charge densities.

See Atkins et al. (2022, Focus 14), Berg (2010, Chapter 6), Cosgrove (2010, Chapter 16), Doi (2013, Chapter 10), Israelachvili (2011, Chapter 14), Lyklema (1995, Vol. 2, Chapter 3), and Ohshima (2024, Chapters 5–7) for additional discussion on Electrostatic Forces.

5.2 van der Waals Forces Are Dominated by Quantum Mechanical Dispersion Forces

5.2.1 An Attractive Dispersion Force of Quantum Mechanical Origin Operates Between Any Two Molecules

Attractive interactions between any pair of molecules arise from the correlation between the motion of the electrons in the two molecules. Even if an atom or molecule has no permanent electric moment, electrons circle the nuclei and create instantaneous dipoles. These dipoles generate a field at the second atom (or molecule) whose electrons respond instantaneously (or more precisely at the speed of light). Because Fritz London first analyzed this mechanism of intermolecular interactions, it is often called the London dispersion interaction. Another name, which refers to the mechanistic description presented earlier, is the induced dipole–induced dipole interaction.

The dispersion interaction emerges naturally from a quantum mechanical description of the interaction between two molecules. It is mechanistically analogous to the classical thermally averaged Keesom interaction, described in Chapter 3. The difference is that for the dispersion interaction, the correlation is between electrons that are described quantum mechanically, rather than the thermally excited molecular rotation. Neglecting molecular anisotropies, we can write the resulting interaction potential $V_{12}^{\text{disp}}(R)$ in a quantum mechanical formalism as

$$V_{12}^{\text{disp}}(R) = -\frac{1}{24(\pi\varepsilon_0)^2}\frac{1}{R_{12}^6}\sum_{n,k}\frac{\left|\langle n|\boldsymbol{m}|0\rangle_1\right|^2\left|\langle k|\boldsymbol{m}|0\rangle_2\right|^2}{\left(E_1^n - E_1^0\right) + \left(E_2^k - E_2^0\right)} \equiv -\frac{C_{12}}{R_{12}^6} \tag{5.2.1}$$

where $\langle n|\boldsymbol{m}|0\rangle_1$ represents the so-called transition dipole moment for molecule 1 going from the ground state $|0\rangle$ with energy E_1^0 to an excited state $|n\rangle$ with energy E_1^n. The transition dipole moment gives the amplitude of the optical transition between the two states. The quantities $\langle k|\boldsymbol{m}|0\rangle_2, E_2^0$ and E_2^k represent the corresponding quantities for molecule 2.

In Eq. (5.2.1), the R^{-6} dependence has the same origin as the dipole–induced dipole interaction in Eq. (3.4.8). The first R^{-3} dipolar factor reflects the strength of the electrical field, and the second reflects the interaction between the spontaneous and the induced dipole. Because E_1^0 and E_2^0 are the ground state energies, the denominators in Eq. (5.2.1) are always positive. The transition energies in the denominator are furthermore much larger than $k_B\mathbf{T}$. It is this feature that fundamentally distinguishes the dispersion interaction from the Keesom interaction involving thermally excited degrees of freedom.

The origin of the name *dispersion interaction* is not obvious, but it stems from the fact that the molecular quantities entering Eq. (5.2.1) are precisely those that determine the optical properties of the molecules. For example, the frequency-dependent isotropic polarizability $\alpha(\omega)$ at the frequency ω is given by

$$\alpha(\omega) = \frac{2}{3}\sum_n \frac{(E^n - E^0)\left|\langle n|\boldsymbol{m}|0\rangle\right|^2}{\left(E^n - E^0\right)^2 - (\hbar\omega)^2} \tag{5.2.2}$$

This equation contains the same quantities as Eq. (5.2.1), but in a different combination ($\hbar = h/2\pi$ is the reduced Planck's constant). This polarizability directly relates to the frequency dependence of the refractive index and thus to the dispersion of light.

London derived an approximate relation for the dispersion interaction in terms of the polarizabilities at zero frequency $\alpha(0)$ and the ionization potentials I of the molecules, that is, the energy required to ionize the molecule,

$$V_{12}^{\text{disp}}(R) \approx -\frac{3}{32(\pi\varepsilon_0)^2}\frac{I_1 I_2}{I_1 + I_2}\frac{\alpha_1(0)\alpha_2(0)}{R_{12}^6} \tag{5.2.3}$$

The dispersion interaction, which is a major reason behind cohesion in condensed states such as liquids and molecular solids, does not vary strongly from molecular pair to molecular pair, resulting in a limited molecular selectivity. However, some general trends determine the magnitude of the dispersion term. An increase in the number of electrons per unit volume in the molecules increases V_{12} by adding terms to the sum in Eq. (5.2.1) and a lowering of the excitation energies also increases V_{12}.

Another important feature of the dispersion interaction that emerges from its quantum mechanical derivation is that it is pairwise additive to a good approximation. Even in a condensed phase, we can obtain the total dispersion energy with reasonable accuracy by simply summing over all molecular pairs. The fact that virtually all molecules are in their electronic ground states at ambient temperatures simplifies this operation. Thus, the total dispersion interaction is independent of temperature in a condensed phase except for the indirect sensitivity to density changes affecting the average separation R_{12}.

5.2.2 We Can Calculate the Dispersion Interaction Between Two Colloidal Particles by Summing Over the Molecules on a Pairwise Basis

If we assume pairwise additivity, we can determine the dispersion force between two colloidal particles by summing over all pairs of molecules, one molecule on each particle. The basic model of a slit separating two surfaces, as in Figure 5.7, still proves useful. We can assume that a low-density gas, which has negligible influence on the interaction, fills the slit.

Using Eq. (5.2.1), the contribution from volume elements $\mathrm{d}V_1 = \mathrm{d}x_1\,\mathrm{d}y_1\,\mathrm{d}z_1$ and $\mathrm{d}V_2 = \mathrm{d}x_2\,\mathrm{d}y_2\,\mathrm{d}z_2$ (one on either side) to the total interaction energy $\mathbf{U}_{12}$ between the two half-spaces 1 and 2 is

$$\mathrm{d}\mathbf{U}_{12} = -\frac{C_{12}\rho_1\rho_2\,\mathrm{d}V_1\,\mathrm{d}V_2}{\{(x_2 - x_1)^2 + (y_2 - y_1)^2 + (z_2 - z_1)^2\}^3} \tag{5.2.4}$$

where ρ is the density. The total interaction energy is calculated by direct integration

$$\mathbf{U}_{12} = -\frac{H_{12}}{\pi^2} \iiiint_{-\infty}^{+\infty} \int_{-\infty}^{0} \int_{h}^{\infty} \frac{\mathrm{d}x_1\,\mathrm{d}x_2\,\mathrm{d}y_1\,\mathrm{d}y_2\,\mathrm{d}z_1\,\mathrm{d}z_2}{\left\{(x_2 - x_1)^2 + (y_2 - y_1)^2 + (z_2 - z_1)^2\right\}^3} \tag{5.2.5}$$

where we have combined $\pi^2 C_{12}\rho_1\rho_2$ to form the Hamaker constant H_{12}, a constant that plays a central role in the description of van der Waals forces between particles. The Hamaker constant has the dimension of an energy, and it depends on both the strength of the intermolecular interaction through C_{12} and the density. Clearly, we expect a weaker interaction between gaseous media than liquid bulk media of the same compounds. As derived here, H_{12} is determined solely

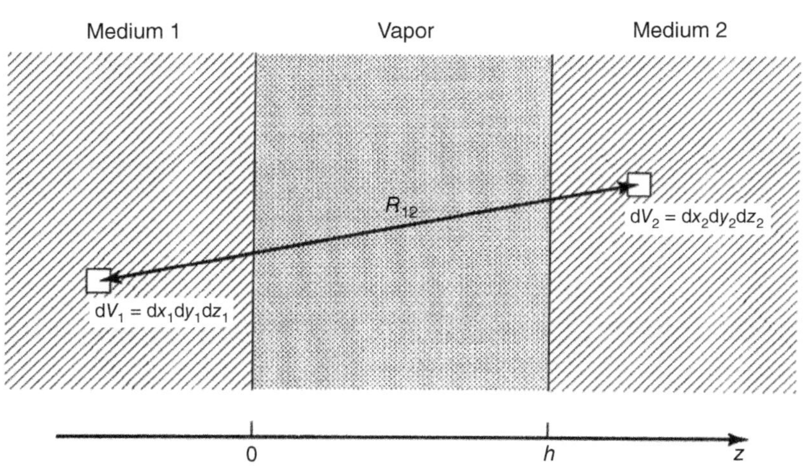

Figure 5.7 Calculation of the London dispersion interaction between two half-spaces separated by a distance h across a vapor. An interaction proportional to R_{12}^{-6} exists between the two volume elements, $\mathrm{d}V_1$ and $\mathrm{d}V_2$. The total interaction is obtained by integrating this interaction over the two half-spaces, respectively.

by the dispersion interaction. Later, we will see that the total Hamaker constant also includes other contributions.

We can solve the integral in Eq. (5.2.5) by first substituting variables $x = (x_2 - x_1); x' = (x_2 + x_1); y = (y_2 - y_1); y' = (y_2 + y_1)$. This substitution brings out an expected proportionality between $\boldsymbol{U}_{12}$ and the area because only the relative position of the two volume elements in the x, y directions matters and

$$\frac{\mathbf{U}_{12}}{\text{area}} = -\frac{H_{12}}{\pi^2} \iint_{-\infty}^{+\infty} \int_{-\infty}^{0} \int_{h}^{\infty} \frac{dxdydz_1 dz_2}{\{x^2 + y^2 + (z_2 - z_1)^2\}^3} \tag{5.2.6}$$

The right-hand side can be integrated most simply by using cylindrical coordinates with $r^2 = x^2 + y^2$, so that

$$\begin{aligned} \frac{\mathbf{U}_{12}}{\text{area}} &= -\frac{H_{12}}{\pi^2} \int_0^\infty \int_{-\infty}^{0} \int_h^\infty \frac{2\pi r dr dz_1 dz_2}{\left\{r^2 + (z_2 - z_1)^2\right\}^3} \\ &= -\frac{H_{12}}{2\pi} \int_{-\infty}^{0} \int_h^\infty \frac{dz_1 dz_2}{(z_2 - z_1)^4} = -\frac{H_{12}}{6\pi} \int_h^\infty \frac{dz_2}{z_2^3} \\ &= -\frac{H_{12}}{12\pi} \frac{1}{h^2} \end{aligned} \tag{5.2.7}$$

By integrating the short-range R^{-6} dispersion term in four directions, we have transformed this interaction into a long-range h^{-2} dependence. (The exponent changes by one for each direction over which the integration is performed.)

At first sight, this result is somewhat surprising. We know that in the bulk phase, the dominant contribution to the energy comes from nearest neighbors as a result of the R^{-6} dependence of the intermolecular interaction. For a narrow slit with h of the order of a molecular diameter, the dominant contribution to $\mathbf{U}_{12}$ comes from the surface layer–surface layer interaction. However, the wider the slit, the less dominant this contribution becomes, so that the total interaction varies less strongly with the separation h.

A similar mechanism operates for a single molecule (2) outside the surface of a bulk phase (1). As in Eq. (5.2.7), we can obtain the interaction potential $V_{12}(h)$ by omitting the integration over dz_2 and dividing out the irrelevant density factor ρ_2, so that

$$V_{12}(h) = -\frac{H_{12}}{6\pi\rho_2} \frac{1}{h^3} \tag{5.2.8}$$

This expression closely resembles the one describing the force F between the two half-planes

$$\frac{F_{12}}{\text{area}} = -\frac{1}{\text{area}} \frac{d\mathbf{U}_{12}}{dh} = -\frac{H_{12}}{6\pi\rho_2} \frac{1}{h^3} \tag{5.2.9}$$

where we have retained the convention that attractive forces are negative. Such analogies between interaction energies and forces can be very useful, as we will demonstrate further in Section 5.4.1

For a liquid held together only by dispersion forces, the surface tension γ clearly relates to the dispersion interaction. To see this quantitatively, we can write the total energy $\mathbf{U}$ of the system in Figure 5.7 as the sum of two bulk terms, $n_1u_1(\text{bulk}) + n_2u_2(\text{bulk})$, in which n_i represents the

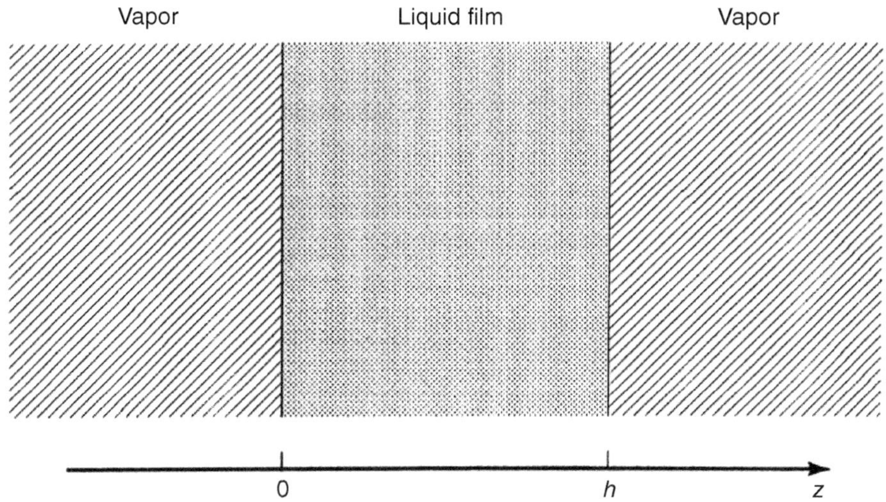

Figure 5.8 A liquid film of thickness *h* in a vapor. The combined systems in Figures 5.7 and 5.8 form a bulk liquid phase if the two media in Figure 5.7 and the liquid in Figure 5.8 are the same.

number of molecules in phase *i*, plus two surface contributions, $\gamma_1 A + \gamma_2 A$, where *A* is the surface area and γ_i the surface energy of phase *i*, and an interaction term to give

$$\mathrm{U} = n_1 u_1(\text{bulk}) + n_2 u_2(\text{bulk}) + \gamma_1 A + \gamma_2 A - \frac{AH_{12}}{12\pi}\frac{1}{h^2} \tag{5.2.10}$$

At contact between the two phases, this equation should reduce to $\mathbf{U} = n_1 u_1(\text{bulk}) + n_2 u_2(\text{bulk}) + \gamma_{12} A$ where γ_{12} equals the interfacial tension and $\gamma_{12} = 0$ if $1 = 2$. A distance of closest approach, δ, exists between the half-planes, and by inserting $h_{\min} = \delta$ we find

$$\gamma_1 + \gamma_2 = \frac{H_{12}}{12\pi}\frac{1}{\delta^2} + \gamma_{12} \tag{5.2.11}$$

When medium 1 is the same as medium 2, the surface tension is

$$\gamma_1 = \frac{H_{11}}{24\pi}\frac{1}{\delta^2} \tag{5.2.12}$$

This distance δ should be on the order of an atomic van der Waals radius. As an example, we can consider a hydrocarbon whose CH_2 group has a van der Waals radius of approximately 0.20 nm. The Hamaker constant for a hydrocarbon is around $H_{11} \approx 5 \times 10^{-20}$ J. Inserting these values into Eq. (5.2.12) leads to an estimate of the surface tension of a hydrocarbon of approximately 17 mJ/m^2, which is in reasonable agreement with the experimental value of approximately 26 mJ/m^2.

A liquid film formed in a vapor phase (such as in a soap bubble), shown in Figure 5.8, provides a situation complementary to the one illustrated in Figure 5.7. In this case, we can obtain the energy of the liquid by integrating over the film. However, because of the complementary geometries in Figures 5.7 and 5.8, we can also use Eq. (5.2.7) to solve the problem algebraically. To illustrate this point, we can subdivide a bulk phase into three parts (1, 2, and 1′) as in Figure 5.9. The total energy is

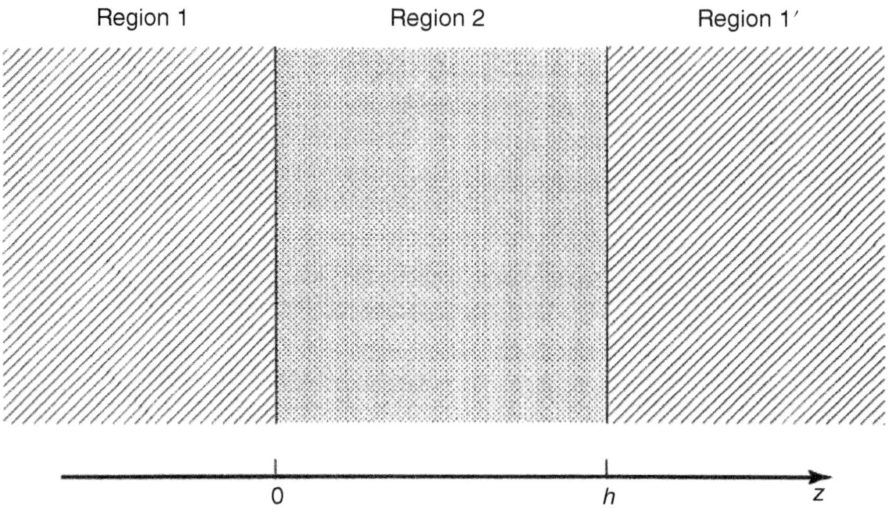

Figure 5.9 Decomposition of a bulk phase into three regions, used to calculate the energy of the liquid film in Figure 5.8. Region 2 corresponds to the liquid film.

$$\mathbf{U} = (n_1 + n_2 + n_1')u(\text{bulk}) = \mathbf{U}_{11} + \mathbf{U}_{12} + \mathbf{U}_{22} + \mathbf{U}_{1'1'} + \mathbf{U}_{1'2} + \mathbf{U}_{11'} \tag{5.2.13}$$

The quantity $\mathbf{U}_{22}$ represents the integration of Eq. (5.2.4) over the film itself, and it constitutes the only contribution to the energy of the film in Figure 5.8. The point is now that we can use the result in Eq. (5.2.10) to obtain the other unknown quantities in Eq. (5.2.13). In Eq. (5.2.10), we determined the energy of two half-spaces separated by a distance h. If 1 and 2 in this equation represent the same material, we could relabel them 1 and 1'. Then $\mathbf{U}$ in Eq. (5.2.10) corresponds to $\mathbf{U}_{11} + \mathbf{U}_{1'1'} + \mathbf{U}_{11'}$ in Eq. (5.2.13). We can obtain the remaining equation needed to calculate $\boldsymbol{U}_{22}$ by noting that

$$\mathbf{U}_{11} + \mathbf{U}_{12} + \mathbf{U}_{22} = \mathbf{U} + n_2 u_2(\text{bulk}) \tag{5.2.14}$$

because both region 1 and region 1 plus 2 represent a bulk phase with one free surface. By combining Eqs. (5.2.13), (5.2.10), and (5.2.14), we find the energy of the film is

$$\mathbf{U}_{22}(h) = n_2 u_2(\text{bulk}) + 2\gamma A - \frac{H_{22}}{12\pi}\frac{1}{h^2}A \tag{5.2.15}$$

since $\mathbf{U}_{12} = \mathbf{U}_{1'2}$ by symmetry. We see from the expression in Eq. (5.2.12) for γ that the energy of the film, $\mathbf{U}_{22}(h)$, is higher than that of a bulk solution. We might have expected this result, but because the h-dependent term is negative, the tendency of the dispersion force to thin the film is not quite so obvious.

As an application of Eq. (5.2.15), we can consider a soap film drawn from an aqueous surfactant solution with varying salt content. The charges on the surfactant molecules create a repulsive double-layer force, which is balanced by an attractive force whose main component stems from the dispersion interaction. Because of the balancing effect of these forces, the film can reach a metastable equilibrium at some distance where the net force is zero. The equilibrium thickness is also affected by the gravitational force as discussed below in Section 5.2.3. By changing the salt concentration in the bulk, we can vary the double-layer force. Figure 5.10 shows how an

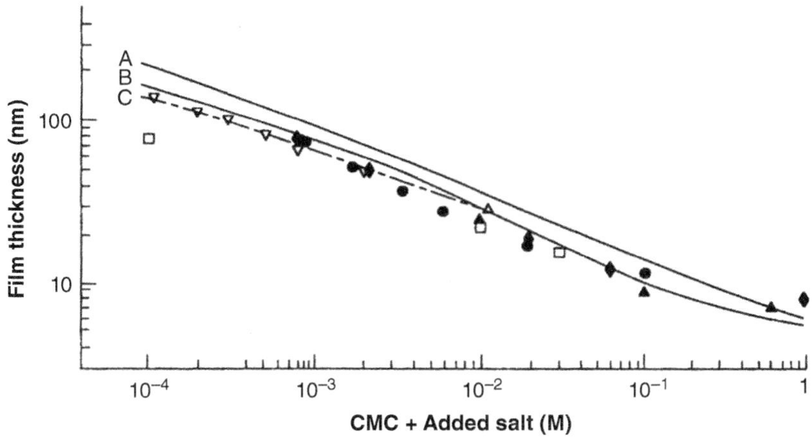

Figure 5.10 Measured thickness of soap films as a function of the electrolyte concentration in the aqueous phase in contact with the film; symbols represent measurements of different researchers. The lines are a theoretical fit, considering that the repulsive double-layer interaction, with surface potential Φ_0 is balanced by a combination of a van der Waals interaction with $H = 5 \times 10^{-20}$ J and a hydrostatic pressure p_h. Curve A; $\Phi_0 = \infty, p_h = 25.5$ Pa; curve B, $\Phi_0 = 65$ mV, $p_h = 0$; curve C, $\Phi_0 = 65$ mV, $p_h = 25.5$ Pa. *Source*: Lyklema and Mysels, 1965/with permission of American Chemical Society.

experimentally determined film thickness varies with the salt concentration in the bulk supporting solution. It also confirms the prediction of how the force varies with h, which we can evaluate by using Eqs. (5.1.31) and (5.2.15). By adjusting the values for the surface potential Φ_0 and/or the Hamaker constant H_{12}, we can get quantitative agreement over a large range of bulk electrolyte concentrations.

5.2.3 The Presence of a Medium Between Two Interacting Particles Modifies the Magnitude of the Hamaker Constant

In most applications of dispersion forces to colloidal problems, the interaction between two particles occurs across a liquid medium; an example such as the soap film in Figure 5.10 is really an exception. In an inhomogeneous situation (particle 1–medium–particle 2), dispersion interactions occur between all the constituents. This substantially reduces the effective interaction relative to the case with a vapor gap shown in Figure 5.7. The net dispersion force may even become repulsive rather than attractive, for reasons that will become clear below.

In keeping with our general approach, we can consider two half-planes of media 1 and 3 separated by a liquid 2, as shown in Figure 5.11. By analogy with Eq. (5.2.13), the total energy becomes a sum of six terms

$$\mathbf{U}_{123}(h) = \mathbf{U}_{11} + \mathbf{U}_{33} + \mathbf{U}_{13}(h) + \mathbf{U}_{22}(h) + \mathbf{U}_{12}(h) + \mathbf{U}_{23}(h) \tag{5.2.16}$$

The sum of the first three terms follows from Eq. (5.2.10) by a suitable change of indices. The fourth term, $\mathbf{U}_{22}(h)$, is the same as in Eq. (5.2.15). We can obtain $\mathbf{U}_{12}(h)$ by applying Eq.(5.2.7) at contact and subtracting its value at separation h so that

$$\mathbf{U}_{12}(h) = -\frac{H_{12}}{12\pi}\left(\frac{1}{\delta^2} - \frac{1}{h^2}\right)A \tag{5.2.17}$$

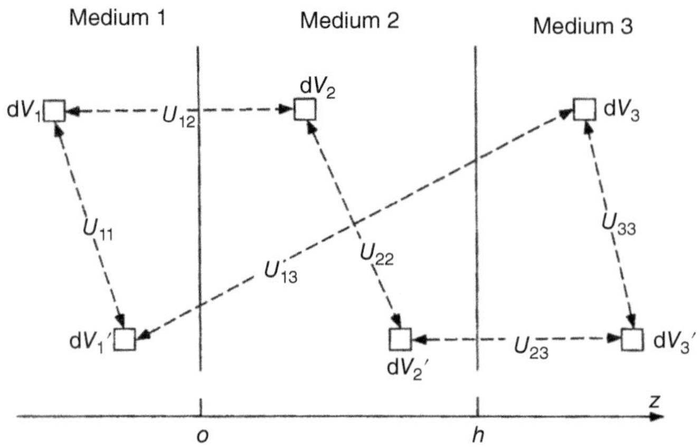

Figure 5.11 Interaction of bulk phases 1 and 3 across a film of phase 2 of thickness h. The interaction energy $\mathbf{U}_{123}(h)$ is obtained by summing the contributions $\mathbf{U}_{ij}(h) = 1, 2, 3)$ as in Eq. (5.2.16).

with an analogous expression for $\mathbf{U}_{23}(h)$. If we combine all these expressions, $\mathbf{U}_{123}(h)$ reduces to

$$\begin{aligned}\mathbf{U}_{123}(h) &= n_1u_1(\text{bulk}) + n_2u_2(\text{bulk}) + n_3u_3(\text{bulk}) + (\gamma_1 + \gamma_3 + 2\gamma_2)A \\ &\quad - \frac{A}{12\pi}(H_{13} + H_{22} - H_{12} - H_{23})\frac{1}{h^2} - \frac{A}{12\pi}(H_{12} + H_{23})\frac{1}{\delta^2} \\ &= \mathbf{U}_{123}(\text{bulk}) + (\gamma_{12} + \gamma_{23})A - \frac{AH_{123}}{12\pi}\frac{1}{h^2}\end{aligned} \tag{5.2.18}$$

In analogy with Eq. (5.2.11), the interfacial tension γ_{ij} is

$$\gamma_{ij} = \frac{1}{12\pi}\frac{1}{\delta^2}\left\{\frac{1}{2}(H_{ii} + H_{jj}) - H_{ij}\right\} \tag{5.2.19}$$

and the effective Hamaker constant H_{123} is

$$H_{123} = H_{13} + H_{22} - H_{12} - H_{23} \tag{5.2.20}$$

This expression can be rearranged to reveal how the effective interaction emerges as a difference between the respective dispersion interactions

$$H_{123} = \left\{H_{13} - \frac{1}{2}(H_{11} + H_{33})\right\} - \left\{H_{12} - \frac{1}{2}(H_{11} + H_{22})\right\} - \left\{H_{23} - \frac{1}{2}(H_{22} + H_{33})\right\} \tag{5.2.21}$$

We see in Eq. (5.2.21) that H_{123} contains contributions from three terms. Each one consists of $H_{ij} - \frac{1}{2}(H_{ii} + H_{jj})$—that is, the Hamaker constant for the interaction between two different media minus the average value of the interaction of the media with themselves. We encountered this combination of interaction constants when we described the regular solution theory in Chapter 1. For a symmetric situation in which media 1 and 3 are the same, Eq. (5.2.21) reduces to

$$H_{121} = H_{11} + H_{22} - 2H_{12} \tag{5.2.22}$$

showing even more clearly the similarity to Eq. (1.1.16). As we show in the following text, the constant H_{121} is always a positive quantity, and an attraction thus always exists between similar particles.

We also note in Eq. (5.2.22) that $H_{121} = H_{212}$ so that the dispersion contribution to the Hamaker constant is the same, for example, for water across hydrocarbon as for hydrocarbon across water.

Instead of providing a general proof that H_{121} is positive, we introduce an assumption that leads us to some useful simplifications. Using the relation (see Eq. 5.2.5),

$$H_{12} = \pi^2 C_{12} \rho_1 \rho_2 \tag{5.2.23}$$

for the Hamaker constant, we see that approximately

$$H_{12} \propto \rho_1 I_1 \alpha_1 \rho_2 I_2 \alpha_2 \tag{5.2.24}$$

where 1 and 2 are either equivalent or different media. When media 1 and 2 are different H_{12} is the geometrical mean of H_{11} and H_{22} so that

$$H_{12}^2 \approx H_{11} H_{22} \tag{5.2.25}$$

and it follows from Eq. (5.2.22) that

$$H_{121} \approx \left(\sqrt{H_{11}} - \sqrt{H_{22}}\right)^2 \tag{5.2.26}$$

and the positive sign of H_{121} becomes apparent.

When all three media illustrated in Figure 5.11 are different, H_{123} is negative if H_{22} is intermediate between H_{11} and H_{33}. In this case, combining Eq. (5.2.21) and the approximation in Eq. (5.2.25) gives

$$H_{123} = \left\{\sqrt{H_{22}} - \sqrt{H_{11}}\right\}\left\{\sqrt{H_{22}} - \sqrt{H_{33}}\right\} \tag{5.2.27}$$

which is negative if $H_{11} > H_{22} > H_{33}$ or equivalently $H_{11} < H_{22} < H_{33}$. This situation occurs most commonly when a liquid (2) wets a solid surface (1) in air (3). In Section 2.2 we discussed the method of measuring the surface tension of liquids using the capillary rise method (see Figure 2.7). When the liquid wets the solid material of the capillary tube a thin liquid film will rise on the walls. We are now in the position to analyze this phenomenon quantitatively for the case when the dispersion interaction dominates.

Consider a thin liquid ring of radius r at height h and the thicknesses Δr and Δh above the bulk liquid with density ρ, as shown in Figure 5.12. The area of the ring is $2\pi r \Delta h$ and its volume is $2\pi r \Delta r \Delta h$. The sum of the gravitational and dispersion energies for a ring at height h is

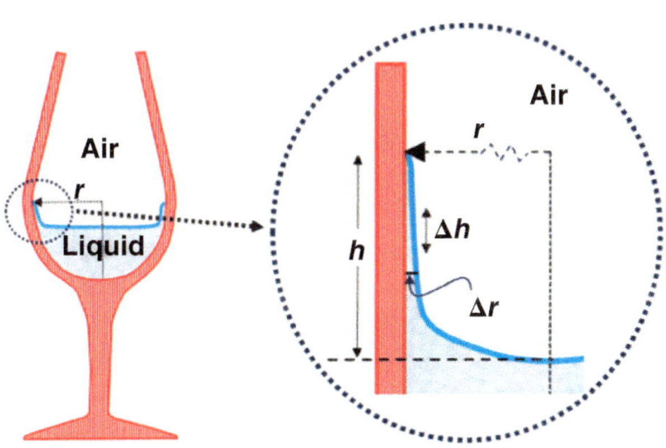

Figure 5.12 For the case when H_{123} is negative a thin film will climb the container surface. The film thickness Δr is determined by a compromise between surface forces and gravitational forces.

$$\Delta E_{\text{ring}}(\Delta r) = \rho g h \times 2\pi r \Delta h \Delta r + \frac{H_{123}}{12\pi \Delta r^2} 2\pi r \Delta h \tag{5.2.28}$$

The first term on the right-hand side increases with increasing thickness Δr while the second term decreases if the Hamaker constant H_{123} is negative. In this case there is an optimal film thickness when $\frac{\mathrm{d}\Delta E_{\text{ring}}(\Delta r)}{\mathrm{d}\Delta r} = 0$ which is satisfied when

$$\Delta r = \left(-\frac{H_{123}}{6\pi \rho g} \right)^{1/3} h^{-1/3} \tag{5.2.29}$$

Problem: Calculate the film thickness at $h = 1$ cm above the bulk surface for $H_{123} = -1 \times 10^{-20}$ J, $\rho = 1 \times 10^3$ kg/m³

Solution: The calculated film thickness is $\Delta r = 1 \times 10^{-20} / \left(6\pi \times 1 \times 10^3 \times 9.81 \times 0.01\right)^{\frac{1}{3}} \approx 18$ nm.

Figure 2.24 shows how such a film of a water-ethanol mixture is developed on a glass surface. Liquid helium represents an extreme case with a low density of $\rho \approx 1.4 \times 10^2$ kg/m³ and for a typical value $H = 6 \times 10^{-21}$ J, the calculated film thickness at a height $h = 1$ m is 6 nm. Since liquid helium has an unusually low viscosity, it has the tendency to flow out of containers through the liquid film formed on the container walls.

5.2.4 Lifshitz Theory Provides a Unified Description of van der Waals Forces Between Colloidal Particles

In his theory of nonideal gases developed at the end of the nineteenth century, van der Waals realized that some attractive interactions operate between atoms or molecules. He also noted that these attractions establish the gas–liquid equilibrium. As we have seen, several mechanistic sources contribute to these attractive forces, which are collectively called van der Waals forces.

Earlier we have discussed the dispersion interaction, which usually provides the main contribution to the van der Waals force. However, in molecular systems, the polar Keesom and Debye interactions treated in Chapter 3 can also make substantial contributions to the net attractive force. Even though we can derive the electrostatic interactions for isolated pairs of molecules, major difficulties arise if we want to calculate the net force in a condensed state.

As noted in Section 3.4, the dipole-induced dipole (Debye) interaction is nonadditive, so we cannot simply integrate it as a pairwise interaction to obtain the total force as we did for the dispersion force in Eq. (5.2.5). Similarly, the dipole–dipole interaction depends on the relative orientation of the two molecules, and in a liquid we must average over the rotational motion. Because we cannot average separately for each pair of molecules, the averaged dipole–dipole (Keesom) interaction is effectively nonadditive as well.

The pronounced many-body character of these two classical components of the van der Waals force between colloidal particles compels us to look for a different approach. For electrolyte solutions, we solved this dilemma by describing the medium as a dielectric continuum. Not surprisingly, the same solution applies in the present case, but going from a dielectric picture to actually calculating the force turns out to be much more difficult.

A group of Russian physicists led by Evgeny Lifshitz solved this problem in the 1950s. To give a detailed account of the Lifshitz theory is far beyond the scope of this text, but in short, it represents a continuum electrodynamics point of view in which each medium is characterized solely by its frequency-dependent relative permittivity $\varepsilon_r(\omega)$ (c.f. Section 3.5).

Although Lifshitz theory and Hamaker theory use very different approaches, the net results are similar, and the underlying molecular interactions are identical. In particular, Lifshitz and Hamaker theories both result in the same distance dependence of the interactions. For example, Eq. (5.2.7) remains valid, although it requires a different interpretation of the Hamaker constant. In Lifshitz theory, one can identify two contributions to the Hamaker constant. One comes from the dispersion interaction and is determined by the dielectric response at frequencies much higher than those associated with the thermal energy $k_B\mathbf{T}/\hbar$. The second, usually smaller, contribution is due to the thermally excited orientational degrees of freedom corresponding to the Keesom and Debye interactions. Here we simply cite an approximate expression for the Hamaker constant for media 1 and 3 separated by medium 2

$$H_{123} \approx \frac{3}{4}k_B\mathbf{T}\left(\frac{\varepsilon_1-\varepsilon_2}{\varepsilon_1+\varepsilon_3}\right)\left(\frac{\varepsilon_3-\varepsilon_2}{\varepsilon_2+\varepsilon_3}\right) + \frac{3\pi\hbar\upsilon_e}{4\sqrt{2}}\frac{(n_1^2-n_2^2)(n_3^2-n_2^2)}{\{(n_1^2+n_2^2)(n_3^2+n_2^2)\}^{\frac{1}{2}}\left\{\left(n_1^2+n_2^2\right)^{\frac{1}{2}}+\left(n_3^2+n_2^2\right)^{\frac{1}{2}}\right\}^{\frac{1}{2}}} \tag{5.2.30}$$

Here υ_e represents an average electronic excitation energy of order 4×10^{-13} s^{-1} and n_i is the refractive index of medium i. Importantly, the magnitude of the first "classical" term cannot exceed (3/4) $k_B\mathbf{T} \approx 3\times10^{-21}$ J. For the symmetric case, medium 1 same as medium 3, Eq. (5.2.30) simplifies to

$$H_{121} \approx \frac{3}{4}k_B\mathbf{T}\left(\frac{\varepsilon_1-\varepsilon_2}{\varepsilon_1+\varepsilon_2}\right)^2 + \frac{3\pi\hbar\upsilon_e}{8\sqrt{2}}\frac{\left(n_1^2-n_2^2\right)^2}{\left(n_1^2+n_2^2\right)^{3/2}} \tag{5.2.31}$$

Equations (5.2.30) and (5.2.31) provide a useful estimate of the Hamaker constant for insulators. For metals and semiconductors, the notion of an average excitation frequency ν_e is, however, not relevant. A metal is conducting for frequencies lower than the so-called plasma frequency. A consequence is that electron–electron correlations are easily established, resulting in a high value of the Hamaker constant. Table 5.1 shows values for H for some illustrative systems.

5.2.5 A Relativistic Correction Weakens the Dispersion Interaction at Long Range

In the expression for the dispersion interaction Eq. (5.2.1), there is a denominator determined by the excitation energies of the two molecules. For molecular separations exceeding the transition wavelength, a correction of relativistic origin appears, resulting in a change in the distance dependence of the interaction from R^{-6} to R^{-7}. This effective weakening of the interaction is

Table 5.1 Values of Hamaker constants in 10^{-20} J for some common combinations of materials *M*.

Material	M|Air|M	M|Water|M	M|Water|Air	M|Air|Water
Water	3.7	0.0	0.0	3.7
Alkanes				
$n = 5$	3.8	0.3	0.15	3.6
$n = 6$	4.1	0.4	0.0	3.8
$n = 10$	4.8	0.5	−0.3	4.1
$n = 14$	5.1	0.5	−0.5	4.2
$n = 16$	5.2	0.5	−0.5	4.3
Fused quartz	6.5	0.8	−1.0	4.8
Fused silica	6.6	0.8	−1.0	4.8
Sapphire	16.0	5.0	−3.8	7.4
Polymethyl methacrylate	7.1	1.1	−1.3	5.0
Polystyrene	6.6	1.0	−1.1	4.8
Polyisoprene	6.0	0.7	−0.8	4.6
Polytetrafluoroethylene	3.8	0.3	0.1	3.7
Mica (green)	10.0	2.1	—	—

commonly called the retardation effect. For molecules, the effect is too weak to be relevant except under extreme conditions. However, when we derived the interaction between two media in Section 5.2, the interaction at a given separation h was found to also have contributions from molecules at separations far exceeding the gap size h. Consequently, the retardation effect becomes significant at separations of only a fraction of the relevant wavelengths, and the effect can have some practical consequences.

Prior to the emergence of Lifshitz theory, in 1948, Casimir derived, in 1948, an expression for the interaction between two metals. Like Lifshitz, he used a field theoretical approach and assumed that the metal was conducting at all frequencies and found an attractive interaction proportional to h^{-3} at all separations. This is called the Casimir force, and the distance dependence is stronger than the h^{-2} dependence of Eq. (5.2.7), which is valid for molecular, nonconducting materials. The consistent h^{-3} dependence is a consequence of the unrealistic assumption of perfect conductivity of the metal at all frequencies, including the high-energy spectrum, leading to retardation of the force at all separations. The Casimir force is thus best seen as a retarded version of the dispersion force.

See Atkins et al. (2022, Focus 14), Berg (2010, Chapter 7), Cosgrove (2010, Chapter 16), Israelachvili (2011, Chapter 13), and Ohshima (2024, Chapter 8) for additional discussion on Van der Waals Forces.

5.3 Electrostatic Interactions Generate Attractions by Correlations

In the two previous sections, we have described condensed phase systems in terms of their average densities or electrolyte concentrations. In reality, there are local variations in these parameters due to the molecular nature of liquids and solids. Charged systems, in particular, can show density correlations on a length scale relevant for colloidal systems. In this section, we describe a number of cases where such correlations influence the force between surfaces in a significant way.

5.3.1 Ion–ion Correlations Can Induce Attraction Between Similarly Charged Surfaces

In Section 5.1, we analyzed the electrostatic double-layer force on the basis of the mean field PB equation and found that a repulsion always exists between similar surfaces. In Section 5.2, on the other hand, we concluded that correlations between electrons of separate molecules always give rise to an attraction between particles made from similar materials. These two basic forces dominate in many colloidal systems, but even within the framework of electrostatic double-layer interactions, more subtle effects can be of major importance under some circumstances. To appreciate the significance of these effects, we must go beyond the description in which we treat electrostatic charge distributions in terms of their mean value.

Charges in colloidal systems are localized to ionic or dipolar groups, and the charge distribution varies strongly between different configurations of ions and molecules. An averaging due to thermal molecular motions leads to a more uniform, mean field, charge distribution. In general, the motion of interacting molecules or ions are, however, correlated, and if we consider only their mean charge distribution, we ignore important physical effects that arise due to these correlations. Figure 5.13 illustrates how, for a specific ion configuration, electric fields penetrate beyond the

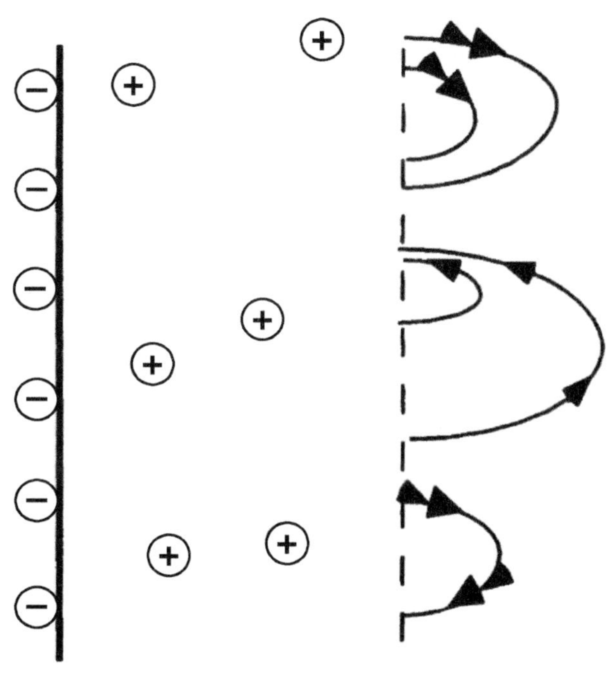

Figure 5.13 The neutralizing counterions are never uniformly distributed outside a charged surface. For each instantaneous configuration of these counterions, there will be a field $\boldsymbol{E}(\boldsymbol{r})$ generated at position $\boldsymbol{r}$ in the opposing half plane. On overage $\langle \boldsymbol{E}(\boldsymbol{r}) \rangle = 0$ but $\left\langle \boldsymbol{E}^2(\boldsymbol{r}) \right\rangle \neq 0$ for positions beyond the midplane. The figure illustrates schematically for one particular configuration the electrical field lines, i.e., lines of constant field magnitude $|\boldsymbol{E}(\boldsymbol{r})|$ with the arrow pointing in the direction of $\boldsymbol{E}$.

average neutral plane. Ions on a nearby surface can, in turn, respond to these fields, potentially inducing an attraction which is absent in the mean field description.

The stronger the interaction between counterions, the larger is their positional correlation. The Coulomb interaction (Eq. 3.5.15) is proportional to the square of the charge and inversely proportional to the relative permittivity, ε_r. Aqueous solutions have a large ε_r, in the range 75–80, and correlation effects are small for systems containing monovalent counterions. However, for highly charged systems with divalent or multivalent counterions correlations can become significant and, as illustrated in Figure 5.14, can even change repulsion to attraction.

The charge density of 0.2 C/m^2 where the force changes from repulsive to attractive in Figure 5.14, can for example be found in lamellar liquid crystals of charged amphiphiles. For aqueous systems with monovalent ions, corrections to the mean-field description occurs at significantly higher charge densities. It is, for example, difficult to account for the properties of black films formed by ionic surfactants without invoking an ion correlation correction to the estimate of the repulsive charge-charge repulsion as will be further discussed in Chapter 11. However, a reversal from repulsion to attraction does not seem to occur in practice for aqueous systems with monovalent counterions.

A very important manifestation of the correlation attraction is the formation of concrete. In the mixture of water, cement powder, and sand particles, the dissolution of the cement leads to a gradual increase in calcium ion concentration. In addition, the *p*H reaches high values in the range *p*H ≈ 13–14, leading to the sand particles becoming strongly negatively charged. The ion correlation effect then induces an attraction between the sand particles and the concrete solidifies. Contamination of the water by monovalent salts has a screening effect on the attraction, while

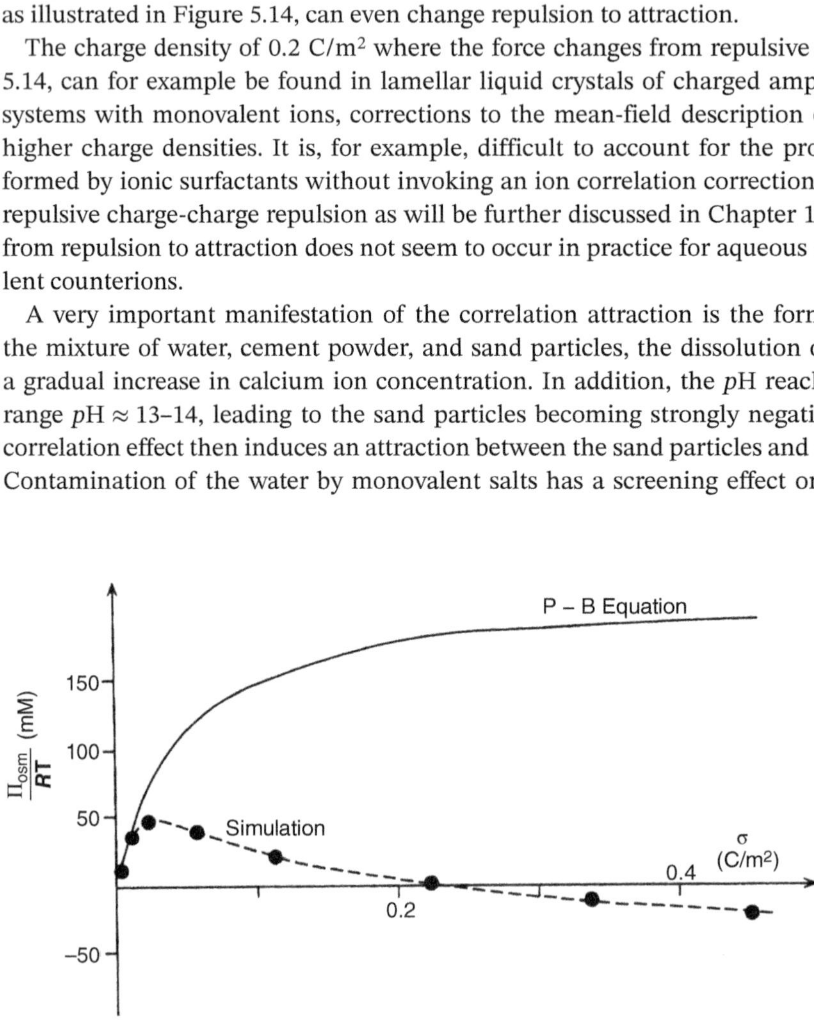

Figure 5.14 Force calculated between two parallel identical charged surfaces separated by a distance $h = 2.1$ nm in a solvent with $\varepsilon_r = 80$ and containing only divalent counterions. The solid line represents the PB calculation and circles the Monte Carlo simulations. The force is expressed in terms of the osmotic pressure divided by ***RT*** so that 1 mM corresponds to 2500 N/m^2. The Monte Carlo simulation explicitly considers the instantaneous ion–ion interactions to account for the additional attractive ion-ion correlation component to the force. For low surface charge densities σ, the attractive component is negligible even for divalent counterions, but becomes substantial for $\sigma > 0.03$ C/m^2. Under the given conditions, the net force is zero at $\sigma = 0.22$ C/m^2 and for even larger charge densities, the attractive component dominates and the PB calculation is qualitatively wrong. *Source*: Guldbrand et al., 1984/with permission of AIP Publishing.

contamination by, for example, sugars induces a buffering effect, which reduces the pH, leading to a reduced surface charge density and thus also reducing the strength of the interaction.

With ions of valency 3 or higher, correlation attractions can appear already at low concentrations. Al^{3+} ions can induce attraction even between moderately charged particles and this effect has been extensively used to induce particle precipitation in waste-water treatment. This effect is also causing the high toxicity to freshwater fish of Al^{3+} ions, which act to clog their gills. Similarly, Al^{3+} ions are efficient in deodorants through the clogging of sweat glands. Oligomeric amines like spermidine (valency +4) furthermore owe their biological function to inducing compaction of single-stranded DNA in the sperm through this electrostatic attraction mechanism.

A subtle correlation effect can appear when the counterion species has a higher charge than that of the individual charged groups on the surface. Counterions can then be attracted to the surface to the extent that they "over-neutralize" the surface charges. Seen from a distance, the overall surface charge is then reversed, leading to a double-layer repulsion at long separations where counterions and coions have exchanged roles.

5.3.2 Surface Dipoles Correlate to Yield an Attraction

In a colloidal system, the chemical composition of the surface layer often differs markedly from the bulk of the particle. Two approaching particles will "see" each other through these surface layers, and the interaction between the surface layers becomes increasingly important as the separation between the particles decreases. Lifshitz theory includes an asymptotically leading electrostatic bulk correlation term. However, strongly polar groups carried by the surface layers can cause an additional dipole-dipole correlation contribution that can dominate at intermediate to short separations.

Aggregates composed of zwitterionic or dipolar amphiphiles have highly polar surface layers where electrostatic interactions play a dominant role. This applies to solvent–headgroup as well as headgroup–headgroup interactions within the aggregate. Consequently, we should also expect headgroup–headgroup interactions between aggregates to be important at close approach. As an example, we can consider an amphiphile with a zwitterionic headgroup. On a planar interface, we can resolve the dipole vector of the head group into two components that point perpendicular and parallel to the surface, which give rise to somewhat different types of correlations.

Let us first consider the parallel component. A dipole oriented along a surface is free to rotate because it has no intrinsic directional preference. Averaged over all the in-plane orientations, the net dipole moment vanishes, so the average electric field component from the in-plane dipole is zero. A second dipole on an opposing surface can, however, interact favorably with the source dipole by preferentially adopting an opposite orientation. The result is an attractive correlation force similar to the Keesom interaction we discussed in Section 3.3. We can obtain an order-of-magnitude estimate of this force by assuming that the dipoles on the same surface are sufficiently separated to be treated as approximately independent. An integration using the Hamaker approach results in

$$\frac{F}{A} = -\frac{\pi \rho^2 m^4}{(4\pi\varepsilon_0\varepsilon_r)^2 k_B \mathbf{T} h^5} \tag{5.3.1}$$

where ρ represents the surface density of dipoles and m the in-plane dipolar component.

From Lifshitz theory, we also obtain the h^{-5} distance-dependence for two interacting thin layers. In fact, for more realistic cases of strong intrasurface correlations the magnitude of the force will change, but the h^{-5} distance-dependence remains valid. In this case, the force is very sensitive to molecular details and the magnitude can vary by an order of magnitude.

For the perpendicular component of the zwitterionic dipole, the only accessible degree of freedom is a translation along the surface. As a result, the thermal averaging is different relative to that of the parallel component. When two surfaces approach, their perpendicular dipolar components are oppositely oriented. At first glance, one might be tempted to conclude that this orientation produces a repulsive interaction between the surfaces. While this is true for two dipoles in juxtaposition, electrostatic interactions are long-ranged, and the dipole–dipole interactions change sign when the dipolar orientation and the vector joining the dipoles form the so-called magic angle of $\theta \approx 54°$ making $3\cos^2(\theta) - 1 = 0$. For a uniform distribution of dipoles, the repulsive interaction between dipoles close to juxtaposition is canceled exactly by an attractive interaction between dipoles that are laterally farther apart see Figure 5.15.

Although the mean-field interaction between the two uniform dipolar distributions is zero, an attractive component appears when we allow for correlations in the lateral positions of the dipoles on the two surfaces. If we fix a dipole at some position on one surface, interactions will affect the dipolar density at the corresponding position on the other surface. For a weakly correlated system in which we can treat the dipoles as independent, the net force is the same as for the parallel dipoles. Equation (5.3.1) then applies, where m now instead denotes the out-of-plane dipolar component.

In the other extreme of very strong intrasurface correlations, as in a lattice of dipoles, we can obtain an attractive interaction by displacing the lattices by half a unit cell. This correlation, which formally occurs in only a single degree of freedom, produces an energy gain that easily exceeds $k_B\mathbf{T}$. The electric field outside an ordered surface decays exponentially on a length scale given by the lattice spacing a. Consequently, the force is short-ranged, but strong. An approximate expression for the force is

$$\frac{F}{A} \approx \text{const} \times \frac{m^2}{\varepsilon_r \varepsilon_0 a^6} \exp\left(-\frac{h}{\lambda}\right) \tag{5.3.2}$$

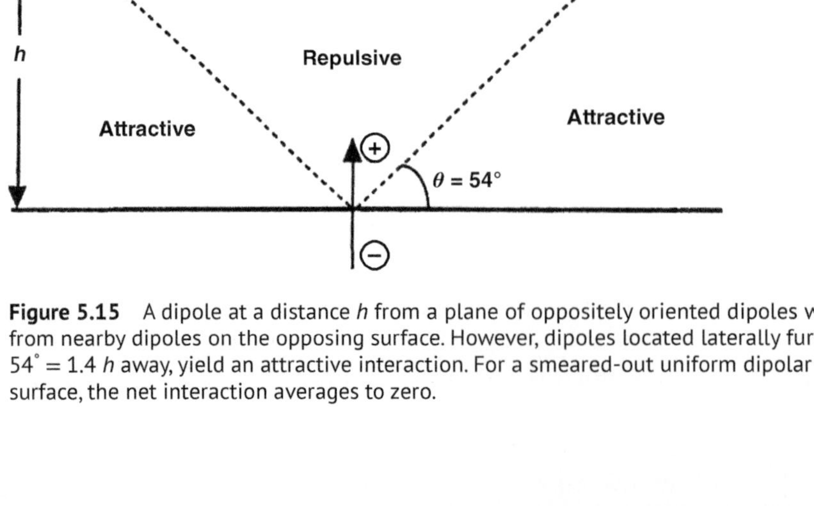

Figure 5.15 A dipole at a distance h from a plane of oppositely oriented dipoles will sense a repulsion from nearby dipoles on the opposing surface. However, dipoles located laterally further, more than h tan $54° = 1.4\,h$ away, yield an attractive interaction. For a smeared-out uniform dipolar distribution on the surface, the net interaction averages to zero.

where the decay length $\lambda \approx a/6.88$ and the constant prefactor is of order 100. This force, which often exceeds the dispersion contribution at short range, provides extra cohesion to polar solids. Ultimately, a strongly correlated system also exhibits the asymptotic h^{-5} dependence of the force, but in this case, the asymptotic regime is attained only at very long separations where the force is insignificant.

5.3.3 Domain Correlations Can Generate Long-range Forces

In the two previous subsections, we discussed attractive interactions that occur due to correlations between individual molecular degrees of freedom; that is, the positions of ions or the orientations/position of dipolar groups. The same basic electrostatic correlations also can operate with entities of colloidal dimensions rather than molecular ones. Some quantitative differences occur because the length scale changes, and with larger entities, the interaction per degree of freedom is more likely to exceed the thermal energy. For such a mechanism to operate, the surfaces must show inhomogeneities on the colloidal scale, which can be generated in several ways.

Surfactants can, for example, adsorb on a surface in the form of a charged micellar aggregate attracted to an oppositely charged surface. (Figure 2.21 shows an example.) The result might be a surface, which is net neutral but still has regions of opposite charge. A similar situation can occur when a highly charged polyelectrolyte adsorbs to yield partial coverage on an oppositely charged surface. Figure 5.16 illustrates a third possibility: lateral phase separation into charged domains. In this case, a perpendicular dipole moment is associated with domains that interact laterally by electrostatic forces. Thus, two such surfaces may interact across a medium. A fourth alternative is for the surface to form a solid structure of low symmetry. In this case, different domains of the same phase can carry an in-plane dipole moment.

In all these cases, a nonuniform charge or dipolar distribution along the surface gives rise to an electric field in the medium outside the surface. This field varies in sign laterally on the length scale of a domain size and decays in amplitude in the perpendicular direction on the same length scale, as illustrated in Figure 5.16. A second surface can respond to this field to generate correlation attraction just as in the case of zwitterionic head groups. However, because larger entities produce stronger interactions, correlations also can develop in the distance regime of 10–100 nm. On the other hand, rearrangement processes are slow, so that correlations may not have time to develop during the timescale of an experiment.

Asymptotically, a formula analogous to Eq. (5.3.1) with an h^{-5} distance dependence can be derived. However, a more interesting regime is the strongly correlated situation arising for short separations, described by (Eq. 5.3.2). Up to this point, our description has neglected the influence of a screening electrolyte. This complication is not a major concern for molecular dipoles because their electric fields decay rapidly. When considering interactions between correlated domains,

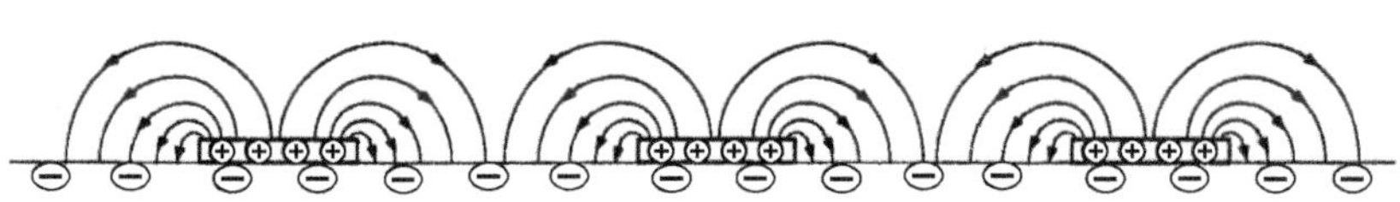

Figure 5.16 A laterally nonuniform charge distribution can arise when a positively charged particle or aggregate adsorbs on a negatively charged surface of a charge density of lower magnitude. Electrical fields are generated outside the surface and these extend out from the surface on the same length scale as the in-plane inhomogeneity as shown by the electrical field lines in the figure.

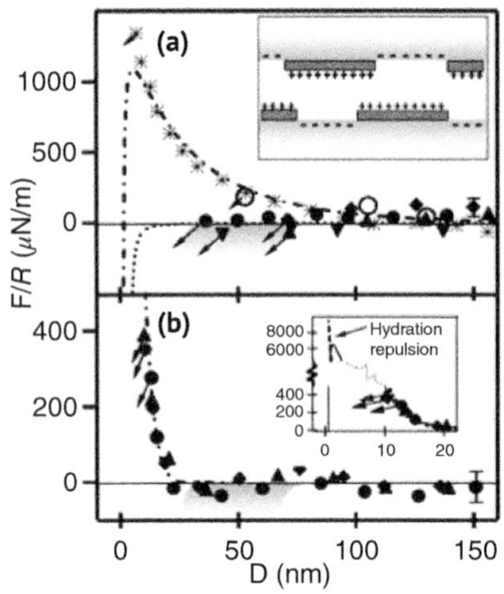

Figure 5.17 Force between curved surfaces measured by the SFA method (see Section 5.4.2). The cationic surfactant CTAB was adsorbed on negative mica surfaces and the force was measured for varying times after CTAB addition. (a) Pure aqueous solvent; (*) bare mica, filled symbols surface partly covered by CTAB. Arrows indicate an attraction instability which occurs at a range that increases with immersion time. The dotted line is an estimate of the van der Waals attraction. (b) 0.01M NaCl solution. A repulsion due to a net surface charge dominates at larger separations, but it is overtaken by an attraction at separations below approximately 10 nm. *Source*: Perkin et al., 2006/with permission of American Physical Society.

however, we are treating much longer length scales where screening should have a strong effect. From a model based on in-plane dipolar domains of radius R and a charge distribution $\sigma = \sigma_0 \cos\phi$, we can solve the linearized PB equation to obtain an explicit equation for the interaction between two similar surfaces. The calculated force is

$$\frac{F}{A} \approx -\frac{\sigma^2}{2\varepsilon_r\varepsilon_0}\left\{\exp\left[\left(\kappa^2 + \frac{3.4}{R^2}\right)^{1/2}\frac{h}{2}\right] + \exp\left[-\left(\kappa^2 + \frac{3.4}{R^2}\right)^{1/2}\frac{h}{2}\right]\right\}^{-2} \tag{5.3.3}$$

This result shows that the range of the force is determined by either the Debye screening length κ^{-1} or the domain radius R—whichever is shorter.

Figure 5.17 demonstrates this competition between R and κ in determining the range of the interaction. In pure water the range is determined by the domain size which grows gradually with increasing immersion time. For a 0.01 M NaCl solution the range of the force is decreased and less sensitive to domain size. Finding a force like the one represented in Eq. (5.3.3) requires rather special conditions, but when these are present, they produce striking effects because the force is very strong and easily dominates over the dispersion force, even at separations that are large on the colloidal scale.

5.3.4 An Electrostatic Repulsion Also Exists in Very Concentrated Salt Solutions and Ionic Liquids

In ocean water, the salt concentration is around 0.5 M and the Debye length (Eq. 3.8.13a) is approximately 0.4 nm and of similar magnitude as one ionic diameter. For still higher salt contents the Debye length becomes gradually smaller, leading to a complete screening of the electrical double layer repulsion. However, it has been found experimentally that for ionic liquids, where the liquid molecules themselves are charged, there is a repulsive force acting between charge surfaces in spite of a negligible Debye length. This repulsive force, which also occurs in very concentrated salt solutions, has an approximately exponential decay but with a decay length which increases with increasing salt concentration (see Figure 5.18), in contrast to the ordinary double layer repulsion.

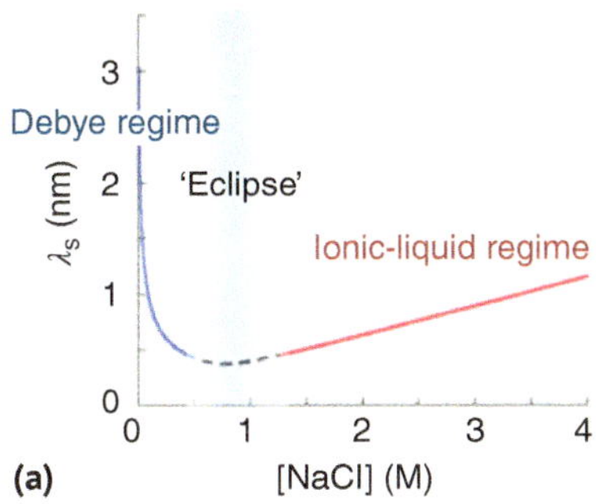

Figure 5.18 Representation of how the decay length of the interaction depends on the electrolyte concentration. Below 0.5 M, there is a Debye screening due to the correlation between a central ion and its counterion cloud. Between 0.5 M and 1.5 M, there is a complete screening of charges, effectively eliminating the electrostatic interaction. Above 1.5 M, strong ion–ion correlations start to reduce the screening so that the screening length increases with increasing salt concentration. *Source*: Wennerström et al., 2020/with permission of National Academy of Sciences.

In the concentrated salt solution, as well as in the ionic liquid, the ions interact strongly resulting in strong positional correlations. A cation is surrounded by anions, which in turn are preferentially surrounded by other cations. The result is that the radial distribution function has the character of a damped oscillation, where the periodicity is determined by the ion size and the decay length by the magnitude of correlations. A similar oscillation in the ion distributions is also present outside charged surfaces. When two similar charged surfaces approach the counterion charge densities start to overlap at separations of order twice the correlation length. In a way similar to what was discussed for the conventional double layer repulsion, the counterion charge will be pushed toward the surface by the presence of the other surface and this generates a repulsion.

The organization of the metabolic processes in a living cell depends on the existence of repulsive electrostatic interactions to avoid molecular aggregation. It is interesting to note that common salt has worked as an effective preservative for thousands of years, most likely through its effect to screen the electrostatic repulsion. Still, there are organisms that can grow in close to saturated salt solutions, such as in the Dead Sea. At such massive salt concentrations, the mechanism discussed above ensures the existence of an electrostatic repulsion between similarly charged species and an attraction between oppositely charged ones, even at these high salt concentrations. An adapted cell can thus control the metabolic processes using the same organizing principles as those of cells growing under low salt conditions.

See Atkins et al. (2022, Focus 14), Berg (2010, Chapter 6), Cosgrove (2010, Chapter 16), Doi (2013, Chapter 10), Israelachvili (2011, Chapter 14), Lyklema (1995, Vol. 2, Chapter 3), and Ohshima (2024, Chapters 5–7) for additional discussion on Electrostatic Forces.

5.4 Measuring Surface Forces

The properties of colloidal systems are to a large extent determined by the forces acting between the particles. It is, however, experimentally challenging to explicitly measure these forces and an equally large challenge to theoretically interpret the observations and disentangle the physical mechanisms resulting in a measured net force. Above, we discussed the most important generic contributions to the total force between two surfaces. Before continuing this overview of forces, we give a brief description of how they can be measured.

Force measurements are in practice based on one of two basic principles. In the most direct methods, one measures a force mechanically, for example through the deflection of a spring, while simultaneously monitoring the separation. A more indirect way to measure interactions is to vary some thermodynamic properties of the system and measure distances by scattering or interference methods. The measurement of double-layer repulsion in soap films shown in Figure 5.10 provides an example of the latter: by varying the electrolyte concentration in the solution, one can induce changes in film thickness that can be measured optically. In the coming chapters we will give further examples of such methods for indirectly measuring colloidal interactions, but in this section, we concentrate on the direct force measurement techniques. To appreciate the versatility of these experimental methods we will first introduce an additional theoretical tool, namely the Derjaguin approximation.

5.4.1 The Derjaguin Approximation Relates Forces Between Curved Surfaces to the Interaction Free Energy Between Planar Surfaces

In 1934, the Russian colloid scientist Boris Derjaguin published a derivation of the relation between the force $F(h)$ between curved surfaces and the (free) energy of interaction per unit area, $\mathbf{G}(h)/A$, acting between two infinite, planar surfaces of the same material(s). The simplest example is the force acting between a spherical particle of radius R and an infinite plane, separated by a smallest distance h, which is given by

$$F(h)=\frac{2\pi R\mathbf{G}(h)}{A} \tag{5.4.1}$$

As will be apparent below, this type of relation between the force between curved surfaces and the interaction free energy between planar surfaces is very general, although the prefactors vary depending on the exact geometry. The various incarnations of the Derjaguin approximation constitute extremely useful relations from both theoretical and experimental points of view. In theoretical treatments based on either analytical or simulation approaches it is typically easier to consider the parallel plate geometry rather than that involving curved surfaces. It also follows that interaction mechanisms involving curved surfaces are also operating between planar surfaces and vice versa. This brings about a conceptual simplification. Chapter 8 is devoted to interactions between colloidal particles and the Derjaguin approximation will be used extensively there to apply the relations derived in Sections 5.1 and 5.2 for planar walls to interactions between spherical particles.

To appreciate the basis of this relation we first consider the case of an infinite plane and a sphere of radius R, both homogeneous and composed of molecules that interact through a pairwise additive interaction $V_{12}(r)$. The plane and the sphere are placed with the sphere center a distance $R + h$ from the plane, as shown in Figure 5.19. When $R \gg h$, we can calculate the force F acting on the sphere by considering the interaction energy $\mathbf{U}$ at two positions displaced by an infinitestimal amount dh,

$$F=\frac{\mathbf{U}(h)-\mathbf{U}(h+\mathrm{d}h)}{\mathrm{d}h} \tag{5.4.2}$$

As h/R is small, we need to consider only the shell of thickness $\mathrm{d}h$ closest to the planar body; the shell on the back side of the sphere makes a negligible contribution to F because the large separation makes the interaction potential negligible.

As a lateral distance r from the point of closest approach, the shell is at a distance h_1 from the surface. A circular strip with a rhombic cross-sectional area of and a volume $2\pi r\mathrm{d}r\mathrm{d}h$ exists in the range h_1 to $h_1 + \mathrm{d}h$. By simple geometry $h_1 = h + r^2/2R$ as long as $R \gg r$ (see Figure 5.19). At a distance h_1 from the surface, Eq. (5.2.8) gives the interaction energy per unit volume $V_{12}(h)$.

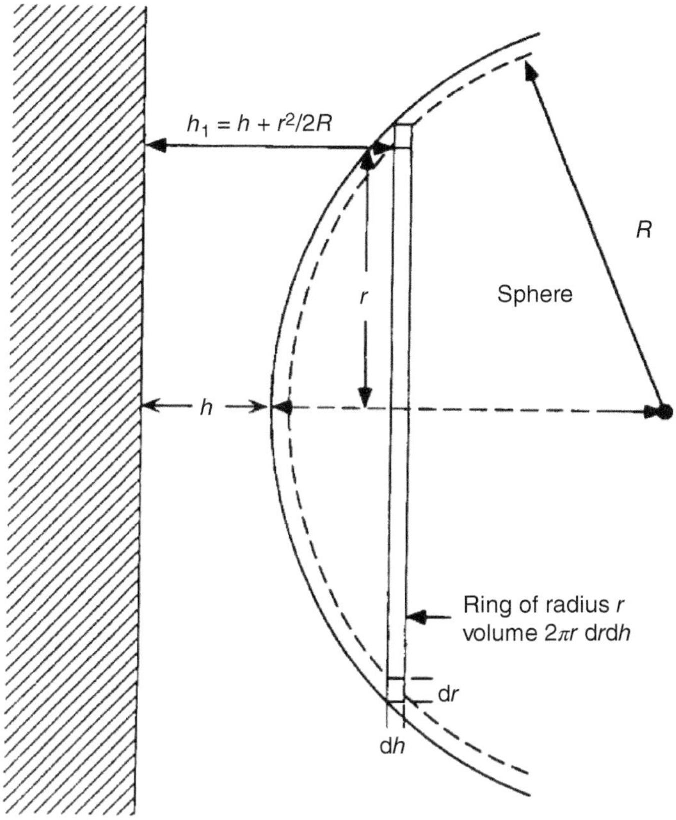

Figure 5.19 Calculation of the force between a half-space and a sphere of radius R. The calculation considers the interaction between the half-space and the shell of thickness dh, obtained as the difference between spheres with centers at $R + h$ and $R + h + \mathrm{d}h$. The thin shell can be divided into rings of radius r, cross-sectional area d$r \times$ dh, volume $2\pi r\mathrm{d}r\mathrm{d}h$ and distance $h_1 = h + r^2/2R$ to the planar surface. Integrating over all radii r gives the net force, assuming that the interaction is non-negligible only when $h_1 << R$. The figure displays a cut through the center of the sphere.

From Eq. (5.4.2) we obtain the total force by summing over all the circular strips—that is, by integrating over r and

$$F_{12} = \int_0^R 2\pi r\rho V_{12}(h_1)\mathrm{d}r \approx 2\pi R\rho \int_h^\infty V_{12}(h_1)\,\mathrm{d}h_1 \tag{5.4.3}$$

using the geometrical relation between h_1 and r. We have extended the upper limit in the integration because we assume that $V_{12}(h)$ is negligible for h_1 of order R.

Let us compare Eq. (5.4.3) with the expression for the interaction $\mathbf{U}_{12}$ between two planar surfaces, as given in Eq. (5.2.5). The final integration step in Eq. (5.2.7) amounts to an integration over planar sheaths of thickness dz_2 and inserting the expression for V_{12} in Eq. (5.2.8) we find that

$$\mathbf{U}_{12}(h)/\text{area} = \rho\int_h^\infty V_{12}(z)\mathrm{d}z \tag{5.4.4}$$

The integrals in Eqs. (5.4.3) and (5.4.4) are the same, except for the integration variable so that

$$F_{12}(h) = 2\pi R \frac{\mathbf{U}_{12}(h)}{\text{area}} \tag{5.4.5}$$

For this relation to be valid the radius R should be larger than both h and the range of the potential $V_{12}(h)$.

Equation (5.4.5) demonstrates how the force operating between a sphere and a plane relates to the interaction energy between two planar surfaces. The Eq. (5.4.4) makes use of an additivity assumption, but we assume nothing particular about the potential $V_{12}(r)$ except that its range is short relative to R. In fact, the result has a far-reaching validity. In a general application of the Derjaguin approximation we should replace the energy $\mathbf{U}(h)$ in Eq. (5.4.5), as well as in Eqs. (5.4.6a) and (5.4.6b), with the free energies $\mathbf{G}(h)$ or $\mathbf{A}(h)$. A relation equivalent to Eq. (5.4.5) exists for the interaction between two crossed cylinders with radii R_1 and R_2

$$F(h) = 2\pi\sqrt{R_1 R_2}\,\frac{\mathbf{G}(h)}{A} \tag{5.4.6a}$$

As we will see below this relation provides the basis for interpreting the results obtained by the surface forces apparatus. For two spheres, a derivation along the same lines results in

$$F(h) = 2\pi \frac{R_1 R_2}{R_1 + R_2} \frac{\mathbf{G}(h)}{A} \tag{5.4.6b}$$

To obtain the dispersion interaction energy between the spheres at short separations we can integrate Eq. (5.4.6b) using

$$\mathbf{G}(h) = \mathbf{U}(h) = -\frac{H_{12}}{12\pi}\frac{1}{h^2} \tag{5.4.7}$$

to yield

$$\mathbf{U}_{\text{sphere}}(h) = -\frac{R_1 R_2}{6(R_1 + R_2)} H_{12} \frac{1}{h} \text{ (}h\text{ small)} \tag{5.4.8}$$

Thus, a slow variation of the interaction with distance emerges in clear contrast to the interaction for spheres at long separations, $h \gg R_1, R_2$, where the R^{-6}dependence of the force enters directly and

$$\mathbf{U}_{\text{sphere}}(h) = -\left(\frac{4}{3}\right)^2 R_1^3 R_2^3 H_{12} (h + R_1 + R_2)^{-6} \text{ (}h\text{ large)} \tag{5.4.9}$$

The Derjaguin approximation is basically of geometric origin, and it is valid under conditions that go beyond those of the derivation above. It can be shown to apply also within Lifshitz theory, eliminating both the condition of pairwise additivity and angular independence as implicitly assumed for $V_{12}(r)$ above. In the following text, we will discuss interaction forces that are influenced by a change in the density or composition of the fluid between the surfaces. The Derjaguin approximation remains valid also for such situations, provided that the density and orientation correlation lengths in the fluid are short relative to R.

5.4.2 Surfaces Forces Can Be Accurately Measured Using the Surface Force Apparatus

The most accurate determinations of surface forces so far have been obtained using the Surface Force Apparatus (SFA), which can be used to measure forces acting over distances up to 10^3 nm with a

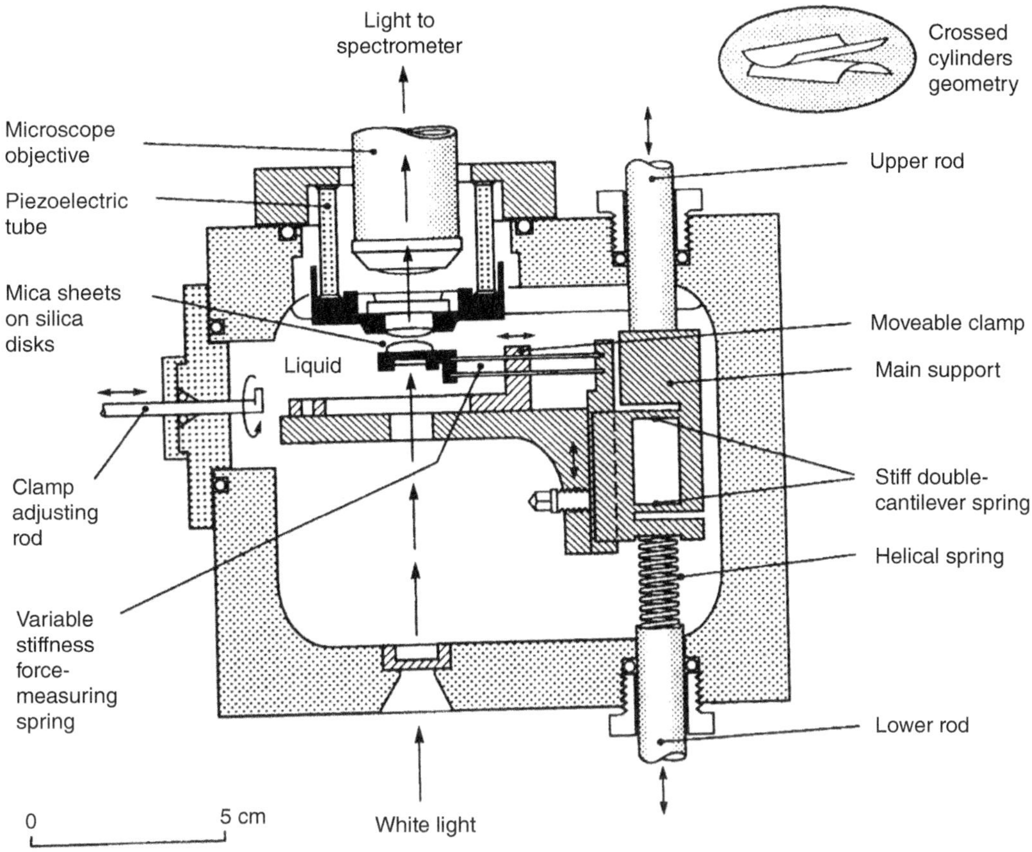

Figure 5.20 Surface forces apparatus (SFA), which permits a direct measurement of forces between molecularly smooth surfaces. *Source*: Israelachvili, 1992/with permission of Elsevier.

resolution of 0.2 nm. Such accurate measurements of colloidal forces between surfaces, however, assume control of their properties on a molecular level. To get reproducible results, the surfaces must be clean, homogeneous, and molecularly smooth; conditions that are generally difficult to achieve on the surface of a bulk material. In the SFA, this is typically solved by preparing surfaces using mica, which is a clay material that can be cleaved into macroscopically large pieces of thin, molecularly smooth planar sheets. In the experimental setup, the sheets are silvered on the back before being glued onto two glass lenses, shaped as cylindrical sections with radii $R \approx 1$ cm. The two lenses are mounted at right angles in the SFA, attaching the lower one to a force-measurement spring and the upper one to a piezoelectric crystal (see Figure 5.20).

A central feature of a force measurement is to control and monitor the separation between the two surfaces. In the SFA this is accomplished by measuring the separation by passing white light through the thin mica surfaces. Partial reflection at the silvered surfaces generates interference patterns of equal chromatic order known as fringes. We can analyze these fringes in a spectrometer and measure the change in separation by the change in the wavelength of a particular fringe as two surfaces approach one another.

A three-stage mechanism controls the separation between the surfaces. The upper rod permits us to move the lower surface over large distances between 1 μm and 1 cm, but it is not used during the actual force measurement. A double-cantilever spring, about 1000 times stiffer than the helical spring attached to the lower lens, operates the lower rod. This differential spring mechanism makes it possible to reduce a 1 μm movement of the lower rod to a 1 nm relative movement of the two mica surfaces. The piezoelectric crystal attached to the upper mica surface then controls small movements of 0.1 to 1 nm.

If we advance the lower micrometer rod when the two surfaces are so far apart that no interaction force can be detected between them, the main support of the double-cantilever spring and the mica surface move at the same rate. However, when we push the mica surfaces so close together that a repulsive force arises, the weak force-measurement spring bends downward. As a consequence, if we optically measure the distance actually moved, it is smaller than we would expect (e.g. 4 rather than 5 nm). The spring, therefore, must have bent 1 nm. If we push the surfaces still closer together, we can record a degree of deformation in the force measurement spring corresponding to each separation h. By measuring the spring constant independently, we can obtain the total force between the mica surfaces as a function of separation distance.

The arrangement with two cylinders oriented at right angles may seem odd but it has two great advantages. First it reduces the problem of aligning the surfaces relative to one another since a slight deviation from exactly 90° orientation has only a negligible effect on the measured force. The second point is that one can apply the Derjaguin approximation so that the measured force F at surface separation h is related to the free energy of interaction $\mathbf{G}$ between planar surfaces through

$$\frac{\mathbf{G}(h)}{A} = \frac{F(h)}{2\pi R} \tag{5.4.10}$$

Native mica crystallizes in layer structures. On cleavage it has a claylike surface with one potassium ion site per 0.50 nm^2. In liquids with low dielectric constant, mica behaves as a neutral surface but in ionizing solvents like water, potassium counterions dissociate from the surface, generating an electrical double layer, and we can measure the interaction forces between the charged surfaces as a function of salt concentration. The theoretical description of double-layer forces given in Chapter 5 has been tested extensively in this way. We obtain a quite different surface in the presence of cationic surfactants, which exhibit a strong Coulomb interaction with the negatively charged mica surface. This interaction leads to a strong surfactant adsorption on the surface. This effect can be used to adsorb an amphiphile bilayer on the surface. In general, the molecular nature of the exposed surface in the SFA can be varied using chemistry, but there are limits to the possibilities. It is furthermore very difficult to create a well-controlled asymmetric system with two chemically different surfaces.

5.4.3 The Atomic Force Microscope Provides an Alternative Way to Mechanically Measure Forces Between Colloidal Particles

In Section 2.6, we described how the atomic force microscope (AFM) can be used to monitor the topography of a surface. This device can also be used to efficiently measure the force between a surface and the probing tip. As in the SFA, the force is monitored by determining the deflection of a cantilever spring. In the absence of a force, the spring moves the same distance as the nominal separation between the tip and the surface. When an interaction appears between them, a deflection results which is measured optically by a laser beam reflected on the cantilever spring whose position is recorded on a detector. As with the SFA method, the AFM measurement relies on the

Derjaguin approximation to convert the measured force to an interaction-free energy between flat surfaces.

The AFM method has some advantages compared to the SFA method. When the tip approaches the surface, this needs to be flat on a lateral length-scale large only with respect to the radius of curvature of the tip, which is a much less stringent condition than for using the SFA. Consequently, one can monitor a heterogeneous surface in the lateral direction and probe how the interaction varies over the surface, as long as the surface is heterogeneous over length scales larger than the tip radius. One can also attach a colloidal particle to the tip, which provides a convenient way to study how forces vary with the chemical nature of the surfaces. The system is furthermore inherently asymmetric, providing rich possibilities for studying asymmetric pair interactions; it is even possible to measure forces involving single macromolecules. However, the distance resolution in an AFM measurement is nearly an order of magnitude smaller than that of the SFA technique, and determining the absolute value of the distance between the surface and the tip is more involved than when using the SFA.

5.5 Density Variations Can Generate Attractive and Oscillatory Forces

When analyzing double-layer and van der Waals forces at long range, we have treated the medium as a continuum and considered the surfaces as flat. In reality, the molecularity of the medium should have an effect, and we expect that this factor becomes more important the shorter the separation between the surfaces. Moreover, surfaces or interfaces are never perfectly flat. Here we distinguish between two major types of surfaces/interfaces: a solid surface that has a fixed geometry, whether it is flat, bumpy, or cracked, while a liquid-like surface adopts whatever shape will minimize the free energy, exhibits thermal fluctuations, and can deform in response to external forces. In this section, we will ignore the complications that arise from the flexibility of liquid-like surfaces and focus on fixed, solid surfaces. However, most of the effects we will discuss can also occur for liquid-like surfaces, although they will be modified by the flexibility of the system.

One specific phenomenon occurring in solid-like systems is what we will call the steric energetic effect. Virtually all solid surfaces are rough at the molecular scale. When two such surfaces approach, their roughness prevents them from establishing close molecular contact over an extended area. Even when only an attractive van der Waals force is operating, the adhesion energy is much smaller than we would estimate on the basis of the surface free energy. This is because the adhesive force decreases due to the roughness of the surface. From Eq. (5.2.7), it follows that the force per unit area between two media is

$$\frac{F}{A} = -\frac{H_{12}}{6\pi h^3} \tag{5.5.1}$$

Assuming that the distance of closest approach between the surfaces is 0.5 nm and using a Hamaker constant of $H_{12} = 4 \times 10^{20}$ J for media 1 and 2 across air, the calculated value of F/A is 17×10^6 Pa. This is a large force that results if one is able to achieve molecular contact between two surfaces over macroscopic areas. Pieces of steel that have been carefully polished adhere strongly through the dispersion interaction, and they can only be separated by sliding the surfaces relative to one another. Such gauge blocks, also called *Jo blocks*, named after the inventor Carl Edvard

Johansson, had a large technical use in precision machining a hundred years ago. Gold foils are sufficiently elastic to be made to adhere on surfaces by the van der Waals force, although in practice one sometimes also applies an adhesive to achieve a lasting adhesion. Some animals, such as the gecko lizard and the common housefly, have a sufficiently flexible surface on their feet so that they can use the van der Waals attraction to walk on vertical walls and even upside down on the ceiling! It is then essential to keep the surface clean from adhering particles and therefore flies can be seen spending time polishing their feet using their mouth.

5.5.1 Molecular Packing Forces Produce Oscillatory Force Curves with a Period Determined by Solvent Size

One assumption made explicitly in the Hamaker theory of dispersion forces and implicitly in the Lifshitz theory is that the density of a liquid remains constant up to the surface or interface. However, a liquid exhibits short-range correlations, and the system is inhomogeneous on the molecular scale. At an interface, similar correlations exist, leading to variations in density on the scale of a few molecular diameters. Such effects are most pronounced at smooth solid surfaces, where the perturbation introduced by the surface is constant in time and also along the surface. Because short-range repulsion effects bear the main responsibility for the structure in a liquid, we can model them with reasonable accuracy by a hard-sphere system. At packing fractions above 40% the packing constraints lead to distinct positional correlations in the bulk liquid. Outside a hard repulsive wall such correlations generate a particle density ρ that approximately follows a damped oscillation $\rho(x) \propto \cos\left(\frac{2\pi x}{\lambda_1}\right)\exp\left(-\frac{x}{\lambda_2}\right)$. Here the wavelength λ_1 corresponds to an effective hard-sphere diameter and is relatively insensitive to density, while the decay length $\lambda_2$varies with density and is not simply related to λ_1.

For a liquid between two solid surfaces, an interference will occur between the two density profiles at sufficiently short separations, as illustrated in Figure 5.21. Separations h that form an integral multiple of the wavelength λ_1—that is, $h = n\lambda_1$ with n an integer—display an optimal arrangement in which the free energy is at a local minimum. We can see this as a partial layering of

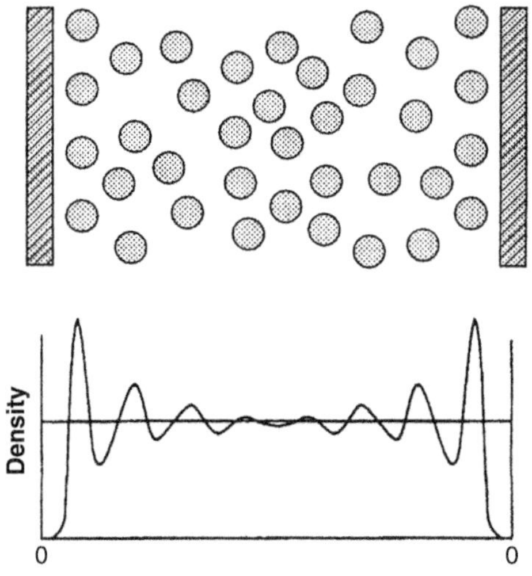

Figure 5.21 Schematic of a hard-sphere fluid contained between two smooth solid surfaces separated by a distance h. Separation by $h = n\lambda$ lead to optimal molecular packing and a local minimum in the free energy; other separations lead to an increase in free energy with a maximum occurring at $h = (n + 1/2) > 1$. *Source*: Israelachvili, 1992/with permission of Elsevier.

the liquid between the surfaces. Packing frustration instead occur for separation values such that $h \approx \left(n + \frac{1}{2}\right) \lambda_1$, where the free energy displays a local maximum. These variations in free energy give rise to an oscillatory force of the approximate form

$$\frac{F_{\text{pack}}(h)}{A} = \text{const} \times \cos\left(\frac{2\pi h}{\lambda_1}\right) \exp\left(\frac{h}{\lambda_3}\right) \tag{5.5.2}$$

where the decay length λ_3 is closely related to λ_2 for the density oscillations and the constant prefactor is of the order of the external pressure. For separations shorter than a molecular diametere, $h < \lambda_1$, solvent molecules can no longer enter the gap, and in addition to direct adhesive forces one has also to initially overcome the external pressure when separating the surfaces.

In a liquid consisting of approximately spherical molecules, an oscillatory force can be measured up to $n = 10$. Nonspherical molecules show a less pronounced but still measurable layering effect between smooth surfaces, and careful experiments show that the effect is present also in aqueous systems. Straight-chain hydrocarbons can display an oscillatory force where the length λ_1 is independent of chain length and corresponds to the diameter of the hydrocarbon chain. Chain branching tends to destroy these oscillations. However, surface roughness due to an uneven solid surface or to capillary waves on a liquid surface tends to eliminate the effect.

Even though the integrated effect of these packing forces is small, they can lead to substantial kinetic barriers. Usually, when two surfaces approach one another, the most sluggish step is to push away the last layer of solvent on approach. Similarly, to pull two surfaces away from adhesive contact, we must overcome a barrier arising from the layering effect as well as the attractive van der Waals force.

5.5.2 A Nonadsorbing Solute Creates an Attractive Depletion Force

In colloidal systems, there is often a solute dissolved in the liquid medium. The solute can either adsorb onto the surface due to a high affinity to it—and consequently alter its surface properties—or it can have a smaller affinity to the surface than the solvent itself. In the latter case, there is a zone close to the surface in which the solute is depleted. Using the terminology of the Gibbs adsorption isotherm (see Section 2.4), we have a negative adsorption. We devote this subsection to an analysis of the consequence of such a negative adsorption on the force between particles.

Even a negative adsorption affects the force between particles in the solution. Outside the surface, the nonuniform concentration profile ultimately approaches the bulk concentration. As the two particles approach, their concentration profiles start to overlap, which will generate a force between the particles, similar to the situation for charged surfaces where a force appears as the diffuse electrical double layers start to overlap.

When the surfaces have come closer than twice the thickness, h_{dep}, of the depletion zone, no solute can be accommodated between them. In this situation, a further decrease of the separation simply results in a transfer of solvent molecules from the now pure solvent between the surfaces to the bulk solution at nonzero osmotic pressure Π_{osm}. This process implies a transfer of solvent from a region of high chemical potential to one of a lower value reducing the free energy. For planar surfaces, the force per unit area equals the negative of the free energy difference $\Delta g_{\beta\alpha}$ per unit volume between the two environments β and α. At total depletion this free energy difference equals the osmotic pressure of the bulk solution and

$$\frac{F}{\text{area}} = -\Delta g_{\beta\alpha} = -\Pi_{\text{osm}} \quad h < 2h_{\text{dep}} \tag{5.5.3}$$

At $h \gg 2h_{\text{dep}}$ the force should go to zero, while the behavior of F/area for distances $h \approx 2h_{\text{dep}}$ depends on the characteristics of the concentration profile outside a single surface. Figure 5.22 illustrates three qualitatively different cases.

To generate a strong depletion interaction, we need a combination of two properties of the solute. First, it should produce a high osmotic pressure so that the force is large, according to Eq. (5.5.3). Second, the repulsive interaction with the surface should be long-ranged so that h_{dep} is large.

Most applications of depletion forces employ a polymer molecule as the active solute. In such cases, the size of the polymer coil determines the range of the force. Due to the polymer's conformational degrees of freedom, it also contributes appreciably to the osmotic pressure, as we will discuss in Chapter 7. For example, polyethylene glycol is used to induce fusion of biological cells. In finding such an effect one could be tempted to conclude that the polymer molecules interact directly with the membrane during the fusion process. In reality, their function is to change the conditions of the bulk solution, by lowering the water chemical potential. Polymers often exhibit a peak in their density profile outside the depletion zone. This density peak results in the schematic force curve shown in Figure 5.22b in which a repulsive barrier occurs prior to the attractive depletion region.

Another possibility is to employ a charged micelle (or polyelectrolyte). In this case, a repulsive electric double layer force can exist between the wall and the micelle while the counterions set up the osmotic pressure. This effect can be important in emulsions. In this case, there is also a strong micelle–micelle repulsion. At higher micelle concentrations, an oscillatory density profile can develop outside the single surface, producing an oscillatory force as in Figure 5.22c. A third alternative occurs for a simple electrolyte solution outside an apolar surface. In this case, a

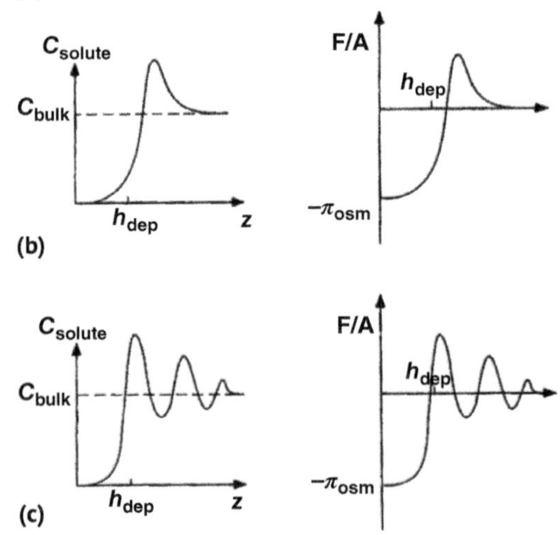

Figure 5.22 The distance dependence of the depletion force depends on the nature of the solute concentration profile outside a single surface. For a monotonically decaying solute concentration as in (a) the force also decays monotonically to the limiting value $-\Pi_{\text{osm}}$. Sometimes the depletion layer is followed by a peak in the concentration at $z \geq h_{\text{dep}}$ as in (b). Then there is a force barrier prior to the attractive regime at $z \geq h_{\text{dep}}$. A third possibility (c) is that there are several oscillations in the solute concentration and then the force also shows oscillations similar to those discussed in Section 5.5.1, but on a different length scale.

so-called image charge interactions repel the ions from the surface (within the range of the Debye screening length). For reasonable electrolyte concentrations, the ion density goes smoothly toward the bulk value, and we have a force profile as shown in Figure 5.22a.

We can obtain an approximate expression for the depletion force acting between two spherical particles of radii R through the Derjaguin approximation (see Eq. 5.4.6b)

$$F(h_0) = \pi R \frac{G_{\text{plane}}(h_0)}{\text{area}} = \pi R \int_{h_0}^{\infty} \frac{F_{\text{plane}}(h)}{\text{area}} dh \approx -2\pi R \Pi_{\text{osm}} h_{\text{dep}} \left(1 - \frac{h_0}{2h_{\text{dep}}}\right) \quad (5.5.4)$$

where we have used Eq. (5.5.3) as an approximate description of the depletion force. The linear proportionality between the force and the separation for $h \ll R$ can be qualitatively understood as arising from the distance-independent force between parallel surfaces, a manifestation of the conceptual simplification provided by the Derjaguin approximation.

5.5.3 Adsorption Introduces an Increased Average Repulsion

Although some solutes are depleted from a surface, adsorption on a surface is far more common. Clearly, adsorption profoundly affects the properties of the surface and thus the forces between particles in a way that depends on the detailed chemistry. Adsorption of a charged surfactant normally leads to the formation of a charged surface that exhibits a double-layer repulsion. However, if a surfactant adsorbs on an oppositely charged surface, it may make the surface electrically neutral or even hydrophobic. Conversely, adsorption of a nonionic surfactant makes a hydrophobic surface more polar, and the force between two such surfaces typically shows a short-range repulsion.

We find a similar scenario with polymers. Their adsorption can result in repulsion or attraction, depending on the balance between the polymer-polymer and polymer–solvent interactions. In the next section—when we discuss liquid interfaces with internal degrees of freedom—we will present a more detailed discussion of the mechanisms behind the forces involved in these cases. Here we concentrate on describing a general relation between force and surface free energy useful for the conceptual understanding of forces in systems containing adsorbing solutes.

In Section 2.4, we discussed how the surface tension of a liquid–vapor or a liquid–liquid interface changed when a solute was added to the system. We found a fundamental relation in the Gibbs adsorption equation stating that changes in surface tension (or surface free energy) γ directly relate to the surface excess of the solute. A solute that shows negative adsorption (surface depletion) increases the surface free energy, while adsorption leads to a decrease in γ.

The same basic relation applies at a solid–liquid interface. Let us now consider the situation where we bring two planar solid surfaces from a large separation up to molecular contact in a solvent containing the solute a. For the separated surfaces, we have an excess surface free energy $2A\gamma_{SL}(c_a)$ where we have explicitly indicated that the interfacial energy depends on the solute concentration c_{a}. The free energy change in bringing the surfaces into contact is the work of cohesion, as expressed in Eq. (2.1.11)

$$\frac{\Delta \mathbf{G}}{\text{area}} = -2\gamma_{SL}(c_a) \quad (5.5.5)$$

Alternatively, we can write the free energy change as an integral of the force as defined in Eq. (5.1.1), using $\Delta \mathbf{G} \approx \Delta \mathbf{A}$

$$\frac{\Delta \mathbf{G}}{\text{area}} = \int_{\text{contact}}^{\infty} \frac{F(h)}{\text{area}} dh = -\ 2\gamma_{SL}(c_a) \quad (5.5.6)$$

For the depletion case, we have that $\gamma_{SL}(c_a) > \gamma_{SL}(c_a) = 0)$ according to the Gibbs equation, and Eq. (5.5.5) thus shows that, on average, the force should be more attractive for $c_a > 0$ than in the pure solvent as discussed for the depletion attraction. For adsorbing solutes on the other hand, γ_{SL} $(c_a) < \gamma_{SL}(c_a) = 0)$ and the average force becomes more repulsive due to the presence of the solute. Note that to obtain direct molecular contact between the solid surfaces, the adsorbed solute must be squeezed off the surface. We thus expect a major extra repulsive contribution to the force at short separations where such forced desorption occurs.

A practical concern in systems with adsorbents is whether there is enough time when two surfaces approach to equilibrate the adsorption. Equation 5.5.5 applies at equilibrium, but another possibility is that the surfaces approach so rapidly that the adsorbed amount remains constant. How does this nonequilibrium force relate to the equilibrium one? We discussed an analogous problem for the electrical double-layer force when we distinguished surfaces approaching at constant charge from those at constant potential. The condition of constant charge is similar to that of constant adsorbed amount. In the case of the electrical double-layer force, we concluded that the forces were the same at large separations where the double-layer overlap is small. A difference appeared only when the constant potential condition implied a substantial change in the surface charge, in which case, the repulsion is weakened. We can generalize this case to the adsorption situation where we conclude that, provided that the individual surfaces at infinite separation have equilibrated with the solution, the force is the same whether one considers a constant adsorbed amount or an h-dependent adsorption equilibrium as long as the interaction is weak.

5.5.4 Capillary Phase Separation Yields an Attractive Force

In Section 2.3, we discussed how surface free energy can trigger the condensation of a liquid in the cracks and grooves on a particle's surface at a vapor pressure below saturation, so-called capillary condensation. Capillary condensation can be seen as a manifestation of a more general phenomenon that we call capillary-induced phase separation (CIPS). When two surfaces approach, they create an arrangement similar to that of a crack in a single particle. The contribution from the surface free energy can be substantial for the medium in the gap and may induce a transition to a phase with lower surface energy. Once the new phase is formed, it will give rise to a strong attractive capillary force. A well-known manifestation of such a capillary force is the sand castle, which cannot be built underwater and collapses when it dries. Surface tension at the air–water interface keeps the castle standing when it is built from moist sand with a bridge of water between the sand grains. Similarly, we know that grains of sugar or salt tend to stick together in a moist atmosphere, since water condenses at the contact area between grains, yielding a force keeping them together. If the atmosphere after some time becomes dry again, it can happen that the collection of grains sticks permanently, forming a cake. Since sugar and salt are water-soluble, there is a dissolution process in the water bridge, and upon drying, the dissolved sugar or salt recrystallizes, forming a solid bridge between the originally separated particles.

Although the transition from gas to liquid is the most familiar form of CIPS, we can also have the reverse, a liquid-to-gas transition, and also, for binary liquids, a liquid-to-liquid phase separation as illustrated in Figure 5.23. Even capillary-induced transitions to liquid crystalline phases have been observed. Let us first analyze the force between two planar surfaces of area A that approach in a bulk medium α, with a surface free energy $\gamma_{s\alpha}$ associated with each of the two surfaces. Now assume that molecules of medium α can transform into a new phase β, which can

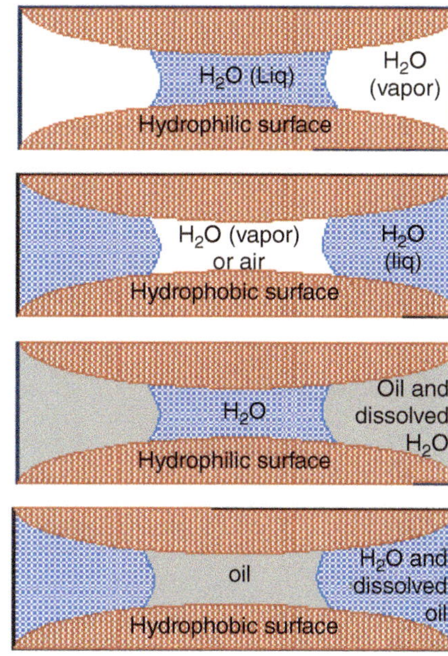

Figure 5.23 Illustration of four different cases of capillary induced phase transition, leading to the formation of a lens of a second phase between curved surfaces in a bulk medium of another phase (from above): capillary condensation of water between hydrophilic surfaces (as in cracks in rock or in a snowflake); capillary vaporization of water between strongly hydrophobic surfaces (such as the material used in boiling chips); separation of water from moist oil between hydrophilic surfaces (common in oil reservoirs); oil separation from oil-contaminated-water (a challenge in cleaning processes).

be a vapor, a liquid, or a liquid crystalline phase. The surface free energy of this phase is $\gamma_{s\beta}$ per unit area. When $\gamma_{s\alpha} > \gamma_{s\beta}$ the surface free energy $\Delta\mathbf{G}_S$ is decreased by the formation of the β in the gap, according to

$$\Delta\mathbf{G}_S = 2A\left(\gamma_{s\beta} - \gamma_{s\alpha}\right) = -2A\gamma_{\alpha\beta}\cos\theta \tag{5.5.7}$$

The second equality follows from the Young equation (2.1.13), where θ is the contact angle. However, because α is the stable bulk phase, creating the new phase also involves an increase $\Delta\mathbf{g}_{\beta\alpha}$ in the bulk free energy per unit volume. The net change in free energy is

$$\Delta G(h) = 2A\left(\gamma_{s\beta} - \gamma_{s\alpha}\right) + hA\Delta\mathbf{g}_{\beta\alpha} \tag{5.5.8}$$

The transition from α to β occurs in the gap when $\Delta\mathbf{G}(h) = 0$, corresponding to a separation

$$h = -2\left(\gamma_{s\beta} - \gamma_{s\alpha}\right)/\Delta\mathbf{g}_{\beta\alpha} \tag{5.5.9}$$

At this separation, we find the onset of an attractive force

$$\frac{F}{\text{area}} = -\frac{1}{\text{area}}\frac{d\Delta G(h)}{dh} = -\Delta\mathbf{g}_{\beta\alpha} \tag{5.5.10}$$

The force is independent of h as long as h is smaller than the critical value given by Eq. (5.5.8). This can be understood from the fact that, once the β phase is formed in the gap, a change in separation dh causes a transformation between β and α of a constant volume Adh irrespective of the value of h.

In the case of a liquid condensing from its vapor α at pressure p, we already calculated the free energy term $-\Delta\mathbf{g}_{\beta\alpha}$ in our discussion of nucleation in Section 2.3.3 as

$$-\Delta\mathbf{g}_{\beta\alpha} = -\frac{k_B\mathbf{T}}{V_L}\ln\left(\frac{p}{p_0}\right) \tag{5.5.11}$$

Here p_0 is the saturation pressure and V_L the molecular volume in the liquid.

Problem: Calculate the force curve between two planar surfaces where $(\gamma_{s\beta} - \gamma_{s\alpha}) = 10$ mJ/m^2 in water vapor at 90% relative humidity and 25 °C.

Solution: For water $V_L = 30 \times 10^{-30}$ m^3 so the force is

$$\frac{F}{A} = \frac{k_B \mathbf{T}}{V_L} \ln 0.9 = \frac{4.1 \times 10^{-21}}{30 \times 10^{-30}} \ln 0.9 \simeq -1.3 \times 10^7 \text{ N/m}^2$$

for

$$h < \frac{2 \times 10^{-2}}{1.3 \times 10^7} \simeq 1.8 \text{ nm}$$

Applications usually involve particles that have curved surfaces. Previously, we dealt with this complication by using the Derjaguin approximation. It is far from evident from the derivation presented in Section 5.4.1 that we can use this approximation when we have two coexisting phases present in the gap. However, as we will show in the following, the Derjaguin approximation nevertheless remains valid even for heterogeneous systems as long as equilibrium is maintained.

Consider a spherical particle of radius R_0 approaching a planar surface of the same material, as illustrated in Figure 5.24. Phase β now forms a lens with radius $R_m$around the point of closest approach, h_0. On forming the lens, we have a change in free energy

$$\Delta \mathbf{G}(h_0) = 2\pi R_m^2 (\gamma_{s\beta} - \gamma_{s\alpha}) + \pi R_m^2 \left(h_0 + \frac{R_m^2}{4R_0} \right) (-\Delta \mathbf{g}_{\beta\alpha}) + 2\pi R_m \left(h_0 + \frac{R_m^2}{4R_0} \right) \gamma_{\alpha\beta} \tag{5.5.12}$$

consisting of a negative surface free energy term, a positive contribution from transforming the volume $\pi R_m^2 \left(h_0 + \frac{R_m^2}{4R_0} \right)$ of the lens from the α to the β phase and a surface term due to the rim of the lens. The last term is only significant for the initial stage of lens formation and we will neglect it below. An optimal value of the radius of the lens exists which varies with separation h_0. Taking the derivative of Eq. (5.5.12) and applying the condition $R_0 \gg R_m \gg h_0$ so that one can neglect the last term on the right-hand side, the optimum value for the radius of the lens is

$$R_m = \left\{ 2R_0 R_K \left(1 - \frac{h_0}{R_K} \right) \right\}^{1/2} \tag{5.5.13}$$

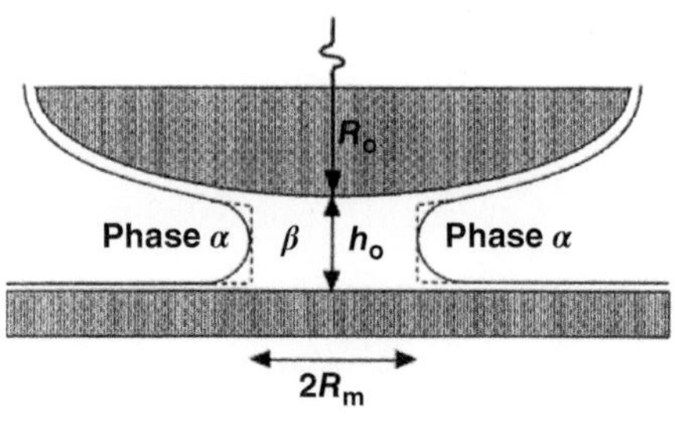

Figure 5.24 A capillary-induced separation of phase β from a bulk phase α when a spherical particle of radius R_0 approaches a planar surface with a distance of closest approach of h_0. The phase β forms a concave lens of radius R_m. Note that the phase β typically also wets the surfaces even in bulk solution.

Where R_K is the Kelvin radius we encountered when discussing nucleation in Section 2.3.

$$R_K \equiv 2(\gamma_{s\beta} - \gamma_{s\alpha})/\Delta \mathbf{g}_{\beta\alpha} \tag{5.5.14}$$

Now we can evaluate the force in a straightforward manner by taking the derivative of Eq. (5.5.12) with respect to h_0 and neglecting the last term of the right-hand side

$$F = -\frac{d\Delta G(h_0)}{dh_0} = 4\pi R_0 \left(\gamma_{s\beta} - \gamma_{s\alpha}\right)\left(1 - \frac{h_0}{R_K}\right) = -2\pi R_0 R_K \Delta \mathbf{g}_{\beta\alpha}\left(1 - \frac{h_0}{R_K}\right) \tag{5.5.15}$$

In fact, we can also obtain this result from Eq. (5.5.8) and the Derjaguin approximation for a sphere and a plane in Eq. (5.4.1), and the free energy change in Eq. (5.5.8) yielding

$$F = 2\pi R_0 \left\{2\left(\gamma_{s\beta} - \gamma_{s\alpha}\right) + h_0 \Delta \mathbf{g}_{\beta\alpha}\right\} = 4\pi R_0 \left(\gamma_{s\beta} - \gamma_{s\alpha}\right)\left(1 - \frac{h_0}{R_K}\right) \tag{5.5.16}$$

exactly as in 5.5.15. In this application of the Derjaguin approximation we have implicitly used the condition that the shortest separation h_0 should be much smaller than the radius of the sphere. The force due to the phase separation is typically very strong, often resulting in direct contact ($h_0 = 0$) between the surfaces. It is however difficult to measure the distance dependence explicitly when one has a capillary condensation where the difference $\gamma_{s\beta} - \gamma_{s\alpha}$ is large. However, Figure 5.25 shows an example of such an attractive force due to a capillary phase transformation from a bicontinuous microemulsion water-oil-AOT to an aqueous phase. Since the mica surface is hydrophilic the mica–water surface free energy is lower than the mica–microemulsion surface free energy.

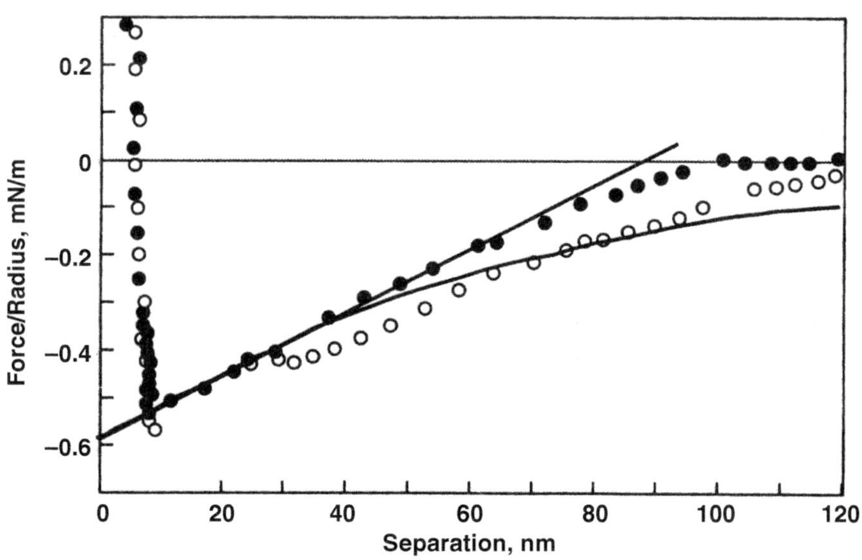

Figure 5.25 Measured force, using the SFA, between two cylindrical mica surfaces in a system with a microemulsion bulk phase. The attractive force is due to a capillary-induced formation of a lens of an aqueous phase between the surfaces. Filled dots represent the force measured on approach while the unfilled dots represent the force measured on separation of the surfaces. The straight line is calculated using Eq. (5.5.15) while the curved line follows Eq. (5.5.17) using the same parameters. The repulsion observed at short separations is due to the double-layer force between the charged mica surfaces. *Source*: From Petrov et al., 1997/with permission of American Chemical Society.

However, the microemulsion contains large aqueous domains that most probably wet the mica surface. Consequently the difference in surface free energy is small so the force is relatively weak.

Our analysis of the force assumed that equilibrium was maintained at all separations. However, some processes can be slow to equilibrate. First, we know that nucleation of a separate phase can be very sluggish. Second, as the surfaces approach one another, material must be transported to and from the slit to establish the difference in concentrations. Thus, we can often observe time-dependent forces. A case of particular importance is when two surfaces that have been in contact for some time are quickly separated. As a first approximation, we can assume that no material is transported on short timescales and that the volume of the lens remains constant. Taking the derivative of the free energy in Eq. (5.5.12) at constant lens volume yields a force

$$F = 4\pi R_0 \left(\gamma_{s\beta} - \gamma_{s\alpha}\right) \left\{ 1 - \left(1 + \frac{R_K^2}{h_0^2}\right)^{-1/2} \right\} \tag{5.5.17}$$

For short separations, this force is the same as in Eq. (5.5.15), but it is more long-ranged. The lower line in Figure 5.25 is a fit to Eq. (5.5.17) to the force measured on separating the surfaces. This nonequilibrium force is clearly more long-ranged than the equilibrium one.

In the discussion above it was assumed that there is a distinct difference between the phases α and β, resulting in a well-defined phase boundary. When the fluid is near a critical point, (see Chapter 10), the situation is more complex. For a binary fluid one of the components has a higher affinity to the surface than the other. The resulting higher density of this component close to the surface propagates into the solution and close to a critical point, density correlations are long-ranged. Between two surfaces there will then be a continuous variation in the density rather than a distinct lens as in Figure 5.24. However, the density variations generate an attraction by basically the same mechanism, where the range is instead determined by the correlation length in the fluid. This kind of attractive force for near critical systems is sometimes referred to as "critical Casimir forces" although the analogy with a quantum Casimir force due to a retarded dispersion interaction (see Section 5.2.5) isn't obvious.

See Cosgrove (2010, Chapter 16), Israelachvili (2011, Chapter 15), and Ohshima (2024, Chapters 10 and 11) for additional discussion on Density-mediated Forces.

5.6 Entropy Effects Influence the Forces Between Liquid-like Surfaces

Emulsion stability depends on the presence of a repulsive force between the droplets. The droplet surfaces are liquid in character, and the interfaces themselves possess thermally excited degrees of freedom that profoundly influence the forces between them. The same argument applies to forces in foams and between self-assembled aggregates, which are similarly controlled to a considerable extent by their liquidlike character. We also find such conditions on solid surfaces covered by a layer of adsorbed surfactant or polymer, which has a liquidlike disorder. Irrespective of molecular details, the net force has an important repulsive component because two interacting surfaces always have reduced configurational freedom relative to free surfaces. This fact constitutes a conceptual difference in comparison to the solid surfaces discussed in the previous subsection. Another complication that we do not address explicitly here is that liquid-like surfaces easily deform under the action of interparticle forces. In Chapter 11, we will see that this property has dramatic consequences for the structure of concentrated emulsions and foams. In fact, due to the deformations it poses fundamental complications to uniquely define a force between two liquid drops or gaseous bubbles.

5.6.1 Reducing Polymer Configurational Freedom Generates a Repulsive Force

A polymer layer can cover a surface either via adsorption from solution or through covalent attachment. If the polymer is soluble in the liquid medium, the parts of the polymer molecule that are not in direct contact with the surface tend to extend out from the surface into the solution in order to maximize the number of polymer configurations. In Chapter 7, we will discuss in more detail the issue of polymers on surfaces and their effects on surface forces. Here we concentrate on the basic mechanism behind the force between surfaces covered by polymers under good solvent conditions. When two such surfaces approach one another, the polymer tails reaching out from one surface start to impede the configurational freedom of the polymers of the other surface, reducing their configurational entropy. Unless direct interaction between the monomer units compensates for this effect, the result is a repulsive force. Because polymer molecules can be large, the range of this force can be significantly longer than that of the van der Waals and double-layer forces.

In the literature, this repulsive force is usually called the *steric force*. This terminology is somewhat misleading because it obscures the insight that the effect is of entropic origin. Although the terminology is established, one should distinguish it from the steric-energetic force operating between solid surfaces discussed in the introduction to Section 5.5. The term overlap repulsion seems more appropriate and it provides a clear distinction to the steric energetic force for solid objects.

5.6.2 Repulsive Undulation Forces Act Between Fluid Bilayers

Lipids often self-assemble into fluid bilayers, as will be discussed in Chapter 6. In a stack of bilayers, there is an additional long-range repulsion due to so-called *undulation forces*. The area of the bilayer is free to adjust to an optimal value, so there is no tension in contrast to the situation at an air–water interface where there are thermally induced capillary waves, whose amplitude is limited by the surface tension. For lipid bilayers, the resistance to deviations from the perfectly flat state is caused by the bending energy determined by the local curvature of the bilayer, which we will discuss in more detail in later chapters. The bending occurs on a length scale long relative to the molecular size, so one can characterize the system with a single parameter called the bending rigidity κ, whose value is determined by the molecular details of the bilayer. Even though the flat bilayer represents the minimum free energy state, its shape will fluctuate due to thermal motion of the molecules, and these bending fluctuations contribute to the free energy of the bilayer. When the average separation between two bilayers decreases, the bending fluctuations will cause occasional short-range contacts between them, as illustrated in Figure 5.26.

If there is a short-range repulsion between molecules on the respective bilayers, such contacts will, to some extent, limit the amplitude of the bending fluctuations. This constraint entropically increases the free energy of the system in qualitatively the same way as for polymer-covered surfaces discussed in the previous section. The German physicist Wolfgang Helfrich derived an explicit equation for this force, typically called the undulation force, which depends on the inverse cube of the bilayer separation h

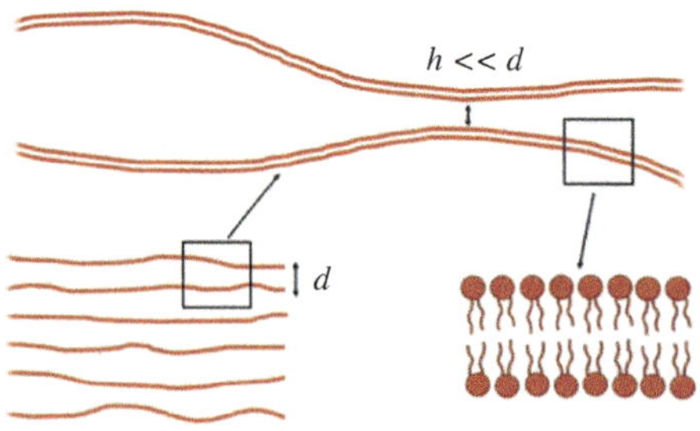

Figure 5.26 The thermally excited bending modes of a bilayer in a stack occasionally result in contacts with adjacent bilayers, even at mean separations that are large relative to the bilayer thickness.

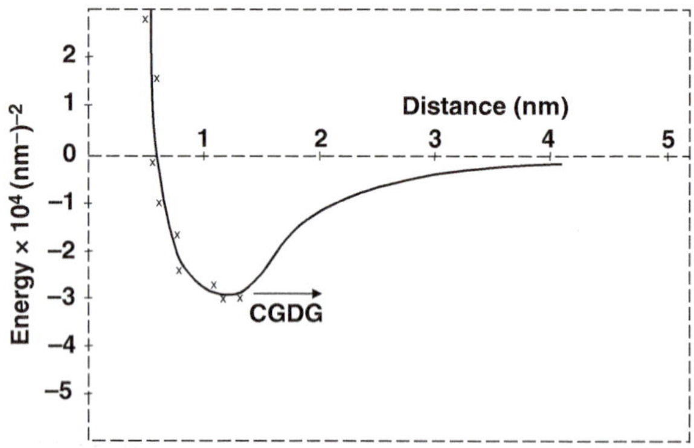

Figure 5.27 Interaction free energy per area, measured using the SFA, between two bilayers of the lipid digalactosyl diglyceride (DGDG) deposited on a mica surface. The curve for distances beyond the minimum of the curve at $d = 1.4$ nm is a reconstruction assuming a dispersion interaction. *Source*: Adapted from Marra, 1985.

$$\frac{F}{A} = \frac{3\pi^2}{64} \frac{(k_B\mathbf{T})^2}{\kappa} \frac{1}{h^3} \tag{5.6.1}$$

This force is more long-ranged than the van der Waals force, which, for finite sheets at separations much larger than their thickness, varies as h^{-5}. The description of bilayer properties in terms of bending ceases to be valid for lengths shorter $a_0/(2l_0)$ where $2l_0$ is the bilayer thickness and a_0 is the area per molecule. It follows that Eq. (5.6.1) is not valid at short separations, where the molecular properties of the bilayer cannot be captured by a single parameter like the bending rigidity. There exists a debate on the precise value of the numerical coefficient in Eq. (5.6.1), but experiments confirm that the value quoted is correct to within a factor of two.

5.6.3 When Layers with Liquid Disorder Come in Direct Molecular Contact a Strong Repulsive Force Can Appear

Consider a liquid layer, consisting of lipids or surfactants, that rests on a planar solid support. In this configuration, the thermally excited bending modes are quenched, and the undulation force is therefore not operating between two such layers. Figure 5.27 shows the measured interaction, using the SFA, between two such supported layers of an uncharged sugar lipid. At long range, there is the expected attraction due to the dispersion interaction and by electrostatic correlation mechanisms discussed in Section 5.3. However, at a surface-surface separation of around 2 nm there is an onset of a repulsive interaction causing a minimum in the interaction free energy. This repulsion increases sharply as the separation becomes smaller. This short-ranged repulsive interaction is generic between bilayers of uncharged lipids, and it is a widespread notion that it is caused by an inadequacy of water to solvate polar groups at short range. However, several of the examples given previously in this section have shown that restrictions on the conformational freedom of molecules generally gives rise to a repulsion of entropic origin. This mechanism should also be operating for lipid systems at short range as it does at long range. The main difference with respect to the undulation force case is that many more degrees of freedom are involved, potentially making the short-range force contribution much larger. A second difference is that, as the surfaces approach, more degrees of freedom are involved in the overlap, indicating a strong distance dependence of the force. In Chapter 6, we will discuss in more detail the quantitative aspects of the repulsive force and how it is related to the nature of the lipid head group and the state of the hydrocarbon chains.

5.7 The Strength of the Hydrophobic Interaction Shows an Unexpected Temperature Dependence

Whether the applications of colloidal science are technological or biological, most processes occur in an aqueous medium. So far in this chapter, we have characterized the medium in terms of simple constants like relative permittivity, refractive index, and molecular size. As we saw in Chapter 3, the strong cohesion that exists in liquid water is due to electrostatic interactions in the form of hydrogen bonds, and they make water an associating liquid. How the structural properties of liquid water influence interactions between colloidal particles has been a classical subject of scientific discussion.

The best-documented case in which water's structural effects are influential is the hydrophobic interaction we introduced in Chapter 1. Because the hydrophobic interaction plays a central role in colloid science, we here return to a more detailed discussion of this phenomenon. Now that we have established the basic intermolecular interactions, we are in a better position to understand some of the subtleties that have generated considerable confusion in the past. To begin with, we can identify two levels of hydrophobic interaction. The first one is the molecular level, where the interaction between individual molecules leads to such effects as micellar self-assembly. The second level relates to the force operating between colloidal entities that possess hydrophobic surfaces.

5.7.1 The Hydrophobic Interaction Is a Special Case of the General Phenomenon of Solvophobic Interactions

Solvents with strong intermolecular interactions of electrostatic origin are in general not effective in accommodating apolar solutes. We can use the regular solution model to illustrate the basis of this property. According to Eq. (1.1.16), the effective interaction parameter w_{AB} for a system of a solvent A and a solute B is given by

$$w_{AB} = \left(W_{AB} - \frac{1}{2}W_{AA} - \frac{1}{2}W_{BB}\right) \tag{5.7.1}$$

If the solvent now has a strong self-interaction $-W_{AA}$ so that $-W_{AA} \gg -W_{AB}, -W_{BB}$, the interaction parameter w_{AB} is clearly positive and the regular solution model predicts a low solubility of the solute (B) in the solvent (A). Ionic liquids have a strong charge-charge interaction between the components, giving a strong cohesion and therefore producing a solvophobic effect. Another, uncharged, example is formamide, which is used as a solvent with strong solvophobic properties due to its high polarity and internal hydrogen bonds.

The regular solution model is transparent but simplistic. For example, it leaves out the fact that intermolecular interactions are orientation dependent. The interaction parameter w_{AB} is treated as pure energy, while in practice, preferential orientations lead to a reduction in entropy, which also affects the net interaction. This shortcoming of the regular solution model is generically present but is particularly apparent for aqueous systems, where it turns out to have some unexpected consequences.

5.7.2 The Interaction Between Apolar Molecules in Water Is Solvophobic

When discussing ion–ion interactions in Section 3.5 we could account for the effect of the solvent by changing the interaction strength by a factor ε_r, the relative permittivity. Even though the strength of the interaction is strongly reduced in polar solvents such as water, its distance dependence and dependence on the charge magnitude is unchanged. When we change the solute from ions to apolar solutes, accounting for the solvent effect becomes more intricate. Consider as an example the case of two methane molecules at separation R in water. In a statistical mechanical description, the free energy of the system is determined by a configurational integral $Z_N(R)$ involving all N water molecules in the system. One can, in principle, evaluate $Z_N(R)$ for all relevant separations. The free energy of interaction is then given by $-k_B\mathbf{T}\ln Z_N(R)$ (see Eq. 5.1.3) and one can define an effective CH_4—CH_4 potential as

$$V_{\text{eff}}(R) = -k_B\mathbf{T}\ln[Z_N(R)] + k_B\mathbf{T}\ln[Z_N(R = \infty)] \tag{5.7.2}$$

The potential that determines the configurational integral Z_N contains, in addition to the direct potential CH_4—CH_4 $V_{\text{Me-Me}}(R)$, also the methane–water interaction as well as the water–water one.

Figure 5.28 shows an evaluation of Eq. (5.7.2) from a molecular dynamics simulation. The simulation demonstrates the existence of a primary and a weaker secondary minimum. Note that the primary minimum becomes deeper with increasing temperature. The hydrophobic interaction is substantially stronger than the direct dispersion interaction, whose effect is reduced by an order of magnitude if one subtracts the corresponding dispersion interaction between the two water molecules that have been displaced by the methane molecules. The depth of the primary minimum demonstrates the strength of the solvophobic effect. If one repeats the calculation for another pair of apolar solute molecules, such as argon, krypton, or CF_4 the result is qualitatively very similar. The position of the primary minimum, however, depends on the size of the solute and its depth also increases with the solute radius.

From this, we conclude that there is a generic hydrophobic pair interaction between apolar solutes in water that dominates over the direct dispersion interaction at molecular contact. It follows from the definition in Eq. (5.7.2) that V_{eff} is, in general, not pairwise additive. However, the solvophobic effect is generic in nature and is qualitatively present also for a cluster of apolar molecules. The nonadditivity of $V_{\text{eff}}(R)$ is more pronounced for the secondary minimum. Its existence depends on the possibility for the water molecule adjacent to apolar groups to engage in hydrogen bonds with neighboring water molecules. Thus, apolar molecules usually aggregate into compact

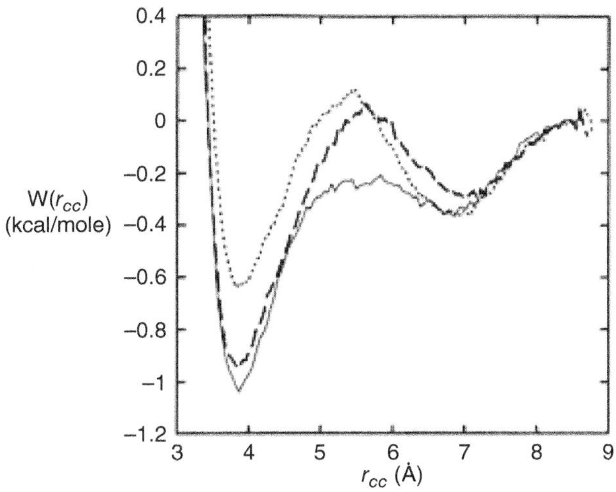

Figure 5.28 Effective methane pair potential $V_{\text{eff}}(R)$ in water showing a primary minimum and a solvent-separated secondary minimum. Dotted line $T = 282$, dashed line $T = 313$, and solid line $T = 298$; $\frac{1\ \text{kcal}}{\text{mole}} = \frac{4.19\ \text{kJ}}{\text{mole}} \approx 1.7 k_B\mathbf{T}$. *Source*: Adapted from Rich and Berne, 1997.

clusters with water molecules on the outside, but there exists a striking exception that we discuss in the next section.

5.7.3 Clathrates Are Crystals with Apolar Molecules in Water Cages

Clathrates are mixed crystals of apolar compounds and water, whose formation is a practically important consequence of the existence of the secondary minimum in the effective interaction between two hydrophobic molecules. They form with a broad variety of apolar substances and Figure 5.29 shows some typical structures.

A common feature of all of them is that water molecules interact to form a hydrogen-bonded network around the apolar molecules that occupy the cavities inside the network, and they act as simple "fillers" without specific structural roles. The stability can be seen as due to the secondary minimum in V_{eff} and clathrates appear under situations where one has the simultaneous presence of water and apolar substances at low temperatures. Clathrates can for example be formed in Nature from natural gas and water. Such precipitates are known to potentially clog natural gas pipelines in very cold environments as in Siberia or Alaska. Large quantities of methane are also stored in the form of clathrates in the permafrost in Arctic regions. When the temperature increases due to climate change, these clathrates melt and release methane into the atmosphere, further increasing the heating effect of the atmosphere. Similarly, clathrates can exist on the ocean floor and melt if the temperature increases.

The observation of solid clathrates inspired Frank and Evans in 1945 to introduce the concept of "iceberg formation" around apolar groups also in liquid aqueous environments. The basic idea is that the cage structure observed in the crystalline structures, as illustrated in Figure 5.29, is to some extent preserved in the liquid state even if the long-range crystalline order disappears. In a more general terminology, the phenomenon is referred to as hydrophobic hydration. Indeed, water molecules that are in direct contact with the apolar solute on average have a larger number of hydrogen bonds with their neighbors than water molecules in the bulk. In addition, the mole-

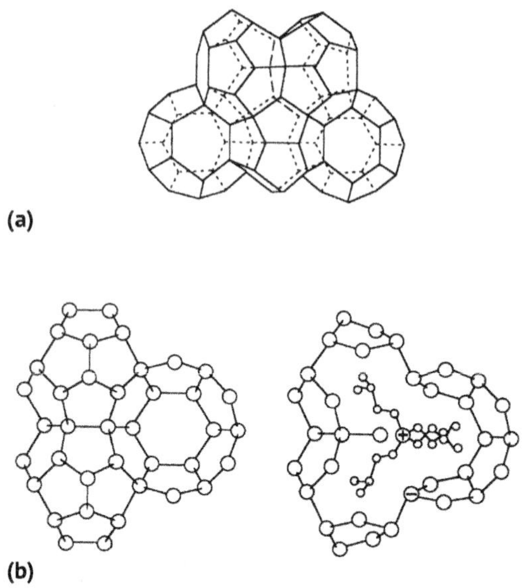

Figure 5.29 Clathrate structures. Clathrates are stoichiometric crystalline solids in which water forms cages around solutes. Each line represents an O—H—O bond. (a) When this clathrate contains Cl_2, the larger tetrakaidecahedral cavities are occupied, but not the dodecahedral ones. With the smaller CH_4 solute, both cavities are filled. *Source:* Pauling, 1960/Cornell University Press. (b) Clathrate structure of $(i\text{-}C_5H_{11})_4NF$ (left) and position of the $(i\text{-}C_5H_{11})_4N^+$ ion (right). The F^- ion replaces one of the water molecules. *Source*: Davidson, 1973/with permission of Springer Nature.

cules surrounding the solute also have longer orientational correlation times a factor of two to four. Taken together, these observations show that there is a significantly increased degree of structural organization of water molecules around hydrophobic solutes.

5.7.4 The Solubility of Apolar Compounds in Water Has an Anomalous Temperature Dependence

In the regular solution model, the solubility of the solute (B) in the solvent (A) is promoted by the entropy of mixing but limited by the unfavorable interaction, as described by the interaction energy w_{AB}. It follows that the solubility of B, denoted by $c_B(\text{sat},\mathbf{T})$, should increase with increasing temperature according to

$$c_B(\text{sat},\mathbf{T}) \propto \exp\left(-\frac{\Delta H^\circ}{\boldsymbol{RT}}\right) = \exp\left(-\frac{w_{AB}}{\boldsymbol{RT}}\right) \tag{5.7.3}$$

In general, one expects interactions to be stronger in the neat phase, leading to $\Delta\mathbf{H}^\circ > 0$, so that the solubility of a substance increases with increasing temperature. For gases like O_2 and N_2, interactions are, however, stronger in solution than in the neat gas, and the solubility instead decreases with increasing temperature. Figure 5.30 shows the temperature dependence of the aqueous solubility of three hydrocarbon benzene derivatives.

In the high-temperature end, $\mathbf{T} > 160$ °C, the solubility is roughly consistent with Eq. (5.7.3), while there are increasingly larger deviations for $\mathbf{T} < 120$ °C. Below 20 °C, the solubility instead increases with decreasing $\mathbf{T}$. It was pointed out by Shinoda that these data reveal an important property of the hydrophobic interaction. To see the connection between the molecular and thermodynamic properties, we recall that the potential V_{eff} as defined in Eq. (5.7.2) has the character of a free energy, since it is obtained from a thermal average. It is possible to separate the effective potential into an enthalpy (energy) and an entropy part

$$V_{\text{eff}}(R) = \mathbf{H}_{\text{eff}}(R) - \mathbf{TS}_{\text{eff}}(R) \tag{5.7.4}$$

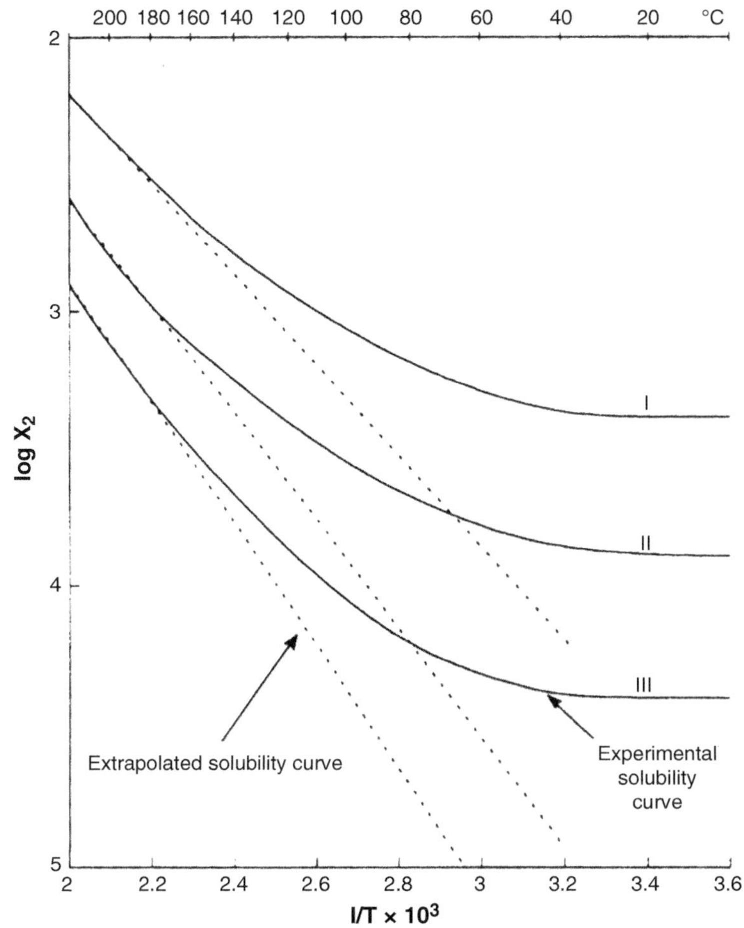

Figure 5.30 Water solubility of alkylbenzenes(I, C_6H_6; II, $CH_3C_6H_5$; III, $C_2H_5C_6H_5$) versus $1/T$. As temperature decreases, the solubility of alkylbenzenes in water initially decreases but then becomes almost constant. For many substances, the solubility displays a minimum around 20 °C ($1/\mathbf{T} = 3.67\ 10^{-3}\ K^{-1}$) and then actually increases with decreasing temperature. The deviation from the straight line dependence at lower temperatures reflects the increased importance of hydrophobic hydration in promoting the solubility of apolar substances in water. *Source*: Shinoda, 1978/with permission of University of Minnesota.

The fact that the solubility curves in Figure 5.30 strongly deviate from a straight line shows that the enthalpy of solution is strongly temperature dependent (see Eq. 5.7.3). The temperature dependence of the enthalpy and entropy is determined by the heat capacity, ΔC_P. The data in Table 1.5 shows that there is a consistently large ΔC_P for the transfer of a hydrocarbon from the neat liquid to water. This table also shows that the corresponding $\Delta\mathbf{H}$ is close to zero at 25 °C. Combined with the ΔC_P data and Eq. (5.7.4), this explains in thermodynamic terms why the solubility curves in Figure 5.30 turn upward around 25 °C. From a molecular perspective, the thermodynamic properties are caused by the fact that the hydrophobic hydration becomes more and more developed as the thermal energy decreases and V_{eff} in Eq. (5.7.4) becomes dominated by the entropic part $\mathbf{TS}_{\mathrm{eff}}(R)$.

The "iceberg" proposition by Frank and Evans unfortunately led to the widespread misconception that the low solubility of apolar compounds in water is due to hydrophobic hydration decreasing the entropy of the system. In fact, the situation is the reverse, as first pointed out by K Shinoda! The hydration structure is there to do the best in an unfavorable situation and provides an aspect of the versatility of water as a solvent.

The temperature dependence of the hydrophobic interaction is determined by

$$\frac{d(\mu^{o}{}_{aq} - \mu^{o}{}_{n})}{dT} = \frac{d\Delta\mu^{o}{}_{hp}}{dT} = -\Delta S^{o}_{hp} \tag{5.7.5}$$

Since ΔS^{o}_{hp} is positive at all temperatures $\mathbf{T} < 100\ °C$, the strength of the hydrophobic interaction decreases monotonically with decreasing temperature over the temperature interval where liquid water is stable. In the data of Figure 5.30, the decrease in the strength of the hydrophobic interaction is compensated by an even stronger decrease in the entropy of mixing term, leading to a decrease in the solubility for $\mathbf{T} > 25\ °C$ but then the solubility has a minimum. At still lower temperatures, the mixing term is not large enough to compensate for the decrease in the interaction, and the solubility increases again.

Globular proteins in general denature at high temperatures but some also on decreasing the temperatures; a phenomenon called cold denaturation. This behavior is a consequence of the weakening of the hydrophobic interaction at lower temperatures. The analogy with the temperature dependence of the solubility of hydrocarbons is striking. A related phenomenon is that cellulose, which is kept in crystalline order by the hydrophobic interaction, increases its aqueous solubility at subzero temperatures.

5.7.5 Interacting Hydrophobic Surfaces Induce Capillary Vaporization

In Sections 5.1–5.5, we have based the description of colloidal forces on the interaction between two parallel surfaces. It turns out that this isn't a fruitful approach for the case of interacting hydrophobic surfaces in water. A pure hydrocarbon surface has an interfacial free energy close to 50 mJ/m^2 and according to Eq. (5.5.6) the free energy change in bringing two surfaces initially far apart into direct contact is $2 \times 50 = 100$ mJ/m^2, which is a very large energy. Many attempts have been made to measure the distance dependence of the attractive hydrophobic force between surfaces and claims have been made that the force is long-ranged. However, the quantitative experimental results are inconsistent, indicating fundamental experimental problems.

The interfacial energy between a hydrocarbon and air is around 27 mJ/m^2, which is substantially lower than the hydrocarbon-water interfacial energy. As two hydrophobic surfaces approach, the surface free energy can thus be lowered by replacing water with a gas phase. The thermodynamic argument shows that capillary vaporization should occur at some separation distance between the surfaces, and this is an example of capillary-induced phase separation (CIPS) discussed in Section 5.5. A quantitative analysis reveals that the replacement of water by a gas should occur at separations exceeding one micron at equilibrium. The occurrence of the CIPS phenomenon makes the question of the range of the molecular hydrophobic force redundant for the case of purely hydrophobic surfaces of macroscopic dimensions.

In practice, surfaces can have limited patches that may be like a hydrocarbon surface but more often they have a less pronounced hydrophobic character. In such cases, surface-induced vaporization is typically not a relevant phenomenon. There are no general relations describing how the strength or the range of the hydrophobic interaction varies with the detailed nature of the surface, although there are a number of empirically based algorithms for estimating the interaction strength. These, however, do not address the question of the range of the force, which has an influence on both equilibrium constants and on the kinetics of association and dissociation. Based on the calculated CH_4—CH_4 potential shown in Figure 5.29, it is most likely that such practically important hydrophobic interactions have a limited range of at most a few times the water molecule diameter.

See Kronberg et al. (2014, Chapter 3) for additional discussion on Hydrophobic Interactions.

5.8 Hydrodynamic Interactions Influence the Dynamic Properties of Colloidal Systems

So far, we have discussed forces between surfaces or particles at rest. These forces determine the thermodynamic properties, and they affect the motion in the system. However, when a particle moves in a solvent, this also induces motion in the surrounding medium, whose collective molecular motions are referred to as *hydrodynamics*. In this picture, the molecular view is replaced by a continuum description, which is then valid on length scales significantly larger than the size of the molecules of the medium. In Section 1.7, we briefly described how the diffusion of a single colloidal particle is affected by the viscosity of the medium, given by Eq. (1.7.15) for the Stokes friction coefficient of a sphere. A moving colloidal particle will give rise to motion in the surrounding liquid and this in turn will pull or push on other particles. The result is that a correlation between the motion of the two particles is generated. This phenomenon is an example of a hydrodynamic interaction, which affects the dynamics of the suspension but not the properties of the equilibrium state.

The central quantity in a hydrodynamic description is the position- and time-dependent velocity field $\boldsymbol{u}(\boldsymbol{r},t)$ of the fluid (liquid or gas) constituting the surrounding medium. In general, the time evolution of this field is dictated by the Navier-Stokes equation. However, for colloidal particles in a liquid medium, we can make two simplifying assumptions and reduce the Navier–Stokes equation to a more manageable form. Firstly, the liquid is considered *incompressible*. The equation of continuity then reduces to

$$\nabla \cdot \boldsymbol{u} = 0 \tag{5.8.1}$$

In the general case, the motion in the medium is affected by inertial, frictional, and external forces. The relative importance of inertial and frictional effects is characterized by the dimensionless *Reynolds number*

$$Re \equiv \frac{\rho U R}{\eta} \tag{5.8.2}$$

Here, ρ is the mass density of the liquid, U is the particle speed, and R is a characteristic size of the particle (such as the radius of a spherical colloid), and η is the viscosity of the medium. For a particle with $R = 10^{-6}$ m settling under gravity in water, with $\eta \approx 10^{-3}$ Ns/m^2, the speed U is of order 10^{-6} m/s resulting in $Re \approx 10^{-6}$, showing that friction totally dominates the effect of inertia in the motion of the particle in the fluid. In this limit, external forces such as gravity are completely balanced by the frictional forces.

When describing the dynamics of a colloidal particle in the fluid, one can adopt two different perspectives. One is to fix the coordinate system on the particle and see the fluid pass by. In the opposite view, the fluid is at rest far from the particle, which moves and affects the velocity of the fluid in its vicinity. The bulk fluid moves as a response to a position-dependent pressure, p^a, while the particle moves as a response to a force $\boldsymbol{F}$ acting on it. One can combine these effects into a dynamic pressure p as

$$p = p^a + \boldsymbol{F} \cdot \boldsymbol{r} \tag{5.8.3}$$

Combined with the equation of continuity (5.8.1), for low Reynolds numbers, the Navier–Stokes equation is simplified to the homogeneous Stokes equation

$$\eta \nabla^2 \boldsymbol{u} = \nabla p \tag{5.8.4}$$

This is a linear differential equation with the property that, if one has two separate solutions, the sum of these is also a solution. There are three vector components in Eq. (5.8.4) and for the case of a particle settling by gravity, there is only a force $F = mg$, where g is the gravitational acceleration, acting in the z-direction, but the motion of the particle will induce velocity components also in the perpendicular direction. We have from Eq. (5.8.4)

$$\nabla^2 \boldsymbol{u} = \frac{g\{\rho(\text{particle}) - \rho(\text{liquid})\}}{\eta} \hat{\boldsymbol{z}} \tag{5.8.5}$$

This is formally similar to the Poisson equation (3.2.1), which provided the basis for the electrical multipole expansion in Section 3.3. This technique can also be used to solve Eq. (5.8.5). The boundary condition is that, at the particle surface, the liquid has the same velocity U as the particle. This is called stick boundary conditions. The solution of Eq. (5.8.5) can then be expressed in terms of a multipole expansion of the flow field. For a sphere, the resulting flow field with components u_α is

$$u_\alpha = U\left[\frac{3R}{4}\left(\frac{\delta_{\alpha z}}{r} + \frac{r_\alpha r_z}{r^3}\right) + \frac{3R^3}{4}\left(\frac{\delta_{\alpha z}}{3r^3} - \frac{r_\alpha r_z}{r^5}\right)\right] \tag{5.8.6}$$

The first term on the right-hand side is the *force monopole* term, often called the *Stokeslet*, decaying as $1/r$. It shows how the particle pushes the liquid in the forward direction. The second is the so-called *source dipole* term, decaying as $1/r^3$, and describes how there must also be counterflow to replace liquid in the wake of the moving particle. This shows that the moving particle affects the flow in the liquid far away. Hydrodynamic interactions are therefore long-ranged.

Perpendicular to the sphere, there is a cancellation between the monopole and the dipolar components, while in the head and tail regions, the components add. The velocity gradients in the medium give rise to a viscous dissipation, which is the source of the friction coefficient

$$\mu = 6\pi\eta R \tag{1.7.15}$$

Equation (5.8.6) describes the flow field around a single moving colloidal particle. This shows that there is a long-range disturbance of the fluid velocity induced by the moving particle, which will be felt by a second particle, whose motion will be affected. It is possible to solve the Stokes equation for two particles settling through gravitation in a liquid. The motions of the particles are coupled due to the hydrodynamic interaction. One can use the multipole picture to describe the coupling and at large particle–particle separations r we can calculate the sedimentation speed U_1 of particle 1 according to

$$U_1 = \frac{1}{8\pi\eta}\left\{\left(\frac{4}{3R} + \frac{1}{r}\right) + \frac{\boldsymbol{r} \cdot \boldsymbol{r}}{r^3}\right\}\{\rho(\text{particle}) - \rho(\text{liquid})\}g \tag{5.8.7}$$

Here the second term in the parenthesis is caused by the coupling between the dynamics of the two particles, at a relative separation $\boldsymbol{r}$, and it gives a positive contribution to the sedimentation rate U_1 of particle 1. Since there is a reciprocal effect on particle 2, the hydrodynamic coupling results in an increased sedimentation rate of the couple as a whole. For three or more settling particles, the situation becomes more complex, but it is a general rule that hydrodynamic interactions enhance the sedimentation rate.

Equation (5.8.7) is valid for particles at separations much larger than their radius. A case of considerable practical importance occurs when particles instead move relative to one another at separations $r \ll R$. For motion along the line between two spherical particles at a relative speed W the force F_h, called the lubrication force, is given by

$$F_h \approx 6\pi\eta RW/h \tag{5.8.8}$$

Here h is the surface–surface separation and the hydrodynamic force thus diverges at close contact. Here the sign of W is such that the force is repulsive on approach, while it is attractive when the particles separate. The divergence for $h = 0$ is a consequence of the hydrodynamic description with a continuum medium. A molecular description is therefore needed to describe the situation when h is of the same order as the size of the molecules in the medium. For motion perpendicular to the center-center line, tangential to the surfaces, there is a weaker logarithmic divergence of the shear force.

In microfluidic devices, colloidal particles typically have a size comparable to the dimensions of the device. This has an influence on the flow pattern. For a liquid flow in a cylindrical pipe of radius R_C, at a pressure drop Δp over a distance L, Poiseuille showed already around 1840 that there is a parabolic flow profile for stick boundary conditions

$$u_z(r) = \frac{\Delta pR^2}{4\eta L}\left\{1 - \left(\frac{r}{R_c}\right)^2\right\} \tag{5.8.9}$$

which is illustrated in Figure 5.31.

A spherical colloidal particle of radius R will travel in the direction of the flow, but if the ratio R/R_C is close to unity the particle will move slower than the liquid by an amount proportional to $(R/R_t)^2$. The velocity gradient generates a shear force on the particle, which responds by rotating around an axis perpendicular to the flow at an angular velocity of $ru_{\max}/R_t^2$. At higher particle concentrations, the hydrodynamic interactions give rise to a complex flow behavior. There is a

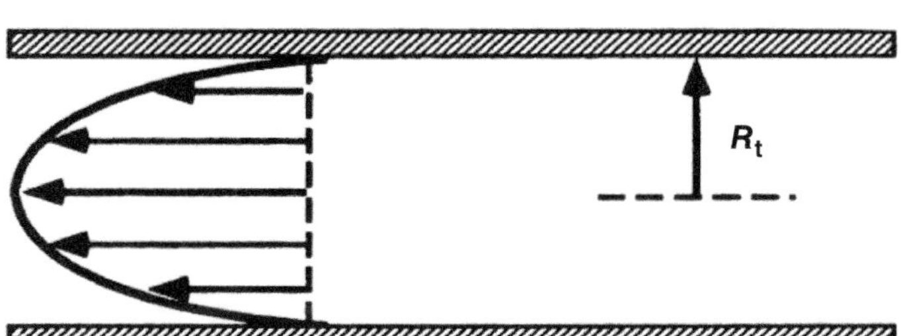

Figure 5.31 The laminar flow profile $u_z(r)$ through a tube of radius R_t length L, a pressure drop Δp, and a radial coordinate $r = (x^2 + y^2)^{\frac{1}{2}}$.

branch of analytical chemistry called field-flow fractionation (FFF) where one makes use of such flow effects to achieve size separation of colloidal particles.

As the particle radius R approaches that of the pipe, particles become concentrated toward the center of the pipe, leaving a thin boundary layer free of particles close to the wall. The velocity gradient is then localized toward the boundary layer due to the lower viscosity of the pure medium relative to that of the suspension. This so-called Fåhraeus effect ultimately leads to a so-called *plug flow* with no velocity gradients away from the boundary layer. A very important example of this type of flow is found in the capillary veins joining the arterial and venous parts of the blood flow. The red blood cells even partly deform when pushed through.

See Berg (2010, Chapter 10) and Israelachvili (2011, Chapter 18) for additional discussion on Hydrodynamic Forces.

Exercises

5.1 A suspension of clay particles is dewatered by applying an external pressure of $1 \times 10^5 N/m^2$ (≈ 1 atm) across a membrane impermeable to the clay particles. At what relative humidity should the piece of clay be stored at room temperature to ensure that it neither dries nor takes up moisture?

5.2 Two identical disk-like clay particles of radius 2×10^{-5} m, thickness 5×10^{-6} m, and density 3×10^3 kg/m^3 move in a stream. One of them settles at the bottom at a calm stretch, while the other goes all the way to the sea. Calculate the thickness of the water layers separating the (highly charged) clay particles from the sediment on the bottom (also carrying a high negative charge) at equilibrium. Assume that the stream is free from salt while the seawater has a NaCl content of 3.5%w/w.

5.3 A film of water can condense from humid air on a quartz surface with a high density of ionized groups. Calculate the equilibrium thickness of the film as a function of the relative humidity. Assume that both the electrical double layer and the van der Waals forces are important. The Hamaker constant quartz–water–air is -1.0×10^{-20} (**T** = 298K).

5.4 Sodium chloride has been deposited on a glass surface by evaporation of a thin electrolyte solution film. The resulting surface density of NaCl is 1/10 nm^2. This surface is placed in a closed container thermostated at 25 °C containing an open beaker with a 1×10^{-5}M NaCl solution. What happens? (a) Neglect the action of van der Waals forces. (b) Consider van der Waals interaction with a Hamaker constant of 1.0 10^{-20} J.

5.5 Israelachvili and Adams measured the interaction energy per unit area between two charged mica surfaces across an aqueous solution of 1.0×10^{-4} MKNO_3. The table summarizes measured values at long range.

Separation (nm)	Interaction free energy (mJ/m^2)
29	141.0
41	89.0

Separation (nm)	Interaction free energy (mJ/m²)
50	67.0
60	45.0
84	28.0
91	15.8
112	8.4
135	3.3

Are these measurements consistent with the theory for weakly overlapping double layers? In such a case, what is the value of Γ_0 (**T** = 298K)? (Neglect the influence of the van der Waals interaction.)

5.6 At the air–water interface of an SDS solution above the CMC (8 mM), there is a highly charged monomolecular film. Micellar aggregates are electrostatically repelled from the surface. Estimate the distance of separation between the monolayer and a micelle at which interaction equals $k_B\mathbf{T}$ ($\mathbf{T}$ = 300 K) by assuming that the Derjaguin approximation applies and both surfaces are highly charged.

5.7 Solve the linear Poisson–Boltzmann equation for two parallel homogeneously charged surfaces with charge densities σ_1 and σ_2 of opposite sign, separated by a distance h, and in equilibrium with a bulk aqueous solution with a concentration c_b of a 1:1 electrolyte. Determine the position and the depth of the minimum in the interaction free energy.

5.8 Using the Hamaker theory, calculate the dispersion interaction between two parallel sheets of thickness D separated by a distance h.

5.9 Derive an approximate expression for the adhesion energy between two rigid spherical particles of radius R, interacting only by van der Waals forces. (Assume a distance of closest approach of δ.)

5.10 Determine the dispersion interaction between a half-plane and a sphere of radius R, using the Hamaker method. (a) At a close separation so that the shortest distance between sphere and plane, h, is much smaller than the radius of the sphere; $h \ll R$. (b) At a separation $h \gg R$. (c) Show by explicit integration the transition between the two limiting cases (a) and (b).

5.11 Two flat surfaces are in adhesive contact in a liquid with molecules of radius R. The surfaces are pulled apart under atmospheric pressure to a separation $h < 2R$ so that no liquid can penetrate the gap. Estimate the force in the range $0 < h < 2R$, assuming that the surfaces only interact by van der Waals forces.

5.12 Two large planar parallel hydrophobic surfaces are placed 0.1 m below the air–water surface. Determine the separation at which it is favorable for air to fill the gap between the surfaces. Assume surface tensions $\gamma_{s,w}$ = 50 mnN/m² and $\gamma_{s,air}$ = 30 mnN/m² (s = surface, w = water, and a = air).

5.13 In the Problem in Section 5.5.4, we found that after the condensation of water between two polar surfaces, the force per area was $-1.3\ 10^{-7}$ N/m^2. Compare this value with the initial van der Waals attraction between the surfaces at a separation 1.8 nm. (Use Hamaker constant 5×10^{-20} J.)

5.14 Calculate the translational diffusion coefficient of a spherical particle with radius 10 nm in water at 25 °C. Determine the average time it takes for such a particle to diffuse from one side to the other of a spherical confinement of radius 10 μm?

5.15 A drop of water of radius 10^{-4} m is falling in air at a temperature of 298 K. Determine the velocity at which the Reynolds number is unity.

5.16 Two identical spheres of radius R are moving toward each other on the center-to-center line. At which velocity is the van der Waals attraction balanced by the repulsive hydrodynamic lubrication force at a surface–surface separation $h = R$?

References

Atkins, P., de Paula, J., and Keeler, J. (2022). *Atkins Physical Chemistry*, 12th ed. Oxford, UK: Oxford University Press.

Berg, J.C. (2010). *An Introduction to Interfaces & Colloids. The Bridge to Nanoscience*. World Scientific.

Cosgrove, T. (2010). *Colloid Science. Principles, Methods and Applications*. Chichester, UK: Wiley.

Davidson, D.W. (1973). *Water, A Comprehensive Treatise*, Vol. 2, F. Franks, ed. Plenum Press.

Doi, M. (2013). *Soft Matter Physics*. Oxford, UK: Oxford University Press.

Guldbrand, L., Jönsson, B., Wennerström, H., and P. Linse. (1984). *J. Phys. Chem*. 80, 2221.

Israelachvili, J.N. (1992). *Intermolecular and Surface Forces*, 2nd ed. London: Academic Press.

Israelachvili, J.N. (2011). *Intermolecular and Surface Forces*, 3rd ed. London, UK: Academic Press.

Kronberg, B., Holmberg, K., and Lindman, B. (2014). *Surface Chemistry of Surfactants and Polymers*, 3rd ed. Chichester, UK: Wiley.

Lyklema, H. ed. (1995). *Fundamentals of Interface and Colloid Science. Solid-Liquid Interfaces*, Vol. 2. Amsterdam, The Netherlands: Elsevier.

Lyklema, K. and Mysels, K.J. (1965). *J. Am. Chem. Soc*. 87, 2539.

Marra, J. (1985). *JCIS*. 107, 446.

Ohshima, H. (2024). *Fundamentals of Soft Interfaces in Colloid and Surface Chemistry*. London, UK: Academic Press.

Pauling, L. (1960). *The Nature of the Chemical Bond*. Ithaca, NY: Cornell University Press.

Perkin, S., Kampf, N., and Klein, J. (2006). *Phys. Rev. Lett*. 96, 038301.

Petrov, P., Olsson, U., and Wennerström, H. (1997). *Langmuir*, 13, 1000.

Rich, S.W. and Berne, B. (1997). *J Phys Chem B*. 101, 10488.

Shinoda, C.K. (1978). *Principles of Solution and Solubility*. New York: Dekker.

Wennerström, H., Vallina Estrada, E., Danielsson, J., and Oliveberg, M. (2020). *Proc. Natl. Acad. Sci*. 117, 10113.

6

Bilayer Systems

Concept Map

Bilayers
Bilayer structures constitute a midpoint between normal and reversed amphiphilic structures and play crucial roles in biological and industrial processes. This chapter focuses on the lipids of biological membranes (Figure 6.1), which constitute the building blocks for many biological structures. Surfactants relevant for industrial applications are also discussed.

Manifestations of Bilayer Structures
Bilayers are the basic structural element of many different global structures, including: • **Lamellar liquid crystals**, which are stacks of bilayers separated by solvent and containing hydrocarbon chains in the melted state. Mechanical forces can induce defects in the bilayers that cause them to close up on themselves to form onion-like shells, called liposomes or vesicles (Figure 6.13). • **Gel phases** contain bilayers with crystallized hydrocarbon chains separated by a liquid-like solvent. Mismatches between the cross-sectional areas of chains and head groups lead to tilted or interdigitated chains or rippled surfaces (Figure 6.10). • **Bicontinuous phases** in which the bilayer is draped on a surface of zero mean curvature (Figure 6.14) or forms a sponge structure, which is a disordered version of the cubic phase. • **Vesicles,** which are usually metastable and whose size depends on the preparation method.

Key Features of Bilayer Systems
The solubility of bilayer-forming amphiphiles is usually low ($10^{-7} - 10^{-10}$ M), and processes that proceed via monomer exchange occur very slowly; attaining equilibrium requires days or even years (Table 6.3).

(Continued)

Key Features of Bilayer Systems (Continued)

When the hydrocarbon chains are melted, amphiphile diffusion in the lateral direction occurs quickly, but molecular exchange between monolayers is slow since it involves motion of the polar head groups through the oil-like center of the bilayer.

Characterization of Bilayer Systems

X-ray diffraction provides information on bilayer structure and thickness, head group area, solvent layer thickness, and whether the chains are melted or frozen (Figure 6.3).

Neutron diffraction using isotopic labeling provides detailed information on head group position and hydrocarbon chain density.

Nuclear magnetic resonance (NMR) yields a structural order parameter (Figure 6.6) reflecting average hydrocarbon chain orientation in the bilayer and how it is changed by membrane additives (Figures 6.7 and 6.8). Lateral diffusion and molecular mobility can also be measured by NMR (Figure 6.11).

Differential scanning calorimetry (DSC) detects changes in phase when an amphiphilic sample is heated and gives values of $\Delta \mathbf{H}$ associated with these transformations (Figure 6.11).

Osmotic stress (OS) and **SFA** techniques provide direct measurements of bilayer interaction forces (Figures 6.12 and 5.27).

Transmission electron microscopy (TEM) gives high-resolution direct images (Figure 6.13) approaching molecular resolution but requires that samples be dehydrated or frozen.

Role of Bilayers in Diffusion Processes

Even though a bilayer hydrocarbon core is only 4 nm thick, it provides an effective barrier to the diffusion of ions and polar compounds.

Black lipid membranes (Figure 6.15) are useful for measuring a solute's membrane permeability, P_S, which is mainly determined by the partition coefficient K_S between bulk solvent and membrane (Eq. 6.3.6). The permeability takes values of order 10^{-6} m/s for water, 10^{-10} m/s for sugars, and 10^{-14} m/s for simple ions.

Selective distribution of ions across membranes creates an electrostatic transmembrane potential called the **Donnan potential** (Eq. 6.3.13). Away from equilibrium, ion fluxes determine the differences in potential according to the **Goldman equation** (Eq. 6.3.15).

Curvature Energies of Vesicles

The curvature of a bilayer is characterized by its mean curvature H and the Gaussian curvature K.

The curvature energy of a vesicle bilayer is determined by three parameters: the bending rigidity κ, the saddle splay constant $\bar{\kappa}$, and the spontaneous curvature H_0.

The curvature energy is an area integral over the local curvatures as in the Helfrich expression (Eq. 6.2.1).

The curvature energy change in a vesicle fusion/fission process is typically many $k_B\mathbf{T}$ and it depends strongly on the value of the spontaneous curvature. Given the circumstances, typically either fusion or fission is thermodynamically allowed.

Role of Bilayer Structures in Biological Processes

A biological membrane is a complex structure containing lipids, proteins, and polysaccharides, in which lipids provide the basic organizing structural unit. Lipid bilayers provide the following:

- A barrier to passive diffusion of small polar solutes (ions and sugars) and for macromolecules (proteins, nucleic acids, etc.).
- A unique two-dimensional solution environment for membrane proteins.
- An internal organizing structure that supports a multitude of chemical processes.
- Transport through membranes by vesicle fusion/fission.

Roles of Bilayers as Solvent to Membrane Proteins

Proteins associate with membranes by Figure (6.24):

- Spanning the width of the bilayer with its polar ends extending into the adjacent aqueous solution.
- Anchoring via hydrophobic interactions to one side of the membrane.
- Anchoring via a hydrocarbon tail.
- Binding electrostatically.

Transmembrane Transport

The selective transport of solutes across membranes is a fundamental physiological process. Transport can occur by:

- A membrane-soluble carrier (Figure 6.18)
- A membrane-spanning channel (Figure 6.19)
- A pump consuming ATP
- Exocytosis/endocytosis (Figure 6.22)

Bilayers formed by amphiphilic molecules occupy a central position in the study of self-assembly for two reasons. First, the bilayer is the basic building unit of the biological membrane. A multitude of processes in the living cell rely on the unique and versatile properties of the membrane lipid bilayer. Second, the bilayer represents the midpoint between normal and reversed structures,

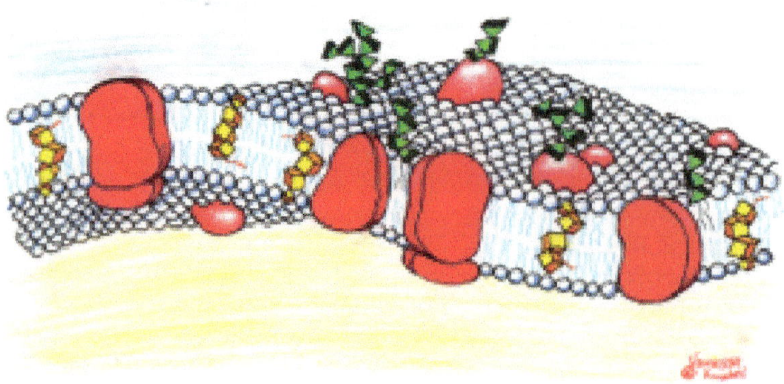

Figure 6.1 Schematic picture of a cell membrane. A number of proteins are dissolved in the lipid bilayer or adsorbed to the surface. *Source*: From Liljas et al., 2017/Vanessa Kunkel.

and under suitable conditions, the majority of amphiphiles aggregate into bilayers. This chapter focuses on the description of bilayer membranes formed by biological lipids. We will also discuss studies of bilayers formed by synthetic surfactants, particularly when illustrating experimental techniques as well as the general principles behind the formation of bilayers.

In the living cell, the lipid membrane (see Figure 6.1) performs four basic functions. First, it serves as a barrier for molecular diffusion processes. Ions and small polar metabolites like sugars penetrate a bilayer very slowly to pass from the inside to the outside or vice versa. Macromolecules, such as proteins, polynucleotides, or polysaccharides, cannot penetrate an intact bilayer by diffusion. The second role of the lipid bilayer is to act as a two-dimensional solvent for hydrophobic membrane proteins. A substantial fraction of the proteins in a cell depends on the lipid membrane for its function. The membrane proteins have adapted to the unique amphiphilic milieu of a 4-nm thick hydrophobic layer with an aqueous environment on both sides. The third function is to generate an intracellular structure that serves to organize the multitude of metabolic processes that occur in the living cell. This is partly accomplished by having a lipid bilayer structure that is laterally inhomogeneous. It is becoming increasingly evident that this is a very important role of the membrane in all but the most primitive organisms. A fourth role is in the communication between cells through the fission and fusion of vesicles.

6.1 Bilayers Show a Rich Variation with Respect to Local Chemical Structure and Global Folding

6.1.1 Many Amphiphiles Form a Bilayer Structure

In Chapter 1, we discussed the self-assembly of amphiphiles and found that for a surfactant number $N_s = v/la_0$ around unity, the preferred aggregate structure is a bilayer. Thus, in relation to a micelle-forming amphiphile, one that produces a bilayer structure should have a bulkier apolar region and/or a smaller head group. Chemically, this structure can be obtained in several ways. One approach for obtaining a bilayer-forming amphiphile involves adding a second apolar chain of approximately the same size to a surfactant molecule. This strategy works irrespective of the nature of the polar head group. Common examples are, in addition to the membrane lipids, the cationic dialkyl dimethyl quaternary ammonium surfactants, which have widespread use in conditioners for hair and textiles, and the anionic double-chain surfactant aerosol OT (di-2-ethylhexylsulphosuccinate) that has been widely used in scientific studies.

Another approach involves reducing head group size. For nonionic surfactants of the alkyl oligooxyethylene type, a shorter ethylene oxide chain promotes bilayer formation. For example, $C_{12}E_4$ forms a lamellar phase at ambient temperatures. Upon adding another ethylene oxide unit, the lamellar state forms first at an increased temperature, around 50 °C. This behavior correlates with the temperature dependence of the hydrophobic interaction, as discussed in Section 5.7. Monoglycerides, which have small head groups, aggregate into bilayers or even inverted structures, suggesting a surfactant number slightly above unity. These amphiphiles have a broad use as food additives and in pharmaceutical formulations.

A micelle-forming surfactant can be induced to self-assemble into a bilayer by adding a long-chain cosurfactant such as an alcohol. It has a small polar head making the average head-group size smaller in the mixture, and ultimately, a bilayer emerges as the most stable structure. The so-called catanionic surfactants, which contain a mixture of cationic and anionic amphiphiles, constitute a more intricate variation on this theme. Since these amphiphiles attract one another electrostatically, the effective head group size becomes much smaller than the sum of the head group areas for the separate systems, and the single-chain surfactants will assemble into a bilayer. In a perfectly matched system, the resulting properties resemble those of a double-chain zwitterionic system.

6.1.2 Membrane Lipids Exhibit Chemical Variations on a Common Theme

For the major fraction of membrane lipids, chemical structure can be seen as variations on a common theme. The apolar group typically consists of two hydrocarbon chains originating from a range of fatty acids. These chains are usually attached via ester or, more rarely, by ether covalent bonds to a linker group, usually a glycerol, which is also bonded to the polar head group. Figure 6.2 summarizes the chemical structure of some common lipids.

From Figure 6.2b, we see that the chemical diversity of polar groups is similar to that found for synthetic surfactants. We find ionic, zwitterionic, and sugar head groups. All common ionic lipids are anionic, a fact that has considerable significance for the electrostatic interactions. On average, typically 10–20% of the lipids of a membrane are charged. The nature of the head group depends on the organism. Glycolipids with mono or disugar polar groups dominate in plants. In animals, phospholipids are most abundant, whereas in bacteria, both types of lipids are common.

The apolar parts of the membrane lipids originate from the fat metabolism, and the chemical diversity in the fatty acid chains is the same in the membrane lipids as in the fat. The hydrocarbon chain length typically ranges from 16 to 22, with the degree of unsaturation increasing with increasing chain length. Table 6.1 shows the chemical formulas and names of some common fatty acids, and Table 6.2 shows the distribution of fatty acids in phospholipids of different organs in some animals.

In addition to the types of lipids shown in Figure 6.2, higher organisms also possess lipids of a quite different chemical nature, where the infamous cholesterol is the prime example. For example, the red blood cell has a very high content of cholesterol ($\approx$ 40 mol% lipid). Cholesterol consists of a rigid, relatively planar steroid skeleton with a single OH as the polar group. By itself, it is not a bilayer-forming lipid at ambient temperatures, and it plays a role similar to that of an alkanol cosurfactant in surfactant systems.

Given the variations in head groups, linkers, and fatty acid chains, and also including substitution patterns, a single cell contains on the order of 100 or more different lipid species. The major enigma of lipid biochemistry is to understand the functional significance of this diversity. Lipid composition

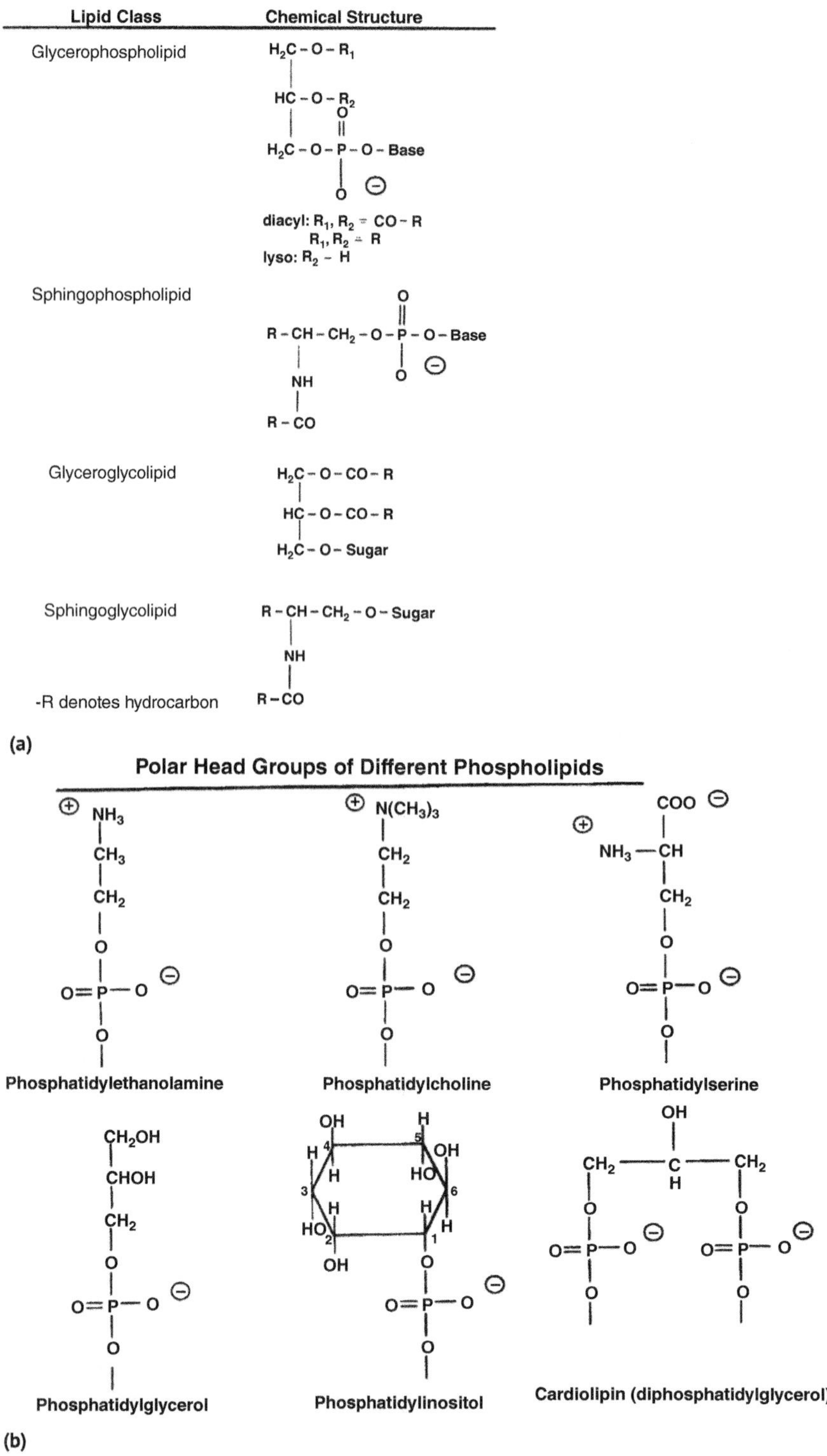

Figure 6.2 (a) Chemical structure of some common classes of lipids. (b) Head group structure of some common classes of phospholipids.

Table 6.1 Common name of acyl chains and melting point of the corresponding fatty acid. Δ denotes positin(s) of double bond(s).

Acyl chain	Common name	Symbol	Melting point of acid (°C)
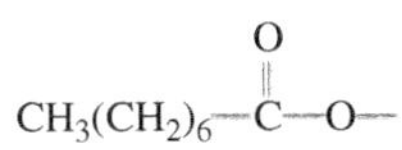	Capryl	8:0	16
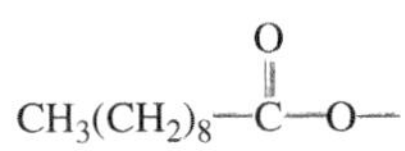	Capric	10:0	32
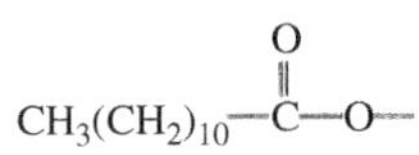	Lauroyl	12:0	44
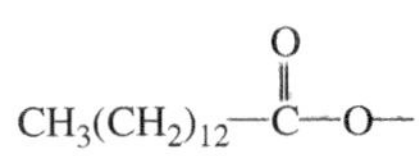	Myristoyl	14:0	54
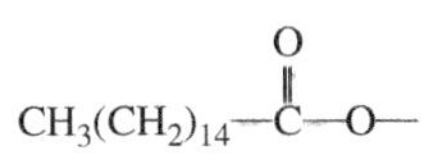	Palmitoyl	16:0	63
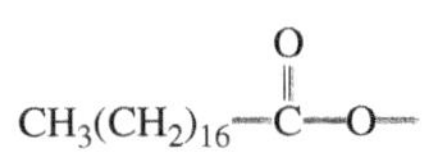	Stearoyl	18:0	70
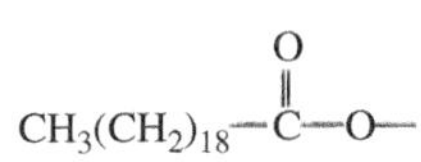	Arachidoyl	20:0	77
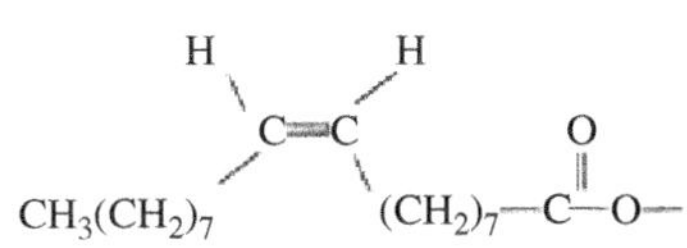	Oleoyl	$18:1^{\Delta 9}$	13
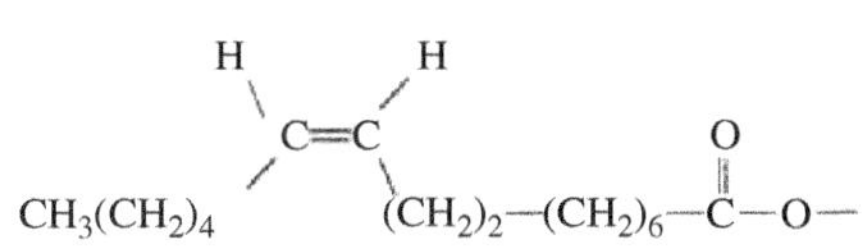	Linoleoyl	$18:2^{\Delta 9,12}$	-5
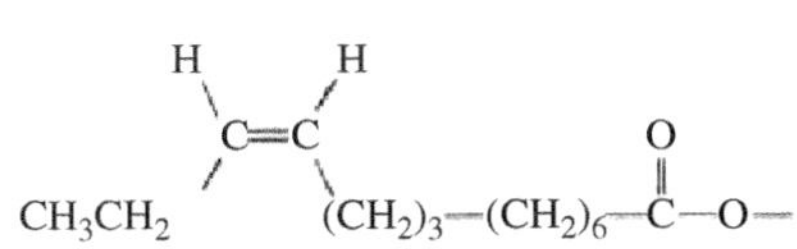	Linolenoyl	$18:3^{\Delta 9,12,15}$	-11
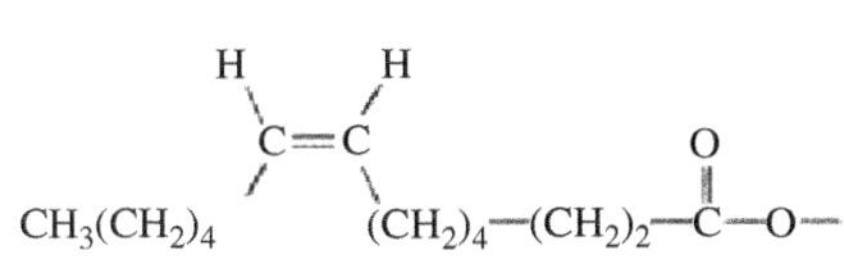	Arachidonoyl	$20:4^{\Delta 5,8,11,14}$	-50

varies between organisms, between cells, between different organelles within the cell, between the two sides of the bilayer, and between planar and curved regions. We can readily understand some trends. The degree of chain unsaturation increases with decreasing growth temperature. The values for tuna muscle given in Table 6.2 exemplify this effect. Saturated alkane chains tend to crystallize at ambient temperatures, and unsaturation is introduced to ensure a liquid-like character of the membrane. As a contrast, Table 6.2 also shows a high fraction of saturated C_{16} chains in the membrane of

Table 6.2 Distribution of the most common fatty acids of phospholipids in different membranes.

	$m:n$[a,b]				
Source	**16:0**	**18:0**	**18:1**	**20:4**	**22:6**
Phosphatidylcholine					
Brain (human)	34	13	45	1	0
Lung (pig)	47	9	26	5	0
Muscle (tuna)	19	5	8	7	55
Liver (rat)	30	17	10	18	3
Phosphatidylethanolamine					
Brain (human)	5	30	9	13	30
Liver (rat)	25	14	6	10	21

Source: Marsh, 1990/with permission of Taylor & Francis Group.
a. m denotes the number of carbons and n the number of cis double bonds.
b. These fatty acids account for ≈90% of the total content in the cases listed.

Table 6.3 Comparison between properties of micellar and bilayer aggregates.

Property	**Micelles**	**Bilayers**
Monomer solubility	10^{-2}–10^{-4}	10^{-7}–10^{-10} M
τ for monomer exchange	10^{-3}–10^{-6} s	10^{2}–10^{-3} s
τ for aggregate lifetime	10^{-1}–10^{-3} s	Days to years
Characteristic temperature	Krafft temperature	Chain-melting temperature
Structural directionality	All directions equivalent	Lateral diffusion rapid, flip-flop slow
Aggregation pattern	Forms a well-defined aggregate at well-defined CMC	Basic structural unit appears in a variety of global structures

pig lungs. Their functional significance may be that unsaturated chains are susceptible to oxidation by O_2. Having saturated chains predominate in cells with the highest oxygen exposure reduces the oxidative chemical degradation.

6.1.3 Comparing the Properties of Spherical Micelles and Bilayers Provides Useful Insight into the Many Distinctive Molecular Properties of Bilayers

Table 6.3 summarizes many of the properties of spherical micelles and bilayer systems. Monomer solubilities of typical zwitterionic and polar surfactants (as measured by their CMCs) are of the order 10^{-3} M, whereas those of phospholipids that form bilayers are much lower, on the order of 10^{-10} M. This difference of a factor of 10^6 in monomer solubilities reflects the difference between having one or two hydrocarbon chains and leads to profound differences in many of the molecular properties of micellar and bilayer systems.

As described in Section 4.2, the formation of micelles has been characterized by a number of experimental techniques. Micelle formation clearly illustrates the characteristics of the aggregation process described in Section 1.4. The CMC defines the "start" of the process, the point at which micelles begin to form spontaneously. Adding more surfactant leads to the formation of more micelles with well-defined aggregation numbers rather than leading to larger micelles. The micelle's radius, which is determined by the length of the fully extended hydrocarbon chain, leads to aggregates with well-defined properties.

Our direct information on the bilayer aggregation process is more limited because most experimental techniques are not sensitive enough to yield useful data at such extremely low monomer concentrations. Monomer solubility still defines the start process, and the bilayer's thickness is a well-defined quantity determined by the length of the hydrocarbon chains, while there is no molecular limit to bilayer growth in the lateral direction. In an ideal situation, bilayers attempt to minimize the degree to which the hydrocarbons at their edges are exposed to the aqueous environment by extending indefinitely in the lateral direction. In reality, finite samples show defects, and bilayers tend to close up on themselves, leading to a variety of macroscopic structures described in the next section.

In micellar solutions, the relaxation times for monomer exchange range from 10^{-3} to 10^{-6} s, and they involve a diffusion-limited process that depends directly on the monomer concentration. In bilayers, monomer exchange is much slower, as a consequence of their much lower monomer solubility.

Lifetimes for typical micelles vary from 10^{-3} to 10^{-1} s. With bilayers, such transformations involve rearranging the entire bilayer structure, and a particular specific macroscopic state often reflects the methods used to prepare the sample. Thus, micelles respond rapidly to changes in their environment, while bilayers are much more sluggish. It can require hours, days, or even years for a bilayer to achieve an equilibrium state.

The transport properties of micelles and bilayers also display major differences. With spherical micelles, all directions are equivalent, and a surfactant molecule can diffuse very rapidly within a micelle. A bilayer, on the other hand, has two distinct diffusion modes. The first involves lateral diffusion within the bilayer plane. It occurs rapidly since it mainly involves the amphiphile chains moving through the oil-like region of the bilayer. The second diffusional mode is slow since it involves moving the head group from one side of the bilayer to the other. This "flip-flop" requires the polar head group to diffuse through the oil-like interior of the bilayer. In pure bilayers, the time required to execute this flip-flop process depends on the nature of the amphiphile, but it is typically at least a minute for a double-chain lipid with a strongly polar head group, while it occurs on a millisecond timescale for cholesterol.

6.1.4 Several Methods Can Be Used to Study a Lamellar Liquid Crystal, Which Is the Natural Reference Phase for Bilayer Systems

Whether it is a synthetic amphiphile or a biological membrane lipid with two chains, it typically forms a lamellar liquid crystalline phase at equilibrium. Bilayers are stacked in a parallel way as shown in Figure 1.6c. We will, in the following, exclusively consider aqueous systems, but a lyotropic liquid crystal can also form in other polar media like ethylene glycol or ionic liquids. One can apply several different experimental methods to unravel structural and dynamic properties of the bilayers in a lamellar phase.

The repeat distance, d, in a lamellar stack is readily observed using X-ray diffraction. Figure 6.3 shows how, for the specific sample of an ionic amphiphile, the repeat distance d increases linearly with the inverse of the amphiphile volume fraction.

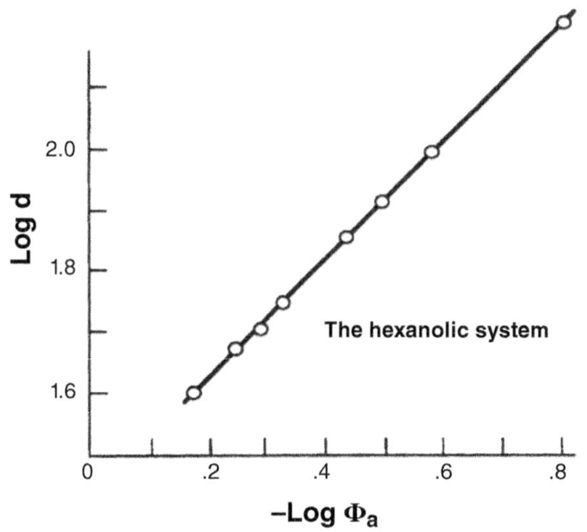

Figure 6.3 Double-logarithmic plot of repeat distance *d* (nm) versus inverse amphiphile volume fraction Φ_a at a fixed molar ratio of 2.38:1 surfactant CTAB to cosurfactant hexanol. The straight lines have a slope of 1.0, showing that $d \propto \Phi_a^{-1}$, which is characteristic of one-dimensional swelling, where added water only increases the thickness of the aqueous layer. *Source*: Fontell et al., 1991/ with permission of Springer Nature.

If one assumes that the amphiphile and water are fully separated, the thickness of the amphiphile bilayer, d_2, is simply given by $d_2 = \Phi_a d$. In Figure 6.3, d is inversely proportional to the volume fraction Φ_a, showing that the bilayer thickness d_2 and thus the area per polar group remain constant during swelling. For such an ionic system, we know from the previous chapter that there is an electrostatic double-layer repulsion between bilayers, and this gives rise to the swelling. One would intuitively expect the bilayer–bilayer repulsion to also affect the area per amphiphile, but this is not observed for the case shown in Figure 6.3. It is a counterintuitive property of the electrostatic interaction in the nonlinear regime to generate a constant lateral pressure (see Eq. 3.9.17), which explains the observation. For zwitterionic amphiphiles, one observes that the area per head group decreases slightly on decreasing bilayer separation. For a given amount of solvent, this reduces the repulsive interaction by increasing bilayer–bilayer separation.

Through neutron diffraction, one can obtain more detailed information about the molecular arrangement in the lamellar stack. By contrast variation (see Section 4.3), different parts of the structure can be highlighted, as illustrated in Figure 6.4 for a phospholipid system. The thickness of the hydrocarbon layer is, as discussed in Section 1.5, significantly smaller than that of two extended chains, indicating that there is a large fraction of gauche conformations in the chain. This is consistent with the liquid character of the bilayer. The double bond of oleic acid is located between carbon 9 and carbon 10 in the chain, and their peaks are separated by 1.8 nm, which is substantially smaller than the length of two stretches of eight carbons as estimated from Eq. (1.5.3). There is a clear dip in density at the bilayer midplane, indicating that there is a limited overlap of chains from one side to the other. Later in this chapter, we will argue that it is a reasonable approximation to consider the two monolayers as approximately independent when discussing bending modes. The sample in Figure 6.4 was equilibrated at a relative humidity of 66%, which corresponds to approximately six water molecules per lipid. In such a case, the water density does not reach the bulk value since there is an overlap between the polar head groups on opposite monolayers. It is important to realize that in contrast to a proper crystalline system, there are substantial thermal fluctuations in the liquid crystal, so the profile in Figure 6.4 represents a thermal average over head group conformations and bending modes.

Computer simulations provide an alternative, powerful route to obtain insight into the molecular organization in a bilayer. Such simulations contain a lot of molecular detail, and, if properly

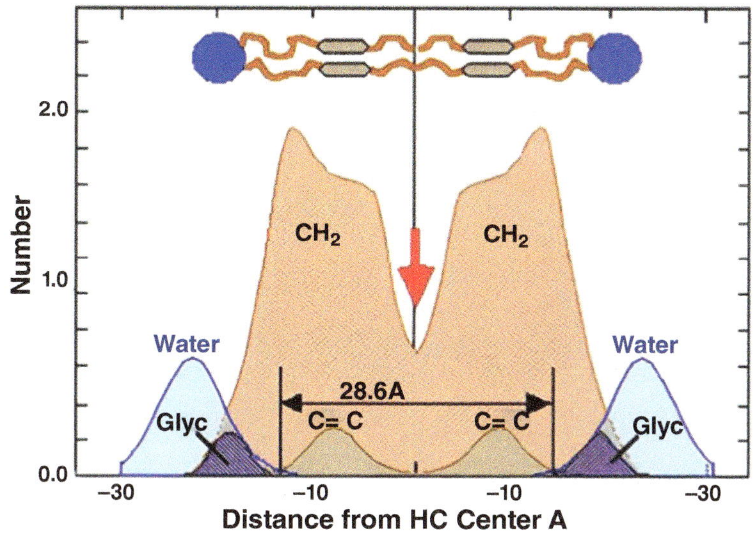

Figure 6.4 Density profiles in a fluid bilayer. The glycerol groups mark the water–methylene boundary for DOPC at 66% relative humidity. The double arrow indicates the thickness of the bilayer, 2.86 nm, taken as the transbilayer separation of the extreme edges of the water distributions. Notice that thermal motion causes a small but significant overlap of the water and double-bond distributions. *Source*: Wiener and White, 1991/with permission of Elsevier.

conducted, they confirm the structural picture shown in Figure 6.4. A quantity that is easily calculated from the simulation is the order parameter profile, $S_{CD}(n)$, representing an average

$$S_{CD}(n) = \frac{1}{2}\langle 3\cos^2\theta_{CD}(n) - 1\rangle \tag{6.1.1}$$

involving the angle θ_{CD} between the bilayer normal and the direction of the C—D bond on carbon n . This average is strongly affected by thermal motions, but it is nonzero due to the anisotropy of the lamellar system. The order parameter is readily measured by nuclear magnetic resonance (NMR) on deuterated lipids; hence, the notation S_{CD}.

Figure 6.5 shows the ^{2}H NMR spectrum of a randomly oriented lamellar system DPPC–sodium cholate–water with deuterated alkyl chains. In this case, the signal from all 16 —CD_2 (—CD_3) groups is resolved. The order parameter S_{CD} is proportional to the observed splitting with a known proportionality constant. Figure 6.6 shows a typical order parameter profile obtained for the palmitoyl chain in a lamellar liquid crystal formed by the lipid palmitoyl-oleoyl phosphatidylcholine (POPC). The order parameter decreases monotonically along the chain, as the distance to the polar head increases. However, if one considers a free single chain with a fixed position of the head group and only confines the chain to a half-plane, the order parameter decreases more rapidly along the chain. This shows that the neighboring chains induce an effective interaction, straightening the chains somewhat. For comparison, Figure 6.7 shows the order parameter profile obtained from the spectrum in Figure 6.5.

When a second compound is added to a lipid bilayer, it affects the chain order. In the system of Figure 6.7, the physiological surfactant sodium cholate has been added to the DOPC lipid. As seen in Figure 6.7, the result is a drastic decrease in the order parameter along the chain. Cholate is a planar molecule with an apolar and a polar side, and it has a tendency to lie flat at the polar–apolar

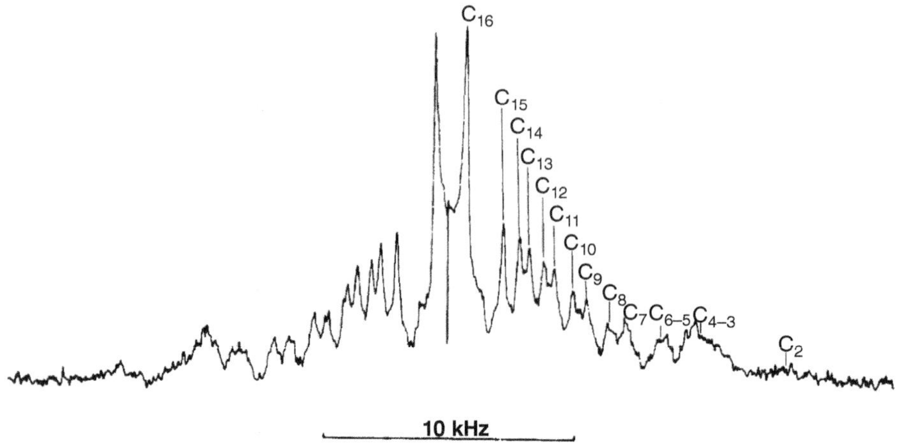

Figure 6.5 Deuterium NMR spectrum of a lamellar liquid crystalline phase in the system d_{31} palmitoyl-oleoylphosphatidylcholine–sodium cholate–water. Signals from all the 15 different deuteriums along the palmitoyl chain are resolved through the difference in the quadrupolar splitting. The numbering refers to the carbon atom to which deuteriums are bound and starts at the carboxylate end of the palmitate. *Source*: Lindblom et al., 1982/with permission of American Chemical Society.

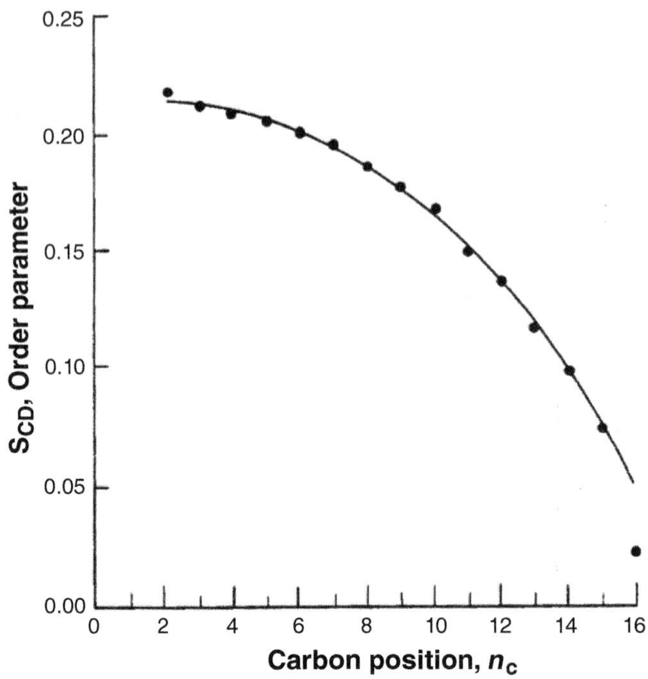

Figure 6.6 Order parameter profile for palmitoyl (d_{31}) chain palmitoyl-oleoylphosphatidylcholine (POPC) in a lamellar liquid crystal at 30 °C. *Source*: Adapted from Lafleur et al., 1990.

interface of the bilayer. When the lipid chains fill the space on the apolar side of the cholate molecule, the result is an induced chain disorder.

Figure 6.8 shows that the cholesterol molecule has an opposite effect on the chain order.

Figure 6.7 The order parameter decreases drastically for the palmitoyl chain of POPC when sodium cholate is added, cholate/DOPC ratio: ▲ 0.084; ■ 0.78. *Source*: Lindblom et al., 1982/with permission of American Chemical Society.

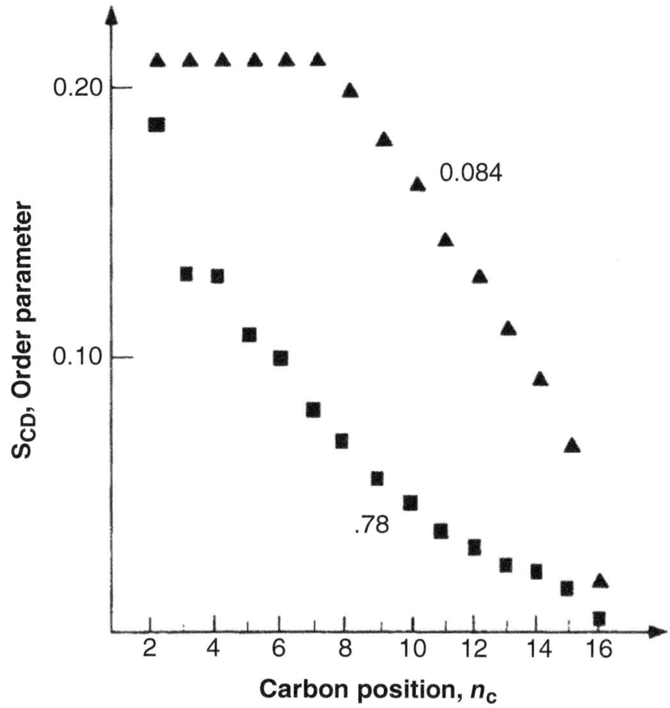

Figure 6.8 Order parameter in a palmitoyl chain increases, as cholesterol is added to the lamellar phase: pure POPC (○); 20 mol% cholesterol (●); 30 mol% (■); 45 mol% (□) at 30°C. *Source*: Adapted from Lafleur et al., 1990.

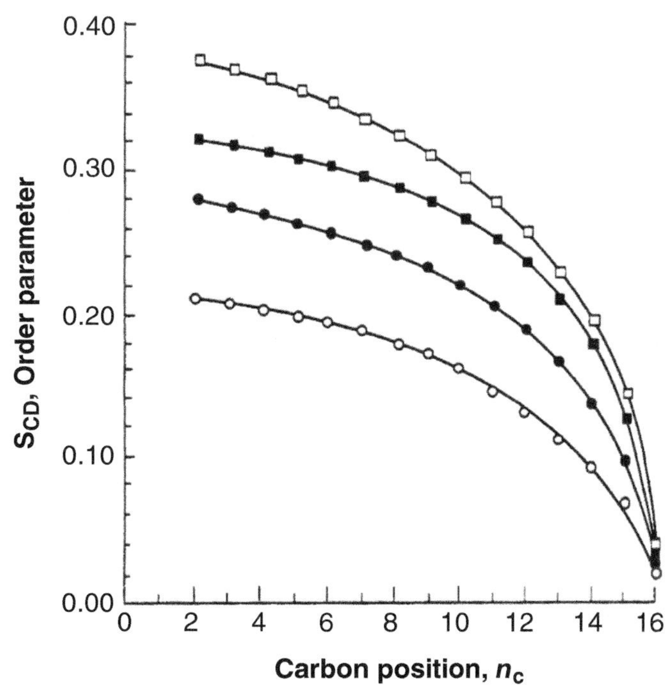

Cholesterol is also partly planar, but both sides are apolar, and the molecule inserts between chains with the hydroxy group at the surface. This results in a straightening of the alkyl chains near the cholesterol molecule, and at 45 mol% cholesterol, the order parameter has nearly doubled relative to the pure phospholipid case. We will discuss the effect of cholesterol further in the following section.

The liquid crystalline character of a lamellar phase is clearly demonstrated by measurements of the self-diffusion coefficient of the lipid/amphiphile. The molecules are confined to the layers, so diffusion in the perpendicular direction is orders of magnitude lower than in the lateral direction. Diffusion along the lamellae is readily measured using pulsed field gradient NMR techniques. The lateral self-diffusion coefficients, D_S, for lipids in a lamellar system are typically in the range 10^{-10} to 10^{-11} m^2/s, which is similar to the values for liquid hydrocarbons of the same chain length.

The water molecules in the aqueous layer also experience a highly anisotropic diffusion. Water molecules can pass through the lipid layer, but this is a very improbable event due to the low local concentration as will be discussed in Section 6.3. In contrast, the lateral water diffusion is rapid, and the measured self-diffusion coefficient approaches that of bulk water as the water content increases.

When a lipid bilayer is bent, there is an induced area difference between the two monolayers with the result that the area per molecule changes. This is a nonequilibrium situation that provides a driving force for lipids changing side in a so-called flip-flop motion. Such a flip-flop requires that the polar group pass through the apolar layer of the lamellae, which makes it a slow process. The rate depends strongly on the chemical nature of the polar group, and it also depends on the bilayer composition. For an ionic lipid, the process is very slow, and it is still slow for zwitterionic and sugar head groups. The processes are so slow that it is difficult to measure the flip-flop rate accurately, but they occur on timescales from seconds to days. An important exception is found for cholesterol with a single —OH polar group so that it has a reasonable solubility in apolar solvents. This ensures a rapid flip-flop time of order milliseconds.

See Israelachvili (2011, Chapter 20) for additional discussion on Lipid Molecular Characteristics.

6.2 Bilayers Can Adopt Many Different Global Structures

6.2.1 Gel Phases Are Formed at Lower Temperatures

When a lamellar lyotropic liquid crystal is cooled to lower temperatures, it is a common finding that there is a transition to another bilayer phase, called the gel phase, with solid ordered hydrocarbon chains but with disorder in the polar part. A direct way to study such transitions is using differential scanning calorimetry (DSC). The DSC experiment monitors the sample temperature as a function of time for a constant heat transfer to the system. This process measures the heat capacity $\mathbf{C}_\mathrm{p}$ of the system, which peaks sharply at phase changes. Ideally, $\mathbf{C}_\mathrm{p}$ should be infinite at a first-order phase transition, but in practice, mechanisms always exist that lead to a finite, but strongly peaked, heat capacity. Figure 6.9 shows this effect for a phospholipid–water system: the peak at 41°C is due to the gel-to-liquid crystal transition, while the bump at 34°C signals a transition between two different gel phases.

By integrating the heat capacity, curve A in Figure 6.9, we can determine the enthalpy change $\Delta\mathbf{H}$ associated with the phase transitions as shown in curve B. Table 6.4 shows the measured values of $\Delta\mathbf{H}$ and $\Delta\mathbf{S}$ for the gel-to-liquid crystal transition for a series of phospholipids. For the saturated chains, $\Delta\mathbf{H}_\mathrm{t}$ increases by around 2.5 kJ/mol per added carbon atom. These $\Delta\mathbf{H}_\mathrm{t}$ values are somewhat lower than the heats of fusion for pure alkanes of similar size, which indicates that although the gel differs from a crystalline phase and the liquid crystal is not quite an isotropic liquid, the process is mainly chain melting.

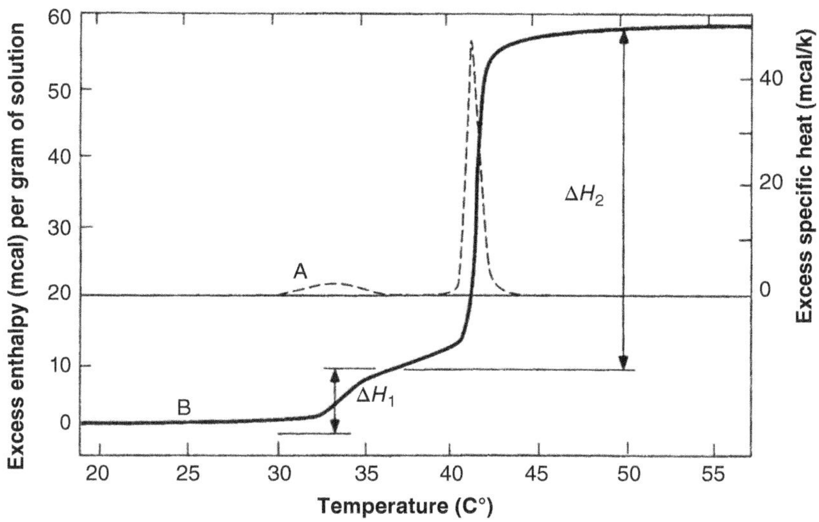

Figure 6.9 The variation with temperature of the excess specific heat (curve A, right-hand ordinates) and the excess enthalpy (curve B, left-hand ordinates) during the gel-to-liquid crystal transition of dipalmitoyl phosphatidyl choline in the presence of excess water. $\Delta \mathbf{H}_1$ is the enthalpy for the transition from $L_\beta{}'$ to P_β at $T \approx 34°C$. $\Delta \mathbf{H}_2$ is the transition enthalpy for P_β to L_α chain melting at 41 °C. *Source*: Hinz and Sturtevant, 1971/with permission of Elsevier.

Table 6.4 Thermodynamic parameters, transition temperature T_t, enthalpy ΔH_t, and entropy ΔS_t for the transition from gel (L_β or P_β) to liquid crystal (L_α) for a number of phospholipids.

Head group	Chain 1	Chain 2	T_t(°C)	ΔH_t (kJ/mol)	ΔS_t (kJ/mol)
Choline	Dodecyl	Dodecyl	−1	12	45
	Tetradecyl	Tetradecyl	23.5	25	83
	Octadecyl	Octadecyl	55	42	129
	Tetradecyl	Hexadecyl	35	33	107
	Tetradecyl	Icosano	40	34	108
	Hexadecyl	*cis*-9-Octadeceno	3	30	100
Ethanolamine	Dodecyl	Dodecyl	30	16	53
	Hexadecyl	Hexadecyl	64	35	105
Serine (Na^+ salt)	Tetradecyl	Tetradecyl	36	31	94
	Hexadecyl	Hexadecyl	54	38	115

Source: Marsh, 1990/with permission of Taylor & Francis Group.

The calorimetric experiments provide a thermodynamic characterization of the system. Diffraction studies reveal the structural characteristics. In the gel phase, chains are ordered and this gives rise to a characteristic wide-angle diffraction peak showing a repeat distance of 0.54 nm. This is due to a hexagonal packing of the chains. Figure 6.10 shows four different structures with ordered chains. In the L_β phase, chains are oriented perpendicular to the lamellae. A similar packing is found in the interdigitated structure, but the lamellar thickness is much smaller and the area per polar group is consequently larger. In the $L_{\beta'}$ phase, chains are tilted relative to

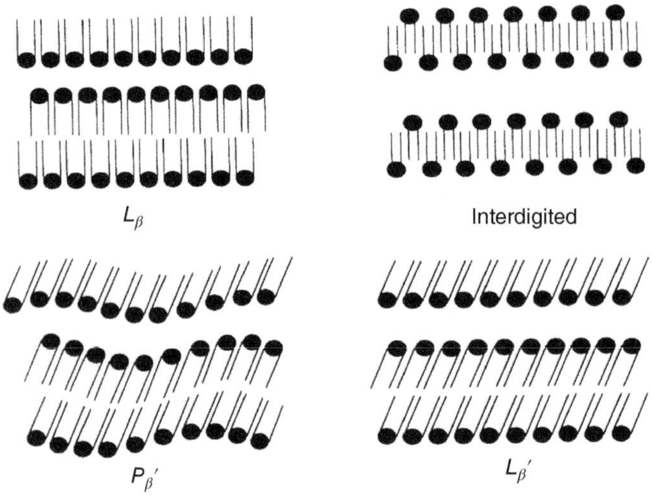

Figure 6.10 When alkyl chains crystallize, they have a cross-sectional area of ≈ 0.20 nm^2 per chain in a plane perpendicular to the direction of the chain. If the area of the polar groups in the bilayer matches that of the chain(s), a structure L_β is adopted. In the more likely case of a mismatch between polar head and chain areas, a tilted structure, as in $L_{\beta'}$, or a rippled structure, as in $P_{\beta'}$, can appear. For single-chain amphiphiles, chains may even interdigitate, making the thickness of the apolar layer similar to the length of a single chain.

the bilayer normal, breaking the cylindrical symmetry of the two previous phases. Finally, in the $P_{\beta'}$ phase, sometimes denoted as P_β, there is an undulation of the apolar–polar interface. In a scattering study, one can determine the characteristic length of the undulation, which typically is of order 10–20 nm. The dimensionless number $g_p = a_0/(0.54^2 \times 3^{1/2})$, representing the ratio between the area of the polar group and the cross-sectional area of the chains, can be seen as the basic factor determining the most favorable packing. For $g_p = 1$, the L_β phase can be formed without internal frustration. For $g_p \approx 2$, the interdigitated phase is optimal. For $1 < g_p < 2$, it is a compromise to tilt the chains to achieve a larger cross-sectional area in the perpendicular direction. The undulation structure adds to the effect of the tilt to increase the effective area, which can be seen from the fact that the midplane area in the P_β phase is, at a given tilt, larger than that of the $L_{\beta'}$ phase.

6.2.2 Addition of Cholesterol to a Lipid Bilayer Can Induce the Formation of a Liquid-ordered State

At the transition from a gel to a liquid crystalline phase, two separate molecular events happen simultaneously. The alkyl chains go from an all-trans conformation to a more disordered state with a substantial fraction of gauche conformations. There is also a loss of the crystalline order in the lateral direction characteristic of a solid-to-liquid transition. For a pure lipid system, there is a sufficient coupling between the conformational and translational order to ensure that a transition occurs simultaneously. As seen in Figure 6.8, the addition of cholesterol induces a straightening of the alkyl chains. A consequence is a weakening of the coupling between conformational and translational order of the alkyl chains. As a result, there can occur a transition from a liquid-disordered, l_d, to a liquid-ordered, l_o, state with mainly straight chains but lacking the translational order typical for a solid state. Figure 6.11 shows how the lateral diffusion coefficient, D_L, of the lipid egg sphingomyelin varies as the cholesterol content is increased. The two-phase coexistence is signaled

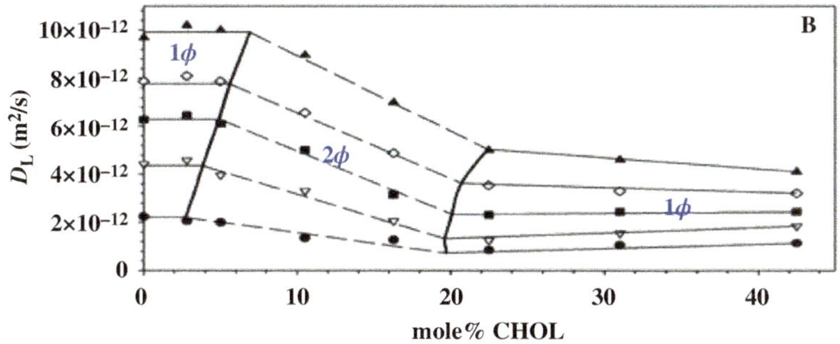

Figure 6.11 Measured lateral self-diffusion coefficients, D_L, of egg sphingomyelin as a function of the cholesterol content at five different temperatures; from bottom to top: 313, 318, 323, 328, 333, and 338 K. There is a two-phase area liquid disordered coexisting with a liquid-ordered phase at higher cholesterol contents of around mole 20%. In the two-phase area, the observed D_L varies linearly with concentration since it is an average over the D_L of the respective phases at the boundary. *Source*: Fillipov et al., 2003/with permission of Elsevier.

by a linear variation of the measured D_L. The diffusion is around a factor of two slower in the more compact l_o phase compared to the l_d phase. However, a value of $D_L > 10^{-12}$m/s^2 is a signature of a liquid character of the l_o phase.

One of many functions of cholesterol in a cell is to facilitate the formation of domains of the l_o state in the membrane. This can act to give a lateral sorting of membrane proteins, as will be further discussed in Section 6.4 and in Chapter 10.

6.2.3 Surface Forces Can Be Readily Measured in a Lamellar Stack

In a stack of bilayers, the polar–apolar interface is spontaneously arranged with a constant mean distance separation between two opposing interfaces. The chemical potential of water, often expressed in terms of the osmotic pressure (see Section 1.2.3), is the thermodynamic parameter that determines the swelling of the aqueous layer. There are several experimental alternatives to control the osmotic pressure in a bilayer system. The most direct method is to equilibrate the system at controlled relative humidity (RH) and measure the water content. One then obtains a direct thermodynamic characterization of the relation between water content and water chemical potential. This works well for high osmotic pressures, or RH $\leq$ 97%, which for typical lipid systems yields a swelling of up to 10 water molecules per lipid. For intermediate and low osmotic pressures, it is more convenient to use the depletion force (see Section 5.5.2) and equilibrate the bilayer system with an aqueous solution of a polymer that is too large to penetrate the polar region between the lipid layers. By simultaneously measuring the repeat distance by X-ray scattering, one obtains a force–distance relation, covering a large range. This is called the osmotic stress method, pioneered by A. Parsegian and coworkers.

Figure 5.17 shows the interaction free energy between two glycolipid bilayers measured by the SFA (see Section 5.4). In the osmotic stress technique, one measures the force, which is the derivative of the interaction free energy, and one can only determine the repulsive part of the interaction curve. This is also the case for measurements based on a controlled RH value. The measurements using the different experimental methods give a consistent pattern of a repulsive force that increases approximately exponentially as bilayers approach. Figure 6.12 shows a compilation of a number of different measurements for phospholipid and glycolipid systems.

The curves show that for all systems, the repulsion depends exponentially on distance except at short separations, where the overlap between head groups makes it difficult to uniquely define

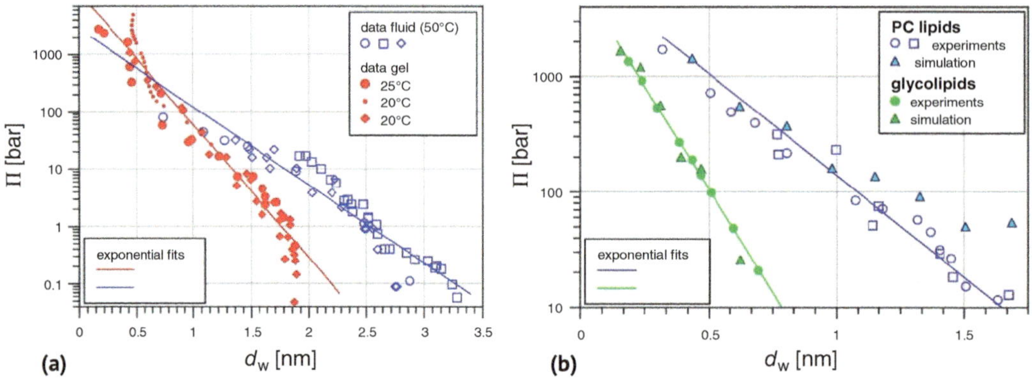

Figure 6.12 Measured repulsive forces, log(Π_{osm}) versus water layer thickness d_w. (a) Forces between phosphatidyl choline bilayers at different temperatures. Red symbols denote gel phase, while blue symbols denote liquid crystalline phase. The solid lines are exponential fits to the data with decay lengths of 0.15 nm (gel) and 0.31 nm (liquid crystal). (b) Comparison between measured forces for the glycolipid, DGDG (green), decay length 018 nm; and the phospholipid, PC (blue), decay length 0.31 nm. The triangles are the results of molecular dynamics simulation determinations of the force. Courtesy of Emanuel Schneck.

a separation distance. The decay length depends on the system with respect to both the head group and the state of the alkyl chains. It was stated in Section 5.6 that the most likely source of the repulsion is the overlap between head groups on opposing bilayers. The data of Figure 6.12 corroborates such a conclusion. In the gel phase, the apolar part of the bilayer is in a solid state with limited fluctuations, while the conformational freedom of the head groups is similar to that in the liquid crystalline state. With fewer degrees of freedom, the repulsion in the gel phase becomes weaker and more short-ranged than for the corresponding liquid crystalline state. A sugar head group has fewer conformational degrees of freedom than a phosphatidyl choline lipid. Thus, one expects a stronger and more long-ranged repulsion for the latter system.

6.2.4 A Lipid Bilayer Can Fold into a Closed Vesicle

A finite piece of a bilayer has a high edge energy, resulting in a line tension. The bilayer readily folds to form a metastable closed shell rather than reverting to the equilibrium lamellar structure. Such aggregates are called liposomes or vesicles, and they can occur both as multilayer and single-layer structures. The latter form is conventionally divided into small unilamellar vesicles (SUVs), with a radius in the nanometer range, and giant unilamellar vesicles (GUVs), with a radius larger than a micron.

The experimental protocol for forming unilamellar vesicles varies widely depending on the desired vesicle size. It is difficult to get narrow size distributions, and there is typically also a fraction of vesicles with more than one bilayer. Once a lipid vesicle has been formed, the number of lipids remains virtually constant due to the low monomer solubility. The vesicle is then free to adapt to an optimal area since there are no external constraints. In addition, water can, on a millisecond timescale, pass through the bilayer, so there is no constraint on the enclosed volume except at rapid shape changes. Therefore, there is no tension in the bilayer as long as there is no difference in osmotic pressure between inside and outside the medium. When stretched, the free energy of a vesicle depends quadratically on the vesicle area. This is in contrast to the case when a bilayer is in equilibrium with a lipid reservoir such that an area change leads to a linear change in the free energy.

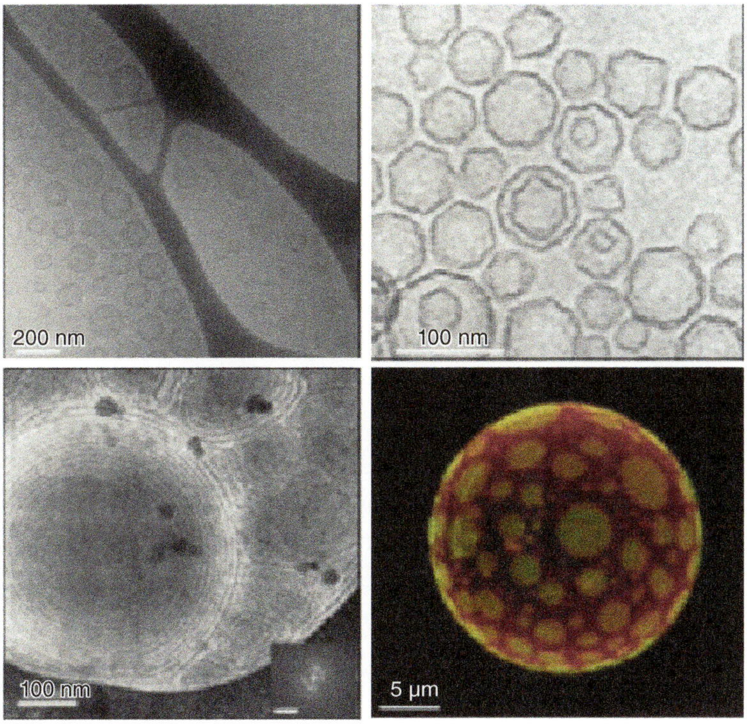

Figure 6.13 Cryo-transmission electron microscope (TEM) pictures of vesicles: Upper panel left: Small unilamellar vesicles formed by fluid lipid bilayers. *Source*: Makasewicz et al., 2022/Cambridge University. Upper panel right: Facetted vesicles of lipids in the gel state. *Source*: Ickenstein et al., 2006/with permission of Elsevier. Lower panel left: Multilamellar vesicles formed by isolated milk fat lipids in the liquid crystalline state. *Source*: Waninge et al., 2004/The Royal Society of Chemistry.) Lower panel right: Confocal fluorescence microscopy of giant unilamellar vesicles composed of phospholipids from pulmonary lung surfactant, coexistence of gel phase, and fluid phase bilayers. *Source*: Bagatolli et al., 2006/with permission of Elsevier.

For a unilamellar vesicle in a state symmetric with respect to inside and outside, the sphere represents the lowest energy state. Due to thermal bending fluctuations, there is at each moment in time a deviation from the spherical shape, but the amplitude of such deviations is typically small relative to the size of the vesicle. However, by breaking the inside–outside symmetry, one can induce major shape changes. Such symmetry breaking can be accomplished by inducing an inside–outside difference in osmotic pressure, by imposing an area mismatch by adding or removing lipids from the outer layer, by creating an inside–outside difference in the monolayer composition, or by external adsorption of proteins. Large vesicles can, under such constraints, adapt several shapes.

There exists a substantial literature discussing vesicle shapes and transitions between different locally stable geometries. The bilayer midplane is conveniently modeled as a mathematical surface. At any point on the surface, there is a tangential plane. If one takes the z-direction along the (positive) normal to this plane, the curvature is determined by the second derivatives of the surface with respect to the x and y directions along the tangential plane. The four second derivatives $-\partial^2 f/\partial x^2$, $\partial^2 f/\partial x\partial y = \partial^2 f/\partial y\partial x, \partial^2 f/\partial y^2$ –form a 2×2 matrix $\boldsymbol{C}$, where $1/2\text{Tr}\boldsymbol{C} = H$, the mean curvature, and $\text{Det}\boldsymbol{C} = K$, the Gaussian curvature. These quantities are independent of the orientation of x,y coordinate system and uniquely determine the curvature of the surface.

The main quantity determining the shape is the curvature free energy, $\mathbf{G}_C$. W. Helfrich showed that the curvature free energy, $\mathbf{G}_C$, to second order in the curvature, is the surface integral of local contributions from the mean and the Gaussian curvatures

$$\mathbf{G}_C = \int \left\{ 2\kappa(H - H_0)^2 + \bar{\kappa}\, K \right\} \mathrm{d}A \tag{6.2.1}$$

This equation provides a theoretical basis for describing vesicle systems. The Helfrich curvature energy expression contains three parameters characterizing the bilayer property. These are the bending rigidity κ, the saddle splay modulus $\bar{\kappa}$, and the spontaneous curvature H_0. (The definition of the spontaneous curvature varies in the literature and sometimes $H_0' = 2H_0$.) For lipid bilayers in the liquid crystalline state, κ is usually in the range 20–100 $k_B\mathbf{T}$. The smaller values occur for bilayers with unsaturated chains, while the larger values occur for thicker bilayers in, for example, the liquid-ordered state. The knowledge about the value for $\bar{\kappa}$ is more limited. The formation of a saddle shape from a plane involves bending the surface in two opposite directions, making the value of the Gaussian curvature K negative. A negative value of $\bar{\kappa}$ shows that the formation of a saddle involves an increase in free energy. For a homogeneous sheet, one can show that $\bar{\kappa} = -\kappa/2$.

To illustrate the use of Eq. (6.2.1), consider the formation of a vesicle with a midplane radius R from a piece of a planar symmetric bilayer. The vesicle area is $4\pi R^2$, the mean curvature is $H = 1/R$, and the Gaussian curvature is $1/R^2$. From Eq. (6.2.1), it follows that

$$\mathbf{G}_C = 8\pi\kappa + 4\pi\bar{\kappa} \tag{6.2.2}$$

Note that $\mathbf{G}_C$ in this case is independent of the radius R. For bending rigidities in the range 20–100 $k_B\mathbf{T}$ and similarly $-\bar{\kappa}$ in the range 10–50 $k_B\mathbf{T}$, the free energy of forming a vesicle is between 400 and 2000 $k_B\mathbf{T}$ according to Eq. (6.2.2). Since $\mathbf{G}_C$ in Eq. (6.2.2) is independent of the radius R, the relation also applies to the fission of one vesicle into two or, with opposite sign, to the fusion of two vesicles. It follows that for symmetrical bilayers with zero spontaneous curvature, fusion is strongly favored relative to fission. In Section 6.4, we will discuss in more detail how, for asymmetric bilayers with nonzero values of the spontaneous curvature, H_0 profoundly influences the energetics of vesicle fusion–fission processes.

A more subtle consequence of Eq. (6.2.2) is that the vesicle bending free energy becomes lower if the value of the elastic constants decreases. This occurs if the bilayers are thinned by stretching. For small vesicles, this balance between bending and stretching free energies can lead to bilayer thinning, which ultimately, for very small values of the radius, results in vesicle instability.

6.2.5 Bilayers Can Fold to Form Bicontinuous Structures

In the expression for the curvature free energy in Eq. (6.2.1), it was stated that the saddle splay constant $\bar{\kappa}$ was typically negative, and for a homogeneous medium, one has that $\bar{\kappa} = -\kappa/2$. A negative value of $\bar{\kappa}$ implies that bending of a plane into a saddle is energetically unfavorable. However, a bilayer is not a homogeneous medium. We have seen repeatedly earlier that aggregate shapes are determined by a balance between head group repulsion and volume constraints due to chain packing. For an amphiphile with surfactant number $N_s \approx 1$, the planar state is optimal also for a monolayer in a bilayer. However, if N_s deviates significantly from unity, the monolayers are frustrated. In the formation of a saddle structure, the area of the midplane is larger than the area at the plane of the polar head groups. For $N_s > 1$, the formation of a saddle shape can relieve the monolayer frustration, and we expect that $\bar{\kappa}$ becomes less negative and even positive as N_s is increased above unity. Conversely, in the formation of a closed vesicle, the area of the midplane is smaller than the area of the polar groups, when averaged over inside and outside. For this case, $\bar{\kappa}$ is expected to become more negative as N_s decreases below unity. Ultimately, $\bar{\kappa}$ becomes so low that the bilayer no longer represents the most stable aggregate geometry as discussed in Chapter 1.

When $\bar{\kappa}$ is positive, it is favorable to form saddle shapes locally. A simple demonstration of the geometrical effect which can be seen when drying a thin slice of potato. When water leaves the surfaces on the two sides, there is an area contraction compared to the central plane. The tension is then released when the planar slice deforms into a saddle-like chip. Now, saddles can combine into a global structure with a complex topology as illustrated in Figure 6.14. The mean curvature can be kept to zero, in which case one is dealing with a so-called minimal surface, of which there is a large family. Such bilayer structures have two physically significant properties. The bilayer separates two aqueous domains between which water or polar solutes can exchange only by passing through the bilayer. Molecules in these domains and in the bilayer can diffuse over macroscopic distances in all three directions with a diffusion coefficient that is two-thirds of the corresponding local value. Such bicontinuous bilayer phases are found for conditions intermediate between those giving lamellar systems and those where the bilayer is no longer stable and the lipids instead pack into reversed hexagonal structures. In terms of a monolayer spontaneous curvature, the corresponding condition is $0 > H_0(\text{mono}) > -1/(2l)^{-1}$. Cubic phases can also occur when $1/(2l)^{-1} > H_0(\text{mono}) > 0$ but in that case, the amphiphile is locally in a rod-like state rather than in a bilayer.

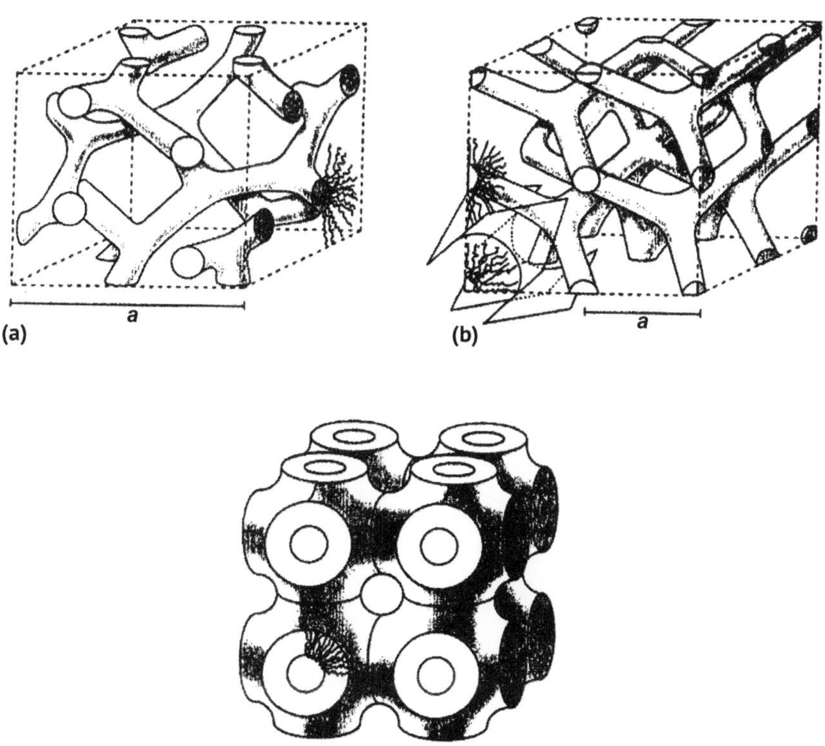

Figure 6.14 Illustration of three different arrangements of bilayers in cubic symmetry: (a) a gyroid structure; (b) a double diamond (as in two interwoven diamond lattices); and (c) "plumbers nightmare." *Source*: Seddon, 1990/with permission of Elsevier.

The topologically complex bicontinuous bilayers can adopt a variety of structures. In the ordered cubic liquid crystalline phases, the structural repeat distance is typically in the range of 10–50 nm. Amphiphiles that have a smaller bending rigidity than typical lipids can also form disordered bicontinuous bilayer structures called sponge phases. Such phases will be discussed further in Chapter 11. The structural length scale in these phases can exceed 100 nm. A third important example of bicontinuous bilayers is found in the so-called "reticulum" membrane structures in eukaryotic cells (reticulum is Latin for network). By having molecular transformations occurring in a reticulum, both polar and apolar substances are free to diffuse within the cell. The reticulum structures have characteristic dimensions of order 100–500 nm, significantly larger than what is found *in vitro* for pure lipid systems.

See Berg (2010, Chapter 3) and Israelachvili (2011, Chapter 20) for additional discussion on Bilayer Structures.

6.3 Transport Across Bilayers Can Be Accomplished in Several Different Ways

A lipid bilayer provides a major barrier for polar molecules to cross from one side to the other. This is a property that is utilized in a profound way in the living cell, and it is also important in other contexts. In this section, we will discuss three different transport mechanisms: diffusion across an intact bilayer, diffusion in defects/holes, and transport by a molecular pump. Transmembrane transport by vesicle fusion/fission is discussed in Section 6.4.1.

6.3.1 Diffusional Processes Are Always Operating in the Living System

In a liquid, small molecules diffuse rapidly on short length scales. A typical value for their translational diffusion coefficient D in water is 10^{-9} to 10^{-10} m/s. According to Eq. (1.7.5), diffusion is effective in eliminating concentration differences over distances $(tD)^{-1/2}$.

Taking 1 s as a typical diffusion time, the corresponding distance is 10 μm. This distance corresponds to the size of a typical cell, which shows that diffusional barriers are necessary to sustain concentration differences on this length scale. The lipid bilayer provides such a diffusional barrier. Although the hydrocarbon core of a bilayer may be only 4 nm thick, it provides an effective barrier to ions and other polar compounds. Solutes dissolved in the aqueous region will quickly establish an equilibrium with the bilayer surface facing the solution. The concentration of polar substances in the bilayer interior is low. Although the mobility of a particular molecule is relatively high in the membrane, this low average concentration results in low net transport.

We can characterize the ability of a molecule A to cross a membrane by the permeability P(m/s) defined by

$$J_A = -P_A \, \Delta c_A \tag{6.3.1}$$

Here, J_A mol/(m^{2}s) represents the flow per unit area of molecule A when a concentration difference Δc_A mol/m^3 exists across the membrane. (For the solvent water, Δc_A should be replaced by $\Delta \Pi_{osm}/\boldsymbol{RT}$.)

One technique for measuring permeability involves forming a black lipid membrane (BLM) in a small orifice separating two aqueous reservoirs, as illustrated in Figure 6.15

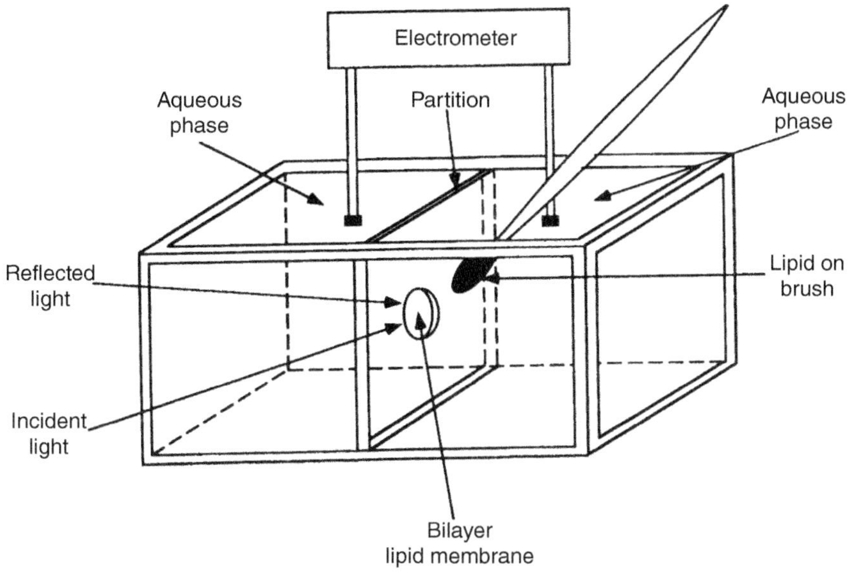

Figure 6.15 Experimental setup for measurements of permeability properties of BLMs. Two aqueous compartments are separated by a partition with a hole (≈ 1 cm diameter). Lipid dissolved in an organic solvent is applied to the hole with a brush. A film is formed, which thins gradually with time. *Source*: Adapted from Fendler, 1987.

We can prepare such a bilayer film by gently brushing an organic solution containing the amphiphile across the orifice. As the organic solvent diffuses into the adjacent aqueous phases, the film thins and exhibits interference colors in reflected light. Ultimately, it appears black in reflected light, when the thickness is much smaller than the wavelength of the light, yet it possesses surprising stability. If BLMs are prepared with meticulous care and scrupulous cleanliness, they can last for hours, and one can measure the permeabilities of polar molecules/ions using such a device.

The permeability coefficient for water in BLMs is $5 - 100 \times 10^{-6}$ m/s. For polar solutes such as glucose, glycerol, and urea, it is much smaller (in the range $5 - 300 \times 10^{-10}$ m/s). For small ions, such as Na^+, K^+, and Cl^-, the values lie in the range $1 - 100 \times 10^{-14}$ m/s, that is, several orders of magnitude smaller.

To understand the molecular basis for these values, Figure 6.16 presents an idealized model of the molecular case of the small permeability of polar molecules in a membrane. Here, a uniform membrane of thickness d_2 separates two bulk phases with a concentration of solute S of c_I and c_{II}. If the bulk phases are properly stirred, their concentration of the solute is uniform. According to Fick's first law (see Section 1.7), the flow of molecules per unit area is

$$J_s(z) = -\frac{D_s(z)}{\boldsymbol{RT}} c_s(z) \frac{\mathrm{d}\mu_s}{\mathrm{d}z} \tag{6.3.2}$$

At steady state—where there is no buildup of the S inside the membrane—the flow, J_s, must be independent of the coordinate z. Similarly, if the membrane has uniform properties, D_s is independent of z and thus $c_s(z)\,\mathrm{d}\mu_s/\mathrm{d}z$ is constant at steady state. If the solute is present in low concentrations and external fields are absent

$$\mu_s = \mu_s^\theta + \boldsymbol{RT} \ln c_s$$

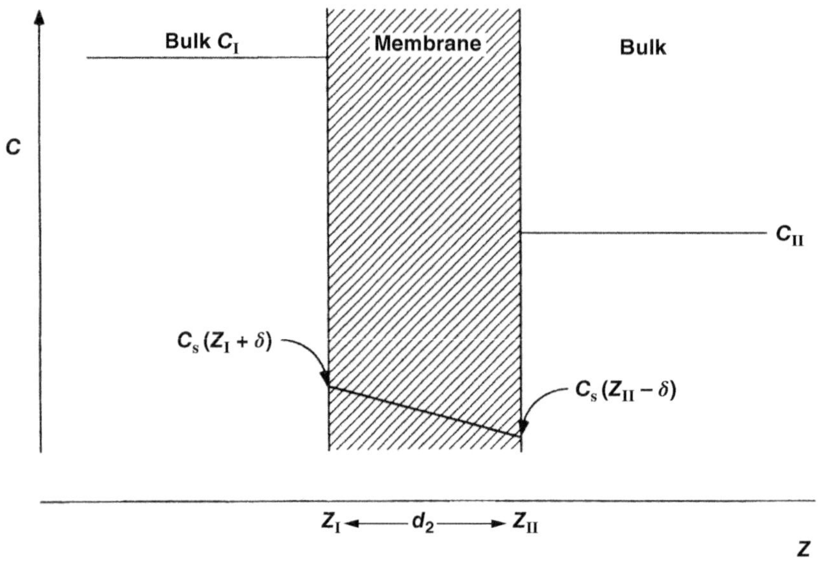

Figure 6.16 The steady-state concentration profile for the diffusion of a solute S across a membrane of thickness d_2, separating two bulk phases with solute concentrations c_I and c_II.

and

$$J_s = -\frac{D_s}{\boldsymbol{RT}} c_S(z) \frac{\mathrm{d}\mu_s}{\mathrm{d}z} = -D_s \frac{\mathrm{d}c_s}{\mathrm{d}z} = -\frac{D_s\{c_s(z_{II} - \delta) - c_s(z_I + \delta)\}}{d_2} \tag{6.3.3}$$

where $c_s(z_{II} - \delta)$ and $c_s(z_I + \delta)$ are the solute concentrations at just a small distance δ into the membrane on either side. At the membrane surface, the solute is in equilibrium with the bulk at the respective sides so that

$$c(z_{II} - \delta) = c_{II}(\text{bulk})K_S$$

$$c(z_I + \delta) = c_I(\text{bulk})K_S \tag{6.3.4}$$

Here, K_S represents the partition coefficient between the membrane and bulk solvent. Thus,

$$c(z_{II} - \delta) - c(z_I + \delta) = K_S\{c_{II}(\text{bulk}) - c_I(\text{bulk})\} \tag{6.3.5}$$

and by combining Eqs. (6.3.3) and (6.3.4) with the definition in Eq. (6.3.1), the permeability coefficient emerges as

$$P_S = \frac{D_S K_S}{d_2} \tag{6.3.6}$$

In a fluid membrane, diffusion coefficients of small molecules have the same magnitude as in bulk solvents, and the important factors determining the permeability are the bilayer thickness and, in particular, the partition coefficient K_S. This fact explains why small ions, whose low solubility in an apolar environment makes the partition coefficient K_S very small, exhibit very low values of P_S.

When ions are transported, an additional effect arises from the electrostatic interactions. We can see the bilayer surrounded by an electrolyte solution on either side as a capacitor (cf. Figure 3.6b). The capacitance C is (cf. Eq. 3.5.5)

$$C = \frac{\varepsilon_r \varepsilon_0}{d_2} \text{ area} \tag{6.3.7}$$

With an estimate of $\varepsilon_r \approx 3$ and $d_2 \approx 4$ nm, $C \approx 7\ 10^{-3}$ As/Vm^2 per area. Thus, to obtain a potential difference of 100 mV, we need to transport a charge Q

$$Q/\text{area} = 7\ 10^{-3} \times 10^{-1} = 7\ 10^{-4}\text{As/m}^2 \leftrightarrow 4.4 \times 10^{15}\ e/\text{m}^2 \leftrightarrow 1e/2 \times 10^2\ \text{nm}^2.$$

For a colloidal system, this corresponds to a low charge density. A potential difference $\Delta\Phi$ across the membrane also implies that an electric field, $-\Delta\Phi/d_2$, exists in the membrane. Ions moving across the membrane will sense this field, which can either facilitate or hinder the motion of the ion. When applying Fick's first law, Eq. (6.3.2), we have to incorporate the electrostatic interaction in the chemical potential, and for an ion i of charge q_i, the flux is

$$J_i = -D_i \left\{ \frac{\mathrm{d}c_i(z)}{\mathrm{d}z} + \frac{q_i}{k_\mathrm{B}\mathbf{T}} c_i(z) \frac{\mathrm{d}\Phi}{\mathrm{d}z} \right\} \tag{6.3.8}$$

When we can neglect the charge inside the membrane, the electrical field is constant and

$$\frac{\mathrm{d}\Phi}{\mathrm{d}z} = \frac{\Phi_{II} - \Phi_I}{d_2} = \frac{\Delta\Phi}{d_2} \tag{6.3.9}$$

At steady state, the flux is constant. As a result of the electrostatic term, however, the concentration gradient is no longer constant as for the diffusing noncharged species in Eq. (6.3.3). By solving the differential Eq. (6.3.8), we find

$$c_i(z) = K_i c_i \exp\left(-\frac{q_i\,\Delta\Phi}{k_\mathrm{B}\mathbf{T}}\frac{z}{d_2}\right) + \frac{J_i k_\mathrm{B}\mathbf{T} d_2}{q_i D_i\,\Delta\Phi}\left\{\exp\left(-\frac{q_i\,\Delta\Phi}{k_\mathrm{B}\mathbf{T}}\frac{z}{d_2}\right) - 1\right\} \tag{6.3.10}$$

where we have fixed the boundary condition at $z = z_I = 0$

$$c_i(0) = K_i c_I \tag{6.3.11}$$

At the other side, $z = z_I = d_2$, the boundary condition is

$$c_i(d_2) = K_i c_{II} \tag{6.3.12}$$

as in Eq. (6.3.4) (see Figure 6.16). When the boundary conditions 6.3.11 and 6.3.12 are applied to Eq. (6.3.10), we obtain a relation between the flux J_i and the potential difference $\Delta\Phi$

$$J_i = \frac{q_i\,\Delta\Phi}{k_B\mathbf{T}} P_i \left\{ \frac{c_{II} - c_I \exp\left(-\frac{q_i\,\Delta\Phi}{k_B\mathbf{T}}\right)}{\exp\left(-\frac{q_i\,\Delta\Phi}{k_B\mathbf{T}}\right) - 1} \right\} \tag{6.3.13}$$

If $c_{II} = c_i \exp\left(-\frac{q_i\,\Delta\Phi}{k_\mathrm{B}\mathbf{T}}\right)$, the flux is zero and the ion i is in equilibrium, and this condition is called a *Donnan equilibrium*. Such a membrane potential arises when there is an asymmetry in the distribution of some charged macromolecular species that has a negligible membrane permeability.

Consider now a more general case in which the individual ion flows are nonzero. At the steady state, the potential difference is constant, and the net electric current must be zero and the individual ion currents cancel so that

$$\sum_i q_i J_i = 0 \tag{6.3.14}$$

This condition can be satisfied only at a unique potential difference. If we consider only the monovalent ions Na^+, K^+, and Cl, we find by combining Eqs. (6.3.13) and (6.3.14)

$$\Delta\Phi = \Phi_{II} - \Phi_I = \frac{k_B\mathbf{T}}{e}\ln\left\{\frac{P_{Na}c_{Na}^{II} + P_K c_K^{II} + P_{Cl}c_{Cl}^{I}}{P_{Na}c_{Na}^{I} + P_K c_K^{I} + P_{Cl}c_{Cl}^{II}}\right\} \qquad (6.3.15)$$

This so-called Goldman equation shows how the electrostatic transmembrane potential depends on ion permeability and concentration. In this nonequilibrium situation, the potential difference is controlled by the permeabilities, so we can regulate $\Delta\Phi$ by making the permeabilities time-dependent. A nerve signal is propagated along the nerve axon by opening and closing sodium and potassium channels, that is, by changing the ion permeabilities.

So far, we have discussed the transport of polar molecules across a single membrane. Some new features appear when we consider the transport in a bulk liquid crystal, although the basic principles remain the same. A lamellar liquid crystal is anisotropic, and we should consider the diffusion parallel and perpendicular to the lamellae separately.

In the parallel direction (see Figure 6.17), the effective diffusion coefficient D_L is simply the weighted average of the diffusion in two domains, which we assume to be isotropic. The average diffusion represents the diffusion in the respective domains weighted by the fraction of molecules there

$$D_L = (d_1 D_1 + K_S d_2 D_2)/(d_1 + K_S d_2) \qquad (6.3.16)$$

where K_S is the partition coefficient between the domains as in Eq. (6.3.4).

To arrive at the expression for the average perpendicular diffusion, it requires a somewhat longer derivation. At steady state, the flow J in the two regions must be the same and

$$-J_\perp = \frac{D_1}{k_B\mathbf{T}}c_1\left(\frac{d\mu}{dz}\right)_1 = \frac{D_2}{k_B\mathbf{T}}c_2\left(\frac{d\mu}{dz}\right)_2 \equiv \frac{D_\perp}{k_B\mathbf{T}}c_{av}\left(\frac{d\mu}{dz}\right)_{av} \qquad (6.3.17)$$

where we have introduced c_{av} and $\left(\frac{d\mu}{dz}\right)_{av}$ as the respective properties seen on an average macroscopic scale and used in Eq. (6.3.2) to identify $D_\perp$. Thus

$$c_{av} = \frac{d_1c_1 + d_2c_2}{d_1 + d_2} \qquad (6.3.18a)$$

$$\left(\frac{d\mu}{dz}\right)_{av} = \frac{d_1\left(\frac{d\mu}{dz}\right)_1 + d_2\left(\frac{d\mu}{dz}\right)_2}{d_1 + d_2} \qquad (6.3.18b)$$

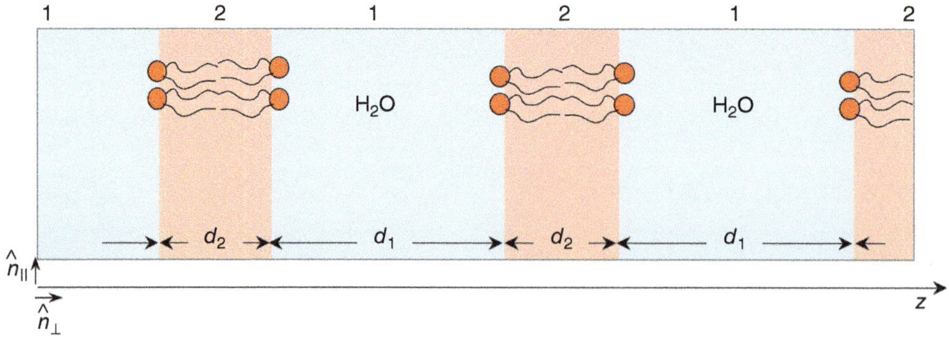

Figure 6.17 Diffusion in a lamellar liquid crystal. In the parallel direction, solutes can diffuse without passing the interface, while in the perpendicular direction, a diffusional process involves repeated transition between regions 1 and 2.

We solve Eq. (6.3.17) for $D_\perp$ by assuming that we can neglect the electrostatic complications. Then $d\mu/dz$ is constant within each layer as in Eq. (6.3.3) and

$$\left(\frac{d\mu}{dz}\right)_2 = \left(\frac{d\mu}{dz}\right)_1 \frac{D_1c_1}{D_2c_2} \tag{6.3.19}$$

using Eq. (6.3.17). If we combine the second equality of this equation with Eq. (6.3.18 a,b), we find

$$k_B\mathbf{T}J_\perp = D_1c_1\left(\frac{d\mu}{dz}\right)_1 = D_\perp\left\{d_1\left(\frac{d\mu}{dz}\right)_1 + \frac{d_2D_1c_1}{D_2c_2}\left(\frac{d\mu}{dz}\right)_1\right\}(d_1+d_2)^{-2}(d_1c_1+d_2c_2) \tag{6.3.20}$$

and the desired average diffusion coefficient is

$$D_\perp = D_1\frac{(d_1+d_2)^2}{(d_1+K_Sd_2)\left[d_1+\left(\frac{D_1}{K_SD_2}\right)d_2\right]} \tag{6.3.21}$$

If we assume for the sake of simplicity that the local dynamics in the different layers are similar, so that $D_1 \approx D_2$, then

$$D_\perp \approx D_1\frac{(d_1+d_2)^2}{(d_1+K_Sd_2)[d_1+K_S^{-1}d_2]} \tag{6.3.22}$$

which has its maximum value $D_\perp = D_1$ when $K_S = 1$. For both $K_S \gg 1$ and $K_S \ll 1$, that is, when the solute strongly prefers one region relative to another, the perpendicular diffusion can be very slow. Thus, an oriented lamellar phase can present a substantial diffusional barrier to most solutes, even though the local molecular motion is as fast as in a liquid.

The outer part of the human skin, the *stratum corneum* or horny layer, has penetration properties relative to small molecules that resemble the properties of $D_\perp$ in Eq. (6.3.22). In particular, the skin protects the organism against penetration from both strongly polar molecules, such as water, and strongly apolar molecules, such as hydrocarbons. However, molecules that are soluble in both water and hydrocarbons—such as dimethyl sulfoxide (DMSO)—readily penetrate the skin. Organic chemists know that a drop of DMSO spilled on the skin partly ends up in the lungs, producing a garlic-like smell and a foul taste in the mouth.

Although the *stratum corneum* has a complex structure, one of its essential elements seems to be bilayers stacked in a lamellar bilayer phase. One way to administer drugs is to apply them on the skin, either in salves or on a bandage. The rate of the diffusive transport of the active component through the skin clearly influences the efficacy and the timescale of the treatment.

6.3.2 Molecules Can Pass a Bilayer Through Holes or by Carriers

In Section 6.3.1, we saw that the permeability P of a solute across a lipid bilayer is determined primarily by the partition coefficient between the bilayer and the aqueous phase. One way to increase P is to increase the partition coefficient. We can accomplish this by introducing a carrier that has a large solubility in the membrane and also associates specifically with a solute. A well-known example is valinomycin, a cyclic peptide that forms a complex with potassium ions, as illustrated in Figure 6.18. With valinomycin present, the effective solubility of potassium ions in the bilayer increases dramatically, while the effective diffusion coefficient is reduced slightly because the valinomycin–potassium complex is now the diffusing entity. The potassium-loaded valinomycin partitions into the bilayer due to a combination of two effects. First, when the ion is bound,

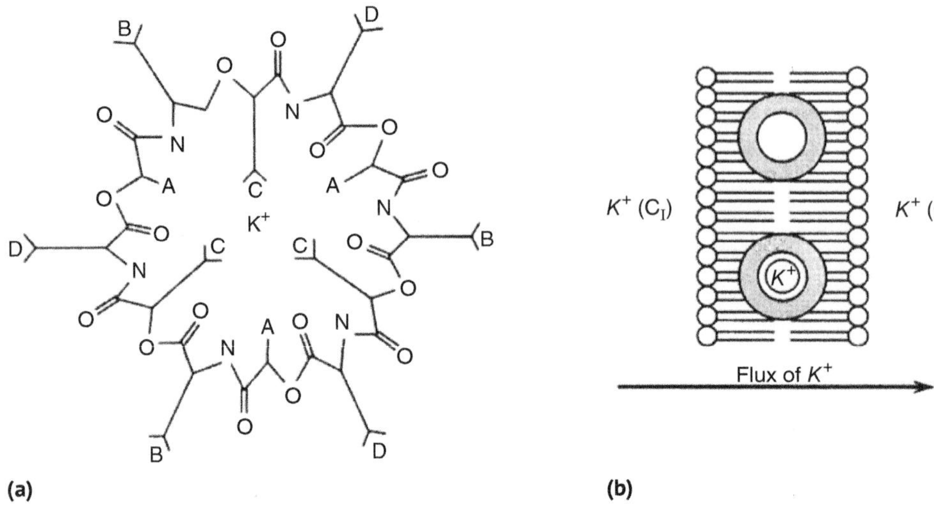

Figure 6.18 Valinomycin is a cyclic decapeptide that gives a strong hydrophobic complex with potassium ions. (a) Molecular structure: A = L-lactate; B = L-valine; C = D-hydroxyisovalerate, and D = D-valine. (b) Potassium transport across a bilayer. Both loaded and unloaded valinomycin diffuse across the bilayer. In the presence of a concentration difference of K^+, $c_1(K^+) > c_2(K^+)$, there is a net flux of K^+ until a compensating electric potential difference is established.

the polar side chains of the cyclic peptide coordinate the charge in the center, leaving an apolar periphery. Second, according to the Born model of ion solvation discussed in Section 3.6, the electrostatic solvation energy of an ion is inversely proportional to its radius. In the present case, the radius should be rather large because the peptide is included, and as a result, it greatly reduces the penalty in electrostatic energy for an ion entering the bilayer. Many ion carriers are based on a cyclic structure like that of valinomycin. Some of them have found a use as antibiotics. The basis for such a physiological effect is that the unspecific ion flux interferes with the formation of an optimal membrane potential.

Protons have unique properties as a cationic species in that they always associate covalently with some other molecular entity. Thus, carriers for proton translocation across the bilayer are based on a covalent binding of the proton. A typical example is 2,4-dinitrophenol, in which the phenolic proton is acidic and both the protonated and acidic forms are present at neutral pH. In the ionized form, the charge is delocalized onto the aromatic ring and the nitro groups. As the Born model predicts, this delocalization gives the charged species a large radius, which enables the charged dinitrophenolate to dissolve into the bilayer. Thus, the molecule can diffuse across the bilayer in both its neutral and charged forms. The net result of a 2,4-dinitrophenol molecule diffusing in one direction in the charged form and returning in its neutral form is the transport of a proton in the direction opposite to that of the negative charge. Compounds that can transport protons through this mechanism are highly toxic to cells. They not only affect the membrane potential but also interfere with the regulation of the pH value in the cell.

A more elaborate strategy for increasing the permeability of small ions across the bilayer is the formation of a channel. In this case, a polypeptide or protein forms a rigid structure consisting of a polar channel that spans the bilayer. This channel prevents the ion from making contact with the apolar interior as it moves across the bilayer. The efficiency per structural unit is much higher for

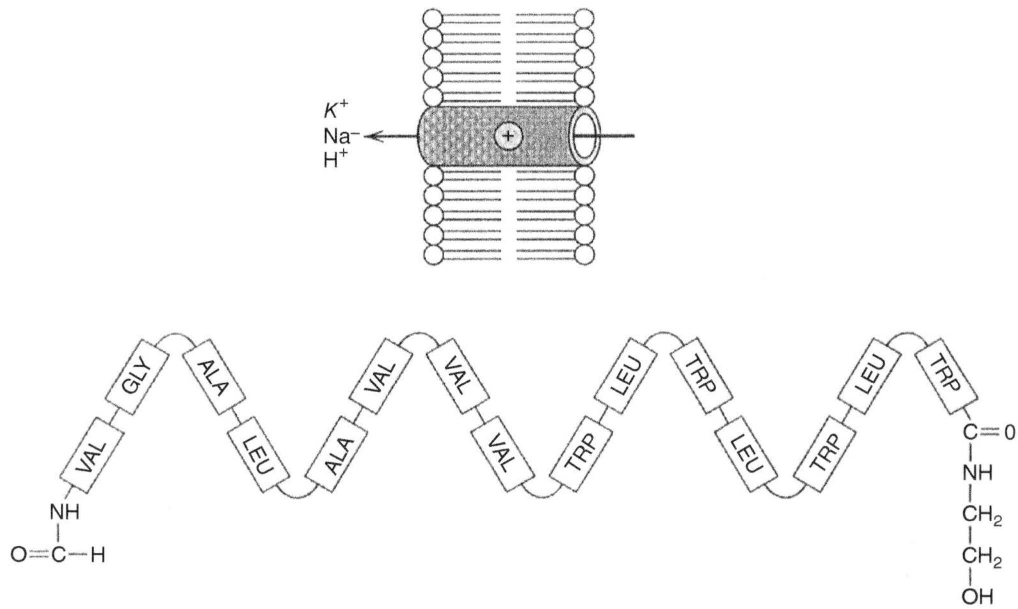

Figure 6.19 Gramidicin A is a polypeptide that forms a helical structure in a bilayer. When two helices anchored in the respective monolayers of the bilayer meet, a polar pore spanning the bilayer is formed. In that case, ions that are small enough can pass freely across the bilayer.

channels than for carriers because the ion is the only moving species, possibly with some hydration water. Gramicidin, shown in Figure 6.19, is a channel-forming polypeptide that has been carefully studied. It adopts a helical structure in the bilayer, but because gramacidin is only 15 amino acids long, two oppositely oriented helices must associate in a dimer in order to form a bilayer-spanning channel. Once a channel is formed, ions diffuse through with such ease that it is possible to measure the conductivity of a single channel. The presence of a nonspecific channel is highly damaging to a cell. For example, bee venom contains the protein melittin, which makes the membrane more permeable, probably through a channel formation. Channel-forming proteins also occur as integral parts of the living cell. However, they usually show high chemical selectivity, and some regulatory mechanism that opens and closes the channels.

In some cells, there is a need for rapid volume changes accomplished by water transport. This applies to red blood cells and cells in the kidney, which both contain protein water channels in the form of aquaporins. These channels are typically narrow enough that only water molecules can pass.

Amphiphiles forming micelles have a general antiseptic effect. In large concentrations, they can simply solubilize the lipids of the cell membrane. In smaller concentrations, the situation is the reverse, where the membrane solubilizes the amphiphile. However, such a bilayer will have a monolayer curvature that is slightly positive, and transiently, there can be holes stabilized by the amphiphile. This leads to a leakage of metabolites, which ultimately generates cell death.

Another way to induce a temporary hole is to apply a strong field across the bilayer. This is called electroporation, and the electrical conductivity is temporarily strongly increased when a hole is formed. Once the local field has been reduced by conduction, the membrane heals until a strong field has been generated again. This is a versatile experimental method to transport larger molecules into vesicles and cells.

6.3.3 Molecular Pumps Can Transport Polar Molecules and Ions Across a Cell Membrane

An even more elaborate way of transporting a small polar molecule/ion C across a bilayer membrane is through an enzymatic pump. In a pump mechanism, translocation requires a concerted chemical transformation of A to B

$$A + C(I) \rightleftarrows B + C(II) \qquad (6.3.23)$$

where I and II denote the two sides of the membrane. Note that the reaction is reversible and can go in either direction, depending on the free energy conditions. Due to the coupling of a reaction to the transport, the membrane-bound enzyme works as a pump. In fact, the reaction in Eq. (6.3.23) represents one of the most fundamental biochemical reaction schemes and one of the simplest of a whole class of reactions coupled to transport. Clearly, the stoichiometric ratio can be different so that more than one C is transported per transformation $A \rightarrow B$. Several species can also be involved in the transport so that when C goes from side I to II, another molecular species (possibly with a different stoichiometric coefficient) goes from side II to I. The physiologically most important example of such a pump is the sodium-potassium ATPase. Using ATP as a free energy source, it translocates three Na^+ ions inside out and two K^+ ions outside in. This involves the net transport of one unit charge per cycle, and a membrane potential is generated in, for example, nerve cells. At a potential difference of around −70 mV, inside negative, the pump has reached its conditional equilibrium, and the reaction proceeds with an equal rate in both directions. When a nerve signal reaches the cell, a channel protein opens, and there is a leakage of Na^+ ions rapidly reducing the membrane potential. This is restored by a flux of K^+ ions in the opposite direction. The ATPase reaction gradually restores the ion concentrations and the potential to their initial resting values. See Berg (2010, Chapter 9), Israelachvili (2011, Chapter 23), and Ohshima (2024, Chapter 6 and 7) for additional discussion on Lipid Molecular Characteristics. See Israelachvili (2011, Chapter 22) and Ohshima (2024, Chapter 22) for additional discussion on Molecular Transport.

6.4 The Lipid Bilayer Supports a Range of Central Metabolic Processes in the Living Cell

In the most primitive organisms like archaea and bacteria, the plasma membrane defines the cell in a molecular sense. Inside these prokaryotic cells, molecules diffuse freely and engage in the molecular transformations typical of life. The cell membrane controls transport to and from the surrounding environment, acting as a general barrier, while allowing for specific molecular species to cross. The evolution toward more complex forms of life leads to the development of a compartmentalization of the cell interior based on internal membranes. This allows for the more effective and complex functions that are found in eukaryotic cells. In this section, we describe some basic features of the membrane cellular function of special relevance from a colloid science perspective.

6.4.1 Vesicle Fusion–Fission Is a Central Process in Living Cells

Vesicle fusion/fission illustrates one of the central themes of this book: the stability–instability dichotomy. Figure 6.20 shows schematic illustrations of fusion/fission for two cases. In the upper panel, two vesicles of equal size merge to an initially deformed larger vesicle, which relaxes to its

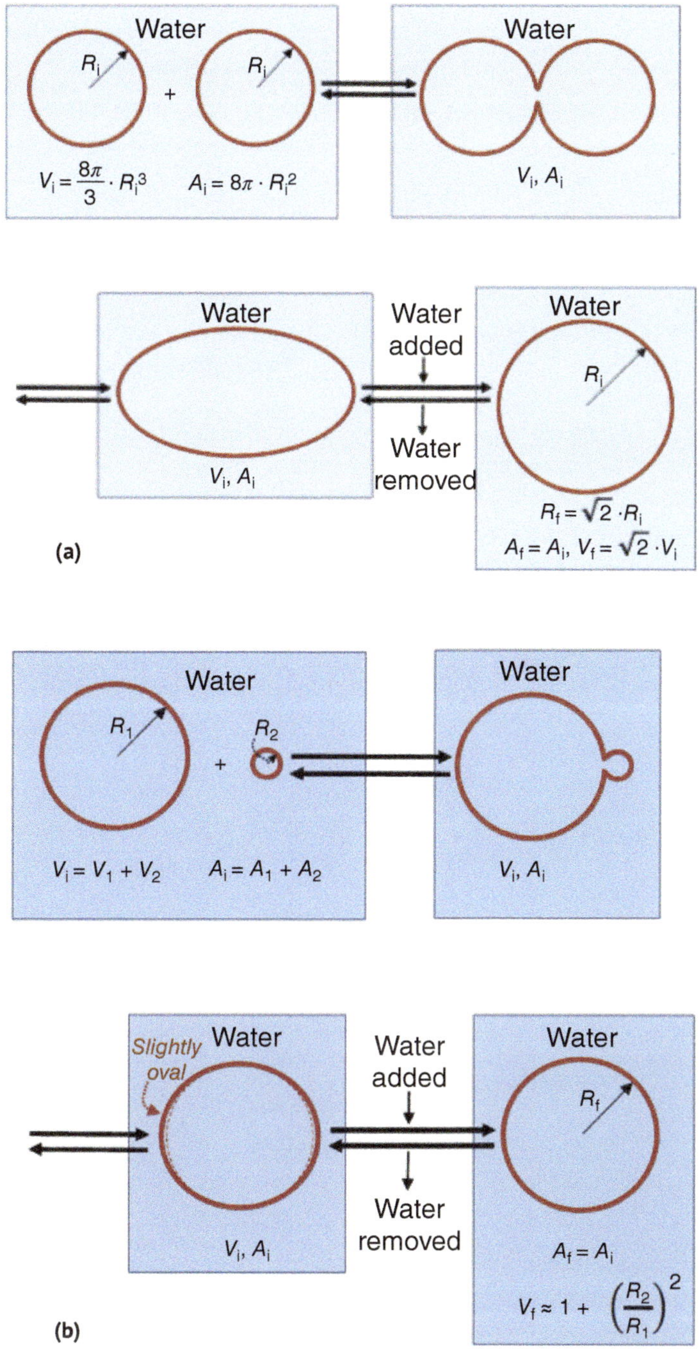

Figure 6.20 (a) Fusion/fission of two equal-sized vesicles. During fusion, internal volume is conserved, and there follows a slow step of water uptake. (b) Fusion/fission of a large and a small vesicle, with only a small volume change in the process.

equilibrium shape by diffusive influx of water. In the lower panel, the fusion is between a small and a large vesicle, more like what is found in cell metabolism. For symmetrical bilayers, the free energy change is, according to Eq. (6.2.2), the same in the two illustrated cases. However, in a typical metabolic situation, the membrane is asymmetric with a nonzero value of the spontaneous curvature H_0. In this case, there can be a large difference in the change in curvature free energy for the two cases.

For a pure lipid bilayer, fusion/fission is associated with a kinetic barrier. Fusion probably occurs in two steps, where in the first stage, there is a hemi-fusion between the two outer monolayers of the respective bilayers, forming a "stalk" connecting the vesicles. Within this "stalk," there is then an additional hemi-fusion resulting in the formation of a water channel connecting the aqueous interiors of the vesicles. Figure 6.21 illustrates this both schematically and in an electron microscopy picture from a cellular system. In the cell, there is an elaborate protein machinery that, in a controlled way, facilitates the kinetics of fusion/fission.

In the stalk mechanism described in Figure 6.21 and in analogous mechanisms of fusion, there are highly curved bilayer regions representing configurations of high free energy. The localization to the highly curved regions of a second component can substantially reduce the free energy barrier. Molecules effective in this way are called fusogens. In the living cell, there are several protein components that promote and control fusion/fission. A common experimental method to induce fusion is to use depletion attraction (see Section 5.5) to force the outer monolayers into contact.

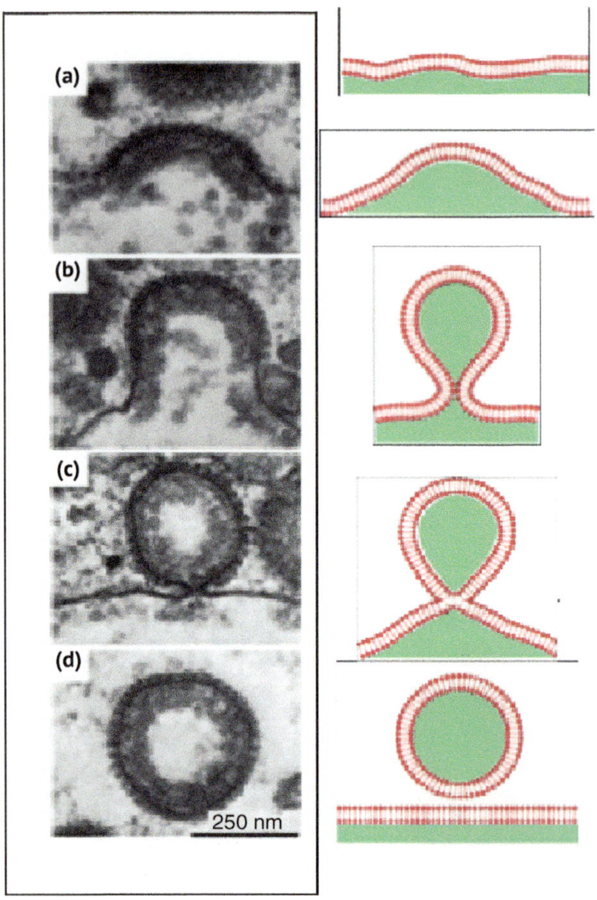

Figure 6.21 Left: Electron microscope pictures the budding of clathrin-coated vesicles. *Source*: Perry et al., 1979/with permission of The Company of Biologists. Right: Schematic picture of the process illustrating the stalk formation as the last step in the vesicle budding.

If vesicle fusion is kinetically allowed, so is the reverse process of fission. The direction of the process is determined by the energetics. The bending free energy provides the main contribution to the overall process, and we can use the Helfrich expression (6.2.1) to obtain a quantitative estimate of the effect. In Section 6.2.4, we found that for a symmetrical bilayer, the fusion process is strongly favored relative to the fission. For an asymmetric membrane separating solutions of different compositions, there is a spontaneous curvature, H_0, which has a strong effect on the bending energy balance. The cell membrane has a radius much larger than a vesicle. In this limit is the change in bending energy when a vesicle of radius R and spontaneous curvature $H_0(v)$ is detached from a practically planar cell membrane with spontaneous curvature $H_0(p)$

$$\Delta \mathbf{G}_b = 8\pi\left\{\kappa_v\left[(1 \mp 2H_0(v)R + R^2H_0^2(v)\right] - \kappa_p R^2 H_0^2(p)\right\} + 4\pi\bar{\kappa}(v) \tag{6.4.1}$$

The sign $\Delta \mathbf{G}_b$ is reversed for the corresponding fusion process. The term $2H_0(v)R$ can have either sign depending on the nature of the process, as will be discussed in the next section. The curvature of a vesicle is defined as positive when the bilayer curves away from the external solvent. It is important to realize that the change in curvature free energy described by Eq. (6.4.1) depends strongly on the value of H_0R, and with only a small variation in this parameter, it can change the sign of $\Delta \mathbf{G}_b$. One can go from a situation where fusion is thermodynamically allowed to the case where it is fission that is favored, with only minor changes to the system.

6.4.2 Cells Can Communicate with the Surrounding and with Other Cells Through Vesicles

Vesicles loaded with a cargo function to exchange molecules between the cell and its surrounding medium or between cells. Figure 6.22 illustrates the four main processes.

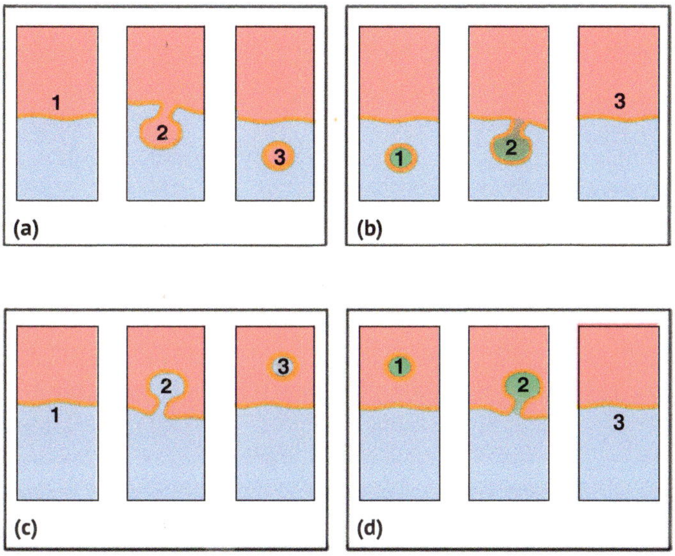

Figure 6.22 Cells (blue) communicate using vesicles in four different ways. (a) The plasma membrane invaginates and buds off a vesicle that reflects the content of the surrounding medium (endocytosis). (b) An internally produced vesicle fuses with the plasma membrane and delivers its content to the medium (exocytosis). (c) The plasma membrane buds off a vesicle, an exosome, whose composition reflects the composition of the cell plasma. (d) An external vesicle from a "green" source fuses with the plasma membrane and delivers its content.

In Figure 6.22a, the membrane invaginates and buds off an intracellular vesicle, which provides a way of importing polar molecules and macromolecules from the medium into the cell. This is called endocytosis. In the reverse process (Figure 6.22b), an internal vesicle fuses with the plasma membrane and releases its content to the extracellular solution. This is called exocytosis. In Figure 6.22c, the membrane forms a bud that is released as an external vesicle, with a content reflecting the composition of the cell plasma. The reverse process is the fusion of an external vesicle with the membrane, bringing external molecules into the cell (Figure 6.22d). These processes serve as an important route for macromolecular communication between cells. They also provide a route to externally deliver molecules to a cell, be it a viral infection or a therapeutic transfer of medicine or genes.

It requires an intricate control system to regulate these vesicle fusion/fission processes. Such a control has both kinetic and thermodynamic aspects. The processes shown in Figure 6.22 typically require special molecular arrangements to make them kinetically feasible. However, when kinetically allowed, the direction of the process is determined by the thermodynamic driving force quantified in Eq. (6.4.1).

For an endocytosis/exocytosis in Figures 6.22a,b, the directions of the monolayers are reversed, and the definition of the curvature is such that the plus sign of the $2H_0(v)R$ term should be used. For the fission process in Figure 6.22c, the outer monolayer of the plasma membrane remains the outer monolayer of the vesicle, and the minus sign of the $2H_0(v)R$ term applies. The situation is the same for the fusion process in Figure 6.22d. In cases (a) and (c), the vesicle membrane has the same composition as the plasma membrane, unless there has been an active molecular sorting in the fusion/fission process. Without such a sorting, the bending rigidity is unaffected so that $\kappa(v) = \kappa(p)$ and the same applies to the spontaneous curvature. Then Eq. (6.4.1) reduces to

$$\Delta \mathbf{G}_b = 8\pi\kappa_v(1 \mp 2H_0R) + 4\pi\bar{\kappa}(v) \tag{6.4.2}$$

This shows that the value of the spontaneous curvature has a large effect on the change in the bending energy of a fusion–fission process. One can note that the factor multiplying H_0R in Eq. (6.4.2) is of order 2000 $k_B\mathbf{T}$ so that only a small, of order 1%, change in H_0R is enough to change $\Delta\mathbf{G}_b$ by ± 20 $k_B\mathbf{T}$ corresponding to a change in the equilibrium constant for fusion/fission by a factor $e^{20} \approx 10^8$! One can go from virtually complete fusion to complete fission at equilibrium through a small change in H_0R. The fission in Figure 6.22c is favored by a positive membrane spontaneous curvature, while to achieve endocytosis as in Figure 6.22a H_0 should be negative. The value of $\bar{\kappa}$ is also relevant, and its magnitude is affected by the monolayer spontaneous curvature as discussed in Section 6.2.5. For the processes in Figure 6.22b,d, the vesicle originates from a membrane system different from the plasma membrane and the more involved Eq. (6.4.1) applies.

In Figure 6.22, we have shown fusion/fission involving vesicles. A related phenomenon occurs when an oil-in-water emulsion droplet (discussed in Section 11.3) or a hydrophobic particle stabilized by an amphiphile monolayer encounters a lipid bilayer. A possible fusion only involves the outer monolayer, and the hydrophobic content of the droplet primarily ends up in the bilayer interior.

The signal transfer across the synaptic cleft in the conduction of nerve signals illustrates the role of vesicle fusion/fission in cell communication. As illustrated in Figure 6.23, vesicles, loaded with a transmitter substance like acetylcholine, are formed in the presynaptic cell. Triggered by a drop in the membrane potential, the vesicles fuse with the plasma membrane, releasing the transmitter substance into the synaptic cleft. Receptors in the postsynaptic membrane associate with the transmitter substance and give rise to a drop in the membrane potential, which is then propagated along the axon of the receiving cell.

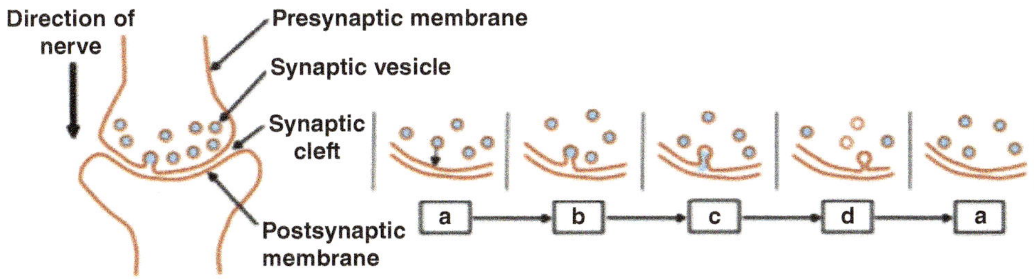

Figure 6.23 Diagram showing the vesicle cycle in a chemical synapse, the junction between two nerve cells. When an electrical signal reaches the presynaptic membrane (a) in the cleft, vesicles fuse with the plasma membrane (b) and deliver the chemical transmitter (c), for example, acetylcholine, into the synaptic cleft. The transmitter diffuses across the cleft and reaches receptors on the postsynaptic membrane. Then there is a fission process generating empty vesicles (d), which are then filled with the transmitter substance regenerating state (a). In an alternative description, the process reverts to the stage between (c) and (d), and the vesicle piece closes in a fission process and comes back (half)-empty into the presynaptic space.

Once the fusion has occurred, the potential across the membrane is restored, and a fission process occurs, which generates unloaded vesicles. Both vesicle fusion and fission occur in the same membrane system, and there is a challenge to have both processes thermodynamically allowed.

6.4.3 The Lipid Membrane Is a Solvent for Membrane Proteins

A biological membrane consists of anywhere between 25% and 75% w/w lipid. The remaining 75–25% is protein, glycoprotein, or lipoprotein. Thus, the protein species appears to be as dominant in the membrane as the lipids. However, proteins penetrate the aqueous media outside the bilayer planes. This fact implies that if one specifies the concentrations by occupied area, the lipid content increases substantially, and lipids emerge as the main component. Indeed, previous discussions in this chapter have been based on this assertion.

Proteins can associate with the membrane in several different ways, as illustrated in Figure 6.24. The archetypal membrane protein is an integral part of the bilayer. It spans the width of the bilayer and is exposed to the aqueous medium on both sides. A protein can also be anchored on one side of the bilayer by a hydrophobic interaction. In this case, the protein often exchanges between a state free in the cytosol with a partly disordered chain conformation and a bound state with an amphipathic α-helix whose apolar side is associated with the membrane. A third possibility is that the protein contains one or a few covalently attached fatty acid chains, which anchor an otherwise polar protein to the membrane. A protein can also associate electrostatically with the membrane.

Proteins are inserted into the membrane with a prescribed orientation, and the reorientation is sufficiently slow that once inserted, they keep their orientation during the biologically relevant time. However, they can undergo chemical transformations. The plasma membrane contains proteins that become heavily glycosylated on the outside, with a substantial fraction of charged sugar units. By this mechanism, a protective shield of entangled sugar units covers the cell. This cover plays an important role in the contact communication between different cells. It is also a target for the adhesion of infectious bacteria.

The bilayer interior provides a very apolar environment for a membrane protein. To fit in, these proteins have a minimum of polar amino acid side chains for the part of the molecule that does not protrude from the bilayer surface. Charged amino acids are very rare in this region; even

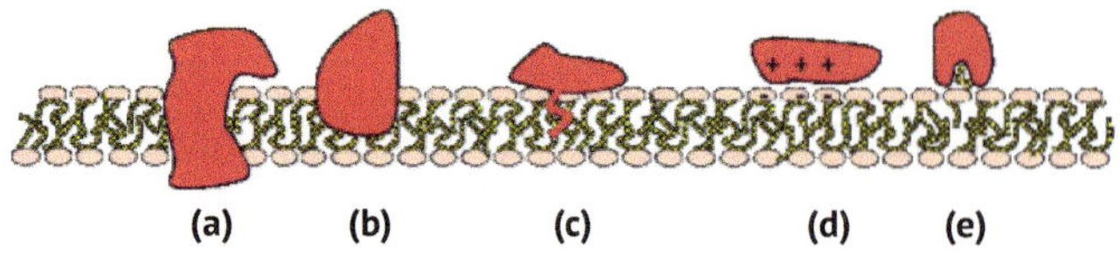

Figure 6.24 Illustrations of different modes of protein–lipid bilayer interactions: (a) integral membrane spanning protein; (b) protein binding to one side of the bilayer through a hydrophobic patch; (c) mainly hydrophilic protein attaching hydrophobically through a covalently bound alkyl chain; (d) hydrophilic protein binding electrostatically; and (e) mainly hydrophilic protein with a hydrophobic pocket that can attract a lipid chain.

exposed OH and NH_2 groups are uncommon. In the backbone, the amide linkage necessarily has a polar character. As a result, the membrane-spanning parts usually form α-helices, and to a lesser extent β-sheets (see Chapter 7) so that the amide and carboxy groups can be engaged in intramolecular hydrogen bonds. The medium provides neither a hydrogen bond donor nor an acceptor. An α-helix elongates 0.15 nm per residue so that a membrane-spanning protein consists of a series of sequences containing 20 to 25 hydrophobic amino acid residues spaced by shorter or longer stretches of more hydrophilic ones. Each such sequence of 20-odd apolar residues forms a membrane-spanning helix, and typically several, 3, 5, and 7, of those fold to form the tertiary structure of a membrane protein.

There is a large range of functions for membrane proteins. A straightforward role is to act passively to allow some specific molecular species to pass through the membrane, as discussed previously for aquaporins In a slightly more sophisticated version, there is a cotransport of two different molecular species in either the same or opposite directions. In this case, an existing concentration difference across the membrane of one component can be utilized to generate a concentration difference of the other species. A more central function occurs for channel proteins that switch between an open and a closed state controlled by some external molecular, electrical, or mechanical signal. Membrane protein pumps use chemical free energy, typically in the form of ATP, to generate concentration differences across the membrane. The pumps often act on simple ions like protons, sodium, potassium, calcium, or chloride ions. A substantial fraction of the free energy used in mammals goes into the sodium–potassium pump. Pumping an ion involves a charge separation, generating an electrical field across the membrane.

In these examples of membrane protein function, the common feature is the translocation of molecules (ions) across the bilayer. A second type of function is found for a class of proteins, called G-proteins, which take part in molecular signaling from outside to inside or vice versa.

From a molecular perspective, the most sophisticated membrane protein function is found in (free) energy transducing systems. In the chloroplasts of green plants, the Photosystem I and, in particular, Photosystem II have complex multiprotein structures designed to convert light into chemical free energy. In the mitochondria, the product of the Krebs cycle NADH is converted into ATP through oxidation by O_2. This is accomplished through a series of electron transfer reactions between proteins in the mitochondrial membrane that occur under isothermal conditions with an efficiency higher than that in a fuel-driven car engine.

In both chloroplasts and mitochondria, the membrane is used to couple spatially separated enzymatic reactions that also involve the pumping of protons across the membrane. The free energy generated by a chemical oxidation step is transiently stored as a combination of proton entropy of mixing and free energy in the transmembrane electrical field. This is called the chemiosmotic mechanism.

6.4.4 Membrane Proteins Interact Laterally in an Inhomogeneous Environment

With membrane proteins serving a range of functions, there is a challenge to organize the system to achieve optimal function. A basic strategy for eukaryotic cells is the compartmentalization into a nucleus, ribosomes, mitochondria, and other organelles. Separated membrane systems have compositions adapted to their functions. Still, there are, in each case, several membrane protein species existing in a high concentration in an effectively two-dimensional solvent. Chapter 5 provides a comprehensive discussion of the interactions between particles in a bulk aqueous solvent. The situation in a lipid bilayer is more complex and much less investigated. Consider first the interaction between two identical membrane-spanning proteins in a bilayer. There is always an attractive dispersion interaction. However, the refractive indices of proteins and lipids are rather similar, so the strength is clearly reduced relative to that found in an aqueous medium. Membrane proteins are usually (net) negatively charged, and there should be a long-range repulsion between the two proteins. There are two difficulties involved in estimating the strength of this effect. In the membrane, a fraction of the lipids are also negatively charged. For a charge density of the protein, σ_p projected to the membrane surface and a charge density of the lipid bilayer, σ_1, the electrostatic interaction is, to first order, proportional to $(\sigma_p - \sigma_1)^2$, so when the charge densities match, there is no net effect. The charge densities differ between inside and outside for both proteins and lipids. There is, in addition, a effect in that the bulk electrostatic screening length is longer on the inside, $\kappa^{-1} \approx 2.2$ nm than on the outside $\kappa^1 \approx 0.8$ nm, so for given charge densities, the electrostatic interaction is stronger and more long-ranged intracellularly (see also Section 8.3).

The dispersion interaction between two membrane proteins is effectively rather weak, and even the electrostatic interaction is not dominating. For membrane proteins, the scenario of a competition between dispersion attraction and electrostatic repulsion as dominating is not as applicable as in the bulk. There is a need to look for other organizing interactions. If the proteins have an exposed polar group like OH or NH_2 in the apolar region, this could give rise to a strong attraction. The formation of an α-helix gives rise to a subtle electrostatic effect. The polarity of hydrogen bonds has a component parallel to the axis of the helix. The net result is that there is a charge separation of magnitude 1/3e between the ends of the helices. In the apolar environment, this can give rise to strong interactions of an electrostatic origin. In addition, there can be other molecular sources of a dipole component perpendicular to the membrane plane, which generates a repulsive interaction. There is a mixture of lipid components, and it is conceivable that some chains fit better than others into the uneven apolar membrane interface. Such a preferential solvation causes a weak repulsion.

There is a complex interplay between molecular interactions at the polar-apolar interface of the bilayer. However, it is difficult to arrive at any general conclusions about the net effect of these interactions. A bilayer spanning α-helix is rather stiff, and if it is longer than the width of the apolar layer, it needs to tilt to fit in. This tilt generates a pulling force on the lipid molecules around, and when two such tilted helices come close, there is a cooperative effect mediated by the lipids, resulting in a protein–protein attraction. Such a mechanism can also be seen as promoting the formation of so-called lipid rafts. These consist of larger regions on the nanometer scale of the lipid-ordered phase l_o. By solubilizing those membrane proteins that prefer lipids with straighter chains, they act to organize the protein distribution in the membrane.

See Israelachvili (2011, Chapter 21) for additional discussion on Bilayers in Cells.

Exercises

6.1 A small unilamellar vesicle is formed at pH 6 in an aqueous solution. How many free protons are there inside the vesicle if the inner radius is 15 nm? Interpret the result.

6.2 A liquid crystalline sample in an open tube is placed in a closed container having a bulk pure water phase at the bottom. What is the osmotic pressure in the liquid crystal if the tube is placed 10 cm above the water surface?

6.3 A characteristic signature of a gel phase is the appearance of a sharp X-ray diffraction peak corresponding to a distance of 0.42 nm between symmetry planes. The chains are packed in a two-dimensional hexagonal array. What is the area per chain?

6.4 For DPPC in the gel $L_{\beta'}$ state, the repeat distance measured at 19 °C by X-ray diffraction is 6.3 nm when the crystal is in equilibrium with excess water. Under these conditions, there are 12 water molecules per lipid. Calculate the angle of tilt of the hydrocarbon chains, given that the chains pack at an area of 0.2 nm^2 measured perpendicularly to the chain direction. The lipid has an effective specific volume of 9.36×10^{-4} m^3/kg.

6.5 The table shows the measured repeat spacing in a lamellar phase of water–sodium octonate (NaC_8) decanol (Dec) for a series of compositions (% w/w).

NaC_8	Dec	H_2O	d(Å)
14.7	42.1	43.2	41.0
13.9	39.9	46.2	43.1
13.0	36.9	50.1	45.8
12.0	32.9	55.1	52.0
10.4	29.4	60.2	58.2
9.0	25.8	64.2	66.9
8.0	22.0	70.0	77.0
7.0	20.0	73.0	84.9

Calculate the thickness of the water and amphiphile layers, assuming that their respective specific volumes are 0.996×10^{-3} and 1.11×10^{-3}. Is the area per head group constant?

6.6 In Exercise 5.8, we found that the van der Waals force between two bilayers of thickness L separated by a solvent layer of thickness d is

$$\frac{F_{vdW}}{\text{area}} = -\frac{H}{6\pi}\left\{\frac{1}{d^3} - \frac{2}{(d+L)^3} + \frac{1}{(d+2L)^3}\right\}$$

At what separation d is this van der Waals force balanced by the repulsive undulation force when $\mathbf{T} = 300$ K, $H = 6 \times 10^{-21}$ J, $L = 4$ nm, and $\kappa_b = 8 \times 10^{-20}$ J? (Assume that the forces are independent.)

6.7 The short-range repulsive force for DPPC in the liquid phase has been measured to be

$$\frac{F}{\text{area}} = 1 \times 10^9 \exp\left(\frac{h}{\lambda}\right)\left(\frac{\text{J}}{\text{m}^3}\right)$$

where $\lambda \approx 022$ nm. Assuming that the Hamaker constant is 5×10^{-21} J, what is the separation h for a phase in equilibrium with pure water? How many water molecules per lipid are there at equilibrium if we assume that the area per lipid is 0.68 nm^2?

6.8 Using the expression for the force between two DPPC bilayers in water, calculate the adhesion energy per unit area between two flat bilayers at equilibrium with pure water.

6.9 Using the results of the preceding exercises, estimate the interfacial free energy of a lamellar DPPC liquid crystal and pure water for an interface parallel to the bilayers.

6.10 Calculate the adhesion energy between two equal-sized DPPC vesicles of radius R. Assume that they do not undulate or deform upon contact and use the constants of the preceding exercises.

6.11 When an ionic surfactant like $C_{16}N(Me)_3Br$ is added to a DPPC lamellar liquid crystal, the surfactant is solubilized in the bilayer. Solubilization introduces an electrostatic double-layer force between bilayers. What ratio of CTAB to lipid is needed to induce an extensive swelling of the lamellar liquid crystal (a) when the medium is pure water with only Br^- counterions and (b) in 1×10^{-2}M NaBr.

6.12 A phospholipid vesicle of inner radius 15 nm is formed in a 1 Mm KCl solution. It is then placed in a pure water solution. Estimate the time required for the ions to leak out into the solution if the permeability coefficient is 1×10^{-13} m/s for both ions and assuming that the vesicle remains intact despite the change in osmotic conditions.

6.13 Consider the vesicle system of Exercise 6.12. Assume that the pure aqueous medium now contains some valinomycin, leading to a drastic increase in the permeability coefficient for K^+. What is the change in membrane potential $\Delta\Phi$ for each transported K^+ion? What is the maximum value of $\Delta\Phi$ that can be generated?

6.14 Estimate the parallel D_P and perpendicular $D_\perp$ diffusion coefficients for water and KCl in a phospholipid lamellar phase with $d_{H_2O} = 3.0$ nm and $d_{\text{lipid}} = 4.0$ nm. Use the bilayer permeability coefficients $P_{H_2O} = 1 \times 10^{-5}$ m/s and $P_K = P_{Cl} \approx 1 \times 10^{-13}$. The diffusion coefficients in water are around 2×10^{-9} m^2/s for all three species.

6.15 To a solution of vesicles with a radius of 25 nm, prepared in pure water, one quickly adds KCl to yield a 1 mM solution. Calculate the time evolution of the vesicle's internal volume using $P_{H_2O} = 1 \times 10^{-5}$ m/s and $P_K = P_{Cl} \approx 1 \times 10^{-13}$. Consider only osmotic effects.

6.16 The squid has unusually large nerve axons, which are suitable for experimental studies. Across the axon membrane, there is an electrostatic potential drop. Inside the axon $[K^+] = 0.41$ M and $[Na^+] = 49$ mM, while on the outside $[K^+] = 22$ mM and $[Na^+] = 0.59$ M. What is the steady-state value of the potential when the ion permeabilities are such that

a. $P_{K+} \gg P_{Na+}$
b. $P_{K+} = P_{Na+}$
c. $P_{K+} \ll P_{Na+}$

Neglect the role of other ions (implying that their permeabilities are smaller).

6.17 Assuming that in Exercise 6.16, $P_{K+} = 5 \times 10^{-10}$ and $P_{Na+} = 1 \times 10^{-12}$. What is the flow of ions at steady state? Calculate the free energy loss per time and area due to ion flux. (The steady state is supported by an Na^+/K^+ion pump protein that consumes ATP.)

6.18 Calculate the capacitance per unit length of a myelin structure in which a central cylindrical (radius R_1) part of a nerve cell is covered by n concentric bilayers as in a liposome. Assume that the bilayers are 4.0 nm thick and separated by water layers of thickness 40 nm. Note that $n(4 + 40)$ can be larger than R_1.

6.19 Using Exercise 6.18, calculate how many monovalent ions per unit length are transferred from the outside of the myelin structure to inside the nerve cell to change the potential from −70 to +30 mV using $R_1 = 1 \times 10^{-6}$ m and $n = 25$.

References

Bagatolli, L. (2006). *Biochimica et Biophysica Acta (BBA) - Biomembranes*, 1758, 1541.
Berg, J.C. (2010). *An Introduction to Interfaces & Colloids. The Bridge to Nanoscience.* World Scientific.
Fendler, J.H. (1987). *Membrane Mimetic Chemistry.* New York: Wiley.
Fillipov, A., Orädd, G., and Lindblom, G. (2003). *Biophys. J.* 84, 3079–3086.
Fontell, K., Khan, A., Maciejewska, D., and Puang, S. (1991). *Colloid Polym Sci.* 269, 727.
Hinz, H.J. and Sturtevant, J.M. (1971). *J. Biol. Chem.* 247, 6071.
Ickenstein, L.M., Sandström, M.C., Mayer, L.D., and Edwards, K. (2006). *Biochimica et Biophysica Acta (BBA) - Biomembranes*, 1758, 171–180.
Israelachvili, J.N. (2011). *Intermolecular and Surface Forces*, 3rd ed. London, UK: Academic Press.
Lafleur, G., Cullis, P., and Bloom, M. (1990). *Eur. Biophys. J.* 19, 55.
Liljas, A., Liljas, L., Ash, M.-R., Lindblom, G., Nissen, P., and Kjeldgaard, M. (2017). *Textbook of Structural Biology* (2nd ed.). World Scientific Publishing Cto. Pte. Ltd.
Lindblom, G., Wennerström, H., Fontell, K., Söderman, O., and Arvidson, G. (1982). *Biochemistry*, 21, 1553.
Makasewicz, K., Wennmalm, S., Linse, S., and Sparr, E. (2022). *QRB Discovery*, 3.
Marsh, D. (1990). *Handbook of Lipid Bilayers*, FL: CRC Press.
Ohshima, H. (2024). *Fundamentals of soft interfaces in colloid and surface chemistry.* London, UK: Academic Press.
Perry, M.M. and Gilbert, A.B. (1979). *J. Cell Sci.* 39, 257.
Seddon, J. (1990). *Biochimica et Biophysica Acta*, 1031, 1.
Waninge, R. (2004). *Phys. Chem. Chem. Phys.* 6, 1518–1523.
Wiener, M.C. and White, S.H. (1991). *Biophys. J.* 61, 434.

7

Polymers in Colloidal Systems

Concept Map

Polymers

- Polymers contain N_p covalently bonded monomers whose collective effect leads to unique and highly useful properties.
- Polymers are synthesized by condensation or by free radical reactions that join monomers together to form a macromolecule. Usually, such reactions produce a range of N_p values, which makes the sample size polydisperse. The ratio M_n/M_w between the number- and weight-average molecular weights provides a measure of the polydispersity. Some biopolymers are monodisperse.

Homopolymers	Block Copolymers	Heteropolymers
N_p identical monomers with uniform interactions along the chains; they often form random coils.	Two or more covalently linked homopolymer chains.	Different monomers joined in a nonregular pattern. Proteins contain 20 different types of monomers.

Polymer Conformations in a Solvent and the Radius of Gyration

The size of a polymer coil is characterized by its radius of gyration R_g (Eq. 7.1.7).

- In dilute solution, polymers can fold (Figure 7.1) in many ways because they possess many internal degrees of freedom. In the compact state $R_g \propto M^{0.33}$, for a linear conformation, $R_g \propto M$, while for a random coil, $R_g \propto M^{\alpha}$, where $0.5 < \alpha < 0.6$.
- The radius of gyration R_g is smaller than the average end-to-end distance R_{1N} (Eqs. 7.1.7 and 7.1.14).
- R_g is determined by (1) bond rigidity and local steric interactions; (2) excluded volume effects resulting from overlap repulsion between monomers distant along the chain; and (3) the balance between solvent–monomer and monomer–monomer interactions as measured by the χ-parameter (Eq. 7.1.10).

(Continued)

Polymer Conformations in a Solvent and the Radius of Gyration (Continued)

- When polymers dissolve in a solvent, many more coil configurations become accessible, which adds a significant entropic contribution to the free energy of solution (Eq. 7.1.19).
- Based on χ, three polymer solution regimes can be defined: $\chi < 0.5$, good solvent where configurational entropy dominates, leading to an open polymer chain; $\chi > 0.5$, bad solvent where monomer–monomer attractions are strong, leading to a dense polymer coil; and $\chi = 0.5$, a theta solvent where polymer configurational entropy and monomer–monomer attractions precisely balance to give a random coil with $R_g \propto M^{0.5}$.
- The persistence length, l_p, describes the stiffness of a polymer chain (Eq. 7.1.15).
- Polyelectrolytes can form rods with large values of l_p (Eq. 7.1.20), while proteins often form compact globular structures in their native conformations.

Polymer Solutions

Three concentration regimes exist (Figure 7.5): dilute–intermolecular interactions are unimportant; semidilute–chain overlap becomes important; and concentrated–chain entanglements dominate.

The transition between dilute and semidilute occurs at the overlap concentration C*, which takes values between 0.1 and 10 wt% (Eq. 7.2.2).

- The Flory–Huggins theory describes the semidilute regime in terms of the regular solution theory and provides a basis for calculating $\Delta \mathbf{G}_{mix}$ (Eq. 7.2.6).

Characterization of Polymer Solutions

- Light, neutron, and X-ray scattering techniques provide R_g and B_2, the second virial coefficient (Figure 7.4).
- In a dilute solution, the self-diffusion coefficient depends on molecular weight as $D \propto M^{-(05-0.6)}$ (Eq. 7.1.23), while for the intrinsic viscosity, $[\eta] \propto M^{-(0.5-0.8)}$ (Eq. 7.1.28). The intrinsic viscosity provides a convenient way to estimate M.
- In the semidilute and concentrated solutions, both D (Eq. 7.2.12) and η (Figure 7.8) show a strong dependence on polymer size as predicted by the reptation model.

Many polymer solutions display a rheology where the response to mechanical perturbation involves both elastic deformation and viscous flow.

Polymer Association

Polymers can associate to form a number of useful structures.

- When both units of a diblock copolymer B_mC_n are in a good or a bad solvent, the copolymer behaves like a homopolymer. When the solvent is good for one unit but bad for the other, the amphiphilic character leads to association and the formation of micelles, bilayers, etc. (Figure 7.10).

- Helices can form when individual monomers possess amphiphilic properties. Transformation between helices and random coils is a cooperative process (Figure 7.11).
- Gels result from self-association or chemical crosslinking of polymers. On a macroscopic level, gels resemble solids, but on a microscopic level, they retain fluid-like properties.

Polymers on Surfaces

Adsorption Processes

Thermodynamics	Kinetics
• Physical adsorption leads to many polymer–surface contacts (Figure 7.12). In practice, the process can be irreversible. • With amphiphilic block copolymers, one block may selectively adsorb onto the surface, while the other extends into the solvent. • Polyions adsorb strongly onto oppositely charged surfaces, but can desorb when an electrolyte is added.	• The rate of adsorption is determined by the slow rate of polymer diffusion to a surface. • Polydispersity can affect the adsorption process. Short polymer chains adsorb first because they diffuse faster than large ones, but the final equilibrium favors adsorption of long chains. • Equilibrium can take a very long time to achieve or may never be reached.

Forces Between Surfaces with Adsorbed Polymers

- Surfaces covered with polymer will first begin to interact even when the surface separation exceeds $2R_g$ owing to occasional polymer tails extending out from the surface. At this distance ($\approx 10 - 100$ nm), van der Waals and double-layer forces are negligible (Figure 7.13).
- In a good solvent ($\chi < 0.5$), a repulsive overlap force (Figure 7.14) arises because the configurational entropy of the interacting chains is reduced, while in a bad solvent ($\chi > 0.5$), the net attraction between adsorbed polymers on opposing surfaces leads to an attraction (Figure 7.15).
- Colloidal particles covered with a polymer will coagulate when the solvent changes from good to bad by variation in temperature, etc. (Table 7.2). This aggregation process is usually reversible.

Polymers are molecules whose size falls within the colloidal range. Historically, polymer science developed as a branch of colloid chemistry, but with the advent of modern polymer chemistry, studies of polymers have become a subject in their own right, with entire textbooks and university courses devoted to them. In this chapter, we focus only on aspects of polymer science that are particularly relevant to the general properties of colloidal systems.

In colloid science, we encounter polymers of several different kinds. **Homopolymers** consist of a monomer unit repeated N_p times, where N_p is called the degree of polymerization. In a **block copolymer**, the polymer chain contains one regular "block" of one repeating monomer unit followed by one or more similar blocks of other repeating unit(s). **Heteropolymers**, in which several different monomer units are joined in a nonregular pattern, are often encountered in biopolymers like proteins and nucleic acids. If the monomer units carry charges, we refer to a polymer as a **polyelectrolyte** composed of a **polyion** and its counterions. Normally, the monomers of a polymer are joined in a linear fashion, but branched polymers also occur. In a **dendrimer**, there is extensive branching so that monomers pack to form a nearly spherical particle with properties intermediate to a typical polymer and a typical colloidal particle or micelle.

In Table 7.1, we list polymers of particular importance for colloidal systems and indicate their use. The difference in chemical complexity of these different polymers is also reflected in their self-assembly properties in solution. The homopolymer, in which the chemical interactions are uniform along the chain, typically forms a random coil conformation. On the other hand, block copolymers resemble amphiphiles, and as we shall see, they show all the different self-assembly structures found with surfactants. Finally, a heteropolymer, like a protein, can organize into a very specific conformation determined by the detailed sequence of amino acids. Clearly, polymer systems show a rich behavior with respect to the conformation of the single molecule and also in the assembly of the different molecules.

Table 7.1 Some polymers of special importance in colloidal systems.

Common name	Repeating unit	Example of use
Synthetic polymers		
Polyethylene oxide (PEO), polyethylene glycol (PEG)	$-CH_2-CH_2-O-$	Steric stabilization in water; inducing depletion forces
Polyacrylamide	$-CH_2-CH(CONH_2)-$	Gels (chromatography, diapers), rheology modifiers
Polystyrene	$-CH\Phi-CH_2-$	Latex particles, rigid plastic containers, thermal insulators (Φ = phenyl)
Polyamines	$-CR=N-$	Flocculant in papermaking
Polyethylene oxide and polypropylene oxide block copolymer	$-CH_2-CH_2-O-$ $-CH(CH_3)-CH_2-O-$	Drug delivery, detergency
Polystyrene sulfonate	$-CH(\Phi SO_3^-)-CH_2-$	Reverse osmosis membranes

Common name	Repeating unit	Source	Example of use
Biopolymers			
DNA	Adenosine, guanosine, cytidine, thymidine nucleotides	Any cell	Identification, gene modification, DNA tagging
Cellulose	Glucose	Plants	Paper making
Starch	Glucose	Plants	Food
Gelatin	Glycerin, valine, alanine	Collagen, bones	Food additive
Agarose	Galactose	Seaweed	Chromatography gel
Hyaluronic acid	Glucuronic acid, N-acetyl glucosamine	Connective tissue	Eye surgery, veterinary medicine
Carrageenans	Galactose, galactose sulfate	Algae	Food additives

7.1 Single Polymer Chains Feature a Variety of Conformations in Solution

7.1.1 We Can Characterize the Polymer Size Distribution with Averages

Polymers are formed by linking monomer units to the growing chain in a stepwise fashion

$$M_n + M \rightarrow M_{n+1} \cdots \tag{7.1.1}$$

In the biological system, the link is formed by enzymatic catalysis. For example, in protein synthesis within the ribosome, there is a precise control of the sequence and number of amino acids, resulting in a protein with a specific molecular weight.

For all synthetic polymers and some biological ones, a partly random event terminates the chain reaction, resulting in a distribution of molecular weights. The size polydispersity of polymer molecules constitutes a complication in all fundamental studies of their properties and also in many applications. However, modern synthesis techniques utilizing chain transfer agents are successfully used to both control the molecular weight and limit the polydispersity of polymer preparations. Post synthesis, chromatographic procedures or fractionated precipitation can be used to achieve relatively narrow size distributions.

Determining the full size distribution $F(N_p)$ of a given polymer sample is demanding, but it is often sufficient to determine specific averages of the distribution. The most accessible values are the degree of polymerization, which is also called the number average:

$$\langle N_p \rangle_{na} = \int N_p F(N_p) \mathrm{d}N_p \tag{7.1.2}$$

together with the so-called weight-averaged polymerization number

$$\langle N_p \rangle_{na} = \int \frac{N_p^2 F(N_p) \mathrm{d}N_p}{\langle N_p \rangle_{na}} \tag{7.1.3}$$

When the size distribution function simply represents a peak at a given N_p value, such as N_p' for a protein,

$$\langle N_p \rangle_{wa} = \frac{N_p'^2}{N_p'} = N_p' = \langle N_p \rangle_{na} \tag{7.1.4}$$

the number average and the weight average are the same. As soon as a true distribution exists, the weight average is always larger than the number average. Longer chains contribute more to the weight than to the number average value.

Different methods for determining molecular weights yield different averages when a distribution in size exists. In Chapter 4, we discussed how to determine micelle sizes by light scattering. The same method can be applied to polymers, as discussed later in this chapter, and this measurement yields a weight-averaged molecular weight M:

$$\langle M \rangle_{wa} = M_1 \langle N_p \rangle_{wa} \tag{7.1.5}$$

where M_1 is the molecular weight of the monomer. In contrast, osmometry counts individual molecules and

$$\langle M \rangle_{na} = M_1 \langle N_p \rangle_{na} \tag{7.1.6}$$

Determining such different average molecular weights provides a useful measure of the width of the size distribution function. The ratio $\langle N_p \rangle_{wa} / \langle N_p \rangle_{na}$ can vary substantially above its minimum of unity. A value of 1.1 is obtained when great care has been taken to obtain a narrow size distribution, while 2 is obtained for a purely random termination reaction. Values significantly larger than 2 have been observed for some biopolymers.

7.1.2 The Interplay Between Chain Entropy and Monomer–Monomer Interactions Determines the Conformation of a Single Polymer Chain

When a polymer is dissolved in a high excess of solvent, we can neglect interactions between polymer molecules. However, many internal degrees of freedom are present in polymers, and the chain can fold in many different ways. How can we characterize the chain conformation? We can distinguish three different types of chain folding (see Figure 7.1). In the compact state, the chain folds back on itself to minimize polymer–solvent contact in a way that resembles a micellar configuration. The linear size of the polymer globule (the radius) depends on the degree of polymerization, N_p, as $N_p^{0,33}$. Soluble proteins in their native state are often found in this globular form.

Some polymers adopt a stiff linear conformation, often dictated by the formation of a helical structure. In this case, the longest linear dimension increases linearly with N_p. A short DNA double helix typifies this behavior.

Most synthetic polymers, as well as proteins and DNA in denatured states, show a much less well-defined structure, which is referred to as a (random) coil. In this case, the characteristic linear dimension R increases with the degree of polymerization as some power, $R \sim N_p^{\alpha}$, where α is typically in the range of 0.5–0.6.

In the coil state, the polymer chain adopts many different configurations, and we can characterize the state only by average properties. The most common measures are the radius of gyration R_g, which for a homopolymer is defined by

$$R_g^2 = \frac{\left\langle \sum_{i=1}^{N_p} \left| \boldsymbol{r}_i - \boldsymbol{r}_{cM} \right|^2 \right\rangle}{N_p} \tag{7.1.7}$$

Figure 7.1 The triangle of Haug, illustrating three extreme types of polymer conformations: the stiff rod $(L \sim N_p)$, the compact globule $(R \sim N_p^{0.33})$, and the coil $(R_g \sim N_p^{\alpha}$ with $0.5 \leq \alpha \leq 0.6)$. Intermediate conformations are possible as indicated by the solid lines.

and the average end-to-end distance R_{1N}

$$R_{1N}^2 = \left\langle |\boldsymbol{r}_N - \boldsymbol{r}_1|^2 \right\rangle \tag{7.1.8}$$

where $\boldsymbol{r}_i$ stands for the position of the ith segment and $\boldsymbol{r}_{\mathrm{cM}}$ is the location of the center of mass. The angular brackets denote a Boltzmann weighted average over all allowed coil configurations. The end-to-end distance is larger than the radius of gyration, but they scale similarly with changes in molecular weight.

Even though the chain conformation is characterized by one length in Eqs. (7.1.7) and (7.1.8), we should not think of the typical chain configuration as a loose homogeneous spherical ball. The density of monomer units decreases out from the center. Furthermore, in a particular configuration, the decrease is more rapid in some directions than in others, so that although the coil is spherical on average, each configuration has a nonspherical distribution of monomers and typically exhibits a distinct asymmetry.

A polymer can adopt a large number Ω of configurations. As an example, we can consider a polymer where the monomer–monomer bond can adopt three basic orientations, as in a carbon–carbon single bond with two gauche and one trans conformations. With a thousand monomers, we have

$$\Omega \approx 3^{999} \approx 10^{477} \tag{7.1.9}$$

configurations of a single molecule! Not all of these configurations have the same energy. The size of the polymer coil, as measured by R_g or R_{1N}, results from an average over all these 10^{477}, weighted by their respective Boltzmann factors. We can distinguish between three types of interactions affecting these weight factors.

First, there are short-range intrachain interactions. For example, in polyethylene, the trans conformation is lower than the gauche by approximately 2.4 kJ/mol, which is similar to the thermal energy $k_B\mathbf{T}$ at ambient conditions. There are two gauche conformations and one trans with the net effect that there is a slight preference to form a locally straight chain. A second, more subtle, effect is that a chain cannot fold back on itself due to the overlap repulsion between monomeric units. Even though this interaction has a very short range in space, it takes on a long-range character when counting along the chain, since monomer units that are well separated along the chain can come close in space as the molecule folds. The net result is to favor more extended configurations of the chain. In the polymer literature, this is referred to as an excluded volume effect.

The third, and most significant, contribution to the energy in the Boltzmann factor comes from the monomer–solvent interactions. Even for a polymer at a high dilution, the local concentration of monomer units may be substantial within a polymer coil. From a molecular point of view, we have a local situation much like that in an ordinary binary liquid mixture, and we can use the same basic model to analyze it that we used for such mixtures in Chapter 1. In the regular solution model, the basic parameter is the interaction constant, w, introduced in Eq. (1.1.16). When we apply it to a polymer system, we only need to reinterpret the bulk coordination number, z_{b}, since each monomer unit is linked covalently to two other monomers.

For the typical case of $w > 0$, monomer–monomer contacts are energetically favorable, and the more compact the coil, the larger the number of such contacts. Balancing this organizing effect is the entropy-promoting randomness, that is, equal weights given to all possible configurations. The crucial quantity determining the equilibrium properties of the coil is in analogy with regular solutions

$$\chi = w/\boldsymbol{RT} \tag{7.1.10}$$

commonly referred to as the χ parameter. When analyzing polymer coil properties at different χ values, we can identify three important regimes:

1. For $\chi < 0.5$, the intermolecular interactions are weak compared to the thermal energy, and chain configurational entropy dominates. This regime is called good solvent conditions.
2. For $\chi > 0.5$, monomer–monomer attractions are strong. To obtain many such contacts, the polymer coil becomes dense, and this regime is called bad solvent conditions.
3. At $\chi = 0.5$, a transition occurs from good solvent to bad solvent behavior. The temperature when $\chi = 0.5$ is called the theta (θ) temperature, and a solvent in which theta conditions apply is called a theta solvent. These have important applications in the fundamental studies of polymer solutions.

7.1.3 Persistence Length Describes the Stiffness of a Polymer Chain

A sufficiently long polymer chain is flexible on a length scale that is smaller than the contour length but may be large compared to the size of a monomer unit. This flexibility leads to a coil conformation. The overall shape of the coil does not depend on the local stiffness of the chain, while the size does. In the same way as a coil of a thin flexible thread is much smaller than a coil of a stiff rope of equal length, the size of a polymer coil depends not only on the length of the chain but also on its stiffness. Moving along a polymer chain in a particular coil configuration is like following the path of a diffusing molecule. The random walk is a mathematical model that describes some of the basic features of both polymer configurations and diffusive motion. In the simplest form, the random walk is a sequence of N_r steps of length l_r each in a random direction. The root mean square distance between the end points $\langle L_r^2 \rangle^{1/2}$ of such a walk is

$$\langle L_r^2 \rangle^{1/2} = N_r^{1/2} l_r \tag{7.1.11}$$

If we impose the additional restriction that the path may not cross itself within a distance of the order of the size of a monomer unit, one has the self-avoiding random walk that is less compact on average and

$$\langle L_r^2 \rangle^{1/2} \propto N_r^{0.6} \tag{7.1.12}$$

When applying the above equations to a polymer, it is tempting to identify the number of steps N_r with the degree of polymerization N_p and the step length with a monomer size. However, this identification does not properly account for the local stiffness of the polymer chain. The stiffness of the chain can be characterized by the persistence length l_p, which is the length over which two parts of the chain keep their orientational correlation. In a simplified picture, we can see the chain as a sequence of freely linked stiff segments, so-called Kuhn segments, whose length corresponds to twice the persistence length. With a length L of the fully stretched chain (L is called the contour length), there are $L/(2l_p)$ Kuhn segments in a chain. Note that a chain will only behave as a flexible chain, long enough to adopt a random coil conformation, when $L \gg l_p$.

We can make use of the random walk properties by identifying the step size in the walk with twice the persistence length of a particular chain. An important characteristic of the θ condition is that the polymer coil has random walk properties and the end-to-end distance

$$R_{1N}^2 = \frac{L}{2l_p}(2l_p)^2 = 2Ll_p \tag{7.1.13}$$

using Eq. (7.1.11). We will see later in this chapter that in scattering experiments, the radius of gyration is experimentally more accessible than the end-to-end distance. It is generally true that for long chains, R_g and R_{1N} are proportional and for the random walk model

$$R_{1N}^2 = 6R_g^2 \tag{7.1.14}$$

When we measure the radius of gyration at θ conditions for a polymer of known contour length, we obtain the persistence length

$$l_p = 3R_g^2/L \tag{7.1.15}$$

using Eqs. (7.1.13) and (7.1.14). For systems of different polymers with the same length L, the size of the coil will increase with increasing persistence length, as we can see from Eq. (7.1.13). The persistence length can vary from values in the range of 0.5–1.0 nm for very flexible polymers, such as $(-CH_2-)_N$, $(-OSi(CH_3)_2-)_N$, and $(-OC_2H_4-)_N$ to more than 100 nm for some polyelectrolytes in low salt conditions.

We can now express how the size of the polymer coil depends on the relevant physical parameters. The monomer–solvent interaction enters implicitly through the solvent quality, while the others enter explicitly. At θ conditions,

$$R_g = \left(\frac{l_M l_p}{3}\right)^{0.5} N_p^{0.5} \propto M^{0.5} \tag{7.1.16}$$

Here, we have used the contour length $L = l_M N_p$, where l_M is the monomer length.

Under good solvent conditions, $\chi < 0.5$, the polymer coil configuration behaves as a self-avoiding random walk and in analogy with Eq. (7.1.12)

$$R_g \sim N_p^{0.6} \sim M^{0.6} \tag{7.1.17}$$

with a proportionality constant that increases with an increasing persistence length. In a bad solvent,

$$R_g \sim N_p^{0.33} \sim M^{0.33} \tag{7.1.18}$$

For very long polymers, the transition from a coil conformation at θ conditions to a compact globule, where Eq. (7.1.18) applies, occurs quite dramatically, as χ increases above 0.5. We emphasize that the proportionality $R_g \sim N^\alpha$ in Eqs. (7.1.16)–(7.1.18) only applies when the contour length L is much larger than the persistence length l_p. For shorter molecules, non-negligible correction terms appear.

7.1.4 When a Polymer Dissolves into a Solvent, Many More Coil Configurations Become Accessible

The solvent quality not only affects the size of the polymer coil but also has a strong effect on the polymer's solubility. Our first question is: Should we expect a polymer to be soluble to any appreciable extent? Today, we know the answer can be yes, but the question was one of major concern in the early days of polymer science. When a polymer is dissolved from a pure state, for example, a crystal, into a dilute solution, there are three contributions to the free energy of the process. As with any solute, there is the entropy of mixing of the polymer, as a molecular entity, with the molecules of the solvent. For dilute systems in which we can neglect interactions between polymer molecules, the ideal entropy of mixing expression, Eq. (1.1.8), applies for polymers as it does for any other solute. This contribution is determined solely by the number of dissolved molecules, and it is, in the polymer case, independent of the degree of polymerization and becomes negligible for large N_p.

A second contribution derives from bringing the polymer from the pure state, where it is surrounded by like polymers, into solution, where it is surrounded by solvent molecules to a large extent. For $w > 0$, this contribution involves an increase in energy on dissolution and thus limits the solubility. Since these interactions involve the monomer units, the energy per molecule increases linearly with the degree of polymerization N_p $\left(\Delta \mathbf{U}_{\text{solution}} \propto wN_p\right)$. Since typically $w > 0$, the energy seems to dominate, provided the molecules are sufficiently long. For this reason, large polymers seemingly cannot be soluble to any appreciable extent. The resolution of this dilemma that faced the polymer scientists around 1930 lies in a third free energy contribution. In the pure state, the configuration of the chain is highly confined by the neighboring chains, while in a solvent, the chain can adopt many more configurations. The change in configurational entropy on the dissolution of a single chain $\Delta \mathbf{S}_{\text{conf}}$ can be written

$$\Delta \mathbf{S}_{\text{conf}} = k_B \mathbf{T} \left\{ \ln(\Omega_{\text{solution}}) - \ln\left(\Omega_{\text{pure}}\right) \right\} \sim k_B \mathbf{T} N_p \tag{7.1.19}$$

since the number of configurations increases exponentially with N_{p} as in Eq. (7.1.9). Since both the positive energy contribution and the negative entropy contribution to the free energy change on dissolution are proportional to N_p, the sign of $\Delta \mathbf{G}$ (solution) is determined by the size of the proportionality factors. As stated in Section 7.1.2, solubility is obtained for $\chi \equiv w/(\boldsymbol{RT}) < 0.5$, while for higher values of $\chi > 0.5$, one has bad solvent conditions with a very low solubility of the polymer. In contrast to colloid particles that lack internal configurational degrees of freedom, a polymer can dissolve even in the limit of a high molecular weight. We can relate the criterion for solubility to the same χ parameter (Eq. 7.1.10) that determines the nature of the polymer coil. It follows that understanding the changes in available polymer configurations provides a key to the conceptual understanding of polymer solutions.

For a polyelectrolyte, the solubility arises from a somewhat different mechanism. In the pure state, the counterions are located in close contact with the charged groups on the polyion. When exposed to a polar solvent like water, the counterions dissolve, increasing their entropy. Combined with the solvation of the charged groups on the polyion, this ensures good solubility of the polyelectrolyte in polar solvents. Even an intrinsically strongly hydrophobic monomer like styrene can give a water-soluble polymer if it is transformed into a polyelectrolyte by sulfonating the monomer. A more complex interplay between hydrophobic attraction and ion dissolution is found in the nucleic acids RNA and DNA. They have a strong hydrophobic attraction between the aromatic rings in the bases and, also to some extent, in the ribose groups. The charged phosphate groups promote aqueous solubility. RNA with its intact ribose group is slightly more hydrophilic than DNA and is rather stable as a single chain in solution, while for DNA, the formation of the double helix allows for hydrophobic attraction between the bases while still leaving the charged phosphates exposed to the aqueous medium. Another case of great practical importance is the dissolution of cellulose. The glucose units of the chain have OH groups that can become charged at high (> 12) values of the pH. This effect can be used to bring cellulose molecules in solution despite the strong cohesion mainly due to hydrophobic interactions in the solid fiber.

7.1.5 Charged Polymer Chains Display a More Extended Conformation

Charged polymers play an important role in colloid chemistry. In addition to the practical aspect that charges on the chain greatly increase the aqueous solubility of polymers, polyelectrolytes strongly influence the electrostatic forces between colloidal particles. The presence of the charged groups influences both intra- and intermolecular interactions. Intramolecular repulsions between the charged groups lead to a more extended conformation of the chain compared to the corresponding neutral polymer. This means that the persistence length is mostly determined by the electrostatic effects. Odijk derived an approximate relation for the persistence length

$$l_p = l_p^0 + l_p^{el} = l_p^0 + \frac{e^2}{16\pi\varepsilon_r\varepsilon_0 k_B \mathbf{T}\kappa^2 b^2} \tag{7.1.20}$$

Here l_p^0 represents the persistence length of the chain in the absence of the long-range electrostatic interactions, l_p^{el} is the electrostatic contribution, and b is the distance between unit charges along the chain. However, the persistence length concept is problematic for polyelectrolytes, and Eq. (7.1.20) should only be used as a qualitative guideline rather than as a quantitative relation.

Clearly, the density of charges along the chain will also influence the role of electrostatics. A common model used to describe a polyelectrolyte starts from a solution of the Poisson–Boltzmann equation for a charged cylinder. This geometry represents an intermediate case between the planar charged surfaces and spherical aggregates discussed in earlier chapters. In the cylindrical geometry, a phenomenon called ion condensation appears as a consequence of the long-range interactions. To a good approximation, the counterion population can be divided into two categories. One category, called condensed ions, is electrostatically associated close to the charged cylinder, while the remaining ions are diffusely bound. The point of this division is that we can easily calculate the fraction of condensed ions. They have the effect of reducing the effective linear charge density, $\xi = \pm e/b$, to a critical value

$$\xi_{\text{crit}} = \pm\frac{4\pi\varepsilon_r\varepsilon_0 k_B \mathbf{T}}{ze} \tag{7.1.21}$$

Thus, with no added salt, the fraction of condensed counterions equals $1 - \xi_{\text{crit}}/\xi$ for $\xi \geq \xi_{\text{crit}}$. For monovalent ions, $z = 1$, in water, $\varepsilon \approx 80$, at room temperature, $\xi_{\text{crit}} = e/(0.71\ \text{nm})$. The distribution of the noncondensed ion can then be calculated using the linearized Poisson–Boltzmann equation. The ion condensation model is appealing since it provides a conceptually simple way of dealing with the electrostatic consequences of ion distribution. However, it should be appreciated that, in the form presented here, it simply represents an approximate way to describe the solution of the nonlinear Poisson–Boltzmann equation in a cylindrical geometry.

Owing to their long persistence length, polyelectrolyte coils are very extended and tend to overlap at very low concentrations. When coils entangle, the viscosity increases. For example, the use of hyaluronic acid in eye surgery is based primarily on its ability to increase the viscosity of the aqueous medium even at low concentrations.

7.1.6 Protein Folding Is the Result of a Delicate Balance Between Hydrophobic and Hydrophilic Interactions and Configurational Entropy

Proteins are heteropolymers in which the monomer building blocks are up to 20 different amino acids connected via amide linkages. The specific amino acid sequence of a protein, which defines its primary structure, is coded by the DNA. The secondary structure refers to the regular arrangement of the chain, as in α-helices or β-sheets, and also in sidechain–sidechain disulfide bonds. The tertiary structure refers to the protein's unique three-dimensional structure, and it is determined by the interactions between individual amino acids.

Using a selection of the 20 different amino acids in specific sequences of a polymer chain, nature has managed to make protein molecules with a remarkable range of functions. Enzymes that occur either as globular water-soluble molecules or bound in the lipid membrane catalyze a multitude of specific biochemical reactions. Proteins like hemoglobin transport and store metabolites. Others like insulin and calmodulin regulate physiological processes. Collagen and silk are used as materials. Gluten in seeds and egg white in eggs store amino acids. Histones control DNA packing in chromosomes. Antibodies act as specific complexing agents. Actin and myosin form a molecular motor in muscles.

It is one of the future challenges of polymer technology and colloid science to, however slightly, approach this degree of sophistication in material design. The protein example demonstrates the great variety of properties that can be obtained by using monomers with different physicochemical characteristics in a controlled sequence. Using the block–copolymer concept, one has made some advances in tailor-making the properties of synthetic polymers. This should only be the beginning.

In the language of polymer chemistry, the compact globular conformation typifies a polymer molecule in a bad solvent, yet the protein remains soluble. The reason is that the different side chains represent a spectrum of χ parameters with respect to water. Thus, the protein chain strives to adopt a conformation that hides the apolar side chains from the aqueous solvent, while retaining the polar side chains at the surface of the folded molecule. However, some compromises have to be made in the folded state. Polar groups, such as the amide linkages of the protein backbone, must enter the apolar interior. To do as well as possible in an intrinsically uncomfortable environment, they tend to link up by hydrogen bonds, as in an α-helix illustrated in Figure 7.2.

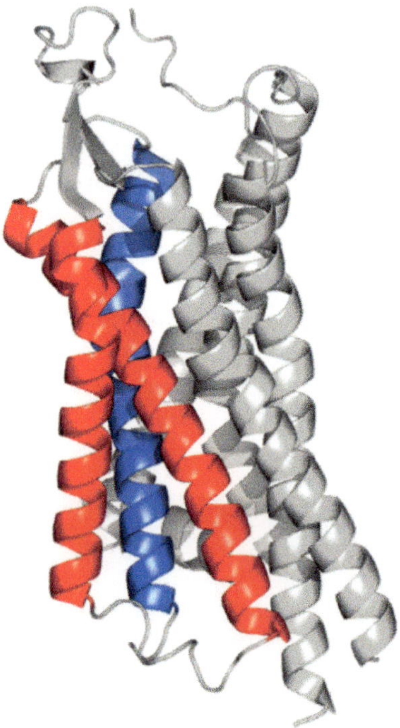

Figure 7.2 Illustration of an α-helix bundle where each turn contains 3.6 residues and a length of 0.15 nm.

Another possibility is the formation of a β-sheet, where sequences of the chain associate in an antiparallel orientation as illustrated in Figure 7.3. A parallel orientation of the chain sequences can also be found, as will be further discussed in Chapter 9.

Thus, the driving force for the formation of a globular protein conformation in water is the hydrophobic interaction and is the same as for the association of surfactants into a micellar aggregate. The packing possibilities are much more constrained in the protein, however, because the units are covalently linked and because polar or semipolar groups are distributed along the chain and in the backbone. Although the globular state does not form because of the hydrogen bonds in the interior of the protein globule, these bonds are necessary to prevent the folded state from becoming too unfavorable.

Heating a globular protein denatures it. Typically, denaturation is completed over a relatively narrow ($\approx$ 10 K) temperature range. Thus, protein folding is a cooperative process, and the cooperativity has the same basic source as in micellization. The alternative to the native folded state is a more or less random coil with a large number of accessible configurations. Denaturation occurs when the temperature is raised because of the increased importance of the **TS** term of the chain configurational entropy in the random coil. Thus, we see how the same principles that operate in a surfactant aggregate also govern the folding of the amino acids in a globular protein.

7.1.7 Polymer Molecules Can Be Effectively Characterized Using Scattering Methods

In Chapter 4, we described how the size of micelles could be determined using light scattering. The molecular weight of a polymer can be measured in a similar way. There is, however, an

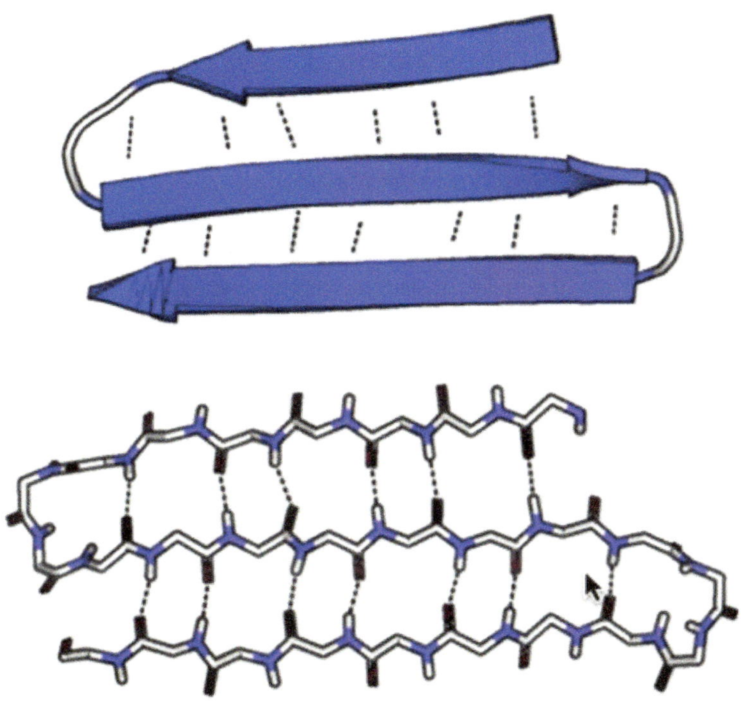

Figure 7.3 In a β-sheet, the amide groups of the protein backbone form hydrogen bonds with other amide groups along the backbone.

additional complication. Polymer coils have a more diffuse structure than micelles, and the extension from the center is such that the dimension of the coil is not negligible compared to the wavelength of visible light, making the scattering intensity dependent on the scattering angle in a nontrivial way. In practice, this has the consequence that to obtain the molecular weight, one has to extrapolate to zero scattering vector to eliminate the angular dependence. Then one can apply Eq. (4.3.17) and extrapolate to zero concentration, to eliminate the effect of coil–coil interactions. The method is called a Zimm plot, after the polymer chemist Bruno Zimm, and an example is shown in Figure 7.4. The molecular weight is obtained at the intersection of the two extrapolated lines representing constant concentration, zero scattering angle ($\theta = 0$) and constant scattering angle, zero concentration ($c = 0$), respectively. From the slope of the $\theta = 0$ line, one obtains the second virial coefficient, B_2. Under good solvent conditions, this is positive due to the repulsion between coils. The value of B_2 is a direct measure of how much excluded volume a polymer coil generates. The slope of the extrapolated $c = 0$ line determines how the form factor $P(q)$ (see Eq. 4.3.6) depends on the magnitude of the scattering vector $\boldsymbol{q}$. In Eq. (4.3.22), we showed how $P(q)$ depends on q for a perfect homogeneous sphere. The density in a polymer coil decreases away from the center, and the first term in an expansion of $P(q)$ defines the radius of gyration according to

$$P(q) = 1 - R_g^2 q^2/3 + \ldots \tag{7.1.22}$$

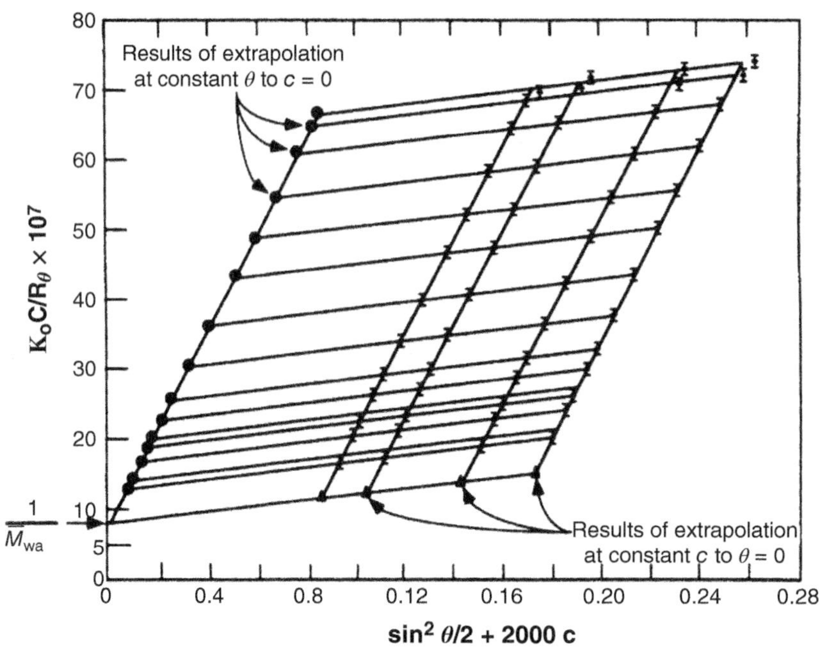

Figure 7.4 Zimm plot for measuring polymer molecular weight. Measurements are performed at a number of scattering angles θ for a series of concentrations. The observed scattering intensity depends both on the scattering angle θ and the polymer concentration c (weight/volume). At the concentration $c = 0$, the structure factor $S(q) = 1$ and the left line in the figure provide a measure of the form factor, and the slope is proportional to the radius of gyration R_g. The form factor at $q = 0$ is inversely proportional to the osmotic compressibility (Eq. 4.3.8), and the lower line representing $\theta = 0$ has a slope proportional to the second virial coefficient and an intercept inversely proportional to the molecular weight M (Eq. 4.3.20). *Source*: Benoit et al., 1954/with permission of American Chemical Society.

7.1.8 Polymer Self-diffusion and Solution Viscosity Reflect the Dynamic and Structural Properties of a Polymer Coil

In solution, a polymer coil has a range of dynamic modes from translation and rotation of the whole molecule to local conformation changes. The monomer units interact locally with the solvent, but when the molecule moves—in part or its entirety—this motion involves solvent molecules. They, in turn, affect other solvent molecules, which can influence a more distant segment of the polymer. We discussed this mechanistic description of hydrodynamic interactions briefly in Section 5.8. Hydrodynamic interactions are important for polymer coil dynamics, and they tend to correlate motions of different segments of the coil.

Let us first consider the translational self-diffusion of a polymer coil. For compact spherical particles of radius R in a solvent of viscosity η, the self-diffusion coefficient is given by the Stokes–Einstein relation

$$D = \frac{k_B \mathbf{T}}{6\pi\eta R} \tag{1.7.17}$$

which is derived by considering a liquid flow past a spherical object with the stick boundary condition that the flow is zero on the surface of the sphere. A polymer coil can be seen as approximately spherical, and we have expressions for the mean radius of the coil, but to use Eq. (1.7.17), we must also justify the stick boundary condition that the liquid velocity is zero at a distance R from the center of the stationary coil. This condition implies that the liquid flow inside the coil is zero. Although individual liquid molecules clearly diffuse relatively freely inside the coil, hydrodynamic interactions could cause the internal solvent to follow the coil. Partial drainage of the solvent in the coil is occurring, but experiments indicate that this effect is relatively small, and in a theta solvent, we find

$$D \propto M^{-0.5} \tag{7.1.23}$$

which is the relation we expected by combining Eq. (1.7.17) and the relation $R_g \propto M^{0.5}$ (Eq. 7.1.16). In a good solvent, the same reasoning suggests an exponent −0.6, and experimentally, one finds values in the range −0.55 to −0.6. Self-diffusion coefficients of polymers are rather demanding to measure experimentally, so while we could in principle use Eq. (7.1.23) to measure molecular weights, this is not done in practice.

A much more easily accessible parameter is the viscosity of a polymer solution. It serves as a simple route for determining molecular weight and particle shape. As the solution is diluted, the viscosity, η, approaches that of the solvent, η_s, linearly with solute volume fraction, ϕ, so that

$$\eta = \eta_s(1 + [\eta]\phi) \tag{7.1.24}$$

where $[\eta]$ is the intrinsic viscosity, also known as the limiting viscosity number. For spheres, Einstein derived that $[\eta] = 2.5$. On the other hand, for stiff rods with length L and radius R and $L \gg R$

$$[\eta] = \frac{L^2}{\pi R^2} \tag{7.1.25}$$

This relationship shows that stiff rods give a much more viscous solution than spherical particles, even in the limit where no direct interaction takes place between the rods. We can understand the molecular origin of the intrinsic viscosity based on the fact that no velocity gradients can exist inside the particles. With stick boundary conditions, a particle in a velocity gradient will sense a torque and rotate. A rotating rod generates more dissipation than a rotating sphere, explaining the larger value of $[\eta]$ for the rods.

For compact spheres, the intrinsic viscosity is, according to the Einstein result, independent of the size and only determined by the volume fraction of particles. Polymer coils are spherical on average, but at a given volume fraction, their intrinsic viscosity does depend on molecular size. To see this, we first adopt the assumption established in our analysis of diffusional motion that we can neglect the drainage of the coil. The radius of the coil varies as

$$R \propto M^{\alpha} \tag{7.1.26}$$

where $\alpha = 0.5 - 0.6$, depending on solvent quality. The volume, V_{coil}, of a coil then varies as $V_{\text{coil}} \propto M^{3\alpha}$. The effective volume fraction of coils is

$$\phi(\text{coil}) = c^* V_{\text{coil}} \tag{7.1.27}$$

where c^* equals the number concentration of polymer molecules.

Now, a coil contains both polymer and solvent so that $\phi(\text{coil})$ is much larger than the volume fraction of polymer ϕ_p. In a measurement, we prepare samples of a known concentration of polymer on a weight basis equal to $c^* M$. Knowing the density of the polymer, this can be converted to a volume fraction of polymer molecules $\phi_p \propto c^* M$. By measuring the viscosity for a series of concentrations, we determine $[\eta]\phi_p$ experimentally. By Einstein's argument (Eq. 7.1.24), this should equal $2.5\phi(\text{coil})$ so that

$$[\eta] = \frac{2.5\phi(\text{coil})}{\phi_p} \propto \frac{c^* M^{3\alpha}}{c^* M} \propto M^{(3\alpha-1)} \tag{7.1.28}$$

with the exponent $3\alpha - 1$ varying between 0.5 and 0.8, as α varies between 0.5 and 0.6. We can use this relation to determine molecular weights, provided that we can calibrate against some known samples. Experimentally, the exponent 0.5 is confirmed for theta conditions, while in good solvents the exponent is distinctly higher; we often find a value around 0.75, but rarely 0.8. This discrepancy could indicate that drainage through the coil is not negligible for these more extended coils.

See Cosgrove (2010, Chapter 7), Doi (2013, Chapter 1), and Lyklema (2005b, Vol. 5, Chapter 2) for additional discussion on Polymer Chains.

7.2 Thermodynamic and Transport Properties of Polymer Solutions Change Dramatically When Coils Overlap at Higher Concentrations

7.2.1 Three Concentration Regimes Can Be Identified in the Description of Polymer Solutions

A polymer molecule with a characteristic radius R_g contains mostly solvent in its random coil domain. The larger the exponent α, the longer the persistence length l_p, and the higher the degree of polymerization N_p, the less dense the coil. The expanded nature of the coils implies that they start to overlap and interact at low polymer concentrations.

Polymer–polymer interactions play somewhat different roles in the distinct concentration regimes illustrated in Figure 7.5. When discussing the polymer chain conformation, we explicitly considered the single chain, that is, the dilute solution. In practice, most polymer solutions

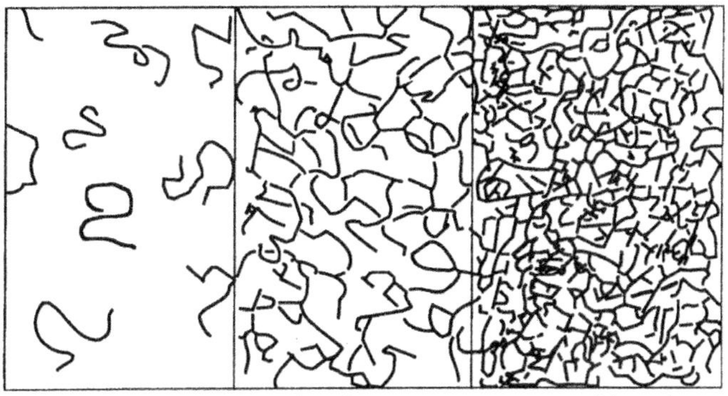

Figure 7.5 The three concentration regimes of a polymer solution. When the concentration is less than the overlap concentration C^*, we have a dilute solution. In the semidilute system, $c > C^*$, but the volume fraction ϕ is still small ($\phi \ll 1$). In the concentrated system, $c \gg C^*$ and ϕ approaches unity.

are in the so-called semidilute regime, where solvent exceeds polymer on a weight-on-weight basis, but the coils of individual molecules overlap so that a monomer in one chain is also likely to make contact with monomers of other chains. In semidilute solutions, the chains form a nonconnected network with a mesh size smaller than the radius of gyration. Furthermore, the mesh size decreases with increasing concentration because the coils are pushed even more into one another.

The crossover from a dilute solution to a semidilute regime is described by the characteristic overlap concentration C^*. Clearly, C^* does not have a precise value, but we can estimate it by assigning a volume V_{coil} to each polymer and calculating when the total number of coils fills the volume of the solution.

If we assume

$$V_{\text{coil}} = \frac{4}{3}\pi R_g^3 \tag{7.2.1}$$

the overlap concentration C^* (molar monomers) is

$$C^* = \frac{3 \times 10^{-3} N_p}{4\pi N_{Av}} \frac{1}{R_g^3} \tag{7.2.2}$$

A more experimentally accessible criterion for the overlap concentration is obtained from Eq. (7.1.28) for the intrinsic viscosity. If we require $\phi(\text{coil}) \approx 0.5$ (the limit for hard-sphere solutions discussed in Chapter 10), the first equality shows that the overlap concentration in volume fraction is

$$\phi_p^* \approx \frac{1.25}{[\eta]} \approx [\eta]^{-1} \tag{7.2.3}$$

simply the inverse of the intrinsic viscosity. The factor of 1.25 is often dropped, as in the second equality, since we are dealing with an order-of-magnitude estimate.

Because R_g varies with N_p with a power 0.5 in a theta solvent and a power 0.6 in a good solvent, $C^* \propto N_p^{-0.5}$ for a theta solvent and $C^* \propto N_p^{-0.8}$ in a good solvent. Thus, the higher the molecular weight of the polymer, the lower the overlap concentration. Estimates of C^* based on measured

$[\eta]$ values show that the overlap concentration typically lies in the range 0.1–10% (w/w). Low values occur for long polymers with stiff chains in a good solvent.

At higher concentrations of polymer, where solvent and polymer are present in approximately the same amount on a weight basis, we encounter yet another concentration regime. In these concentrated solutions, the picture of a loose polymer network is no longer relevant. Instead, the system consists of highly entangled chains, with properties similar to those of a polymer melt. This concentration regime is relevant, for example, in describing the latex particles so important for modern paints. In this regime, the polymer–polymer interaction also affects the coil conformation, which tends to become more compact at higher concentrations in good solvents, so that one finds $R_g \propto N_p^{0.5}$ irrespective of solvent quality in this concentration regime.

7.2.2 The Thermodynamic Properties of the Semidilute Regime Is Well Described by the Flory–Huggins Theory

Mixing polymer and solvent produces two types of entropy effects. As with any binary mixture, entropy increases from mixing two different molecular species. However, few polymer molecules are present on a molar basis, and this entropy contribution is small. The more important effect is that the entropy of the polymer configuration increases as more space is available in the solution. Having small solvent molecules as neighbors allows for many more chain configurations than being surrounded by other chains.

By accounting for the connectedness between molecules, Paul Flory and Maurice Huggins in 1953 independently succeeded in generalizing the regular solution theory to semidilute polymer solutions. Using a lattice model, as in Figure 1.2, and assuming a random distribution of the monomer units on the lattice sites, with the constraint that the polymer chains remain intact, they were able to calculate the entropy of mixing, $\mathbf{S}_{\text{mix}}$. It adopts the deceptively simple form

$$\mathbf{S}_{\text{mix}} = \boldsymbol{R}\left(n_s \ln \phi_s + n_p \ln \phi_p\right) \tag{7.2.4}$$

where s and p denote solvent and polymer, respectively. The number of moles of species i is denoted by n_i, and ϕ_i represents the volume fraction they occupy. A solvent molecule and a monomer unit are assumed to occupy the same volume so that

$$\phi_s = \frac{n_s}{n_s + n_p N_p};\ \phi_p = \frac{n_p N_p}{n_s + n_p N_p} \tag{7.2.5}$$

It is important to realize that although Eq. (7.2.4) is formally closely analogous to the ideal entropy of mixing Eq. (1.1.8), the main contribution to $\mathbf{S}_{\text{mix}}$ comes from the change in configurational entropy of the polymer chain, rather than from the mixing of molecular units.

We can derive the interaction contribution to the free energy of mixing exactly as we did for the regular solution theory in Section 1.1. The only difference is that the number of nearest neighbors is reduced by two because the monomer units in the chain are covalently linked to two other monomers. The resulting expression for the free energy of mixing is

$$\Delta\mathbf{G}_{\text{mix}} = \boldsymbol{RT}\left(n_s \ln \phi_s + n_p \ln \phi_p\right) + \left(n_s + N_p n_p\right) w\, \phi_s \phi_p \tag{7.2.6}$$

in close analogy with Eq. (1.1.18) for a regular solution. A notable feature of $\Delta\mathbf{G}_{\text{mix}}$ is that for a given value of the polymer volume fraction ϕ_p, it varies only slightly with the degree of polymerization N_p. It follows from Eqs. (7.2.5) and (7.2.6) that only the small term, $n_p \ln \phi_p$, changes with N_p at constant ϕ_p and $N_p n_p$.

Physically, therefore, the relevant quantity in the solution is the mesh size in the network of chains, and this size is determined solely by the polymer volume fraction. Thus, in practice, it is very difficult to determine polymer molecular weights in the semidilute regime by measuring a thermodynamic quantity, while a number of transport properties are sensitive to the size of the molecules.

We can obtain the chemical potential of the solvent by differentiating Eq. (7.2.6).

$$\mu_s = \mu_s^\theta + \boldsymbol{RT}\ln\left(1 - \phi_p\right) + \boldsymbol{RT}\left(\phi_p - \frac{\phi_p}{N_p}\right) + w\phi_p^2 \tag{7.2.7}$$

When the volume fraction of polymer is small, we can write

$$\ln\left(1 - \phi_p\right) \approx -\phi_p - \left(\frac{1}{2}\right)\phi_p^2 \tag{7.2.8}$$

so that the chemical potential μ_s equals

$$\mu_s = \mu_s^\theta - \boldsymbol{RT}\left[\frac{\phi_p}{N_p} - \left(\chi - \frac{1}{2}\right)\phi_p^2\right] \tag{7.2.9}$$

Here $\chi \equiv w/\boldsymbol{RT}$ as in Section 7.1. The term that is linear in ϕ_p vanishes for long polymers (large N_p) while the term quadratic in ϕ_p becomes positive when $w > 0.5$ **RT**, or $\chi > 0.5$. Then we have a bad solvent, and in the limit $N_p \to \infty$, the polymer is insoluble since the chemical potential μ_s increases with increasing polymer concentration. For $\chi < 0.5$, the polymer is soluble over the whole semidilute regime, regardless of its length, and we have a good solvent. In the regular solution theory of a binary mixture, demixing occurs when $\chi > 2$ so that polymers are intrinsically less soluble than monomers. Because of the internal entropy of the chain, however, a sizable positive value of the interaction parameter w is required to achieve phase separation.

When using the Flory–Huggins theory and the free energy expression (Eq. 7.2.6), we should remember that from a molecular point of view, the theory is based on the same simplified picture as the regular solution theory. The theory is conceptually appealing, but inaccuracies can arise when it comes to quantitative predictions.

7.2.3 In a Semidilute or Concentrated Solution, Polymer Diffusion Can Occur Through Reptation

The Brownian motion of the solvent molecules is only moderately affected by the presence of the polymer molecules in the semidilute regime. Their self-diffusion coefficient is reduced by an obstruction factor approximately proportional to the polymer volume fraction. But even though there is a rapid local molecular dynamics, the diffusion of a polymer molecule is strongly inhibited. There are local changes of chain conformation, but they rarely lead to a translation of the center of mass of a polymer molecule. In Section 7.1.8, we analyzed how an intact single coil moved through a bare solvent. This mode of motion is completely quenched in a more concentrated system in which extensive entanglements exist between coils. So, which motional mode makes the dominant contributions to the translation motion under these circumstances?

One mode that is operative even at very high concentrations is polymer sliding along the position of its own chain, as illustrated in Figure 7.6.

Steric interference with this motion from other polymer molecules can occur only at the chain ends. The polymer motion must occur perfectly along the chain only at the highest polymer concentrations. In a semidilute system, the chain can make lateral excursions on the scale of the mesh

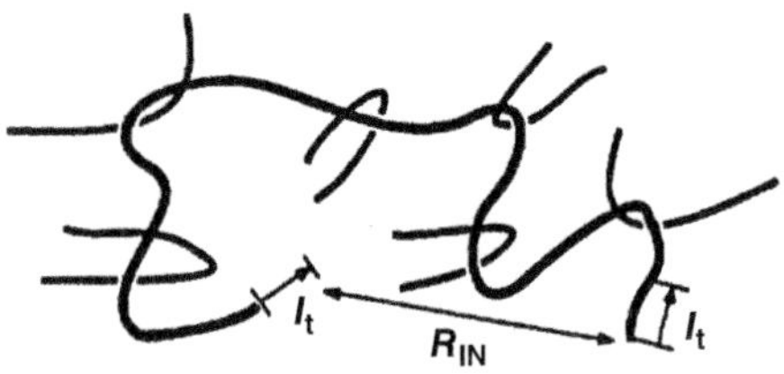

Figure 7.6 Illustration of reptation motion in a concentrated polymer solution (or polymer melt). Due to the constraints from other chains, shown as thin lines, the illustrated chain, thick line, can only slide along its own axis. A movement of length l_t causes on average a change in the center of gravity $\Delta r_M \propto l_t R_{1N}/L$.

size, and we can envisage the chain moving along in a snake-like motion. This movement forms the conceptual basis of the reptation model of diffusion and viscosity of entangled polymer solutions.

To arrive at the molecular weight dependence of the self-diffusion coefficient, we first note that the friction factor μ for sliding the polymer in the tube increases linearly with the length of the chain, so we can write

$$\mu = \mu_{\text{mon}} N_p \propto M \tag{7.2.10}$$

Furthermore, a unit displacement, l_t, along the tube does not cause a similar displacement of the center of mass, r_M of the polymer since the polymer is in a coil. If the two ends of the coil are located close together in space, moving material from one end to the other hardly affects the location of the center of mass, while if they are located on opposite sides of the coil, there is a substantial translational component to the motion. The average mean square displacement of r_M is related to the average end-to-end distance through

$$\langle \Delta r_M^2 \rangle = l_t^2 \frac{\langle R_{1N}^2 \rangle}{L^2} \propto \frac{M^{2\alpha}}{M^2} \propto M^{-2(1-\alpha)} \tag{7.2.11}$$

which can be derived using the random walk model. Combining Eqs. (1.7.5) and (7.2.11), we find that in this reptation model the self-diffusion coefficient is

$$D \propto M^{-2} \tag{7.2.12}$$

since α tends toward 0.5 for concentrated polymer systems. Comparing Eq. (7.2.12) with Eq. (7.1.23), we see that the diffusion coefficient depends much more strongly on molecular size in concentrated systems than in dilute ones.

Polymer self-diffusion is strongly concentration-dependent. In Eq. (7.1.23), we gave a value relevant for infinite dilution. As concentration increases, other polymer coils initially exclude volume, resulting in a decrease of D linear with concentration in analogy with Eq. (7.1.24) for the viscosity. As the solution approaches the overlap concentration, the concentration dependence increases, and in the semidilute regime, we find a relation

$$D \propto \phi_p^{-\beta} \tag{7.2.13}$$

where the exponent β has been found to be 1.75 for a good solvent and 3 for a theta solvent. At high volume fractions, all systems approach theta conditions.

7.2.4 Polymer Solutions Show a Wide Range of Rheological Properties

One widespread use of polymers is to control the rheological properties of a solution or colloidal dispersion. Flow properties are crucial in food processing (jam, ice cream, sauces, etc.), as well as in paints and in diverse medical applications. Rheological properties are essential also from a fundamental point of view because they can be used to characterize a system.

In general, rheology deals with how a system responds to a mechanical perturbation in terms of elastic deformations and viscous flow. Normally, we associate the elastic response with solids and the viscous response with liquids, although both systems show both types of behavior, albeit on very different timescales. The typical colloidal system shows a behavior intermediate between the solid and the simple liquid, and not surprisingly, we find in these systems characteristic elastic and viscous behaviors. In a simple rheological experiment, the sample is put between two parallel walls at a separation y, as shown in Figure 7.7. One of the walls is stationary, while the other can be displaced. Displacing this wall a distance x will generate a shear force F on the stationary wall. This force is proportional to the area A of the walls, and we can define a shear stress σ as

$$\sigma = \frac{F}{\text{area}} \tag{7.2.14}$$

For a perfect elastic solid, Hooke's law applies, and the shear stress is proportional to the displacement x divided by the thickness y, that is, the shear strain γ

$$\gamma = \frac{x}{y} = \frac{\mathrm{d}x}{\mathrm{d}y} \tag{7.2.15}$$

where in the last equality we have considered both x and y to be infinitesimal quantities. For an ideal (Hookean) solid, we can calculate the shear strain from macroscopic distances x and y because it is constant through the sample, but in general, we will have to use the differential form. For a Hookean solid

$$\sigma = G_0 \gamma \tag{7.2.16}$$

where G_0 is called the static shear modulus.

On the other hand, if the sample is a simple liquid, a displacement x will induce only a transient force on the stationary plate. A shear stress constant in time is observed only if the second plate is moved at a constant rate, $\mathrm{d}x/\mathrm{d}t$, which implies a constant shear rate

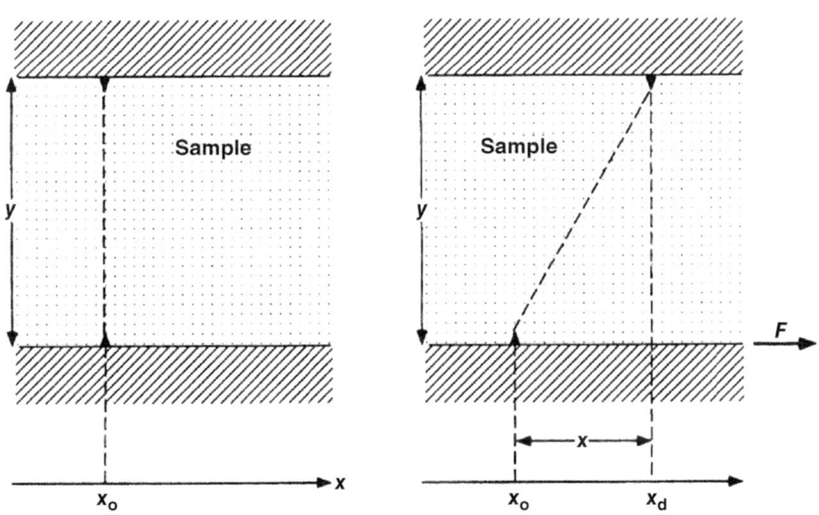

Figure 7.7 Schematic representation of a shear cell. Two parallel walls are separated by a sample of thickness y. When the upper surface is displaced a distance x from x_0 to x_d, a force F is generated on the lower surface.

$$\dot{\gamma} = \frac{\mathrm{d}}{\mathrm{d}t}\left(\frac{\mathrm{d}x}{\mathrm{d}y}\right) \tag{7.2.17}$$

in the sample. For simple fluids, a linear relation exists between the shear stress σ (i.e., the force per area on the stationary surface) and the shear rate $\dot{\gamma}$ (i.e., the rate at which the other surface is moved) so that

$$\sigma = \eta\dot{\gamma} \tag{7.2.18}$$

Here η stands for the viscosity of the system. Liquids that obey this linear relation over a large range of shear rates $\dot{\gamma}$ are called Newtonian.

Concentrated polymer solutions typically show non-Newtonian behavior under high shear rates, but it is possible to study the Newtonian range at low shear rates to obtain the (Newtonian) viscosity. In the semidilute regime, a polymer coil must constantly disentangle (and entangle) in a velocity gradient to sustain a different velocity from its neighboring coils. In a concentrated system, this process becomes more sluggish as the molecular weight increases, as illustrated in Figure 7.8. A straightforward application of the reptation theory yields an exponent of 3.0, which is in reasonable, but not perfect, agreement with the experimental observation. In this theoretical description, the crucial molecular factor is the time it takes for a polymer coil to reptate completely out of its initial tube.

In the Newtonian regime, the viscosity alone determines the rheological behavior. However, the most relevant rheological properties in applications of polymer solutions and colloidal dispersions relate to their non-Newtonian behavior. To describe these properties, we need a conceptual framework that brings us continuously from the ideal Hookean solid to the Newtonian liquid.

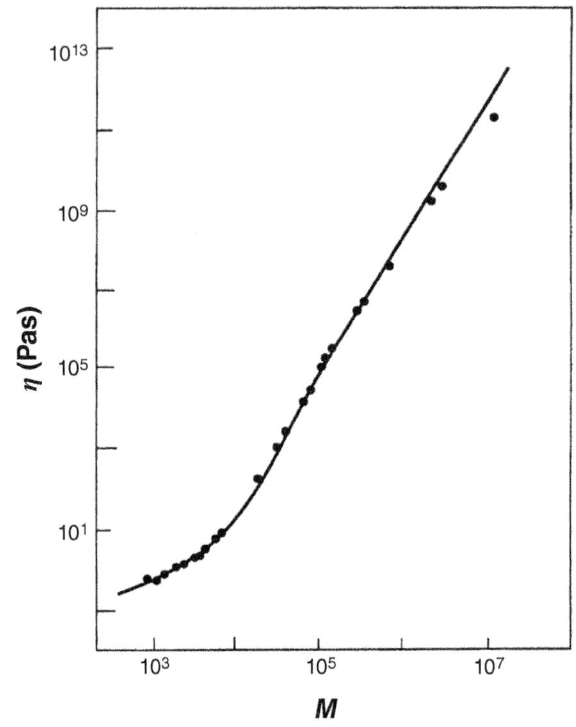

Figure 7.8 The molecular weight dependence of the viscosity of a polybutadiene melt at 25 °C. Above $M = 10^4$, the viscosity depends on the molecular weight to the power 3.4, as shown by the straight line. *Source:* Colby et al., 1987/with permission of American Chemical Society.

With the setup of Figure 7.7, we can consider an experiment in which the upper wall is moved periodically with a frequency ω and an amplitude x_0. Under these conditions, the shear strain γ and the shear rate $\dot{\gamma}$ will vary as

$$\gamma = \frac{x_0}{y}\sin(\omega t) = \gamma_0 \sin(\omega t) \tag{7.2.19a}$$

$$\dot{\gamma} = \frac{x_0}{y}\omega\cos(\omega t) = \gamma_0\,\omega\cos(\omega t) \tag{7.2.19b}$$

The oscillatory motion will give rise to a time-varying shear stress at the stationary surface with the same frequency. In general, however, a phase shift δ takes place so that

$$\sigma = \sigma_0 \sin(\omega t - \delta) \tag{7.2.20}$$

For sufficiently small amplitudes γ_0, the response in shear stress will depend linearly on γ_0, and we can write

$$\sigma = \gamma_0\{G'(\omega)\sin(\omega t) + G''(\omega)\cos(\omega t)\} \tag{7.2.21}$$

Here

$$G' = \left(\frac{\sigma_0}{\gamma_0}\right)\cos\delta \tag{7.2.22a}$$

and

$$G'' = \left(\frac{\sigma_0}{\gamma_0}\right)\sin\delta \tag{7.2.22b}$$

are called the storage modulus and the loss modulus (in N/m^2), respectively. These properties depend on the frequency ω. For an ideal solid, the storage modulus is $G'(\omega) = G'(0)$ and the loss modulus is $G''(\omega) = 0$, while for a Newtonian liquid, $G'(\omega) = 0$ and $G''(\omega) = \eta\omega$.

By measuring the frequency dependence of the loss and the storage moduli, we obtain information on the timescales for mechanical relaxation processes in the sample. The method is of general use in colloid science. Here we use the semidilute polymer solution to illustrate some aspects of the rich rheological behavior found in colloidal systems.

For polymers in the semidilute solution, shear will create constant disentanglements/entanglements to accommodate a velocity gradient. The perturbation from the shear will cause a change in the local structure of the solution, which in turn will cause deviations from the linear Newtonian behavior. To somewhat decrease the disentanglement, the polymer coils will, on average, orient and deform to extend along the lines of flow. This facilitates the flow and leads to a shear thinning and the loss modulus $G''(\omega)$ has a weaker than linear dependence on the frequency ω. For example, in a paint there should be facile liquid flow under the action of the brush while, once applied, the paint should remain on a wall and not form drops that could flow under the action of gravity.

Shear thickening is qualitatively opposite to shear thinning. For polymers in the semidilute regime, it appears that the effect only appears when there is some chemical heterogeneity along the chain. As the coils deform under shear, transient attractive interactions can appear between such heterogeneous units. Such contacts can then hinder the development of velocity gradients.

Figure 7.9 compares the rheological properties of a polymer solution and a polymer gel. At low frequencies, the storage modulus remains substantial for the gel but not for the solution.

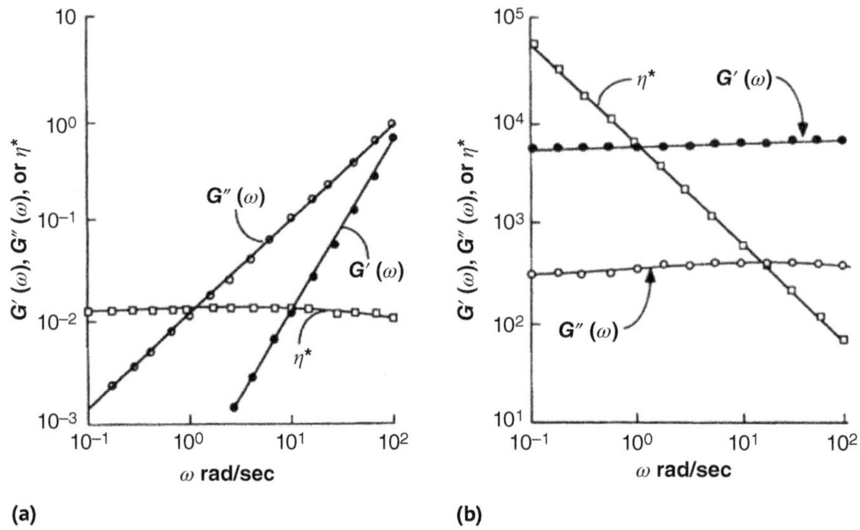

Figure 7.9 Rheological parameters: □, dynamic viscosity $\eta^* = \left(G'^2 + G''^2\right)/\omega$ (Ns/m²); ●, storage modulus $G'(\omega)$; and ○, loss modulus $G''(\omega)$ for (a) a polymer dextrane aqueous solution and (b) an aqueous agarose gel. Note the nearly perfect Newtonian behavior of the solution and the strongly elastic properties of the gel. *Source*: Morris and Ross-Murphy, 1981.

Furthermore, the polymer solution has a Newtonian viscosity, while gel viscosity decreases dramatically with increasing frequency. Some systems show viscoelastic properties intermediate between a gel and a viscous solution. A bread dough responds elastically to rapid deformations, but responds with viscous changes under sustained shear. These examples indicate how we can use rheological measurement as a sensitive indicator for association phenomena in polymer systems.

Another example of non-Newtonian behavior is found for systems where the polymer chains crosslink reversibly, for example, through the formation of a chemical complex. On time scales short relative to the average lifetime of the crosslink, the system behaves as a connected rubber and responds elastically, while for longer times, or low frequencies, the system flows as a liquid. This is the principle behind the properties of silly putty, which can be made from polyvinyl alcohol and water by mixing in some borate to make labile crosslinks.

See Cosgrove (2010, Chapter 7), Doi (2013, Chapter 2), Israelachvili (2011, Chapter 16), Kronberg et al. (2014, Chapter 9), and Lyklema (2005a, Vol. 4, Chapter 6) for additional discussion on Polymer Solutions.

7.3 Polymers May Associate to Form a Variety of Structures

So far, we have focused on polymer properties that can be associated with the intrinsic structures of random coils, rods, and compact globules. Interaction between polymer molecules can also promote cooperative processes that lead to self-assembly of polymeric systems.

7.3.1 Block Copolymers Show the Same Self-assembly Properties as Surfactants

A diblock copolymer composed of monomer units *B* and *C* has the general formula

$$\underbrace{B\cdots B}_{m} - \underbrace{C\cdots C}_{n}$$

that is, B_mC_n. In a solvent, the conformation of a single isolated chain depends strongly on the solvent quality with respect to the two types of monomer units. If it is a good solvent for both units, the conformation will resemble that of a homopolymer in a good solvent. If it is a bad solvent with respect to both units, it will not dissolve. However, when the solvent is good relative to only one of the monomers, say *B*, the molecule displays amphiphilicity. In the single molecule, the *C* part of the chain is in a bad solvent, and the *C* chain adopts a compact structure to which a random coil *B* chain is attached (see Figure 7.10). A strong attraction exists between the compact *C* parts. Indeed, when the concentration is increased sufficiently, an association that is closely analogous to the micelle self-assembly process results. In practice, one often has a distribution in the degrees of polymerization *m* and *n* with the consequence that the CMC is less well-defined (see Section 4.4.2 for an illustration). In addition, the aggregation typically occurs at very low concentrations and is thus difficult to detect.

The self-assembly process starts with the attraction between *C* units, while the stop occurs because *B* coils of different molecules overlap more and more as the aggregation number increases. At some point, the effective repulsion between the *B* units balances the attraction between the *C* parts. The geometry of the resulting aggregate depends on the relative length of the two blocks as well as on the persistence length of the *B* chain and the goodness of the solvent for the *B* chain.

The ratio n/m between *C* and *B* units plays a similar role as the surfactant number N_s of amphiphiles, and by varying n/m, one can observe the same sequence of preferred structures as when varying N_s of surfactants as illustrated in Figure 7.10. This will be discussed further in Chapter 10.

Poloxamers are *B–C–B* block copolymers with *B* as ethylene oxide and *C* as propylene oxide, which have a wide use in aqueous systems. The extra methyl group on the propylene oxide makes it sufficiently hydrophobic to experience water as a bad solvent. In addition, the ethylene oxide–water interaction is temperature-dependent (section 5.7), so that the aggregation properties of the system can be controlled by temperature as well as by the values of the segment lengths *m* and *n*. Poloxamers will be discussed further in later chapters.

7.3.2 Polymers with Amphiphilic Monomer Units Often Form Ordered Helix Structures

When discussing solvent quality, we have considered the monomer as a single unit with a given interaction parameter relative to the solvent. For molecularly complex monomers, this is clearly a simplification, and one group of the monomer may interact favorably with the solvent while the rest of the molecule dislikes the solvent, as in an amphiphilic monomer unit. How can we build a structure that solves the dilemma of having solvent exposure for the solvophilic parts while protecting the solvophobic parts of the monomer from solvent contact?

We discussed one possibility for the globular proteins, but this solution to the problem really requires a distribution of different monomer units. In the case of homopolymers, but also for some

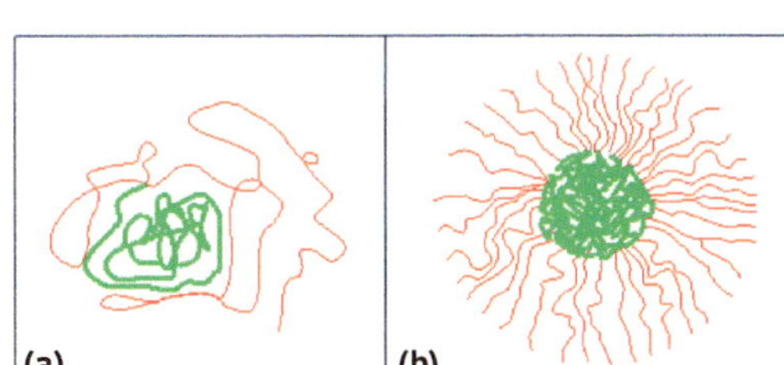

Figure 7.10 (a) Single block copolymer *BC* in a solvent that is bad for *C* but good for *B*. The *C* part (green) adopts a compact state, while the *B* part (red) is in a coil conformation. (b) A micelle formed by the self-assembly of *BC* block–copolymer molecules. Crowding caused by the aggregation results in the stretching of the *B* chains.

heteropolymers, an alternative is the formation of a helix. There are rather strict conditions to obtain optimal packing in a helical structure. In some cases, like the α-helix in a protein, it can be realized by a single strand, while for DNA, there is a double helix, and in collagen, the native form of gelatin, a triple helix is preferred. In a protein, the α-helix is formed to shield the polar amide backbone from an apolar environment. It is the dominant protein structure in the part of an integral membrane protein that resides within the lipid bilayer, but it also occurs in globular proteins.

The force that drives the formation of the double helix of DNA is the hydrophobic interaction between base pairs. The primary force acting against the dimerization is the electrostatic free energy due to the charged phosphate groups. Association also involves a loss of solvation energy of the polar hydroxy and nitrogens of the purine and pyrimidine rings. However, to a large extent, the specific hydrogen bond matching in the helix reduces this free energy cost. It is a widespread misconception that the hydrogen bonding, rather than the hydrophobic interaction, provides the main driving force for helix formation in DNA.

Agarose, widely used in protein and nucleic acid electrophoresis, is a polysaccharide obtained from a red algae source. It is partly charged and associates into double helices by hydrophobic attraction, not hydrogen bonding. The carrageenans used as food additives in, for example, ice cream and yogurts are also polysaccharides obtained from red algae. They are charged by sulphate groups providing the polar parts of the monomer units, while the sugar part is partly hydrophobic, resulting in a double-helical structure that often appears in an aqueous environment.

Helix formation is a cooperative process, which is signaled by the occurrence of the transition from a random coil to a helix over a narrow temperature region. The helix formation can be seen as a nucleation and growth process, where the cooperativity is found in the nucleation step. The molecular characteristics of this step depend on the type of helix to be formed. For a single-stranded helix like an α-helix, the nucleation is largely completed once the chain has adopted a conformation with a single turn. This implies only a weak cooperativity and a broad transition from coil to helix for an isolated chain. In practice, most α-helices interact with adjacent structures, giving further stability to the helix and making the association process more cooperative.

The formation of a DNA double helix is more complex. The first step is the encounter between two single-stranded molecules. These are highly charged and experience an electrostatic repulsion, decreasing the probability for direct molecular encounter between monomer units on separate strands. However, this electrostatic interaction provides a bias for encounters between residues at the chain ends. Once two such monomers have made molecular contact, initial formation of a helix requires appropriate orientation of the respective chains. When approximately one turn has been formed, the helix formation enters the growth phase. Since the initial nucleation step involves more molecular states with a higher free energy, the cooperativity is larger than for the formation of a single helix structure. The cooperativity ensures that a transition from helix to random coil occurs over a narrow temperature range, as illustrated in Figure 7.11, which also demonstrates that the transition involves a strong electrostatic component, as shown in panel C.

Collagen, one of the most common proteins in mammals by weight, forms a triple helix already during biosynthesis. The triple helix has a structure-forming role in cartilage, skin, and some organs. Extracted collagen is the main component in gelatine, which is used to control the rheological properties of food in the kitchen. On heating an aqueous suspension, the collagen adopts a random coil conformation. When the system is cooled, triple helices are reformed in a nucleation and growth process. The formation of the initial helical nucleus is a random process, and the protein has, in addition, typically a molecular weight polydispersity. The consequence is that there can appear a mismatch between the length of the different polymer strands in the triple helix, resulting in loose ends that can occasionally join some other triple helix. The net result is a

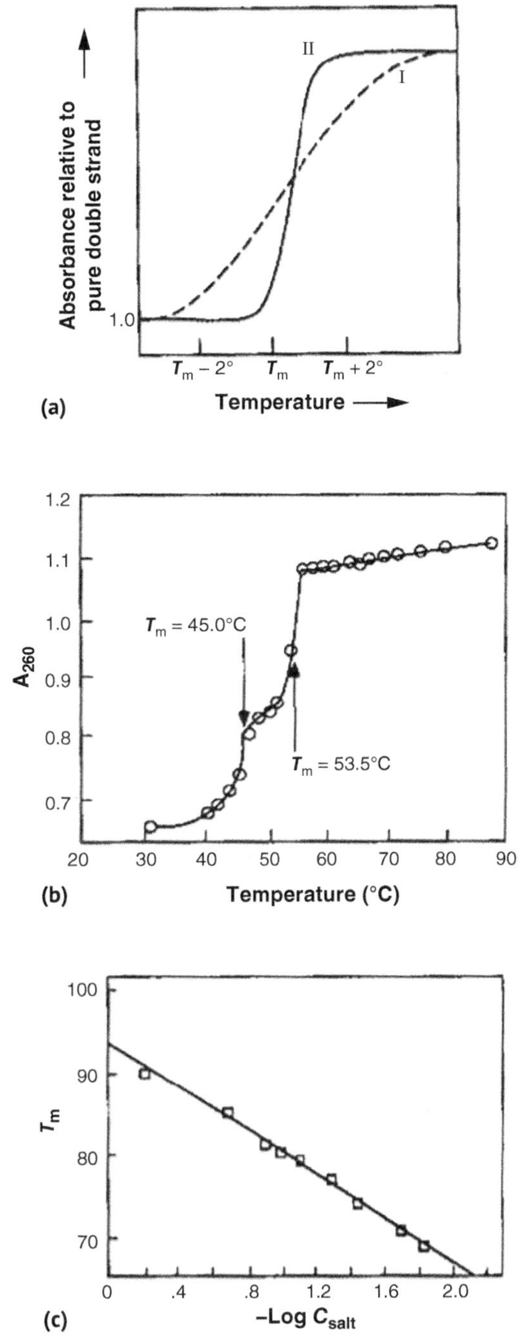

Figure 7.11 Helix–coil transitions of bacterial DNA in a KCl solution. (a) Schematic representation of the transition for a mixture of two polynucleotides with complementary monomers (solid line) and natural DNA (dashed line). The transition can be monitored by measuring light absorbance at 260 nm. It is complete within 2 °C for the pure system, while in the native state, it occurs over 5 °C. (b) Two-step melting for DNA with uracil (U) and adenine (A) in a 2:1 ratio. An intact double helix exists below 45 °C, and above 54 °C, we find only single strands. Between the two transitions is a A–U double helix with U single strands. *Source*: Stevens and Felsenfeld, 1964/John Wiley & Sons. (c) The salt dependence of the helix–coil transition. Varying the KCl concentration from 0.7 to 0.01 M changes the transition point by 20 °C. *Source*: Adapted from Schildkraut and Lifson, 1965.

nonequilibrium disordered helix network covering the entire system. By increasing the temperature, the system can be brought back to the random coil solution, but there is a clear hysteresis in the helix formation–helix melting processes, indicating the nonequilibrium character of the helix network. A second indication of this property is that if one lets a seemingly stable system rest for a long time at constant temperature, there can be a separation in a denser network and excess solvent. This process is called syneresis.

Helices formed by polymers of a biological origin are always chiral. When two such helices interact, the chirality puts constraints on the arrangement in the axial direction. A common manifestation of this effect is that on attraction, the helices tend to arrange into bundles. In a next stage, these chiral bundles in turn aggregate into bundles in a hierarchical structure as, for example, four helix bundles of α-helices and actin–myosine helices in a muscle fiber.

When a polymer goes from a random coil conformation to one containing helices, there is a loss of conformational freedom, and the polymer potentially becomes less soluble. When the polymer contains charged groups, the electrostatic interactions promote solubility, but for uncharged polymers, the solubility is decreased by the formation of helices.

7.3.3 Polymers Form Gels Through Chemical Crosslinking and by Self-association

Above the overlap concentration C^*, the polymer chains are entangled, and the mesh size of the entanglements changes with concentration. However, entire chains can move relative to one another. For example, even if the dynamics of chain diffusion are slow, the system behaves as a liquid. In many applications, it is desirable to create a system in which the local dynamics are rapid, yet on a larger scale, the system is stationary. This goal can be achieved by creating a polymer gel.

In a chemically crosslinked gel, overlapping polymer coils are covalently connected at a few positions. As the crosslinking progresses, larger and larger polymers are created until macroscopic molecular weights are reached, in which single molecules extend through the whole macroscopic sample. Once the crosslinked polymer has been formed, the solvent can be evaporated, and the polymer will retain a memory of its macroscopic shape that reappears when it is redissolved. The polymer and the solvent form a gel (see Chapter 10 for a general discussion of the term) in which both elastic and viscous properties remain at low frequencies, that is, $G'(\omega) \neq 0$ for small ω.

Crosslinked polymer gels have many practical uses. One is in the formation of filter membranes for purification processes. The active part of a reverse osmosis system for water purification is a crosslinked polyelectrolyte gel membrane. By controlling the pore size, only the solvent and low molecular weight compounds can be made to pass through the membrane. Disposable diapers contain small granules of the polyelectrolyte sodium polyacrylate. These granules can swell extensively (up to 1000 times) to incorporate an aqueous solution by electrostatic double-layer forces, so the degree of swelling depends on the electrolyte content in the water.

A number of polymers can form gels without chemical crosslinking, so-called physical gels. In these cases, the gelling of a polymer solution can be induced by changes in temperature or salt concentration, or by adding a specific chemical that can adsorb on the polymer. The most widely known gel former is gelatin, a protein obtained from collagen, but a number of polysaccharides also form gels. These include agarose, widely used in chromatography, pectin, the basic gelling component of natural fruit jelly, and carrageenan, a common food additive obtained from seaweed.

Many of these polymers can form helices, and the molecular event that triggers gel formation is a coil–helix transition. As we remarked in Section 7.3.2, helical polymers have intrinsically low solubility because they possess little configurational entropy. Thus, we can expect precipitation at the formation of helices. However, the size and rigidity of the helices, and possibly also chemical heterogeneity, create a possibility that an ordered equilibrium packing of helices is prevented, and kinetically trapped configurations appear. In one such scenario for polymers that form double or triple helices, a given strand is involved in several helical sections of limited length with different partners, leading to a branched network structure. However, stiff rods of helices or bundles of helices lacking such branches can also be trapped in nonequilibrium arrangements by weaker local interactions, resulting in gel-like properties of the system.

7.3.4 Oppositely Charged Polyions Associate to Form Coacervates with Liquid-like Properties

When one, in an aqueous solution, mixes polyelectrolytes of opposite sign, the respective polyions attract strongly electrostatically. One possible outcome is that the association results in a solid precipitate. However, it is also common that there is instead a formation of a concentrated separate liquid phase named coacervate by Bungenberg de Jong, who was the first to study such systems systematically in the 1920s. In practice, the mixing of the two components easily leads to the formation of metastable coacervate droplets dispersed in the aqueous solution. Formally, such a state represents an emulsion discussed in Chapter 11.

To understand the mechanisms leading to the liquid-like behavior of the associate, we recall the results from Section 5.1.6 dealing with the interaction between two planar surfaces with opposite charge, but with different values of the magnitude of the charge densities. Due to the lack of charge matching, the electrostatic attraction changes to a repulsion at short range, leaving a liquid film separating the surfaces. Similarly, when there is no structural matching between the polyions, there appears a repulsion at short range between the oppositely charged chains. The result is that in the aggregate, there is some solvent, typically water, and also the occasional small counterion with respect to the polyion with the highest charge density. In the same way as electrostatic repulsion is decreased in strength by the addition of salt, so is the strength of an electrostatic attraction also reduced by the addition of electrolyte. Thus, coacervates tend to dissolve at higher electrolyte concentrations.

During the association process, polyions successively add to a separated nucleus. Typically, one of the polyions is in excess relative to the other and then separate aggregates repel electrostatically. As will be discussed in detail in Chapter 8, this repulsion might be sufficiently strong to kinetically prevent the aggregates to associate and grow indefinitely in size to a separate phase.

The term coacervate is sometimes also used for the case of a polymer or protein separating from a solution in a bad solvent in a situation where a solid precipitate is not formed. Within the framework of this book, we consider this as one route to the formation of a colloidal particle, as discussed in Chapter 9.

See Berg (2010, Chapter 3), Cosgrove (2010, Chapters 8 and 12), Doi (2013, Chapters 3 and 9), Israelachvili (2011, Chapter 16), and Kronberg et al. (2014, Chapters 11 and 19) for additional discussion on Polymer Association.

7.4 Polymers at Surfaces Play an Important Role in Colloidal Systems

7.4.1 Polymers Can Be Attached to a Surface by Spontaneous Adsorption or by Grafting

A molecule in solution will adsorb to a surface, provided the adsorption energy is substantially larger than $k_B\mathbf{T}$. An adsorbing polymer molecule typically makes many contacts with the surface, as illustrated in Figure 7.12b. Even if the energy per monomer contact is relatively small, the number of contacts results in a high adsorption energy. To some extent, this high adsorption energy is compensated for by a loss of configurational entropy, but if the monomer has some affinity for a surface, the polymer usually adsorbs strongly and in practice often irreversibly.

We can also promote polymer adsorption through more specific effects. A block copolymer with one block experiencing bad solvent conditions will be very surface active and will have a strong tendency for adsorption (Figure 7.12a). A polyion will adsorb strongly on an oppositely charged surface. Finally, the polymer chain can be grafted to a surface by a covalent bond (Figure 7.12a).

The chain conformation at the surface depends on the forces that bring the polymer down to the surface. For a grafted polymer in a good solvent, in which the monomer has no affinity for the surface, the chain will adopt a coil conformation, stretching out into the solution (see Figure 7.12). Neighboring chains will repel one another, and this repulsion will influence the coil conformation when the grafting density is higher than R_g^{-2}, the inverse second power of the radius of gyration. For this dense so-called polymer brush, the monomer concentration c_M decays parabolically out from the surface at $z = z_0$

$$c_M(z) = c_M(z_0)\left\{1 - \left(\frac{z - z_0}{z_{\max} - z_0}\right)^2\right\}; z < z_{\max} \tag{7.4.1}$$

Similarly, for a block copolymer B_mC_n, where only the C part adsorbs strongly on the surface, the conformation of the B coil depends on the surface density of polymer chains. In a series of polymers with constant n and increasing m, the coil–coil repulsion becomes more and more significant as m increases. For small m values, the polymer density is determined solely by the packing of the C chains on the surface and the B coil extends a distance $2R_g$ in both the parallel and perpendicular directions adopting a so-called mushroom conformation. When the C coils become so small that the surface density of polymer chains exceeds R_g^{-2}, the packing density at the surface will decrease, and the B coil will stretch away from the surface. When $m \gg n$, the chain conformation will be the same as in the dense polymer brush, and the distribution function of Eq. (7.4.1) will apply, assuming that the polymer still adsorbs on the surface.

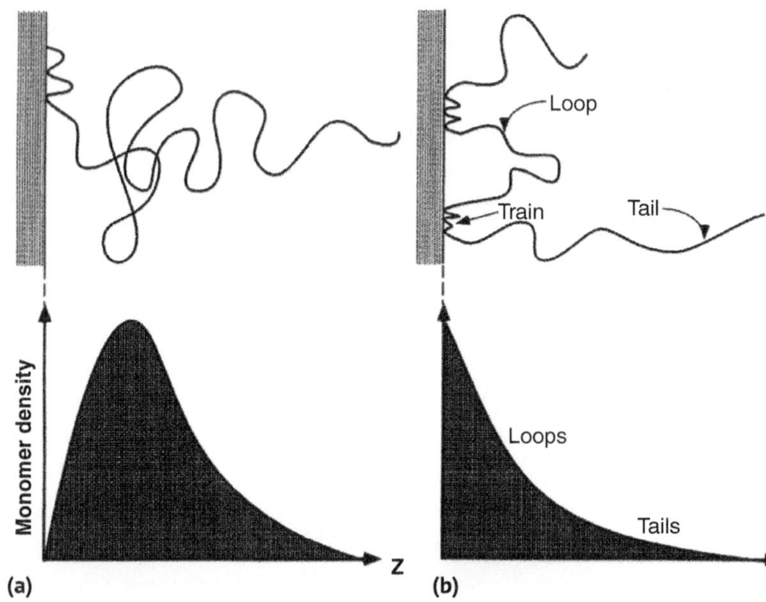

Figure 7.12 Polymer configurations at a surface. (a) Grafted chain or block copolymer with one highly adsorbing segment at a low surface coverage. The main chain is assumed to be in a good solvent with no affinity for the surface. Thus, it strives to avoid the surface. (b) Adsorbing a homopolymer in a good solvent, indicating a loop, two trains, and the two tails that extend farthest into the solution. *Source*: Israelachvili, 1992/with permission of Elsevier.

Electrostatic forces cause a polyion to adsorb on an oppositely charged surface. The polyion acts as a multivalent counterion to the surface, and entropy increases greatly when it replaces multiple small counterions at the surface and simultaneously releases its own counterions. Rinsing with pure water will never remove the polyion from the surface, but the entropy effects can be greatly reduced in a concentrated electrolyte solution, so that the polyion can readily come off the surface.

Because the adsorption and the conformation of the polyelectrolyte in solution are governed mainly by electrostatic effects, the surface conformation depends strongly on salt concentration in an aqueous medium. In pure water, adsorption is very strong, and the polyion adopts a conformation that provides an optimal match between polymer and surface charges. This process can lead to a chain that essentially lies flat on the surface. The surface density of chains will then be close to $-\sigma/Z_p e$, where σ stands for the surface charge density of the bare surface and $Z_p e$ the charge of the polyelectrolyte. When the intrinsic charge density is higher on the polyion than on the surface, the adsorption leads to a reversal of the effective surface charge. This can also be accomplished by an excess of polyelectrolyte in the solution.

As the salt concentration is increased, the effective strength of the electrostatic interactions is decreased. The polymer can gain configurational entropy by extending somewhat into the solution, and the ions of the electrolyte act as counterions. Increasing the salt concentration gradually loosens the electrostatic binding, and ultimately, the polymer reaches a state in which it no longer binds to the surface or in which more chemical forces determine the adsorption behavior.

A noncharged homopolymer that adsorbs from a good solvent onto a surface will not change its coil conformation dramatically. Typically, the thickness of the adsorbed layer, as well as the lateral extension of an individual chain, is of the same magnitude as the radius of gyration R_g. The chain conformation can be described in terms of two tails (i.e., the two ends of the chain sticking out from the surface), a number of trains (sequences of the chain that have contacts with the surface) and a number of loops in which the chain makes long excursions into the solution between two trains (see Figure 7.12b).

7.4.2 Kinetics Often Determines the Outcome of a Polymer Adsorption Process

Exposing a clean surface to a solution containing an adsorbing polymer creates a nonequilibrium situation. The kinetics of the adsorption is typically transport-controlled. In the unstirred boundary layer outside the surface, diffusion determines the transport to the surface. Owing to the size of the polymer molecules, the diffusion coefficients are so small that the adsorbed layer builds up on a timescale, which can be on the order of seconds and minutes, even for this diffusion-controlled process.

In practice, the distribution of polymer molecular weights usually plays a role. Thermodynamically, the longest chains have a higher affinity for the surface because they make more surface contacts per molecule. However, the shorter chains have a higher diffusion coefficient, and they will be the first to reach the surface. Thus, in the initial state, the shorter chains are preferentially concentrated at the surfaces, while at equilibrium, the longer chains are the ones that reside at the surface. To reach equilibrium, longer chains must replace the shorter ones. This process is slow because it requires desorption of the shorter chains. Depending on the circumstances, changes can be observed over hours and days, or the exchange may be so slow that for practical purposes, the initial state remains stable. The adsorption of proteins on surfaces exposed to blood serum is an important phenomenon in a medical context. The serum contains many different proteins, and one often finds that initially the smaller ones adsorb on the surface, while at longer exposures, the small proteins are replaced by larger ones. This is called the Vroman effect.

Because kinetics is so slow, the performance of the adsorption process becomes an issue of great practical importance. Not only the transport to the surface but also surface rearrangements can constitute very slow processes, so surface polymers may take a long time to respond to changes made in the solution. For this reason, polymer adsorption processes are delicate to study in the laboratory, and in practical applications, kinetic effects must be carefully watched and turned to an advantage.

7.4.3 Forces Between Surfaces Change Drastically When Polymers Adsorb

Surfaces covered with polymers will "see" one another through the outer part of the polymer coil. These coils will overlap substantially at surface–surface separations of order $2R_g$, but individual chains make excursions beyond the radius of gyration, and a surface–surface interaction arises at separations clearly larger than $2R_g$. In practice, separations in the range 10–100 nm are so large that the effect of van der Waals and double-layer forces between the bare surfaces is negligible, and polymer–polymer overlap provides the dominant contribution to the force.

In a good solvent, the polymer coil expands away from the surface to gain configurational entropy. When it encounters a coil emanating from the second surface, its allowed configurations are restricted and the free energy increases. This mechanism, illustrated in Figure 7.13, generates a repulsive force between the surfaces as discussed in Section 5.6. In older literature, this is called steric stabilization. The force is entropic in origin, and a better term is overlap repulsion. The dimensions of the polymer coil determine the range of the force, and configurations that extend far from the surface are particularly important in determining the onset of the repulsive force.

Figure 7.14 shows an experimentally measured force curve for mica surfaces covered with polyethylene oxide (PEO) in water at 20 °C, where water is a good solvent for PEO. The force is indeed

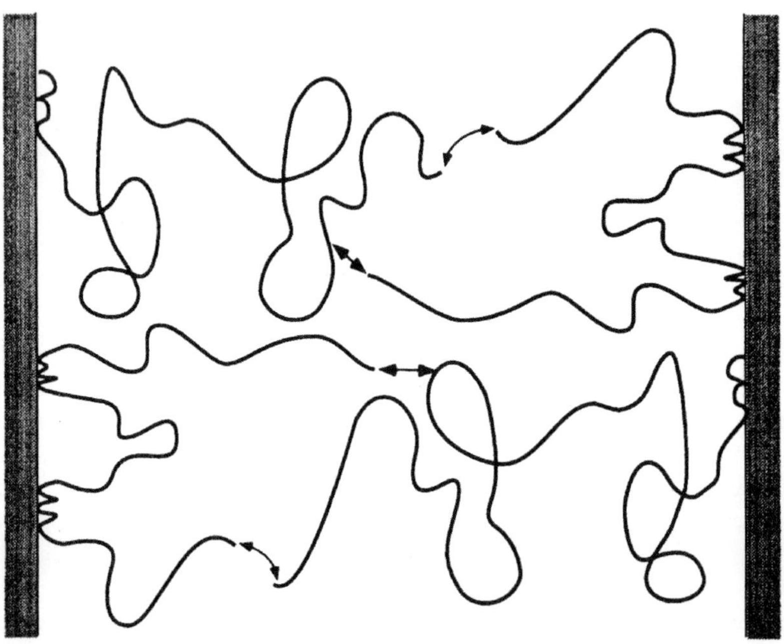

Figure 7.13 Schematic illustration of conformations of polymers on two opposing surfaces restricting each other, resulting in a decrease in entropy and normally in a repulsion between the surfaces.

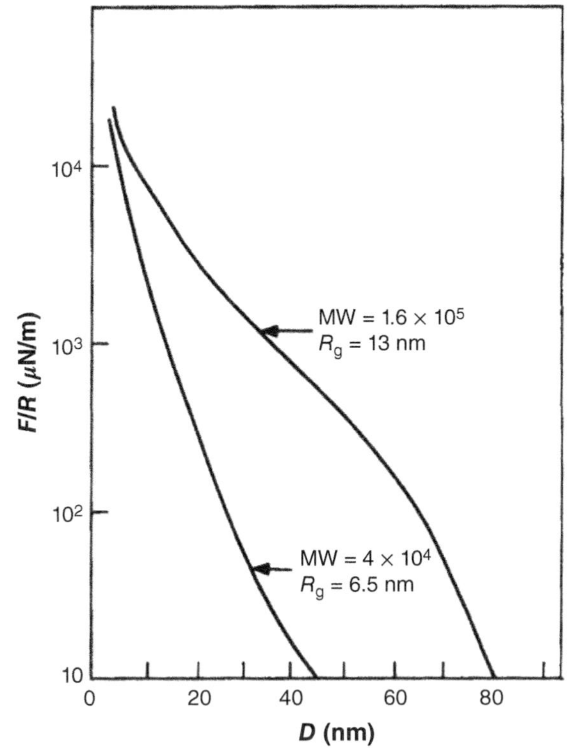

Figure 7.14 Force between the two curved mica surfaces with adsorbed polyethylene oxide in water, which is a good solvent at room temperature. A monotonic repulsion is observed, and the range of the force scales with the radius of gyration. Measurable forces are obtained at distances corresponding to several multiples of R_g showing the importance of the odd tails sticking far into the solution. *Source*: Adapted from Klein and Luckham, 1982.

repulsive, increasing with increasing molecular weight as R_g does. The force is strong enough to be appreciable out to several multiples of the radius of gyration, R_g.

In a bad solvent, that is, $\chi > 0.5$, the polymer wants to contract at the surface to avoid contact with the solvent. As a second surface approaches, a chain anchored at one surface can establish contact with a chain of the opposite surface. This leads to an increase in monomer–monomer contact as well as to an increased configurational entropy. These mechanisms add up to a strongly attractive force sensed at shorter separations than the repulsion in a good solvent for polymers of the same molecular weight (see Figure 7.15). Furthermore, the attraction turns at short range into a repulsion because the polymer chains start to squeeze each other.

Table 7.2 demonstrates the crucial role played by solvent quality. It shows that suspensions of colloidal particles covered by adsorbed polymers became kinetically unstable at a temperature very close to the theta temperature for the polymer–solvent pair.

A slightly different version of the polymer overlap interaction occurs when the surfaces are covered with a dense polymer brush. Figure 7.16 shows a series of experiments with surfaces covered by a short block copolymer of the general formula $C_m E_n$ (see Section 4.4), where m and n range from block copolymer values to values typical of nonionic surfactants. Shlomo Alexander and Pierre-Gilles de Gennes (1987) derived an expression for the force per unit area for a chain density S^{-2}

$$\frac{F}{\text{area}} = \text{const}\frac{k_B T}{S^3}\left\{\left(\frac{2L}{h}\right)^{\frac{9}{4}} - \left(\frac{h}{L}\right)^{3/4}\right\};\ h < 2L \tag{7.4.2}$$

which fits the data if the thickness L of the brush is treated as a parameter, as illustrated in Figure 7.16.

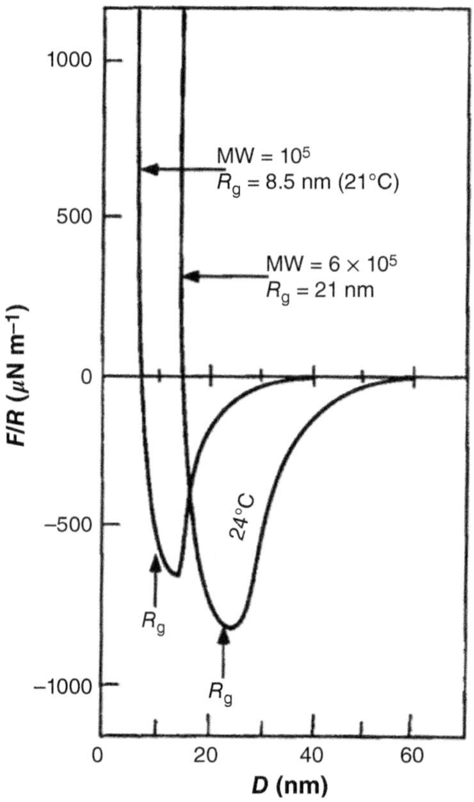

Figure 7.15 Force between curved surfaces with adsorbed polystyrene in cyclohexane at 24 °C, which is below the theta temperature. In the bad solvent, there is a strong attraction, with a maximum strength at a separation around R_g, and appreciable up to separations around $2R_g$. *Source*: Adapted from Klein, 1980.

Table 7.2 Comparison of theta temperature with critical flocculation temperature (CFT).

			Temperature (K)	
Stabilizer	**Molecular weight**	**Dispersion medium**	**CFT**	θ
Polyethylene oxide	10 000	0.39 M $MgSO_4$	318 ± 2	315 ± 3
Polyethylene oxide	1000 000	0.39 M $MgSO_4$	317 ± 2	315 ± 3
Polyacrylic acid	9800	0.2 M HCl	287 ± 2	287 ± 5
Polyacrylic acid	89 700	0.2 M HCl	281 ± 1	287 ± 5
Polyvinyl alcohol	26 000	0.2 M HCl	302 ± 3	300 ± 3
Polyvinyl alcohol	270 000	0.2 M HCl	312 ± 3	300 ± 3
Polyisobutylene	23 000	2-Methyl butane	325 ± 1	325 ± 2
Polyisobutylene	150 000	2-Methyl butane	325 + 1	325 + 2

If we create a system lacking enough polymers to saturate the surfaces, an attraction can occur even in a good solvent. In this case, a polymer coil can extend from one surface to the other and make molecular contact on both sides. One surface then provides the missing polymer to the other, resulting in an attractive interaction usually called *bridging*.

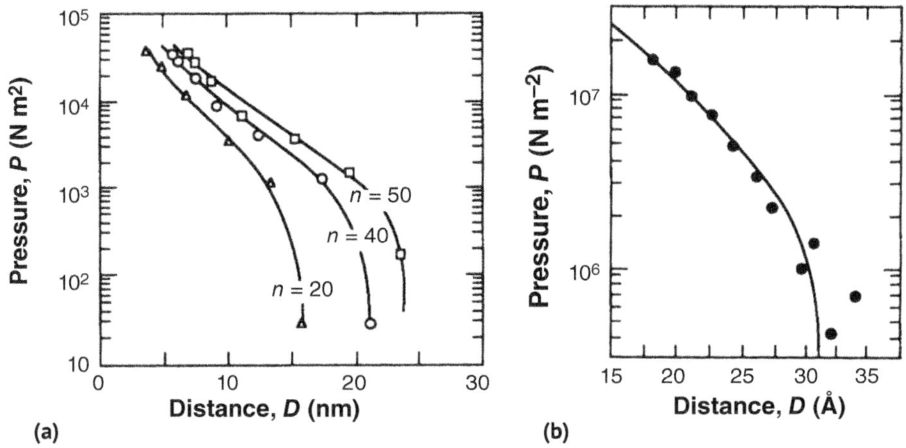

Figure 7.16 (a) Measured forces between adsorbed monolayers of the polyethylene oxide surfactant $C_{18}E_n$ in water. The solid lines are fits to Eq. (7.4.2) with overlapping head groups modeled as a polymer brush layers, using $L = 7.8$ nm for $n = 20$, $L = 10.5$ nm for $n = 40$, and and $L = 12$ nm for $n = 50$ (all fitted for $S = 13$ nm). (b) Measured forces (dots) between bilayers of $C_{12}E_4$ in water. Solid line fit to Eq (7.4.2) using $L = 1.6$ nm corresponding to 0.4 nm/EO group, $S = 0.93$ nm and $\mathbf{T} = 25$ °C. *Source*: Israelachvili and Wennerström, 1992/with permission of American Chemical Society.

A similar attractive mechanism can operate in a nonequilibrium state in which two surfaces have been pushed toward each other and left to equilibrate for some time. In the compressed state, the polymer coil from one side can adsorb on the opposing surface or become entangled in loops of the polymer adsorbed on the opposing surface. At this stage, any attempt to separate the surfaces on a timescale shorter than that required for disentangling the chains causes an attractive nonequilibrium force to appear. This force is one of many sources of hysteresis effects with polymer-covered surfaces.

7.4.4 Polyelectrolytes Can Be Used Both to Flocculate and to Stabilize Colloidal Dispersions

Polyelectrolytes have a large practical use in controlling the stability of colloidal dispersions. They combine the effects of electrostatic and polymer-mediated interactions. By a careful design, one can generate stabilizing/destabilizing effects that are difficult to accomplish by other means.

When added to a suspension of charged colloidal particles, the polyion acts as a multivalent counterion. In Section 5.3, we discussed how multivalent coions could induce an electrostatic attraction between oppositely charged surfaces through a correlation effect. The mechanism leading to an attraction also operates for polycounterions with an additional complication due to the configurational freedom in the polyion. On the addition of a 1:1 electrolyte, monovalent counterions partly replace polyion charges close to the surface, reducing the attraction, and the surface–surface interaction can turn repulsive again.

For a sufficiently highly charged polyion, there is charge reversal leading to a long-range repulsive double-layer force. However, the sensitivity to the addition of electrolyte might change. If the polymer adsorbs on the surface solely by electrostatic forces, then a high salt concentration will make it desorb. If, on the other hand, there is also some chemical interaction between the surface and the monomer unit, the polyelectrolyte remains adsorbed, and one has a situation similar to a noncharged polymer in a good solvent. Thus, the overlap force will be operating, and there is a repulsive force also under high salt conditions.

We observe dramatic effects when only small amounts of the polyelectrolyte are added to the system. Then only a fraction of the surface is covered with adsorbed polyions. The surface is then alternatively positive and negative, and a dipolar field extends out into the solution. When two surfaces approach one another, correlations between the charged domains then generate a strong attractive interaction as described in Section 5.3. In practice, this behavior is very useful because adding only small amounts of the polyelectrolyte can have a major effect on the stability of colloidal solutions. For example, positively charged polyelectrolytes are used extensively in papermaking to flocculate cellulose fibers together with clay fillers and other negatively charged additives. Practical experience affirms the importance of the order in which the different components are mixed, again illustrating the important role of the nonequilibrium states in polymer adsorption.

See Berg (2010, Chapter 7), Cosgrove (2010, Chapter 9), Kronberg et al. (2014, Chapter 18), and Lyklema (2005b, Vol. 5, Chapter 1) for additional discussion on Polymer Stabilization.

Exercises

7.1 A polymer sample has a Gaussian size distribution function $F(N_p) = C\exp\left\{-a\left(N_p - N_{av}\right)^2\right\}$. Calculate how the ratio between the weight-averaged and number-averaged aggregation numbers depends on a.

7.2 Using the random walk model, calculate the radius of gyration for a polyethylene coil at theta conditions. Assume that the Kuhn length is approximately 1 nm, corresponding to seven CH_2 units. Do the calculation for $(CH_2)_{1000}$ and $(CH_2)_{10,000}$.

7.3 Estimate the overlap concentration C^* for the polymers in Exercise 7.2.

7.4 Estimate the overlap concentration of an aqueous DNA solution with a persistence length of 15 nm and a contour length of the double helix of 200 nm.

7.5 Using the Flory–Huggins theory, calculate the structure factor $S(Q)$ at $Q = 0$ for a semidilute solution with volume fraction $\Phi = 0.05$ of a polymer in a solvent with $\chi = 0.3$. What is the difference between $N_p = 1000$ and $N_p = 10\,000$?

7.6 A protein with a molecular weight of 40 000 has 250 amino acid residues. It adopts a globular conformation in water. Estimate the number of residues in contact with the solvent.

7.7 Two polymers of equal size are mixed in a melt. What is the free energy of mixing in the framework of the regular solution theory?

7.8 What is the maximum depletion force obtained between two phospholipid bilayers in a 10% (w/w) polyoxyethylene aqueous solution? Use the Flory–Huggins theory with $\chi = 0.4$.

7.9 Calculate how the depletion force contribution to the adhesion between two planar surfaces varies with the molecular weight of the polymer. Assume that full depletion occurs for separations smaller than $2R_g$ and neglect it for larger separations. The polymer is in a good solvent with $\chi = 0.4$.

7.10 Colloidal particles of radius 10^{-5} m are dispersed in water and stabilized by the adsorption of $C_{18}\,EO_{20}$. Estimate the adhesion energy when the Hamaker constant is 1×10^{-20} J. (*Hint*: Use the data of Figure 7.16.)

7.11 Calculate the electrostatic contribution to the double-helix, single-coil transition of DNA at 1×10^{-2} and 1×10^{-1}M KCl (**T** = 300 K). The double helix has a radius of approximately 12 Å and one negative charge per 1.7 Å projected on the axis. For simplicity, assume that the denatured state is a single helix with the same radius but half the charge density. (*Hint*: Use Eq. (4.1.26) and note that the radius of curvature is half that of a sphere with the same radius.)

References

Benoit, H., Holtzer, R.M., and Doty, P. (1954). *J. Phys. Chem.* 58, 635.

Berg, J.C. (2010). *An Introduction to Interfaces & Colloids. The Bridge to Nanoscience.* World Scientific.

Colby, R.H., Felters, L.J., and Graessley, W.W. (1987). *Macromolecules* 20, 2226.

Cosgrove, T. (2010). *Colloid Science. Principles, Methods and Applications.* Chichester, UK: Wiley.

Doi, M. (2013). *Soft Matter Physics.* Oxford, UK: Oxford University Press.

Israelachvili, J. (1992). *Intermolecular and Surface Forces*, 2nd ed. London: Academic Press p. 291.

Israelachvili, J.N. (2011). *Intermolecular and Surface Forces*, 3rd ed. London, UK: Academic Press.

Israelachvili, J. and Wennerström, H. (1992). *J. Phys. Chem.* 96, 520.

Klein, J. (1980). *Nature* 288, 248.

Klein, J. and Luckham, P.F. (1982). *Nature* 300, 429.

Kronberg, B., Holmberg, K., and Lindman, B. (2014). *Surface Chemistry of Surfactants and Polymers*, 3rd ed. Chichester, UK: Wiley.

Lyklema, H. ed. (2005a). *Fundamentals of Interface and Colloid Science. Particulate Colloids*, Vol. 4. Amsterdam, The Netherlands: Elsevier.

Lyklema, H. ed. (2005b). *Fundamentals of Interface and Colloid Science. Soft Colloids*, Vol. 5. Amsterdam, The Netherlands: Elsevier.

Morris, E.R. and Ross-Murphy, S.B. (1981). *Techniques in the Life Sciences*, Vol. B3, Elsevier/North-Holland, Amsterdam, 1981, Chap. B310.

Stevens, C. and Felsenfeld, G. (1964). *Biopolymers* 2, 293.

Schildkraut, C. and Lifson, S. (1965). *Biopolymers* 3, 195.

8

Colloidal Stability

Concept Map

Colloidal Stability
When two colloidal particles undergoing Brownian motion approach each other, they experience two types of interaction: the static forces described in Chapter 5, which we consider first, and dynamic effects as described in Sections 1.7 and 5.8. Figure 8.1 depicts two important types of static interaction. When no repulsive energy barrier exists, particles come together into a primary minimum. A barrier creates a potential energy curve with a maximum separating a deep primary minimum at short separations and often a shallow minimum further out. Many colloids are thermodynamically unstable and remain in solution only if a potential barrier $\gg k_B\mathbf{T}$ prevents them from moving into the primary minimum. Flocculation involves reversible particle association. Coagulation is irreversible.

DLVO Theory
In the DLVO theory, colloidal stability (Eq. 8.2.1a) is determined by a balance between double-layer repulsion $V_R(h)$, which increases exponentially with decreasing distance, h, and van der Waals attraction $V_A(h)$. The DLVO expressions for parallel plates (Eq. 8.2.4a) and identical spheres (Eq. 8.2.9) show that $V_R(h)$ depends on the screening length $(\kappa)^{-1}$ and the surface potential Φ_0. The critical coagulation concentration (CCC) is a useful measure of colloidal stability. It approximately corresponds to the electrolyte concentration at which the maximum value of $V(h) = 0$. The CCC becomes independent of potential when $ze\Phi_0/k_B\mathbf{T} \gg 1$ and then it is very sensitive to counterion valency z because CCC $\propto z^{-6}$ (Eq. 8.2.11), verifying the century-old empirical Schulze–Hardy rule. When $ze\Phi/k_B\mathbf{T} < 1$, the CCC $\propto \Phi_0^4/z^2$ (Eq. 8.2.12) and thus becomes extremely sensitive to the potential, but less sensitive to counterion valency.

The Colloidal Domain: Where Physics, Chemistry, Biology, and Technology Meet, Third Edition. Håkan Wennerström.
 Published 2026 by John Wiley & Sons, Inc.

Kinetics of Aggregation: Rapid Coagulation

The initial rate of coagulation involves dimer formation between identical particles P and is expressed as $-d[P]/dt = k_r[P]^2$ where k_r is a second-order rate constant.

We obtain an expression for k_r by calculating the flux resulting from particles diffusing toward a central particle, and the diffusion-limited rate constant is $k_D = (16/3)k_B\mathbf{T}/\eta$ (Eq. 8.3.1b) for identical spheres. The rate is independent of R because the increased probability that large particles will collide as a result of their size (as measured by R) is compensated by the increased probability that small particles will collide because of their greater mobility (as measured by D).

We can generalize the expression for k_D to include the formation of trimers and larger aggregates and obtain an expression (Eq. 8.3.8), giving the change in the total number of particles $\sum_j [P_j]$ with time (Figure 8.5).

Slow Coagulation

When $V(h)$ has a barrier much larger than $k_B\mathbf{T}$, coagulation rates can become orders of magnitude slower. By including a potential-dependent term in Fick's first law, we obtain $k_r = k_D/W$ (Eq. 8.3.18); W represents the stability ratio and is proportional to $\exp(V_{\max}/k_B\mathbf{T})$ (Eq. 8.3.21).

Figure 8.6 shows how k_r increases as the particle radius decreases, ultimately eliminating kinetic colloidal stability for small particles.

When coagulating particles become locked into a configuration set by their initial contact, a loosely packed, less dense precipitate forms, which displays fractal properties (Figure 8.8).

Electrokinetics: Measurement of the Zeta Potential

Application of the DLVO theory requires knowledge of the surface potential Φ_0, which is often difficult to obtain in practice. Instead, we can correlate the zeta potential ζ obtained from electrokinetic measurements on the stability of colloidal systems (Table 8.4).

In an experimental determination of the zeta potential, we induce flow between a colloidal surface and the adjacent fluid and measure the potential at the position in the double layer where fluid begins to flow.

There are four electrokinetic effects: **electrophoresis** and **sedimentation potential** measurements involve the motion of charged particles in a stationary liquid; **streaming potential** and **electro-osmosis** measurements involve the flow of fluid past a stationary charged surface.

Electrophoresis

In an electrophoresis experiment, we impose an electrical field E and measure the colloidal particle's velocity v, which is more conveniently expressed in terms of the mobility $u = v/E$. The relation between u and ζ depends on the ratio between particle radius and Debye length. In the limit $\kappa R \ll 1$, we obtain the Hückel equation (8.4.6), while the condition $\kappa R \gg 1$ results in the Helmholtz–Smoluchowski equation (Eq. 8.4.12).

Streaming Potentials, Electro-osmosis, and Sedimentation Potential

In a streaming potential measurement (Figure 8.10), we cause a solution with a specific conductance of κ_b to flow through a capillary by applying a pressure gradient Δp. The zeta potential (Eq. 8.4.18) arises because the flow of counterions induces a potential Φ_{st}.

In an electro-osmosis experiment, we apply a potential Φ_{eo} across two electrodes connected by a capillary of length L. The field exerts a force on the counterions in the double layer, and they, in turn, carry the solution in the tube along with them. The relation between volumetric flow rate and the zeta potential is given by Eq. (8.4.19). A sedimentation potential is generated when a charged particle settles in a gravitational field. For dilute systems, the relation between the induced electrical field and the zeta potential is given by Eq. (8.4.20).

One of the key issues underlying our ability to apply colloidal science to technical problems, biological questions, or scientific experimental methods is the stability of the colloidal state. This question fascinated Faraday 200 years ago. He was able to prepare a sol of gold particles with seemingly indefinite stability. The dichotomy between colloidal stability–instability remains a relevant question, although with an awareness of more sophisticated implications.

The observant eye discovers many commonplace manifestations of the interplay between stable and unstable colloidal systems. Ordinary milk, for example, displays an amazingly rich range of colloidal behavior. It owes its white color to multiple light scattering from colloidal fat globules and protein aggregates. Milk fresh from the cow shows colloidal stability except for gravitational effects, leading to cream formation at the surface. If we use a centrifuge to separate the cream from the milk, we create two stable systems, both containing fat and protein. Whip the cream, and it becomes a metastable foam. Whip it a bit more, the foam collapses; an aqueous phase, buttermilk, separates out; and the rest forms butter. This is a metastable colloidal system of water droplets in fat. We can keep butter for a very long time in the refrigerator, but when heated, the fat and water separate, forming clarified butter. Starting as milkfat in water, the fat finally has separated completely from the water.

Milk's protein component shows a similar richness. Around 20% of the protein fractions are soluble whey proteins, while approximately 80% of milk proteins are caseins present as particles called casein micelles (not to be confused with surfactant micelles) of around 100 nm in radius. The casein micelles also contain small ($\approx$ 2 nm) calcium phosphate crystals. The casein micelles are stabilized by a charged κ-casein polymer brush, and they are colloidally stable in ordinary milk. Even at higher temperatures, as in a cup of milky tea, stability prevails, but try to add a few drops of lemon juice! The skin we find on milk that has been heated to boiling is surface-precipitated β-lactoglobulin, and the protein adsorption makes the saucepan hard to wash. The latter effect causes great problems in the dairy industry, where surfaces foul during pasteurization. However, colloidal instability can be turned to an advantage. Delicate creamy yoghurt results from microbial activity, producing lactic acid, lower pH, and colloidal coagulation. Alternatively, proteolytic enzymes in rennet transform the surface properties of the casein micelles so that the fat and the caseins coprecipitate to form cheese curd, leaving the whey behind.

Because of their considerable economic importance, the properties of milk systems have been studied in great detail, and we will not pursue any of these special phenomena. However, they represent a rich spectrum of colloidal stabilization–destabilization processes, such as lipid stabilization of fat globules, foam formation in whipped cream, emulsion inversion in the formation of butter, emulsion breaking by heating, and polymer brush degradation by enzymes.

The protein component shows stabilization through a polyelectrolyte brush, coagulation by charge neutralization using lemon juice, surface/interface adsorption by heating, coagulation by partial neutralization in yogurt, and coagulation by surface modification through the action of rennet. We will encounter more detailed discussions of these types of processes in this and subsequent chapters.

8.1 Colloidal Stability Involves Both Thermodynamic and Kinetic Factors

8.1.1 The Interaction Potential Between Particles Determines Kinetic Behavior

A colloidal particle experiences a Brownian motion in solution. On approaching another particle, it senses not only the surrounding solvent but also the effective interaction with this other particle. Particle–particle interaction has two components. One is a result of the static forces discussed in detail in Chapter 5. The other is a dynamic effect that leads to a correlation of the motions of the two particles mediated by the motions of the solvent molecules.

Before we enter into a description of the dynamics, let us qualitatively discuss the consequences of the static interaction potential. For a colloidal system, we can identify interaction curves of the two general types shown in Figure 8.1. In Figure 8.1a, a monotonically decreasing free energy persists up to a primary minimum. No energy barrier prevents the system from reaching this minimum, and it easily achieves this bound state. The extent to which the association between the particles is, in practice, reversible depends on the depth of the potential well and on possible further processes in the aggregated state.

A reversible association is often called flocculation, while an irreversible association is called coagulation. The distinction between the two terms is a qualitative rather than a quantitative one. Nonionic micellar solutions are examples of a system whose potential energy minimum is particularly shallow. With some exceptions, micellar systems show little tendency toward aggregation, and they are thermodynamically stable in the dispersed state.

On the other hand, oil droplets in water manifest a strong adhesion force. Furthermore, once the two droplets come into contact with one another, they merge irreversibly into one bigger drop. In this case, coalescence follows coagulation. For vesicles of neutral phospholipids, the energy minimum can be so deep that a relatively stable dimer is formed. The dimer can grow to form a trimer, and so on, by the addition of more vesicles. Mild mechanical agitation can redisperse such a floc, but if left unperturbed, it will sediment, and ultimately, irreversible fusion will occur.

The nonmonotonic interaction potential of Figure 8.1b allows for more intricate behavior. In this example, the potential curve displays two minima: a deep primary minimum at short separations and a shallow secondary minimum farther out. A potential energy barrier separates the two. If the system reaches the primary minimum, the coagulation process is often irreversible. In addition, the secondary minimum may be deep enough to support the type of flocculation we have already discussed. If the energy barrier exceeds $k_B\mathbf{T}$ by far, the primary minimum becomes inaccessible, and the system is kinetically stable. With a barrier on the order of only a few $k_B\mathbf{T}$, a relatively rapid approach is possible. Stable aggregates are formed for sufficiently deep or wide interaction wells at sufficiently high particle concentrations.

To obtain a semiquantitative understanding of the conditions necessary for aggregation, let us analyze the first step, the formation of a dimer P_2 from two particles P

$$P + P \rightarrow P_2 \tag{8.1.1}$$

The degree of association at equilibrium is determined by the equilibrium constant

$$K = \frac{[P_2]}{[P]^2} \tag{8.1.2}$$

K is closely related to the second virial coefficient B_2 (Eq. 1.2.15). For spherical particles of radius R, and with a distance of closest approach of $2R$, the expression for K is

$$K = \int_{2R}^{\infty} \left\{ \exp\left(-\frac{V(r)}{k_B\mathbf{T}}\right) - 1 \right\} 4\pi r^2 \mathrm{d}r = -\left(2B_2 - \frac{32\pi R^3}{3}\right) \tag{8.1.3}$$

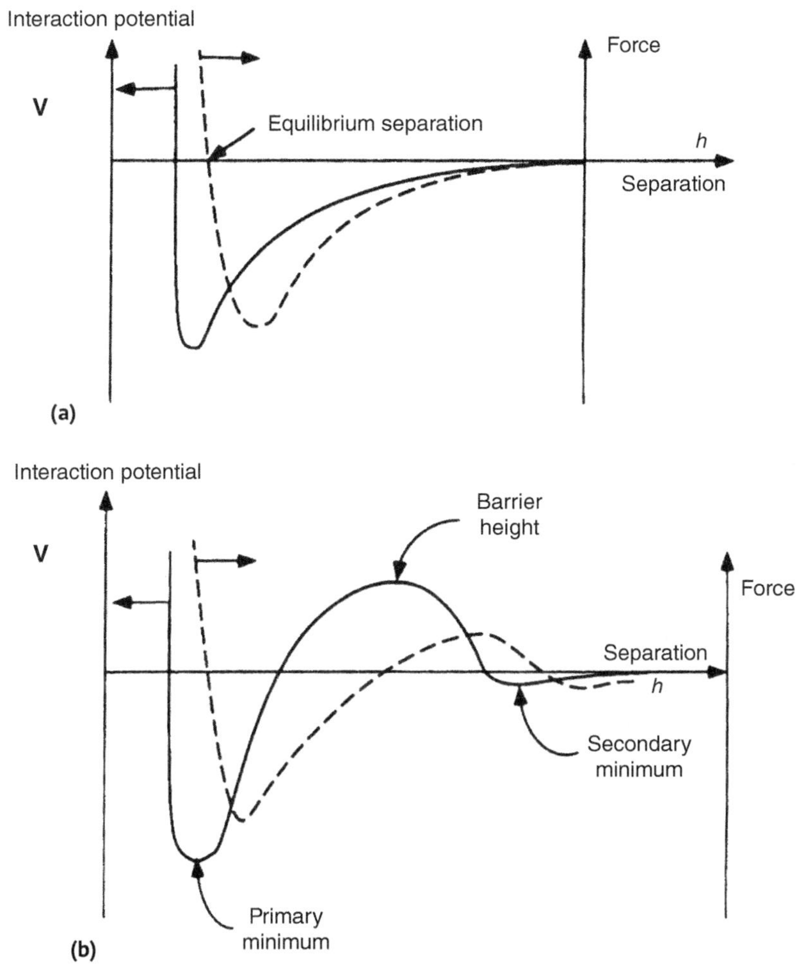

Figure 8.1 (a) Monotonically decreasing potential energy leads to reversible association or coagulation, depending on the depth of the potential energy well. Dashed line shows the corresponding force curve. Note that the force is zero at the equilibrium separation and is attractive for larger h. (b) Potential energy curve showing two minima, a shallow secondary minimum at large distance and a primary minimum at shorter distances, separated by a maximum. If the maximum is sufficiently high, it serves as a barrier to particle coagulation. Generally, the primary minimum is so deep that once particles have passed over the barrier, association becomes irreversible.

This expression for the equilibrium constant is valid when the pair formation only involves thermally excited, classical degrees of freedom. When a covalent bond is formed, the expression for K is more complex.

When the particle–particle interaction potential $V(r)$ is strongly negative over an appreciable range, K can become large, and dimerization can occur at low concentrations. The fraction of dimerized particles relative to monomers is

$$\frac{2[P_2]}{[P]} = 2K[P] \tag{8.1.4}$$

using Eq. (8.1.2). Thus, aggregation is not a major event if

$$2K[P]_{\text{tot}} < 1 \tag{8.1.5}$$

where $[P]_{\text{tot}}$ stands for the total particle concentration. On the other hand, if $2K[P]_{\text{tot}} > 1$, substantial aggregation occurs in the system, and the aggregates are likely to grow. Thus, for the solution to be thermodynamically stable, we require that the equilibrium constant K, or equivalently the second virial coefficient B_2, should be substantially smaller in magnitude than the inverse particle concentration so that $[P]_{\text{tot}}^{-1} > K, -B_2$.

8.1.2 Particles Deformed upon Aggregation Change Their Effective Interaction Potential

In the interaction potentials shown in Figure 8.1, the independent coordinate represents the distance between the particles. We implicitly assumed that no major internal rearrangements had altered the particle structure. However, at short range, particularly in the primary minimum, the interparticle interactions conceivably might be strong enough to affect the shape and nature of the particle. Whether or not such changes of the particles occur provides a demarcation line between the sol systems to be discussed in Chapter 9 and the emulsions and foams discussed in Chapter 11. A solid particle rearranges at high temperatures only in a sintering process. On the other hand, liquid drops or bubbles deform easily. They flatten on contact and may coalesce.

A more intricate question of particle integrity arises when a surfactant or polymer molecular film covers the surface. The final outcome of an aggregation process essentially depends on how the film responds to the perturbation of the interparticle contact. In vesicle systems, the result may be fusion or flocculation; for emulsions, the response determines whether coalescence accompanies coagulation; for adsorption-stabilized sols, the properties of the surface film regulate the degree of coagulation versus coalescence.

Clearly, the result of the colloidal aggregation process in these cases depends on a crucial balance between interparticle and intraparticle interactions. If we seek a detailed understanding, we must consider both types of interaction and the interplay between them. However, in this chapter, we will concentrate on the first step in the aggregation process and ignore the complications associated with the internal rearrangements of the particles until we discuss emulsions in Chapter 11.

See Berg (2010, Chapter 7), Cosgrove (2010, Chapter 3), Kronberg et al. (2014, Chapter 18), and Ohshima (2024, Chapter 9) for additional discussion on Colloidal Stability.

8.2 The DLVO Theory Provides a Basic Framework for Thinking About Kinetic Colloidal Stability

8.2.1 Competition Between Attractive van der Waals and Repulsive Double-layer Forces Determines the Stability or Instability of Many Colloidal Systems

A major advance in colloidal science occurred during the 1940s when two groups of scientists—the colloidal chemist Boris Derjaguin and the physicist Lev Landau, Nobel laureate in 1962, in the Soviet Union, and Evert Verwey and Theo Overbeek in the Netherlands—independently published a quantitative theoretical analysis of the problem of colloidal stability. The theory they proposed became known by the initial letters of their names: DLVO.

DLVO theory assumes that the more long-range interparticle interactions mainly control colloidal stability. Two types of force are considered. A long-range van der Waals force operates irrespective of the chemical nature of the particles or the medium. If the particles are similar, this force is always attractive (see Section 5.2). Furthermore, most colloidal particles acquire a charge either from surface charge groups or by specific ion adsorption from the solution. For similar particles, this charge leads to a repulsive double-layer force.

Thus, we can write the total interaction potential as

$$V(h) = V_A(h) + V_R(h) \tag{8.2.1a}$$

and the corresponding force equals

$$F = -\frac{\mathrm{d}V}{\mathrm{d}h} = -\frac{\mathrm{d}V_A}{\mathrm{d}h} - \frac{\mathrm{d}V_R}{\mathrm{d}h} \tag{8.2.1b}$$

Depending on the relative strength of the attractive and repulsive terms, we can generate the interaction potential versus distance curves shown in Figure 8.1, which also illustrates the corresponding force curves. One should realize that interaction potentials have the character of free energies and contain both energetic and entropic contributions.

Let us illustrate the typical properties of the interaction potentials and the corresponding force by considering the interaction between two identical half-planes separated by a distance h. According to Eqs. (5.2.7) and (5.2.20) the attractive van der Waals interaction V_A is

$$\frac{V_A(h)}{\text{area}} = -\frac{H_{121}}{12\pi h^2} \tag{8.2.2}$$

where H_{121} represents the Hamaker constant for medium 1 separated by a liquid 2.

At large separations, $h \geq \kappa^{-1}$, the double-layer interaction force is

$$\frac{F}{\text{area}} \approx 32(k_B\mathbf{T})^2 \frac{\varepsilon_r\varepsilon_0\kappa^2}{z^2e^2} \Gamma_0^2 \exp(-\kappa h) = 64k_B\mathbf{T}c_0^*\Gamma_0^2 \exp(-\kappa h) \tag{5.1.33}$$

as derived in Section 5.1. We obtain the corresponding potential by integrating Eq. (5.1.35) and

$$\frac{V_R}{\text{area}} = \frac{64k_B\mathbf{T}c_0^*\Gamma_0^2}{\kappa} \exp(-\kappa h) \tag{8.2.3}$$

Here, $\Gamma_0 = \tanh(ze\Phi_0/4k_B\mathbf{T})$ (Eq. 3.8.14) depends on the potential at the surface, but the distance dependence is solely determined by the Debye screening in the solution.

Now we can write Eq. (8.2.1) as

$$\frac{V(h)}{\text{area}} = -\frac{H_{121}}{12\pi h^2} + \frac{64k_B\mathbf{T}c_0^*\Gamma_0^2}{\kappa}\exp(-\kappa h) \tag{8.2.4a}$$

and the corresponding force

$$\frac{F}{\text{area}} = -\frac{H_{121}}{6\pi h^3} + 64k_B\mathbf{T}c_0^*\Gamma_0^2\exp(-\kappa h) \tag{8.2.4b}$$

The attractive term $V_A(h)$ dominates the repulsive term $V_R(h)$ when h is very large or very small. At intermediate separations, the double-layer force gives rise to a potential energy barrier if the surface is sufficiently charged and if the electrolyte ions do not screen too much. Figure 8.2 illustrates the resulting total potential for two choices of the screening length κ^{-1}.

When the low electrolyte concentration makes κ^{-1} large, a substantial energy barrier exists in the total $V(h)$ and the colloidal system is typically stable. On the other hand, for a short screening length κ^{-1}, the barrier is small or nonexistent, and the colloidal system is unstable.

To understand the approximate location of the barrier, let us consider the special case when the barrier is so small that the maximum value of $V(h)$ is zero (see Figure 8.2, curve $V(2)$). At the maximum

$$\frac{dV(h)}{dh} = \frac{dV_A}{dh} + \frac{dV_R}{dh} = 0 \tag{8.2.5}$$

and differentiating Eq. (8.2.4a) yields

$$\frac{dV(h)}{dh} = -\kappa V_R - \frac{2}{h}V_A = 0 \tag{8.2.6}$$

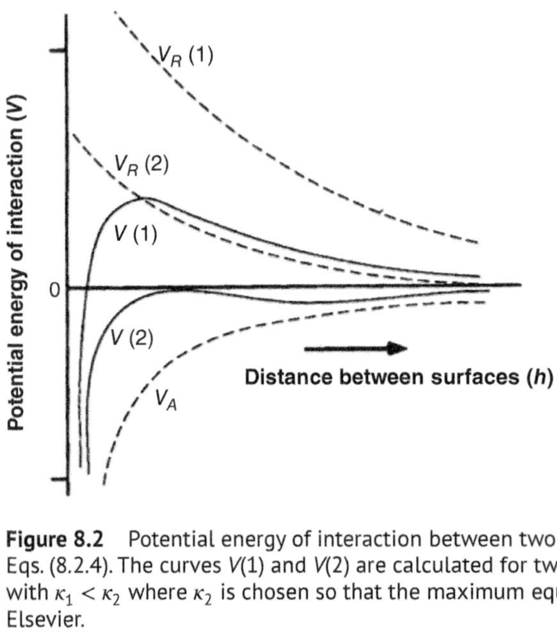

Figure 8.2 Potential energy of interaction between two planar surfaces according to the DLVO theory, Eqs. (8.2.4). The curves $V(1)$ and $V(2)$ are calculated for two different values of the inverse Debye length with $\kappa_1 < \kappa_2$ where κ_2 is chosen so that the maximum equals zero. *Source*: Shaw, 1980/with permission of Elsevier.

Combined with the condition $V(h) = 0$, that is, $V_A = -V_R$, we simply obtain

$$h_0 = \frac{2}{\kappa} \tag{8.2.7}$$

where h_0 is the separation at the top of the barrier.

When the top of the barrier occurs for $V(h_0) > 0$, the separation h_0 is larger than the one given in Eq. (8.2.7), and aggregation becomes slower. In the opposite case when $V(h_0) < 0$, the colloidal system coagulates rapidly. Since the asymptotic expressions in Eqs. (8.2.2) and (8.2.3) are valid for separations, $h > \kappa^{-1}$, it can be used to predict stability/instability.

8.2.2 The Critical Coagulation Concentration Is Sensitive to Counterion Valency

In the DLVO theory, colloidal stability is caused by a balance between double-layer repulsions and van der Waals attraction. The simplest way of controlling this balance in practice is to change the nature and concentration of the electrolyte. Experimentally, we observe a rather abrupt change from stability to instability on changing the salt concentration. Such studies measure the critical coagulation (electrolyte) concentration (CCC).

Starting from a DLVO potential and viscosities of the solvent, we can, in principle, calculate the expected association rate for any electrolyte concentration. However, a simpler, more qualitative approach is to specify a potential $V(h)$, which approximately represents the transition from stability to instability in practical terms. A mathematically convenient choice is to let $V(h) = 0$ at the barrier maximum. In this case, increasing the electrolyte concentration would yield an unstable system, but by slightly decreasing the electrolyte concentration, one would achieve long-term stability. Later in the chapter, we will analyze the relation between the kinetics of coagulation and interaction potential more closely.

Let us apply the model to a system of identical spherical particles, in which the radii R are large enough for the Derjaguin approximation (see Section 5.4.1) to hold. We can obtain the force by combining Eq. (5.4.6b), which relates the force between spherical particles to the interaction potential between flat surfaces, with Eqs. (8.2.2) and (8.2.3):

$$F(h) = \pi R \left\{ -\frac{H_{121}}{12\pi h^2} + \frac{64 k_B \mathbf{T} c_0^* \Gamma_0^2}{\kappa} \exp(-\kappa h) \right\} \tag{8.2.8}$$

When integrated using $dV(h) = -F(h)\,dh$, this equation yields

$$V(h) = \pi R \left\{ -\frac{H_{121}}{12\pi h} + \frac{64 k_B \mathbf{T} c_0^* \Gamma_0^2}{\kappa^2} \exp(-\kappa h) \right\} \tag{8.2.9}$$

Figure 8.3 shows how the total interaction potential $V(h)$ varies with the screening length κ^{-1}. At short κ^{-1}, $V(h)$ is monotonically decreasing, while for longer screening lengths, a barrier appears that becomes higher as κ^{-1} increases. It follows from Eq. (8.2.9) that the shape of $V(h)$ is independent of the radius R while the height of the barrier is proportional to the radius.

We now seek the relation between κ and h_0 where the barrier occurs for a value zero of the potential, $V(h_0) = 0$. Solving Eqs. (8.2.8) and (8.2.9) for the distance h_0 at the top of the barrier, we find in close analogy with Eq. (8.2.7)

$$h_0 = \frac{1}{\kappa} \tag{8.2.10}$$

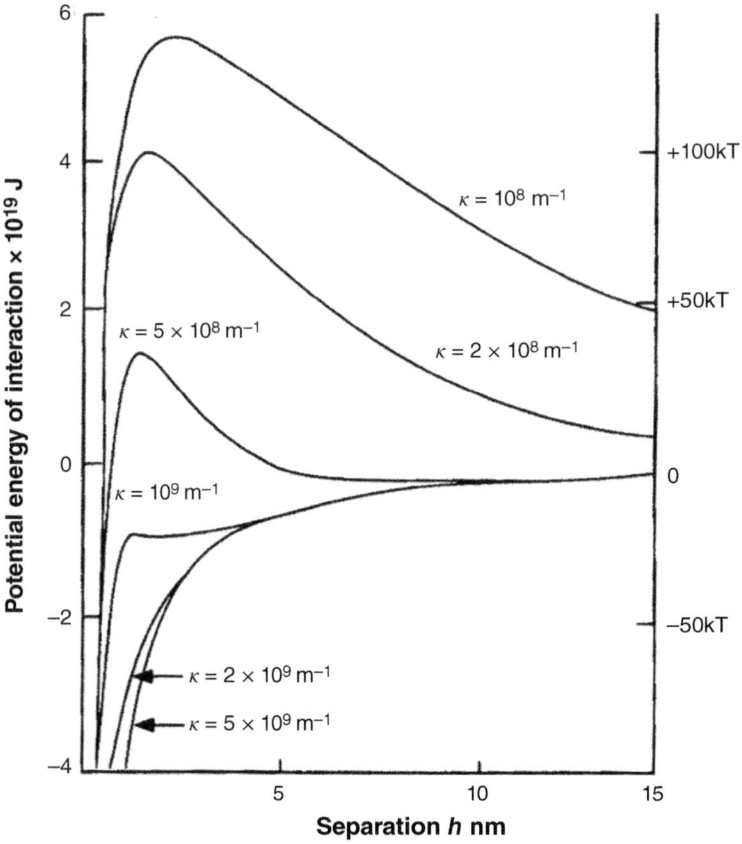

Figure 8.3 The influence of inverse Debye length κ on the total potential energy of interaction of two spherical particles calculated using Eq. (8.2.9). Parameters: $R = 10^{-7}$m; $\mathbf{T} = 298$K; $z = 1$; $H_{121} = 6 \times 10^{-20}$ J; $\varepsilon_r = 78.5$; $\Phi_0 = 50$ mV. *Source*: Shaw, 1980/with permission of Elsevier.

For spherical particles, this maximum occurs at half the distance required for flat surfaces.

We can estimate the electrolyte concentration that results in an onset of coagulation, called the CCC, by inserting $h_0 = 1/\kappa$ into Eq. (8.2.9) and solving for c_0^*, noting that the screening length also depends on c_0^*, as in Eq. (3.8.13). This yields

$$\text{CCC} \approx 1 \times 10^5 \frac{\left(\varepsilon_r\varepsilon_0\right)^3 \left(k_B\mathbf{T}\right)^5}{z^6 e^6 H_{121}^2} \Gamma_0^4 \tag{8.2.11}$$

Although this formula is numerically approximate, it clarifies that the CCC depends strongly on the parameters.

For a highly charged surface, Γ_0 approaches a constant value of unity (see Section 3.8) and the CCC depends on the sixth power of the ion valency. The strong valency dependence of the CCC was observed empirically around 1900 and became known as the Schulze–Hardy rule. Providing a transparent theoretical derivation of this rule was one of the successes of the DLVO theory.

At lower charge densities, where we can linearize the Poisson–Boltzmann equation, Γ_0 takes the value $ze\Phi_0/(4k_B\mathbf{T})$ and Eq. (8.2.11) becomes

$$\text{CCC} \approx 4 \times 10^2 \frac{\left(\varepsilon_r\varepsilon_0\right)^3 k_B\mathbf{T}}{\left(zeH_{121}\right)^2} \Phi_0^4 \tag{8.2.12}$$

Thus, for low surface potentials Φ_0, the CCC depends on the ion valency to the inverse second power.

Problem: Derive the CCC for constant surface charge density in the linear regime of the Poisson–Boltzmann equation.

Solution: In the linear regime, $\sigma = \varepsilon_r\varepsilon_0\kappa\Phi_0$ (Eq. 3.8.20) and $\kappa^{-1} = \left(\varepsilon_r\varepsilon_0 k_B\mathbf{T}/2z^2e^2c_0^*\right)^{1/2}$ (Eq. 3.8.13). From Eq. (8.2.12)

$$\mathrm{CCC} \approx 4\times 10^2 \frac{(\varepsilon_r\varepsilon_0)^3 k_B\mathbf{T}}{(zeH_{121})^2}\left(\frac{\sigma}{\varepsilon_r\varepsilon_0\kappa}\right)^4 = 4\times 10^2 \frac{(\sigma)^4 k_BT}{(zeH_{121})^2\varepsilon_r\varepsilon_0}\frac{8(\varepsilon_r\varepsilon_0 k_B\mathbf{T})^2}{4(ze)^4 CCC^2} \rightarrow$$

$$\mathrm{CCC}^3 \approx 10^2\frac{\sigma^4(k_B\mathbf{T})^3\varepsilon_r\varepsilon_0}{(ze)^6H_{121}^2} \rightarrow \mathrm{CCC} \approx \frac{4.6k_B\mathbf{T}(\varepsilon_r\varepsilon_0)^{1/3}\sigma^{3/4}}{(ze)^2H_{121}^{2/3}}$$

In Eqs. (8.2.11)–(8.2.13), we consistently use SI units, and the CCC is given in molecules per volume.

For mathematical convenience, the derivations have been made for a symmetrical electrolyte. In practice, this turns out not to be a major problem, provided we obey some simple rules when applying the results to unsymmetrical electrolyte systems. For example, in highly charged systems, Eq. (8.2.11) still applies if we interpret z as the valency of the counterion. In the linear regime, the true parameter in the problem is κ, and by using the general formula for κ given in Eq. (3.8.13), we can remake the derivation for an arbitrary mixture of electrolytes. The result will deviate somewhat from those given in Eq. (8.2.12).

A notable feature of the equations for the CCC is that they do not depend on the particle radius. This independence arises from our somewhat arbitrary condition that $V(h_0) = 0$ at the CCC. In reality, a slightly higher $V(h_0)$ is required to obtain long-term stability, and a R-dependence enters the calculation through a nonzero left-hand side of Eq. (8.2.9). This effect will be discussed further in Section 8.3.

For many colloidal systems, we know neither the surface potential Φ_0 nor the surface charge density σ_0. In practice, we often must resort to indirect methods to characterize the electrical properties of a particle. Perhaps the most convenient way to do this is to measure the migration of single particles in an electric field, a method that enables us to deduce the so-called zeta potential. This concept will be discussed in more detail in Section 8.4 on electrokinetic phenomena.

Until this point, we have discussed colloidal stability simply as a question of reaching the primary minimum. The further fate of the aggregated particles depends on the interaction at short range, where chemical specificity enters the problem. The usefulness of the DLVO theory is based on the fact that the van der Waals and electrical double-layer forces are long-ranged, with a known asymptotic form. Thus, we can hope that the DLVO potential is accurate up to the location of the energy barrier, even without a good theoretical or experimental basis for assuming that this potential is accurate up to contact between the surfaces. At short range, all the complications discussed in Sections 5.3–5.6 can enter the problem. Thus, the properties of the aggregated state depend in a sensitive way on the nature of these short-range forces as well as on the detailed geometry of the particles. In general, the DLVO potential cannot be used to discuss this aspect of the problem of colloidal particle aggregation.

8.2.3 A Colloidal Suspension Can Be Stabilized by Adsorbing Surfactants or Polymers

In the DLVO theory description, colloidal stability emerges for charged particles in a medium with low electrolyte concentration. In practice, there are many other circumstances when we want to achieve the stability of a suspension. The most efficient way to accomplish this is to add an additional component like an amphiphile or a polymer, which adsorbs on the colloidal particles and changes their surface properties. With a judicious choice of adsorbent, one can obtain stability for uncharged systems, such as particles in an apolar medium, in aqueous systems with a high salt concentration, and also for small particles for which stability is not achievable using the DLVO concept (see Section 8.3). There are many applications where we want to be able to control the properties of the suspension so that it is stable up to a certain point, at which instability may be introduced by some simple means. Polymer and surfactant-stabilized systems provide a greater flexibility in this respect.

Surfactants adsorb readily from an aqueous solution on the surface of a (slightly) hydrophobic particle and also on hydrophilic surfaces in an apolar medium. They self-assemble on the surface to form either micellar-type aggregates or a monolayer film. An adsorbed film of surfactant typically eliminates the charges on the particle itself; the resulting repulsive force between the particles is mainly determined by the interaction between the surfactant monolayers. There is a contribution to the force from the dispersion attraction between the particles themselves, but due to the substantial thickness of the adsorbed layers, this dispersion interaction is only operating at a relatively long range, substantially reducing its strength.

Problem: Calculate the dispersion interaction between two similar spherical particles of radius $R = 1 \times 10^{-7}$ m separated by two monolayers of phospholipids (2.0 nm per monolayer) and water region of thickness 2.0 nm. Assume a Hamaker constant of 6×10^{-21} J and neglect the contribution from the lipid layers.

Solution: From Eq. (5.4.8)

$$U_{\text{sphere}} = -\frac{RH_{121}}{12h} = -\frac{1 \times 10^{-7} \times 6 \times 10^{-21}}{12(2+2+2) \times 10^{-9}} = 6.25 \times 10^{-21}\ \text{J}$$

corresponding to $1.5k_B\mathbf{T}$.

In an apolar medium, the apolar chains of the surfactant will be exposed to the medium and provide stability like a short polymer brush, provided the medium is a good solvent for the chains. In an aqueous suspension, the stability with repulsive force will depend on the nature of the polar head group. For charged surfactants, the long-range electrostatic repulsion will operate effectively, and even at high salt concentrations, stability can prevail under favorable conditions through a more short-range interaction since the strength of the van der Waals force has been reduced. One disadvantage of using surfactants to stabilize a dispersion is that the required monomer concentration in the bulk is approximately equal to the CMC. For dilute suspensions, in particular, this can consume large amounts counted per particle.

It was noted in the previous chapter that polymers readily adsorb on surfaces and that this can occur even at very low bulk concentrations. In a good solvent, the adsorbed polymers generate a repulsive entropic overlap force with a range determined by the radius of gyration of the polymer (cf. Figure 7.14). Since for a typical polymer, R_g can exceed 10 nm, the attractive dispersion force between the particles is virtually eliminated. However, an attraction can be introduced by changing to bad solvent conditions (cf. Table 7.2 and Figure 7.15).

With a polyelectrolyte on the particle surface, one has the combined effect of an electrostatic and a steric entropic repulsion. As long as the polymer remains on the surface, a suspension is stable even for high electrolyte concentrations due to the overlap repulsion. In this limit, the charges on the polymer have the effect of ensuring good solvent conditions in an aqueous medium. For example, adding salt to milk does not induce coagulation of casein micelles since the overlap repulsion remains operative.

It is an interesting question whether the stabilizing effect of an adsorbing surfactant or polymer is kinetic or thermodynamic. As we will discuss in Chapter 11, both possibilities are found in emulsion systems. With colloidal particles of a solid character, the shapes are given by the preparation, and the question of thermodynamic stability seems to depend strongly on particle shape. For perfect spheres, there is not much direct molecular contact in the adhered state, and the particle–particle interaction might not be strong enough to push away an adsorbed surfactant or polymer layer. If the particles, on the other hand, have large planar faces, which easily occur for particles with crystalline atomic order, the direct particle–particle contact is more effective in giving a strong adhesion, and the adsorbed layers can be pushed out. Molecularly rough particles represent an intermediate case between the perfect sphere and the locally planar particles.

8.2.4 Interactions in Highly Concentrated Systems

In the previous sections, we have assumed that interactions between the particles could be described using the infinite dilution conditions. This remains a reasonable approximation as long as the particle concentration is so low that the mean surface–surface separation, h(mean), between particles is larger than the screening length $1/\kappa$. For a system where h(mean) is smaller than $1/\kappa$, the quantitative expressions for the electrostatic interaction used above are no longer valid. There is a gradual change from an exponentially varying interaction in the presence of screening to an algebraic distance dependence typical of a counterion-only system.

We can use an ionic micellar system to illustrate some of the complications that arise in the description at high particle (micellar) concentrations. Particles as small as micelles cannot be stabilized kinetically since the barrier height is too low, as we will analyze further in Section 8.3. Instead, stability against aggregation must be due to interactions at short range. Micellar solutions are usually stable up to volume fractions of order 30–40%. If one considers a uniformly charged sphere, the dispersion interaction at short range varies as h^{-1}. The electrostatic interaction between planes varies at short range as h^{-1} (Section 5.1.3), and using the Derjaguin approximation, this implies a divergence as $\ln(h)$ at contact. Since h^{-1} diverges more strongly than $\ln(h)$, as h goes to zero, we have the formal result that the uniform spheres should aggregate, even though there is a competition between dispersion attraction and electrostatic repulsion. This result is relevant for a particle lacking internal degrees of freedom. A micelle is liquid-like, and the internal degrees of freedom tend to lose entropy on tight association, and this extra free energy contribution is preventing association through mechanisms discussed in Section 5.6.

8.2.5 Protein–Protein Interactions Are Strongly Orientation-dependent

A living cell can be seen as an example of a concentrated colloidal system. Proteins, nucleic acids, and some lipids are predominantly negatively charged, and the stability relative to association is primarily due to an electrostatic repulsion. As in the micellar system discussed earlier, the average separation h(mean) is smaller than the screening length. In a prokaryotic cell, without internal compartmentalization, proteins at a concentration of around 35% (V/V) diffuse continuously, encountering other proteins. It is found that the charge density of the proteins has a narrow charge distribution centered around the value of -9×10^{-3} C/m^2. The implication is that at longer range $h \geq 0.5$ nm, there is a generic weak electrostatic repulsive protein–protein interaction common to all protein pairs, as illustrated in Figure 8.4.

This allows for short-range protein–protein encounters to occur. At such separations, short-range interactions like hydrogen bonds and hydrophobic contacts can be sampled in the search for a specific binding partner. If such a fit is not operating, the proteins separate in a diffusive motion in their further search.

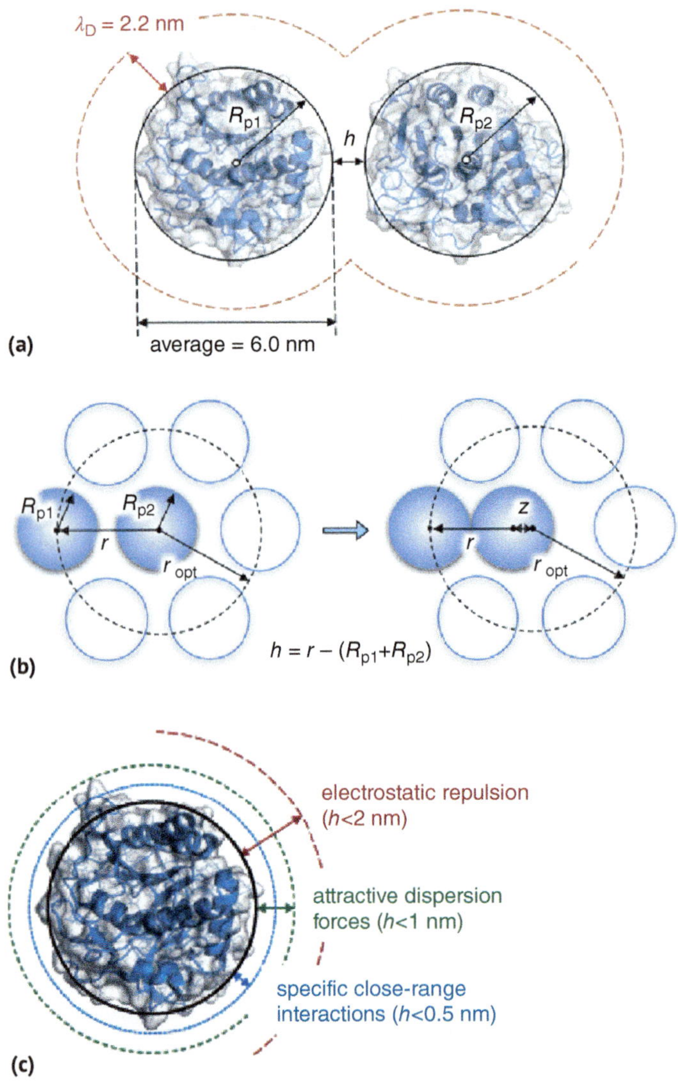

Figure 8.4 Illustration of protein–protein interactions in a cell. The mean separation h is around 2.0 nm, while the Debye screening length κ^{-1} is slightly larger. At separations $h > 1.0$ nm, the electrostatic repulsion dominates. For 0.5 nm $< h <$ 1.0 nm, the electrostatic and dispersion interactions largely compensate one another, while at shorter separations, more specific interactions dominate. *Source*: Wennerström et al., 2020/National Academy of Sciences/CC BY-NC-ND 4.0.

Protein–protein interactions depend strongly on the relative orientation, and this has important consequences for the association behavior. Previously, we have analyzed the association properties of colloidal particles using the simplifying assumption of orientation-independent interaction potentials. In such a scenario, adding a monomer to a dimer gives a stronger interaction free energy than for the formation of the dimer itself. This is a cooperative mechanism, and it promotes the formation of a separate precipitate. For a system with a highly anisotropic interaction, it can happen that the association can stop at the dimer or multimer stage since attractive patches have been covered.

Protein–protein association provides an important illustration of the complexity of the self-association of colloidal particles with strongly anisotropic interactions. Equation (8.1.3) gives the relation between the interaction potential and the dimerization equilibrium constant K. The orientation of an anisotropic body is specified by three angles, and it requires five angles, $\{\alpha\}$, to specify the relative orientation in a dimer and denote the interaction potential $V_d\{r, \alpha\}$. The equilibrium constant is then

$$K \approx \int\!\!\int\!\!\int\!\!\int\!\!\int\!\!\int \left[\exp\left\{-\frac{V_d\{r,\{\alpha\}\}}{k_B\mathbf{T}}\right\} - 1\right] d\Omega 4\pi r^2 dr \Big/ \int d\Omega \tag{8.2.13}$$

It is in practice hard to evaluate this expression for the dimerization constant, and the generalization to a trimer is impractical. However, such relations illustrate the complexity one can encounter for protein self-association. There are many examples of how smaller proteins assemble into functional entities as dimers, tetramers, and other specific multimers. In Chapter 9, we will discuss the formation of amyloid aggregates, which can be associated with many cases of physiological malfunction.

See Berg (2010, Chapter 7), Cosgrove (2010, Chapter 16), Israelachvili (2011, Chapter 14), Kronberg et al. (2014, Chapter 18), and Lyklema (2005a, Vol. 4, Chapter 3) for additional discussion on DLVO Theory.

8.3 Kinetics of Aggregation Allow Us to Predict How Fast Colloidal Systems Will Coagulate

8.3.1 We Can Determine the Maximum Rate of Particle–Particle Association Knowing the Diffusion Constant

In Section 1.7, we showed that the rate constant, k_d, for a binary diffusion-controlled association between A and B is given by

$$k_d = 4\pi R_r D_{AB} \tag{1.7.11}$$

where R_r is the center to center separation at which the association occurs, and below, we will assume that R_r is simply the sum of the particle radii. Both species A and B diffuse independently and $D_{AB} = D_A + D_B$. We are in colloidal applications usually interested in homoassociation so that $A = B$. For that case, one has to add a factor of 2 in the right-hand side of Eq. (1.7.11) since the monomer concentration refers to one species rather than two. For spherical particles, the diffusion constant is given by the Stokes–Einstein relation

$$D = k_B\mathbf{T}/(6\pi\,\eta R) \tag{1.7.16}$$

Combining Eqs. (1.7.11) and (1.7.16), we find a universal rate constant for the diffusion-controlled association of two particles of equal size

$$k_d = \frac{8k_B\mathbf{T}}{3\eta} \tag{8.3.1a}$$

And if they are identical

$$k_d = \frac{16k_B\mathbf{T}}{3\eta} \tag{8.3.1b}$$

For water, the viscosity $\eta = 1.0 \times 10^{-3}$ Ns/m^2 at 20 °C, so that at this temperature $k_d \approx 2.1 \times 10^{-17}$ m^3/s $\approx 6.5 \times 10^9$ M^{-1}s^{-1} for identical particles.

For the dimerization $P + P \rightarrow P_2$, we have the general rate equation

$$-\frac{\mathrm{d}[P]}{\mathrm{d}t} = 2\frac{\mathrm{d}[P_2]}{\mathrm{d}t} = k_r\,[P]^2 \tag{8.3.2}$$

With the initial conditions at $t = 0$; $[P] = [P(0)]$ and $[P_2(0)] = 0$, the equation has the solution

$$[P(t)] = \{k_r t + [P(0)]^{-1}\}^{-1} = \frac{[P(0)]}{1 + \frac{t}{\tau}} \tag{8.3.3}$$

Here, $\tau = \{[P(0)]k_r\}^{-1}$ is the time when half of the particles have associated so that $[P(\tau)] = [P(0)]/2$.

Problem: Determine the initial concentration of particles for the case where the half-life τ is 1 s in water at 20 °C for a diffusion-controlled reaction.

Solution: $[P(0)] = \frac{1}{\tau k_d} = \frac{3}{16}\frac{1 \times 10^{-3}}{k_B\mathbf{T}} = 5 \times 10^{16}\frac{\text{particles}}{\text{m}^3} \approx 1 \times 10^{-10}\text{M}.$
Diffusion-controlled reactions are rapid.

8.3.2 We Can Calculate Complete Aggregation Kinetics if We Assume That Rate Constants Are Practically Independent of Particle Size

In a situation involving rapid coagulation, the association process does not stop with dimers but continues to trimers, tetramers, and so forth until macroscopic particles precipitate. Following the argument of the preceding section, we can obtain the rate constant k_{ij} for the diffusion-controlled association of an i-mer P_i with a j-mer P_j by combining Eqs. (1.7.9) and (1.7.16)

$$k_{ij} = 4\pi\left(D_i + D_j\right)\left(R_i + R_j\right) = \frac{2}{3}\frac{k_B T}{\eta}\left(\frac{1}{R_i} + \frac{1}{R_j}\right)\left(R_i + R_j\right) \tag{8.3.5}$$

The rate constant carries only a weak dependence on the size of the particles because the factor $\left(1/R_i + 1/R_j\right)\left(R_i + R_j\right)$ varies between 4 (when $R_i = R_j$) and R_i/R_j (when $R_i \gg R_j$). Thus, we can create a reasonable approximation of reality by making all rate constants equal and setting $\left(1/R_i + 1/R_j\right)\left(R_i + R_j\right) \approx 4$ so that

$$k_{ij} \approx \frac{8k_B T}{3\eta} \equiv k \tag{8.3.6}$$

becomes a universal rate constant, which we can insert in the kinetic scheme of the total aggregation process. Because $k_{ij} \geq k$, we will underestimate the rate somewhat.

For the concentration of species P_j, the rate equation is

$$\frac{d[P_j]}{dt} = k\left\{\sum_{i\le(j-i)}[P_i][P_{j-i}] - [P_j]\left(\sum_i [P_i] + [P_j]\right)\right\} \tag{8.3.7}$$

The first sum of the right-hand side accounts for the formation of aggregate j by association of smaller aggregates, while the second sum accounts for the consumption of aggregates j by association. When two j aggregates combine, there is a double loss of these, which accounts for the extra term $[P_j]$. When j goes from 1 to j_{max}, Eq. (8.3.7) represents j_{max} coupled differential rate equations, which were first solved by von Smoluchowski.

Let us start by considering the change in the total particle concentration $\sum_j [P_j]$. This quantity is changed by all associations, and because the rate constant is the same, we can, in an analogy with Eq. (8.3.2), write the rate equation

$$-\frac{d}{dt}\sum_j [P_j] = k\left(\sum_j [P_j]\right)^2 \tag{8.3.8}$$

Each association eliminates two aggregates but produces one new. With the initial condition of total particle concentration

$$[P]_0^{tot} = \sum_j [P_j]_0 \text{ at } t = 0$$

Eq. (8.3.8) has, as in Eq. (8.3.5), the solution

$$\sum [P_j] = \frac{[P]_0^{tot}}{1 + t/\tau};\ \tau = \left(k[P]_0^{tot}\right)^{-1} \tag{8.3.9}$$

This solution shows that for short times, $t < \tau$, the decrease in total number of particles is linear in time, while for long times, $t > \tau$, the decrease is inversely proportional to t. The characteristic time τ separating the two regimes decreases with increasing total particle concentration. Note that τ equals the time required to reduce the total number of particles in solution to one-half their initial number.

Given the expression for the total concentration of particles in Eq. (8.3.9), we can decouple the differential equations in Eq. (8.3.7) by successively solving for $j = 1$, $j = 2$, and so on, since for $j = 1$, the first term on the right-hand side vanishes. A simple analytical solution is obtained if one in Eq. (8.3.7) neglects the extra term $[P_j]^2$, which is only substantial for short times. For the initial condition of only monomers at $t = 0$, the approximate solution is

$$[P_j] \approx [P]_0^{tot}\left(\frac{t}{\tau}\right)^{j-1}\left(1 + \frac{t}{\tau}\right)^{-j-1} \tag{8.3.10}$$

Figure 8.5 shows the time evolution of the concentrations of the monomer, dimer, trimer, and tetramer. The monomer concentration decays monotonically, while for the others, an initial increase is followed by a decline in the concentration at longer times. Equation (8.3.9) provides predictions that can be simply tested:

1. If Eq. (8.3.9) describes the total course of coagulation, then the value of τ should be independent of the stage of coagulation. Table 8.1 provides an example of this.
2. According to Eq. (8.3.9), the product $\tau[P]_0^{tot}$ should be independent of the initial concentration, as shown in Table 8.2.

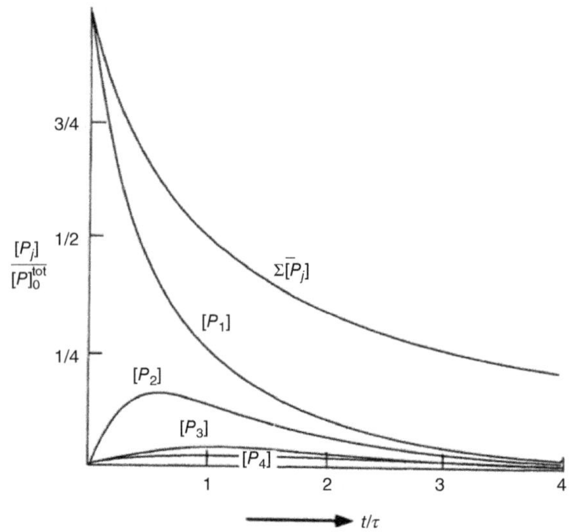

Figure 8.5 The predicted change in concentration of monomers, dimers, and so forth, where $[P_j]/[P]_0^{tot}$ is plotted versus t/τ. The concentration of monomers decreases continuously, but the concentration of all other species initially increases, then goes through a maximum, and finally decreases. *Source*: Kruyt, 1952/ with permission of Elsevier.

Table 8.1 Illustration of the validity of Eq. (8.3.9) for the coagulation of a gold sol.

/J	Gold Sol (number of particles × 10^{-14} m^{-3})		
t	**Obs.**	**Calc ($\tau = 79$ s)**	**τ(s)**
0	20.20	20.20	
30	14.70	14.40	80
60	10.80	11.19	69
120	8.25	7.74	83
240	4.89	4.78	76
490	3.03	2.71	85
			$\tau_{av} = 79$

Source: Adapted from Tuorila, 1926.

Table 8.2 Illustration of the independence of the rate constant on initial conditions, where $\tau^{-1} \equiv k[P]_0^{tot}$.

Dilution solution	Mean τ(s)	$\tau[P]_0^{tot}$(sm^{-3})
1:1	22.7	7.1×10^{17}
1:2	44.0	7.0×10^{17}
1:10	213.4	6.9×10^{17}

Source: Adapted from Kruyt and van Arkel, 1920.

Several features of the coagulation process deserve additional comments. Fluid flow can have a pronounced effect on the rate at which a process involving mass transfer occurs. We can express the ratio for the rates of coagulation with (J_S) and without (J_B) forced convection as

$$\frac{J_S}{J_B} = \frac{\eta R_{ij}^3 \left(\frac{dv}{dx}\right)}{2k_B\mathbf{T}} \tag{8.3.11}$$

where dv/dx is the velocity gradient in the fluid. The ratio of the rates involves the sum of the particle radii, $R_{ij} = R_i + R_j$ to the third power. For water, $\eta \approx 1 \times 10^{-3}$ Ns/m^2, and if the shear rate dv/dx = 1 s^{-1}, then $J_S/J_B \approx 1$ when $R_{ij} = 2$ μm and 125 when $R_{ij} = 10$ μm. Thus, stirring has no or little effect at the beginning of a coagulation process when the particles are small. However, stirring a quiescent, still-clear solution after coagulation has begun and the particle size has enlarged immediately results in increased turbidity.

All the discussion and experimental results cited earlier involve spherical particles. The product of *RD* of the effective reaction radius R_{eff} and the diffusion coefficient *D* is always larger for any nonspherical shape than it is for a sphere of the identical mass. This is the case because the asymmetric particles can also meet through rotational diffusion. Thus, nonspherical particles flocculate more rapidly than spherical ones.

In fact, stirring solutions with large aggregates or particles can have unexpected consequences. This applies particularly to systems with elongated particles where the mechanical tensions caused by stirring easily lead to breaking the internal association. It is, for example, difficult to handle samples with DNA extended helices without breaking the covalent bonds in the strands. A similar effect is found for amyloid aggregates discussed in Section 8.3.5 where it is necessary to handle samples with extreme care to avoid shear-induced aggregate fragmentation.

8.3.3 As the Barrier Height Increases, There Is a Gradual Transition from Fast to Slow Flocculation Followed by Long-term Stability

When the potential energy curve *V* has a barrier appreciably larger than $k_B\mathbf{T}$, the rate of coagulation can become orders of magnitude slower than the diffusion-limited rate Eq. (1.7.9). We can derive an expression for the rate at the initial stages of coagulation by using Eq. (1.7.7). In the more general form, Fick's first law says that the flux $\boldsymbol{J}$ is proportional to a friction factor $D/k_B\mathbf{T}$, a concentration term $[P]$, and a driving force, the gradient in chemical potential $\nabla\mu$, so that

$$\boldsymbol{J} = -\frac{D}{k_B\mathbf{T}}[P]\nabla\mu \tag{8.3.12}$$

In an external potential, $V(r)$, the chemical potential is

$$\mu = \mu^\theta + k_B\mathbf{T}ln[\boldsymbol{P}] + V(r) \tag{8.3.13}$$

Combining Eqs. (8.3.12) and (8.3.13) leads to

$$\boldsymbol{J} = -D\nabla[P] - \frac{D}{k_B\mathbf{T}}[P]\nabla V \tag{8.3.14}$$

We can solve Eq. (8.3.14) using the same procedure we used to calculate rapid coagulation in Section 1.7.3. If the central particle is located at the origin of the coordinate system and the particle is spherically symmetrical, the total flux through a shell of radius *r* is

$$(JA)_i = -8\pi r^2 D\left\{\frac{\mathrm{d}[P]}{dr} + [P]\frac{\mathrm{d}\left(\frac{V}{k_B\mathbf{T}}\right)}{dr}\right\} = \text{constant} \tag{8.3.15}$$

at steady state. In this equation, we have invoked the approximation that the relative diffusion coefficient is twice the individual D for similar particles.

We can integrate Eq. (8.3.15) by noting that $\exp(V/k_B\mathbf{T})$ is an integrating factor. The concentration profile around the central particle is

$$[P] = -\frac{(JA)_i}{8\pi D}\int_r^{\infty} r'^{-2}\exp\left(\frac{V(r')}{k_B\mathbf{T}}\right)\mathrm{d}r'\exp\left(-\frac{V}{k_B\mathbf{T}}\right) + [P]_b\exp\left(-\frac{V}{k_B\mathbf{T}}\right) \tag{8.3.16}$$

Here we have used the boundary condition that $[P]$ should tend to the bulk value $[P]_b$ for large r. At a radius $r < 2R$, we consider a dimer to have formed. Thus, a second boundary condition is $[P] = 0$ for $r = 2R$, as used earlier.

Problem: Show that Eq. (8.3.16) satisfies Eq. (8.3.15).

Solution: Evaluating the derivative of Eq. (8.3.17), we find that

$$\frac{\mathrm{d}[P(r)]}{dr} = \frac{JA}{8\pi D}\left\{-r^{-2}\exp\left[\frac{V(r)}{k_B\mathbf{T}}\right]\exp\left[-\frac{V(r)}{k_B\mathbf{T}}\right] + \int_r^{\infty}(r')^{-2}\exp\left[\frac{V(r')}{k_B\mathbf{T}}\right]dr'(-1)\frac{\mathrm{d}\left(\frac{V}{k_B\mathbf{T}}\right)}{dr}\exp\left[-\frac{V(r)}{k_B\mathbf{T}}\right]\right\} - [P]_b\frac{\mathrm{d}\left(\frac{V}{k_B\mathbf{T}}\right)}{dr}\exp\left(-\frac{V}{k_B\mathbf{T}}\right).$$

Inserted into the right-hand side of Eq. (8.3.15) we have

$$\text{RHS} = -8\pi r^2\left\{\frac{JA}{8\pi D}\left[r^{-2} + \int_r^{\infty}(r')^{-2}\exp\left[\frac{V(r')}{k_B\mathbf{T}}\right]dr'(-1)\frac{\mathrm{d}\left(\frac{V}{k_B\mathbf{T}}\right)}{dr}\exp\left[-\frac{V(r)}{k_B\mathbf{T}}\right]\right]\right.$$
$$-[P(b)]\frac{\mathrm{d}\left(\frac{V}{k_B\mathbf{T}}\right)}{dr}\exp\left(-\frac{V}{k_B\mathbf{T}}\right) + \int_r^{\infty}(r')^{-2}\exp\left[\frac{V(r')}{k_B\mathbf{T}}\right]dr'(-1)\frac{\mathrm{d}\left(\frac{V}{k_B\mathbf{T}}\right)}{dr}$$
$$\left.\exp\left[-\frac{V(r)}{k_B\mathbf{T}}\right] - [P(b)]\frac{\mathrm{d}\left(\frac{V}{k_B\mathbf{T}}\right)}{dr}\exp\left(-\frac{V}{k_B\mathbf{T}}\right)\right\} = JA = \text{LHS}.$$

We can now solve for the flux $(JA)_i$ in Eq. (8.3.15) and

$$JA = -8\pi D[P]_b\left\{\int_{2R}^{\infty} r'^{-2}\exp\left(\frac{V(r')}{k_B\mathbf{T}}\right)\mathrm{d}r'\right\}^{-1} \tag{8.3.17}$$

Using the same argument that leads from Eq. (1.7.10) to Eq. (1.7.11), the rate constant k_r, in the presence of a barrier, becomes

$$k_r = 8\pi D\left\{\int_{2R}^{\infty} r'^{-2}\exp\left(\frac{V(r')}{k_B\mathbf{T}}\right)\mathrm{d}r'\right\}^{-1} \equiv \frac{k_d}{W} \tag{8.3.18}$$

Thus, an energy barrier decreases the rate of coagulation by a factor W^{-1}, where the stability ratio W is given by

$$W = 2R\int_{2R}^{\infty} \exp\left(\frac{V(r)}{k_BT}\right)r^{-2}\mathrm{d}r \tag{8.3.19}$$

If the barrier $V(r)$ is only a few $k_B\mathbf{T}$, the association is in practice still rapid, and the system is seen as unstable. Previously, we have estimated the half-life, $t_{1/2}$, on the order of a second for a dilute sol in water when the association is diffusion-controlled. To extend the half-life to 24 h ($\approx 10^5$ s), the parameter W must reach a value of 10^5 or larger. In such a case, the value of the integral in Eq. (8.3.19) is dominated by r-values in the vicinity of the maximum of $V(r)$ at r_{max}. To have accordance with the notation for the interaction potential in Section 8.2, we transform to the variable $h = r - 2R$ and expand the potential $V(h)$ around its maximum at h_0, where the first derivative of $V(h)$ is zero

$$V(h) = V_{\mathrm{max}} + \frac{1}{2}(h - h_0)^2\left(\frac{\mathrm{d}^2V}{\mathrm{d}h^2}\right)_{\mathrm{max}} \tag{8.3.20}$$

Since $V(h)$ has a marked peak around h_{max}, we assume that h is much smaller than R, and we can replace r^{-2} in the integrand of Eq. (8.3.19) by $(2R)^{-2}$ so that

$$W \approx \frac{1}{2R}\exp\left(\frac{V_{\mathrm{max}}}{k_B\mathbf{T}}\right)\int_{-h_0}^{\infty}\exp(-b^2x^2)\,\mathrm{d}x \approx \frac{\pi^{1/2}}{2Rb}\exp\left(\frac{V_{\mathrm{max}}}{k_B\mathbf{T}}\right) \tag{8.3.21}$$

Here we have used the notation $b^2 \equiv -\left(\frac{\mathrm{d}^2V}{\mathrm{d}h^2}\right)_{\mathrm{max}}/(2k_B\mathbf{T})$, the variable substitution $x = r - 2R - h_0$, and the assumption that $h_0 \gg b^{-1}$. Using the DLVO potential of Eq. (8.2.9), we can evaluate the parameter b using the condition that the first derivative of $V(h)$ is zero at the maximum, which gives the relation

$$\exp(-\kappa h_0) = \frac{H_{121}\kappa}{768\pi k_B\mathbf{T}c_0^*\Gamma_0^2}\frac{1}{h_0^2} \tag{8.3.22}$$

Equation (8.3.22) has two solutions for h_0 and the maximum occurs for the smaller value. Taking the second derivative of $V(h)$ yields

$$b^2 = \frac{RH_{121}}{24k_B\mathbf{T}h_0^2}\left\{\frac{2}{h_0} - \kappa\right\} \tag{8.3.23}$$

Our final expression for the stability ratio W is

$$W = \left(\frac{H_{121}R^3}{6\pi k_B\mathbf{T}}\right)^{-\frac{1}{2}} h_0\left(\frac{2}{h_0} - \kappa\right)^{\frac{-1}{2}}\exp\left(\frac{V_{\mathrm{max}}}{k_B\mathbf{T}}\right) \tag{8.3.24}$$

Combining Eqs. (8.2.9) and (8.3.22), we find that the value of V at the maximum is

$$V_{\mathrm{max}} = \frac{RH_{121}}{12h_0}\left\{\frac{1}{\kappa h_0} - 1\right\} \tag{8.3.25}$$

From Eq. (8.3.22), we have that the position of the maximum h_0 is independent of the particle radius, R, which implies according to Eq. (8.3.25) that the barrier height increases linearly with R. Since this R dependence enters into the exponent in Eq. (8.3.24), it follows that the stability ratio increases strongly with increasing particle radius as shown in Figure 8.6.

8.3.4 For Particles of the Same Sign of the Charge, Homoassociation Usually Dominates over Heteroassociation

Many colloidal systems contain particles with different sizes, shapes, Hamaker constants, and surface potentials. Practical examples include paint formulation, water purification, and filtration. In these systems, particles of different types can associate, leading to heteroassociation rather than homoassociation. However, heteroassociation is not limited to colloidal sols. This effect also occurs in flotation, which involves particle–bubble interactions, and adhesion, in which particles attach to planar surfaces. The description of the area is complex because we must simultaneously consider both homoassociation between like particles and heteroassociation between unlike particles. It is clearly a more demanding task to design a system of several types of colloidal particles, which is nevertheless stable. When should we expect heteroassociation to be a significant process? Within the framework of the DLVO theory, stability is determined by a balance between attractive van der Waals and repulsive double-layer forces.

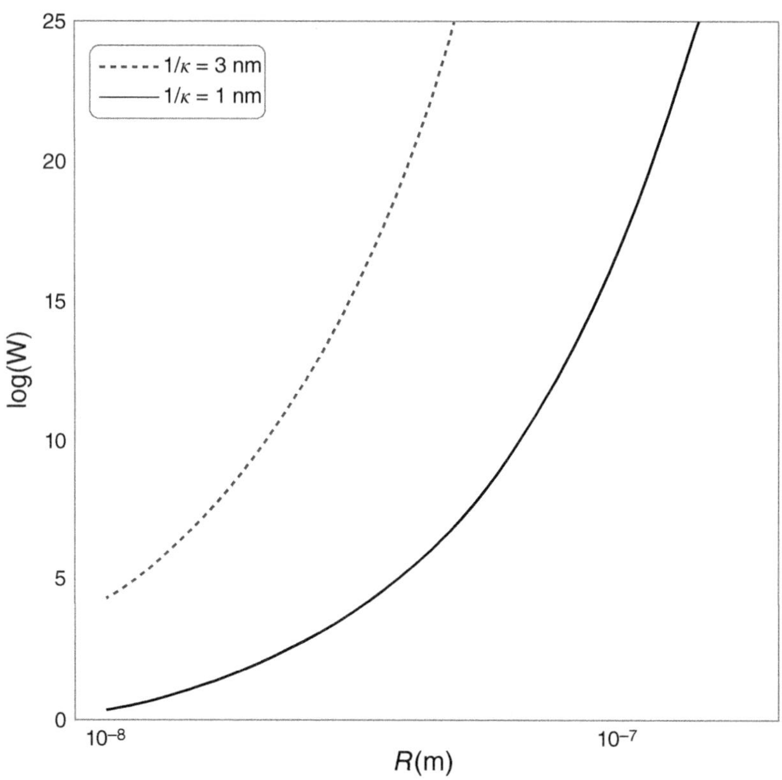

Figure 8.6 The calculated stability ratio W as a function of particle radius for highly charged systems with a surface potential $\Phi_0 = 60$ mV and for two values of the screening length. For small particles, $R \leq 10$ nm, W is smaller than 10^5, indicating that one cannot achieve electrostatically induced colloidal stability for small particles.

If the different types of particles have opposite charge, both the van der Waals and the electrostatic forces are attractive at long range, and flocculation based on heteroassociation should occur. Note that the electrostatic force can turn repulsive at short range, as discussed in Sections 5.1 and 7.3.

When all types of particles have the same sign of the charge (although with different magnitude), a delicate situation can arise when the difference in the attractive force between different pairs of particles is nearly balanced by a similar difference in the repulsive force. The three parameters that mainly govern this balance are the surface potentials, the Hamaker constants, and the particle sizes. For a system containing particles of types *A* and *B*, we can have three types of associations: *A-A*, *B-B*, and *A-B*. Consider a system that is initially colloidally stable where we slowly increase the electrolyte concentration or change the pH. According to the previous section, we expect that the association that occurs first is the one with the lowest barrier, $V_{\max}$.

The generalization of Eq. (8.2.9) to the interaction between two different particles *A* and *B* is

$$V(h) = \pi \frac{2R_A R_B}{R_A + R_b} \left\{ -\frac{H_{A2B}}{12\pi h} + \frac{64k\mathbf{T}c_0^* \Gamma_{0A} \Gamma_{0B}}{\kappa^2} \exp(-\kappa h) \right\} \tag{8.3.26}$$

The position of the maximum, h_0', is as in Eq. (8.3.22) but with Γ_0^2 replaced by $\Gamma_{0A}\Gamma_{0B}$ and with the proper value of the Hamaker constant. Analogously in Eq. (8.3.25) for $V_{\max}$, R is replaced by $2R_A R_B/(R_A + R_b)$ and the same substitution applies in Eq. (8.3.24_ for the stability ratio W.

The association with the lowest stability ratio W will dominate. The decisive factor in W is the barrier height in the exponential function. For homoassociation, the barrier is given by Eq. (8.3.25), and the corresponding expression for heteroassociation is

$$V'_{\max} = \frac{R_A R_B H_{A2B}}{12(R_A + R_B)h_0'} \left\{ \frac{1}{\kappa h_0'} - 1 \right\} \tag{8.3.27}$$

From Eq. (5.2.25), we have that the Hamaker constant for a mixed system is approximately the geometrical mean of the Hamaker constants of the respective pure systems. The pair to form first when the electrolyte concentration is slowly varied is the one with the lowest barrier $V_{\max}$. The relative size of the barriers depends on the ratio of the particle radii, the ratio of the Hamaker constants H_{A2A} and H_{B2B}, and on the parameters Γ_{0A} and Γ_{0B} determined by the surface potentials, which decrease in magnitude as the electrolyte is added. It turns out that, due to the fact that the quantities entering Eq. (8.3.27) for $V'_{\max}$ are intermediate between those of the pure systems, the lowest barrier is obtained for the A – A or B – B association for a dominating part of the parameter space. Thus, heteroassociation is an infrequent event when the instability conditions are approached gradually.

8.3.5 Highly Anisotropic Interactions Promote the Formation of Linear Aggregates

There are several cases of serious diseases whose molecular origin is associated with the formation of protein or polypeptide aggregates. These often take the form of amyloids, which typically are elongated precipitates held together by hydrophobic interactions involving β-sheets. Figure 8.7 shows an electron microscopy picture of fibers formed by a fragment of the peptide involved in Alzheimer's disease.

These diseases include Parkinson's and Alzheimer's diseases as well as ALS (amyotrophic lateral sclerosis), diabetes type II, and BSE (bovine spongiform encephalopathy), and each of these has been associated with the aggregation of a specific protein or polypeptide.

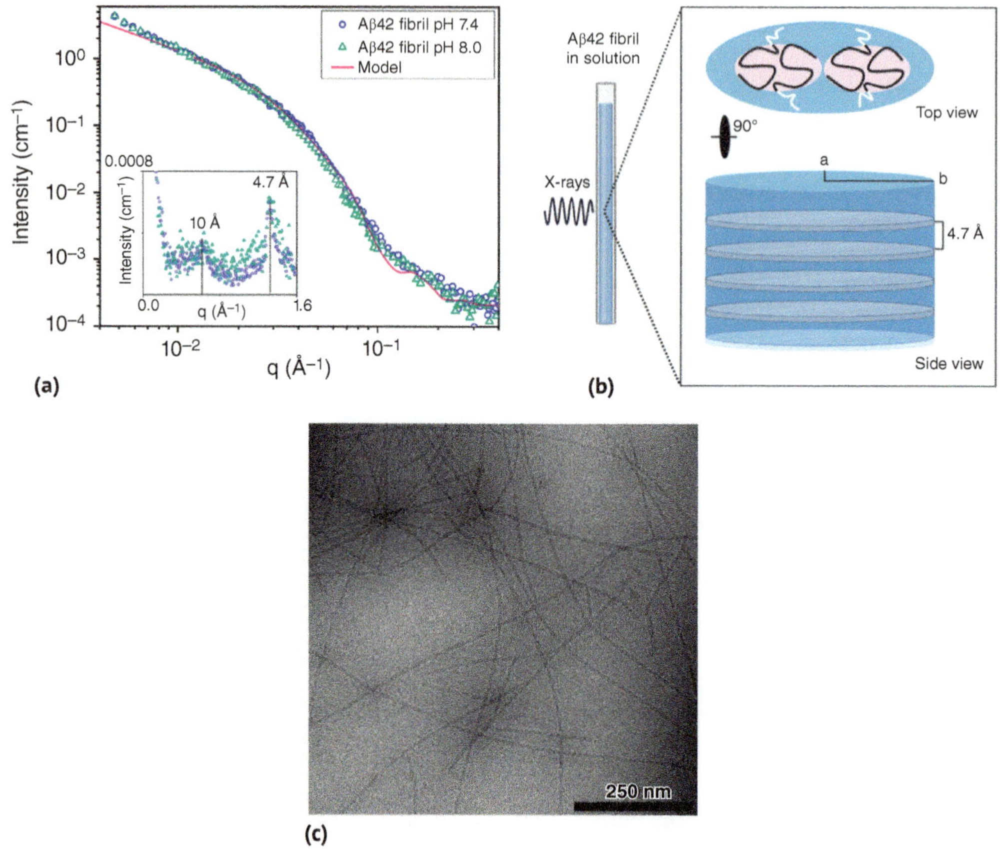

Figure 8.7 The amyloid precipitation of the polypeptide Aβ (A-beta) containing 42 amino acids is involved in the pathology of Alzheimer's disease. Panel (a) shows the SAXS scattering pattern for samples precipitated at two different pH values. The inset displays the wide-angle (WAXS) pattern, with sharp peaks corresponding to a separation of 0.47 nm typical for a beta-sheet packing along the fiber. In addition, there is a peak corresponding to a packing of the chains in the perpendicular direction. Panel (b) shows a schematic representation in the scale of A*β*42 fibril elliptical cross section with an ordered core (red) and a less ordered surface region (blue). *Source*: Lattanzi et al., 2021/National Academy of Sciences/CC BY-NC-ND 4.0. Panel (c): A cryo-transmission electron microscope picture of the amyloid fibrils. *Source*: Törnqvist et al., 2020/The Royal Society of Chemistry/CC BY 3.0.

Even though these proteins/polypeptides are chemically unrelated, they are often present in solution with an intrinsically disordered conformation. In contrast, the protein adopts a partly ordered conformation in the aggregate. The association thus involves a step with a conformational change, where hydrophobic residues fold into a conformation where intermolecular beta sheets (see Figure 7.3) can form. For a single protein, the conformation with the exposed hydrophobic patch(es) is of high free energy and thus occurs only rarely. The observation of large aggregates shows that the addition of a monomer to such a structure is thermodynamically favorable. However, such a state is reached first after a nucleation phase.

In Section 2.3, we analyzed the nucleation of a phase that grows in three dimensions and showed that a nucleation barrier is caused by an increasing volume-to-area ratio as molecules are added to the nucleus. This cooperative mechanism is not present in amyloid formation, where aggregates grow linearly. This latter situation was discussed in Chapter 7 in connection with the formation of double and triple helices. When amyloid aggregates are formed by proteins with more than a hundred amino acid residues, homogeneous nucleation is rarely observed. It appears that one protein molecule needs to bind to a surface, a membrane, or a preexisting protein aggregate before an amyloid aggregate is formed in a heterogeneous nucleation process. There is a low probability that two protein molecules meet in solution, both having a conformation allowing association.

8.3.6 Aggregates of Colloidal Particles Can Show Fractal Properties

The final outcome of an aggregation of colloidal particles also depends on processes occurring after the initial association. As discussed in Section 8.1, rearrangements are likely to occur once two particles have come into molecular contact. However, for an irreversible association of solid particles, the system is often kinetically trapped in its initial configuration. As the association progresses, this fact has important consequences for the nature of the aggregates and also for the kinetics of the association.

The equilibrium state of the aggregate should always be compact, but when the aggregate is locked into a configuration set by the initial contact, a much less dense system is formed. The analysis of the kinetics made in Section 8.3.2 no longer applies. It has been shown that aggregates grow in size as a so-called fractal object. In contrast to a compact aggregate, in which the mass M and the radius R are related as $M \propto R^3$, the radius of a fractal object depends on M with an exponent $d_f < 3$. If M and R are related by

$$M \propto R^{d_f} \tag{8.3.31}$$

we are dealing with a fractal object, where the exponent d_f represents the fractal dimension, which is a rational number less than 3. The smaller the value of d_f, the less compact the aggregate is. In the previous chapter, we discussed the polymer coil, which, in fact, shows fractal properties. By inverting Eqs. (7.1.16) and (7.1.17), we find $d_f = 2$ for a coil in a theta solvent, while $d_f = 1.67$ in a good solvent.

Figure 8.8 shows a transmission electron micrograph of an aggregate of colloidal gold particles. The open nature of the aggregate is apparent, and we can see clearly how the density of particles decreases away from the center. This effect is a necessary consequence of Eq. (8.3.31).

Fractal dimensionality depends on the nature of the aggregation process. The least dense structure appears for a diffusion-limited aggregation in which particles stick irreversibly and permanently at first contact. Computer simulations have shown that $d_f = 1.8$ in this case. For the experiments with colloidal gold shown in Figure 8.8, the estimated value of $d_f = 1.75 \pm 0.05$ agrees well with theory.

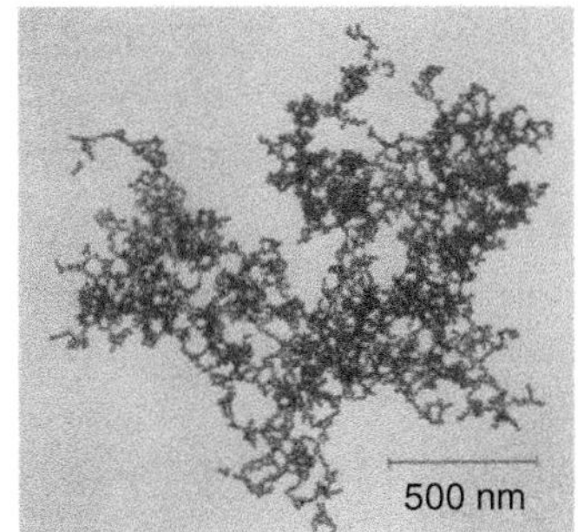

Figure 8.8 Fractal cluster of gold particles prepared by diffusion-limited aggregation. *Source*: Weitz et al./U.S. Department of Energy/Public Domain.

When a potential barrier must be overcome so that the stability ratio $W > 1$, an incoming particle might bounce against the aggregate several times. In that case, it is more likely to reach the interior of the aggregate and stick there, leading to a more compact structure and a larger d_f is observed experimentally.

See Berg (2010, Chapter 7), Cosgrove (2010, Chapter 3), and Lyklema (2005a, Vol. 4, Chapter 4) for additional discussion on Association Kinetics.

8.4 Electrokinetic Phenomena Are Used to Determine Zeta Potentials of Charged Surfaces and Particles

To use the DLVO theory, we need to know the potential energy of attraction and repulsion between colloidal particles. As we have seen, we can estimate the attraction contributions in terms of known values of the Hamaker constant. However, determining the repulsive double-layer force is more problematic. In many practical situations, it is difficult to obtain a reliable estimate of a colloidal particle's surface potential or surface charge density. Under these circumstances, a possible strategy is to use electrokinetic measurements, which we can interpret in terms of the zeta potential, ζ. Experience has shown that we can correlate colloidal stability with this readily accessible experimental quantity: Table 8.3 shows how the CCC depends in a systematic way on the measured zeta potentials. Consequently, it is important to see how electrokinetic techniques can be used to determine the zeta potential.

The four electrokinetic phenomena shown in Table 8.4 all play a role in colloidal science. Electrophoresis and sedimentation potential involve the motion of charged particles in a quiescent liquid. Streaming potential and electro-osmosis involve the flow of fluid past a stationary charged surface. We can also classify these phenomena in terms of driving force and response. In

Table 8.3 Comparison of CCC and critical zeta potential.

As_2S_3 Sol(−)			Fe_2O_3 Sol(+)		
Electrolyte	**CCC (mM)**	**Critical zeta potential (mV)**	**Electrolyte**	**CCC (mM)**	**Critical zeta potential (mV)**
KCl	40.0	44	KCl	100.0	33.7
$BaCl_2$	1.0	26	NaOH	7.5	31.5
$AlCl_3$	0.15	25	$CaSO_4$	6.6	32.5
$Th(NO_3)_4$	0.20	27	K_2CrO_4	6.6	32.5
			$K_2Fe(CN)_6$	0.65	30.2

Table 8.4 Summary of electrokinetic effects.

Potential	Stationary (wall)	Moving (colloidal particles)
Applied	Electro-osmosis	Electrophoresis
Induced	Streaming potential	Sedimentation potential

electrophoresis and electro-osmosis experiments, we apply an electrical field and generate particle or fluid flow. In sedimentation potential or streaming potential experiments, we impose an external pressure gradient or a gravitational force and generate an electrical potential. Our main interest lies in electrophoresis, but we will also briefly consider the other electrokinetic effects.

There are many examples of electrokinetic effects in technical processes and biological systems. The flow of liquids containing even a small amount of charge can result in an induced streaming potential. This, in turn, can trigger corrosion processes in pumps and valves. It can even lead to damaging electrical discharges. Electrophoresis is used in biochemical purification processes and surface coating technologies, while electro-osmosis provides a mechanism for inducing flow in thin capillary tubes. Ocean water desalination by reverse osmosis is a technical process of paramount practical and political importance. A pressure is applied across a polymer membrane having channels with negatively charged walls. Anions pass the membrane only with difficulty, and an electrostatic streaming potential develops across the membrane so that the cations are also left behind while water can pass.

8.4.1 We Can Relate the Electrophoretic Velocity of a Colloidal Particle to the Electrical Potential at the Slip Plane

In an electrophoresis experiment, we impose an electrical field across a colloidal solution and measure the colloidal particle velocity, $\boldsymbol{v}$. Often, colloidal particles are large enough to permit the use of a light or dark field microscope to follow their motion directly. We then measure distance traveled versus time to obtain $\boldsymbol{v}$.

The term *electrophoresis* implies the interplay between electrical phenomena and motion. We know that a viscous fluid adjacent to a solid surface moves with the same velocity as the surface, yet at a distance, the surface does not affect the fluid's velocity. Therefore, if we induce motion either of a charged solid particle or in the adjacent solution, we expect to find relative motion between the particle and the counterions. The position δ outside the charged surface, where the liquid begins to flow, defines the location of the zeta potential. Although we expect the distance δ to be on the order of a few molecular diameters, we do not know its exact position.

When we first apply an electrical field, the charged particles experience a force that causes them to accelerate. However, the opposing viscous force increases linearly with velocity so that the particle almost instantaneously reaches a steady-state velocity, $\boldsymbol{v}$, in which electrical force and viscous force are equal. In that situation, we can estimate the velocity using Stokes's law (cf. Eqs 1.7.15–1.7.17) to yield

$$\boldsymbol{v} = \frac{\boldsymbol{F}}{6\pi\eta R} \tag{8.4.1}$$

in which $\boldsymbol{F}$ is the external force experienced by the particle. In general, we do not know the force acting on the particle caused by the external electrical field $\boldsymbol{E}$. For small fields, the force varies linearly with the field, so that $\boldsymbol{F} \propto \boldsymbol{E}$. We can then define the electrophoretic mobility u as

$$u = \frac{\boldsymbol{v}}{\boldsymbol{E}} \tag{8.4.2}$$

This quantity is the primary value determined in an electrophoretic experiment.

For the simple case of an isolated particle with charge q in a field $\boldsymbol{E}$, the force is $q\boldsymbol{E}$, and by combining Eqs. (8.4.1) and (8.4.2), we have

$$u = \frac{q}{6\pi\eta R} \tag{8.4.3}$$

Equation (8.4.3) is of limited use for a colloidal system because we cannot know the extent to which the particle moves independently of the surrounding electrolyte ions. However, we can rephrase the problem and consider the (spherical) particle plus some counterions as one unit with an effective charge q' and effective radius R'.

According to Eq. (3.8.25), in a system for which the Debye–Hückel theory is valid, the potential is

$$\Phi(r) = \frac{q'}{4\pi\varepsilon_r\varepsilon_0} \frac{e^{-\kappa r}}{r} \frac{e^{\kappa R'}}{1+\kappa R'} \tag{8.4.4}$$

Here we can identify the factor q'/R' as in Eq. (8.4.3) by choosing $r = R'$ and we define the zeta potential, ζ, as

$$\zeta \equiv \Phi(R') = \frac{q'/R'}{4\pi\varepsilon_r\varepsilon_0} \frac{1}{1+\kappa R'} \tag{8.4.5}$$

Furthermore, if we assume $\kappa R' \ll 1$, Eq. (8.4.3) provides a simple relation between the electrophoretic mobility and the zeta potential:

$$\zeta = \frac{3}{2} \frac{\eta u}{\varepsilon_r\varepsilon_0} \tag{8.4.6}$$

This relationship is known as the Hückel equation. We have subsumed the uncertainty in both charge q' and radius R' into one quantity, the zeta potential. Even though absolute numbers are uncertain, we should expect the measured ζ to increase if we increase the true charge q, or if we decrease the true radius R. As a result, we can use Eq. (8.4.6) for comparative studies.

In the electrophoretic measurement, we primarily measure the balance between a viscous drag force and an electric pulling force. Our discussion used the Stokes equation for the viscous part and the Debye–Hückel theory for the electric force. Furthermore, we restricted our treatment to small particles under the condition $\kappa R \ll 1$. Now we approach the problem from a slightly different angle. Consider instead the limit of large particles so that $\kappa R \gg 1$. In such a case, it suffices to consider a planar charged surface exposed to a liquid and to balance the viscous and electrical forces in this system, as illustrated in Figure 8.9.

Consider a slice of the liquid with a thickness dx, at a distance x from the surface. The viscous force on either side of the slice is proportional to the gradient dv/dx in the velocity of the liquid. The net viscous force F_v on the slice is proportional to the difference in the gradients and

$$\frac{\mathrm{d}F_v}{\text{area}} = \eta \frac{\mathrm{d}^2 v}{\mathrm{d}x^2} \mathrm{d}x \tag{8.4.7}$$

The electrical force F_E equals the field E times the charge in the slice

$$\frac{\mathrm{d}F_E}{\text{area}} = E\rho \mathrm{d}x \tag{8.4.8}$$

in which we can obtain the charge density ρ from the Poisson equation (Eq. 3.5.13) so that

$$\rho = -\varepsilon_r\varepsilon_0 \frac{\mathrm{d}^2\Phi}{\mathrm{d}x^2} \tag{8.4.9}$$

At steady state, the forces should balance at all x and

$$\eta \frac{\mathrm{d}^2 v}{\mathrm{d}x^2} = \varepsilon_r\varepsilon_0 E \frac{\mathrm{d}^2\Phi}{\mathrm{d}x^2} \tag{8.4.10}$$

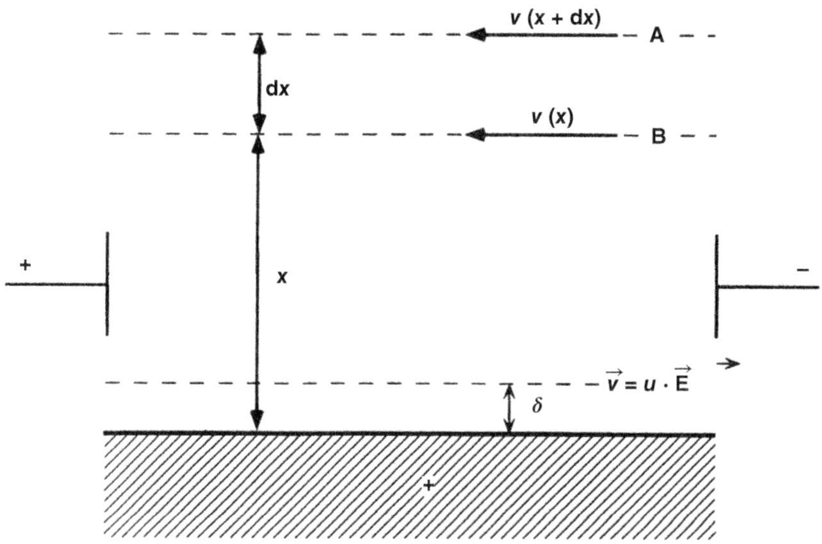

Figure 8.9 Electrophoresis of a large particle $\kappa R \gg 1$. At a distance δ from the surface, the liquid velocity equals that of the particle, while a velocity gradient exists for larger x values. Across the AB sheet, the viscous force balances the electric force.

The general solution of Eq. (8.4.10) is

$$\eta v(x) = \varepsilon_r \varepsilon_0 E\Phi(x) + C_1 x + C_2 \tag{8.4.11}$$

because $v(x)$ and $\Phi(x)$ have proportional second derivatives. Now $v(x)$ and $\Phi(x)$ cannot increase without limit as x increases, so $C_1 = 0$. Furthermore, far from the surface, both $v(x)$ and $\Phi(x)$ should remain constant, and we assign zero for both quantities so that $C_2 = 0$.

In the electrical field, the particle will have a velocity $\mathbf{v} = u\mathbf{E}$. At a small distance δ from the surface, we apply the hydrodynamic stick boundary condition, so that the velocity is $\mathbf{v}(\delta) = u\mathbf{E}$. Writing $\Phi(\delta) = \zeta$, we can express Eq. (8.4.11) as

$$\zeta = \frac{\eta u}{\varepsilon_r \varepsilon_0} \tag{8.4.12}$$

which is called the Helmholtz–Smoluchowski equation. The Hückel equation (Eq. 8.4.6) and the Helmholtz–Smoluchowski equation (Eq. 8.4.12), which are valid in the two limits $\kappa R \ll 1$ and $\kappa R \gg 1$, respectively, differ by a factor of 1.5. These two limits seldom apply strictly in colloidal applications.

Our analysis assumed that the liquid flow or the external electric field scarcely perturbs the electrical double layer outside a charged particle. Full analysis of the dynamic effects occurring in an electrophoresis experiment of a colloidal system presents a major challenge. For example, a more detailed description of the electrophoretic rate of the particles also depends on the mobilities of the counterions and on local viscosities close to the particle. However, in many cases, the simple equations (8.4.6) and (8.4.12) provide a reasonable estimate of the zeta potential.

8.4.2 We Can Determine the Zeta Potential for a Surface by Measuring the Streaming Potential

In a streaming potential measurement, we cause a solution to flow through a capillary tube by applying a pressure gradient across it, as illustrated in Figure 8.10.

This movement displaces part of the double layer in the mobile phase so that a charge accumulates in one flask relative to the other flask. The flow of counterions constitutes a current that generates a potential difference between the two flasks. A steady state is reached at a potential difference Φ_{st}, where the counterion current near the charge surface is compensated by a countercurrent in the bulk liquid.

The fluid flow through the capillary follows Poiseuille's law (Eq. 5.8.9), and the fluid velocity at radius r in a capillary of radius R, length l and with a pressure drop Δp is

$$v = \frac{\Delta p}{4\eta l}(R^2 - r^2) \tag{8.4.13}$$

The electrical current $\mathrm{d}I$ associated with the fluid flow in a shell of area $\mathrm{d}A = 2\pi r\mathrm{d}r$ is

$$\mathrm{d}I = \rho v \mathrm{d}A = \rho \frac{\Delta p}{4\eta l}(R^2 - r^2)\,2\pi r\mathrm{d}r \tag{8.4.14}$$

Here ρ stands for the volume charge density. If we replace r by a distance measured from the surface of shear, $x = R - r$, and assume $x \ll R$, we can write

$$\mathrm{d}I = \frac{\rho\Delta p}{4\eta l}(2Rx - x^2)2\pi(R - x)\,\mathrm{d}x \approx \frac{\rho\Delta p}{\eta l}\pi R^2 x\mathrm{d}x \tag{8.4.15}$$

From the point of view of the electrical double layer, the surface of the capillary is flat ($\kappa R \gg 1$). Using the Poisson equation (Eq. 3.5.13) for ρ yields

$$\mathrm{d}I = \frac{\pi\varepsilon_r\varepsilon_0\Delta p R^2}{\eta l}\frac{\mathrm{d}^2\Phi}{\mathrm{d}x^2}x\mathrm{d}x \tag{8.4.16}$$

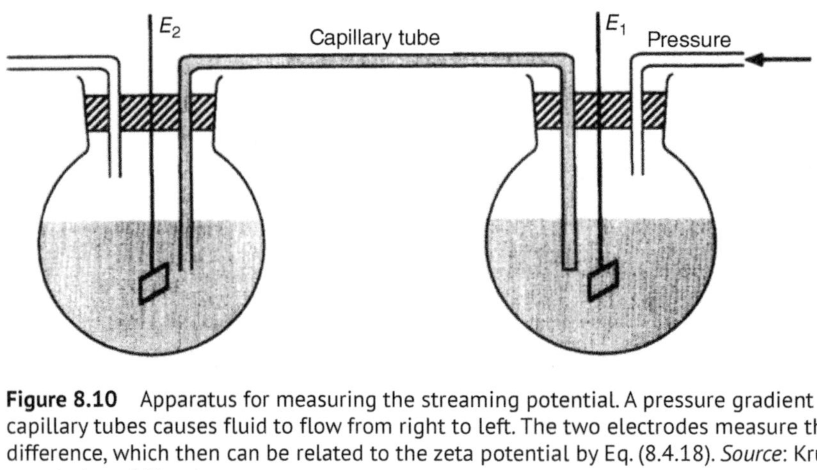

Figure 8.10 Apparatus for measuring the streaming potential. A pressure gradient applied across the capillary tubes causes fluid to flow from right to left. The two electrodes measure the resulting potential difference, which then can be related to the zeta potential by Eq. (8.4.18). *Source*: Kruyt, 1952/with permission of Elsevier.

Through an integration by parts

$$I = \frac{\pi \varepsilon_r \varepsilon_0 \, \Delta p R^2}{\eta l} \left(x \frac{d\Phi}{dx} \bigg|_0^R - \int_0^R \frac{d\Phi}{dx} dx \right) = \frac{\pi \varepsilon_r \varepsilon_0 \, \Delta p R^2}{\eta l} \zeta \tag{8.4.17}$$

using $\Phi = \zeta$ at $x = 0$ and $\Phi = \frac{d\Phi}{dx} = 0$ at $x = R$.

The current in Eq. (8.4.17) is balanced by a current in the bulk liquid of conductivity, k_b. The resistance in the tube is $(\pi R^2 k_b)^{-1}$ per unit length, and the counterion current of Eq. (8.4.17) is balanced by the current in the bulk at the potential difference

$$\Phi_{st} = \frac{\varepsilon_r \varepsilon_0 \, \zeta}{\eta} \frac{\Delta p}{k_b} \tag{8.4.18}$$

We can determine the zeta potential by measuring Φ_{st} versus Δp and knowing ε_r, η, and k_b of the bulk. If the solid surface is also conducting, a current can occur in the substrate and Eq. (8.4.18) is no longer valid. In practical situations, like in hydraulic systems, currents in the solid material can lead to corrosion problems. Thus, hydraulic oils are often made weakly conducting to prevent corrosion.

8.4.3 Electro-osmosis Provides Another Way to Measure the Zeta Potential

In an electro-osmosis experiment, we apply a potential Φ_{eo} across two electrodes connected by a capillary tube of length L. For charged surfaces, the electrical field exerts a force on the counterions in the double layer. This force causes the counterions to move, and they carry the solution in the tube along with them. Because the tube's radius R is on the order of millimeters, the double layer is virtually planar, and $\kappa R \gg 1$.

In electrophoresis, we considered a moving particle in a fixed solvent, but in electro-osmosis, the converse is true. However, the relative motion between surface and bulk solution is the same, and, because $\kappa R \gg 1$, the Helmholtz–Smoluchowski equation applies (Eq. 8.4.12). We obtain the volume flow of liquid dV/dt as the product of the velocity and the area A of the tube. The velocity is related to the mobility u and the electric field, $E = \Phi_{eo}/L$ by Eq. (8.4.2) so that we can relate the volume flow to the zeta potential through

$$\frac{dV}{dt} = vA = uEA = \frac{\pi \varepsilon_r \varepsilon_0 \zeta R^2 \Phi_{eo}}{\eta L} \tag{8.4.19}$$

Electrophoretic measurements by the microscope method are complicated by the simultaneous occurrence of electro-osmosis. The internal glass surfaces of the cell are usually charged, which causes an electro-osmotic flow of liquid near the tube walls. Because the cell is closed, the charged glass surfaces also cause a compensating return flow of liquid with maximum velocity at the center of the tube. This combination results in a parabolic distribution of liquid speeds with depth, and we can observe the true electrophoretic velocity only at locations in the tube where the electro-osmotic flow and the return flow of the liquid cancel each other. For a cylindrical cell, this "stationary level" is located at 0.146 of the internal diameter from the cell wall.

8.4.4 A Sedimentation Potential Is Generated by Settling Charged Particles

When a charged particle settles in a gravitational field, one has a situation that is opposite to electrophoresis. When the particles move in the liquid, it will pull on the counterions, which will also move but lag behind, as illustrated in Figure 8.11. The resulting charge separation is

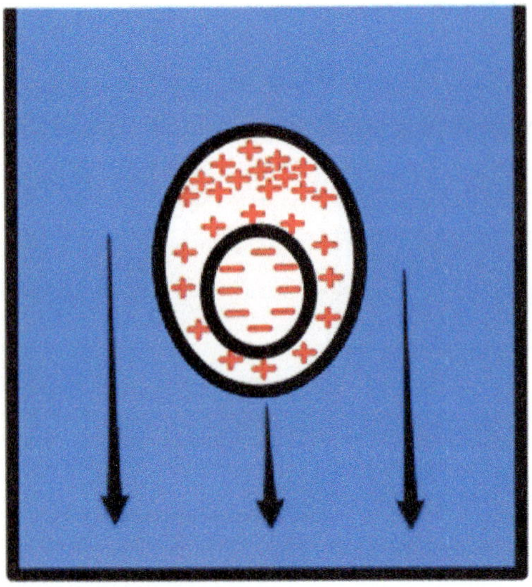

Figure 8.11 For a negatively charged particle settling by gravity in a fluid, the lagging counterions generate a dipolar field.

equivalent to a dipole whose electrical field develops around the particle. The longer the Debye screening length, the larger the dipole. For a single particle, this dipolar field gives a local perturbation in the liquid. However, for a solution of particles of volume fraction ϕ_p, there is a cooperative phenomenon. Consider a situation where the particle concentration is low so that the average particle separation is larger than the screening length, and direct particle–particle interaction is negligible. The perpendicular components of the dipolar fields from separate particles tend to cancel, while the components in the velocity direction add. From an electrostatic point of view, one has a situation similar to what is present in a capacitor, as illustrated in Figure 3.6. The net result is an electric field, and consequently a position-dependent potential, called the sedimentation potential.

More than a 100 years ago, von Smoluchowski derived an expression for the field E_s in a system with sedimenting particles with a radius larger than the screening length.

$$E_s = \varepsilon_r \varepsilon_0 g \Delta\rho \phi_p \zeta/(\eta k_b). \tag{8.4.20}$$

Here $\Delta\rho$ is the difference between the mass densities of the particles and the medium. At higher concentrations, there are effects from the direct interaction between the particles. In addition, the hydrodynamic coupling, discussed in Section 5.8, affects the sedimentation rate.

Thunder provides a dramatic manifestation of a sedimentation potential. In a cloud that is dominantly negatively charged, water droplets settle toward the lower end of the cloud, leaving a positively charged upper side of the cloud. Due to the very low conductivity of the atmosphere, this charge separation can result in very high differences in electrostatic potential. However, the situation within a cloud in a developing thunderstorm is much more complex than what is described by Eq. (8.4.20).

See Berg (2010, Chapter 6), Cosgrove (2010, Chapter 2), Doi (2013, Chapter 10), Lyklema (1995, Vol. 2, Chapter 4), and Ohshima (2024, Chapters 12–15) for additional discussion on Electrokinetics.

Exercises

8.1 Consider two identically charged spherical colloidal particles with a surface potential of 25 mV dispersed in an aqueous medium with $\kappa^{-1} = 1$ nm. The effective Hamaker constant is 7×10^{-21}J. Determine the separation h_0 at which the interaction potential goes through zero between the barrier and the primary minimum.

8.2 Calculate the interaction potential between two similar charged spheres of radius 10^{-7} m and a surface potential of −50 mV. The NaCl concentration of the aqueous medium is 10 mM and the Hamaker constant is 5×10^{-20} J.

8.3 Calculate the quantity

$$-\left(\frac{\partial^2 V}{\partial h^2}\right)_{h=h_{max}}$$

for a system of spherical colloidal particles of radius 100 nm and surface potential 35 mV in a 2.0 mM KCl solution using the Hamaker constant $H = 5 \times 10^{-20}$ J and $\mathbf{T} = 300$ K. The second derivative of the interaction potential should be evaluated at the primary maximum.

8.4 Tuorila (1926) measured the rapid coagulation of a gold sol, radius 50 nm, and obtained the following values:

t(*s*)	Number of particles/m^3×10^{-14}
0	20.2
30	14.7
60	11.2
120	7.74
240	4.78
480	2.71

Determine the association rate constant and compare it with the theoretical value for a diffusion-controlled reaction.

8.5 Estimate the CCC in water at $\mathbf{T} = 300$ K for highly charged spherical colloidal particles for counterion valencies $z = 1$, 2, and 3. $(H_{121} = 1.0 \times 10^{-20}\ \text{J})$.

8.6 Calculate the expected value of the stability ratio W for two spheres of radius 100 nm in a 2.0 mM KCl aqueous solution at $\mathbf{T} = 300$ K with surface potential 35 mV and Hamaker constant $H = 5 \times 10^{-20}$ J.

8.7 Polymer chains in a solvent can be viewed as fractal objects. What is the fractal dimension (a) at theta conditions and (b) in a good solvent?

8.8 Spherical particles of radius 100 nm aggregate by a diffusion-limited process. Estimate the size of a cluster of (a) 100 particles and (b) 1000 particles, assuming a fractal aggregate of dimensionality 1.8.

8.9 Dunstan and Saville (1992) measured the electrophoretic mobility of intrinsically uncharged alkane particles, average radius 500 nm, in aqueous salt solutions and found the following electrophoretic mobilities, u:

Electrolyte	Concentration (M)	$U \times 10^8$ ($m^2s^{-1}V^{-1}$)
$AlCl_3$	10^{-3}	3.1
	10^{-2}	3.0
HCl	10^{-3}	−0.9
	10^{-2}	±0
KCl	10^{-3}	−2.7
	10^{-2}	−2.1
$CaCl_2$	10^{-3}	−2.1
	10^{-2}	−1.8

Determine the zeta potential for these cases. Estimate the minimum magnitude of the surface charge. Discuss the molecular origin of the charge. If the surface charge arises from impurities, what is the lower value of the concentration of the impurity if the particles are present at a volume fraction of 0.001? Using a more general theory, Dunstan and Saville estimated the following zeta potentials (mV): 40, 35, −15, 0, −45, −30, −35, −25 in the order given in the table. How do these values deviate from the ones you have calculated?

8.10 Stigter and Mysels (1955) measured the electrophoretic mobility of SDS micelles as a function of added NaCl. Their results are summarized as follows.

NaCl (m/L)	CMC (mM/L)	N	10^8u ($m^2V^{-1}s^{-1}$)	$10^{10}R_{mic}$(m)	κR_{mic}
0	7.12	80	4.55	21.5	0.61
0.01	5.29	89	4.26	22.1	0.86
0.03	3.13	100	3.84	23.0	1.32
0.05	2.27	105	3.63	23.4	1.69
0.1	1.46	112	3.42	24.0	2.40

Calculate the zeta potential for the five cases using the Smoluchowski equation.

8.11 A spherical colloid with a radius of 50 nm has a mobility of 4.0×10^{-8} $m^2V^{-1}s^{-1}$ in a 0.001 M KCl solution. Calculate the zeta potential using the Smoluchowski equation and then the Henry correction.

8.12 Calculate the electro-osmotic transport of an aqueous solution through a glass capillary 5 cm long with a 0.5 mm internal diameter under a potential difference of 100 V when the zeta potential is 80 mV. The viscosity of the solution is 1×10^{-3} kg $m^{-1}s^{-1}$ and its dielectric constant $\varepsilon_r = 79$.

8.13 Calculate the expected streaming potential and streaming current (i.e., charge transported by convection) if the zeta potential is –100 mV, and the solution conductance is 1×10^{-3} $\Omega^{-1}m^{-1}$. The pressure difference is 0.1 atm across the capillary tube, which is 30 cm long and 1 mm in diameter. Also, calculate the resistance of the capillary using the parameters of the previous problem. Repeat the calculation for the flow of a hydrocarbon (gasoline) through the capillary under the same conditions; specific conductance equals 1×10^{-6} Ω^{-1} m^{-1} and $\varepsilon_r = 2$. Do you see why the first thing that is done when delivering fuel to an airplane via a meter-long fueling line is to connect a grounding wire between the airplane and the pumping track?

8.14 In an electro-osmosis experiment, the flow rate was 2×10^{-5} $m^3 s^{-1}$, the current through the cell was 0.5 mA, and the conductance of the solution was 1×10^{-3} $\Omega^{-1}m^{-1}$. The viscosity of the solution is 1×10^{-3} kg, $m^{-1}s^{-1}$ and its dielectric constant is $\varepsilon_r = 79$. Calculate the zeta potential of the porous plug. (You can neglect the pressure difference on both sides of the plug.) If the effective radius of the porous plug is 0.5 mm and its length is 15 cm, what pressure difference across the plug would result in zero net flow of solution?

8.15 In an electrophoresis experiment, we visually measure the distance traveled by individual particles over a set period of time and calculate an average displacement from a number of observations. The distance an individual particle moves reflects a combination of its motion under the influence of an electrical field and the random movement resulting from Brownian motion. Suppose that you determine the electrophoretic mobility of 4.0×10^{-7} $m^2V^{-1}s^{-1}$ for a colloidal sol in water.

(a) How far does the particle move under an electric field of 4000 V/m in 20 s?

(b) Calculate the maximum and minimum distance value you might expect when the effect of Brownian motion is taken into account for particles with a radius of 500 nm. Repeat the calculation for particles $R = 5$ nm.

References

Berg, J.C. (2010). *An Introduction to Interfaces & Colloids. The Bridge to Nanoscience.* World Scientific.

Cosgrove, T. (2010). *Colloid Science. Principles, Methods and Applications.* Chichester, UK: Wiley.

Doi, M. (2013). *Soft Matter Physics.* Oxford, UK: Oxford University Press.

Dunstan, D.E. and Saville, D.A. (1992). *J. Chem. Soc. Faraday Trans.* 88, 2031. Israelachvili, J. N. (2011). *Intermolecular and Surface Forces*, 3rd ed. London, UK: Academic Press.

Kronberg, B., Holmberg, K., and Lindman, B. (2014). *Surface Chemistry of Surfactants and Polymers*, 3rd ed. Chichester, UK: Wiley.

Kruyt, H.R. (1952). *Colloid Science*, Vol. 1, New York: Elsevier.

Kruyt H.R. and van Arkel, A.E. (1920). *Rec. Trav. Chim.* 39, 656.

Lattanzi, V., André, I., Gasser, U., Dubackic, M., Olsson, U., and Linse, S. (2021). *Proc. Natl. Acad. Sci.* 118, e2112783118.

Lyklema, H. ed. (1995). *Fundamentals of Interface and Colloid Science. Solid-Liquid Interfaces, Vol.* 2. Amsterdam, The Netherlands: Elsevier.

Lyklema, H. ed. (2005a). *Fundamentals of Interface and Colloid Science. Particulate Colloids, Vol.* 4. Amsterdam, The Netherlands: Elsevier.

Ohshima, H. (2024). *Fundamentals of Soft Interfaces in Colloid and Surface Chemistry*. London, UK: Academic Press.

Shaw, D.J. (1980). *Introduction to Colloid and Surface Chemistry*, 3rd ed., London: Butterworth-Heinemann.

Stigter, D. and Mysels, K.J. (1955). *J. Phys. Chem*. 59, 45.

Törnqvist, M., Cukalevski, R., Weininger, U., Meisl, G., Knowles, T.P.J., Leiding, T., Malmendal, A., Akke, M., Linse, S. (2020). *Proc. Natl. Acad. Sci*. 117, 11265.

Tuorila, T. (1926). *Kolloid Chem. Beih*. 22, 191.

Weitz, D.A., Lin, M.Y., and Huang, J.S. (1987). *Physics of Complex and Supramolecular Fluids*, Safran, S.A. and Clark, N.A. eds., p. 518, New York: Wiley.

Wennerström, H., Vallina, E., Danielsson, J., and Oliveberg, M. (2020). *Proc. Natl. Acad. Sci*. 117, 10113.

9

Colloidal Sols

Concept Map

Preparation of Colloidal Particles

In Chapter 8, we described the general principles that govern colloidal stability; here we consider specific colloidal sols that demonstrate typical properties and illustrate some of the complexities encountered in practical applications.

Three basic strategies for forming colloidal particles involve (1) dispersion in which large particles are broken into small ones, (2) condensation, in which exceeding an equilibrium condition, such as a solubility limit, leads to the formation of particles of a new phase, and (3) chemical synthesis where monomers are attached covalently to a growing particle. Usually, these processes lead to heterogeneous sols that possess a wide range of particle sizes.

One way of preparing oligodisperse colloids is to separate and control nucleation and growth, as illustrated in Figure 9.1. The idea is to induce nucleation in a short burst by briefly exceeding the nucleation concentration and then promote growth by maintaining the reactant concentration low enough to avoid nucleation but high enough to promote particle growth.

Potential-determining Ions

We prepare AgI sols by exceeding the solubility product $c(Ag^+)c(I^-) = 7.5 \times 10^{-17} M^2$.

Iodide ions adsorb preferentially on AgI particles so that the concentration of I^- (or Ag^+) sets the potential on the AgI particles.

We divide ions into two categories: (1) indifferent ions such as KNO_3, which affect interaction by changing the screening length κ^{-1}, and (2) potential-determining ions Ag^+ and I^-, which directly change Φ_0 and thus can have a profound effect on colloidal stability.

With proteins and latex spheres containing carboxyl and/or amino groups, H^+ ions are potential determining, while in a sol like $Al(OH)_3$, the ions Al^{3+} and OH^- play the same role; in either case, the surface potential depends on pH. Some of the complexities encountered in colloidal systems are illustrated by the pH dependence of the Al^{3+} system, which forms a variety of complexes whose electrical charge is set by concentration and pH (Figure 9.3).

The Colloidal Domain: Where Physics, Chemistry, Biology, and Technology Meet, Third Edition. Håkan Wennerström.

Clays

Clay minerals are silicon–aluminum systems that crystallize in sheets kept together in multilayers by electrostatic and dispersion forces. The individual layers can be uncharged as in talc or kaolinite or carry an electrostatically coordinated ion like K^+ in mica or Na^+ in montmorillonite.

Talc and kaolinite are easily cleaved along the sheet plane, but they don't swell in water and have a partly hydrophobic character.

Mica, on the other hand, swells in water once cleaved and can be used to accurately measure surface forces.

Montmorillonite swells spontaneously in water and has a use as an ion exchange material.

Finite clay particles can associate edge-to-plane and form a "house of cards" extended structure. In the presence of Ca^{2+}, particles can associate to tactoids, consisting of parallel sheets separated by a thin, 1 nm, aqueous layer.

Latex Particles Can Be Synthesized into Many Different Structures

Emulsion polymerization of hydrophobic monomers yields latex particles. Straightforward polymerization leads to size polydispersity of the product. With care, one can obtain nearly uniform particle size. Other reaction conditions can lead to "janus" particles, core-shell particles, shell particles, or thermo-responsive particles (see Figure 9.8).

Using Self-assembly Features of Surfactants and Block-copolymers, One Can Make Particles with Internal Structure

The spontaneous self-assembly of surfactants and block-copolymers can be utilized to make a number of particles with uniquely designed properties. Mesoporous silica can be made from a templating phase of a block-copolymer. Quantum dots can be produced by using a water-in-oil microemulsion as a templating structure. Lipid nanoparticles have been formed from chemically modified lipids. Dispersion of a liquid crystalline phase can result in polymer-stabilized particles for controlled release applications (see Figure 9.9).

Aerosols

Aerosols containing solid or liquid particles dispersed in gas (air) play important roles in many medical, technological, and environmental contexts.

In a gaseous medium, van der Waals forces dominate particle interactions, and the systems are not colloidally stable, even when particles are charged. Hetero- and homocoagulation are equally likely to occur, often leading to the formation of open fractal-like aggregates.

Aerosols are often charged, for example by cosmic radiation, and they migrate in an electrical field, which can be used for air purification.

Particles settle in a gravitational field and can generate a gravitational potential. There are three motional regimes for settling. When colloidal particles move in a medium, one is usually in the low Reynolds number regime, where Stokes' law applies. This regime is also valid for smaller particles, but when they grow, motion is faster as for raindrops, and one approaches the turbulent high Reynolds number regime. At low gas densities, as in the upper atmosphere, the mean free path length of gas atoms can be of the same magnitude as the particle diameter, and the medium can no longer be considered as homogeneous.

Water drops are in a dynamic exchange with water molecules in the gas phase and the equilibrium drop–gas is affected by surface free energy, drop composition, temperature gradients, electric charge effects, as well as the presence of a solid particle in the drop.

During the nineteenth and early twentieth century, the study of small solid particles dispersed in liquid played a pivotal role in establishing the atomic nature of matter and defining the colloidal domain. In addition, analysis of sols has helped to establish fundamental relationships in nature. In the early twentieth century, there were important contributions from von Schmulokowski, Perrin, and the immensely insightful Einstein. For example, both the kinetic nature of matter as manifested by Brownian motion and the first determination of Avogadro's number involved colloidal sols. More recently, characterization by neutron scattering and modern microscopy visualizations of colloids such as latex spheres has provided insights into phase transformations.

Colloidal sols provide the basis of many industrial processes and products and there is currently a large interest in developing applications of colloidal particles. Processing of clay–water dispersions produces ceramics ranging from delicate china figurines to toilet bowls and masonry bricks. Applying modern colloid technology to silica sols yields tough, fracture-resistant ceramics that eventually find application ranging from household devices such as knives and non-stick saucepans to high-temperature automobile engine parts, rocket nose cones, and long-lasting medical prostheses.

Virtually all coating systems employ colloidal materials, and most industrial products are coated for either protective or decorative reasons. Paints are colloidal suspensions that may contain titanium dioxide or zinc oxide and latex particles. Drilling muds, which are indispensable as lubricants and rheology control agents in oil exploration, are complex colloidal materials. Papermaking involves producing a meshwork from colloidal cellulose fibers in which clay particles can be used as filler to improve opacity and to produce a pleasing surface texture. Inks used in ordinary ballpoint pens, xerography, and high-speed printing presses are colloidal liquids or pastes. Toothpaste, scouring powders, and other heavy-duty cleaning agents contain colloidal materials such as pumice.

Colloidal particles are increasingly used in drug and vaccine formulation. Inhalation of aerosol particles provides a method to administer a drug specifically to the lungs. Incorporating a drug into a polymer or lipid matrix particle can result in a controlled release. A recent major success was achieved when RNA molecules could be incorporated into lipid nanoparticles, allowing for entry into cells producing an effective COVID vaccine.

Control of colloidal systems is also a key issue in dealing with a host of environmental problems associated with our heavy use of technology. Smog consists of colloidal-sized particles generated by an atmospheric photochemical reaction involving petroleum as well as natural products. Controlling the fine powders and other colloidal debris associated with processing at wood pulping plants, mineral flotation sites, coal grinding processing units, and asbestos plants requires the application of colloidal chemical techniques. Some of the key steps in the purification of water and the treatment of sewage also depend on the adsorptive capacity of colloidal materials.

In the previous chapter, we discussed the basic features of colloidal stability and instability. In applications, the stability versus association is a general theme. In this chapter, we aim to illustrate how the stability/instability dichotomy can be handled in practice, both in the production of particles and in the application. The field is very rich, and we merely illustrate some general themes based on specific examples that provide useful insight into the world we live in.

9.1 Colloidal Sols Can Be Formed by Dispersion, Precipitation, or Chemical Synthesis

Three basic strategies for preparing colloidal sols are dispersion, which involves breaking large particles into small ones, condensation, in which exceeding the solubility limit leads to the formation of a new phase, and chemical synthesis in which a particle is formed by successive condensation chemical bond formation.

9.1.1 Particles from Ground Solids

In the dispersion method, a mechanical process breaks bulk material down to colloidal dimensions. For example, the titanium dioxide used in paints is processed in a ball mill. We can view clays as colloidal materials that originally were crystallized from silicon, aluminum, and oxygen compounds and subsequently dispersed into colloidal-size particles by natural forces. Iron powder made from a porous iron source is a major industrial product used to make metal parts by sintering the powder at high temperature and pressure.

It was shown by the Russian mathematician A. Kolmogorov that such a random fragmentation results in a log-normal size distribution

$$F(x) = \frac{1}{x\sigma\sqrt{2\pi}} \exp\left\{ -\frac{[\ln(x) - \mu]^2}{2\sigma^2} \right\} \tag{9.1.1}$$

Here x denotes the particle size, μ the mean value of $\ln(x)$ and σ^2 the variance. This is a very broad distribution and to have a system for practical use, the fragmentation is typically combined with some size-selective fractionation. Another practical problem is that in a grinding process, small particles tend to agglomerate during grinding and to resist redispersion in subsequent processing steps. Use of a suitable dispersing agent can control such an agglomeration. For example, if we want to grind sulfur in a ball mill, we can add sucrose to coat the surface of the particles and prevent them from adhering. We can remove the sugar later by dialysis. Another dispersion method involves passing an electric arc between two plates submerged in a liquid. This process tears small particles from the metal surfaces. If we perform the dispersion in the absence of oxygen, we obtain a metal sol. Otherwise, the sol's surface is covered by metal oxide.

9.1.2 Particles Can Be Formed by Precipitation

Condensation methods are much more widely used and more versatile than dispersion methods. They include chemical reaction, condensation from the vapor, and dissolution and reprecipitation.

The reaction

$$KI + AgNO_3 \rightarrow AgI(sol) + K^+ + NO_3^- \tag{9.1.2}$$

exemplifies how we can form a sol by exceeding the solubility limit of AgI. To stabilize the sol, we can reduce the concentration of unprecipitated ions by dialysis.

Condensation processes can form new colloidal phases of a variety of materials. For example, reduction of the gold chloride complex by hydrogen peroxide

$$2Au(Cl^-)_4 + 3H_2O_2 \rightarrow 2Au(sol) + 6HCl + 3O_2 + 2Cl^- \tag{9.1.3}$$

forms colloidal gold. Oxidation of hydrogen sulfide with oxygen

$$2H_2S + O_2 \rightarrow 2S(sol) + 2H_2O \tag{9.1.4}$$

forms the sulfur sol. Heating an aqueous solution of $FeCl_3$

$$FeCl_3 \rightarrow Fe(OH)_3(sol) + 3HCl \tag{9.1.5}$$

forms the azeotrope of HCl and simultaneously the $Fe(OH)_3$ sol.

The dissolution–reprecipitation method involves, as an example, rapidly injecting a solution of steric acid in ethanol into a large volume of water

$$C_{17}H_{35}COOH_{EtOH} + H_2O \rightarrow \text{fatty acid(sol)}$$

Because all the processes just described involve concurrent nucleation and growth, they lead to colloidal particles with a range of particle sizes. For example, when we simply mix together a solution of $AgNO_3$ and KI, we create small regions of solution in which the concentration of Ag^+ and I^- ions exceeds the solubility product. This condition initiates the nucleation process and growth follows. As mixing continues, nucleation begins in other regions of the solution. Sequential nucleation, accompanied by continuous growth, results in sols with a wide distribution of particle sizes. Although such sols have many uses, monodisperse colloids are desirable in scientific investigations and also in some refined applications.

9.1.3 Controlling Nucleation and Growth Steps Can Produce Oligodisperse Sols

The control of the nucleation and growth processes is a key to the formation of a narrow particle size distribution, oligodisperse sols. As we saw in Chapter 2, homogeneous nucleation occurs at concentrations or vapor pressures that exceed the equilibrium value by a considerable amount. Thus in the absence of seeding particles, precipitation of colloidal material begins at a concentration of reactants that considerably exceeds the solubility product.

Figure 9.1 illustrates a strategy used to produce oligodisperse colloidal sols. Two important concentrations are involved: the nucleation threshold and the solubility limit or saturation

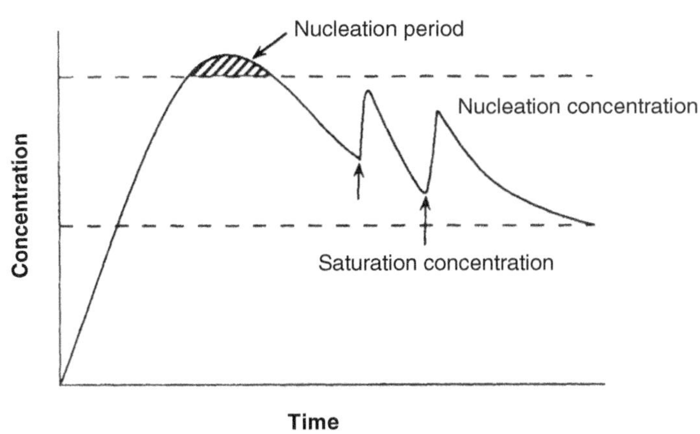

Figure 9.1 Producing a monodisperse sol requires control and separation of nucleation and growth processes. The number of nuclei is set by increasing the concentration of reactants above the nucleation concentration for a brief period. Particle growth is controlled by maintaining the reactant concentration below the nucleation threshold but above the equilibrium concentration long enough to obtain the desired particle size. Arrows indicate small additions of reactants after the nucleation step.

concentration. The formation process is initiated by adjusting the temperature or concentration so that the solute concentration uniformly exceeds the nucleation threshold. The result is that nucleation occurs in a single short burst that causes the concentration of reactants to fall below the nucleation threshold. We can control subsequent growth by adjusting reaction conditions to remain below the nucleation threshold, but above the equilibrium line, by consecutive small additions of reactants or by changing the temperature. The number of nuclei formed in the initial stage determines the number of particles; the length of the growth period determines their size.

Formation of gold sols with a narrow size distribution illustrates how we can separate nucleation and growth entirely. In the nucleation step, small gold particles are formed by the reaction

$$\mathrm{AuCl_4^-} + \mathrm{P(red)} \rightarrow \mathrm{Au(nuclei)} + \text{reduction products} \tag{9.1.6}$$

The growth step involves adding a mild reducing agent (such as formaldehyde) along with more $AuCl_4^-$ producing gold atoms in solution. Under conditions that allow no new nuclei to be formed, the gold atoms associate with the existing gold particles in a growth process

$$\mathrm{Au(N)} + \mathrm{Au} \rightarrow \mathrm{Au(N+1)}. \tag{9.1.7}$$

Typically, in such a reaction the association rate is diffusion-controlled and we can use this circumstance to analyze growth dynamics. The growth rate is then

$$\frac{\mathrm{d}N}{\mathrm{d}t} = k_d(N)\{c_{Au} - c_{Au}(eq) - c_{Au}(OR)\}. \tag{9.1.8}$$

Here $c_{Au}(eq)$ is the saturation concentration of gold atoms and $c_{Au}(OR)$ the correction term due to the particle surface free energy causing Ostwald ripening. These two concentrations determine the off rate, which must be deducted from the on-rate to describe the net growth. For the case of a small reactant reaching a larger particle, the diffusion-limited rate constant $k_d(N)$ is proportional to the particle radius R and the diffusion coefficient of the small reactant (see Eq. 1.7.11)

Assume that we can add reactants continuously in such a way that the concentration of gold atoms is kept constant and ignore the small effect of the Ostwald ripening term. Then Eq. (9.1.8) simplifies to

$$\frac{\mathrm{d}N}{\mathrm{d}t} = a(\Delta c)N^{1/3} \tag{9.1.9}$$

Here $a(\Delta c)$ is a constant and we have used the fact that the particle radius is proportional to the third root of the volume, which in turn is determined by the number of atoms N in the particle. It is straightforward to integrate Eq. (9.1.9) and for a particle of size $N(0)$ at $t = 0$ the size at a time t later is

$$N(t) = \left\{ \mathrm{N}(0)^{\frac{2}{3}} + \frac{2}{3}\,\mathrm{a}(\Delta c)t \right\}^{3/2} \tag{9.1.10}$$

Problem: Compare the growth of two particles A and B where A has twice the volume of B at $t = 0$ ($N_{A0} = 2N_{B0}$). Show that the difference in size increases, while the relative size decreases with increasing t.

Solution: From Eq. (9.1.10) follows $\frac{N_A(t)}{N_B(t)} = 2 \times \frac{\{1 + \sqrt[3]{2}\, a(\Delta c)t/(3N_{B0}^{2/3})\}^{3/2}}{\{1 + 2\, a(\Delta c)t/(3N_{B0}^{2/3})\}^{3/2}} < 1$, which shows that the relative size decreases. The difference in size is

$$N_A(t) - N_B(t) = 2N_{B0}\left\{1 + \sqrt[3]{2}\frac{a(\Delta c)t}{3N_{B0}^{2/3}}\right\}^{3/2} - N_{B0}\left\{1 + 2\frac{a(\Delta c)t}{3N_{B0}^{2/3}}\right\}^{3/2}$$
$$\approx N_B\left\{1 + \frac{2\left(\sqrt[3]{2} - 1\right)}{N_{B0}^{2/3}}a(\Delta c)t\right\} > N_B$$

showing that the difference in size, measured on an absolute scale, increases.

As the problem shows, the growth step increases the homogeneous nature of the sol when diffusion of material to the particle surface is the rate-determining step. A similar result is obtained for the case when a surface reaction is rate-limiting and the rate is proportional to the particle area. Only in the unlikely event that the rate is proportional to the particle's volume does narrowing of the distribution fail to occur. For this case, the exponent is one instead of 1/3 as in the growth Eq. (9.1.9). With unity as the exponent, a particle will grow exponentially in size with time, while the size ratio

$$\frac{N_A(t)}{N_B(t)} = \frac{N_A(0)\exp\{a(\Delta C)t\}}{N_B(0)\exp\{a(\Delta C)t\}} = \frac{N_A(0)}{N_B(0)} \tag{9.1.11}$$

remains constant. Thus, the growth process usually leads to a narrowing of the particle size distribution. One exception is the Ostwald ripening (see Chapter 2), where large particles grow and small ones decrease in size.

9.1.4 Chemical Synthesis Can Be Used to Make Particles with a Range of Properties

The most versatile ways of making colloidal particles make use of a chemical synthesis, where monomer molecules are coupled to the growing particle through the formation of a covalent bond. It is an art to choose reaction conditions that give the particles the desired properties with respect to composition, size, and shape, as well as size distribution. Here we illustrate the principles of this way of particle preparation by two examples.

The synthesis of silica particles is often accomplished through the so-called Stöber process. The starting material is a tetraethyl orthosilicate or an analogous compound. This can be dissolved in a water–alcohol mixture. On addition of base or acid, the ethanol–silica bond is hydrolyzed, exposing Si—O^- or Si—OH groups, which react to form a siloxane Si—O—Si bridge. Since the Si is tetravalent, the condensation results in a compact solid structure.

The synthesis of latex particles relies on one of the standard polymer synthesis methods, primarily yielding a linear polymer. The reaction can be controlled by letting the polymerization occur in an aqueous medium, where the monomer has a low but finite aqueous solubility, but the ionic initiator is highly soluble. As the monomer is consumed by the polymerization, the solution is replenished by monomers stored in dispersed emulsion drops and the method is called emulsion polymerization. Due to the ionic initiator, the emerging polymer chain carries a charge. In the initial stage of the process, short growing chains associate to form small particles. At some point, particles have acquired such a charge and size that they repel one another and further growth in size occurs by monomer addition. One has then growth conditions similar to those discussed for gold particles in Section 9.1.2.

See Berg (2010, Chapter 5), Kronberg et al. (2014, Chapter 21), and Lyklema (2005, Vol. 4, Chapter 2) for additional discussion on Particle Synthesis.

9.2 Colloidal Particles Acquire Surface Charges by Specific Ion Adsorption

9.2.1 The Concentration of Silver and Iodide Ions Determine the Surface Potential of Silver Iodide Sols

The AgI-sol, one of the most thoroughly characterized colloidal systems, has played an important role in developing our understanding of colloidal suspensions. The solubility product of AgI is 7.5×10^{-17} M^2 at 25 °C. Under conditions in which $c(Ag^+) = c(I^-) = 8.7 \times 10^{-9}$ M, electrophoretic measurements on AgI colloidal particles show that the particles carry a negative charge. By varying the concentration of Ag^+ (or I^-) by adding an appropriate salt, we can determine that when $c(Ag^+) > 3.0 \times 10^{-6}$ M, the particles are positively charged, while below this value they are negative. Thus, the point of zero charge pzc occurs when $c(Ag^+) = 3.0 \times 10^{-6}$ M. Furthermore, the potential on the particles, as determined by electrophoretic measurements, varies linearly as a function of the logarithm of added Ag^+ (or I^-) ions. This shows that the relative concentrations of Ag^+ and I^- ions in solution determine the potential on the surface of the AgI particles.

We can understand this behavior in terms of the ion binding model of Section 3.7. At equilibrium, the chemical potentials of the Ag^+ and I^- ions in solution equal those on the colloidal particle, and on the surface, the chemical potential of the ions varies with the electrostatic surface potential. The point of zero charge occurs when the ions are adsorbed equally, and the fact that $c_{pzc}(I^-) = 2.5 \times 10^{-11}$ M is much smaller than $c_{pzc}(Ag^+) = 3.0 \times 10^{-6}$ M simply reflects the readier adsorption of I^- ions on the AgI surface.

Away from the point of zero charge, a nonzero potential Φ_0 exists at the surface of the sol particle. For $c(Ag^+) > 3.0 \times 10^{-6}$ M, this potential is positive, while it is negative at lower concentrations of Ag^+. We can understand the quantitative relation between ion concentration and surface potential by considering the sol particle as a giant ion, P^{m+}, of charge me, where m can have either sign.

In the solution, we have the equilibria

$$P^{m+} + Ag^+ \rightleftharpoons P^{(m+1)+} \tag{9.2.1a}$$

$$P^{m+} + I^- \rightleftharpoons P^{(m-1)+} \tag{9.2.1b}$$

The free energy change of adsorbing a silver ion is

$$\Delta \mathbf{G}_{Ag} = \Delta\mu^{\theta}_{Ag} + (m+1)e\Phi_0 - k_B\mathbf{T} \ln c(Ag^+) - me\Phi_0 \tag{9.2.2}$$

neglecting the small entropy of mixing terms for the particles. Similarly, for the addition of an iodide ion

$$\Delta \mathbf{G}_{I} = \Delta\mu^{\theta}_{I} + (m-1)e\Phi_0 - k_B\mathbf{T} \ln c(I^-) - me\Phi_0 \tag{9.2.3}$$

The free energy changes $\Delta\mathbf{G}_{Ag}$ and $\Delta\mathbf{G}_{I}$ are both zero at equilibrium. In particular, at the point of zero charge (pzc), where $\Phi_0 = 0$, we find

$$\Delta\mu^{\theta}_{Ag} - \Delta\mu^{\theta}_{I} = k_B\mathbf{T}\ln\left\{\frac{c_{pzc}(Ag^+)}{c_{pzc}(I^-)}\right\} \tag{9.2.4}$$

By combining Eqs. (9.2.2)–(9.2.4), we can eliminate $\Delta\mu_{Ag}^{\theta}$ and $\Delta\mu_{I}^{\theta}$ to find the surface potential

$$\begin{aligned}\Phi_0 &= \frac{k_B\mathbf{T}}{2e}\left(\ln\left[\frac{c(Ag^+)}{c_{pzc}(Ag^+)}\right] + \ln\left[\frac{c_{pzc}(I^-)}{c(I^-)}\right]\right)\\ &= 0.059\log\left(\frac{c(Ag^+)}{3.0\times10^{-6}}\right) = 0.059\log\left(\frac{c(I^-)}{2.5\times10^{-11}}\right).\end{aligned} \tag{9.2.5}$$

This is sometimes called the Nernst equation. The second equality refers to $T = 298$ K.

The concentrations, or more specifically the chemical potentials, of the Ag^+ and I^- in the solution set the surface potential on the AgI particle. Figure 9.2 shows how the zeta potential varies with Ag^+ concentration:

We see two types of deviation from the ideal behavior described by Eq. (9.2.5). Around the point of zero charge, a linear relation exists between the zeta potential and log[c(Ag^+)], but the slope is smaller than in Eq. (9.2.5). Furthermore, at high and low values of log[c(Ag^+)], the zeta potential seems to saturate. These deviations could be explained by the difference between the surface potential Φ_0 and the zeta potential.

In a system like the AgI sol, we divide ions into two categories, potential-determining ions and indifferent ions. For example, KNO_3 functions as an indifferent electrolyte. These affect the interaction between colloidal particles mainly by changing the Debye length, $1/\kappa$. They also change the activity of potential-determining ions through the Debye—Hückel screening, but that effect is generally small. Potential-determining ions directly change Φ_0. Consequently, minute changes in the concentration of potential-determining ions can produce dramatic effects on the stability of colloidal systems.

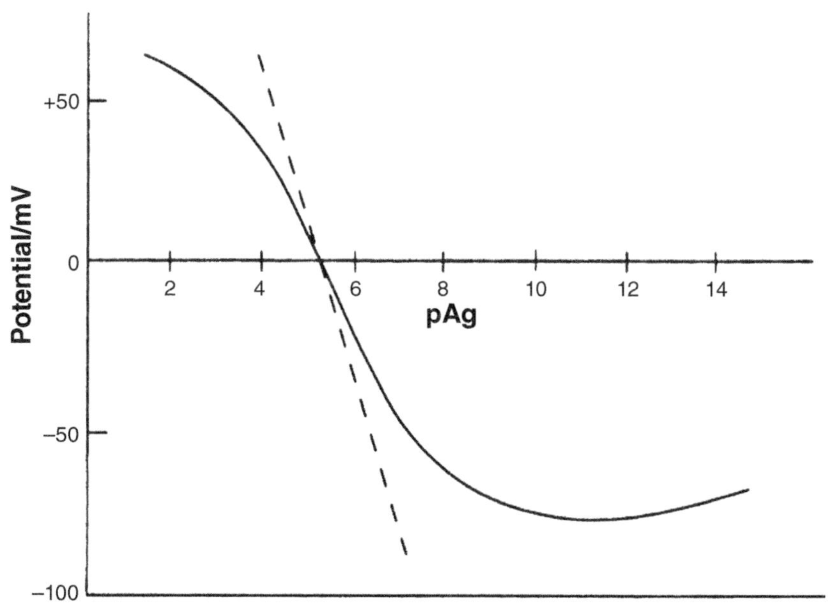

Figure 9.2 Zeta potentials ζ for silver iodide sol as a function of −log[c(Ag^+)]. As the silver ion concentration decreases, ζ changes from a positive to a negative value; the point of zero charge occurs at 3×10^{-6}. The prediction of the Nernst Eq. (9.2.5) (dashed line) is shown for comparison purposes. *Source*: Adapted from Kruyt, 1952.

9.2.2 Ion Complexation Plays an Important Role in Controlling Colloidal Stability

Potential-determining ions play a key role in a large number of colloidal systems. For example, hydroxide and hydrogen ions determine the surface potential for colloidal molecules or particles containing weak acidic or basic groups or both. This category includes some inorganic oxides, polymers, latex spheres, and amphiphilic aggregates containing carboxyl and amino groups, as well as almost all proteins. Chloride ion is a potential-determining ion for the gold sol because it forms a particularly strong complex. Trivalent ions like Al^{3+} and Fe^{3+} form a variety of complexes whose electrical charge is set by the ion concentrations and the pH. These two ions are important in water and sewage treatment, and here we will consider the behavior of an aluminum system in some detail as an illustration of the complex properties.

Like many other multivalent cations, aluminum can form a bewildering variety of complexes in water, whose stability depends on metal ion concentration and pH. At low pH(<4) $Al^{3+}(H_2O)_6$ is the dominating species while at high pH (≥10) $Al(OH)_4^-(H_2O)_2$ dominates. Around neutral pH the aluminum hydroxide $Al(OH)_3$ precipitates at equilibrium, but due to the loose network structure, the kinetics is slow. In intermediate pH regimes, the aluminum ions associate into oligomeric and polymeric structures, which can be seen as precursors of the precipitating hydroxide. Below pH 7, these oligomers/polymers should be positively charged, while at higher pH values, they are predominantly negatively charged. We know that the complexes are important in controlling colloid stability in the aluminum system, and we illustrate this by considering the effect of $Al(NO_3)_3$ on the stability of the AgI sol.

Figure 9.3 displays the pH dependence of the critical coagulation concentration (CCC) for an AgI sol, prepared by mixing 0.1 mM $AgNO_3$ and 0.4 mM KI solutions.

In the absence of aluminum, these sols will be highly negatively charged according to Eq. (9.2.5), and this ensures colloidal stability. The figure shows that the presence of aluminum has a dramatic effect on the stability in a way that depends on the pH value. It is possible to understand this variation based on the solution properties of the aluminum ions:

1. When the pH is less than 4, the CCC is constant and equal to 2.5×10^{-5} M, a value characteristic of destabilization of the colloid by the trivalent Al^{3+} species inducing correlation attraction as discussed in Section 5.3.

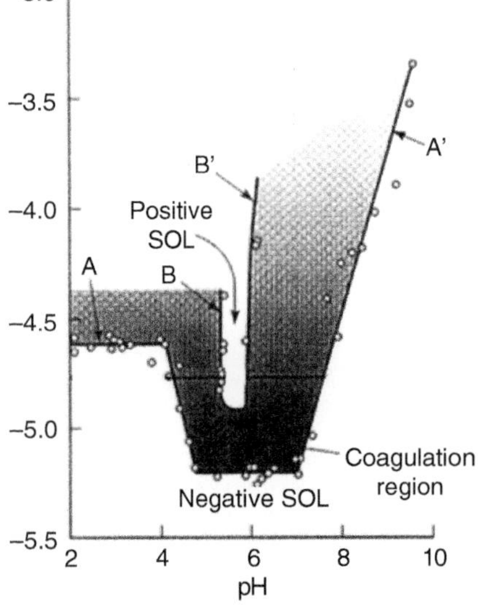

Figure 9.3 Coagulation behavior of a silver iodide sol as a function of pH and aluminum nitrate concentration (y-axis, M). The AA′ curve represents the CCC and the BB′ curve the stability limit due to charge reversal resulting from the adsorption of positively charged aluminum complexes. *Source*: Adapted from Matijević et al., 1961.

2. At pH 4, the CCC drops sharply to 7×10^{-6} M and remains almost constant up to pH 7. This fact implies the formation of a hydrolyzed species with a charge greater than +3. In this regime, we have a range of oligomeric and even polymeric aluminum species, which have a charge per Al of less than $+3e$ but where the total charge can reach higher values. The highly charged species will act like a polyelectrolyte and induce attraction due to correlations as discussed for polyions in Section 7.4.3
3. When the pH increases above 7, the CCC rises strongly. Oligomers and polymers are progressively more and more negatively charged, and they also decrease in size. This leads to a transition toward the behavior of a 1:1 electrolyte solution.
4. Following the *BB′* path: between pH 5 and 6 and at a sufficiently high concentration of $Al(NO_3)_3$, zeta-potential measurements show that the AgI sol becomes positively charged. This charge reversal results from the adsorption of a multinuclear Al complex and reflects the preferential adsorption of highly charged complexes rather than the hydrated metal ions themselves. In Section 7.4, we discussed a similar recharging of surfaces by adsorption of a polyelectrolyte on a surface.

9.2.3 Apolar Particles Can Acquire a Charge by Adsorbing Long-chain Amphiphiles from Solution

It is a common finding that oil drops, bubbles or latex particles dispersed in water carry a negative charge. Such observations have sometimes been rationalized as a manifestation of a specific surface adsorption of hydroxide ions. A much more likely explanation is that there is a weak contamination of highly hydrophobic molecules carrying a carboxylic group, such as a long-chain fatty acid. The adsorption of such molecules to the surface of an oily particle or an air–water interface can occur at sub-millimolar bulk concentrations. Figure 9.4 shows how the electrophoretic mobility of nominally noncharged emulsion drops depends on the pH. Below pH 4, the mobility is virtually zero, but

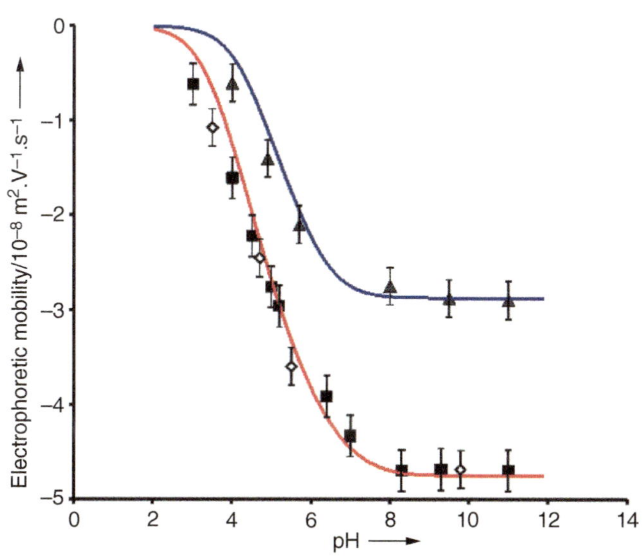

Figure 9.4 The pH dependence of the electrophoretic mobility of an emulsion of hexadecane droplets. Red line (■) commercially available oil of 99.8% purity; blue line (▲) purified sample; red line (◇) purified sample with 5.8 mM oleic acid. *Source*: Roger and Cabane, 2012/with permission of John Wiley & Sons.

it increases with increasing values of the pH reaching a constant value of 5×10^{-8} $m^2V^{-1}s^{-1}$ above pH 7. Purifying the system meticulously reduces the mobility to less than 3×10^{-8} $m^2V^{-1}s^{-1}$ and the mobility of the original sample could be restored by adding a small amount of a long-chain fatty acid. The pH dependence of the mobility follows the titration behavior of a fatty acid.

See Berg (2010, Chapter 6), Cosgrove (2010, Chapter 2), Israelachvili (2011, Chapter 14), and Lyklema (2005, Vol. 4, Chapter 5) for additional discussion on Particle Charge.

9.3 Clays Are Colloidal Sols Whose Surface Charge Density Reflects the Chemistry of Their Crystal Structure

Clay is a material composed of colloidal particles commonly found in nature. Clays have typically been formed by particle sedimentation from rivers or oceans. Clay is an important component of agricultural soil, where its capacity to resist drying and capacity to buffer pH are important features. The clay structure can also cause a major hazard when persistent rain can lead to extensive swelling of the soil with a resulting danger for landslides. Clay materials in the form of marl have historically been the first mineral fertilizer used in agriculture, mainly to make the soil less acidic. Clays have been used by man since ancient times to make ceramics and later porcelain and to make bricks. In current technology, clays are, for example, used as drilling muds, as fillers in glossy paper, as cosmetic products, and as ion exchangers. Due to the sheet structure, diffusion in the perpendicular direction is hindered (see Section 6.3), and barriers of clay materials are used to reduce leakage from old garbage and nuclear waste deposits.

9.3.1 Clay Are Silicates with a Sheet Structure Having Covalent Bonds Extending in Two Dimensions Only

The building blocks of clays are two-dimensional arrays of silicon and aluminum tetrahedrally coordinated with oxygen sometimes combined with aluminum and magnesium octahedrally coordinated. There are no covalent bonds between sheets so that the forces between sheets are much weaker than within them. As a consequence, clay minerals easily break up into platelets.

The few basic atomic components can be arranged in many different ways, giving rise to a rich variation in net chemical composition accompanied by a variation in properties. When an aluminum ion replaces a silicon atom in a tetrahedral structure, a local negative charge arises. By substituting an aluminum atom by a magnesium, one similarly creates a local charge. The system is then made neutral by externally coordinated Na^+, K^+, or Ca^{2+} ions. Oxygen atoms are engaged in bridging silicon or aluminum atoms, but it also occurs that the oxygen atom is only coordinated to one such atom, while the second one binds a hydrogen atom on the surface of the sheet. The sheets have, for most clay minerals, a mirror image plane along the sheet, so that they retain a planar structure when it is cleaved. However, the mineral krysotil, which is the main component of asbestos, lacks such a perfect symmetry plane, with the result that the sheets have a spontaneous curvature (see Section 6.2). As a result, such sheets roll up into cigar-shaped particles, and when these are packed, one obtains a porous material, where the air in the void acts to give the material a very low thermal conductivity. Unfortunately, the ends of the "cigars" can act as atomically sharp knives and damage the lungs when inhaled.

Due to their sheet structure, clays are easily cleaved, compared to other solids. However, there are large differences between different clays depending on their surface properties. Talc is a silicon–magnesium clay with no net charge in the covalent structure and with only bridging oxygens exposed on the surface, which has a slight hydrophobic character. These properties makes the material useful for many purposes such as solid lubricants, fillers in paper and in cosmetic products.

Kaolin, a main ingredient in porcelain, has a composition similar to talc, but it has a —OH group on the surface producing a stronger coupling between sheets, resulting in a larger resistance to cleavage. In clay minerals like mica or montmorillonite, the covalent network carries a net negative charge that is neutralized by a coordinated cation. This coordination can occur between sheets, making these materials harder to cleave. But once cleaved, they tend to acquire a net surface charge in an aqueous environment by dissociation of the surface-coordinated ion.

9.3.2 Experimentally Measured Forces Between Two Mica Surfaces Provide Insight into Fundamental Properties of Colloidal Systems

Mica is a clay mineral where neighboring sheets are coupled by coordinated potassium ions. It can, with care, be cleaved into sheets where the surface can be atomically smooth over cm^3 sized areas. Glued to glass cylinders, one obtains a substrate ideal for measuring interaction forces in the SFA apparatus (see Section 5.4). A mica sheet coordinates one K^+ ion per 0.5 nm^2 and when immersed into pure water, dissolution of all ions would create a high charge density of 0.32 C/m^2.

The force between two bare mica surfaces is readily measured, and as shown in Figure 9.5, the observed interaction at long range is in quantitative agreement with the DLVO theory. The electrostatic repulsive component is consistent with $\Gamma_0 = 1$ (see Section 5.1), which is the high charge limit and thus independent of the precise magnitude of the charge density. The agreement between theory and experiment is obtained without any adjustable parameters! There is a large free energy penalty associated with generating such a charge density at low electrolyte concentrations in the bulk. As a result, hydrogen ions in the solution bind to oxygen atoms exposed on the surface to reduce the charge density somewhat.

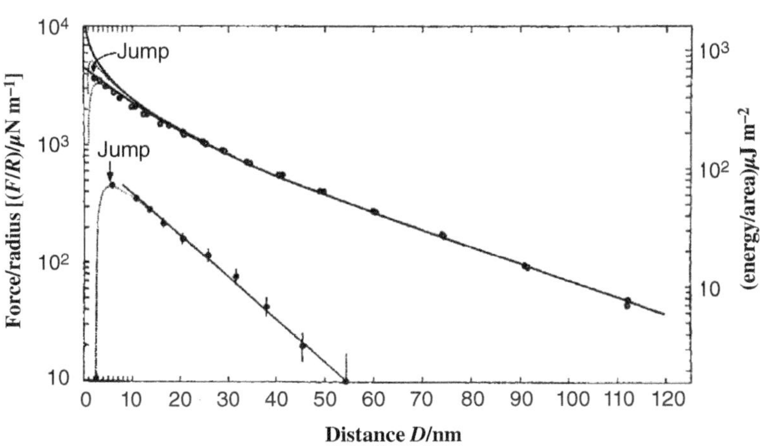

Figure 9.5 Measured interaction free energy between two mica surfaces in a 0.0107 mM KNO_3 aqueous solution (upper curve) and in a 0.0094 mM $Ca(NO_3)_2$ aqueous solution (lower curve). Dots measured points, line fitted to the DLVO potential for a highly charged surface. (Hamaker constant 2×10^{-20}J). *Source*: Israelachvili and Adams, 1978/with permission of Royal Society of Chemistry.

At short separations, we don't expect the DLVO theory to accurately represent the force. Measurements at different pH values and with different alkali ions in the electrolyte confirm this expectation. The force measurements show that there is a competition between covalent hydrogen ion binding and surface coordination of the larger cations. There is a specificity in the ion coordination, and it is significant that the native counterion K^+ has the highest affinity to the surface among the alkali ions. However, it binds significantly less effectively than the hydrogen ion.

9.3.3 Coagulated Structures Influence the Colloidal Stability of Clay Sols

Mica needs to be cleaved mechanically to expose the charged surface. There are clays that swell spontaneously in water. The most studied mineral is montmorillonite, which has Na^+ as its natural counterion. It has a lower surface charge density than mica, with the consequence that the cohesion is less in the dry state, making swelling more probable. When montmorillonite swells under controlled relative humidity, one can observe a stepwise "crystalline" swelling. This is a manifestation of the existence of the oscillatory solvent packing force discussed in Section 5.5.

Clay from natural sources often has defects in their crystal structure, which can give rise to a complex swelling behavior in practice. Another feature of practical importance is that clay leaflets easily break mechanically. As a result, the edges usually acquire a positive electrical charge.

As illustrated in Figure 9.6, negatively charged particles readily adhere to the positively charged edges of a leaflet. A specific manifestation of this effect is that the edge of one leaflet can act as a giant counterion to the flat face of another leaflet. In this way, the system can aggregate into a "house of cards" structure, illustrated in Figure 9.7b. Such structures can incorporate large quantities of water and break and collapse, for example, under the pressure from a tractor wheel.

Highly concentrated clay suspensions with low viscosities are desirable in many practical applications, such as drilling muds. We can achieve such properties by adding a so-called peptization agent to reverse the charge on the edges of the clay particles. For example, polymetaphosphates chemisorb onto the clay's edges, forming complexes with the aluminum ions. By causing the charge on the edges to become negative, small amounts of such material can transform rigid gels into freely flowing solutions as shown in Figure 9.7a.

When Ca^{2+} is available, it preferentially exchanges with monovalent alkali ions, and it can become the dominant counterion in clays with charged surfaces. These are highly charged and

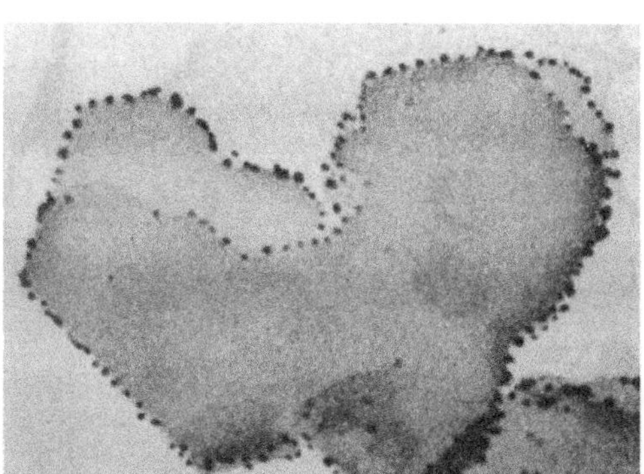

Figure 9.6 Transmission electron microscope image of clay particles mixed with a negatively charged gold sol. Gold particles adsorb on the positively charged edges of the clay particles. *Source*: Thiessen, Luc Mathieu Maria (1947).

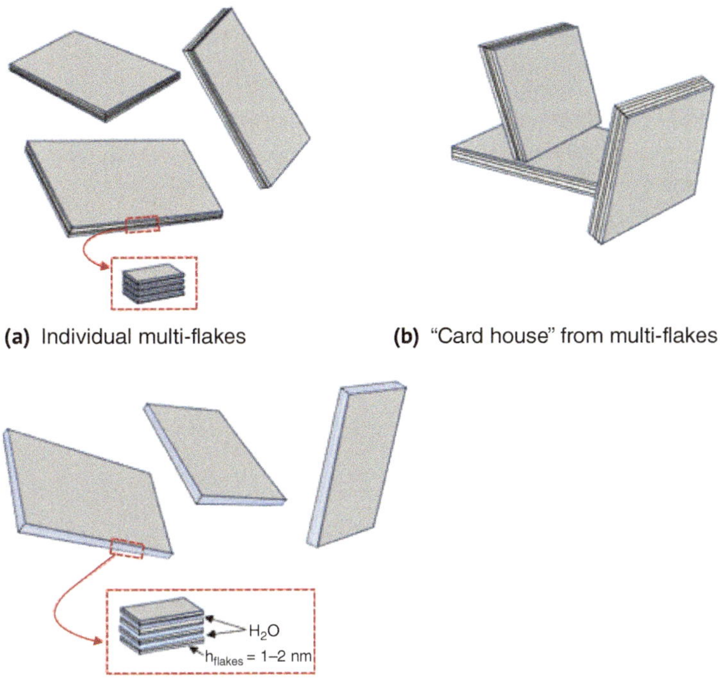

(a) Individual multi-flakes **(b)** "Card house" from multi-flakes **(c)** Stack of single flakes

Figure 9.7 Modes of particle association in clay suspensions: (a) dispersed multisheets state, (b) edge-to-face "house of cards" association of sheets, and (c) tactoid stacks of single sheets, separated by a thin aqueous layer.

one has in this case the attractive correlation forces discussed in Section 5.3. Clay sheets can then aggregate side-by-side forming the so-called tactoid structures with a thin aqueous layer separating the sheets, as illustrated in Figure 9.7c.

An opposite effect can be observed in clay soils initially rich in Ca^{2+} and held together by electrostatic correlation forces. After extensive rains, it can happen that the Ca^{2+}-ions are exchanged for sodium, resulting in swelling. An effect that can trigger a landslide.

See Berg (2010, Chapter 6), Cosgrove (2010, Chapter 9), and Israelachvili (2011, Chapter 15) for additional discussion on Clays.

9.4 Polymer and Lipid-based Particles Can Be Made To Serve a Number of Purposes

9.4.1 Latex Particles Can Be Designed to Have a Number of Special Features

When an apolar polymer is synthesized in a polar medium, the bad solvent conditions lead to the formation of a separate polymer phase. When polymerization occurs in a finite aggregate, for example, a swelling micelle, the product consists of separated latex particles, which are surfactant and/or charge stabilized. There are rich possibilities to vary the synthesis conditions to control the properties of the particles. Figure 9.8 illustrates some of the numerous possibilities.

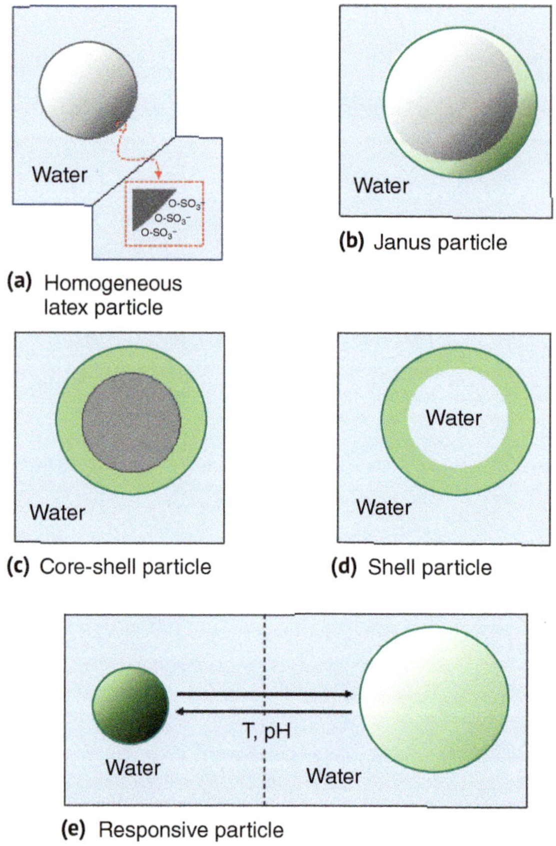

Figure 9.8 (a) Homogeneous latex particle. (b) Janus latex particle from two polymers, grey and green. (c) Core-shell particle; outside layer of green polymer, inside grey. (d) Shell particle with water inside. (e) Responsive particles that swell/deswell on external stimuli.

Figure 9.8a shows the product of a standard emulsion polymerization, where the average particle size can be easily controlled, but one obtains a rather wide size distribution. Such particles are, for example, a common ingredient in paints. In the so-called Ugelstad method, practically monodisperse particles are obtained by adding an inert oil to the surfactant micelles at the start of the polymerization. This has the effect that during the growth process, the monomer chemical potential is controlled by particle size, which is then kept uniform. Such particles of uniform size have an important use in medical applications as, for example, carriers of diagnostics.

Figure 9.8b shows a Janus particle where one has switched monomer during the synthesis so that the particle consists of two different polymer species. Since polymers typically do not mix in a melted state (see Chapter 10) one has an internal phase separation in the particle. The detailed geometry of the particle depends on the interfacial free energies of polymer A–polymer B and polymer–solvent.

Figure 9.8c shows a core-shell particle, where the inner latex particle is covered with a shell of a different polymer. Such a structure can result from a polymer mixture where the shell part has a much lower surface free energy toward the solvent than the core material. It can also be made by polymerization on a preformed latex particle.

Figure 9.8d shows a polymer shell particle. This can be made from a core-shell particle by selective hydrolysis of the inner component. Such a polymer shell has a porosity on the molecular scale so that small molecules can penetrate through, but not macromolecules. Provided that one can solve the loading problem, shell particles can be used as carriers of, for example, drugs, which are then protected from degrading enzymes in the environment.

Figure 9.8e shows a responsive particle, where the monomer is chosen so that one, by external changes, in temperature, pH, or electrolyte concentration, can change from bad to good solvent conditions and thus control swelling properties. In this case, crosslinking is necessary to keep the particle identity.

9.4.2 Templating by Amphiphiles or Block Copolymers Can Be Utilized to Make Particles with Internal Structure

One can affect the shape and internal structure of a particle by utilizing self-assembly processes of amphiphiles and block copolymers. Figure 9.9 shows particles formed by mono- and di-olein stabilized by a PEO-PPO block copolymer. Panel A displays an internal bicontinuous structure. The symmetry of the internal structure is also reflected in near cubic shape of the particles. This effect is even more obvious in the particle in panel B, which is built from a reversed hexagonal phase. These types of particles can be loaded with pharmaceutical or other active substances and serve as a source for controlled release.

Figure 9.10 illustrates how self-assembled structures can be used for templating a porous silica particle. The synthesis starts from an aqueous solution of a block copolymer PEO-PPO-PEO forming micelles together with a silica source like $(C_2H_5O)_4Si$. The process is then initiated by adding either base or acid according to the Stöber synthesis protocol (see Section 9.2). The growing silica clusters adsorb on the block copolymer micelles, with the consequence that their effective surfactant number increases and they elongate, and the adsorbed silica clusters induce a micelle–micelle attraction, ultimately resulting in the formation of particles as in panel B of Figure 9.9. One can achieve other ordered structures with another choice of the block copolymer. In the final stage, the sample is heated so that both the water and the block copolymer are removed, while the silica network is further reinforced. The final product is porous solid particles with several potential possible uses.

Figure 9.11 shows the structure of a virus, λ-DNA phage, particle. The outer capsid is made of proteins, while the interior is loaded with DNA, which during formation is forced into the capsid

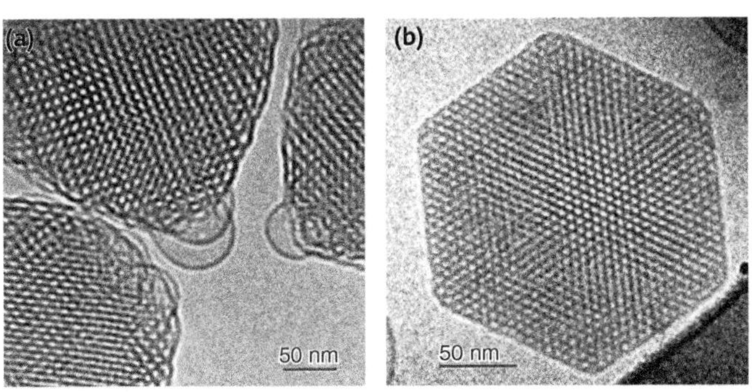

Figure 9.9 (a) Cubosome particles with an internal bicontinuous structure of water–glycerol monooleate stabilized by PEO-PPO block copolymer with an internal bicontinuous structure viewed along the 001 direction. (b) Hexosome particle of water–diglycerol monooleate–glycerol dioleate stabilized by POE-PPO block copolymer with an internal reversed hexagonal structure. *Source*: Barauskas et al., 2005/with permission of American Chemical Society.

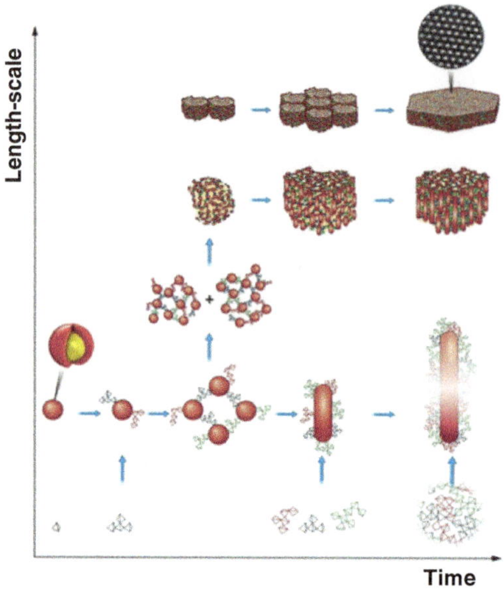

Figure 9.10 Illustration of the formation process of mesoporous disc-like porous silica particles consisting of hexagonally packed channels in a frame of polymerized silica. Channels are normal to the plane of the disc. The structure is templated on a liquid crystalline PEO-PPO block copolymer framework. The polymer is removed by heating the sample. *Source*: Adapted from Alfredsson and Wennerström, 2017.

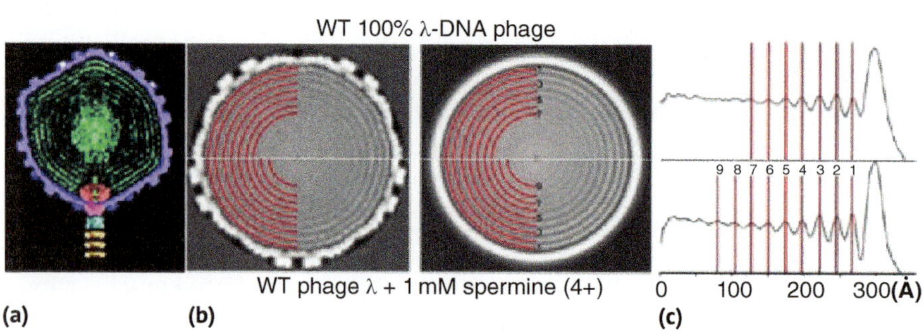

Figure 9.11 Lambda-DNA phage. (a) TEM reconstruction of cut through center. *Source*: Liu et al., 2014/ National Academy of Sciences. (b) Identification of layers. *Source*: Liu et al., 2014/with permission of National Academy of Sciences. (c) Radial DNA density profile.

by a motor-like enzyme illustrated at the lower end of the capsid in panel A. The DNA double helix enters a process that consumes free energy in the form of ATP. The electrostatic repulsion between the helix segments leads to the formation of an ordered structure analogous to a hexagonal phase. In the final virus, the DNA occupies the majority of the available space, and the electrostatic forces result in a pressure on the capsid wall. When infecting, an opening in the capsid results in an effective ejection of the DNA into the host cell due to the electrostatically generated repulsion.

There is a large incentive to deliver DNA or RNA into cells for therapeutic purposes. This technique came into the limelight during the COVID pandemic in the early 2020s. An efficient way of producing a vaccine was developed based on RNA-induced synthesis of an efficient virus antigen. A central challenge was to deliver the RNA molecules to a proper destination in the cells. RNA-degrading enzymes are abundant, and one had to achieve two partly conflicting goals: protection and delivery.

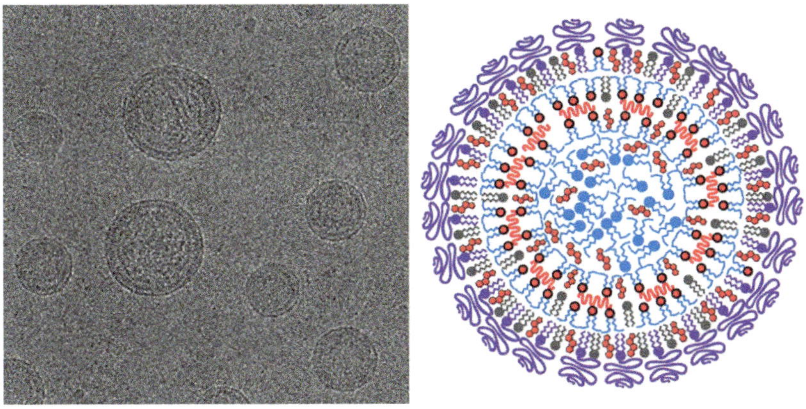

Figure 9.12 Lipid nanoparticle having an outer layer of PEO layer grafted to phospholipid and an interior containing siRNA and cationic lipid. *Source*: Kulkarni et al., 2018/with permission of American Chemical Society.

This was accomplished using so-called lipid nanoparticles (LNPs) illustrated in Figure 9.12. They have an outer monolayer of phospholipids where some have grafted PEO chains.

This outer layer provides protection against enzymatic degradation during transport from vaccine injection to the target cells. The RNA molecules inside the LNP are electrostatically associated with a cationic lipid, forming a less-ordered structure than the one inside the virus in Figure 9.11. Even though the efficiency of RNA delivery to the cells was limited, it was enough to give rise to a therapeutically successful effect. Never before in history has applied colloid science had such a direct impact on the well-being of mankind.

See Kronberg et al. (2014, Chapter 22) for additional discussion on Structured Particles.

9.5 Aerosols Involve Particles in the Gas Phase

9.5.1 Some Aerosols Occur Naturally, but Many Others Are Produced in Technical Processes

Aerosols are systems in which colloidal solid particles or liquid drops are dispersed in a gaseous medium, typically air. They are used in a number of technical processes, ranging from the synthesis of sulfuric acid to aerosol coating processes and spray drying. Both liquid and solid aerosols are used in nose sprays and to administer drugs to and through the lungs. However, a majority of aerosols occur naturally or are inadvertently created.

Clouds are magnificently shaped collections of small water droplets. Trees, grasses, and some other plants cross-fertilize through airborne pollen particles. The Smoky Mountains get their name from aerosols formed by organic compounds emitted by trees growing on the mountain slopes. Aerosols formed by oxidized terpenes appear in the air of northern boreal forests. During wildfires, the afternoon sky can appear reddish-gray to viewers more than a thousand kilometers from the scene due to the scattering from ash and soot particles. Volcanic eruptions can generate large quantities of aerosol ash, which cause major disturbances to aeroplane traffic. In extreme cases, ash clouds also lead to such a screening of sunlight that the climate is affected.

These days, we often connect aerosols with air pollution problems. The combustion in car engines can lead to the formation of air-borne particles, resulting in contaminated air in cities. Urban smog is a complex mixture of particles containing water as well as inorganic and organic compounds.

The chemical composition of these particles undergoes a continuous change. Through the action of sunlight, reactive pollutants are oxidized by O_2, possibly in its singlet form or by ozone. Aerosols are involved in another major environmental hazard. The depletion of the ozone layer over the polar caps during the dark period of the year is catalyzed not only by freons that were used (among other things) to create aerosols for coating processes, but also by a chain of reactions that also depends on the presence of ice particles contaminated by inorganic compounds.

Drugs delivered in the form of aerosols help asthma patients and other allergy sufferers, but in fact, aerosols in the form of pollens, mold spores, and other particles also cause the ailment. Virus infections are often transmitted by aerosol drops. This mechanism applies to the common cold, to influenza, and became a center of interest during the COVID pandemic of 2020 and onward.

9.5.2 Aerosol Properties Differ Qualitatively from Those of Other Colloidal Dispersions in Three Respects

Because aerosols are dispersed in a gas, their colloidal properties differ from those of the dispersions discussed thus far in this chapter. Virtually, no compensating buoyancy exists in the medium, so substantial gravitational force is always acting on aerosol particles. Gravitational potential energy is $m_p gh$ and changes by 1 $k_B\mathbf{T}$ per meter for a particle with a mass

$$m_p = k_B\mathbf{T}/g = 4 \times 10^{-22}\ \text{kg}$$

which corresponds to a water droplet radius ≈ 5 nm. At equilibrium, therefore, all but the very smallest aerosol particles should settle to the bottom of a container.

A second difference between aerosols and dispersions in liquids emerges when we consider the kinetics of particle motion, for example, in gravitational settling. At steady state, the velocity of a drop, v, with radius R and mass density ρ, is determined by the balance between the gravitational force $F = mg$ and the frictional force according to the Stokes Eq. (1.7.15) so that

$$v = \frac{mg}{6\pi R\eta} = \frac{2R^2 g\rho}{9\eta} \tag{9.5.1}$$

At $\mathbf{p} = 1$ atm and $\mathbf{T} = 20\,°\text{C}$, the viscosity, η, of air is 1.82×10^{-5} kg/ms, which is more than 50 times lower than the value for water. Thus, the settling rate is much higher in air than in a liquid at the same gravitational force.

Problem: Calculate the settling rate in air at $\mathbf{T} = 20\,°\text{C}$ and $\mathbf{p} = 1$ atm for a water droplet of radius $R = 1\,\mu\text{m}$.

Solution: From Eq. (9.5.1), it follows that

$$v = \frac{2\left(10^{-6}\text{m}\right)^2 9.81(\text{ms}^2)1 \times 10^3\left(\frac{\text{kg}}{\text{m}^3}\right)}{9 \times 1.82 \times 10^{-5}\left(\frac{\text{kg}}{\text{ms}}\right)} = 1.2 \times 10^{-4}\text{m/s}$$

A velocity of 1×10^{-4} m/s found in the problem is low in relation to a macroscopic scale and it takes a substantial time for micron-sized particles to settle in a gravitational field. More importantly, air velocities of 10^{-4} m/s occur readily by convection so that, in practice, gravitational equilibrium is hardly ever reached for small aerosol particles.

Interparticle interactions constitute a third special feature of aerosols. The gaseous medium does not effectively screen interactions, and for practical purposes, we can neglect the effect of the medium. Thus, Hamaker constants are large, electrostatic interactions are as derived in the first sections of Chapter 3, and we need not account for screening from a dielectric medium nor from free charges. The Debye screening length is effectively infinite, except for the contribution from the aerosol particles themselves.

In the examples given above, an aerosol dispersion has the same qualitative properties as any other colloidal dispersion. The quantitative differences are substantial, however, and their combined effect causes aerosols to show qualitatively different colloidal behavior. For example, they can be studied only under conditions that are dilute from a colloidal dispersion point of view. Even then, stability is a major concern in the laboratory. Particles not only coagulate and settle, but they also combine with the container walls. Particle trajectories are strongly influenced by convective motions, which occur much more readily in a gas than in a liquid.

9.5.3 Aerosol Particles Interact by van der Waals Forces as well as Electrostatically and Hydrodynamically

Aerosol particles always interact through an attractive van der Waals force. The screening effect exerted by the intervening medium is negligible, and Hamaker constants are large, with typical values in the range of $5 - 20 \times 10^{-20}$ J (see Table 5.1). Whatever the nature of the particles, the van der Waals force is strong, so heterocoagulation occurs as readily as homocoagulation. Van der Waals attraction also operates between an aerosol particle and macroscopic bodies such as container walls.

When liquid aerosol droplets associate, they also coalesce to form a single larger droplet. Solid aerosol particles will stick at the point of contact. When a third particle is added to the aggregate, it will also stick at the contact. If no subsequent rearrangements take place, this aggregation process can lead to the formation of loose aggregates whose fractal appearance resembles that shown in Figure 8.8. Household dust provides one example of such aggregated aerosol particles.

The van der Waals interaction energy U_{vdW} between two spheres of radii R_1 and R_2 is (Eq. 5.4.9)

$$U_{vdW} = -\left(\frac{4}{3}\right)^2 \frac{H_{12} R_1^3 R_2^3}{R^6} \tag{9.5.2}$$

when the separation R is much larger than the radii. At short distances, the interaction changes to (Eq. 5.4.8)

$$U_{vdW} = -H_{12} \frac{R_1 R_2}{6(R_1 + R_2)} (R - R_1 - R_2)^{-1} \tag{9.5.3}$$

If we consider the ambient atmosphere, $k_B\mathbf{T} \approx 4 \times 10^{-21}$ J, and typical Hamaker constants are a factor of 10–50 larger than k_BT. For Eq. (9.5.2) to apply, the separation R must be at least three times larger than the radii R_1 and R_2, and it then follows that $U_{vdW} \leq k_B\mathbf{T}$ at these separations. On the other hand, for small separations R, in which Eq. (9.5.3) is valid, the interaction energy is always larger than $k_B\mathbf{T}$. Thus, the van der Waals interaction has the same magnitude as the average kinetic energy at particle surface–surface separations of order of the radii of the particles.

In practice, most aerosol particles acquire a net electrical charge, so they also interact by electrostatic forces. A range of mechanisms results in charged particles. When they are detached from a bulk phase, their electrolyte ions may not obey total charge neutralization or the particle can acquire excess electrons. It is an everyday experience that friction between two different fabric materials can induce such an electron transfer. Once formed, aerosol particles collide with ions and electrons generated in the atmosphere by, for example, ionizing radiation.

In Section 3.6, we used the Born model to calculate the solvation energy for an ion in a bulk phase. This energy is smaller in magnitude in a spherical drop of radius R_1 because the electrical field is that of a vacuum for distances from the ion larger than R_1. For a single charge q in the drop, the solvation energy is (see Eq. 3.6.3)

$$A_{\text{solv}} = -\frac{q^2}{8\pi\varepsilon_0}\frac{1}{R_{\text{ion}}}\left[1 - \frac{1}{\varepsilon_r}\left(1 - \frac{R_{\text{ion}}}{R_1}\right) - \frac{R_{\text{ion}}}{R_1}\right] = A_{\text{solv}}(\text{bulk}) + \frac{q^2}{8\pi\varepsilon_0 R_1}\left(1 - \frac{1}{\varepsilon_r}\right) \tag{9.5.4}$$

This solvation energy is dominated by the term representing the bulk term, which is large even for an apolar medium. For example, charges created by cosmic radiation will associate strongly with a liquid aerosol drop, which is a mechanism resulting in charged aerosol drops. Ambient urban air contains on the order of $1 - 4 \times 10^8$ small ions of either sign per cubic meter. When these ions collide with aerosol particles, they stick and change the particle's net charge. Such charging mechanisms have the consequence that aerosol particles are typically charged. They then interact through an unscreened Coulomb interaction

$$V_{12}(r) = \frac{q_1 q_2}{4\pi\varepsilon_0 r} \tag{3.2.6}$$

Here r is the center-to-center distance and this interaction reaches above $k_B\mathbf{T}$ only at rather short surface-to-surface separations. Electrostatic interactions between aerosol drops typically do not dominate the van der Waals interaction.

An important consequence of the particle charge is that particles migrate in an electrical field as they do in an electrophoresis experiment. The low conductivity of air makes it possible to achieve a high electrical field over macroscopic distances. By applying a voltage between two large electrode surfaces and letting air slowly flow by, we can collect aerosol particles on the electrodes.

The electrical force on a particle, radius R, is the charge, ze, times the field $\boldsymbol{E}$. At steady state, the particle has a drift velocity v, determined by a balance between the electric force and the viscous drag force $6\pi R\eta v$, assuming that Stokes' law (Eq. 1.7.15) applies. For air at room temperature, the viscosity is $\eta = 1.8 \times 10^{-5}$ kg/(ms). Taking $R = 10^{-6}$ m and field of $E = 10^6$ V/m, the velocity of a particle is

$$v = \frac{zeE}{6\pi R\eta} = z \times 5 \times 10^{-4}\,\text{m/s} \tag{9.5.5}$$

If a particle with $z = 1$ spends a minute in an electrode gap 2 cm wide, it will reach the electrode surface. Such a deposition of aerosol particles on electrodes is widely used. Charged aerosol pigment particles can selectively coat metallic surfaces. Smoke from a combustion process can be purified with respect to soot particles, or aerosol particles can be removed from ambient air in rooms occupied by allergic persons.

9.5.4 Aerosol Particles Possess Three Motion Regimes

Several times in the two preceding sections, we have used Stokes' equation when describing the dynamics of aerosol particles. This equation is derived from the Navier–Stokes' hydrodynamic equation, assuming that viscous forces predominate over inertial ones. This assumption covers most cases of interest for a colloidal particle in a liquid, but we need to be cautious for aerosols.

The viscosity of air is much lower than that of liquids. Consequently, viscous forces are small, and we are likely to encounter conditions under which inertial forces play a role. We can measure the relative importance of inertial and viscous effects by the dimensionless Reynolds number Re, which we encountered in Section 5.8. For a sphere of radius R_1 and density ρ, moving with a velocity v in a medium of viscosity η,

$$Re = \frac{\text{inertial force}}{\text{viscous force}} = \frac{vR_1\rho}{\eta} \tag{5.8.2}$$

When $Re < 1$ the fluid flow around the moving particle is laminar, and the Stokes' equation applies. For $Re > 10^3$, on the other hand, the fluid flow is fully turbulent. In this regime, the resistance coming from the medium is the drag force

$$F_D \approx 0.2\pi\rho_m R^2 v^2 \tag{9.5.6}$$

proportional to the velocity squared, the medium mass density, ρ_m, and the cross-sectional area πR^2 of the particle. In the regime $1 < Re < 10^3$, there is a gradual transition from the Stokes regime, in which the drag force is proportional to the velocity, and the turbulent regime, where F_D is proportional to v^2. Raindrops, for example, move under turbulent flow conditions so that Eq. (9.5.6) describes the drag force.

A second regime in which the Stokes' equation fails is found when the gas can no longer be described as a homogeneous medium. This regime is characterized by the Knudsen number, which is the ratio between the mean free path length λ and the particle radius R, $K_n = \lambda/R$. In deriving the hydrodynamic equations, we assume that all relaxation processes in the medium occur rapidly relative to the motion of the particle. The local equilibrium is established through molecular collisions in the medium.

The mean free path length λ measures the average distance traveled between collisions. For hard-sphere particles of radius R_1 at a number density ρ_N, we have (cf. Eq. 1.7.4)

$$\lambda = \left(\sqrt{8}\pi\rho_N R_1^2\right)^{-1} \tag{9.5.7}$$

For a nitrogen molecule in air at 20 °C and $\mathbf{p} = 1$ atm, the estimated mean free path length is 70 nm. We can expect Stokes' law only to apply to aerosol particles longer than this size.

We can generalize Stokes' law to include a so-called Cunningham correction factor C_c, which has the effect of reducing the drag force

$$F_D = \frac{6\pi\eta R_1 v}{C_c} \approx 6\pi\eta R_1 v\left\{1 + \frac{\lambda}{R_1}\left[1.26 + 0.40\exp\left(\frac{-1.1R_1}{\lambda}\right)\right]\right\}^{-1} \tag{9.5.8}$$

Equation (9.5.8) should be used for particles in ambient air with a diameter less than 1×10^{-6} m. This correction becomes even more important in the upper atmosphere, where the mean free path length is longer.

9.5.5 The Fate of a Water Droplet Depends on the Relative Humidity of the Surrounding Atmosphere

Water drops occur naturally in the form of clouds, mist, fog, and ocean spray. We produce drops when coughing or sneezing. Spray drying is used in the food industry. In all these cases, there is a dynamic exchange of water molecules between the drops and the gas phase. Consider first cloud formation. When warm humid air flows to larger heights, it meets a colder environment, and the humidity can exceed the dew point. Water has a high surface tension, resulting in a high barrier for homogeneous nucleation (see Section 2.3.2). As a consequence, heterogeneous nucleation is common. This occurs on solid aerosol particles, but even charges generated by cosmic radiation have a nucleating effect, which is utilized in the so-called Wilson cloud-chamber.

Once water droplets are nucleated in a cloud, there are several factors that influence the future development of the drops. The condensation of water from the gas phase creates heat, and due to the low heat capacity of the air, the result is a significant temperature increase. In Section 2.7, we saw that even vaporization from an air–water surface can yield a measurable temperature change of the surface water.

In the cloud, there is a dynamic quasi-equilibrium between water in drops and water molecules in the surrounding air. Neglecting the complication of temperature gradients, we expect that the local relative humidity slightly exceeds 100%. In Section 2.3.2, we concluded that the surface free energy has the effect that locally the chemical potential of a condensing compound is increased relative to the bulk equilibrium value. This has the consequence that larger aggregates grow smaller, shrink, and dissolve (see Figure 2.11). This mechanism is also operating in clouds, but complicated by several factors, in addition to the presence of temperature gradients, as discussed in the following section.

An additional general colloidal effect operating in clouds is the gravitational settling. As discussed in Section 8.4, the settling of charged particles generates an electrostatic gravitational potential. Once established, it affects the settling rate of charged drops. Hydrodynamic interactions also affect the settling rate. The hydrodynamic pair interaction is, as discussed in Section 5.8, at long range, inversely proportional to the particle separation and proportional to the viscosity of the medium. The viscosity of air is 50 times smaller than that of water, but this is compensated by the fact that, for basically the same reason, the frictional resistance is also proportional to viscosity, indicating that the hydrodynamic coupling between settling particles has the same magnitude in the two media. In practice, the settling of water drops in clouds is strongly affected by the local wind. However, when drops become large enough, the settling dominates, resulting in rainfall.

When water drops are generated in air with a low relative humidity, in a spray dryer or by a person sneezing in ambient air, drops shrink by evaporation. In analogy with the discussion of evaporation from macroscopic surfaces in Section 2.7, the evaporation rate is limited either by heat transfer or by molecular transport, in air or within the drop. The latter possibility is only relevant for a multicomponent drop. In addition to the kinetic transport argument, there is also a question about the thermodynamic driving force. For pure water drops, we showed in Section 2.3 that the surface free energy gives an extra term to the chemical potential in the drop. When the drop is an aqueous solution of an involatile solute, water evaporation will increase the concentration of the solute. The result is that the chemical potential of water gradually decreases, lowering the tendency for evaporation, and this mechanism generates drops with equal concentration of solutes. A more substantial reduction of the evaporation rate can occur if the solute can self-assemble and generate an outer shell on the drop as described in Figure 2.26.

When the drops contain solid particles, there is yet another mechanism that can reduce the evaporation rate. A solid normally has a Hamaker constant larger than water, and water wets the solid surface. In a drop, we have the situation discussed in Sections 2.2.1 and 5.2.3. There is an effectively repulsive van der Waals interaction tending to thicken the film and thus resisting evaporation.

For charged drops, there is yet another effect changing the driving force for evaporation. Drops typically carry a charge, and water is a good solvent. Equation (9.5.4) shows that the solvation energy in a drop is slightly higher than in a bulk phase. The second term on the right-hand side of Eq. (9.5.4) is much smaller than the first, but for small droplets it still has a significant magnitude.

Problem: Calculate the radius R_1 for which the solvation free energy of a charge $q = e$ in water is 1 $k_B\mathbf{T}$ above the bulk value at $\mathbf{T} = 300$ K.

Solution: $$A_{solv} - A_{solv}(\text{bulk}) = \frac{q^2}{8\pi\varepsilon_0 R_1}\left(1 - \frac{1}{\varepsilon_r}\right) = \frac{1.6^2 \times 10^{-38}}{8\pi \times 8.85 \times 10^{-12} R_1} \times 0.987$$
$$= 1.38 \times 10^{-23} \times 300 \rightarrow R_1 = 18\,\text{nm}$$

The solvation energy is proportional to the square of the charge, making it proportionally more important as the charge increases. It is inversely proportional to the radius of the drop. This implies that the free energy gain from adding a water molecule is larger for smaller drops.

When a charged water drop shrinks by evaporation, one encounters a somewhat surprising phenomenon. The drop can become unstable relative to fission into two separate smaller drops, as illustrated in Figure 9.13.

Consider a drop with $2n$ unit charges e and radius R_0. If it is divided into two drops of equal radius R_1 at constant total volume, one has $R_0 = 2^{-1/3}R_1$ and the increase in area is $2A_1 - A_0 = 4\pi R_0^2\left(2^{\frac{1}{3}} - 1\right)$. The combined change in solvation and surface free energy on fission is then

$$\Delta A = \frac{2^{4/3}(ne)^2}{8\pi\varepsilon_0 R_0}\left(1 - \frac{1}{\varepsilon_r}\right) - \frac{4(ne)^2}{8\pi\varepsilon_0 R_0}\left(1 - \frac{1}{\varepsilon_r}\right) + 4\pi R_0^2\left(2^{\frac{1}{3}} - 1\right)\gamma_{H_2O} \tag{9.5.9}$$

For large values of R_0, the surface energy term dominates, but as a drop shrinks by evaporation, the solvation term can become larger, and fission occurs—an effect that is of some importance in the mass spectrometry of macromolecules.

See Cosgrove (2010, Chapter 11) for additional discussion on Aerosols.

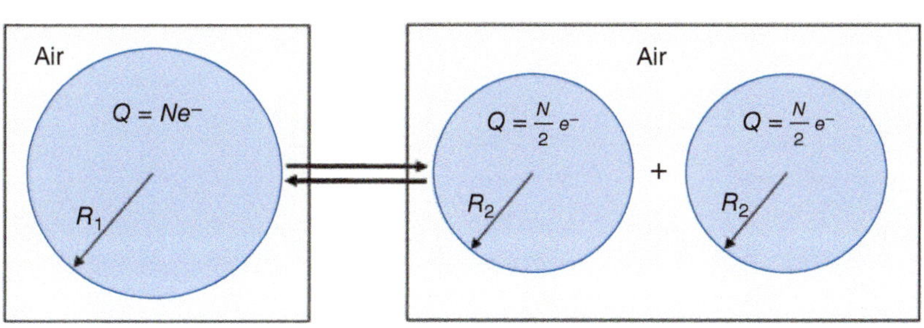

Figure 9.13 An aqueous drop of radius R_1 and charge Q can be unstable to fission into two drops of smaller radii and charge $Q/2$.

Exercises

9.1 The following critical coagulation concentrations (CCC) were found for an $Fe(OH)_3$ sol:

Electrolyte	CCC (mM)
NaCl	9.25
KBr	12.2
$BaCl_2$	4.8
K_2SO_4	0.20
$MgSO_4$	0.22

What qualitative conclusions can you draw from these observations? Try also to estimate the surface potential.

9.2 The CCC for a 0.25% (w/w) suspension of a montmorillonite clay was found to be 14 mM in a NaCl solution and 2.8 mM in a $CaCl_2$ solution. Interpret this observation.

9.3 Homogeneous AgI particles of radius 1.0×10^{-6} m in a 2×10^{-2}M $NaNO_3$ solution are at a concentration of 10^{18} particles per cubic meter. The silver ion concentration is adjusted to 3×10^{-7} M.

(a) What is the surface potential?
(b) Estimate the net charge on one colloidal AgI particle.
(c) How much $AgNO_3$ must be added to the sol to cause rapid coagulation? Assume a Hamaker constant of 5×10^{-5} J.
(d) How much KI must be added to cause rapid coagulation?
(e) Estimate the second-order rate constant for rapid coagulation.
(f) When rapid coagulation occurs, how much time is needed to reduce the number of particles by (i) a factor of 2 and (ii) a factor of 8?

9.4 Horn et al. (1988) measured the interaction between two sapphire surfaces in 1×10^{-3} M NaCl at different pH values as follows:

Separation h (nm)	Energy ($\times 10^5$ J/m^2)	
	pH 6.7	pH 11
30	≈0	0.40
25	−0.15	0.55
20	−0.23	1.27
17.5	−0.32	1.8
15	−0.40	2.4
12.5	−0.55	2.9
10	<−0.55	4.3
8.0	<−0.55	4.9
6.0		<4.9

The value of the Hamakar constant is 6.7×10^{-20} J. Do the data fit the DLVO theory? Estimate the magnitude of the surface potential for the two cases. What is its sign?

9.5 de Kruif et al. (1985, p. 315) measured the structure factor $S(0)$ by light scattering in a sterically stabilized solution of monodispersed silica particles:

S(0)	Concentration (kg/m^3)
0.91	25
0.83	50
0.67	75
0.59	100
0.50	125
0.43	150
0.31	200

The density of the particles is 1.35×10^3 kg/m^3 and their radius 34 nm. Determine the second virial coefficient. Make a quantitative interpretation of the value.

9.6 Kihira et al. (1992) determined the stability ratio W of polystyrene latex spheres $R = 65$ nm as a function of KNO_3 concentration. The following stability ratios and zeta potentials were found:

$C(KNO_3)$ mM	W	ξ(mV)
10	1000	-55
20	500	-40
40	100	-33
100	10	-29
300	1	-22
500	1	18

Determine the CCC and the zeta potential at this point. Equating the surface potential and zeta potential and using a Hamaker constant of 1.0×10^{-20}J, does the CCC agree with the estimate of Eq. (8.2.11)? Estimate theoretically the value of W at KNO_3 concentrations 0.1 M and 0.01 M. How do these estimates agree with the observations?

9.7 If equal numbers of equal-sized AgI particles and Au particles are mixed, a rapid coagulation is observed. However, if the relative concentrations differ by a factor of 4, extensive coagulation does not occur. Suggest a mechanism for this behavior.

9.8 A water droplet contains four monovalent positive ions. The water is gradually evaporated so the drop is shrinking. At what size is it favorable for the drop to split into two parts containing two ions?

9.9 A water drop contains a single monovalent ion. What is the equilibrium drop size in a relative humidity of 95% at $\mathbf{T} = 25\,°\text{C}$?

9.10 Calculate the electrostatic contribution to the free energy of adding a water molecule of volume V_{H_2O} to a drop of radius R having a net charge q.

9.11 Calculate the Debye screening length for air with a density of monovalent ions of 2×10^8 per cubic meter.

9.12 A raindrop with a radius of 1 mm is falling. What is the steady-state velocity?

9.13 Water droplets are formed as the vapor from a cup of hot coffee meets the cool morning air. The heat from the cup also generates a convective upward flow of air, which we assume is 10^{-2}m/s. Which water drops will rise and which will fall?

9.14 A water drop of radius 1.0×10^{-4} is falling in air (20 °C, $\mathbf{p} = 1$ atm.) with a rate of 0.25 m/s. What is the Reynolds number?

9.15 A silica sphere of radius 2.0×10^{-7}m is settling in air (20 °C, $\mathbf{p} = 1$ atm.). Determine the velocity at steady state. How long does it take to fall 1 m?

References

Alfredsson, V. and Wennerström, H. (2017). *Acc. Chem. Res.* 50.

Barauskas, J., Johnsson, M. and Tiberg, F. (2005). *Nano Lett.* 5, 1615.

Berg, J.C. (2010). *An Introduction to Interfaces & Colloids. The Bridge to Nanoscience.* World Scientific.

Cosgrove, T. (2010). *Colloid Science. Principles, Methods and Applications.* Chichester, UK: Wiley.

de Kruif, G.C., Jansen, J.W. and Vrij, A. (1985). *Physics of Complex and Supermolecular Fluids*, Safran, S. and Clark, N., eds. Wiley, New York.

Horn, R.G., Clarke, D.R. and Clarkson, M.T. (1988). *J. Mater. Res.* 3, 413.

Israelachvili, J.N. (2011). *Intermolecular and Surface Forces*, 3rd ed. London, UK: Academic Press.

Israelachvili, J. and Adams, G.E. (1978). *J. Chem. Soc. Faraday Trans. I*, 74, 975–1001.

Kihira, H., Ryde, N. and Matijević, E. (1992). *J. Chem. Soc. Faraday Trans.* 88, 2379.

Kronberg, B., Holmberg, K., and Lindman, B. (2014). *Surface Chemistry of Surfactants and Polymers*, 3rd ed. Chichester, UK: Wiley.

Kruyt, H.R. (ed.) (1952). *Colloid Science*, Vol. 1, Elsevier, New York.

Kulkarni, J.A., Darjuan, M.M., Mercer, J.E., Chen, S., van der Meel, R., Thewalt, J.L., Tam, Y.Y.C., and Cullis, P.R. (2018). *ACS Nano*, 12, 4787.

Liu, T., Sae-Ueng, U., Li, D., Lander, G.C., Zuo, X., Jönsson, B., Raue, D., Shefer, I., and Evilevitch, A. (2014). *Proc. Natl. Acad. Sci. U.S.A.* 111, 14675.

Lyklema, H. ed. (2005). *Fundamentals of Interface and Colloid Science. Particulate Colloids, Vol. 4.* Amsterdam, The Netherlands: Elsevier.

Matijević, E., Mittal, K.L., Ottewill, R.H., and Kerker, M. (1961). *J. Phys. Chem.* 65, 826.

Roger, K. and Cabane, B. (2012). *Angew. Chem. Int. Ed.* 51, 5625.

Thiessen, P.A. (1947). *Z. Anorg. Chem.*, 253, 162.

10

Phase Equilibria, Phases, and Their Applications

Concept Map

Phase Diagrams
Phase equilibria constitute the most essential part of the thermodynamic description of a colloidal system. Analysis of their features provides links between macroscopic behavior and intermolecular interactions. Phase diagrams for colloidal systems display a rich morphology, which includes the following: • The usual states of matter—**crystalline solid, isotropic liquid**, and **gas**. • **Plastic crystals** in which molecules remain organized as a lattice but rotate—characteristic of long-chain alkanes. • **Amorphous solids** and **glasses** in which molecules simultaneously display liquid-like molecular arrangements and local dynamics of solids—typical of polymers. • **Liquid crystals** and **gels**, some of which were described in Chapter 6.

Phase Equilibria
The Gibbs phase rule (Eq. 10.1.3) gives the number of intensive variables {**T**, **p**, μ, …} that can be varied independently, given the number of components c and phases p in a system. In a multicomponent system with two phases in equilibrium, the chemical potential μ_i of each component is the same in both phases. In a two-component system, the Gibbs–Duhem equation permits us to calculate the chemical potential of one component by measuring the concentration dependence of the other. Phase diagrams provide powerful ways to visualize phase behavior: • Two-component systems are often represented by plotting temperature versus composition (Figure 10.5) at fixed pressure. In two-phase regions, the relative amount of each phase is given by the lever rule (Eq. 10.1.12). • Three-component systems use triangular diagrams (Figure 10.6) by fixing two variables (**T** and **p**).

(Continued)

The Colloidal Domain: Where Physics, Chemistry, Biology, and Technology Meet, Third Edition. Håkan Wennerström.

Phase Equilibria (Continued)

Constructing phase diagrams involves preparing samples and determining the number, nature, and composition of the phases by combining visual examination with calorimetric, scattering, and spectroscopic measurements.

Examples of Multiple-phase Colloidal Systems

If two phases are in equilibrium, one is usually more ordered—possessing low energy and entropy—while the other is less ordered, with higher energy and entropy.

In colloidal systems, interparticle interaction energies are typically larger than $k_B\mathbf{T}$, while $\Delta\mathbf{S}_{mix}$ depends only on mole fractions X_A and X_B. This fact promotes the formation of multiple ordered phases, particularly in self-assembled systems that involve an intricate interplay between inter- and intramolecular interactions. Examples illustrating this include the following:

- Monodisperse hard spheres display a liquid–solid coexistence at volume fractions of ϕ between 0.49 and 0.54.
- Ionic surfactants in which inter- and intra-aggregate electrostatic interactions lead to multiple phases (Figures 10.7 and 10.10) that are almost independent of temperature, yet can be changed by the addition of a cosurfactant such as an alcohol.
- Nonionic surfactants in which phase equilibria are very sensitive to changes in temperature that change solvent–head group interactions.
- Block copolymers whose amphiphilic behavior (micelles–inverted micelles) in an appropriate solvent is set by the relative volumes of the two subblocks.
- Monomolecular films whose rich phase behavior reflects the multiple ways of packing parallel hydrocarbon chains at the air–water interface.

Calculation of Phase Diagrams

By calculating phase equilibria, we can relate microscopic interactions to macroscopic phase behavior. The strategy involves developing expressions for free energy or chemical potentials and requires careful manipulation of thermodynamic expressions, as illustrated by the following examples:

- Liquid–liquid separations modeled using regular solution theory show that when $w/\mathbf{RT} > 2$, phase separation occurs, but $w/\mathbf{RT} = 2$ corresponds to a critical point at which two coexisting phases have just merged.
- Two-component systems can be modeled in which A and B form an ideal solution in the liquid phase but not in the solid. Phase coexistence regions can be established using the intercept method (Figure 10.4).
- Impurities create a special type of two-component system. Unequal distributions of the impurity between two phases lead to the creation of a two-phase area in a phase diagram, as illustrated in Figure 10.21.
- The calculated phase diagram (Figure 10.24) shows how a short-range overlap force determines lamellar to gel equilibria in lipid–H_2O systems.

Continuous Versus Discontinuous Phase Transformations

First-order phase transformations involve a discontinuous change in all dependent thermodynamic variables except free energy, and at a molecular level, they reflect transformations from an ordered to a disordered state (or some other drastic structural change) or vice versa. Melting of a crystalline solid is an example.

In continuous phase transformations, dependent variables—such as heat capacity, compressibility, and surface tension—diverge/vanish as the independent variable, for example, $\mathbf{T}$ approaches a critical value, $\mathbf{T}_c$.

The transformation of a gas–liquid two-phase system to a supercritical fluid at the critical point provides an example. As the physical differences between the two phases diminish and merge at $\mathbf{T}_c$, long-range molecular correlations become important.

Continuous transformations can be characterized by power law expressions with characteristic exponents (Eqs. 10.4.2–10.4.5) that can be predicted from statistical mechanics. Surprisingly, the same set of critical exponents describes many seemingly diverse physical systems.

We can illustrate many features of continuous phase transitions using regular solution theory. The critical exponents are inaccurate, reflecting the failure of mean field theory to correctly account for long-range molecular correlations.

Other Sources of Continuous Transformations

Thus far, we have focused on macroscopic systems possessing a few well-defined components. Other sources of apparently continuous phase transformations include the following:

- Systems in which finite-size effects play a dominant role, such as in the formation of micelles leading to a broad critical micelle concentration (CMC).
- Systems containing impurities or inhomogeneities, which broaden phase transformations.
- Systems in which equilibrium times are long compared to observation times, as in glass transitions for polymers.

In colloidal systems, interparticle or interaggregate interactions can easily be strong enough to promote macroscopic ordering, which can lead to the formation of a sequence of phases. Slight variations in conditions may be enough to trigger phase changes, which in turn induce major changes in macroscopic properties. Surfactants and biological lipids display particularly rich behavior; but the phenomenon is also important in other colloidal systems.

The study of phase equilibria fulfills a dual purpose. As the most important factor in the thermodynamic characterization of a system, phase behavior often provides an essential clue to understanding the macroscopic behavior. In addition, we can link phase behavior in an intimate, but nontrivial way, with molecular or particle interactions. If we understand this connection properly, we can control phase behavior by changing interactions on a molecular level. Conversely, we can study interaggregate interactions by observing phase behavior.

10.1 Phase Diagrams Depicting Colloidal Systems Are Generally Richer Than Those for Molecular Systems

10.1.1 Several Uncommon Aggregation States Appear in Colloidal Systems

Colloidal systems show a greater diversity in aggregation states than those found in molecular systems. Reviewing the general terminology used to describe different states and phases can help us appreciate this fact, but we must remember that whenever we attempt to systematize a complex physical reality, we encounter borderline cases that may not fit a given classification system.

At ambient temperatures, matter exists in a solid, liquid, or gaseous state, or in a mixture of these. Dynamic properties at the molecular level provide the basis for characterizing aggregation states. In a **solid**, a molecule's translational motion is slow to very slow. In a **liquid**, molecules (or colloidal particles) diffuse quite rapidly in space and interact constantly with their neighbors. In a **gas**, molecules may travel unperturbed over distances large compared to their size before colliding with another molecule.

These three main aggregation states display a range of phase types, each with its own specific characteristics. For example, the most common form of solid is a **crystal**, in which molecules are arranged with long-range positional and orientational order. Closer to the melting point, a crystal can undergo a phase transition, becoming noticeably softer. It maintains full positional order in this **plastic crystal** state, but it loses some orientational order, and molecules rotate in their lattice sites. This behavior commonly occurs for compounds with long hydrocarbon chains.

In **amorphous solids**, molecules (or atoms) are not located in precisely determined lattice sites, as they are in crystals. An important example—found commonly in solid polymers but also seen in other colloidal systems—is a **glass**. The molecular arrangements we observe in glasses are very similar to those found in liquids, while the local dynamic properties are typical of solids. Amorphous solids often represent nonequilibrium structures, but because of the slow dynamics in the solid state, they can remain unchanged for indefinite times.

Although molecules diffuse rapidly in a liquid, a long-range order might be maintained. However, for ordinary **isotropic liquids**, only a short-range order exists between neighboring molecules, and no correlations exist between molecules farther away. Most substances melt from a crystalline solid to an isotropic liquid. This change implies a transition from a very ordered state to a totally disordered one. However, a number of compounds show a more gradual melting behavior, with one or more transitions between the crystalline solid and the isotropic liquid. These intermediate phases or **liquid crystals** retain some, but not all, of the macroscopic order present in the crystal, while there is a high molecular mobility.

Liquid crystalline phases induced by temperature changes are denoted as thermotropic liquid crystals. There is a considerable chemical diversity among the substances forming thermotropic liquid crystals. The most well-known type is the one containing a stiff, elongated aromatic backbone with a short alkyl chain at the ends. However, compounds like cholesterol esters, pure soaps, and some polymers can also form thermotropic liquid crystals. In colloidal systems, we are normally dealing with two or multicomponent systems, with particles or aggregates in a solvent. In this case, one uses the term lyotropic liquid crystals. The thermotropic and lyotropic liquid crystals differ somewhat in the chemical interactions causing the formation of the ordered phases, but the thermodynamic properties are basically the same.

In a **smectic** liquid crystal, molecules show long-range (macroscopic) orientational order and also long-range translational order in one or two directions. The term **smectic**, which is derived from the Greek word for soap, sometimes refers specifically to a layered structure like those found in the lamellar liquid crystals discussed in Chapter 6, but other forms of positional ordering are also possible like the hexagonal or cubic phases mentioned in Chapters 1 and 6. Figure 10.1 illustrates a typical molecular ordering in a thermotropic smectic phase.

Columnar phases formed by stacked molecules form a special version of smectic phases.

When a smectic phase is heated, it can melt into an isotropic liquid. More often, however, it transforms into a phase in which the long-range positional order has disappeared, but long-range orientational order remains. The nature of this orientational order depends on whether or not the molecules are optically active (chiral).

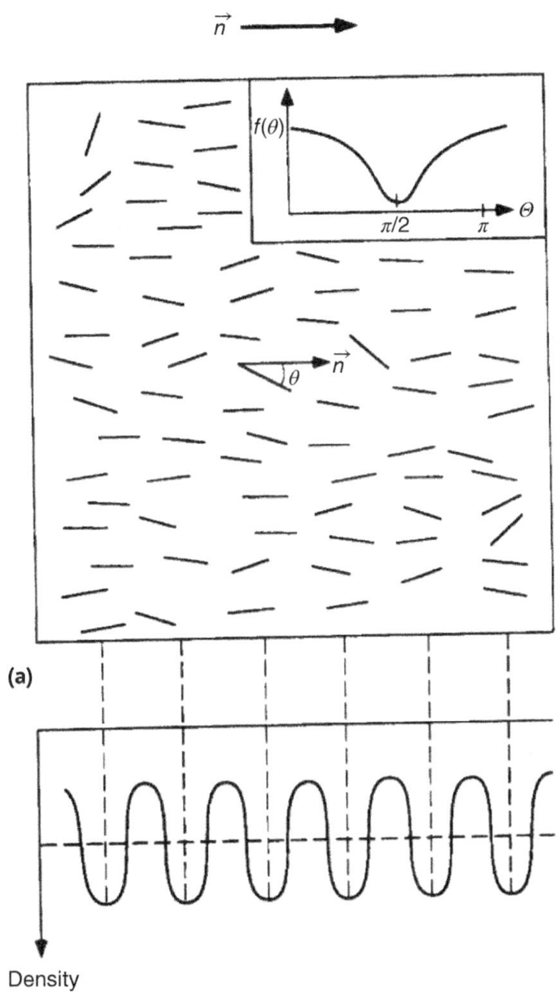

Figure 10.1 Illustration of molecular order in a smectic phase. As shown in panel (a), the molecules are preferentially oriented along the direction *n*. The inset shows the orientational distribution function $f(\theta)$. A careful examination also reveals a layering in the direction of *n*, as demonstrated by the density distribution function in panel (b). When the layering and the orientation lie along the same direction, *n*, the phase is called a smectic *A* phase; otherwise, it is called a smectic *C* phase.

A nonchiral molecule produces a **nematic** liquid crystalline phase, as illustrated in Figure 10.2, in which the preferred orientation is homogeneous throughout the system. Because nematic phases readily orient in magnetic and electric fields, they can be used to control the optical properties of a macroscopic sample. This property is one reason for the use of liquid crystals in optical displays.

In chiral systems, the molecules have a nonuniform orientational order, and the preferred orientation changes direction in a screw-like fashion (Figure 10.3).

This behavior leads to a **cholesteric** phase, also called a **chiral nematic** phase. The pitch of the screw typically lies in the range 300 nm–30 μm depending on the nature of the molecules and on temperature. When this screw-like order is present on length scales matching the wavelength of visible light, it leads to spectacular optical properties.

One application of cholesteric liquid crystals is in thermometers with color indicators. The screw pitch is very temperature-sensitive, being determined by a compromise between energy and entropy; as a result, changing light-scattering properties creates an apparent color change. Cholesteric liquid crystals are transparent to light except for wavelengths that are integer

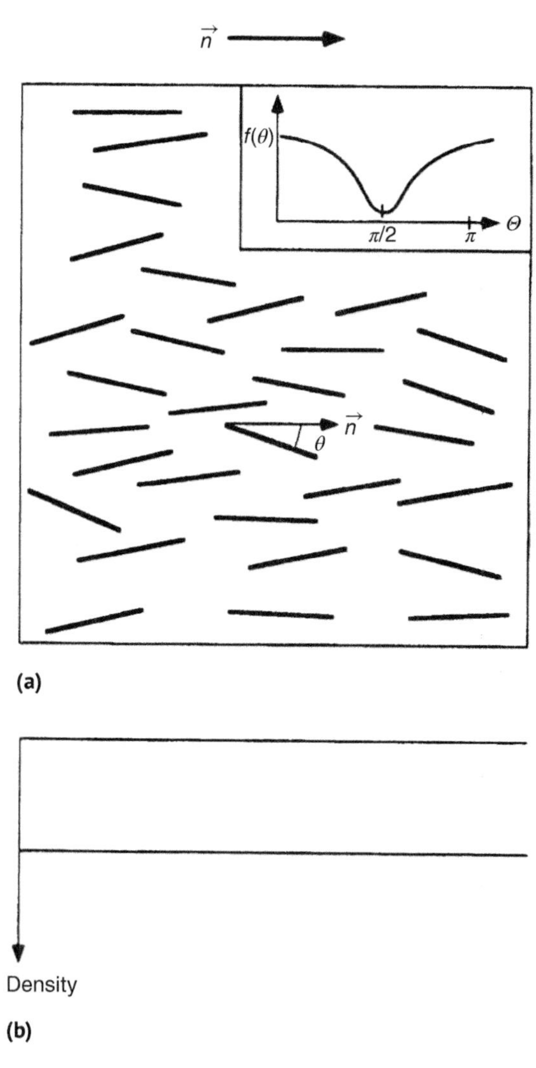

Figure 10.2 Molecular order in a nematic phase. (a) The orientational distribution function $f(\theta)$ having the same qualitative properties as in the smectic phase. (b) The center-of-mass density, which is uniform.

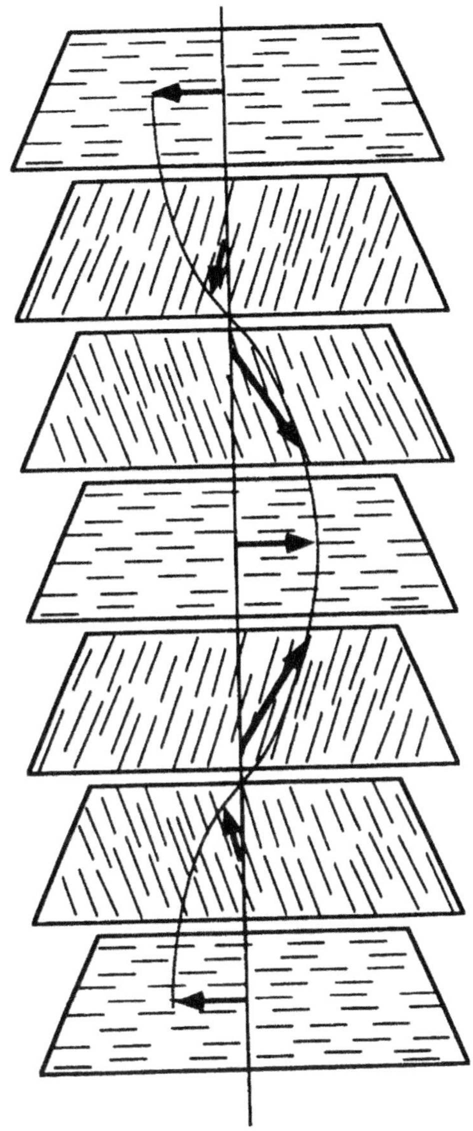

Figure 10.3 In a cholesteric phase, the preferred orientation rotates through the sample. Each plane shown corresponds to the situation in the nematic phase illustrated in Figure 10.2.

multiples of the pitch multiplied by the refractive index, so the observed color is due to a wavelength-dependent scattering.

In systems with two or more components, phases may show solid-like behavior with respect to some components, but liquid-like behavior to others. Usually, we call such phases **gels**, although some authors attach a broader meaning to this term. In a typical gel, one component—a crystalline lipid bilayer or a crosslinked polymer—has low molecular mobility, while the other—usually a solvent—moves as in a liquid. Both components should form domains of infinite extension.

The term gel has been used rather nebulously to describe anything with a jelly-like consistency, sometimes with the specification that there should be a nonzero yield stress. This is a phenomenological classification criterion, which, in our opinion, should be avoided when possible. Thus, liquid crystalline phases are sometimes denoted as gels, and emulsions and foams could be classified as gels using rheology as a criterion. Such a nomenclature is bound to cause confusion.

The weakly interacting molecules found in the gas phase do not show the complex behavior displayed in condensed phases. In the colloidal literature, the term **gas** sometimes designates a dilute solution of particles. This terminology is based on a close analogy between the equilibrium statistical mechanics of a two-component incompressible liquid and a one-component compressible gas. We also used the term for a dilute surface film in Chapter 2, where the gaseous state implies that the area per molecule is so large that surface molecules rarely meet. From a molecular point of view, both these cases involve a liquid system, and the use of the term gas can be misleading since the analogy applies only to equilibrium properties, while the dynamics differ widely. Aerosols, on the other hand, are true gaseous colloidal systems.

10.1.2 The Gibbs Phase Rule Guides the Thermodynamic Description of Phase Equilibria

Thermodynamics provides some general and very useful rules for phase equilibria. The most important of these is the Gibbs phase rule. To derive this rule, let us first recall some basic thermodynamic concepts.

It is a most fundamental assertion of thermodynamics that we can write the equation of state for a one-component system in the absence of an external field as $F(V, n, \mathbf{p}, \mathbf{T}) = 0$; for example, in the ideal gas law, $\mathbf{p}V/(n\mathbf{R}\mathbf{T}) - 1 = 0$. This equation shows that three independent variables uniquely determine the equilibrium state. We can now distinguish between extensive variables like volume V and amount n, which are proportional to the size of the system, and intensive variables like pressure $\mathbf{p}$ and temperature $\mathbf{T}$, which are independent of system size.

Of the three independent variables, we can choose one as extensive, characterizing the size of the system, and two as intensive. If we now take a dependent intensive variable—for example, the density $\rho = n/V$—we can generally write

$$\rho = f_\rho(n, \mathbf{p}, \mathbf{T}) = f_\rho(\mathbf{p}, \mathbf{T}) \tag{10.1.1}$$

where the second equality follows from our conclusion that intensive variables are independent of the system size. Thus, by this formal argument, we have shown the intuitively expected fact that the density of a gas depends only on temperature and pressure; the amount of the gas is, in the thermodynamic limit, irrelevant. The general statement of Eq. (10.1.1) is that all intensive quantities in a one-component system are a function solely of two independent intensive variables.

We can now introduce the concept of the number of degrees of freedom, f, representing the number of independent intensive variables that remain after all possible constraints are taken into account. For a one-component system with no additional constraints, $f = 2$. If we add the condition that the two phases, α and β, are at equilibrium, an additional equation $\mathbf{G}^0_\alpha(\mathbf{T}, \mathbf{p}) = \mathbf{G}^0_\beta(\mathbf{T}, \mathbf{p})$ states that the two phases possess equal molar free energies. This equation reduces the number of degrees of freedom f to one, since $\mathbf{T}$ and $\mathbf{p}$ are no longer independent and, in fact,

$$\frac{\Delta \mathbf{p}}{\Delta \mathbf{T}} = \frac{\Delta \mathbf{S}}{\Delta \mathbf{V}} \tag{10.1.2}$$

for two phases in equilibrium according to the Clausius–Clapeyron equation. If three phases are in equilibrium, there is no degree of freedom, and all intensive variables have a unique value, for example, at the triple point of water ($\mathbf{T} = 270.16$ K; $\mathbf{p} = 611.66$ Pa).

For a two-component system, we must add another intensive variable, which we can represent conveniently as the mole fraction of the second component, $X_2 = \frac{n2}{n1 + n2}$. Thus, with three independent intensive variables and no constraints, $f = 3$.

We can generalize these observations to apply to systems with an arbitrary number of components c. The **Gibbs phase rule** states that

$$f + p = c + 2 \tag{10.1.3}$$

where p is the number of phases.

Determining the number of components sometimes presents a nontrivial task. For example, we usually consider H_2O to be a one-component system, but this is true only as long as dissociation into H_2 and O_2 remains negligible. In the presence of dissociation, water becomes a two-component system. The constraint that H_2 and O_2 occur in the molar ratio of 2:1 in the total system affects an extensive variable, not an intensive one.

The mixture of two salts, such as NaCl and KBr, in which all four ions are different, is a three-component system. For example, a particular phase may well be rich in NaBr. The requirement that the total system contains equal molar amounts of Na and Cl again constrains an extensive variable, not an intensive one.

Polydispersity in the molecular weight always characterizes a polymer system, and according to strict thermodynamics, the number of components is large. However, the components are so similar that we can often consider the polymer as one component if it is chemically homogeneous. However, in a phase separation, there will be a size sensitivity resulting in some fractionation of the polymer component, which can have substantial consequences for the properties of the system.

10.1.3 In a Multicomponent System with Two Phases in Equilibrium, the Chemical Potential of Each Component Is the Same in Both Phases

Equilibrium at constant temperature and pressure occurs when the Gibbs free energy **G** has a minimum value. For a system with two components, 1 and 2, with two phases, α and β,

$$\mathbf{G} = \mathbf{G}^\alpha + \mathbf{G}^\beta \rightarrow \mathrm{d}\mathbf{G} = \mathrm{d}\mathbf{G}^\alpha + \mathrm{d}\mathbf{G}^\beta \tag{10.1.4}$$

but the change in free energy as molecules are added to or subtracted from the phase is determined by the chemical potentials μ_1 and μ_2, and

$$\mathrm{d}\mathbf{G}^\alpha = \mu_1^\alpha \mathrm{d}n_1^\alpha + \mu_2^\alpha \mathrm{d}n_2^\alpha \text{ and } \mathrm{d}\mathbf{G}^\beta = \mu_1^\beta \mathrm{d}n_1^\beta + \mu_2^\beta \mathrm{d}n_2^\beta \tag{10.1.5}$$

The total amount of matter is constant, so that

$$\mathrm{d}n_1^\beta = -\mathrm{d}n_1^\alpha \text{ and } \mathrm{d}n_2^\beta = -\mathrm{d}n_2^\alpha \tag{10.1.6}$$

Thus,

$$\mathrm{d}\mathbf{G} = \left(\mu_1^\alpha - \mu_1^\beta\right) \mathrm{d}n_1^\alpha + \left(\mu_2^\alpha - \mu_2^\beta\right) \mathrm{d}n_2^\alpha \tag{10.1.7}$$

and at the minimum, $\mathrm{d}\mathbf{G} = 0$, which can only occur for

$$\mu_1^\alpha = \mu_1^\beta \text{ and } \mu_2^\alpha = \mu_2^\beta \tag{10.1.8}$$

because n_1^α and n_2^α can vary independently. In the general case,

$$\mu_i^\alpha(X_i^\alpha) = \mu_i^\beta(X_i^\beta) \tag{10.1.9}$$

for all components i, where X_i^α and X_i^β denote the compositions of the two phases α and β, respectively.

By definition

$$\mu_i = \left(\frac{\partial \mathbf{G}}{\partial n_i}\right)_{\mathbf{T},\mathbf{p},n_j} \tag{10.1.10}$$

Thus, the chemical potential can be calculated if **G** is known as a function of the composition.

The intercept method provides a convenient way to graphically determine μ_1 and μ_2 for two-component systems. It is particularly useful for determining qualitative features. Figure 10.4 illustrates the method through a plot of $\mathbf{G}/(n_1 + n_2)$ versus X_1 (or X_2).

With two components, there are three independent intensive variables. At fixed **T** and **p**, only one is left. The chemical potentials μ_1 and μ_2 are both intensive variables. Thus, they cannot be varied independently at a constant **T** and **p**. The interdependence between the two is expressed in the Gibbs–Duhem equation (see Eq. 1.2.22):

$$n_1\,\mathrm{d}\mu_1 = -n_2\,\mathrm{d}\mu_2 \tag{10.1.11}$$

When we, for a two-component system, measure the concentration dependence of the chemical potential of one component at a given **T** and **p**, we can calculate the concentration dependence of the other by using Eq. (10.1.11). This explains why we can use a measurement of the osmotic pressure—that is, the chemical potential of the solvent—to study interactions between solute molecules or colloidal particles as, for example, when obtaining the second virial coefficient using light scattering.

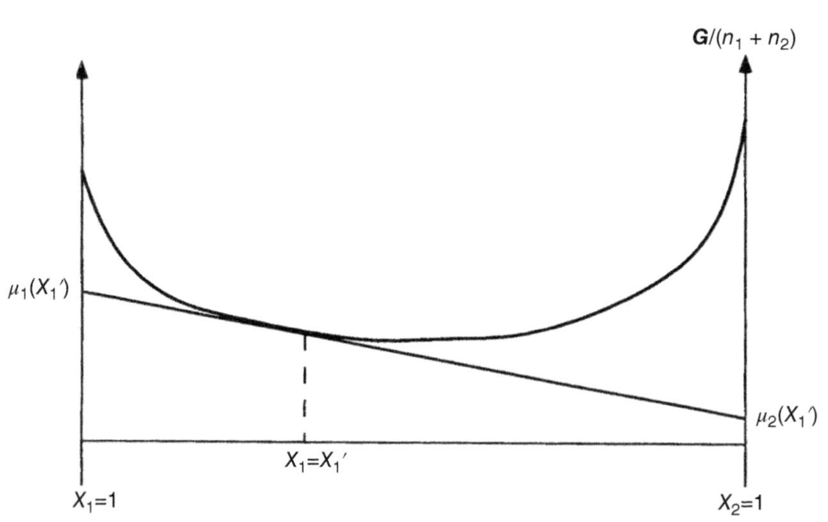

Figure 10.4 The intercept method for determining chemical potentials. Plot free energy **G** over total amount $(n_1 + n_2)$ versus concentration. The chemical potentials at a composition $X_1{}'$ are obtained by constructing the tangent to the $\mathbf{G}/(n_1 + n_2)$ curve at the particular concentration. The two chemical potentials $\mu_1(X_1{}')$ and $\mu_2(X_1{}')$ are then obtained as the intercepts at $X_1 = 1$ and $X_2 = 1$, respectively.

10.1.4 Phase Diagrams Conveniently Represent Phase Equilibria

Phase diagrams provide us with a useful visual way to represent phase behavior. In one diagram, we can summarize a large body of experimental data, and we can also use the phase diagram for predictive as well as cognitive purposes.

A two-component system may have up to three degrees of freedom, but we can display only two dimensions conveniently in a graphic representation. Pressure usually influences condensed phases only slightly. In such a case, we decrease the number of degrees of freedom by one by holding the pressure constant (e.g., at 1 atm), while we vary the temperature **T** and composition X. Figure 10.5 illustrates some general features of $\mathbf{T} - X$ ($\mathbf{p}$ fixed) diagrams through a generic case of two components A and B, which form separate solid phases, yet show miscibility in the liquid state. For one-phase regions with two degrees of freedom ($f = 2$), **T** and X vary independently, and the phase remains the same, although it exhibits continuously varying properties.

When two phases are in equilibrium, $f = 1$. In the diagram, these regions also have a two-dimensional character, but changes in total composition do not change the individual compositions of the two phases in equilibrium. Their only effect is to change the relative amount of each phase. For example, a system containing one mole at composition X' and temperature $\mathbf{T} < \mathbf{T}_t$ (as illustrated in Figure 10.5) will separate a moles of a solid rich in A and b moles of a solid rich in B in such a way that $a + b = 1$. Conservation of component B implies

$$X' = aX(A') + bX(B') \rightarrow a\left[X' - X(A')\right] = b\left[X(B') - X'\right] \tag{10.1.12}$$

which is called the **lever rule**.

The horizontal straight line at temperature $\mathbf{T}_t$ is called a three-phase line. The three phases, A, B, and l, are in equilibrium at this temperature, and consequently $f = 0$. No degree of freedom exists, and both the temperature and the compositions of the phases in equilibrium are unique.

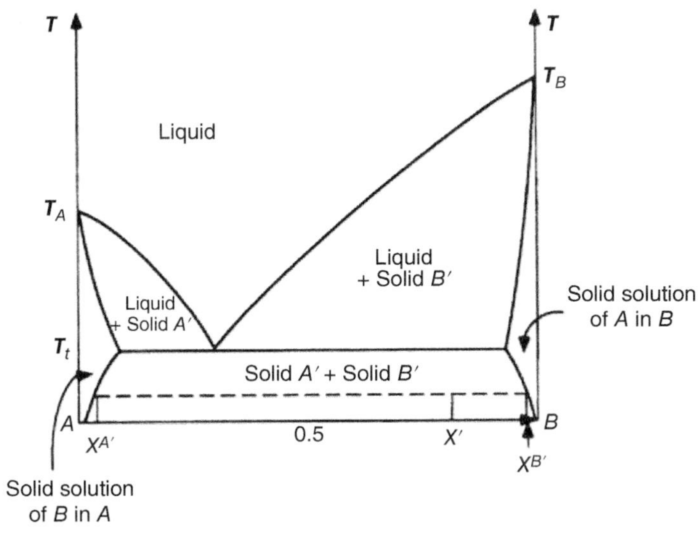

Figure 10.5 Binary $\mathbf{T} - X$ phase diagram at constant $\mathbf{p}$ showing one-phase and two-phase areas and a three-phase line at $\mathbf{T} = \mathbf{T}_t$. The diagram represents the system Ag(A)–Cu(B) and the melting temperature $\mathbf{T}_A$ of silver is 961 °C and for copper $\mathbf{T}_B = 1083$ °C. The dotted line illustrates the use of the lever rule, and a sample with total composition X' separates into solid solutions of composition $X(A')$ and $X(B')$.

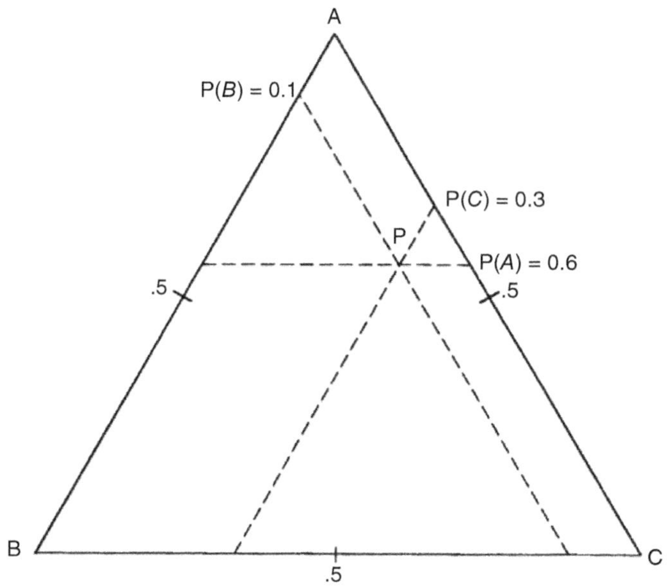

Figure 10.6 Evaluating the composition at a point P in a triangular diagram. The relative amounts of *A*, *B*, and *C* are 0.6, 0.1, and 0.3, respectively.

Systems with three components may have as many as four degrees of freedom. To represent the phase behavior of such a system in a two-dimensional diagram, we must fix two degrees of freedom, usually **T** and **p**. Then we can represent the composition in a triangular diagram. Figure 10.6 demonstrates how to represent the composition using such a diagram.

Each corner represents a pure component, while the opposite base of the triangle represents the binary system of the other two components. We arrive at the relative amount of component *A* at an arbitrary point P by drawing a line parallel to the A base. The intersection of this line with the sides of the triangle determines the fraction of A. The fractions of B and C are determined in an analogous way.

As an example, Figure 10.7 shows the phase equilibria for a water–potassium decanoate ($C_9H_{19}COOK$)–octanol system.

The phase diagram divides the triangle into three types of areas. One-phase areas have two degrees of freedom and can take any shape. Two-phase areas may have four corners joined by two opposing straight lines and two opposing curved lines. In these areas, there is only one degree of freedom, which shows that variations in one direction do not change intensive variables. In a complete phase diagram, we mark the invariant direction by so-called tie lines, which connect points on the two opposing curved lines as exemplified in Figure 10.7. Tie lines connect one-phase points in equilibrium. Only the proportions of the two phases change for points along the same tie line, and the lever rule (Eq. 10.1.12) applies. Finally, for three phases in equilibrium, $f = 0$. This situation generates so-called three-phase triangles. For a point within such a triangle, the three phases represented by the corners of the triangle coexist in equilibrium. The relative amounts are determined in the same way as explained in connection with Figure 10.6, except that we are no longer dealing with an equilateral triangle.

10.1.5 Determining Phase Equilibria Is a Demanding Task

Determining the phase equilibria or phase diagram is usually the most important characterization of the macroscopic properties of a colloidal system. In general, the determination of a complete

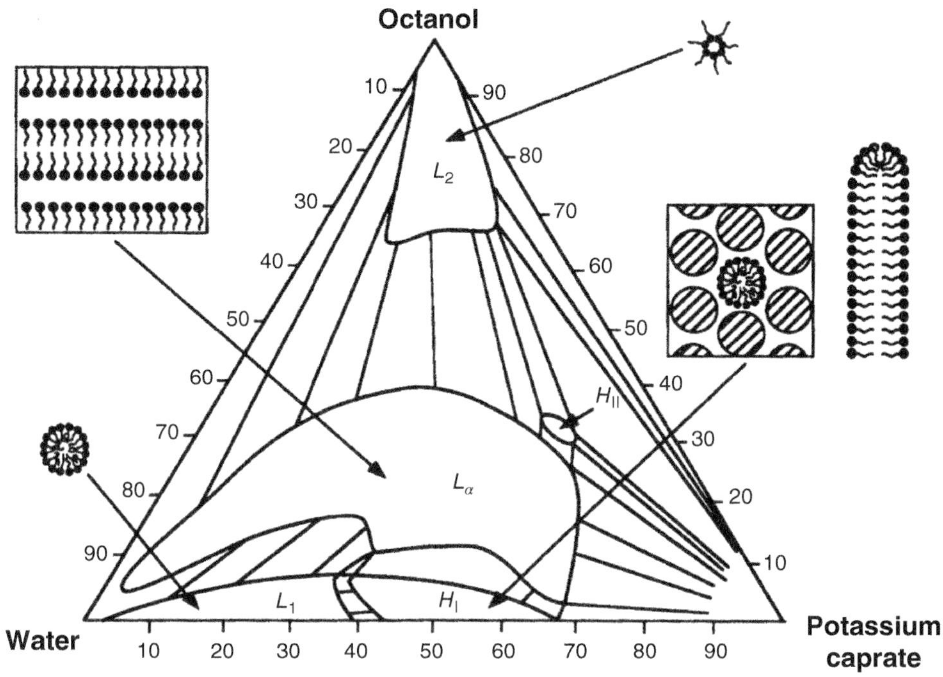

Figure 10.7 Phase diagram of the ternary system: water–potassium decanoate (caprate)–octanol at constant pressure (1 atm.) and constant temperature (20 °C). L_1 indicates a micellar and L_2 a reversed micellar solution. H_I is a normal hexagonal and H_{II} a reversed hexagonal liquid crystalline phase; L_α is a lamellar phase. *Source*: Jönsson and Wennerström, 1987/with permission of American Chemical Society.

phase diagram involves a lot of work. The basic principle is to mix the components and observe the number and nature of the phases. However, this can be done in a number of ways, depending on the particular characteristics of the sample.

The most obvious, direct, and often the most versatile method is direct visual observation, be it by the naked eye or with a polarizing microscope. Isotropic liquid phases typically phase-separate readily, and mixing at a given composition and analyzing the amount and composition of the different phases in equilibrium make it possible to determine a tie line. Thus, when phases readily separate macroscopically, determining the phase diagram is a straightforward task.

However, colloidal systems that abound in liquid crystals and other viscous phases require a long time to achieve a macroscopic separation of the sample into two distinct regions with a single planar interface. It is more common that phases in equilibrium are dispersed within one another, with domain sizes of tens or hundreds of micrometers. For each prepared sample, we face three questions:

- How many phases are present?
- What is the nature of each phase (liquid, liquid crystal, or solid), and what is the symmetry?
- What are the compositions of the phases (if more than one phase is present)?

If macroscopic phase separation does not occur, we are forced to perform measurements on the whole sample and obtain some sort of macroscopic average. The three most common approaches are calorimetry, (X-ray) scattering, and spectroscopic (NMR, electron spin resonance, fluorescence) measurements.

Calorimetric measurements are useful for making a $\mathbf{T} - X$ diagram. Differential scanning calorimetry (DSC) readily measures the heat capacity as a function of temperature, $\mathbf{T}$. By repeating

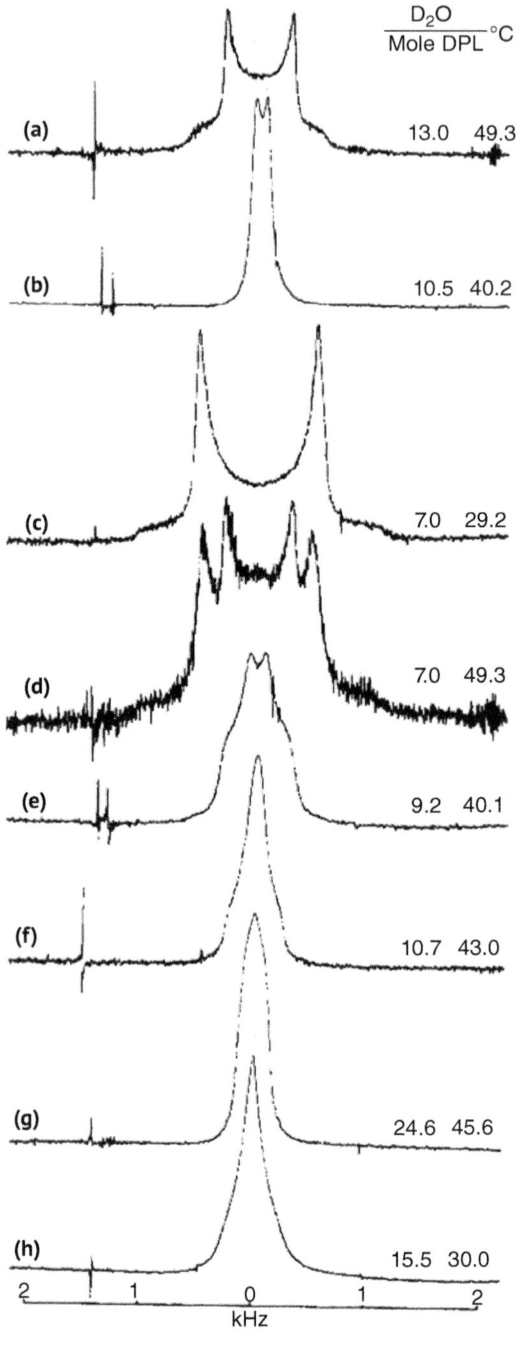

Figure 10.8 Representative deuterium NMR spectra of the binary DPPC–D_2O system. (a) Molar ratio D_2O/DPP = 13, T = 49.3 °C; lamellar liquid crystalline phase L_α. (b) Molar ratio 10.5, T = 40.2 °C; rippled phase P_β. (c) Molar ratio 7.0, T = 29.2 °C; gel phase $L_{\beta'}$. (d) Molar ratio 7.0, T = 49.3 °C; two-phase $L_\alpha + L_\beta$ since this spectrum is a superposition of spectra (a) and (c). (e) Molar ratio 9.2, T = 40.1 °C; two-phase $L_\alpha + P_{\beta'}$ since the spectrum is a superposition of (a) and (b). (f) Molar ratio 10.7, T = 43.0 °C; $P_{\beta'}$ phase. (g) Molar ratio 24.6, T = 45.6 °C; $L_\alpha + D_2O$. (h) Molar ratio 15.5, T = 30.0 °C; $L_{\beta'} + D_2O$. *Source*: Ulmius et al., 1977/with permission of American Chemical Society.

the measurement for a range of compositions, we can localize the temperatures of three-phase lines and of melting transitions (see Figure 6.9). However, phase changes that involve only small changes in enthalpic properties may remain undetected.

Small-angle X-ray or neutron scattering is sensitive to the structure of the phase. Thus, it is very useful for determining the nature of a phase, although it is more difficult, but still possible, to use the method when several phases are present.

Spectroscopic methods rely on the fact that spectroscopic properties are often sensitive to the local molecular environment. Thus, the same molecule in different phases will give different spectroscopic responses. In a multiphase sample, this behavior can be very useful, making it possible to count the number of phases in a straightforward way. Figure 10.8 shows, as an example, a series of deuterium NMR spectra used to construct the phase diagram of the system dipalmitoylphosphatidylcholine (DPPC)–water, shown later (Figure 10.23). By measuring the number of signals, their intensity, and the magnitude of the quadrupole splittings, the combined information is more than sufficient to construct the phase diagram.

We conclude this section by noting that some occasions require a rapid method for determining phase behavior. It is no coincidence that such methods have been developed at the scientific laboratories of the leading detergent producers in the world. A common principle is to take a surfactant crystal, place it in a microscope between coverslips, and let water penetrate the crystal from one side by a diffusion process. During the diffusion process, a concentration gradient is generated and, in principle, the whole compositional range can be sampled across the coverslip. By focusing the (polarizing) microscope at different positions along the concentration gradient, one can determine the sequence of phases. Phase boundaries appear as sharp lines, making it relatively easy to detect a change in phase structure. The situation is similar to the one shown in Figure 2.27 for a sample where the gradient is caused by water evaporation rather than water penetration.

See Atkins et al. (2022, Focus 6), Doi. (2013, Chapter 2), Israelachvili (2011, Chapter 6), and Kronberg et al. (2014, Chapter 5) for more information on Phase Equilibria.

10.2 Examples Illustrate the Importance of Phase Equilibria for Colloidal Systems

A phase transition or an equilibrium between two phases typically occurs when a balance exists between two macroscopic states: one more ordered (with a lower energy and entropy) and one less ordered (with a higher energy and entropy). In colloidal systems, interactions between particles scale with size, and they can easily be of the order $k_B\mathbf{T}$ and higher, yet entropies of mixing N_A and N_B objects are the same irrespective of the size of the objects. Consequently, ordered phases occur more readily in colloidal than in small-molecule systems. This observation is particularly true for self-assembling systems, whose intricate interplay between inter- and intra-aggregate interactions results in a particularly rich phase behavior. We will discuss a number of colloidal systems in which phase behavior constitutes an important aspect of the properties of the system in general.

10.2.1 Purely Repulsive Interactions Can Promote the Formation of Ordered Phases

One of the early successes in computer simulation studies of liquids was the demonstration that even for a hard-sphere system consisting of equal-sized spheres that cannot overlap but otherwise are noninteracting, an ordered (solid) phase can coexist with a disordered (fluid) phase.

Since particle overlap cannot occur, the system only has kinetic energy. As a consequence, the configurational integral Z is independent of temperature, and the thermodynamics is completely determined by entropy except for the contribution from the kinetic energy. Yet the system shows an ordered phase over a concentration range from the close packing limit of volume fraction $\phi = 0.74$ down to $\phi = 0.54$, while the fluid state is stable below $\phi = 0.49$. Ordering occurs at high values of ϕ because the increased entropy created by the rattling of the hard spheres at their sites in the lattice more than compensates for the decrease in entropy associated with changing from long-range disorder to long-range order.

The hard-sphere order–disorder transition would remain an interesting theoretical observation if we could not design colloidal particles that behave like hard spheres. A method for doing so is to first make spherical silica or polymer particles that are as monodisperse as possible. Second, the chemistry of the particle surfaces is controlled to avoid charges. Then the particles are dispersed in a solvent whose refractive index nearly matches that of the particle. This match is important because, according to Eq. (5.2.31), it reduces the effective Hamaker constant to a negligibly low value. The resulting particle system exhibits neither double-layer repulsion nor a van der Waals attraction at long range. At molecular contact, some soft repulsive interaction might occur, but because the range of this interaction is much smaller than the particle radius, the system is effectively a hard-sphere one.

Such systems have been prepared, and as illustrated in Figure 10.9, the observations agree strikingly well with the hard-sphere model. Because the lattice spacings in the ordered phase are of the same magnitude as the wavelength of visible light, we can see spectacular optical effects caused by Bragg diffraction. The optical properties of opals, which are formed by crystallized spherical silica particles of colloidal size, have this origin.

If we prepare the silica or polymer latex particles under such conditions that they acquire an electrical charge, a dominant repulsive electrostatic interaction will result. When reducing the level of electrolyte, an ordered phase can appear at low volume fractions. Here, the ordering is also caused by

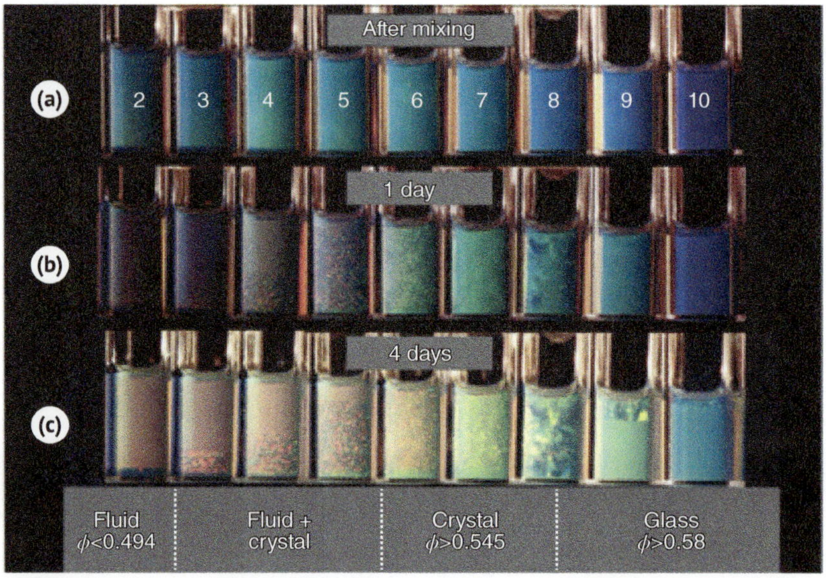

Figure 10.9 Phase behavior of hard-sphere colloids. (a) Immediately after mixing, (b) after 1 day, and (c) after 4 days. At $\phi < 0.494$, there is a dense, disordered liquid. At $\phi > 0.545$, a crystal is formed. At still high volume fractions $\phi > 0.58$, a glass easily forms. *Source*: Pusey et al., 2009/with permission of The Royal Society.

a repulsive interaction. Repulsion is minimized by having the particles localized on sites, and this can compensate for the entropy decrease in the ordering process. The system responds very sensitively to the addition of electrolyte, showing that the repulsion is caused by electrostatic double-layer forces.

10.2.2 Ionic Surfactants Self-assemble into a Multiplicity of Isotropic and Liquid Crystalline Phases

When surfactant molecules self-assemble in a dilute system, local interactions determine the preferred aggregate shape as measured by the surfactant number discussed in Section 1.5. However, when the concentration increases, the interaction between aggregates becomes important, and this effect can lead either to aggregates ordering relative to one another or interactions becoming strong enough to change the aggregate shape. The compromise between these two effects leads to a very rich phase behavior. Figure 10.10 shows the phase diagram for a binary cationic soap–water system.

In addition to the micellar solution, it exhibits a number of different liquid crystalline phases. The most abundant is the normal hexagonal phase H_1 formed by infinite cylindrical aggregates packed in a two-dimensional hexagonal array. A schematic picture of this phase is shown in Figure 10.7. The hexagonal phase was called the middle phase by the soap boilers. The other common phase is the lamellar phase L_α, which the soap boilers called the neat phase. In Figure 10.10, there are two different cubic

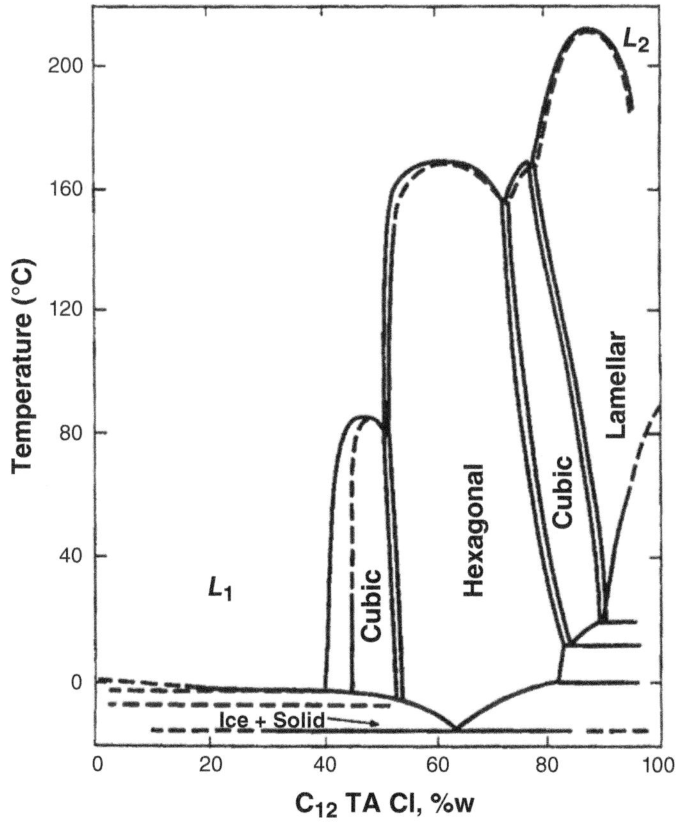

Figure 10.10 T–X phase diagram of the binary system water–dodecyltrimethylammonium chloride. Recent studies show that intermediate phases are present between the hexagonal and cubic phases. *Source*: Balmbra et al., 1969/with permission of Springer Nature.

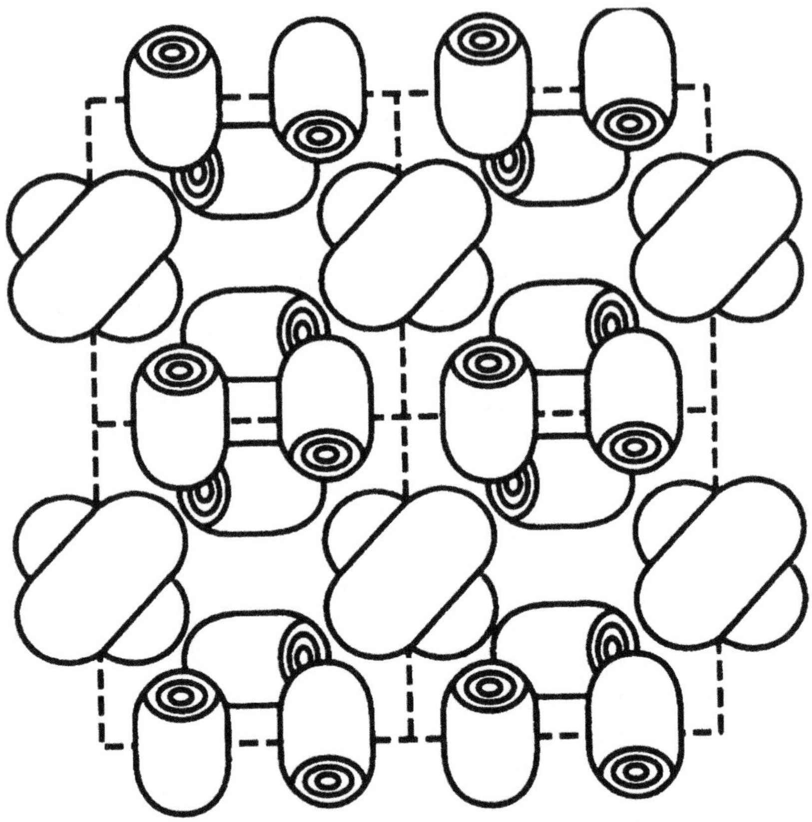

Figure 10.11 Schematized structure in a cubic phase containing ordered ellipsoidal micellar aggregates. Two inequivalent positions exist, as well as a certain degree of orientational disorder. Dashed lines represent the projections of unit cell boundaries. *Source*: Fontell et al., 1985/with permission of Taylor & Francis.

phases. The one between the micellar solution and the hexagonal phase contains finite, short, rod-like aggregates packed in a cubic array, as shown in Figure 10.11.

The other is a bicontinuous phase whose structure resembles the cubic structure shown in Figure 6.14. In fact, a range of different structures of cubic phases have been found for other ionic amphiphiles, as well as so-called intermediate phases with symmetries lower than those of the two-dimensional hexagonal and three-dimensional cubic phases.

Another noticeable feature of the phase diagram in Figure 10.10 is that the phase boundaries between the different liquid crystalline phases are surprisingly insensitive to temperature changes. This behavior is typical for ionic surfactant systems, and we can trace it back to the fact that all important free energy contributions have approximately the same temperature dependence.

Figure 10.7 showed a ternary phase diagram for ionic surfactant–alcohol–water. In such systems, the alcohol is often called a cosurfactant. Because this is not charged, it reduces the charge density on aggregate surfaces and also the optimal area per head group, thus changing the effective surfactant number N_S. These effects explain why adding octanol induces the formation of a lamellar phase. Other phases in this diagram are the reversed micellar solution L_2, which consists of small water surfactant aggregates in a continuous liquid octanol medium, and the reversed hexagonal phase H_2, which contains small water channels packed in a two-dimensional hexagonal array.

10.2.3 Temperature Changes Dramatically Affect Phase Equilibria for Nonionic Surfactants

Surfactants, C_nE_m, with a polar group consisting of an oligoethylene oxide chain with the general formula $CH_3(CH_2)_{n-1}(OC_2H_4)_mOH$ also display a range of different isotropic and liquid crystalline phases. In spite of profound differences in the inter- and intra-aggregate forces relative to those of ionic surfactant systems, phases of the same types still occur.

Figure 10.12 shows a $\mathbf{T} - X$ diagram of the binary system water—$C_{12}E_5$. At temperatures below 20 °C, the sequence of phases resembles the one in Figure 10.10. However, increasing the temperature has a profound effect that is qualitatively similar to adding octanol to a potassium decanoate–water mixture as in Figure 10.7. Increase in temperature does not primarily lead to the melting of aggregate order, but rather a decrease in the area per polar group, which results in an increase in the surfactant number. The basic molecular reason for the decrease in area of the polar group with increasing **T** has not been unequivocally established. However, water becomes a relatively less-efficient solvent for apolar groups as temperature increases due to the increase in the hydrophobic interaction as discussed in Section 5.7. In the language of polymer solutions, water becomes a less good solvent for the C_2H_4O—groups at higher **T**.

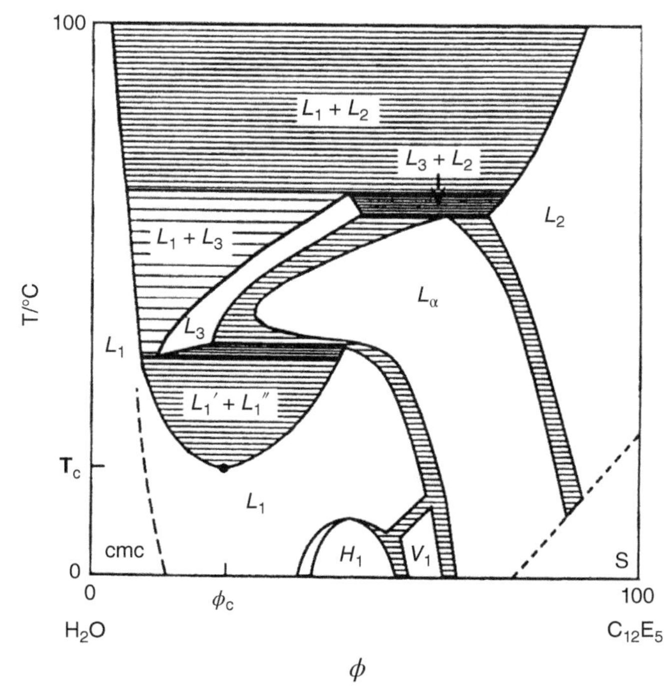

Figure 10.12 $T - X$ phase diagram of the binary system H_2O—$C_{12}H_{25}(OC_2H_5)_5OH$. To show more clearly the behavior at low surfactant concentration, the composition scale is nonlinear. L_1 and L_2 denote normal and reversed micellar solutions, respectively. H_1 is a hexagonal, V_1 a cubic, and L_α a lamellar liquid crystal. L_3 is a sponge phase containing bilayers in an isotropic solution, and S denotes the solid surfactant. Horizontal lines are tie lines. A critical demixing occurs at a volume fraction $\phi_c = 0.013$ and temperature $\mathbf{T}_c = 31.9$ °C. *Source*: Strey et al., 1990/with permission of Royal Society of Chemistry.

10.2.4 Block Copolymers Show as Rich a Phase Behavior as Surfactants

In Chapter 7, we showed that block copolymers self-assemble into micellar aggregates in dilute solution, according to the same principles that apply to surfactants. It should come as no surprise that when the polymer concentration is increased, ordered phases readily appear. Depending on the selectivity of the solvent and on the relative size of the two blocks, hexagonal, lamellar, and bicontinuous cubic phases are formed. Even the pure polymer can show this range of phases as the relative size of the blocks changes.

Let us illustrate the general behavior through the polystyrene (PS)–polyisoprene (PI) system:

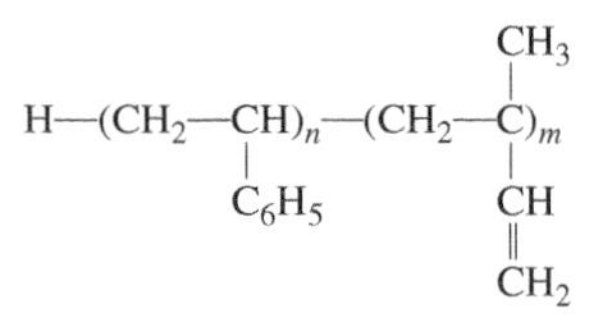

As the ratio between PS and PI is changed, the polymer melt goes through a sequence of structures that is very similar to the ones found for surfactant–water systems. Four of these are illustrated in Figure 10.13. To rationalize the trend in the structural changes, we could clearly use a parameter similar to the surfactant number introduced in Chapter 1.

For sufficiently long blocks, the PS and PI subchains virtually do not mix at all, and the samples consist of pure PS and PI regions. Because PS and PI parts are joined by covalent bonds, some sort of regular pattern must arise. One important factor in determining the structure is the relative volume of each of the two subblocks. In a homogeneous polymer melt, the chains fold as ideal

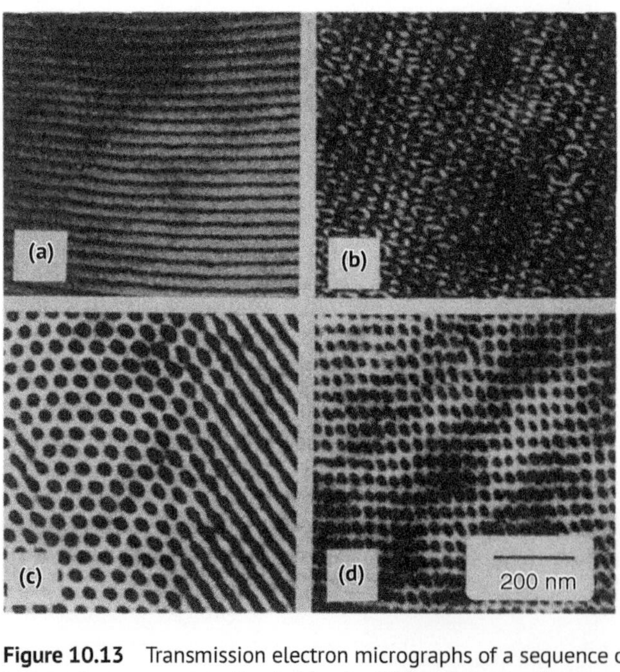

Figure 10.13 Transmission electron micrographs of a sequence of morphologies with increasing homopolymer concentration in blends of PS and styrene-*b*-isoprene diblock copolymer. (a) 10% PS, lamellae; (b) 30% PS, the ordered bicontinuous cubic morphology; (c) 50% PS, cylinders on a hexagonal lattice; and (d) 70% PS, spheres on a cubic lattice. *Source:* Winey et al., 1991/with permission of AIP Publishing LLC.

random coils, and the coil dimension is proportional to $N^{1/2}l_p$ where l_p stands for the persistence length and N the degree of polymerization. Because the persistence lengths of PS and polybutadiene are of similar magnitude the phase behavior is symmetrical with respect to the volume fraction of the two components. Thus, PI forms spherical centers in PS up to 18%, while above 80% PS forms spheres in a PI matrix. A lamellar state is preferred around equal volume fractions, and a two-dimensional hexagonal packing occupies the space between the lamellar packing and the spheres. Between the hexagonal and the lamellar phase, a bicontinuous cubic phase appears as shown in the phase diagram of Figure 10.14.

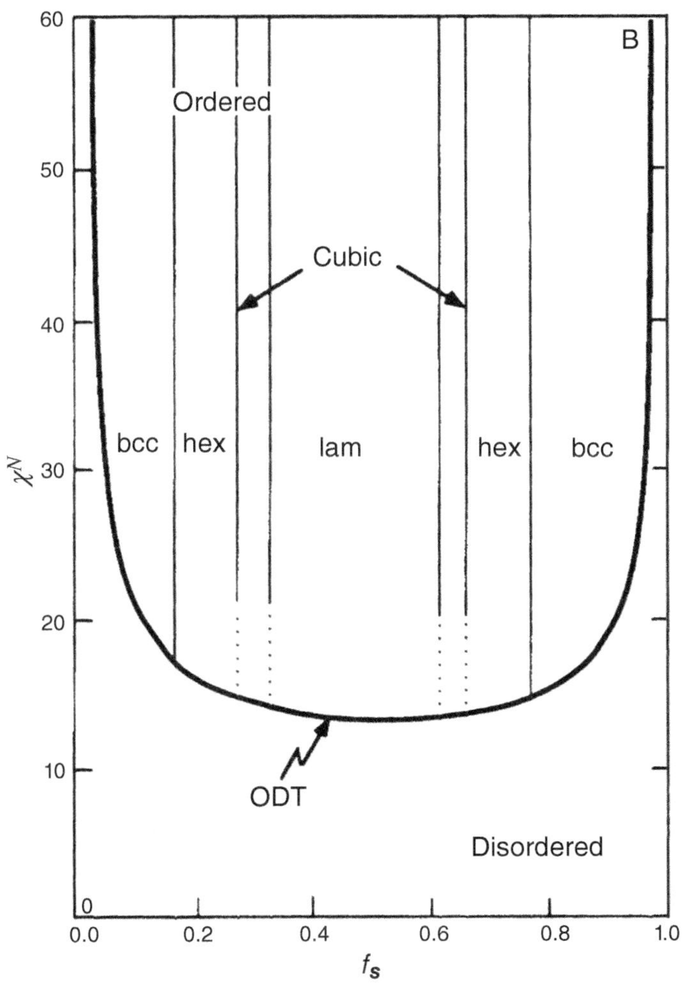

Figure 10.14 Experimentally determined phase diagram for the block copolymer system PI–PS. The fraction of styrene is represented by f_S, and χN is the product of the χ interaction parameter and the number N of monomer units per block. bcc denotes a body-centered cubic phase of finite aggregates of blocks of the less abundant component in a matrix of the more abundant component. If the blocks are too short, ordered phases do not form, and ODT denotes the curve for the order–disorder transition. *Source:* Bates, 1991/with permission of American Association for the Advancement of Science.

10.2.5 Monomolecular Films Show a Rich Phase Behavior

In Section 2.5.2, we briefly discussed the relationship between surface pressure Π_s of a monolayer and the area per molecule a as measured in a Langmuir balance. This relationship corresponds to a $\boldsymbol{p} - V$ diagram for a bulk system, but typically the $\Pi_s - a$ curves show much more structure (Figure 2.18). One reason is that the change from three-dimensional to two-dimensional molecular packing profoundly influences the cooperativity in a molecular ordering process. The number of nearest-neighbor molecules is lower in the film than in bulk.

In the $\Pi_s - a$ curves, we can observe two types of transition. At a certain pressure, the curve appears flat, showing that the area can be changed without affecting the pressure. This is typical of a normal first-order phase transition. In other cases, we observe a distinct kink in the $\Pi_s - a$ curve with different but nonzero slopes on either side of it. This kink indicates a continuous phase transition whose properties will be discussed in Section 10.4. As an illustration of the complexity found for high surface pressures or small areas per molecule, Figure 10.15 shows the phase $\mathbf{T} - \Pi_s$

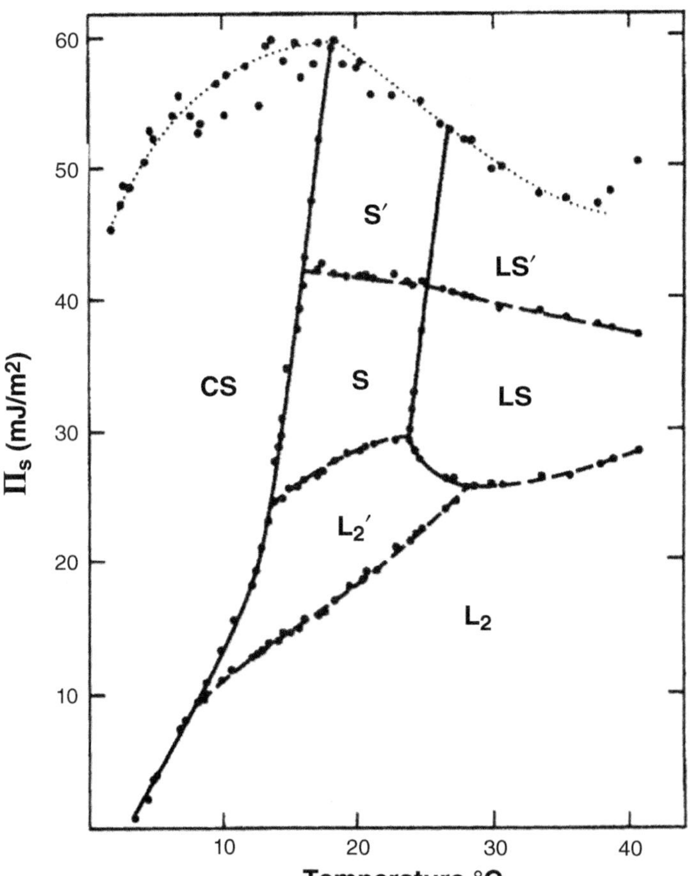

Figure 10.15 Phase diagram showing the different phases formed in a two-dimensional film of *n*-docosanoic acid $C_{21}COOH$ at the air–water interface. Depending on surface pressure and temperature, seven phases are identified. The molecular arrangement in the different phases has not been established, but the higher the pressure and the lower the temperature, the more the film becomes like a solid. Solid lines denote first-order transitions and dashed lines continuous transitions. The dotted line at pressures around 50 mJ/m^2 denotes a mechanical instability of the film. *Source:* Lundquist, 1978.

diagram for the saturated *n*-docosanoic acid (22 carbon atoms) at the interface between air and an aqueous low pH hydrochloric acid solution so that headgroups are uncharged.

In the temperature range of 10–40°C, this diagram displays a sequence of two to four transitions of both the first-order type and the continuous type. All the different phases shown represent dense films whose area varies between 0.25 nm^2 per molecule at lower pressures to approximately 0.18 nm^2 at the collapse of the film.

Detailed molecular organizations present in the individual phases in this example are not fully understood, but the observed rich phase behavior (which is not particularly sensitive to the nature of the polar head group) indicates the existence of many possibilities for varying the packing of approximately parallel hydrocarbon chains. In a bulk phase, we usually see a distinct transition from a solid state with low molecular mobility to a liquid state with high molecular mobility. In the surface film, the transition between these two general states is more gradual and involves a series of phase transitions.

At the highest values of π_s and at lower temperatures, the molecules are packed with a long-range positional and orientational order along the monolayer. Decreasing π_s or increasing **T** results in less ordered structures. Several of these show long-range positional order and molecules in a hexagonal arrangement, but with some disorder in alkyl chains. Furthermore, these chains can show a long-range tilt order. The combination of these effects leads to a multitude of possible phases, many of which are observed. It is in general easier to pack molecules in two dimensions rather than in three. This leads to fewer constraints on possible structures with the result that one can observe more variants of ordered structures.

See Atkins et al. (2022, Focus 6) and Kronberg et al. (2014, Chapters 6, 9, 14, and 17) for more information on Phase Diagrams.

10.3 We Obtain an Understanding of the Factors That Determine Phase Equilibria by Calculating Phase Diagrams

To understand the relationship between microscopic interactions and macroscopic phase behavior, we need to develop methods for calculating phase equilibria. The general principle requires us to find an expression of the free energy or its derivative, the chemical potential, and calculate the phase behavior from this. Such a calculation can be accomplished in many ways, and the choice depends on purpose and circumstances. Thermodynamic phase behavior shows a certain degree of universality, and different molecular mechanisms can lead to essentially the same macroscopic properties. In the following section, we will illustrate several ways to calculate phase diagrams using different methods.

10.3.1 The Regular Solution Model Illustrates Liquid–Liquid Phase Separation

In Section 1.1, we showed that we can calculate the chemical potential for a binary mixture of liquids A and B using the regular solution model (see Eq. 1.2.7) as

$$\mu_A = \mu_A^\theta + \boldsymbol{RT} \ln X_A + wX_B^2$$

$$\mu_B = \mu_B^\theta + \boldsymbol{RT} \ln X_B + wX_A^2 \tag{10.3.1}$$

where w represents the effective interaction parameter between molecules A and B (Eq. 1.1.16). If w is positive, the $A - B$ interaction is unfavorable, and from this point of view, the system would prefer to separate into a pure A and a pure B phase. However, the entropy of mixing, represented by the term $\boldsymbol{RT}\ln X$ in Eq. (10.3.1), promotes mixing of the liquids. Clearly, when w is sufficiently large, the entropy of mixing is not strong enough to prevent phase separation.

Let us assume that the system separates into two liquid phases, α and β, with compositions X_A^α and X_A^β. The chemical potentials in these two liquids should be equal, and combining Eqs. (10.1.9) and (10.3.1), we have the two equations

$$\boldsymbol{RT}\ln X_A^\alpha + w(1 - X_A^\alpha)^2 = \boldsymbol{RT}\ln X_A^\beta + w\left(1 - X_A^\beta\right)^2$$

$$\boldsymbol{RT}\ln(1 - X_A^\alpha) + w(X_A^\alpha)^2 = \boldsymbol{RT}\ln\left(1 - X_A^\beta\right) + w\left(X_A^\beta\right)^2 \tag{10.3.2}$$

which we can solve for the two unknowns, X_A^α and X_A^β. For what values of $w/\boldsymbol{RT}$ can we obtain a solution where $X_A^\alpha \neq X_A^\beta$? We can easily solve the system of Eq. (10.3.2) numerically, but to illustrate a graphical method, let us use Eq. (10.3.1) directly. The parameter that determines whether or not a phase separation occurs is $w/\boldsymbol{RT}$. Now we can solve for equal chemical potentials for both components by plotting $\mu_B - \mu_B^\theta$ versus $\mu_A - \mu_A^\theta$ for a given $w/\boldsymbol{RT}$ and a range of compositions X_A.

In Figure 10.16, we show such plots for the three cases $w/\boldsymbol{RT} = 1$, 2, and 3. When $w/\boldsymbol{RT} = 1$, panel (a), the μ_B versus μ_A plot simply provides a monotonic curve, and the chemical potentials are never equal at different compositions. On the other hand, for $w/\boldsymbol{RT} = 3$, panel (c), the curve displays a loop, and at its intersection point, the chemical potentials are equal for both components at two different compositions. This means that phase separation has occurred. By going back and determining the compositions at the crossing, we can obtain the phase boundaries. For the case $w/\boldsymbol{RT} = 2$, panel (b), there is a cusp in the curve at $X_1 = 1/2$ and this case represents the transition point between monotonic and loop behavior. The cusp signals that the two coexisting phases have just merged into one phase and we have reached the critical point.

Figure 10.17 shows the $\mathbf{T} - X$ phase diagram calculated using the regular solution model.

The critical point occurs at $\mathbf{T} = w/(2\boldsymbol{R})$ and $X_B = X_A = 0.5$. In the coexistence region, the dashed line signifies the compositions at the cusps in the loop in Figure 10.16c, and it follows that between the two cusps, both $\partial\mu_A/\partial n_A$ and $\partial\mu_B/\partial n_B$ are negative, so that when a component is added, its chemical potential decreases. Such a system is locally unstable so that spontaneous fluctuations in the concentration lead to a decrease in free energy and they will tend to grow in amplitude. On the other side of the dashed line in Figure 10.17, which is called the spinodal line, the system is locally stable and requires a nucleation process for the equilibrium phases to form.

Particularly for polymer and colloidal systems, in which equilibration processes can be inherently slow, we can observe intricate domain formations if a system is quickly brought from the isotropic solution into the unstable region. In such a spinodal decomposition, characteristic metastable concentration patterns can develop.

10.3.2 Liquid State Miscibility and Solid State Demixing Lead to a Characteristic Phase Diagram

Figure 10.5 showed a typical phase diagram for a binary A–B system with a common liquid phase and separated solid phases. We can rationalize this phase behavior within a simple model, again using the regular solution model. If we assume the melted (liquid) phases of A and B mix readily,

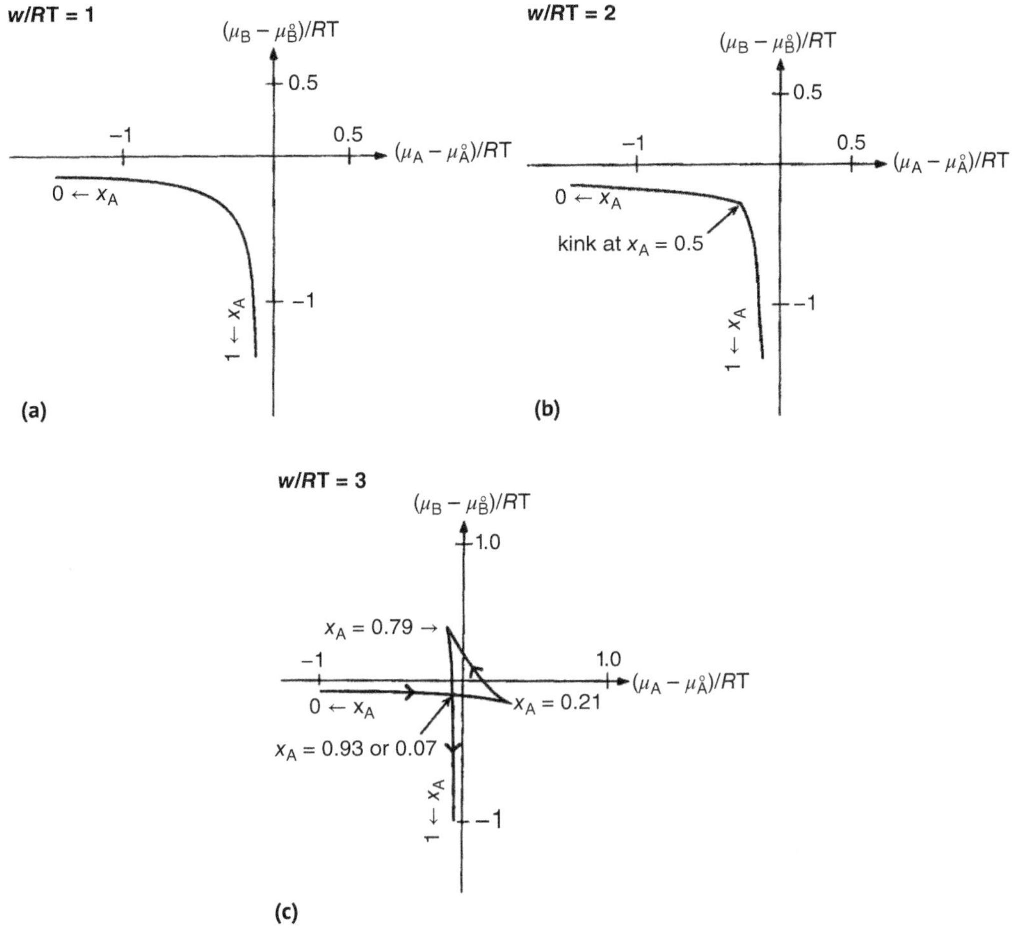

Figure 10.16 Plots of μ_B versus μ_A for the regular solution model. (a) At $w/\mathbf{RT} = 1$, the curve is monotonic, implying miscibility at all compositions. (b) At $w/\mathbf{RT} = 2$, a kink in the curve signals a critical point. (c) At $w/\mathbf{RT} = 3$, the curve makes a closed loop. At the self-intersection, μ_A and μ_B are equal at two different compositions, $X_A = 0.93$ and $X_A = 0.07$. For compositions intermediate between these two, corresponding to the closed loop shown, the system phase separates. Between the kinks at $X_A = 0.79$ and $X_A = 0.21$, the liquid is locally unstable and would show spinodal decomposition. Arrows indicate increasing X_A.

and for the sake of simplicity that they form an ideal solution, the free energy per mole of substance becomes

$$\frac{\mathbf{G}_l}{n_A + n_B} = X_A\left(\mu_A^\theta(l) + \mathbf{RT}\ln X_A\right) + X_B\left(\mu_B^\theta(l) + \mathbf{RT}\ln X_B\right) \tag{10.3.3}$$

In the solid phases, which place much higher constraints on the packing of a B molecule in an A lattice and vice versa, the effective interaction parameter in the regular solution theory is high and positive. A solid solution does not form readily. For the solid phase

$$\frac{\mathbf{G}_s}{n_A + n_B} = X_A\left(\mu_A^\theta(s) + \mathbf{RT}\ln X_A + wX_B^2\right) + X_B\left(\mu_B^\theta(s) + \mathbf{RT}\ln X_B + wX_A^2\right) \tag{10.3.4}$$

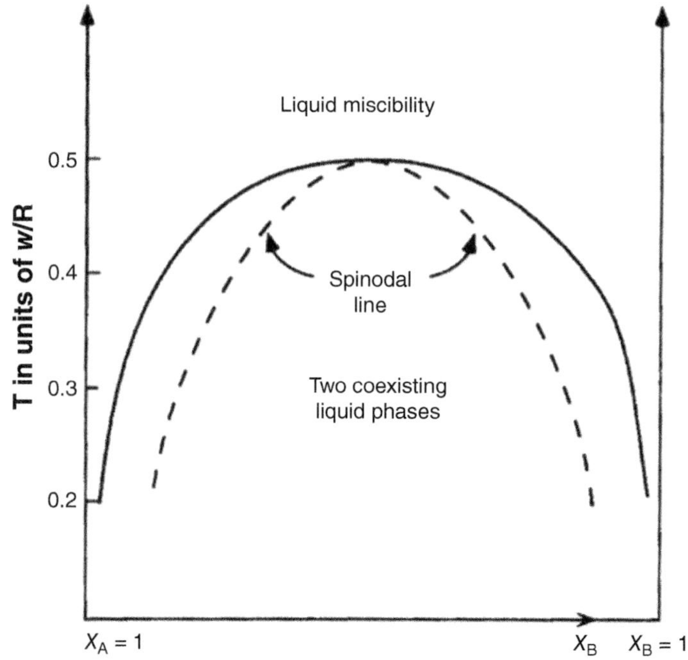

Figure 10.17 Phase diagram calculated for two liquids using the regular solution theory. The dashed line is the calculated spinodal line. A phase quenched to below this line is locally unstable, and the components segregate spontaneously without requiring a nucleation process.

where we expect $w/(\boldsymbol{RT})>>2$. The standard chemical potentials are temperature-dependent, but at the melting point $\mathbf{T}_A$ of A, we have

$$\mu_B^\theta(\mathbf{T}_A, s) = \mu_B^\theta(\mathbf{T}_A, l)$$

but for $\mathbf{T} < \mathbf{T}_A$

$$\mu_B^\theta(\mathbf{T}_A, s) < \mu_A^\theta(\mathbf{T}_B, l)$$

and vice versa for $\mathbf{T} > \mathbf{T}_A$. The same behavior occurs for B and $\mu_B^\theta(\mathbf{T}_B, s) = \mu_B^\theta(\mathbf{T}_B, l)$ at its melting point $\mathbf{T}_B$. Furthermore, $\mu_B^\theta(\mathbf{T}_B, l)$ decreases more rapidly with increasing $\mathbf{T}$ than $\mu_B^\theta(\mathbf{T}_B, s)$. To find the general phase behavior, we need not specify the parameters quantitatively. It suffices to note that the standard chemical potentials display the most significant temperature dependence.

Let us again use a graphical method for determining the phase diagram. In Figure 10.4, we saw how the intercept method can be used to find the chemical potentials once the free energy is known. To solve Eq. (10.1.9), for the coexistence of two phases, we need to find a common tangent at two points. The two compositions X_A^α and X_B^α, at which a common tangent touches the free energy curve $\mathbf{G}_m = \mathbf{G}/(n_A + n_B)$, represent the coexisting phases.

As Figure 10.18 shows, we have two free energy curves: one for the solid phase and the other for the liquid. Above the melting point of both compounds at temperatures $\mathbf{T} > \mathbf{T}_A$, $\mathbf{T}_B$, the free energy of the liquid is lowest at all compositions, and the liquid is the stable phase irrespective of the value of X_A (Figure 10.18a). At low temperatures $\mathbf{T} < \mathbf{T}_A$, $\mathbf{T}_B$, the solid phases are more stable than the liquid ones. As a result of the high $w/\boldsymbol{RT}$ value, however, a solid–solid phase separation occurs, which can be determined by finding the common tangent in Figure 10.18b. Now,

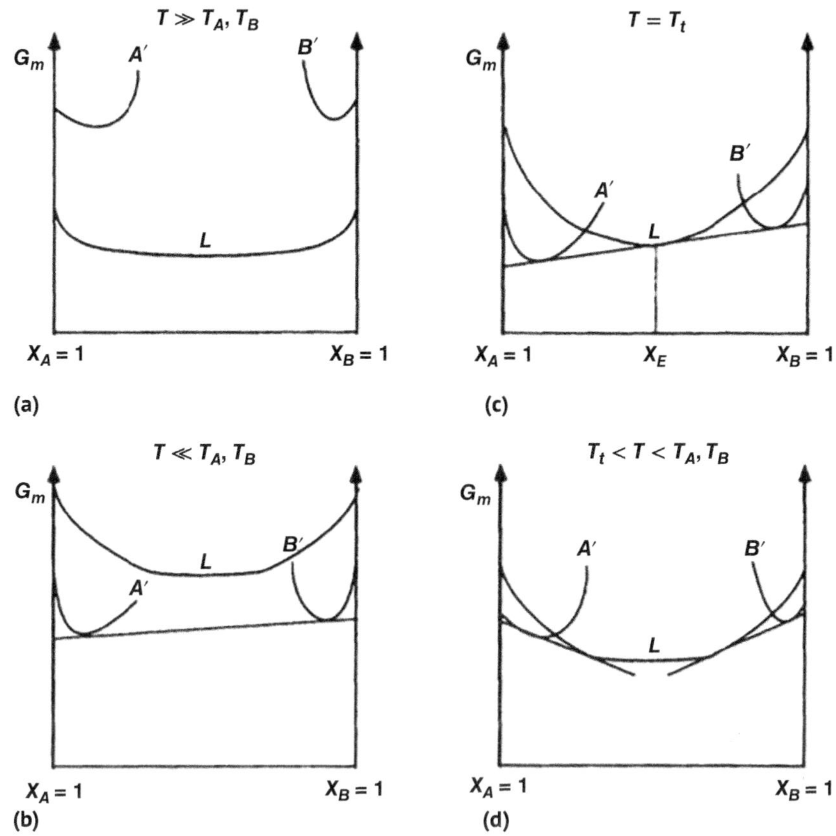

Figure 10.18 Plots of concentration dependences of the free energy $\mathbf{G}_m = \mathbf{G}/(n_A + n_B)$ per mole of a binary *A–B* mixture with one liquid *L* and two solid phases, *A* and *B*. (a) The temperature $\mathbf{T}$ is higher than both the melting temperatures $\mathbf{T}_A$ and $\mathbf{T}_B$ of the pure solids, A and B. (b) Temperature much lower than the melting temperatures, $\mathbf{T}_A$, $\mathbf{T}_B >> \mathbf{T}$. (c) Temperature $\mathbf{T} = \mathbf{T}_t$, where a common tangent to the free energy curves of all three phases exists. (d) Temperature intermediate between $\mathbf{T}_t < \mathbf{T} < \mathbf{T}_A$, $\mathbf{T}_B$ and the melting temperatures and three-phase line.

as $\mathbf{T}$ increases from a low value, the free energy of the liquid phase comes closer and closer to that of the solid system. At a particular temperature $\mathbf{T}_t$, the tangent joining the two solid phases also touches the liquid free energy curve as shown in Figure 10.18c. Here three phases manifest equal chemical potential of the two components. Thus, at $\mathbf{T}_t$, we have a three-phase line. For $\mathbf{T}_t < \mathbf{T} < \mathbf{T}_A$, $\mathbf{T}_B$, we can find two common tangents: one joining the solid *A* phase and the liquid and the other joining the liquid and the solid *B* phase (Figure 10.18d).

By collecting the behavior we have found in the different temperature regimes, we arrive at the resulting phase diagram shown in Figure 10.19.

Figure 10.5 shows that the phase behavior of a silver-copper binary system has the same characteristics as that of Figure 10.19 derived on the basis of the regular solution model. The generality of the thermodynamic model is demonstrated by the fact that the same basic model also describes the phase behavior of some colloidal systems. For example, this type of reasoning is very useful for understanding the phase behavior of mixtures of phospholipids that form bilayer structures. Here, the "liquid" phase is in reality a liquid crystalline phase and the solid is the gel phase with solid

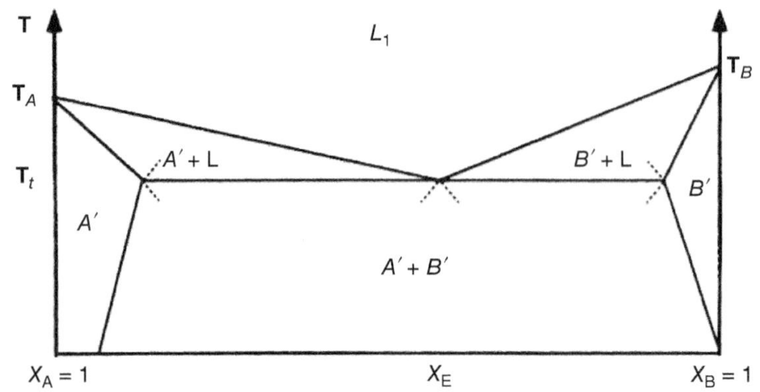

Figure 10.19 Phase diagram deduced from the curves of Figure 10.18. Dotted boundaries show metastable equilibria and X_E is the composition at the eutectic point (i.e., the composition of the liquid phase that emerges first when a two-phase sample is heated).

alkyl chains. Depending on the similarity between the two components, a range of behavior can be observed, from that shown in Figure 10.19 to that shown in Figure 10.20.

10.3.3 Two Lipids That Exhibit Different Melting Points but Mix Ideally in Both the Gel and Liquid Crystalline Phases Produce a Simple Phase Diagram

Figure 10.20 shows an experimentally determined phase diagram for a mixture of two phosphatidylcholines with slightly different alkyl chain lengths in excess water.

We can generate a model that reproduces the qualitative features of Figure 10.20 simply by assuming that the two components mix ideally in both lamellar phases: gel (g) and liquid crystal (lc). The chemical potentials for the two components in the gel are

$$\mu_1^g\left(\mathbf{T}, X^g\right) = \mu_1^{\theta, g}(\mathbf{T}) + \boldsymbol{RT}\ln X^g \tag{10.3.5a}$$

$$\mu_2^g\left(\mathbf{T}, X^g\right) = \mu_2^{\theta, g}(\mathbf{T}) + \boldsymbol{RT}\ln\left(1 - X^g\right) \tag{10.3.5b}$$

where X stands for the mole fraction of component 1. The same expressions apply for the liquid crystal with a change of superscript g to lc. Equilibrium at equal chemical potentials implies

$$\mu_i^g\left(\mathbf{T}, X^g\right) = \mu_i^{lc}\left(\mathbf{T}, X^{lc}\right) \quad i = 1, 2 \tag{10.3.6}$$

We can solve for the two unknowns, $X^g(\mathbf{T})$ and $X^{lc}(\mathbf{T})$, in terms of the difference in standard chemical potential

$$\Delta\mu_1^\theta(\mathbf{T}) = \mu_1^{\theta, g}(\mathbf{T}) - \mu_1^{\theta, lc}(\mathbf{T}) \tag{10.3.7}$$

Denote $\Delta\mu_1^\theta(T)/\boldsymbol{RT} \equiv A$; $\Delta\mu_2^\theta(T)/\boldsymbol{RT} \equiv B$. Then we can write the solution to Eq. (10.3.6) as

$$X^g(\mathbf{T}) = \frac{1 - e^B}{e^A - e^B} \tag{10.3.8a}$$

$$X^{lc}(\mathbf{T}) = \frac{e^A(1 - e^B)}{e^A - e^B} \tag{10.3.8b}$$

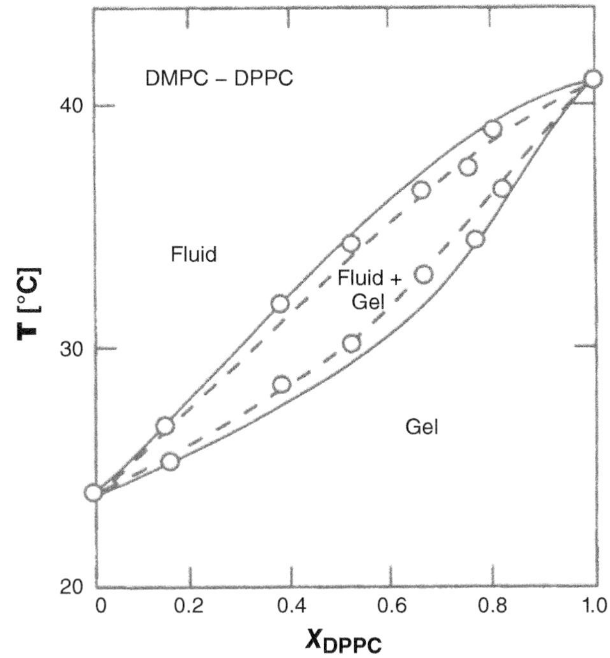

Figure 10.20 Experimental phase diagram for the binary mixture of dimyristoylphosphatidylcholine (DMPC) and DPPC in the presence of excess water. A liquid crystalline, L_α, and a gel, $P_{\beta''}$ lamellar phase (see Chapter 6), or both, are present in the system. The dashed line shows the calculated phase diagram. *Source*: Marley and Sturtevant, 1976/with permission of National Academy of Sciences.

Problem: Show that $X^i(T)$ of Eq. (10.3.8) solves Eq. (10.3.6) with μ_i given by Eq. (10.3.5).

Solution: The equality of the chemical potentials of component 1 yields

$$A = \ln\left(X^{lc}/X^{g}\right)$$

The RHS equals using Eq. (10.3.8)

$$\ln\left\{\frac{e^A(1-e^B)}{e^A-e^B}\frac{(e^A-e^B)}{1-e^B}\right\} = \ln\, e^A = A = \text{LHS}$$

For the second component

$$B = \ln\left\{\frac{\left(1-X^{lc}\right)}{\left(1-X^{g}\right)}\right\}$$

the RHS equals

$$\ln\left\{\frac{\left[1-\frac{e^A(1-e^B)}{e^A-e^B}\right]}{\left[1-\frac{1-e^B}{e^A-e^B}\right]}\right\} = \ln\left\{\left[e^A-e^B-e^A+e^{A+B}\right]/\left[e^A-e^B-1+e^B\right]\right\}$$

$$= \ln\left\{e^B(e^A-1)/(e^A-1)\right\} = B = \text{LHS}$$

The mole fractions $X^g(\mathbf{T})$ and $X^{lc}(\mathbf{T})$ are temperature-dependent through the temperature dependence of the standard chemical potentials $\mu_i^\theta(\mathbf{T})$, which are properties of the pure systems. At the melting points $\mathbf{T}_{0i}$, we have an equilibrium between the two phases, and for the pure compounds, $\Delta\mu_i^\theta(\mathbf{T}_{0i}) = \Delta\mathbf{H}_i^\theta - \mathbf{T}_{0i}\,\Delta\mathbf{S}_i^\theta = 0$. We can expand $\Delta\mu_i^\theta(\mathbf{T})$ around the point $\mathbf{T}_{0i}$ in a Taylor series, using the relation $\partial\mu/\partial\mathbf{T} = -\mathbf{S}$, to obtain

$$\Delta\mu_i^\theta(\mathbf{T}) \approx (\mathbf{T} - \mathbf{T}_{0i})\left\{\frac{\partial\Delta\mu_i^\theta(\mathbf{T})}{\partial\mathbf{T}}\right\}_{\mathrm{T}0i} = (\mathbf{T} - \mathbf{T}_{0i})\,\Delta\mathbf{S}_i^\theta = \left(\frac{\mathbf{T}}{\mathbf{T}_{0i}} - \mathbf{1}\right)\Delta\mathbf{H}_i^\theta \tag{10.3.9}$$

where $\Delta\mathbf{S}_i^\theta$ and $\Delta\mathbf{H}_i^\theta$ now refer to the melting transitions ($g \to lc$) of the pure lipids. From this equation, we find the temperature dependence of the two parameters A and B to be

$$A = -\frac{\Delta\mathbf{H}_1^\theta}{\boldsymbol{R}}\left(\frac{\mathbf{1}}{\mathbf{T}_{01} - \mathbf{T}}\right) \text{ and } B = -\frac{\Delta\mathbf{H}_2^\theta}{\boldsymbol{R}}\left(\frac{\mathbf{1}}{\mathbf{T}_{02} - \mathbf{T}}\right)$$

Figure 10.20 also shows the calculated phase diagram as a dashed line. It has the same shape as the experimental one, but its two-phase area is narrower. A simple extension of this model would introduce a small nonideality correction using a regular solution model.

10.3.4 The Presence of an Impurity Broadens a Phase Transition by Introducing a Two-phase Area

For a system where one component is in large excess, one has some simplifying properties. Understanding how an impurity influences the phase behavior relative to that of a pure one-component system is often of practical importance. Consider the case in which the pure, abundant component (1) exhibits a phase transition $\alpha \to \beta$ on heating at a temperature $\mathbf{T}_c$. The presence of impurity (2) will change the properties of this phase transition.

Since we are specifically concerned with low concentrations of the impurity, we naturally choose the infinite dilution as its standard state (Section 1.5.2). That is, we utilize that $w(1 - X_2)^2 \approx w$ and absorb the interaction parameter w in the standard chemical potential. Thus, in phase i, the chemical potential of the impurity is

$$\mu_2^i = \mu_2^\theta(\mathbf{T}, i) + \boldsymbol{R}\mathbf{T}\ln X_2^i \tag{10.3.10}$$

where the standard chemical potential refers to the infinite dilution state. If both phases α and β are present, the impurity will be distributed between the phases according to the distribution equilibrium

$$\frac{X_2^\beta}{X_2^\alpha} = \exp\left[\frac{\mu_2^\theta(\mathbf{T}, \alpha) - \mu_2^\theta(\mathbf{T}, \beta)}{\boldsymbol{R}\mathbf{T}}\right] = K_d \tag{10.3.11}$$

where K_d represents the distribution coefficient of component 2 between phases α and β.

Equation (10.3.11) gives the ratio of the concentrations. If we also want to obtain the absolute value of X_2, we must consider the chemical potential of component 1. For small X_2, this equals

$$\mu_1^i \approx \mu_1^\theta(\mathbf{T}, i) + \boldsymbol{R}\mathbf{T}\ln(1 - X_2^i) \approx \mu_1^\theta(\mathbf{T}, i) - \boldsymbol{R}\mathbf{T}\,X_2^i \tag{10.3.12}$$

As in Eq. (10.3.9), we can relate the temperature dependence of the chemical potential to the transition enthalpy $\Delta\boldsymbol{H}$ and

$$\mu_1^\theta(\mathbf{T},\alpha) - \mu_1^\theta(\mathbf{T},\beta) = -\Delta\mu^\theta(\mathbf{T}) \approx \left(\frac{\mathbf{T}}{\mathbf{T}_c} - 1\right)\Delta\mathbf{H} \tag{10.3.13}$$

Solving for equal chemical potential of component 1, we find

$$X_2^\alpha - X_2^\beta = \left(\frac{\mathbf{T}}{\mathbf{T}_c} - 1\right)\frac{\Delta\mathbf{H}}{\mathbf{RT}} \tag{10.3.14}$$

and using Eq. (10.3.11),

$$X_2^\beta = -\left(\frac{\mathbf{T}}{\mathbf{T}_c} - 1\right)\frac{\Delta\mathbf{H}}{\mathbf{RT}(1 - K_d^{-1})} \tag{10.3.15}$$

If $K_d > 1$ implying that the impurity dissolves better in the high-temperature phase, which is normally the case, a solution exists for $\mathbf{T} < \mathbf{T}_c$. The composition in the α phase is

$$X_2^\alpha = -\left(\frac{\mathbf{T}}{\mathbf{T}_c} - 1\right)\frac{\Delta\mathbf{H}}{\mathbf{RT}(K_d - 1)} \tag{10.3.16}$$

The constructed phase diagram of Figure 10.21 shows that straight lines initially border the two-phase area and that an impurity introduces a two-phase area into the diagram.

Increasing the amount of impurity widens the two-phase area, but the width also depends on the magnitude of the distribution coefficient K_d. The more K_d deviates from unity, the wider the two-phase area. In a calorimetric experiment, one observes that the peak in heat capacity at the transition temperature (see Figure 6.9) is spread out over an interval $\mathbf{T}_c - \mathbf{T}$, which is determined by the amount of the impurity—an effect that is sometimes misinterpreted as a change in the cooperativity of the transition (see also Figure 10.28).

The old method of determining a compound's molecular weight by measuring the freezing point depression is based on the assumption that $K_d \gg 1$ so that it can be neglected in the denominator of Eq. (10.3.15). Then measuring $\mathbf{T}_c - \mathbf{T}$ and knowing the transition enthalpy $\Delta\mathbf{H}$ gives the mole fraction X_2. From the added amount of component 2, one then obtains its molecular weight.

Normally, a high-temperature phase provides a better solvent than a low-temperature phase, so the two-phase area slopes downward with temperature, as we see in Figure 10.21. However,

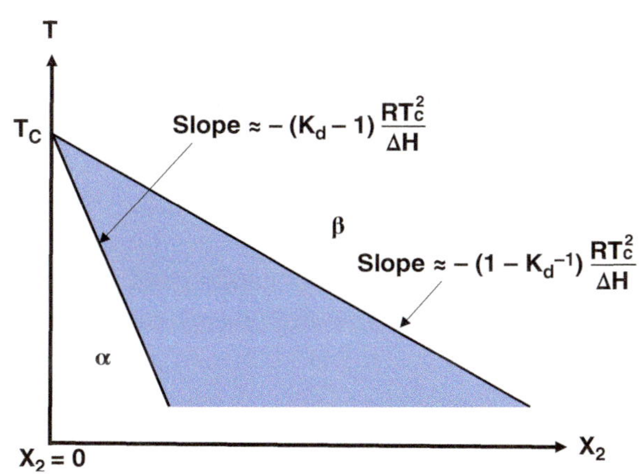

Figure 10.21 Calculated effect of an impurity (2) on the equilibrium between phases α and β. For the pure component 1, the transition temperature occurs at $\mathbf{T} = \mathbf{T}_c$. The slope of the lines is determined by the transition enthalpy $\Delta\mathbf{H}$ and the partition coefficient K_d for the impurity between the two phases.

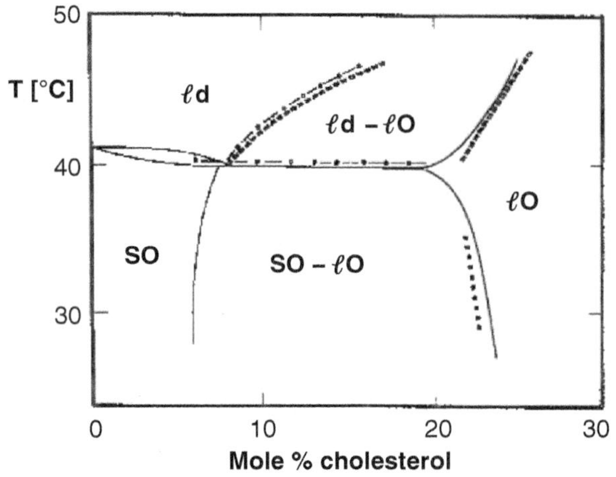

Figure 10.22 Phase diagram DPPC–cholesterol in excess water. *Source*: Ipsen et al., 1987/with permission of Elsevier.

exceptions to the rule clearly exist. Figure 10.20 shows one such exception, where ideal mixing was assumed for both gel and liquid crystalline phases. Another important exception occurs when cholesterol is added to a pure lipid, such as DPPC, as the phase diagram in Figure 10.22 shows.

As discussed in Section 6.1, the planar cholesterol molecule has a preference for straight conformations of the lipid alkyl chains, while, in contrast, the disorder in the liquid crystalline phase makes it a more versatile solvent than the gel phase. At low cholesterol contents, these two effects nearly cancel each other, and a two-phase area is hardly observed. At higher cholesterol contents, the straightening of the chains leads to the formation of the liquid-ordered state, which is then stable down to low temperatures.

10.3.5 The Short-range Repulsive Overlap Force Influences the Equilibrium Between Liquid Crystalline and Gel Phases in Lipid–Water Systems

We can study another example of phase equilibria between lamellar liquid crystal and gel phases of lecithins in the phase diagram of DPPC and water, as illustrated in Figure 10.23.

In contrast to the example discussed in Section 10.3.3, water is not necessarily present in excess, and the figure shows that lowering the water content triggers phase changes between liquid crystal and gel. In Figure 6.9, we saw that in excess water, there is a pretransition $L_{\beta'}$ to P_β at 34 °C and a main transition P_β to L_α at 41 °C. In Figure 10.23, three-phase lines appear at these two temperatures, where pure water represents the third phase. There is an additional three-phase line at $\mathbf{T} \approx 44$ °C associated with the disappearance of the P_β phase at high temperature and low water content. Above $\mathbf{T} \approx 44$ °C, there is a transition from a L_α phase with a low water content to a $L_{\beta'}$ phase of even less water. This phase change occurs since the overlap force is stronger in the more flexible liquid crystalline phase than in the gel. Based on measured forces (see Section 6.3) and structural X-ray characterizations of the lipid phases, it is possible to model the phase diagram in some detail, as shown in Figure 10.24.

This diagram also shows how the respective phases swell in excess water. That the L_α phase has a more extensive swelling is another manifestation that the overlap force is stronger for this phase than for the gel phases, as discussed in Section 5.5.

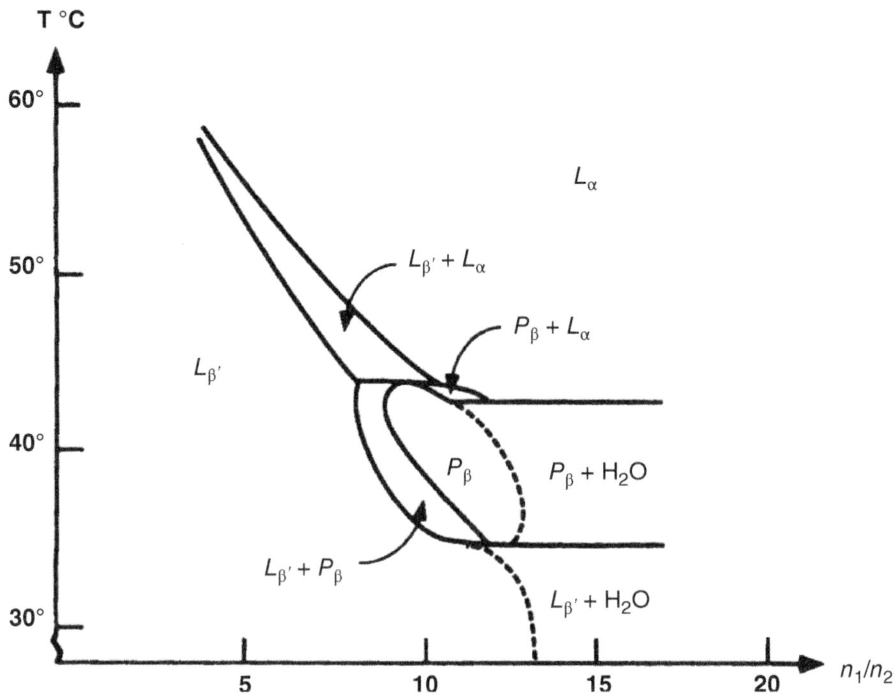

Figure 10.23 Experimental phase diagram for the system water(1)–DPPC(2). *Source*: Ulmius et al., 1977/with permission of American Chemical Society.

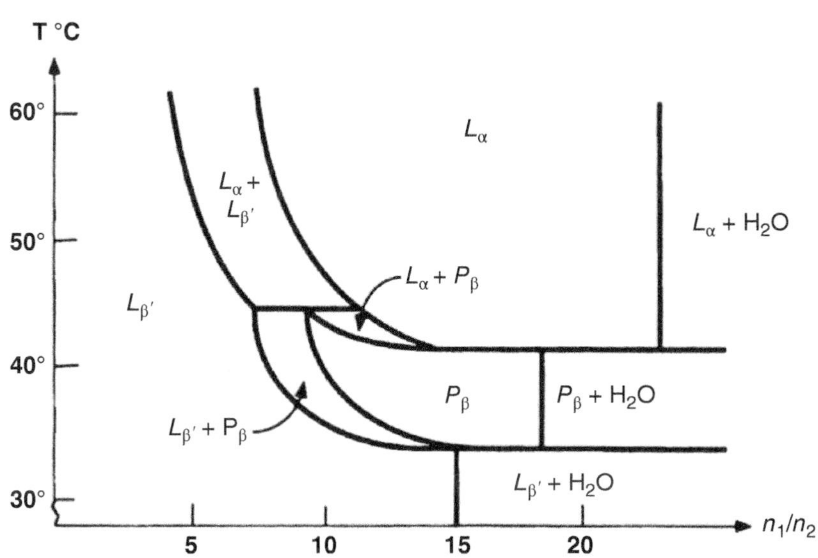

Figure 10.24 Calculated phase diagram for the system shown in Figure 10.23. *Source*: Guldbrand et al., 1982/with permission of Elsevier.

10.3.6 The Isotropic to Nematic Transition Can Be Caused by an Orientation-dependent Excluded Volume

A nematic phase with long-range orientational correlations is typically found in one-component systems when the molecules are rather stiff and elongated. In colloidal systems, nematic phases occur with stiff rod-like particles that do not attract one another. In fact, the first theoretical description of an isotropic to nematic transition was developed by Lars Onsager in 1949 to explain the observation of a nematic phase in suspensions of Tobacco Mosaic Virus (TMV). He modeled the TMV particles as stiff rods of length L and diameter D, and they were supposed to interact by a hard-rod potential stating that rods cannot overlap. Onsager realized that there was also an electrostatic interaction but left that complication for future work.

In Section 10.2.1, we discussed the formation of an ordered phase in a hard-sphere system. In the hard-rod system, the isotropic to nematic transition has a similar cause involving a balance between two competing entropy contributions. In the ordered phase, the entropy of the orientational degrees of freedom is decreased. What could compensate for that loss of entropy?

In the isotropic phase, the particles adopt all orientations, and neglecting correlations, the rotating rods exclude a volume of order L^2D, which for large L/D ratios is substantially larger than the actual particle volume. On the other hand, for orientationally correlated particles, the excluded volume is a multiple of the particle volume and of order LD^2. The total excluded volume increases linearly with particle concentration, so in the oriented phase, the particles have more accessible volume than in the isotropic phase, and the relative difference is higher at increased concentration. Thus, for the oriented phase, there is a higher value of the entropy associated with the translational degrees of freedom. A dilute solution should be isotropic, while there is a possibility that there is a transition to an ordered phase as the concentration is increased. Onsager was able to capture these qualitative arguments in a quantitative description. Instead of giving a detailed account of quantitative estimates, we here cite the result of a numerical evaluation of the model. For spherically capped cylinders, the isotropic phase is stable up to a volume fraction of rods $\phi_{\text{iso}} \approx 3.3D/L$ and then it is coexisting with a nematic phase up to a volume fraction $\phi_{\text{nem}} \approx 4.5D/L$, and at higher concentrations, there is a single nematic phase. These estimates are reliable for aspect ratios $L/D > 10$.

The competition between nematic and smectic orders is a common feature in practice. By elaborating the rod model by making a distinction between cap–cap and cap–center interactions between two rods, we could introduce a mechanism for the formation of a smectic system. When the cap–cap interaction is more favorable than the cap–center, the oriented system will also have a tendency to form layers as illustrated in Figure 10.1. The free energy balance between the nematic and smectic phases is determined by the strength of the interaction relative to $k_B\mathbf{T}$. Depending on such details of the interaction between the particles, one can have a sequence isotropic solution–nematic liquid crystal–smectic liquid crystal–solid crystal. In this sequence, either of the two middle stages, or both, can be absent.

10.3.7 The Flory–Huggins Theory Accounts for the Liquid–Liquid Phase Separation in Solutions of Polymer Mixtures

It is a common observation that solutions of two different polymers in a solvent separate into two coexisting liquid phases at higher polymer concentrations. We can use a direct generalization of the Flory–Huggins theory of polymer solutions to see the source of this phase separation. For two polymer species 1 and 2, the generalization of Eq. (7.2.6) for the free energy of the system becomes

$$\begin{aligned}\Delta G_{\text{mix}} = \boldsymbol{RT}\{ n_s \ln \phi_s + (n_{p1} + n_{p2}) \ln(\phi_{p1} + \phi_{p2}) \} \\ + (n_s + N_{p1} n_{p1}) w_1 \phi_s \phi_{p1} + (n_s + N_{p2} n_{p2}) w_2 \phi_s \phi_{p2} \\ + (N_{p1} n_1 + N_{p2} n_{p2}) w_{12} \phi_{p1} \phi_{p2}\end{aligned} \tag{10.3.17}$$

It is a considerable challenge to analytically solve for phase equilibrium in this ternary system, but it is readily addressed numerically. Compared to the two-component solvent–polymer case, there is a polymer 1–polymer 2 interaction term w_{12}, which is typically positive. There is also a small contribution from mixing the polymer species, and this goes to zero for long polymers. To qualitatively understand why phase separation can occur, compare a homogeneous solution with a phase-separated system of equal volumes of the separated phases. These two parts are rich in polymer 1 and polymer 2, respectively, but at the same total polymer concentration. Since the constraint on polymer conformations depends only on the total polymer content and not on the chemical nature of the polymer, this part of the free energy is the same in the homogeneous solution and in the phase-separated case. The difference between the two cases lies in the entropy of mixing of entire polymer molecules, which promotes mixing, and the polymer 1–polymer 2 contacts, which are proportional to the product of the two volume fractions. As in the regular solution model, such hetero contacts are typically less favorable than contacts between the same monomers, leading to $w_{12} > 0$. The conclusion is that the polymer 1–polymer 2 interaction term favors polymer segregation in solution and melt phases. At higher polymer concentrations, this term tends to dominate, resulting in a liquid–liquid phase separation.

Figure 10.25 shows calculated and experimental phase diagrams for the system water, polyethylene glycol–dextran. This two-phase system has been widely used in protein purification by extraction.

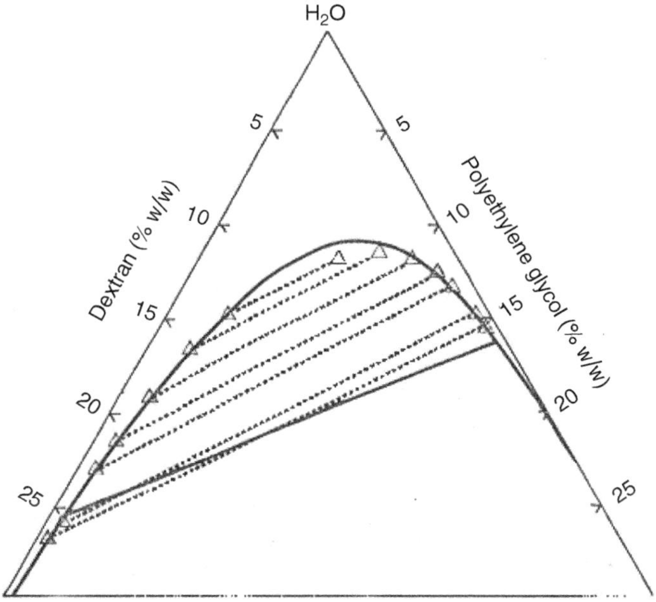

Figure 10.25 Partial ternary phase diagram of the system water–dextran (1–6,1–4 polyglucose)–polyethylene glycol (polyethylene oxide). Triangles show measured coexistence, and the solid line represents calculated values using Eq. (10.3.17). *Source:* Gustavsson et al., 1986/with permission of Elsevier.

10.3.8 Multicomponent Protein Solutions Can be Stable up to High Concentrations: The Eye Lens Protein Case

The lens in the eye of vertebrates is built from elongated cells with their long axis preferentially oriented along the optical axis. These cells contain three classes of proteins: α-, β-, and γ-crystallins. To enhance the effect of the lens, the cells have a protein concentration that varies concentrically from close to 50% w/w along the optical axis to around 15% at the periphery. This results in a variation of the refractive index as in a Fresnel lens. The proteins have a compact globular character so that in contrast to the polymer case discussed in the previous section, the proteins have no or small internal conformational freedom. A further remarkable feature is that once formed, the cells die and are depleted of all content except the lens proteins, their counterions, and water. They are then designed to remain intact during the lifetime of the organism. The lens should be transparent with a low ability to scatter light, and it should also be flexible to allow for accommodation, changing the refraction. In such a concentrated system, the challenge is to avoid phase separation and aggregation, which induces light scattering. Aggregation can be induced by chemical modifications of the proteins that occur during the span of a lifetime. It is the molecular event causing cataract formation, which ultimately causes blindness unless the lens is replaced. A related effect is the formation of cold cataracts, which is observed when the lens is brought to temperatures clearly below the body temperature. In this case, there is a liquid–liquid phase separation causing the scattering, but the lens returns to its original properties when the temperature is raised again.

With three classes of proteins, each encompassing individual variations, the cells in the lens contain a complex protein solution. We can illustrate some of the intricate properties of such a solution by analyzing the properties of the solutions of the main component, α-crystallin. It exists in the cell as a multi-subunit assembly of a small, $\approx$ 20 kD, protein. There is a size polydispersity in the range of 30–40 units per α-crystallin unit. Studies by X-ray and light scattering reveal that solutions of α-crystallin behave nearly as a hard-sphere system, indicating that the protein–protein interaction has been tuned so that van der Waals, electrostatic, and other short-range interactions nearly cancel. As a consequence, the scattering of light determined by structure factor $S(q = 0)$ decreases with increasing concentration and becomes negligibly small at typical *in vivo* concentrations, where the solution is transparent, as shown in Figure 10.26.

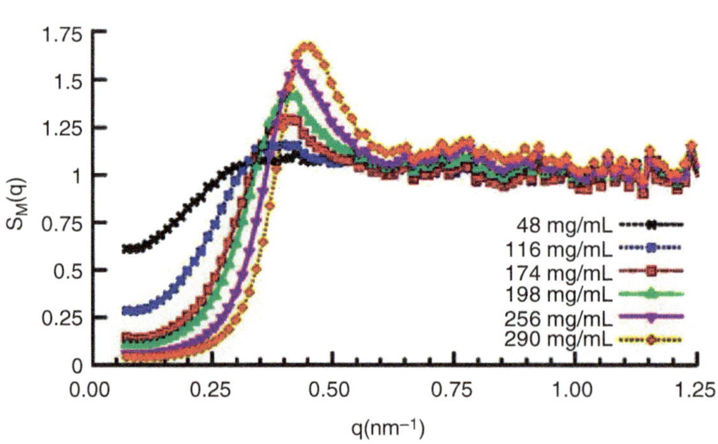

Figure 10.26 Measured structure factor for aqueous solutions of α-crystallin at different concentrations. *Source*: Foffi et al., 2014/with permission of National Academy of Sciences.

Another feature of the system is that the size polydispersity quenches the liquid to solid transition starting at a volume fraction of 0.49 for uniform hard spheres. The solution is intact up to a glass transition occurring around a volume fraction of 0.58. At this point, the viscosity diverges, and for concentrations above 0.5, the solution becomes very viscous, which in the lens case would imply that accommodation is hindered.

10.3.9 Mixing Colloidal Entities of Opposite Charge Often Results in an Associative Phase Separation

When we add a polyelectrolyte to a solution containing another polyion or micelle of opposite charge, there is a strong association leading to one of the three possibilities. One is that there is a precipitation forming a compact solid-like phase. A second possibility is the formation of a complex, often carrying an excess charge, which remains in solution. A third possibility is that there is a liquid–liquid phase separation, where the charged molecular components remain in solution also in the concentrated phase. In this case, there is a limited short-range charge matching so that the concentrated phase contains substantial amounts of water and also some small electrolyte ions. When dealing with two polyelectrolytes, this type of phase is called a coacervate, as discussed in Section 7.3. In a more general context, this can be referred to as associative phase separation, in contrast to the segregative phase separation in polymer mixtures discussed in the previous Section 10.3.7.

The case of local charge mismatch can be illustrated by the phenomenon of DNA compaction. The native DNA double helix is by itself long and stiff and *in vitro* it is easily broken by shear forces. *In vivo*, DNA is packed in a compact form by the action of histones, which are highly positively charged proteins. *In vitro*, one can induce a similar compaction using cationic amphiphiles. Even for a concentration that is only a tenth of the CMC, a cationic amphiphile like CTAB (see Table 1.2) tends to form micellar aggregates associated with the negatively charged DNA helix. The micelle has a higher charge density than the helix, so one has a local charge reversal, which potentially attracts other helices. However, if there is a low concentration of helices, the only available negatively charged helix patches are located on the same long DNA. The attraction to the excess micelle charge induces a folding of the helix, which in turn induces the formation of more micelles, and the end result is a compact DNA-amphiphile aggregate as shown in Figure 10.27.

See Doi (2013, Chapter 5) for more information on Calculations of Phase Equilibria.

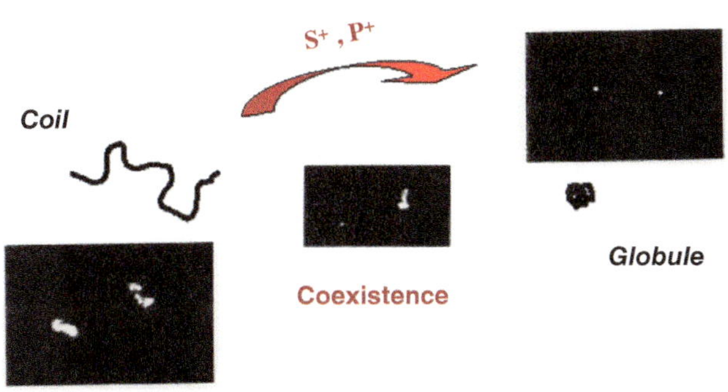

Figure 10.27 The coil–globule transition of a single DNA molecule observed by confocal fluorescence microscopy. The addition of a cationic amphiphile, or a positively charged polyion like spermidine, makes the extended DNA molecule fold up onto itself as seen as a small dot in the upper right image. Courtesy of Björn Lindman.

10.4 Continuous Phase Transitions Can Be Described by Critical Exponents

10.4.1 Phase Changes Can Be Continuous

The phase changes discussed in this chapter have been almost exclusively of a discrete character. At a given temperature or composition, a new phase appears whose properties distinctly differ from those present earlier. Such first-order phase changes are characterized by discontinuous changes in the dependent thermodynamic variables, except for the free energy. They represent indeed the most common, but not the only, type of phase change. The alternative, briefly discussed in connection with Figure 10.15, involves transitions that are collectively known as second-order transitions, phase transitions of the second kind, or continuous phase transitions. We will adopt the latter term, since it is the most descriptive.

In a continuous phase transition, some of the dependent thermodynamic variables diverge, as the independent variable, say the temperature **T**, reaches the critical or transition value, $\mathbf{T}_c$. As an example, Figure 10.28 shows how heat capacity, $\mathbf{C}_p$, can vary at a continuous phase transition. The enthalpy varies strongly but smoothly around $\mathbf{T}_c$. In a first-order transition, on the other hand, enthalpy changes discontinuously at $\mathbf{T}_c$, and consequently, $\mathbf{C}_p$ is unlimited at $\mathbf{T}_c$.

The different thermodynamic characteristics of first-order and continuous transitions have their source in the molecular events that occur as the transition point is approached. Melting of a crystal to a liquid is a typical example of a first-order transition. From a molecular point of view, an abrupt

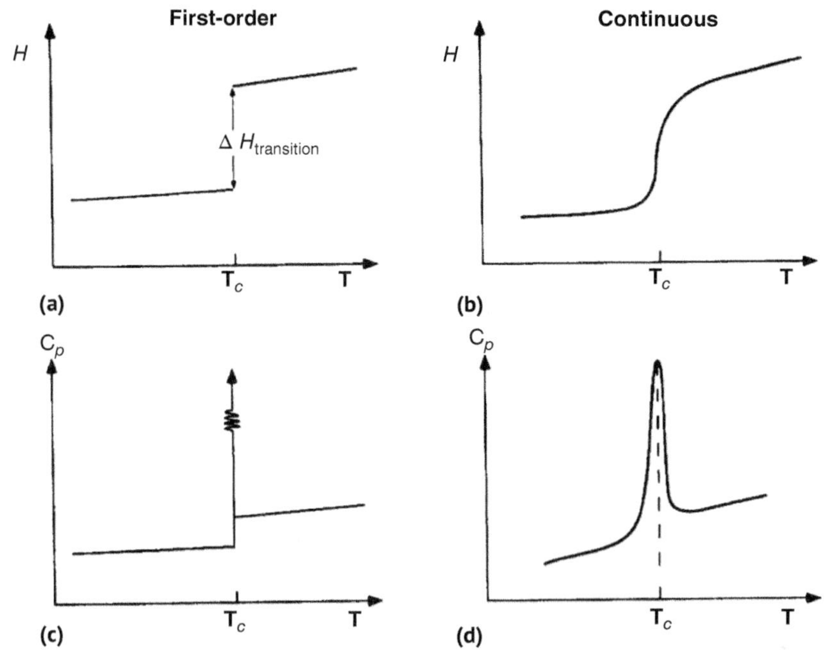

Figure 10.28 Schematic comparison of a first-order (a), (c), and a continuous (b, d) phase transition. In the continuous transition, enthalpy varies continuously but strongly in the region of the transition temperature $\mathbf{T}_c$, while for a first-order transition, a jump in **H** occurs. This jump also implies an infinite heat capacity for the first-order transition, while in a continuous transition, the heat capacity can diverge gradually. For some types of continuous transition, it can also stay finite.

change occurs at the melting point from a well-ordered crystalline state to the much more disordered liquid state. Clearly, discontinuity in the thermodynamic variables reflects a discontinuous change at the molecular level.

The change from gas–liquid coexistence to a supercritical fluid at the critical point serves as an illustration of a continuous transition. At temperatures below the critical point, a gas and a liquid phase coexist. As the temperature is raised at a net concentration corresponding to the composition at the critical point, the physical difference between the two coexisting phases diminishes until they merge at the critical point. Above the critical temperature, there is a single fluid phase. In this transition, the molecular conditions change in a smooth way, and discontinuities in the thermodynamic variables are not necessarily present. However, in the example we are dealing with, there is a change from a two-phase to a one-phase situation, so we should expect the thermodynamics to exhibit some signs of this change.

The Russian physicist L. Landau, the person behind the "L" in the DLVO theory, constructed a model that in simple form could illustrate the difference between a first-order and a continuous phase transition. The key concept is the order parameter η, which for a fluid around the critical point is the difference between the liquid and gas densities, while for an Ising model (see next section) of a magnet, it is the net polarization. The free energy, **A**, is assumed to be a function of temperature at constant pressure, and we can write

$$\mathbf{A}(\mathbf{T},\eta) - \mathbf{A}(\mathbf{T},0) = a(\mathbf{T})\eta^2 + b(\mathbf{T})\eta^3 + c(\mathbf{T})\eta^4 \tag{10.4.1}$$

The linear term only affects the baseline and can be incorporated in $\mathbf{A}(\mathbf{T},0)$. If now the third-order term $b(\mathbf{T})$ is zero, either by symmetry as for the Ising model or by choice of density, as for a critical point of a fluid, we have two terms of the free energy. For a stable system, $c(\mathbf{T})$ is necessarily positive, and consider the case when $a(\mathbf{T})$ changes sign as $\mathbf{T}$ passes a temperature $\mathbf{T}_0$. As the coefficient $a(\mathbf{T})$ changes sign, the minimum in the free energy goes continuously from zero to small values of the magnitude of the order parameter, so that $\eta \sim |\mathbf{T} - \mathbf{T}_0|^{1/2}$. This relation provides an example of a critical exponent discussed in the following section.

10.4.2 Continuous Transitions Are Characterized by the Values of Critical Exponents

For a continuous transition, a limited number of thermodynamic variables diverge or alternatively become zero. For these variables, power laws with characteristic critical exponents can describe the behavior around the transition point. Determining the values of these critical exponents for a particular transition has challenged both theory and experiment.

The heat capacity around the transition temperature can be described by

$$\mathbf{C}_\mathrm{p} \propto \left(|\mathbf{T} - \mathbf{T}_c|\right)^{-\alpha} \tag{10.4.2}$$

where the same critical exponent α has been found to apply on either side of the transition point.

For the gas–liquid system approaching the critical point, the difference in density, $\rho(\text{liquid}) - \rho(\text{gas})$, approaches zero in a characteristic way described by a critical exponent β:

$$\rho(\text{liquid}) - \rho(\text{gas}) \propto (\mathbf{T}_c - \mathbf{T})^{\beta} \tag{10.4.3}$$

The density difference in this example is a particular case of a more general concept called the order parameter of the transition. For a general continuous transition, the exponent β describes how the order parameter approaches a constant value (zero in Eq. 10.4.3). The first step in any analysis of a continuous phase transition is to identify the relevant order parameter(s). For a liquid–liquid phase separation, the density difference is the order parameter, while in a smectic to nematic transition, the order parameter is the amplitude of the density variation illustrated in Figure 10.1.

Large fluctuations in density or, more generally, large fluctuations in the order parameter are found above the critical point. We can relate the long-wavelength part of these fluctuations to the compressibility $\frac{1}{V}\left(\frac{\partial V}{\partial \mathbf{p}}\right)_{\mathbf{T}}$ for the fluid. Compressibility diverges at the critical point as

$$-\frac{1}{V}\left(\frac{\partial V}{\partial \mathbf{p}}\right)_{\mathbf{T}} \propto (\mathbf{T}-\mathbf{T}_c)^{-\gamma} \tag{10.4.4}$$

Equations (10.4.2)–(10.4.4) summarize the thermodynamic behavior around a critical point. What we observe at the molecular level is an increase in the range of molecular correlations in the system as we approach the critical point. The correlation length diverges at the critical point, and in our example, the correlations have resulted in the formation of separate phases, liquid and gas, below the critical point. Above the critical point, we can picture either small drops of liquid in a gas or vice versa. These drops become larger and larger as we approach the critical point, but the density contrast relative to the bulk becomes smaller and smaller. On the other side of the critical point, the drop has reached a macroscopic size, while the density difference relative to the other phase is negligibly small.

For applications to colloid systems, it is particularly important to realize that the surface tension between two phases also goes to zero at the critical point, and the width of the interface diverges. The decrease in surface tension follows naturally from the diminution of the molecular difference between the two coexisting phases on approaching the critical point. The surface tension γ follows the general equation

$$\gamma = \gamma_0\left(1-\frac{\mathbf{T}}{\mathbf{T}_c}\right)^{\mu} \tag{10.4.5}$$

where μ is yet another critical exponent.

For a long time, the theoretical calculation of critical exponents was a major challenge in statistical mechanics. The basic problem was to account properly for the long-range correlations. This problem was solved by introducing the so-called renormalization group theory. To describe even the essence of this theory would lead far outside the scope of this text, but one important qualitative result is the explanation of the so-called universality of critical behavior. Systems that superficially appear to be physically very different can have the same set of critical exponents. In fact, only a limited number of sets of critical exponents have been observed. The exponents for both gas–liquid one-component systems and liquid–liquid two-component systems display so-called Ising-like values, with $\alpha \approx 0.1$; $\beta \approx 0.32$; and $\gamma \approx 1.25$, while the surface tension exponent is $\mu \approx 1.26$.

10.4.3 We Can Use the Regular Solution Theory to Illustrate the Ising Model and to Calculate Mean Field Critical Exponents

The Ising model has been of fundamental importance in building up our understanding of continuous phase transitions. Originally, the model was introduced to describe magnetic systems with two discrete states of each spin. However, we can establish the model by referring to the regular solution model of liquid mixtures, as discussed in Chapter 1.

In a lattice model of a liquid mixture of A and B (see Figure 1.2), each site can be occupied by either A or B. The energy associated with the site depends on the occupancy of the z near-neighbor sites, where an $A-B$ pair has an effective interaction energy of w/z (see Eq. 1.4.16). We can solve for the most probable configurations and the thermodynamic variables in his model using exact statistical mechanics. Near the critical point of the mixture, this approach ultimately would lead to the complexities of the renormalization group theory. As an illustrative alternative, we adopt the simplifying assumptions of the regular solution theory, which leads to the so-called mean field exponents. In the calculations of Section 10.3.1, we found that for sufficiently high temperatures, $w/\mathbf{RT} < 2$, A and B will mix at all proportions. For $w/\mathbf{RT} = 2$, a critical point occurs at $X_A = X_B = 1/2$. The transition at $X_A = X_B = 1/2$ and temperature $\mathbf{T}$ from slightly above to slightly below $w/2\mathbf{R}$ is continuous.

Using the regular solution theory, we can calculate the critical exponents. For example, the width of the coexistence region is described by the exponent β. Combining Eq. (10.3.1) for the chemical potential with the condition for equilibrium between two phases, Eq. (10.3.2) implies that

$$\mathbf{RT}\ln\left(\frac{1}{2}-\Delta X\right)+w\left(\frac{1}{2}+\Delta X\right)^2=\mathbf{RT}\ln\left(\frac{1}{2}+\Delta X\right)+w\left(\frac{1}{2}-\Delta X\right)^2 \tag{10.4.6}$$

Here, the compositions of the two phases are denoted by $X_A^\alpha \equiv \frac{1}{2}-\Delta X$ and $X_A^\beta \equiv \frac{1}{2}+\Delta X$, using the fact that the system is symmetric with respect to the exchange of A and B. Writing $\mathbf{T} = (\mathbf{T}-\mathbf{T}_c)+\frac{w}{2\mathbf{R}}$, we find after expanding the logarithm to the third order that

$$\Delta X=\left(\frac{3}{4\mathbf{T}_c}\right)^{1/2}(\mathbf{T}_c-\mathbf{T})^{1/2} \tag{10.4.7}$$

Comparing this result with Eq. (10.4.3), we find that the mean field critical exponent β is 0.5 rather than the experimentally found value of 0.32.

In Eq. (10.4.4), we saw that compressibility diverges at the critical point of a fluid. For a binary mixture, osmotic compressibility $(\partial\Pi_{\text{osm}}/\partial c_2)^{-1}$ exhibits a corresponding divergence. In Chapter 4, we saw that this quality determines the intensity of light scattering, and we explained the critical opalescence as a "cloud point" behavior observed close to critical points. To determine the mean field value of the critical exponent γ, we can again use the regular solution theory. Because there is no basis for designating one component as solvent, we replace $-V_A\Pi_{\text{osm}}$ with the chemical potential of the A component

$$\left(\frac{\partial\Pi_{\text{osm}}}{\partial c_2}\right)^{-1}\cong-\left(\frac{\partial\mu_A}{\partial c_B}\right)^{-1}V_A\cong-\left(\frac{\partial\mu_A}{\partial X_B}\right)^{-1} \tag{10.4.8}$$

where V_A represents the molar volume of A (and B). Using Eq. (10.3.1), we find

$$\frac{\partial\mu_A}{\partial X_B}=-\frac{\mathbf{RT}}{1-X_B}+2wX_B \tag{10.4.9}$$

Near the critical point, $\mathbf{T}_c = w/(2R)$ and $X_B = 0.5$

$$-\left(\frac{\partial\mu_A}{\partial X_B}\right)^{-1}=\frac{1}{2\mathbf{R}}(\mathbf{T}-\mathbf{T}_c)^{-1} \tag{10.4.10}$$

Thus, the mean field value of the exponent γ becomes 1 instead of the true value $\approx$1.25.

Mean field values of the exponents differ from the exact values because the critical fluctuations have not been properly accounted for. Theories much more elaborate than the regular solution theory also misrepresent the critical exponents if they do not treat the critical correlations properly.

10.4.4 The Term Continuous Phase Transition Sometimes Characterizes the Less Well-defined Phase Changes

We have discussed first-order and continuous phase changes from a thermodynamic point of view. The systems have been considered to consist of a few (one, two, or three) well-defined components. In principle, we assumed that samples were indefinitely large, and true equilibrium was always implied. Many real physical systems provide examples that deviate from the idealized models, although we can identify many features of a phase transition in them.

In a system of finite size, the variables cannot show true discontinuities or divergences. The phase change occurs more smoothly. However, for the finite-size effect to be observable with typical experimental methods, we must go down to a colloidal scale. As an example, consider the formation of micelles. With some effort, we can see that the micellization of a cooperative aggregation of 100 monomers takes place over a small but finite concentration range. However, if we increase the aggregation number to 1000 monomers, the transition is so sharp that its width lies below the typical detection limit.

Sample inhomogeneity also broadens a phase transition. In a solution of colloidal particles or in a synthetic polymer system, it is virtually impossible to obtain a system of perfect homogeneous composition. We always find a size distribution and usually also a certain chemical heterogeneity. Thermodynamically, we are dealing with a truly multicomponent system, and in any phase transition, the different components will prefer, however slightly, one phase over another relative to the mean behavior. This preference leads to a certain broadening of the transition.

Another common effect in colloidal systems is slow equilibration. Two particles may easily be stuck in a particular configuration for times longer than those experimentally available for obtaining equilibrium. Even though thermodynamics would predict a sharp phase transition, slow kinetics precludes this from happening. Instead, we observe a somewhat more gradual change into a partly nonequilibrium state. The glass transition often seen in pure polymer systems provides an example of such nonequilibrium behavior.

See Doi (2013, Chapter 5) for more information on Continuous Phase Transformations.

Exercises

10.1 The phase diagram of Figure 10.10 for the water–dodecyltrimethylammonium chloride (C_{12}TAC) system is a **T**–X diagram. Consider a composition of 20% H_2O and 80% C_{12}TAC. Give the sequence of phases when such a sample is heated from 0 °C to 180 °C with the stability range for each phase.

10.2 Using the phase diagram of Figure 10.7, determine the nature of phases and their composition for the following total composition in grams:

H_2O	Octanol	Potassium caprate
4	2	4
1	12	2
16	1.2	2.8
2.5	10	7.5
1	5	4.5

10.3 Figure 10.20 shows the phase diagram of the DMPC–DPPC binary system. Calculate the temperature dependence of the heat capacity for an equimolar mixture. Assume that the enthalpy change in going from gel to liquid crystal is the same as in the pure compounds ($\Delta\mathbf{H} \approx 25$ kJ/mol and $\Delta\mathbf{H} \approx 35$ kJ/mol for DMPC and DPPC, respectively). The heat capacities of the pure compounds are 1.3 kJ/mol/K for DMPC and 1.6 kJ/mol/K for DPPC.

10.4 Calculate the ternary phase diagram for a mixture of three liquids, A, B, and C, at 25 °C. Use regular solution theory with the parameters
a. $w_{AB} = 7$ kJ/mol, $w_{AC} = w_{BC} = 0$
b. $w_{AB} = 7$ kJ/mol, $w_{AC} = 6$ kmol, $w_{BC} = 0$

10.5 For Exercise 7.8, we found that the free energy of mixing of two polymers of equal size is

$$\mathbf{G}_{\text{mix}} = k_B\mathbf{T}\left\{n_A \ln \phi_A + n_B \ln \phi_B + N_p(n_A + n_B)\chi\phi_A\phi_B\right\}$$

Calculate the combinations of χ and of polymerization, N_p, for which the polymers are miscible at all compositions.

10.6 A small membrane protein is added to a DPPC bilayer system. The protein is much more soluble in the liquid crystalline phase than in the gel phase. Calculate the heat capacity C_p for the temperature range of 35–45 °C for protein concentrations of 0.01 and 0.02 in mole fraction units. Plot the result as C_p versus $\mathbf{T}$ as it would be measured in a DSC experiment. Assume that the contribution to the heat capacity from the protein is constant.

10.7 Extend the model calculations leading to the calculated phase diagram of Figure 10.20 by allowing for a nonideal mixing in the gel state. Assuming regular solution theory, use (a) $w = 1.0$ kJ/mol and (b) $w = 10$ kJ/mol.

10.8 Calculate the phase diagram of Figure 10.5 using the regular solution model. Use the following parameters: $\mathbf{T}_A = 40$ °C, $\mathbf{T}_B = 60$ °C with melting transition enthalpies $\Delta\mathbf{H}_A = 40$ kJ/mol and $\Delta\mathbf{H}_B = 50$ kJ/mol; heat capacities $C_p^A(l) = C_p^B(l) = 1.6$ kJ/mol/K and $C_p^A(s) = C_p^B(s) = 1.3$ kJ/mol/K, and interaction parameters $w_{AB}(l) = 0$, $w_{AB}(s) = 10$ kJ/mol.

10.9 Use the Flory–Huggins equation to calculate how the critical point for a polymer of N_p units depends on the solvent quality as measured by the χ parameter.

References

Atkins, P., de Paula, J., and Keeler, J. (2022). *Atkins Physical Chemistry*, 12th ed. Oxford, UK: Oxford University Press.

Balmbra, R.R., Clunie, J.S., and Goodman, J.F. (1969). *Nature*. 222, 1159.

Bates, F.S. (1991). *Science*, 251, 898.

Doi, M. (2013). *Soft Matter Physics*. Oxford, UK: Oxford University Press.

Foffi, G. Savin, G., Bucciarelli, S., Dorsaz, N., Thurston, G.M., Stradner, A., and Schurtenberger, P. (2014). *Proc. Natl. Acad. Sci. U.S.A*. 111, 16748.

Fontell, K., Fox, C., and Hansson, E. (1985). *Mol. Cryst. Liq. Cryst*. 1, 9.

Guldbrand, L., Jönsson, B., and Wennerström, H. (1982). *J. Colloid Interface Sci*. 89, 532.

Gustavsson, Å., Tjerneld, F., and Wennerström, H. (1986). *Polymer*. 27, 1768–1770.

Ipsen, J.H., Karlström, G., Mouritsen, O.G., Wennerström, H., and Zuckermann, M.J. (1987). *Biochim. Biophys. Acta*. 905, 162.

Israelachvili, J.N. (2011). *Intermolecular and Surface Forces*, 3rd ed. London, UK: Academic Press.

Jönsson, B. and Wennerström, H. (1987). *J. Phys. Chem*. 91, 338.

Kronberg, B., Holmberg, K., and Lindman, B. (2014). *Surface Chemistry of Surfactants and Polymers*, 3rd ed. Chichester, UK: Wiley.

Lundquist, M. (1978). *New Concepts in Lipid Research*, R.T. Holman ed. New York: Pergamon Press.

Marley, S. and Sturtevant, J.M. (1976). *Proc. Natl. Acad. Sci. U.S.A*. 73, 3862.

Pusey, P.N., Zaccarelli, E., Valeriani, C., Sanz, E., Poon, W.C., and Cates, M.E. (2009). *Philos. Trans. R. Soc. A*. 367, 4993.

Strey, R., Schomäcker, R., Roux, D., Nallet, F., and Olsson, U. (1990). *J. Chem. Soc. Faraday Trans*. 86, 2253.

Ulmius, J., Wennerström, H., Lindblom, B., and Arvidson, G. (1977). *Biochemistry*. 16, 5742.

Winey, K.I., Thomas, E.L., and Fetter, L.J. (1991). *J. Chem. Phys*. 95, 9367.

11

Microemulsions, Emulsions, and Foams

Concept Map

Characterization of Surfactant Films
The energetic properties of a surfactant film at an oil–water interface can be characterized by five phenomenological parameters • γ—the surface tension at a macroscopic interface. • K_A—the area expansion elastic modulus, which for phospholipid bilayers is 0.1–0.4 J/m^2. • H_0—the spontaneous curvature, giving the curvature that an unconstrained film would adopt. • κ—the bending rigidity, typically $3 - 50\ k_B\mathbf{T}$. • $\bar{\kappa}$—the saddle splay, modulus, which is usually negative and of order $-\kappa/2$. The Helfrich curvature free energy $\mathbf{G}_C$ provides a relation between these parameters and the properties of surfactant films.

Microemulsions
Microemulsions are thermodynamically stable isotropic solutions containing oil, water, and surfactants, which are composed of either droplets, $R \approx \pm H_0^{-1}$ (Figure 11.3), or bicontinuous structures (Figure 11.4), which occur when $H_0 \cong 0$ corresponding to no preferred curvature of the surfactant film. Triangular phase diagrams (Figure 11.5), at constant **T** and **p**, are widely used; in addition, the χ, constant surfactant, (Figure 11.6) and fish, constant water-to-oil ratio, (Figure 11.8) plots provide useful phase representations. For nonionic surfactants of the C_nE_m type, temperature controls microemulsion structure and stability; H_0 decreases strongly with increasing temperature, so that O/W forms when $\mathbf{T} < \text{PIT}$ (phase inversion temperature, $H_0 = 0$). Around the PIT, we find a bicontinuous structure and for $\mathbf{T} > \text{PIT}$, a W/O microemulsion. With ionic surfactants, we can add salt and cosurfactants to obtain the same type of phase behavior obtainable with nonionic surfactants via variations in temperature.

Emulsions

Emulsions are thermodynamically unstable dispersions containing two immiscible liquids and an emulsifier. Emulsions can be stabilized by amphiphiles, polymers, proteins, or colloidal particles (Figure 11.16).

Macroemulsions contain either drops of oil in water (O/W) or water in oil (W/O).

The emulsifier protects the drops. An emulsion's properties depend crucially on the static and dynamic properties of the protecting film.

We prepare emulsions by **dispersion**—that is, injecting the mechanical energy required to break up bulk phases and create a large interfacial area, or by **condensation**—that is, using chemical energy stored in reactants to transform bulk phases via surface-driven instabilities into emulsion droplets.

Macroemulsion Stability

Figure 11.15 depicts the sequence of steps that determine the time evolution of an emulsion:

- Creaming—differences in density between the drop and the medium result in concentration of drops in the top or bottom of the container.
- Flocculation—drops move into the secondary minimum, a process promoted by large drop size.
- Coagulation—drops move into the primary minimum.
- Ostwald ripening—diffusion of droplet molecules through the medium causes small drops to decrease in size and disappear while large drops grow.
- Coalescence—two drops merge into one.

Stabilizing an emulsion involves minimizing these steps; destabilization involves promoting them.

Surfactant Film Properties Play Key Roles in the Formation and Destabilization of Emulsions

During emulsion formation, the processes of drop formation and coalescence compete, and the fastest-formed proper surfactant film leads to the longest-lived drop.

- While new O/W interface is being created, surfactant interfacial density is nonuniform. Nonuniformity induces Marangoni surface-driven flows, promoting the formation of a stabilizing surfactant film.
- The phase in which the emulsifier is most soluble usually emerges as the continuous phase. This phase correlates with the surfactant number because $N_s < 1$ corresponds to greater water solubility and O/W emulsions, while $N_s > 1$ leads to W/O emulsions.
- When two emulsion drops meet, fluctuations lead to a local neck formation (Figure 11.19). For an unstable emulsion, this neck grows spontaneously, leading to coalescence. For more stable emulsions, it needs to pass a critical size before coalescence occurs. Such a stability is obtained when $H_0 > 0$ for O/W emulsions and when $H_0 < 0$ for the W/O case.

Destabilization Strategies

We can add salt to break an electrostatically stabilized O/W emulsion; centrifuge to promote creaming; agitate to promote coalescence; or heat if the surfactant is in a gel state.

We can change the curvature by adding a cosurfactant or an antagonistic surfactant.

Foams

A foam is a dispersion of gas bubbles in a solid or liquid. It is never thermodynamically stable.

A foam is formed in a liquid through:

- Forcing gas through a nozzle
- Using a supersaturated solution of gas
- Superheating the liquid
- Mechanical agitation of a liquid surface

A solid foam forms when the liquid of a foam solidifies.

A gas bubble in a liquid rises to the surface by the action of gravitation, where it can form a concentrated foam or froth if the liquid film that separates the bubbles is stabilized by surfactants or polymers. In the concentrated foam, the bubbles deform to polyhedral shapes with flat contact areas and Plateau borders at the corners.

The properties of the liquid film in a foam can be studied in the surface film balance, which measures film thickness as a function of bulk osmotic pressure. This technique also allows for reproducible measurements of film rupture under osmotic stress. The film in a foam ruptures by qualitatively the same mechanisms as for an emulsion, but there are substantial quantitative differences.

Oil and water do not mix. This observation is equally relevant in the chemical laboratory, as in the oil field, and the kitchen. However, on many occasions, we actually do want oil and water to mix, and we often accomplish this by means of emulsification. For example, the art of making a good sauce can include a crucial moment of oil and water mixing. In the food industry, emulsions have numerous uses in products such as (low-fat) margarines, milk substitutes, soups, and chocolate drinks. The active components of many drugs are insoluble in water, and emulsification can be used in the formulation of medications in the form of pastes or salves. Such formulations have a wide application in skin care products. Emulsions have found an increasing use in paints and in other situations calling for the replacement of organic solvents for health and conservation reasons. They are also used in formulating machining fluids, in surface coating processes, in making the bitumen composite used in surfacing highways, and in emulsion polymerization and numerous other processes.

Sometimes we need to create a low surface tension between water and oil rather than obtaining a mix. The prime example is tertiary oil recovery, in which an aqueous phase is used to push out the oil remaining in porous rock.

Many living organisms process fatty materials in a polar environment. Examples include emulsified fat droplets in milk or solubilization of fats by lipoprotein particles in blood and egg yolk or by bile in the intestine. Also, in these cases, the procedure employs surface-active components.

In foams, we disperse a gas in a liquid or solid medium. As we will see in the last section of this chapter, foams show many important analogies with emulsions when it comes to the central questions of structure and stability.

From a thermodynamic point of view, proper emulsions are metastable and eventually separate into a polar and an apolar phase. In contrast, microemulsions contain large fractions of both water and oil while still being thermodynamically stable. Although the systems have many structural features in common, microemulsions typically have polar and apolar domains on a smaller length scale than macroemulsions, as the prefixes suggest.

To ensure (meta)stability, the water–oil or air–liquid interface must be protected by some third component. This chapter focuses, for the most part, on amphiphile-stabilized interfaces, keeping in mind that polymer or particle adsorption to the interface can also stabilize emulsions and foams.

11.1 Amphiphiles Form a Semiflexible Elastic Film at Interfaces

11.1.1 We Can Characterize the Elastic Properties of a Film Through Five Phenomenological Constants

Micro- and macroemulsions, as well as foams, are in most cases stabilized by a molecular film separating the aqueous and oil (gas) domains. The properties of these systems are largely determined by those of the film. A fruitful approach to describing the system is to view the molecular layer as a film characterized by a few phenomenological parameters that depend on intensive thermodynamic variables: temperature, pressure, and electrolyte concentration. In the emulsions and foams, we have domains of bulk phases (water, oil, air), separated by the film. The free energy of the system can then be seen as the sum of the free energy of the respective bulk phase domains plus an extra contribution that can be solely attributed to the film. The free energy of the film can, in turn, be seen as a sum of three different contributions. For a given configuration of film, a free energy can be attributed to the number of molecules per unit area. Furthermore, there is an additional contribution from the curvature of the film. A third free energy contribution comes from the entropy associated with fluctuations of the film configurations. In the discussion of the undulation force in Section 5.6, we already encountered one manifestation of the entropic contribution to the free energy of a bilayer film.

When we consider a film that can stretch and bend, there are, to second order, five characteristic parameters. We can define two of these with reference to the simplest geometry of the film: the planar state. The surface tension or surface free energy per unit area, γ, which was introduced at the beginning of this book, represents the first derivative of the film free energy $\mathbf{G}_f$ with respect to area. The area expansion modulus K_A is a second-order derivative

$$K_A = \left(\frac{\partial^2 \mathbf{G}_f}{\partial A^2}\right)_{n_s} \frac{A}{2} \left(\frac{\mathrm{J}}{\mathrm{m}^2}\right) \tag{11.1.1}$$

evaluated at a constant number n_s of surfactants in the film. Normally, when we change the area of the film, the surfactants will equilibrate between the film and the bulk, and in this case $\mathbf{G}_f$ is a truly linear function of A. However, in some situations, the constraint of a constant amount is relevant. Such cases exist when there is no reservoir of surfactants, when transport of surfactants to the film is slow relative to the timescale of area changes, or when we want to consider local concentration fluctuations. For macroscopic interfaces, the area expansion modulus K_A is called the Gibbs elasticity.

In Section 6.2.4, we introduced the three parameters H_0, κ, and $\bar{\kappa}$, which in the Helfrich equation determine the free energy change contributions associated with a curved bilayer.

$$\mathbf{G}_C = \int \left\{ 2\kappa (H - H_0)^2 + \bar{\kappa} K \right\} \mathrm{d}A \tag{6.2.1}$$

The same characterization is valid for any film, while the values of the parameters differ from system to system. In contrast to a symmetric bilayer system, the spontaneous curvature, H_0, is not zero due to symmetry, and it varies with temperature and bulk phase composition. The bending rigidity κ is, for a film of a homogeneous material, inversely proportional to the third power of the thickness. In contrast, one has for a bilayer with a slipping plane in the center, the value of κ is approximately the sum of those of the monolayers. The saddle splay rigidity $\bar{\kappa}$ is typically negative for a film separating oil and water. For a saddle-shaped surface, the two radii of curvature have opposite signs, making the Gaussian curvature K negative. A negative value of $\bar{\kappa}$ then implies that it is energetically unfavorable to deform a planar surface into a saddle shape. It is a challenge to measure $\bar{\kappa}$ since it affects the free energy of a film only at topology changes according to the Gauss–Bonnet theorem. We found in Section 6.2.5 that the value of $\bar{\kappa}$ for a bilayer is $-\kappa/2$ for a homogeneous film.

11.1.2 The Spontaneous Curvature of a Film Depends on Temperature and Compositions of the Bulk Media

We concluded in Section 6.2.4 that the spontaneous curvature H_0 had a strong influence on the free energy change in vesicle fusion/fission. We will later in this chapter see that H_0 plays an equally important role in determining the properties of microemulsions, emulsions, and foams. In a film consisting of a single molecular layer, there is an apolar side exposed to an apolar medium: hydrocarbon, fluorocarbon, or air. On the other side, the polar groups are exposed to a polar medium, usually water, possibly containing solutes. In the film, the driving force for eliminating direct molecular contact between the polar and apolar media is, at equilibrium, compensated by the repulsive force between polar head groups and between alkyl chains. We can use the alkyl ethylene oxide surfactants as an illustration in Figure 11.1.

At a macroscopic interface, the film is forced to be planar by gravitational forces, but on a mesoscopic scale, the molecular forces dominate the gravitation and the film adopts, in general,

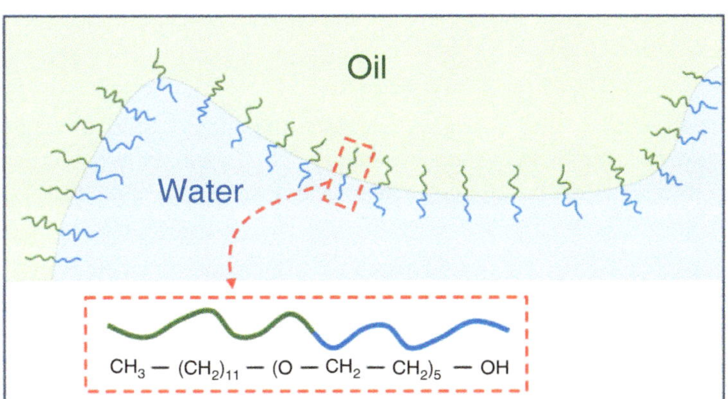

Figure 11.1 A film of the surfactant $C_{12}E_5$ at an oil–water interface.

a curvature determined by minimizing the bending energy of Eq. (6.2.1). Since there are no symmetry restrictions, the spontaneous curvature H_0 can adopt any value. For the particular system in Figure 11.1, it is possible to use temperature to tune to a state $H_0 = 0$. For a given surfactant, the specific temperature, which is fulfilled, depends on the alkane of the apolar medium and on the presence of solutes in the aqueous and oil domains.

Temperature changes affect both sides of the film. The fraction of gauche conformations increases with increasing temperature, and this has the effect that there is a slight increase in the cross-sectional area of the chains. The consequence is that there is an increase in the curvature toward the polar medium (decrease of H_0). The polar groups also containing methylene groups have some hydrophobic interaction with the water, and this positive component to the free energy increases with increasing temperature (see Section 5.7). As a consequence, the water content in the head group region decreases, decreasing the head group area a_0, leading to a decrease in H_0. For C_nE_m systems in general, the temperature effects on H_0 on the apolar and polar sides of the film cooperate to give a distinct decrease of H_0 with increasing temperature, and for these systems, one has for n, m in the measurable range

$$H_0(\mathbf{T}) = C_H \times (\mathbf{T} - \mathbf{T}_0);\ C_H \approx -5 \times 10^6 \mathrm{m}^{-1}\mathrm{K}^{-1} \tag{11.1.2}$$

The value of $\mathbf{T}_0$, called the phase inversion temperature (PIT), varies the values of n, m and also the properties of the surrounding alkane and aqueous media, while the coefficient C_H remains the same within experimental uncertainty.

On the apolar side, there is some mixing of the surfactant hydrocarbon chains and the hydrocarbon molecules of the medium, as illustrated in Figure 11.2. The shorter the alkyl chains of the medium, the larger the mixing ratio. This leads to an expansion of the apolar side of the film, pushing the spontaneous curvature to more negative values. The result is that the PIT for a given C_nE_m system, the balanced temperature $\mathbf{T}_0$, is higher by 2–3 °C for each additional CH_2 group in the alkyl chain of the oil medium.

The conditions in the aqueous domain also affect the spontaneous curvature. An added cosurfactant like an alcohol associates with the film and has a large effect on H_0. Polar solutes like electrolytes or sugars are depleted from the polar region of the film. With a mechanism analogous to the depletion attraction discussed in Section 5.5, the absence of such solutes affects the properties of the polar region, by changing the chemical potential of the water. In a simplified model with a stepwise layer of full depletion of thickness δ and bulk osmotic pressure Π_{osm}, the resulting change in spontaneous curvature is

$$\Delta H_0 \approx -\frac{\Pi_{\mathrm{osm}}\delta^2}{4\kappa} \tag{11.1.3}$$

For films formed by ionic surfactants or other charged molecules, the spontaneous curvature depends strongly on the electrolyte concentration in the aqueous medium. In Section 4.1, we analyzed how the electrostatic free energy of a spherical micelle depends on the micelle radius, Eq. (4.1.26).

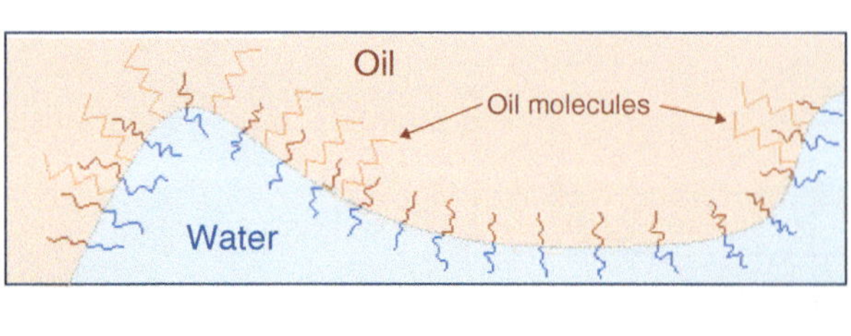

Figure 11.2 Penetration of oil molecules between the apolar chains in the surfactant film. Note the greater extent of penetration when the film curves toward water, which changes the spontaneous curvature of the film.

By generalizing the micelle radius R to a curvature $H = 1/R$, we can interpret Eq. (4.1.26) as the electrostatic contribution to the curvature free energy

$$\Delta \mathbf{G}_C(el) = -\frac{4k_b\mathbf{T}H}{s\kappa}\ln\left\{\frac{1+[1+s^2]^{\frac{1}{2}}}{2}\right\} \tag{11.1.4}$$

Here κ^{-1} is the Debye screening length and s is the dimensionless charge density:

$$s \equiv \sigma(8k_B\mathbf{T}c_0\varepsilon_r\varepsilon_0)^{-1/2} \tag{3.9.14}$$

The free energy $\Delta\mathbf{G}_C(el)$ is balanced by other contributions to give the spontaneous curvature H_0.

As discussed in Section 3.9, the electrostatic free energies in aqueous systems are dominated by the entropy and increase in magnitude with increasing temperature. Thus, when a constant electrolyte concentration increases the temperature, the electrostatic free energy promotes increased curvature toward the solvent. This trend is opposite to that found for the nonionic C_nE_m systems, and it is only partly compensated by the packing effects of the alkyl chains. In the following Section 11.2, we will show how the knowledge about the spontaneous curvature provides a most essential component in the understanding of microemulsions.

Surfactant films are also extensively discussed in Lyklema (2005, Vol. 5, Chapter 6).

11.2 Microemulsions Are Thermodynamically Stable Isotropic Solutions That Display a Range of Self-assembly Structures

11.2.1 Microemulsions Can Consist of Finite Domains in a Medium or Have a Bicontinuous Structure

A microemulsion is an isotropic solution containing substantial amounts of both a strongly polar component (usually water) and a strongly apolar component (usually oil) that are stabilized thermodynamically by an amphiphilic additive. As we will see, there is a substantial structural diversity in microemulsions meeting this thermodynamic criterion.

The most clear-cut examples are the oil-in-water (O/W) and water-in-oil (W/O) droplet microemulsions illustrated in Figure 11.3.

In these cases, the dispersed medium forms spherical droplets. We can calculate their size using a simple geometrical argument. It is easy to determine the volume fractions ϕ_w, ϕ_o, and ϕ_s (water, oil, and surfactant) from the given composition and densities of the components. We can view the oil-in-water microemulsion droplet as composed of a hydrocarbon core of radius R_c covered by a layer of surfactant head groups, making the overall radius R_H. By simple geometry, the radius R_c is related to the ratio between the hydrocarbon volume V_{hc} and the area A_{hc} of the core

$$R_c = 3\frac{V_{hc}}{A_{hc}} \tag{11.2.1}$$

The hydrocarbon volume is the sum of the oil and a fraction β of the surfactant volume so that

$$V_{hc} = V_{\text{tot}}(\phi_o + \beta\phi_s) \tag{11.2.2}$$

We now assume that the surfactant contributes a fixed area a_0 per molecule, making the total core area

$$A_{hc} = \frac{V_{\text{tot}}\,\phi_s}{v_s}\,a_0 \tag{11.2.3}$$

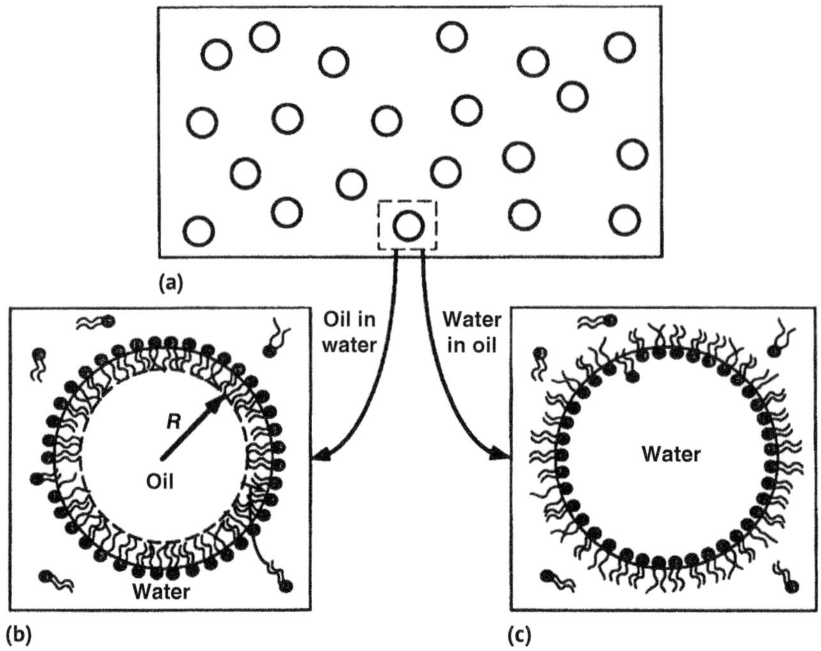

Figure 11.3 A droplet microemulsion phase (a) can be of the oil-in-water (b) or the water-in-oil type (c). Typically, but not exclusively, the drops are nearly spherical and have a small size polydispersity. The radii R of the drops are usually one to four times the thickness l_0 of the surfactant film. The excess surfactant is almost exclusively in the continuous phase.

where v_s is the molecular volume of the surfactant. We can now, combining Eqs. (11.2.1)–(11.2.3), obtain an expression for the core radius in terms of experimentally accessible quantities

$$R_c = 3\left(\frac{\phi_0 + \beta}{\phi_s}\right)\left(\frac{v_s}{a_0}\right) \tag{11.2.4}$$

Here β is typically on the order of 0.5, while v_s/a_0 lies in the range 1.5–2 nm so that microemulsion droplets mostly have radii in the interval 5–20 nm.

We could apply the same argument equally well to the W/O microemulsion. An important feature of Eq. (11.2.4) is that the droplet radius R_c is solely determined by the volume fractions. When the spontaneous curvature H_0 of the film is close to the inverse radius R_c^{-1} of the drop, we can expect a relatively stable situation with a practically unfrustrated surfactant film. However, when $|H_0| \gg R_c^{-1}$, that is, for large values of the spontaneous curvature, the film can relax toward a lower free energy state by decreasing the drop size, expelling the emulsified oil (or water for negative H_0) to a bulk phase. When $|H_0| \ll R_c^{-1}$, the film can relax by forming larger nonspherical shapes.

Ultimately, when $H_0 \approx 0$, there is no preference for a O/W relative to a W/O structure. Instead, the surfactant film separates two macroscopic oil and water regions in a bicontinuous structure, which represents a disordered version of the bicontinuous cubic phases discussed in Chapters 1 and 6.

Figure 11.4 shows a series of freeze-fracture electron micrographs of a bicontinuous microemulsion phase. Figures 11.3 and 11.4 illustrate several types of microemulsion structures, but there are also other possibilities.

Figure 11.4 Scanning electron micrographs of a bicontinuous microemulsion containing 7 wt% $C_{12}E_5$ and equal amounts of water and octane. The sample was prepared by freeze-fracture. The panels show how the structure changes with the initial equilibration temperature (shown in insets). At low temperatures (30 °C), there are oil drops in a water medium, and at high temperatures (90 °C), there are water drops in oil. In between, there is a gradual transition with bicontinuous structures. Contrast between the oil and water phases occurs because the coating forms a mottled pattern on the oil regions. *Source*: W. Jahn et al. (1988)/ with permission of American Chemical Society.

11.2.2 Temperature Controls the Structure and Stability of Nonionic Surfactant Microemulsions

Nonionic surfactants with oligoethylene oxide head groups form stable microemulsions. These types of surfactants are both technically important and fundamentally interesting. As we discussed in the preceding section, their behavior is very temperature-dependent.

Most microemulsion systems that have been studied contain four or more components. In addition to oil, water, and surfactant, we usually must add a cosurfactant (alcohol) and/or an electrolyte to tune the system to the rather special conditions necessary for microemulsion stability. However, we can tune the nonionic surfactants in a ternary system by adjusting the temperature. Furthermore, the weak interbilayer forces in these systems also favor the stability of the microemulsion relative to the lamellar phase.

Working with a strict three-component system has two virtues. First, it simplifies the representation of the phase diagram. For each temperature, we can use a triangular representation similar to that shown in Figure 10.6. A second virtue of a ternary system becomes apparent in applications. For example, to employ a microemulsion in tertiary oil recovery, the surfactant film's properties must remain intact as the process goes on in time. With more than one amphiphilic component, the loss of surfactant due to surface adsorption will inevitably be selective. In a sense, the process resembles large-scale silica gel chromatography. Changing composition results in changed properties, which is usually an undesirable feature.

Figure 11.5 shows a series of triangular diagrams for the system $C_{12}E_5$–water–tetradecane at three different temperatures.

Determining one such diagram is rather tedious work, and the creation of a consistent series represents a major effort. Conversely, the diagrams summarize a great deal of experimental observation, although extracting this information from the diagrams requires some training.

The diagrams show a liquid crystalline lamellar phase, and, for a given temperature, several isotropic liquid phases. These can be identified as a dilute micellar solution in water, a dilute inverted micellar solution in oil also containing monomeric surfactant, a microemulsion phase, and an L_3 or a sponge phase. As the temperature changes, these isotropic one-phase regions split into two and merge with others. Through a concomitant change of concentration and temperature, one can accomplish passing from one of these isotropic phases to all the others without entering a two-phase area.

Although the thermodynamic nature of the phase does not change in such a process, major structural changes can occur on the microscopic scale. For example, in the water corner, $\phi_w \gg \phi_o, \phi_s$, we expect surfactant micelles to be slightly swollen with oil. Similarly, in the oil corner, $\phi_o \gg \phi_w, \phi_s$, the typical structure is an inverted micelle in oil. In this case, we also find some surfactant monomers in the oil medium.

For $\phi_w \approx \phi_o \gg \phi_s$, provided ϕ_s is present in nonnegligible amounts, a three-phase equilibrium exists between an aqueous solution, an oil phase, and a microemulsion phase containing the majority of the surfactant. At a specific temperature, $\mathbf{T} = 47.8\ °C$ in Figure 11.5a, the microemulsion in the three-phase triangle contains equal amounts of oil and water. This represents the PIT discussed in the previous section.

By increasing the temperature above the PIT, the microemulsion of the three-phase triangle becomes increasingly rich in oil, as seen in Figure 11.5b for $\mathbf{T} = 51.2\ °C$. Slightly above 51.2°C (at 51.6°C), the "microemulsion" merges with the inverted micellar solution, and the three-phase triangle disappears. Similarly, by lowering the temperature from the PIT, the microemulsion becomes richer in water (see Figure 11.5c) and ultimately merges with the normal micellar solution at a temperature slightly below 44.5°C (at 43.8°C). Thus, in the present example, the three-phase triangle appears only in a limited temperature range of 43.8 to 51.6°C.

This series of ternary phase diagrams provides a nearly complete picture of the phase equilibria in the system. To aid visual analysis, we can make constant composition cuts so that a single diagram can convey various temperature effects. Cuts can be made in many ways. Here we illustrate the two most common ones: the χ cut and the fish cut. The rationale for the names will become apparent.

In the χ cut, we keep the surfactant concentration constant at a value slightly higher than that for the microemulsion in the three-phase triangle. In Figure 11.6, the surfactant concentration is fixed

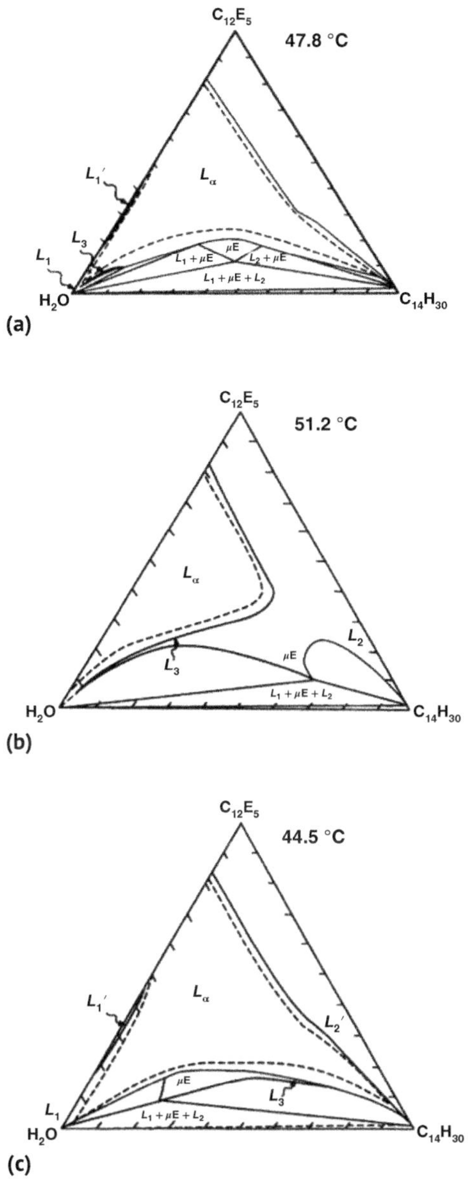

Figure 11.5 Phase diagrams of the ternary system water–tetradecane–$C_{12}E_5$. (a) At 47.8 °C, five distinct isotropic phases can be observed: the dilute normal micellar solution L_1 in the left-hand corner; the concentrated micellar solution L_1' close to the binary water–$C_{12}E_5$ axes (cf. Figure 10.12); the sponge or L_3 phase in the left-hand corner; the inverted micellar solution L_2, which encompasses the pure oil and pure surfactant phases; and finally, the microemulsion μE in the center of the diagram. In addition, there is a large lamellar phase region L_α. The three corners of the fan-shaped μE region connect to three three-phase triangles. At low surfactant concentrations, the microemulsion is in equilibrium with the L_1 and L_2 phases. (b) At 51.2 °C, a temperature slightly higher than that shown in (a), the μE corner of the three-phase triangle $L_1 + \mu E + L_2$ has moved toward the oil corner. The μE region now has a L_3 "nose" pointing toward the water corner. Furthermore, the μE region merges smoothly into the oil-rich L_2 phase. (c) At a lower temperature (44.5 °C), the three-phase triangle defined by $L_1 + \mu E + L_2$ has its μE apex moved toward higher water contents. In addition, the upper right-hand corner of the μE fan has been drawn out to a long "nose" encompassing the inverted L_3 phase containing an aqueous sheet in an apolar environment. *Source*: Kunieda et al., 1982/with permission of Taylor & Francis.

at 16.6% (w/w). The oil/(oil + water) (w/w) ratio, α, and the temperature are the two independent variables. (Using the volume/volume ratio is more consistent from a conceptual point of view, but experimentally, weights are the measured quantities.)

A limitation in using these cuts is that the multiphase areas could contain phases that are not present as pure phases in the cut, while the extent of the one-phase areas emerges clearly. The cut at constant surfactant concentration means a cut parallel to the base of the triangular diagrams in Figure 11.5.

In Figure 11.6, a microemulsion one-phase "channel" goes from the lower left to the upper right corner, representing one leg of a Greek χ. In addition, a narrower one-phase channel goes from the lower right to the upper left corner, representing the other leg of the χ. This channel is a sponge or

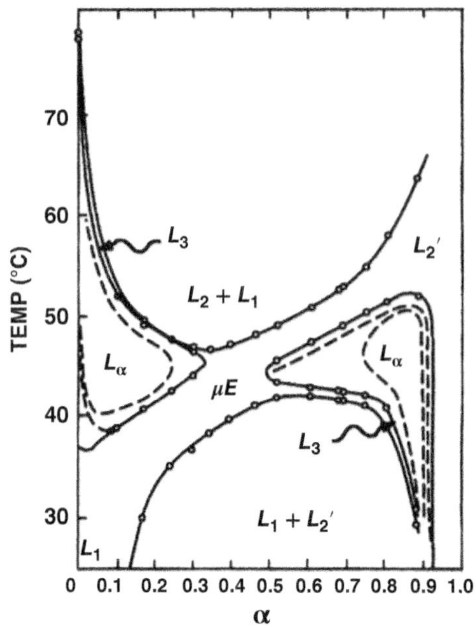

Figure 11.6 Phase equilibria in the χ cut at a constant amount, 16.6% (w/w), of the surfactant $C_{12}E_5$. The two independent variables are the temperature and the weight fraction oil, $\alpha = (C_{14}H_{30})/[(H_2O) + (C_{14}H_{30})]$. The channel running from the lower left-hand corner to the upper right defines the isotropic microemulsion phase, which passes from normal surfactant micelles in water-to-oil droplets in water (O/W) via a bicontinuous structure to water droplets in oil (W/O) and finally reversed micelles in oil. In addition, a narrow L_3 channel goes from the upper left to the lower right-hand corner. Phase notation as in Figure 11.5. *Source*: Olsson et al., 1986/with permission of American Chemical Society.

L_3 phase, and its location at $\alpha = 0$ corresponds—as it should—to the sponge phase in the binary surfactant–water system shown in Figure 10.12.

The PIT is located where the microemulsion and the sponge phase meet at the center of the χ. If we follow the microemulsion channel, we find an uninterrupted transition from a water-continuous medium at low **T** and α to a bicontinuous one at intermediate **T** and α to an oil-continuous one at high **T** and α. Figure 11.7 illustrates this transition and shows the variation of self-diffusion coefficients with α plotted as ratios relative to a neat phase of either component.

At α around 0.5, both oil and water have reduced diffusion to approximately 60% of the bulk value, while the surfactant has its maximum value. This condition constitutes the signature of a bicontinuous system symmetric with respect to the two continuous media. At either side of this point, the surfactant diffusion decreases, while if $\alpha > 0.5$ oil diffusion increases and water diffusion decreases or vice versa if $\alpha < 0.5$.

We can obtain another perspective on the phase equilibria from the fish plot of Figure 11.8.

Here the oil/(oil + water) ratio remains constant at 0.5, while the surfactant concentration varies. This plot reveals a striking symmetry with an approximate mirror plane at **T** = PIT. The three-phase triangle has its largest extension at the PIT. The area where the three phases are in equilibrium adopts the shape of a fish. At higher concentrations, the lamellar phase appears. For both the microemulsion and the lamellar phase, the one-phase regions extend to the lowest surfactant concentration at the PIT.

Figures 11.5–11.8 summarize experimental observations of a particular three-component microemulsion system, but other nonionic surfactant systems qualitatively behave in the same way. Quantitative changes occur in the PIT and in the surfactant concentration of the Winsor III microemulsion when the surfactant or the oil is changed. However, the characteristic patterns of the χ plot and the fish plot remain unchanged. How can we understand what molecular factors govern this generic phase behavior?

In Figures 11.5, 11.6, and 11.8, **T** = PIT represents a symmetric state in which water and oil seemingly are equivalent in the sense that the same structure is obtained for α and $1 - \alpha$. Now a microemulsion can be seen as a system composed of two media (polar and apolar) separated by a surfactant

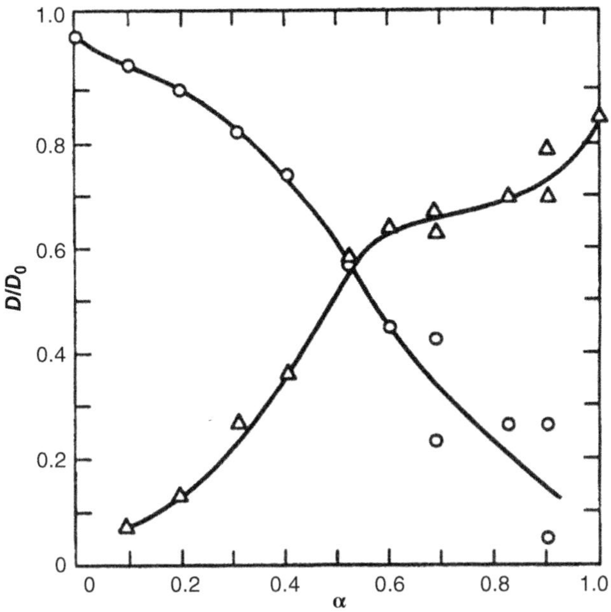

Figure 11.7 Variation of the self-diffusion coefficient of water (○) and tetradecane (Δ) as a function of weight fraction α for the water–oil–surfactant microemulsion system of Figure 11.6. As evident from Figure 11.6, composition and temperature must be varied simultaneously. Diffusion coefficients are plotted relative to the observed values D_0 in the respective neat liquids. At low α values corresponding to a O/W microemulsion, D/D_0 is close to unity for water and small for tetradecane. At high α values, where a W/O microemulsion occurs, the D/D_0 roles are reversed. For α values around 0.5, the water and the oil diffusion are equally affected by the structure, which is the signature of a bicontinuous phase. *Source*: Olsson et al., 1986/with permission of American Chemical Society.

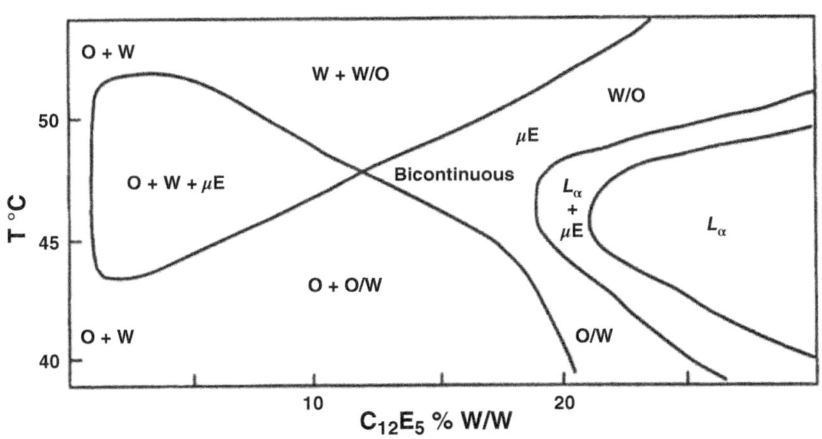

Figure 11.8 The fish cut. For an equal ratio (W/W) of water and tetradecane, $\alpha = 0.5$, the sequence of phase varies with increasing surfactant concentration at different temperatures around the phase inversion temperature of 47.8 °C. The symbol O denotes a tetradecane phase with some surfactant and W an aqueous phase with small amounts of surfactant. Other symbols are as in Figure 11.6. The region of the three phases joining the one-phase microemulsion region resembles the tail of a fish.

film with some spontaneous curvature, H_0. When $H_0 = 0$, the film curves as readily toward the oil as toward the water, equating the two media even though chemically they are very different. This equivalence leads to the conclusion that the PIT is the temperature where $H_0 = 0$.

At $H_0 = 0$, we would expect maximum stability of the lamellar phase because the film then prefers the planar state, and indeed, this can be observed. The microemulsion occurring at the PIT and thus at $H_0 = 0$ should not be of either a W/O or an O/W type, but it should display a structure where the two media have equivalent roles. As the diffusion experiments show, the microemulsion is indeed bicontinuous with equivalent D/D_0 values for oil and water.

If we assume that H_0 is the most essential parameter of the system, the symmetry property of Figure 11.8 follows from Eq. (11.1.3), implying that

$$H_0(\mathbf{T}_0 + \Delta\mathbf{T}) = -H_0(\mathbf{T}_0 - \Delta\mathbf{T}) \tag{11.2.5}$$

At $\mathbf{T}_0 + \Delta\mathbf{T}$, the film wants to curve toward water as much as it wants to curve toward oil at $\mathbf{T}_0 - \Delta\mathbf{T}$. When the film curvature energy provides a dominant contribution to the free energy, we should, in addition, expect the same phase structure at temperature $\mathbf{T}_0 + \Delta\mathbf{T}$ and composition α as we find for temperature $\mathbf{T}_0 - \Delta\mathbf{T}$ and composition $1 - \alpha$. Thus, at a change

$$(\alpha, \mathbf{T}_0 + \Delta\mathbf{T}) \leftrightarrow (1 - \alpha, \mathbf{T}_0 + \Delta\mathbf{T}) \tag{11.2.6}$$

the aggregate structure remains the same, but the oil and the water exchange places. Thus, at both $\mathbf{T}_0 + \Delta\mathbf{T}$ and $\mathbf{T}_0 - \Delta\mathbf{T}$, the spontaneous curvature is H_0 is toward the medium with the volume fraction α. The film does not notice that the oil and water have exchanged places because it sees only a spontaneous curvature and a fraction.

A number of predictions follow from Eqs. (11.2.5) and (11.2.6). In the fish diagram $\alpha = 1 - \alpha$ by choice, so that the temperatures PIT + $\Delta\mathbf{T}$ and PIT − $\Delta\mathbf{T}$ should give equivalent phase structures according to Eq. (11.2.6). The symmetry apparent in Figure 11.8 supports our assumption.

In addition, the χ-plot implies that the partial phase diagram should be invariant with respect to a 180 ° rotation around the point $\mathbf{T}$ = PIT; $\alpha = 0.5$, which is indeed the case with an amazing accuracy.

The simple rationalization of the apparently complex series of phase diagrams represented in Figure 11.5 in terms of a surfactant film with a temperature-dependent spontaneous curvature separating two media of volume fractions ϕ_w and ϕ_o implies that spontaneous curvature is the main control parameter. A further manifestation of the pivotal role of spontaneous curvature is found in the value of the surface free energy between a microemulsion phase and excess solvent, oil or water. At a macroscopic interface, the surfactant film is planar, and the free energy cost in going from an optimal curvature of H_0 to a plane is the dominant contribution to the surface free energy and one has

$$\gamma(\mathbf{T}) \propto \kappa H_0^2(\mathbf{T}) = 3.3\kappa C_H^2 (\mathbf{T} - \mathbf{T}_0)^2 + \gamma_0 \tag{11.2.7}$$

Here γ_0 represents the very low value of the surface tension $10^{-6} < \gamma_0 < 10^{-5}$ N/M at the balanced point $\mathbf{T}_0$. As shown in Figure 11.9, Eq. (11.2.7) gives, with C_H from Eq. (11.1.4), an excellent representation of the measured surface tension for the $C_{12}E_5$–water–oil systems.

Equation (11.2.7) represents one of the few cases where we have a simple quantitative interpretation of the value of a surface tension in terms of the microscopic parameters.

When the surfactant volume fraction ϕ_s exceeds either ϕ_w or ϕ_o, we must refine the discussion. For the case $\phi_s/\phi_o \geq 1$, the surfactant film is not exposed to an excess of oil, and the penetration of oil molecules in the film changes. This will affect the value of H_0. Similarly, for $\phi_s/\phi_w \geq 1$, the polar groups experience an interaction with other polar groups across a thin aqueous region. This affects the spontaneous curvature so that H_0 will show some concentration dependence under these circumstances.

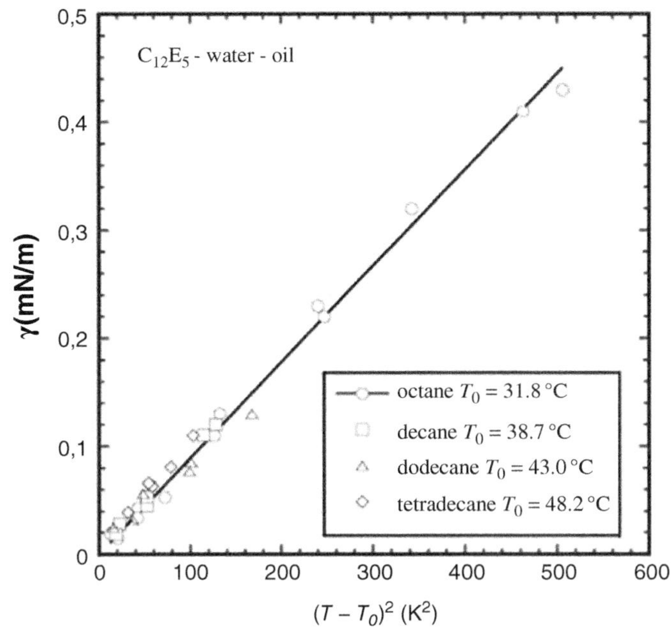

Figure 11.9 Measured surface tension for systems $C_{12}E_5$–water–oil for four different oils. Solid line: Eq. (11.2.7). *Source*: Data from Sottmann et al., 1997; Balogh et al., 2006.

We have used the nonionic C_nE_m systems to illustrate the generic properties of microemulsions, with a focus on the properties of the film separating the apolar and polar domains. There is a remaining question of why microemulsions don't appear in other systems showing films of similar qualitative properties. The nonionic systems have two properties necessary for microemulsion formation. The first is that the film has a low bending rigidity κ of the order of a few $k_B\mathbf{T}$. As a consequence, there are large bending fluctuations in the film, which ultimately favor disordered structures relative to ordered ones. For example, block-copolymers are too stiff to form an isotropic microemulsion, and when $H_0 \approx 0$, there is a formation of a lamellar phase. The same observation has been made with amphiphilic lipids like lecithin, which forms microemulsions first when large amounts of alcohol are added to make the monolayer softer. The second important property of the C_nE_m systems is that the repulsive overlap interaction is of short range between polar groups in opposing monolayers. Strong interactions promote ordered structures. In the fish plot of Figure 11.8, we see that at high surfactant concentrations, the lamellar phase is formed around the PIT temperature. Under these conditions, the film–film separation is so short that there is a repulsive force quenching bending fluctuations, and an ordered phase has a lower free energy.

11.2.3 We Often Need Electrolytes to Obtain Microemulsions for Ionic Surfactants

In Section 11.2.1, we found that bicontinuous microemulsions occur when the spontaneous curvature of the surfactant film is around zero. For nonionic surfactants, we tuned such conditions by changing the temperature. How can we generalize these concepts so that they apply to ionic systems? For single-chain ionic surfactants, the surfactant number N_s is of the order 0.3 or $H_0 \approx 1/l_0$ in terms of curvature, which is far from the optimal condition $H_0 \approx 0(N_s \approx 1)$. Adding an electrolyte can decrease H_0 by decreasing the importance of the electrostatic contribution, but this decrease is insufficient to yield the curvature needed for microemulsion formation.

One solution to the problem of obtaining a balanced microemulsion is to use a cosurfactant, usually a suitable alcohol such as butanol, pentanol, or hexanol. This addition increases the effective surfactant number because the alkyl chains of the alcohols contribute to the hydrocarbon volume, while the alcohol group replaces apolar groups at the polar–apolar interface. The cosurfactant also has the effect of reducing the bending rigidity. However, only adding cosurfactant is not sufficient. Even though the charge density of the film is decreased, there is still an appreciable long-range, double-layer repulsion across polar regions. Microemulsions are first formed when electrolyte is added to the aqueous phase to reduce the repulsion.

Thus, we can use both cosurfactant/surfactant ratios and electrolyte concentration to control the spontaneous curvature. In the previous section, we also concluded that H_0 is a function of temperature as it was for the nonionics, but with a weaker temperature dependence and of opposite sign. In practice, we can thus use temperature only for fine-tuning H_0 in ionic surfactant systems.

An obvious alternative to adding cosurfactants to achieve $H_0 \approx 0$ conditions is to introduce double-chain surfactants. This approach has been widely used with both anionic and cationic surfactants. Here, we concentrate on microemulsions formed by the anionic double-chain surfactant sodium di(ethyl hexyl)sulfosuccinate (AOT). Figure 11.10 shows the ternary phase diagram AOT–water–isooctane. The system shows lamellar and reversed hexagonal liquid crystalline phases over a large concentration range. These phases incorporate some amount of

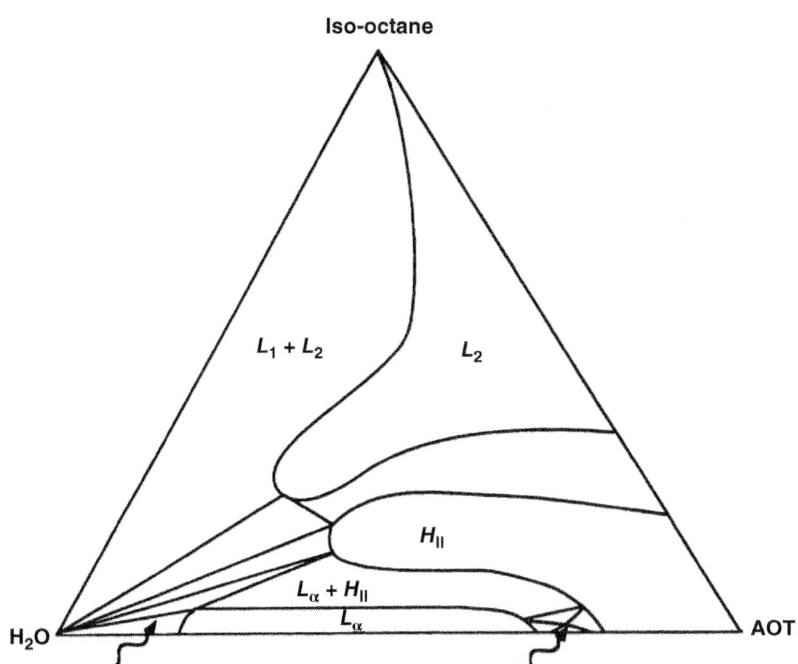

Figure 11.10 Phase diagram of the ternary system H_2O–AOT–isooctane at 20 °C. Three liquid crystalline phases are shown: a lamellar phase L_α, a reversed hexagonal phase H_{II}, and a bicontinuous cubic phase I. The two isotropic phases are a dilute aqueous solution L_1 and a W/O microemulsion phase–oil surfactant solution denoted L_2. Fontell, private communication.

oil, and an isotropic solution of the W/O microemulsion type appears at high surfactant and low water contents.

To extend the range of the microemulsion, we must add an electrolyte. Its role is twofold: we have already discussed how the electrolyte decreases H_0, but its effect on the film–film interaction is an even more important factor. As we saw explicitly in the preceding section, both the microemulsion and the lamellar phase are most stable when $H_0 \approx 0$. For the salt-free AOT system, a lamellar phase rather than a microemulsion forms at $H_0 \approx 0$ conditions.

Isotropic phase regions expand when we add electrolyte, and bicontinuous microemulsions appear. The analogy with the nonionic system can be seen in the fish plot shown in Figure 11.11.

That the spontaneous curvature increases with increasing temperature for ionic systems is shown clearly in Figure 11.11. Increasing the salt content of the aqueous phase leads to a decrease in H_0. Nevertheless, a state of $H_0 \approx 0$ can be found by increasing the temperature. Thus, the three-phase microemulsion equilibrium is found in the range 22–25 °C for 0.40% NaCl, 30–37 °C for 0.58% NaCl, and 50–65 °C for 1.0% NaCl. We see also in Figure 11.11 that the tail of the fish tilts, and the tilt

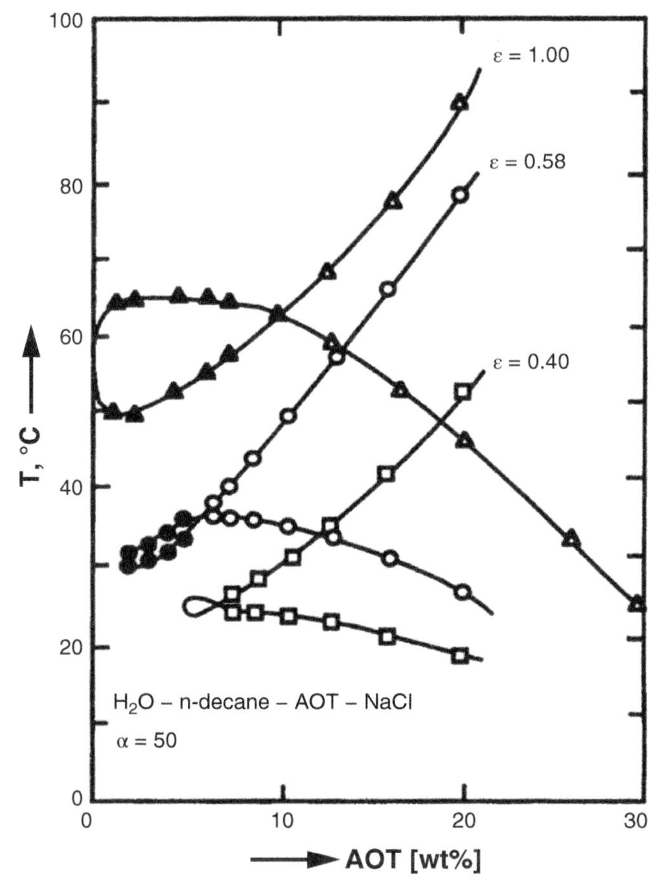

Figure 11.11 Fish cut for the system *n*-decane–AOT–H_2O–NaCl for a mixture of a constant 1:1 ratio of decane to brine. Curves are shown for three different concentrations of NaCl in water: $\varepsilon = 1.0, 0.58$, and 0.40% w/w. The higher the salt content, the higher the temperature at which the three-phase triangle occurs and thus the higher the point at which $H_0 = 0$. Solid symbols denote the boundary of the three-phase equilibrium; open symbols denote the boundary of the microemulsion. *Source*: Kahlweit and Strey, 1986/with permission of American Chemical Society.

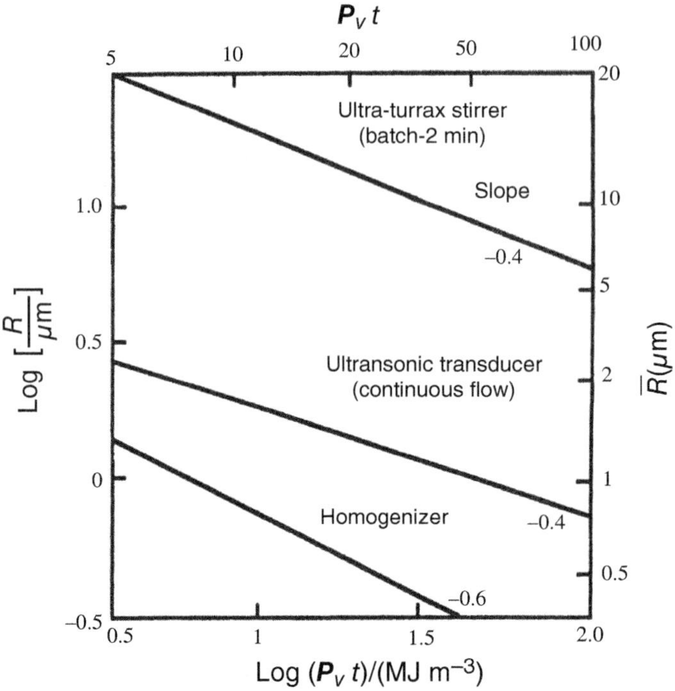

Figure 11.12 Average droplet size $\bar{R}$ as a function of energy input P_V t per volume for diluted paraffin oil-in-water emulsions produced in various machines. *Source*: Walstra, 1983/with permission of Taylor & Francis Group.

becomes more pronounced at lower the salt contents. The reason is that the electrostatic interactions are affected by all ions in the aqueous phase, including both the added electrolyte and the counterions of the surfactant. Thus, as the AOT is added, the Na^+ ion concentration in the aqueous phase increases, leading to a decrease of H_0. The effect of the counterions becomes relatively more important as the NaCl concentration lowers.

In this section, we have demonstrated that we can describe and understand the properties of microemulsions using the Helfrich curvature energy concept combined with a knowledge about surface forces. This insight can now be used to treat the properties of emulsions and foams.

See Berg (2010, Chapter 9), Cosgrove (2010, Chapter 5), Kronberg et al. (2014, Chapters 17 and 25), and Lyklema (2005, Vol. 5, Chapter 5) for more information on Microemulsions.

11.3 Macroemulsions Consist of Drops of One Liquid in Another

Macroemulsions are thermodynamically unstable dispersions containing two immiscible liquids and an emulsifier. They find wide commercial use because in many instances we want to process a macroscopically homogeneous system containing both water and oil. At times, however, a technical process results in an emulsion, but we desire a separation into two bulk phases. This problem is commonly found in oil recovery.

Emulsions combine properties associated with colloidal sols and amphiphilic aggregates. Because macroemulsions and sols are thermodynamically unstable, we use similar strategies to prepare them. Their kinetic stability as colloidal entities also has a common basis. However, emulsion droplets can transform in size, shape, and number, and the systems exhibit many properties associated with amphiphilic systems. This combination of features makes emulsions complex and challenging to understand.

We can state our general goals for forming, stabilizing, and breaking of an emulsion in a simple way. Preparation requires the creation of a metastable state containing either oil-in-water (O/W) or water-in-oil (W/O) droplets. Bicontinuous macroemulsions have not been observed, presumably because any perturbation in the system immediately leads to separation into macroscopic domains. Stabilization demands that we keep the droplet shape and size intact as long as possible, while the strategy for breaking an emulsion requires opening up a pathway for destabilization.

The final fate of an emulsion is clear; it will separate into two or more equilibrium phases. The way we think about emulsion stability depends in a large part on our final goal. If long shelf-life or durability during a process is necessary, we want to stabilize the phases. In other cases, emulsions are unwanted products in a manufacturing process. Also, we sometimes want emulsions to be stable up to a certain point in a process, after which phase separation becomes crucial. Thus, the dichotomy between stability and instability typical of colloidal systems in general also characterizes emulsions.

11.3.1 Forming Macroemulsions Requires Input of Mechanical or Chemical Energy

Two basic kinetic schemes are used for preparing emulsions just as they are for colloidal sols. We can use mechanical energy to mix all the emulsion ingredients (dispersion) or use the chemical energy that is stored in the ingredients and liberated when they come into contact.

The free energy $\Delta\mathbf{G}_{em}$ needed to disperse a liquid of volume V into drops of radius R in a solvent is

$$\Delta\mathbf{G}_{em} = \gamma \frac{3V}{R} \tag{11.3.1}$$

where γ represents the interfacial tensions. When making a salad dressing just by shaking a vinegar–oil mixture, it is difficult to reach a drop size $R < 1$ mm since the high surface tension makes $\Delta\mathbf{G}_{em}$ relatively large. Addition of a surface active component has the dual effect of lower $\Delta\mathbf{G}_{em}$ and also hindering drop coalescence.

In practice, we must use considerably more energy than what is given in Eq. (11.3.1) to reach the dispersed state for two reasons. Directing the input energy solely to forming droplets is technically difficult because any mechanical agitation converts much of the energy to heat through viscous dissipation. Additionally, the energy barrier against the coalescence of drops that stabilizes the emulsion implies the existence of a barrier against forming two droplets from a bigger one. External perturbation must be large enough to overcome this barrier.

Because an emulsion is a nonequilibrium state, its properties depend not only on the variables of the state, such as temperature and composition, but also on the method of preparation. This fact lends an extra level of complexity to these systems when it comes to scientific experimental studies as well as to practical applications.

Although establishing consistently valid rules for the preparation of emulsions presents real difficulties, we can identify a number of important factors that influence the final properties of the mixed system. Both oil and water droplets generally form in a mechanical agitation process. What determines whether the final product is a W/O or O/W emulsion? The fate of the droplets depends on two competing processes: migration of the emulsifier to the droplet's interface (which stabilizes them) and coalescence (which destroys them). The phase that coalesces most rapidly—for whatever reason—will become the continuous phase.

Another factor that determines the final form of the emulsion is the order in which we mix the ingredients. We can dissolve the emulsifier directly in either the water or the oil phase. Alternatively, we can dissolve components of the emulsifier in both phases. We can carry out the emulsification by simply pouring all the ingredients into a container and mixing them, or we can add the aqueous phase bit by bit to the oil-emulsifier phase (or vice versa).

11.3.2 Turbulent Flow During the Mixing Process Governs Droplet Size

Figure 11.12 illustrates how average droplet size varies with energy input in the case of several emulsification techniques. We can obtain a qualitative description of the droplet formation process using arguments drawn mainly from dimensional analysis.

As we discussed in Chapter 2, a spherical drop of radius R exhibits a Laplace pressure

$$\mathbf{p}(\text{Laplace}) = 2\gamma/R \tag{11.3.2}$$

To tear such a drop into two parts, we need a pressure gradient of magnitude

$$\frac{\Delta \mathbf{p}}{\Delta x} \approx \frac{2\gamma}{R^2} \tag{11.3.3}$$

The smaller the drop size, the larger the pressure gradient we need. Moreover, Eq. (11.3.3) shows that lowering the surface tension γ decreases not only the minimum energy input given by Eq. (11.3.1) but also the actual energy input, because a smaller pressure gradient is needed to achieve a certain drop size.

In mixing processes, the pressure gradients, $\Delta\mathbf{p}/\Delta x$, that exert shearing forces on liquid droplets are associated with the eddies present in turbulent flow. These eddies are local regions of chaotic flow characterized by instantaneous velocities that fluctuate around an average velocity. Eddies have a characteristic size l_e, which measures the distance over which the instantaneous velocity changes appreciably. While we can develop equations that relate eddy velocity, energy, and size to the pressure gradients required for emulsification, they are difficult to apply in a quantitative way. Although our alternative approach is a qualitative one based on dimensional analysis, it provides many of the salient features we need to understand emulsification.

First, we consider a pressure difference associated with an eddy of size l_e where a difference in velocity, Δv, exists. According to Bernoulli's equation, the difference in energy density $\rho(\Delta v)^2/2$ relates to the difference in pressure $\Delta\mathbf{p}$. If this pressure difference exceeds the Laplace pressure, then a droplet of diameter $\mathrm{d} \geq \mathrm{d}_{\mathrm{min}}$ contained in the eddy region will break up. If we assume that this breakup diameter depends on the power input per unit volume P_v, density ρ, and surface tension, we can use dimensional analysis to argue that with a method based on turbulent flow, the size distribution function is uniquely determined by the dimensionless quantity

$$\mathcal{R} = R\left(\frac{P_v}{\rho}\right)^{2/5}\left(\frac{\rho}{\gamma}\right)^{3/5}$$

where R is the radius of the emulsion drop.

Problem: Show that $\mathcal{R}$ is indeed dimensionless.

Solution: R(m); P_v ($\text{Jm}^{-3}\text{s}^{-1}$); ρ($\text{kgm}^{-3} = \text{Js}^2\text{m}^{-5}$); γ(Jm^{-2})

$$\rightarrow \dim\{R\} = \text{m}\left(\frac{\text{Jm}^5}{\text{m}^3\text{sJs}^2}\right)^{2/5}\left(\frac{\text{Js}^2\text{m}^2}{\text{m}^5\text{J}}\right)^{3/5} = \text{m}\frac{\text{m}^{4/5}}{\text{s}^{6/5}}\frac{\text{s}^{6/5}}{\text{m}^{9/5}} = \dim\{1\}$$

We thus expect the average radius R_{av} to depend on the power input P_v as

$$R_{av} \propto P_v^{-0.4} \tag{11.3.4}$$

which is verified for the upper curve in Figure 11.12. Analysis of the pressure gradients generated by cavitation in ultrasonic irradiation also leads to an energy dependence of −0.4, while for a high-pressure homogenizer, the dependence is closer to −0.6, as shown in Figure 11.12.

A whole spectrum of eddy sizes exists in turbulent flow, and so we expect a spectrum of droplet sizes. Mechanically prepared emulsions are generally characterized by a wide particle distribution. In the analogous process of grinding a solid into small particles discussed in Section 9.1, one also finds for emulsions a log-normal size distribution

$$F(R) = \frac{1}{(2\pi)^{1/2} R\sigma} \exp\left\{-\frac{(\ln R - \ln R_0)^2}{2\sigma^2}\right\} \tag{11.3.5}$$

R itself is considered the independent variable, and $F(R)$ is a distribution skewed toward large R-values but still specified by only two parameters, R_0 and σ. Such a broad distribution can, however, be made much more homogeneous by repeatedly passing the liquid sample through a frit with a suitable pore size or by similar treatments.

11.3.3 A Chemical Nonequilibrium State Can Induce Emulsification

A simple example of a chemically induced emulsification is observed when mixing ouzo, anisette, or raki with water. The initial ca. 40% ethanol–water liquid contains apolar oils with very low solubility in pure water. When the original mixture is diluted with water, the oil component becomes insoluble and forms emulsion droplets with ethanol and some water. The drink becomes white due to the light scattering from the emulsion drops. This emulsion is unstable since there is no emulsifier, and the system goes toward phase separation. For proper taste, there is a need to consume the drink before this occurs! The ouzo effect provides an example of the general principle of creating an emulsion by nucleation and growth from a supersaturated solution.

Another emulsification mechanism can be realized when, for example, pure water contacts an oil phase containing a solute that is also soluble in water. When we bring a solution of ethanol in toluene saturated with water into contact with water, the interface becomes locally unstable and emulsions of toluene in water and of water in toluene form. As the ethanol diffuses, it leaves behind the toluene, which coalesces into unstable emulsion drops. The free energy for this process comes from the mixing of ethanol with water.

11.3.4 Emulsion Drops Can Change from a Spherical to a Polyhedral Shape as the Concentration Is Increased

We often picture an emulsion in terms of separate spherical drops of the dispersed phase moving by Brownian motion in an excess of the solvent. However, in practice, emulsions can become more concentrated, by direct drop–drop attraction, by creaming, or simply because the emulsion was prepared with a high volume of the dispersed phase. For an emulsion with drops of approximately equal size, drop–drop interactions necessarily become substantial at dispersed phase volume fractions above 50%. When the concentration is increased further, there are two possible scenarios for the response of the drops. The most common is that the size is preserved, and the increased drop–drop contact leads to a flattening of the contact area and the drops tend toward a polyhedral shape as illustrated in Figure 11.13. In fact, it is possible to prepare such emulsions in which the dispersed phase volume fraction is 0.99. In Section 11.4, we will discuss the structure of such emulsions in more detail since the same scenario is commonly found in foams.

A second possibility to achieve a high volume fraction of the dispersed phase is that there is a packing-induced change in the size distribution of the drops. One can fill space by having spheres of all sizes, as illustrated in Figure 11.14, first considered by the ancient Greek mathematician Apolonius.

For surfactant-stabilized emulsions, the smallest sphere radius is that of micelles, which puts an upper limit to the possible packing ratio. Bernard Cabane and coworkers have been able to prepare emulsions of this type with a volume fraction dispersed phase exceeding 95%!

11.3.5 A Number of Different Mechanisms Affect the Evolution of an Emulsion

A freshly prepared emulsion changes its properties with time owing to a series of events that occur on a microscopic scale. Depending on the particular molecular characteristics involved, these

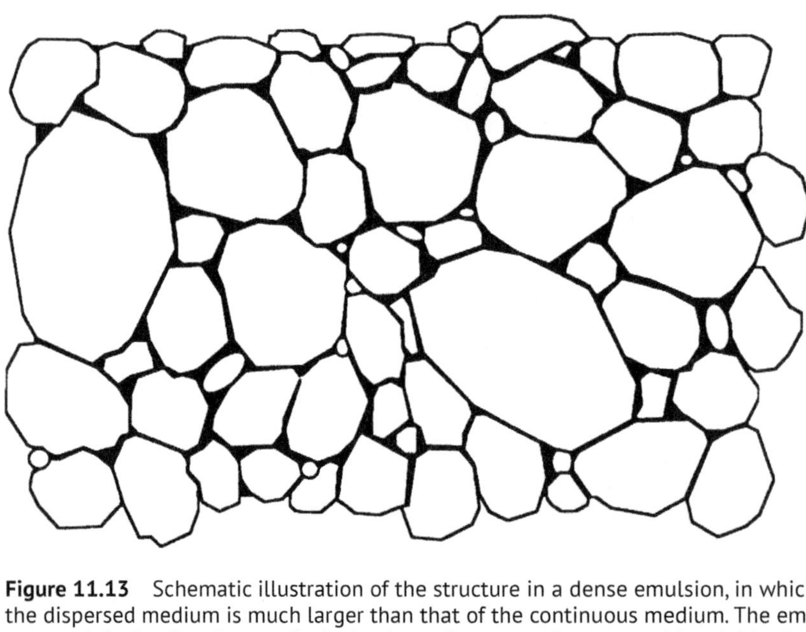

Figure 11.13 Schematic illustration of the structure in a dense emulsion, in which the volume fraction of the dispersed medium is much larger than that of the continuous medium. The emulsion droplets adopt more polyhedra-like shapes, displaying large flat areas of contact with neighboring drops, which are separated by Plateau regions containing small reservoirs of solvent and surfactant.

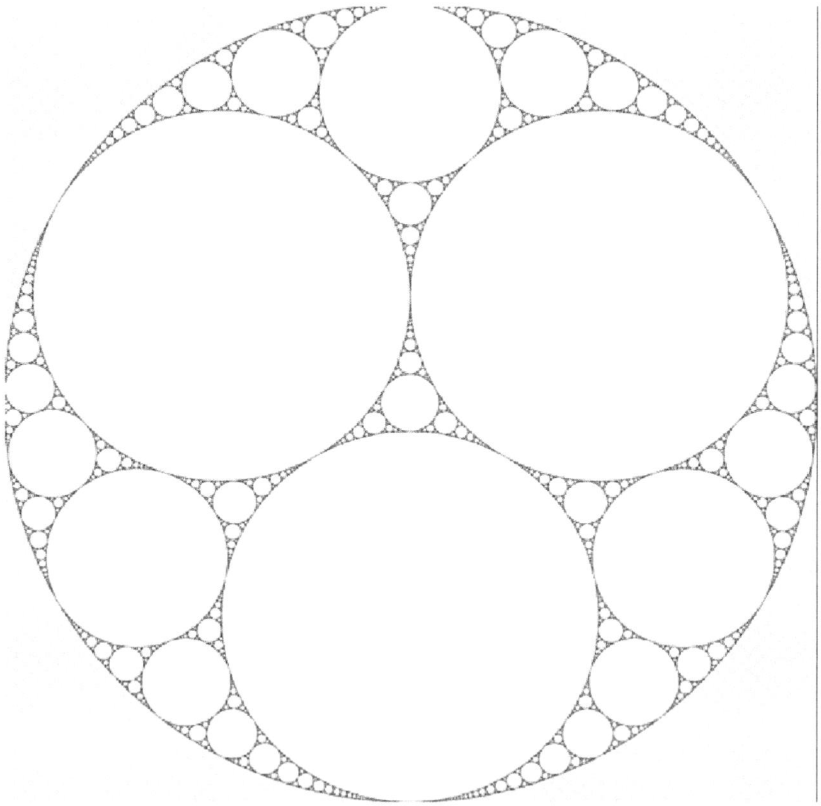

Figure 11.14 Apolonian packing of spheres. By having spheres of all sizes, one can fill space to 100%. For emulsions, there is a lower limit on the size of the sphere and in practice also an upper limit.

initial events rapidly lead either to the overall equilibrium state with macroscopically separated phases, "unstable emulsion," or to a metastable state, "stable emulsion." Clearly, the distinction between stable and unstable emulsions is qualitative and depends partly on the context.

A number of mechanisms contribute to the destabilization of an emulsion. As with the sol systems discussed in Chapter 9, the basis for stability is the existence of a repulsive regime for the force between drops. For oil-in-water, O/W, emulsions stabilized by ionic surfactants, we can use the DLVO terminology from Chapter 8 to describe the features of the droplet–droplet interaction. In the water-in-oil, W/O, case, the continuous medium is an oil, and the dominant stabilizing force is typically the entropic overlap interaction.

Emulsion droplets rather easily **flocculate** into the secondary minimum (see Figure 8.1). This effect is more significant for dilute emulsions, <25% w/w, than for more concentrated ones. Since when the mean distance between drops is less than the separation at the secondary minimum, the system experiences no initial energy gain in bringing drops closer together.

Whether flocculated or not, drops can migrate in the gravitational field to increase the concentration at either the top or the bottom of the vessel. The best-known manifestation of the phenomenon gives this effect the name **creaming**.

Once droplets have been brought into proximity by flocculation and/or creaming, they can **coagulate** into the primary minimum while still keeping their identity as separate droplets. This stage of destabilization is final for colloidal sols, but a further process occurs for emulsions.

Coalescence, in which two droplets merge into a bigger one, constitutes the final stage in the life of an emulsion droplet. It requires reorganization of the stabilizing surface layer surrounding the drops. The surfactant film changes its topology during the coalescence process, and this change usually is connected with passing a free energy barrier.

The general DLVO picture only applies to certain emulsion systems. In other cases, the interaction curve for two intact droplets exhibits a single energy minimum. In such a case, the distinction between flocculation and coagulation becomes tenuous, and the appropriate term depends on the location and depth of the minimum in the interaction curve. Figure 11.15 summarizes the different steps in the time evolution of an emulsion.

In addition to the sequence of flocculation/creaming, coagulation, and coalescence, a more molecular path contributes to destabilizing the emulsion. When the dispersed substance has a non-negligible solubility in the solvent, droplet sizes can change by molecular diffusion. In this Ostwald ripening process, small drops decrease in size and finally disappear, while large drops incorporate material through diffusional transfer. The influence of surface free energy causes local concentration to be slightly higher adjacent to small drops than around larger ones. Thus, concentration gradients exist in the medium with concomitant diffusional flows as shown in Figure 2.11.

We can use the Ostwald ripening process to generate monodisperse drops in an elegant way. If we initially produce an emulsion in the presence of monodisperse polymer so that no drops contain more than one polymer molecule, the presence of the polymer affects the chemical potential of the dispersed medium. Ostwald ripening leads to a quasi-equilibrium state in which all droplets containing a polymer are of a uniform size. Droplets without polymer will disappear irrespective of their size. This process has been used to produce large monodisperse latex particles by emulsion polymerization as discussed in Chapter 9.

This example suggests an effective way to stabilize an emulsion by quenching the Ostwald ripening mechanism. The basic strategy is to include a solute whose solubility in the continuous phase is lower than that of the dispersed solvent itself. Obvious choices of solutes are polymers for oil, and electrolyte

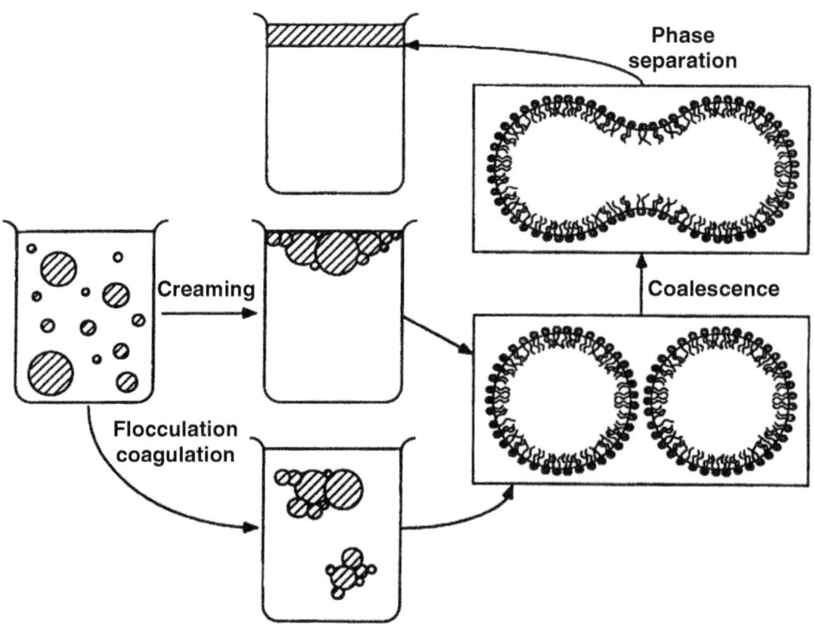

Figure 11.15 Schematic illustration of the different steps in the development/evolution from a freshly prepared emulsion to a final separation into two macroscopic phases.

or sugars for water. As the molecules of the dispersed phase diffuse from the smaller droplets to the larger droplets the solute concentrations increase in the smaller droplets and decrease in the larger ones. Thus, a situation is rapidly established in which the surface tension balances the direct osmotic contributions to give a uniform chemical potential of the dispersed solvent in drops of different sizes.

In Section 11.3.4, we described concentrated emulsions where the drops are deformed. In this state, the drop–drop interaction provides a large contribution to the free energy and the description of the ripening process based solely on surface free energies no longer applies. In the concentrated system, the diffusion path from drop to drop is small so from a molecular transport perspective, one would expect a rapid ripening process. However, it is complex to quantitatively identify a driving force and it is uncertain to what extent ripening caused by molecular diffusion contributes to the long-term stability of concentrated emulsions and also for foams discussed in Section 11.4.

11.3.6 Emulsion Can Be Stabilized by Preventing Contact Between the Dispersed Phase in Different Drops

In practice, there are several ways of achieving emulsion stability, some of which are illustrated in Figure 11.16.

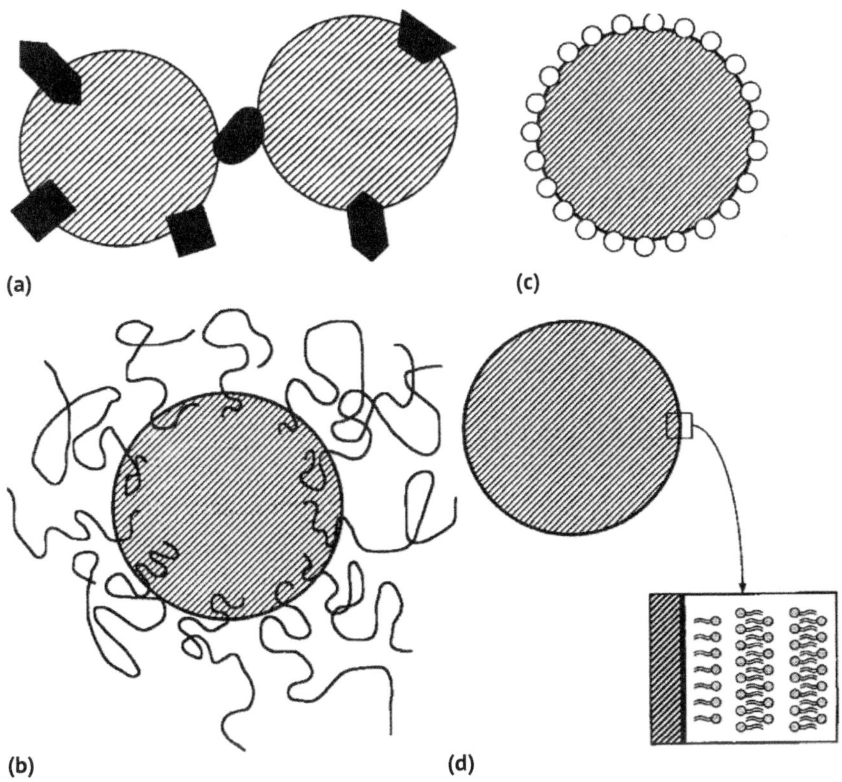

Figure 11.16 (a) Particles can adsorb at the interface between the emulsion drops and the medium and particles will stick out into the medium and prevent drop–drop contact. (b) A polymer protects an emulsion droplet most efficiently if the continuous medium is a better solvent than the emulsified liquid. (c) A dense film of adsorbed protein molecules can also stabilize an emulsion drop. (d) If an emulsion droplet is prepared in a three-phase system, oil–water–lamellar liquid crystal, it can be protected by one or a few bilayers in addition to a monolayer film.

One possibility is to use small solid particles to stabilize the emulsion drops (Figure 11.16a). Such systems are called Pickering emulsions. As shown in Figure 2.15, solid particles often adsorb at interfaces, reducing the combined surface free energy. Once adsorbed, the particle will stick out from the surface to an extent that depends on the relative magnitudes of the surface free energies. When drops are covered with particles, direct contact is prevented between the dispersed liquid phases of two drops. This mechanism contributes, for example, to stabilize the W/O drops in butter. In this important case, water droplets are covered by small fat crystals that adsorb on the drops but penetrate into the apolar continuous medium. When these crystals melt, the emulsion becomes unstable and an aqueous phase appears at the bottom of the saucepan with clarified butter as the end product.

A second method of generating emulsion stability is by adding a polymer. If the monomer unit is slightly surface active, the polymer will adsorb on the polar–apolar interface and the chain will make excursions into both media (Figure 11.16b). To be effective as a stabilizer, the polymer should prefer the continuous medium, because having the largest fraction of the chain on the outside of the drop generates the strongest repulsive force between drops.

A variation of these two stabilization mechanisms can be observed for some protein systems. Globular proteins, like ovalbumin in egg white, can be surface active even in their native state and they adsorb on a polar–polar interface. If the major part of the molecule remains in the continuous medium, a surface film of protein molecules can stabilize an emulsion drop, as illustrated in Figure 11.16c.

The most common method for obtaining a stable emulsion is by using a surfactant or an amphiphile. Typically, they will form a dense monolayer at the polar–apolar interface, which acts to prevent direct contact between the dispersed medium of two drops. How successfully the film prevents this contact depends in a sensitive way on the properties of the film itself and on the interaction between films.

For amphiphiles, a monolayer film is not the only possible arrangement that may occur at the polar–apolar interface. Stig Friberg and coworkers convincingly demonstrated that easily prepared and yet stable emulsions can be obtained if the equilibrium state of the emulsified system is a three-phase sample composed of oil, water, and a lamellar liquid crystal containing the majority of the surfactant. In this three-phase equilibrium, the liquid crystal phase will wet the oil–water interface; that is, the interfacial tensions between oil–liquid crystal and water–liquid crystal are much smaller than the oil–water tension. Thus, when water is dispersed in oil, or vice versa, the lamellar liquid crystal can remain as a protective sheath around drops, as illustrated in Figure 11.16d. Such a multibilayer arrangement clearly can provide a very strong protection for the emulsion drops. A particular example of this principle is found in natural milk. The fat drops produced in the milk glands are stabilized by first a monolayer and then a bilyer of phospholipids (see Figure 11.16d).

The commercial milk is usually "homogenized," which means that the fat drops have been reduced in size by roughly a factor of 3 in radius so that they are stabilized by only a monolayer of phospholipids, which is enough to prevent drop coalescence. The convection, always present in practice, is then effective in preventing these smaller drops from creaming.

Problem: Calculate the reduction in drop size in going from a bilayer + monolayer to only a monolayer at constant fat volume and area per lipid.

Solution: One initial drop radius R_i with volume $(4\pi/3)R_i^3$ and monolayer area $12\pi R_i^2$ is split into N drops of total volume $(N4\pi/3)R_f^3$ of area $N4\pi R_f^2$. Conservation of volume yields $R_i/R_f = (N)^{1/3}$ and constant area then implies $3R_f^2(N)^{2/3} = NR_f^2$ and $N = 27$.

11.3.7 Emulsion Stability Depends on the Properties of the Amphiphilic Film

In Section 11.1, we discussed in detail the elastic properties of molecular films formed by amphiphiles. This knowledge can now be applied to understand how the film responds in an emulsion.

The first question is: What type of emulsion will we obtain if we mix oil, water, and surfactant and emulsify the mixture mechanically? At the early stage of the emulsification process when a drop has just been formed, there is a critical moment. An optimal amphiphile film has not had time to develop at the newly formed surface, and coalescence is very likely to occur if a second drop is encountered. Thus, the more rapidly a proper film forms, the more long-lived the drop. When a new drop emerges by budding from a larger one, it has some surfactant at the interface, but because its area has been increased, the mean molecular density is low and initially nonuniform. At this stage, an important phenomenon called the Marangoni effect comes into play. Nonuniform distribution of surfactants creates nonuniform interfacial tension. This nonuniformity results in a mechanical instability in the surface, and surfactant will flow along the surface toward the area of low surfactant concentration and high surface tension as illustrated in Figure 11.17.

This process is faster than a surface diffusional equilibration of the concentration. Therefore, the action of the Marangoni effect quickly establishes a uniform but thin surfactant film around the drop.

The second stage of the process leading to a relaxed state of the drop is the transport of surfactant to the surface. In this step, the preferential location of the surfactant becomes important. If it concentrates in the dispersed medium inside the drop, transport mainly occurs by diffusion; but if it concentrates in the continuous medium, which is mechanically agitated, convection dominates and diffusion is less important. As a result, the newly formed film heals more rapidly if the surfactant is found in the continuous phase, and rapid healing leads to a final outcome of the emulsification process in which the phase that preferentially dissolves the surfactant becomes a continuous medium. This empirical observation is called the Bancroft rule.

If we apply the knowledge about surfactants that we have developed earlier, we know that normal aggregates are preferred for surfactant numbers $N_s < 1$. Those surfactants have a higher solubility in water than in oil, while surfactants with $N_s > 1$ prefer oil to water. Phrased in terms of the spontaneous curvature of the surfactant film H_0, we expect oil in water (O/W) emulsions when $H_0 > 0$ and water-in-oil emulsions (W/O) when $H_0 < 0$.

What makes one surfactant form a more stable emulsion than another? To discuss this question in a coherent way, we must realize that the path from emulsion droplets dispersed in an excess solvent to a macroscopically phase-separated system contains several steps, as illustrated in Figure 11.15. Let us now concentrate on the coalescence step, which is peculiar to emulsions and foams.

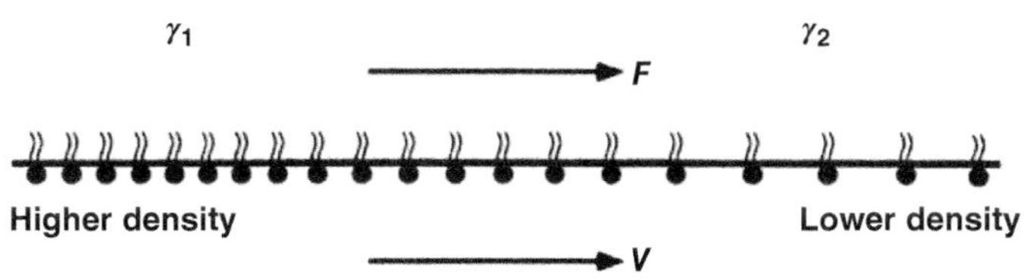

Figure 11.17 Surfactants are distributed nonuniformly in a freshly created interface. The relaxation to a homogeneous state does not occur primarily through diffusion, but by a more rapid process called the (Gibbs)–Marangoni effect. The difference in the concentration of surfactants in the interface induces gradients in surface tension, which generate a flow driven by a force F per unit length of $\gamma_2 - \gamma_1$.

We can identify three sequential crucial events in the total process that lead from two separate emulsion droplets to one combined droplet:

1. For a coalescence to be possible, the two surfactant films must come into molecular contact.
2. The two surfactant films must fuse, forming a neck with a direct contact between the dispersed liquid in the two drops.
3. The neck must grow in size so that the two drops merge completely.

The rate-determining step for the coalescence process can involve any of these three events, depending on the nature of the system. The same factors that control the stability of colloidal sols also control the first step involving the mutual approach of the two drops. For example, if we have an oil-in-water emulsion stabilized by ionic surfactants and with a low electrolyte concentration in the aqueous medium, a strong repulsive double-layer force will prevent the drops from establishing molecular contact, and the emulsion should have long-term stability relative to coalescence. On the other hand, if the potential barrier is small or nonexistent, or if the concentration of the dispersed phase is high, drops approach one another. Then, one has a thin gap of the continuous medium between the two surfactant films on the surface of the drops. Due to the presence of an attractive force, either due to the van der Waals force or due to a high osmotic pressure in the bulk reservoir (see Section 5.1), the liquid droplets deform, leading to a flat contact area as shown in Figure 11.18.

Now we have reached Step 2 in the coalescence process. Clearly, if the surfactant film is poorly developed, it will rupture easily, while if the surfactant has reached the CMC in the bulk, a well-developed film could resist the fusion of the two drops. In Section 6.4.1, we discussed the analogous problem of vesicle fusion/fission involving bilayers rather than monolayers. The breaking of a bilayer or an emulsion monolayer involves a topology change of the respective films. A rough estimate would give that, other conditions being equal, a bilayer should give a barrier roughly twice that of the monolayer. A tentative conclusion is then that thermal fluctuations in the monolayer are much more likely to result in a topology change than in the bilayer case. Still, for a very stiff and dense monolayer, for example, with an amphiphile in the gel phase, local fusion can be prevented. However, for the more typical case of surfactant monolayers with a liquid character thermally generated, local fusion seems to occur.

In the third and final step of the coalescence process, the connecting neck of dispersed liquid that formed in Step 2 needs to grow and eliminate the layer of continuous phase liquid between the drops, as shown in Figure 11.19.

During the growth process, the surfactant film remains intact but changes area and curvature. In the first section of this chapter, we developed a formalism that allows us to quantitatively analyze the free energy changes associated with the growth of the neck. Figure 11.20 illustrates a geometrical description of the neck of radius a and a semicircular rim of radius b, implying a liquid film of thickness $2b$ separating the flat sections of the drops.

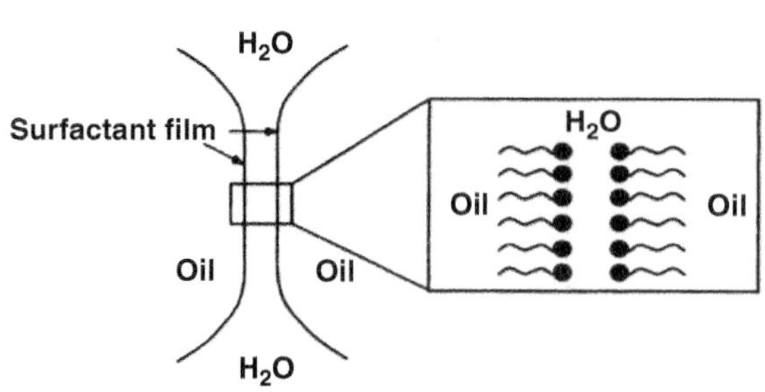

Figure 11.18 Two emulsion droplets in contact will deform, giving flat surfaces separated by a film of the continuous medium, water in this case. The thickness of the film is determined by the forces between drops.

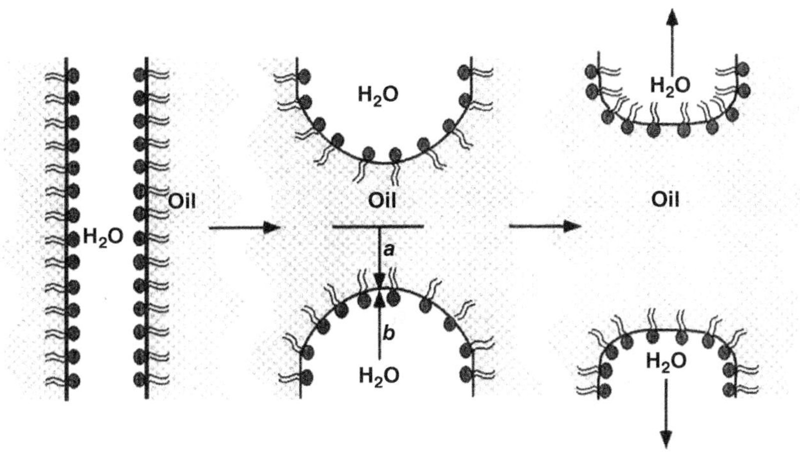

Figure 11.19 When the liquid film between two emulsion droplets ruptures, a neck of the dispersed liquid connects the two drops. We describe the neck in terms of the radius *a* in the plane of the film and the thickness *b* of the film. The growth of the neck involves an increase in *a* and could also lead to a change in the film thickness *b*.

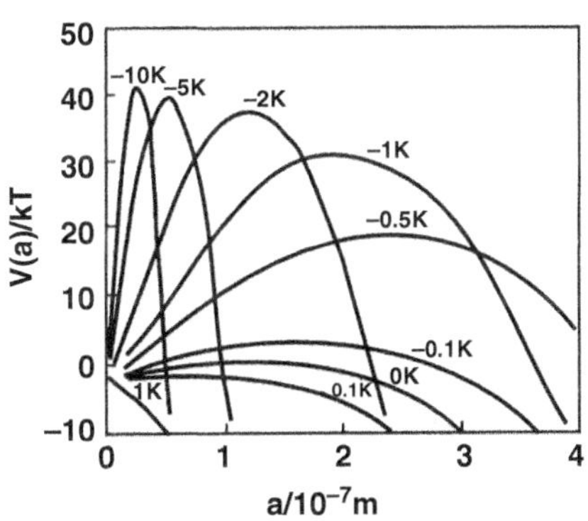

Figure 11.20 The calculated free energy barrier for the growth of a neck as a function of the neck radius *a*. The calculation is for O/W or W/O emulsions stabilized by the nonionic surfactant $C_{12}E_5$. The curves show the barrier at different temperatures in (K) below (O/W) or above (W/O) the balanced temperature $\mathbf{T}_0$, where the spontaneous curvature $H_0 = 0$. Note that the barrier starts to decrease significantly only one degree below/above $\mathbf{T}_0$. *Source*: Kabalnov and Wennerström, 1996/with permission of American Chemical Society.

We can write the excess free energy of the surfactant film $\mathbf{G}_f$ relative to an intact flat film as a sum of four terms

$$\mathbf{G}_f = W_1 + W_2 + W_3 + W_4 \tag{11.3.5}$$

The first two terms are due to changes in area. The flat part of the film decreases in area, resulting in a change in interfacial energy.

$$W_1 = -2\pi\gamma(a+b)^2 \tag{11.3.6}$$

The curved part of the film is assumed to have the geometry of a revolving semicircle, and we can write the surface free energy as

$$W_2 = 2\pi\gamma\left[\pi b(a+b) - 2b^2\right] \tag{11.3.7}$$

which is always positive. In addition, there is a bending free energy

$$W_3 = 2\kappa \int \left\{(H - H_0)^2 - H_0^2\right\} \mathrm{d}A \tag{11.3.8}$$

where the last term in the integral comes from the bending energy of a flat monolayer. Evaluating the integral in Eq. (11.3.8) requires algebra, and we simply state the final result

$$W_3 = 2\pi\kappa \left\{ 2\pi H_0 a + \frac{2(a+b)^2}{b\left[a(a+2b)\right]^{1/2}} \arctan\left(1 + 2\frac{b}{a}\right)^{1/2} + 2(\pi - 4)bH_0 - 4 \right\} \tag{11.3.9}$$

The fourth component is due to the contribution from a Gaussian curvature term, since a saddle bend occurs in the neck but not in the original flat films. This term takes the simple form

$$W_4 = -4\pi\bar{\kappa} \tag{11.3.10}$$

As discussed in Section 11.1, the Gaussian curvature is determined solely by the topology of the film, which does not change during the growth of the neck.

To find the free energy barrier, we optimize the value $\mathbf{G}_f$ of Eq. (11.3.5) for each a and define a potential energy

$$V(a) = \mathbf{G}_f(a, b_0) \text{ with } \frac{\partial \mathbf{G}_f(a,b)}{\partial b} = 0 \text{ at } b = b_0 \tag{11.3.11}$$

As discussed previously in this chapter, the microemulsions of nonionic surfactants have been carefully studied, so we have experimental estimates of $\gamma(\mathbf{T})$, $H_0(\mathbf{T})$, and κ. Therefore, $\mathbf{G}_f(a, b)$ becomes a known function of the two geometrical parameters.

Figure 11.20 shows an evaluation of the potential barrier $V(a)$ for the $C_{12}E_5$ system at different temperatures.

The first striking feature is that a barrier exists for temperatures when the spontaneous curvature is toward the dispersed phase, but for temperatures on the other side of the temperature $\mathbf{T}_0$ where the spontaneous curvature is zero, the barrier is absent. This example provides an explanation of the Bancroft rule. The second striking feature is that the barrier height is high $\approx 40k_B\mathbf{T}$ well away from $\mathbf{T}_0$ but decreases dramatically as the temperature approaches $\mathbf{T}_0$.

The conclusion emerges that we can obtain long-term stability with respect to coalescence due to the barrier for the growth of a neck joining two emulsion droplets. Such a barrier exists when the spontaneous curvature of the film is toward the dispersed phase, but there is no barrier by this mechanism when the spontaneous curvature is in the opposite direction. We conclude this discussion by showing experimental results confirming these theoretical predictions. Figure 11.21 shows the stability of an emulsion containing $C_{12}E_5$ as stabilizing surfactant.

Stability was characterized by the time τ required to separate a certain amount of the initially dispersed phase. The spontaneous curvature was varied both by temperature and the salt content of the aqueous phase. We can determine these variations of H_0 independently. The experiments show the following:

1. The emulsions are highly unstable when the spontaneous curvature is toward the continuous phase for both O/W and W/O emulsions, irrespective of whether the H_0 is tuned by temperature or salt content.
2. The stability time τ is the same irrespective of how we tune H_0 to a given value.
3. The stability time varies dramatically with H_0 and thus with temperature and salt content in the vicinity of $H_0 = 0$. The temperature needs to be controlled to less than 0.1°C to obtain reproducibility.

Emulsifiers are often also classified on the basis of an essentially empirical scale called the hydrophilic–lipophilic balance (HLB) number. Typical W/O emulsifiers have an HLB in the

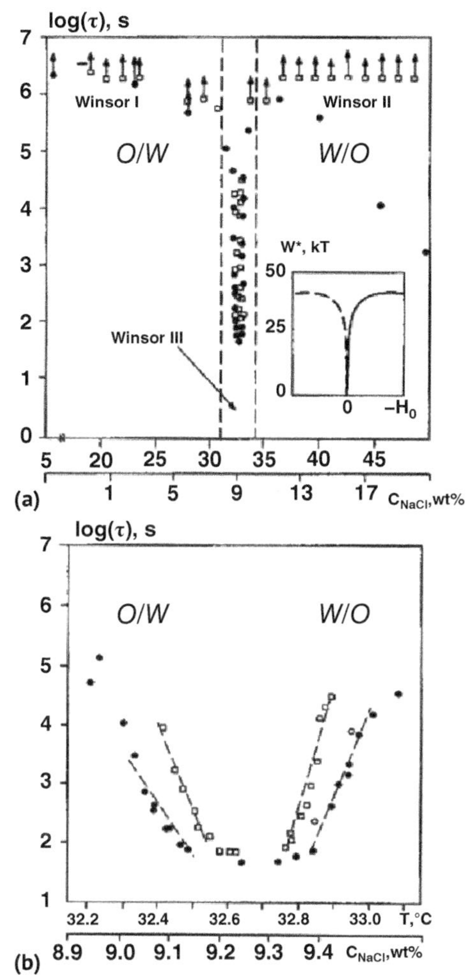

Figure 11.21 Measured stability of emulsions stabilized by the surfactant $C_{12}E_5$. The stability is measured by the time, τ, taken to form 1/4 of the separating bulk phase oil for O/W and water for W/O emulsions. The spontaneous curvature is tuned by either a temperature change or a change in salt content. Open square refers to experiments with a temperature change and the filled circles to changes in salt content. Panel (b) shows a detail of panel (a). Note that the stability time can change by two orders of magnitude with a change in temperature of only 0.2 K! *Source*: Kabalnov and Weers, 1996/with permission of American Chemical Society.

range of 4–8 and O/W emulsifiers are in the range of 12–16. The HLB number originally was developed empirically on the basis of nonionic surfactant properties, but a comparison with established properties of spontaneous film curvature shows that the balanced state $H_0 \approx 0$ corresponds to HLB $\approx$ 10. Thus, we have interpreted the HLB number as a measure of surfactant film spontaneous curvature. In Section 11.1, we saw that the spontaneous curvature of a surfactant film is not determined by the surfactant alone, but also depends on factors such as temperature, salt content, and the extent of oil penetration. Thus, the HLB number, which is a characterization of the surfactant only, provides an approximate measure of the expected emulsion stability.

11.3.8 The Phase Diagram Is a Useful Guide for Selecting an Emulsifier

We can obtain another useful rule for predicting emulsion stability by starting from the equilibrium phase behavior of the ternary system water–oil–amphiphile. Lecithin, or dialkylphosphatidylcholine, is the main lipid component of egg yolk. It is commonly used to emulsify fat in food, as for example, in mayonnaise, bearnaise sauce, milk, and milk replacement products. As we discussed in Chapter 6, lecithins form a lamellar phase in water that incorporates about 40% H_2O (w/w). Very little oil dissolves in this lamellar state. Therefore, a large three-phase triangle with a slightly swollen lamellar phase is in equilibrium with virtually pure water, and pure oil dominates the phase diagram. Empirical observations consistently demonstrate that a sample in this three-phrase triangle readily forms a stable O/W emulsion.

The stability of the lamellar phase shows that a monolayer's spontaneous curvature is close to zero and, in fact, slightly positive toward oil. To obtain a W/O microemulsion, the equilibrium phase for the surfactant should have a curvature slightly more toward water. A commonly used emulsifier in this context is the monoglyceride monoolein. It has a small head group and the monolayer has a negative spontaneous curvature, and the phase diagram shows that an inverted cubic phase is in equilibrium with excess water.

Knowledge of equilibrium phase behavior enables us to understand and control emulsion stability in yet another way. Whatever the spontaneous curvature of the surfactant film, coalescence involves a local rearrangement and change in film topology. For surfactants in the liquid state, we can associate the energy barrier with the formation of a local high-energy structure. By freezing the chains, we can slow down the dynamics on the molecular scale by orders of magnitude. Sometimes, we can take advantage of this phenomenon by preparing an emulsion at a slightly elevated temperature with fluid chains, and then cooling the system through a temperature at which a gel phase forms. The emulsion drops are then covered with a rather rigid shell, making the system very long-lived.

11.3.9 We Can Catalyze Coalescence by Changing the Spontaneous Curvature and by Inducing Depletion Attraction

In many applications, the goal is to destabilize an emulsion that is no longer useful or one that has been formed inadvertently. The optimal strategy first identifies the main mechanism behind the stability and then selectively changes the conditions that affect this critical step in the sequence of events that lead to macroscopic phase separation.

For an electrostatically stabilized O/W emulsion, we can add electrolytes to facilitate contact between drops. Centrifugation can induce creaming. Gentle mechanical agitation facilitates the coalescence of drops. If the drops are stabilized by surfactant in the gel state, heating the mixture above the melting temperature produces a dramatic effect.

A slightly more subtle method involves adding surfactant or cosurfactant to affect the spontaneous curvature of the film. It may seem puzzling that adding a cosurfactant, such as a long-chain alcohol, can either stabilize or destabilize the system depending on the circumstances. However, we can explain this finding if we realize that the emulsion has an optimal stability at a particular value of H_0.

Changes in temperature also affect the value of H_0. As discussed in Section 11.2, ionic and nonionic surfactants show opposite dependence of the spontaneous curvature H_0 on the temperature **T**. Thus, moderate heating of a W/O emulsion stabilized by ethylene oxide as a nonionic surfactant does not have an effect, but destabilization can very well occur if ionic surfactants are used as stabilizers. For cooling, the situation is the reverse.

Adding electrolyte will have a moderate, but measurable, effect on the stability of the emulsion based on nonionic surfactants, as shown in Figure 11.21. We tend to think of adding electrolyte to a system containing ionic surfactants primarily in terms of changing the double-layer repulsion, but clearly, there is also an effect on the spontaneous curvature. For O/W emulsions, the two mechanisms have qualitatively the same effect, and it may be difficult to distinguish which is most important. For W/O emulsions, on the other hand, the double-layer repulsion is usually insignificant, but the electrolyte effect on H_0 still operates to the same extent. Thus, changing the electrolyte content of the dispersed water phase can be an effective way to tune the spontaneous curvature of the stabilizing ionic surfactant film.

Adding surfactant can lead to another destabilizing mechanism. Typically, excess surfactant is present in the continuous phase as micellar aggregates. These aggregates are excluded from the regions where two surfactant films approach distances on the order of a micellar size or the Debye length for charged surfactants. The result is an attractive depletion force, as discussed in Section 5.5, which pushes the films even closer together, making film rupture a more likely event. The depletion attraction can also be induced by polymers that dissolve in the continuous phase and do not adsorb on the surfactant film.

See Berg (2010, Chapter 9), Cosgrove (2010, Chapter 6), Kronberg et al, (2014, Chapter 24), and Lyklema (2005, Vol. 5, Chapter 5) for additional discussion on Emulsions.

11.4 Foams Consist of Gas Bubbles Dispersed in a Liquid or Solid Medium

A foam is the reverse of an aerosol. It is a dispersion of a gas, usually air, in a liquid or solid medium. We know of no example of a thermodynamically stable foam. In analogy with emulsions, foam is used to advantage in a wide range of applications, but foams also may form inadvertently and cause severe problems.

Solid foams have important applications as useful materials. The presence of gas bubbles results in a substantial reduction in the average density of the material. It also reduces the thermal conductivity, making solid foams very suitable for insulation materials such as polystyrene foam. Solid foams also find applications as lightweight materials in mechanical gadgets. In this case, the challenge is to achieve an effective compromise between weight and strength. In foods like bread, ice cream, souffles, and even marshmallows, solid foams give the food an agreeable texture.

In preparing a solid foam, we start from a liquid foam and induce solidification by a chemical polymerization reaction, by lowering the temperature through the melting point, or even by increasing the temperature to induce a structural transition as in the baking of bread.

Porous solids in their dry state can superficially appear as solid foams since they can show the same ratio between gas and solid in the material. However, the porous media have a bicontinuous structure, are less insulating, and tend to adsorb liquid by capillary forces. The simplest way to prepare a porous solid is through the sintering of solid particles as in ceramics.

By far the largest technological application of liquid foams is in mineral froth flotation, and this process uses a substantial fraction of the world production of amphiphiles. Froth flotation is also used in other separation processes, such as the de-inking of recycled paper and in wastewater purification. Other uses of liquid foams include fire-fighting, in cleaning processes (shampoo, shaving cream), in foods such as whipped cream or egg whites, and in the preparation of solid foams as discussed earlier. The foam that inspires scientists most is probably the beer froth!

In the household, foams can also be a nuisance, for example, when they form in boiling liquids such as milk; or if we, by mistake, use a liquid detergent designed for cleaning by hand in a dishwasher or laundry machine. The same type of phenomenon can readily occur in industrial processes with rather more troublesome consequences. Controlling foam stability is an important art of considerable practical importance. Thus, with foams, we encounter the same dichotomy between stability and instability as for sols and emulsions.

11.4.1 A Surface Film Develops on Bubbles as They Rise

The first step in the formation of a foam is the generation of a gas bubble in the bulk liquid. We can accomplish this most simply by forcing the gas through a suitably designed nozzle, but we can also nucleate the bubble from a supersaturated solution, as in beer, or from a superheated liquid, as in ordinary boiling. Another way to form bubbles is by mechanical agitation of the liquid surface as occurs in breaking waves on the sea or when whipping cream and eggwhite. In a bread dough, the foam is generated by the CO_2 production of the yeast or by the decomposition of bicarbonate in baking soda.

After the bubble forms, it will migrate toward the liquid surface through the action of gravitational forces. Unless the bubbles are very small, gravitational energy exceeds thermal energy and the bubbles travel to the surface.

Problem: Determine the radius of an air bubble for which the gravitational energy differs by 1 $k_B\mathbf{T}(\mathbf{T} = 300\text{ K})$ over 1 cm in water. The gravitational energy is $m_{\text{eff}}\, gz$.

Solution: The density of air is negligible compared to that of water $(\simeq 1 \times 10^3\text{ kg/m}^3)$. The effective mass of the bubble is then

$$m_{\text{eff}} = -\frac{4\pi 10^3 R^3}{3}$$

and over 1 cm, the gravitational energy changes

$$\frac{4\pi 10^3 R^3}{3} \times 9.81 \times 10^{-2} = 4.1 \times 10^2 R^3$$

Equating this to the thermal energy $k_B\mathbf{T} = 4.1\ 10^{-21}$ yields $R = \left(10^{-23}\right)^{1/3} \approx 2 \times 10^{-8}$ m.

Note that the buoyancy force acting on a bubble is the same as the gravitational force acting on a liquid drop of the same size in an aerosol, except for its direction. In Section 9.5, we analyzed the settling rate of such a particle, and the same analysis applies here. We only need to multiply the velocities by $\eta_{\text{air}}/\eta_{\text{liq}}$ to get the relevant values for gas bubbles in the liquid. For water and air at atmospheric pressure, the viscosity ratio is 1/50, showing that gas bubble rises considerably slower than an aerosol particle settles, neglecting the effect of convection.

As the bubble rises, the hydrostatic pressure decreases, resulting in a slight expansion of its size. A more significant effect is that if surface-active molecules are present in the liquid, they can adsorb on the bubble liquid interface during the rise phase. There is then a decrease in the surface tension leading to a decrease in the Laplace pressure (see Section 2.1.5), and the bubble expands.

If a colloidal particle collides with a bubble, the chance is that the particle sticks to the gas–liquid interface. This is the equilibrium state if the contact angle is finite; that is, if the liquid does not completely wet the surface of the particle. This is the principle of flotation separation of colloidal particles based on their surface properties. For example, particles of elemental gold can be extracted from a slurry of ground ore even when the gold content is as low as 10 g per 10^3 kg. The trick is to use a surface-active alkane thiol, called a collector, whose SH head group specifically adsorbs onto the gold surfaces, while showing a low affinity for the abundant sulfide and silicate minerals in the ore. After attaching to the bubbles, the gold particles become part of the froth at the surface of the slurry, where they can be skimmed off.

Ultimately, the rising bubble reaches the liquid–air (gas) interface. Let us first consider a virgin interface with no foam present. Assume that the bubble has traveled far enough to reach an equilibrium between its surface and the bulk liquid with respect to surface-active solutes. At the bulk surface, which is in equilibrium with the same bulk phase, the curved surface film of the bubble meets the flat film of the bulk surface. The molecular features of such an encounter should be quite analogous to those described in our discussion of an encounter between two emulsion droplets.

For a pure liquid, the bubble will readily break the film and join the bulk gaseous phase, as for ordinary boiling water. For a surfactant solution, on the other hand, the liquid film is stable enough to rise with the bubble. If this bubble is followed by others, a froth develops that rises in the container until it overflows, the primary bubble generation stops, or bubble coalescence and foam drainage balance the foam formation to achieve a steady state.

11.4.2 Concentrated Foams Consist of Polyhedral Gas Compartments Packed in an Intriguing Way

The bubbles of a concentrated foam, sometimes referred to as a froth, rapidly establish a mechanical equilibrium that affects the bubble's shape. When the bubbles enter the concentrated foam, they drag bulk liquid with them through the action of hydrodynamic forces. As the bubbles come to rest in the foam, this liquid starts to drain back into the remaining bulk liquid. If we assume that coalescence does not occur on the timescale needed for drainage, a local equilibrium is established, in which the force acting across the surface film is close to zero and a balance exists between the attractive van der Waals force and a repulsion originating from surface–surface interactions. In foams, gravitational forces also contribute to the stationary thickness of the liquid film.

Thus, in this conditional equilibrium state, the gas bubbles deform into polyhedra with rounded corners when the force per area between bubbles exceeds the Laplace pressure γ/R as in concentrated emulsions. Let us now consider the packing of these polyhedra in closer detail. The Belgian scientist Joseph Plateau performed pioneering studies of this problem in the second half of the nineteenth century. By careful experimental studies, Plateau established a number of rules for the behavior of soap films. Two of them, shown in Figure 11.22, are particularly relevant for the structure of concentrated foams:

1. Three flat sides of a polyhedra meet at 120°, but if four or more sides come together along a line, they form an unstable configuration.
2. At the corners of the polyhedra, four edges meet in a regular tetrahedral arrangement. The body of the tetrahedron is called the Plateau border, and it serves as a reservoir of the medium.

When the number and volume of the polyhedral compartments are given, the optimal structure of the foam is the one that creates the smallest total film area. This condition constitutes a well-defined mathematical optimization problem of considerable difficulty. The best present solution is that optimally the polyhedra should have 13.39 sides (!) if one disregards the constraint that we expect an integral number of sides in a real system. Careful experimental studies reveal that the

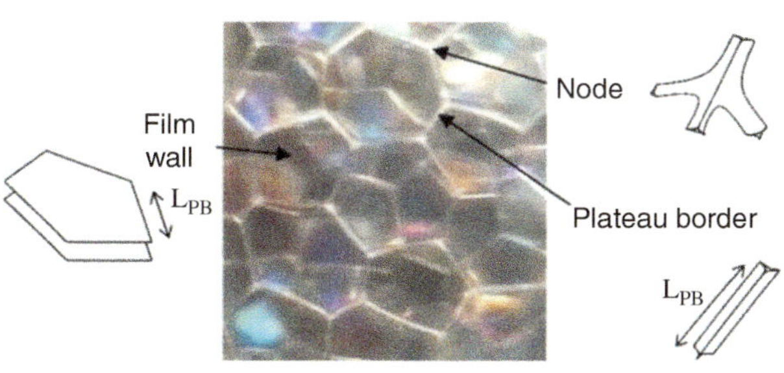

Figure 11.22 In a foam, three compartments meet at an angle of 120° along a straight line, forming a Plateau border of length L_{PB}. Four Plateau borders meet at a node in a tetrahedral geometry at ideally 109.5°. *Source*: S. Abbott/ CC BY ND.

polyhedra most commonly found in foams have 14 sides, followed by 12 sides as the second choice. This description of the structure of foams is equally valid for the highly concentrated emulsions discussed in Section 11.3.5.

11.4.3 The Foam Can Reach a Metastable State Where the Liquid Film Thickness Is Determined by the Balance of Forces

A foam with a well-developed surface film can show long-term stability while resting on a liquid surface. During the formation process, the bubbles are dragged by hydrodynamic forces, drawing some liquid into the foam. Consider a state when the foam production has stopped and a foam layer of height L_0 has been formed. There is then a gravitational force acting on the liquid in the foam, and the strength of this force increases linearly with height. The excess liquid then flows back into the bulk phase, thinning the liquid layer separating the surfaces of the drops. The drainage stops when the gravitational force pushing on the liquid is balanced by a repulsive surface force. There is then a uniform chemical potential of the liquid medium, usually water, across the foam. The situation is analogous to the case of a liquid film forming on the wall of the container analyzed in Section 5.2.3.

A practical complication, which we will neglect below, arises at the upper end of the foam, which is exposed to the ambient air, where the relative humidity usually does not match the chemical potential of the bulk medium. This leads to an extra drying of the surface region (see Section 2.8). In soap bubbles produced by children, this drying is a major cause of film rupture and can, in that case, be partly compensated by mixing in a nonvolatile compound like glycerol into the liquid phase.

In addition to the gravitational force, we have in the film the same surface forces acting as between solid particles and emulsion drops. The Hamaker constant determining the strength of the van der Waals force is higher than for typical emulsions since one of the media is a gas. For a film stabilized by an ionic surfactant, there is a strong repulsive component, and the metastable equilibrium thickness of the film can be of order 100 nm as in Figure 5.10. This state is called a common black film (CBF), since it is so thin that there is only a weak reflection of light (see Section 6.3). By adding electrolytes or increasing the osmotic pressure of the bulk by other means, the film can be forced to thin. Then, one can reach an instability and a transition to a state where the film is only marginally thicker than the width of two surfactant monolayers. Then, one can reach high charge densities and a situation with only counterions in the liquid layer. This is a situation where ion–ion correlations can quantitatively change the surface force from strong repulsion to weak repulsion or attraction and cause a transition from a CBF to a thin Newton black film. Figure 11.23 shows a different example of a transition from a CBF to a Newton black film.

In this case, the film is mainly formed by a nonionic sugar surfactant, but it is contaminated by an ionic species, probably a long-chain fatty acid (see Section 9.2.3). The presence of charges gives a long-range repulsion, and in accordance with the DLVO theory, the van der Waals interaction takes over at intermediate separations. At this stage, the thickness of the film jumps into near molecular contact, forming a Newton black film. This is stabilized by the short-range overlap forces shown in Figures 5.27 and 6.12. When the surfactant concentration is increased above the CMC of the surfactant, the apolar charged impurity can be solubilized in solution, and its concentration in the film decreases and the double-layer repulsion decreases in magnitude.

11.4.4 Foams Disintegrate by Film Rupture

The factors leading to the relaxation of a foam to separate bulk liquid and gaseous phases are essentially the same as those discussed for the separation of an emulsion into oil and water phases.

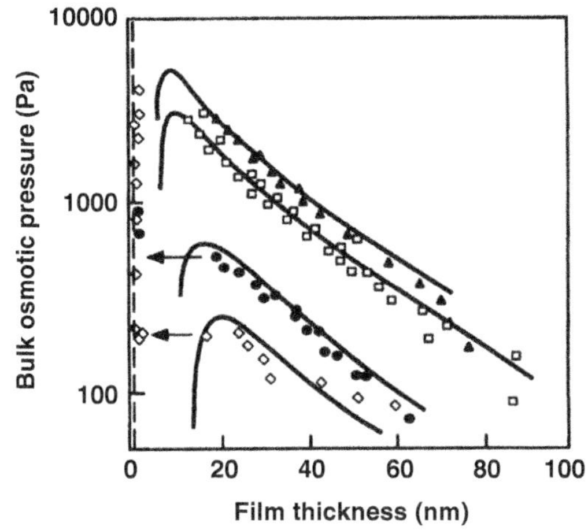

Figure 11.23 Experimentally determined relation between bulk osmotic pressure and film thickness for aqueous 0.1 mM KBr β-octyl glucoside solutions (CMC 20 mM); ▲, 3 mM; □, 10 mM; ● 21 mM ◇, 25 mM surfactant concentration. The solid lines are calculated according to the DLVO theory. The arrows indicate a transition to a Newton black film. *Source*: Bergeron et al., 1996/with permission of American Chemical Society.

The gravitational drainage in a foam is analogous to the creaming in an emulsion. These processes lead to a concentration of the foam/emulsion but not to a phase separation as such. Ostwald ripening of a normal emulsion results in a gradual decrease in the number of drops. Although the same basic driving force is also operating in concentrated emulsions and foams, it appears to be of lesser importance in these circumstances. In dilute emulsions, the increase in the chemical potential of the molecules in the drops is determined by the surface free energy only. In concentrated emulsions and in foams, there is a strong interaction between the compartments, and it is a challenge to determine the free energy change in moving a molecule from one compartment to another. In a foam, one has the additional complication of a change in pressure as material is exchanged between compartments. The net effect of the compartment–compartment interaction seems to slow down the Ostwald ripening process.

Bubble–bubble coalescence through film rupture is the primary route to foam destabilization. The molecular processes are fundamentally the same as those observed in the coalescence of emulsion droplets. Again, we can identify three crucial steps: (1) approach of the two surfaces; (2) local breakthrough; and (3) growth of the hole in the film. Here, we will briefly discuss the quantitative changes in the three steps that occur in the case of foams.

For some foams, the decrease in film thickness due to gravitational drainage is sufficient to induce film rupture. For such systems, a long-range electrostatic repulsion preventing close contact is a possible way to induce stability. However, in many cases, surfactant-stabilized films can thin even to the state of a Newton black film without immediate rupture. We argued in the previous section that emulsion stability was, only in the case of very stiff monolayers, caused by the resistance to the topology change involved in the formation of a local neck combining two separate compartments. The same argument should then apply to the case of foams since the stabilizing monolayers have the same molecular characteristics. Thus, solid foams show long-term stability. The stability of a Newton black film, and the long-term stability of a liquid foam in general, should then be determined by the free energy changes when a small neck grows to join the Plateau borders at the edges of the compartment. We showed in the previous section that this mechanism is important for explaining emulsion stability.

A way to consider the film rupture is to treat the problem as a two-dimensional nucleation process in analogy with the corresponding three-dimensional case discussed in Section 2.3.2. In 2D, it is the line tension, σ, whose associated energy is proportional to the length of the periphery of the hole, which takes the role of surface tension in the 3D case. As the hole radius r grows the area of the film decreases. There is an interface at both sides of the film the change in surface free energy is

$$\Delta G_A \approx -2\pi r^2 \gamma. \tag{11.4.1}$$

In the hole there is also a line tension free energy contribution

$$\Delta G_L = 2\pi r\sigma \tag{11.4.2}$$

The total free energy of the hole is then

$$\Delta G_H \approx 2\pi r(-r\gamma + \sigma) \tag{11.4.3}$$

For small r the line tension dominates, while for larger r the surface free energy takes over. The maximum of ΔG_H occurs for

$$r(m) = \frac{\sigma}{2\gamma} \tag{11.4.4}$$

where it takes the value

$$\Delta G_H(m) = \frac{\sigma^2}{2\gamma} \tag{11.4.5}$$

We now assume that the line tension is determined by the curvature energy of the film and adopt the description of emulsion stability in Section 11.3.7. The mean curvature is approximately

$$H \approx \frac{1}{2}\left(-\frac{1}{r} + \frac{1}{b}\right) \approx \frac{1}{2b} \tag{11.4.6}$$

Here b is the radius of the rim and we have in the second equality assumed that for relevant cases the barrier occurs for r>>b. Based on Eq.(11.3.8) the curvature energy is

$$G_C = 2\kappa \int \left\{ \left(\frac{1}{2b} - H_0 \right)^2 - H_0^2 \right\} dA = 2\kappa \left(\frac{1}{4b^2} - \frac{H_0}{b} \right) 2\pi r \pi b \tag{11.4.7}$$

Comparing this expression with the expression for the line tension energy in Eq.(11.4.2) we have

$$\sigma = 2\pi\kappa \left(\frac{1}{4b} - H_0 \right) \tag{11.4.8}$$

Inserted into Eq.(11.4.5) We find an estimated value of the nucleation barrier to be

$$\Delta G_H(m) \approx \frac{\pi^2 \kappa^2}{\gamma} \left(\frac{1}{b} - 4H_0 \right)^2$$

The larger the line tension, the more stable the film. The bending rigidity κ increases with the monolayer thickness to the third power, and it follows that the longer the hydrocarbon chain of the stabilizing surfactant, the more stable the film. It follows similarly that a larger positive value of the spontaneous curvature promotes film instability.

A consequence of describing the film rupture in terms of a 2D homogeneous nucleation event is that it can be sensitive to a heterogeneous nucleation caused by impurities in the film. For example, one fails to whip eggwhite into a foam if there are traces of egg yolk present. Presumably, the lipoprotein particles of the egg yolk act as inhomogeneous nucleation sites for film rupture. It can also be difficult to create a film with a large area since the probability of the presence of a single impurity particle is high.

Even though a foam based on a liquid film can show stability over a sizeable time, it is normally not long-term stable. To achieve foams that can be stored, a practical way is to convert the film to a solid state. This can be achieved by changing the temperature. To make a meringue or soufflé, one paradoxically heats the eggwhite foam to form a solid. There is denaturation of the egg albumin, and the highly concentrated protein solution turns into a glass-like solid, as in the white part of a boiled egg. In the baking of bread, the CO_2 bubbles formed by the action of the yeast are stabilized by lipids and gluten proteins. In the heat of the oven, the starch granules of the flour swell, converting the concentrated particle solution to a glass-like polymer network. This is, however, only metastable, and the starch reorganizes during storage to a more ordered state, and the bread loses its elasticity.

The more obvious way of forming a solid film is by decreasing the temperature. Pumice is formed in a volcanic process, where magma containing gas bubbles is cooled rapidly to a solid state. The ice cream foam is stable as long as it is stored well below 0 °C, but it quickly loses its agreeable texture at somewhat higher temperatures. In a polystyren-based foam, the polymer is below its glass transition, and the foam is indefinitely stable at ambient temperatures and can tolerate the heat of tea or coffee when used as the material of a cup.

11.4.5 Liquid Foams Show Varied Rheological Properties

When whipping cream, we gradually introduce air bubbles. As the process continues, the apparent volume of the liquid increases, as does the viscosity. For volume fractions of air less than 40%, the bubbles are spherical, except for deformations under shear. The bubbles are stabilized by surface active components of the cream, and one can use stick boundary conditions when evaluating the viscosity. To first order, this is given by the Einstein formula Eq. (7.1.24). Above 40% volume fraction, bubble–bubble interactions start to be significant, resulting in a faceting of the bubbles and non-Newtonian rheological behavior. It is difficult to reach a gas volume fraction larger than 50% in whipped cream, but in this state, the deformation of the bubbles is large enough to prevent the easy flow of the liquid phase, and the whipped cream has a finite yield stress, enabling nice decoration of a cake! In this state, one has a partitioning of the colloidal components between the film sections and the Plateau borders, where there is an accumulation of fat globules and casein micelles. A consequence of the repulsion between gas compartments is that the osmotic pressure is increased. Excessive whipping produces an instability of the foam and the formation of the water-in-oil emulsion called butter.

See Berg (2010, Chapter 9), Kronberg et al (2014, Chapter 23), and Lyklema (2005, Vol. 5, Chapters 6 and 7) for additional discussion on Foams.

Exercises

11.1 Calculate the elastic curvature energy associated with the formation of a vesicle from a planar symmetrical bilayer. What is the dependence on the vesicle radius?

11.2 Repeat the calculation of Exercise 11.1, but consider the two monolayers separately. Let the monolayer be 20 Å thick and associate the curvature energy with the polar–apolar interface. Do the calculation for a vesicle radius of 20 nm; let the monolayer bending constant be 5×10^{-20} J, and use each of three monolayer spontaneous curvatures:

a. $H_0 = 0$

b. $H_0 = 5 \times 10^7\ \text{m}^{-1}$

c. $H_0 = -5 \times 10^7\ \text{m}^{-1}$

11.3 Vesicles are prepared in a 1×10^{-3} M NaCl solution. To the newly formed vesicle solution, one adds NaCl to yield a 1×10^{-2} M solution. Discuss what happens qualitatively.

11.4 Vesicles are prepared in a 1.0 mM NaCl solution. This solution is then mixed with an equal volume of pure water. Calculate the expansion of the initially spherical vesicles, using an area expansion modulus of 0.2 Jm^{-2}.

11.5 For a dilute O/W microemulsion, one finds that the self-diffusion coefficient is $D = 2 \times 10^{-11}$ for both the oil and the surfactant at $\mathbf{T} = 300$ K. What qualitative and quantitative conclusions can you draw from this observation?

11.6 Tetradecane is added to a micellar solution of 10% (w/w) $C_{12}E_5$ in water to yield a system of equal amounts (by weight) of surfactant and oil. The density of the oil is 0.84×10^3 kg m^{-3} and $\approx 1.0 \times 10^3$ kg m^{-3} for the surfactant. Calculate the resulting drop size, assuming spheres and a monolayer 3.5 nm thick formed by the surfactant, half of which is hydrocarbon.

11.7 A sample is prepared at 47.8 °C by taking the microemulsion phase of the three-phase triangle in Figure 11.5a.

a. The sample is then cooled to 44.5 °C. Give the composition and relative amounts of the respective phases.

b. The sample is then heated to 51.2 °C. Give the composition and amount of the equilibrium phases.

11.8 In the phase diagram of Figure 11.6, an equilibrium O/W microemulsion occurs at lower temperatures and high water contents. Assume that the lower phase boundary forms when the radius of the microemulsion droplet corresponds to the inverse spontaneous curvature of the film. Calculate $H_0(\mathbf{T})$ in the range $25°C < \mathbf{T} < 35°C$.

11.9 Use the result of Exercise 11.8 to estimate (a) the temperature $\mathbf{T}_0$, where $H_0(\mathbf{T}) = 0$, and (b) the microemulsion oil–surface tension at 30 °C, given that the bending constants are $\kappa + \frac{\bar{\kappa}}{2} = 1.5 \times 10^{-20}$ J.

11.10 Determine the rate of creaming of a fat droplet of radius 2.5×10^{-6} m in milk. Assume a density of 930 kg/m^3 neglect droplet–droplet interactions, and use the viscosity of pure water.

11.11 An oil-in-water emulsion is made from a 0.3 M solution of decyltrimethylammonium bromide. When two emulsion drops get in close proximity, micelles are excluded from the aqueous gap between them. Exclusion of micelles gives rise to an attractive depletion force. Estimate the magnitude of the force, using the data of Figure 4.7.

11.12 A surfactant film is stretched laterally with pressure Π. This pressure can be relieved by breaking the film. Such a rupture, however, requires the generation of a local hole. Assume that the cost of forming a hole can be described by a line tension χ. At what size does the growth of the circular hole become spontaneous?

11.13 A bubble of radius 10 μm is generated with a clean air–water interface 0.1 m below the surface in a beaker. Calculate the size of the bubble when it reaches the surface. The aqueous phase contains 10 mM SDS, and an equilibrium surfactant film has time to develop at the water–bubble interface during the rise of the bubble. Neglect a possible dissolution of the air in the water.

11.14 Determine the gravitational pressure across an aqueous film at a position 0.1 m above the liquid surface.

11.15 An aqueous film stabilized by a surfactant has an electrical surface potential of 25 mV. At what film thickness does the repulsive double-layer force balance the hydrostatic pressure at a position 0.1 m above the bulk solution containing 10 mM NaCl?

References

Balogh, J., Kaper, H., Olsson, U., and Wennerström, H. (2006). *Phys Rev E*. 73, 041506.

Berg, J.C. (2010). *An Introduction to Interfaces & Colloids. The Bridge to Nanoscience*. World Scientific.

Bergeron, V., Waltermo, Å., and Claesson, P.M. (1996). *Langmuir*. 12, 1336.

Cosgrove, T. (2010). *Colloid Science. Principles, Methods and Applications*. Chichester, UK: Wiley.

Jahn, W. and Strey. R. (1988). *J. Phys. Chem*. 92, 2294.

Kabalnov, A. and Weers, J. (1996). *Langmuir*. 12, 1931.

Kabalnov, A. and Wennerström, H. (1996). *Langmuir*. 12, 276.

Kahlweit, M. and Strey, R. (1986). *J. Phys. Chem*. 90, 5239.

Kronberg, B., Holmberg, K., and Lindman, B. (2014). *Surface Chemistry of Surfactants and Polymers*, 3rd ed. Chichester, UK: Wiley.

Kunieda, K. and Shinoda, K. (1982). *J. Dispension Sci. Technol*. 3, 233.

Lyklema, H. ed. (2005). *Fundamentals of Interface and Colloid Science. Soft Colloids*, Vol. 5. Amsterdam, The Netherlands: Elsevier.

Olsson, U., Shinoda, K., and Lindman, B. (1986). *J. Phys. Chem*. 90, 4083.

Walstra, P. (1983). *Encyclopedia of Emulsion Technology*, Vol. I, P. Becher, ed. New York: Dekker, Chap. 2.

12

Epilogue

Throughout this book, we have emphasized themes that provide a basis for understanding the vast number of topics encompassed within the colloidal domain. These themes, like those in a symphony, are usually stated with prominence and then subsumed to form an integral part of the score. New themes emerge as the power and beauty arising from interconnectedness develop. The trained ear can follow each theme as it re-emerges—often in a somewhat different form appropriate to the individual movement—and trace it back to its original source. To the untrained ear, however, this is a difficult undertaking because the perspective that provides an integrating framework is incomplete.

By its very nature, the colloidal domain encompasses entities that contain many molecules. Thus, it occupies an intermediate position between the molecular and the macroscopic worlds. A major goal in this book has been to show how molecular interactions are manifested in the colloidal domain, as well as how colloidal phenomena are manifested in the macroscopic world, which we perceive with our senses. In establishing these links, we were guided by the need to address four questions:

- How molecular, mesoscale, and macroscopic structures merge in the colloidal domain.
- How molecular interaction forces determine colloidal behavior.
- How the interplay between entropy and energy organizes colloidal entities.
- How dynamic, kinetic, and transport properties influence colloidal processes.

The answers to these questions sometimes appeared quite different in the context of the three main types of colloidal systems: amphiphilic aggregates, polymers, and colloidal sols. Although comparisons among these systems are implicit in the text, often in the hurly-burly of dealing with immediate concepts and issues, they were not made explicit or given a major focus. Our goal in this epilogue is to develop an integrating framework to clarify the unifying features as well as the major differences among the three colloidal systems. We begin by giving a short description of the development of colloid science and technology during the twentieth century.

12.1 Colloid Science Has Changed from a Reductionistic to a Holistic Perspective During the Twentieth Century

At the start of the twentieth century, colloid science represented an intellectual frontier in both physics and chemistry. In those days, the colloidal domain contained the smallest entities amenable to direct observation. Milliken used charged oil droplets to determine the elementary charge.

The Colloidal Domain: Where Physics, Chemistry, Biology, and Technology Meet, Third Edition. Håkan Wennerström.

Lorentz, Mie, and Debye analyzed the light scattering of objects of the same linear size as the wavelength of light. These theoretical studies were followed by experimental work by Svedberg and Szygmondy, among others. The magnitude of Avogadro's number was determined by studies of colloidal systems. For example, Perrin made careful measurements of sedimentation equilibria in the Earth's gravitational field. Knowing the size and weight of the particles, he could determine the value of Boltzmann's constant from the measured concentration profile and obtain Avogadro's number as the ratio $\boldsymbol{R}/k_B$.

The dynamic properties of colloidal systems were also addressed from a fundamental perspective. Einstein analyzed Brownian motion and established the relation between the diffusion coefficient and molecular friction (Section 1.7). He also derived the basic relation between the viscosity of a solution and the concentration of colloidal particles (Eq. 7.1.15). von Schmoluchowski considered the diffusion-controlled association rate of colloidal particles and derived the basic equation for the rate constant (Eq. 8.3.9).

At the turn of the century, the final clashes occurred in the controversy between the atomists (represented by Boltzmann) and those inspired by the success of thermodynamics (like Poincaré, the monumental proponent of this position), who insisted on seeing matter as a continuum. The battle was fought mainly by theoretical arguments, but the atomists typically had to refer to the colloidal domain for an experimental justification of their views. Evidence for the molecular nature of matter became apparent in molecular films of the amphiphilic compounds that were studied initially by Benjamin Franklin and then by Agnes Pockels and Irving Langmuir. These simple measurements provided one of the first clear demonstrations of molecular sizes.

In the 1920s, quantum mechanics entered the scene, and the atomistic view of matter was established beyond any doubt. Now, the means were available to press further into the molecular world. Following the path of scientific reductionism, the focus of urgent fundamental scientific questions moved from the colloidal domain to molecules, atoms, and elementary particles. As a sign of this shift in scientific attention, we note that six of the scientists discussed earlier, who played key roles in defining the colloidal domain, were Nobel laureates. After 1930, only three prizes were given in this area: to the polymer scientists Staudinger and Flory, and more recently to the condensed matter physicist de Gennes.

After the 1930s, colloid science was relegated to the intellectual backwaters of science where it survived through the actions of scientific enthusiasts and researchers with a more applied perspective. However, the last decades have seen the emergence of a new realization that the fundamental motivation for the reductionist strategy in science is to synthesize knowledge about parts into an understanding of the whole. Thus, in the new century, the scientific frontier is passing through the colloidal domain once more. This time, the emphasis is on understanding and controlling the macroscopic manifestations of matter through knowledge of atomic and molecular interactions. During the more than 100 years that have passed since the fundamental contributions from Einstein, Debye, and others, scientific understanding of the colloidal domain has increased tremendously. A holistic approach has been adopted in many disciplines such as molecular biology, biochemistry, biophysics, solid-state physics, mesoscopic physics, and soft condensed matter physics. Colloid science can be seen to provide the integrating intellectual framework.

The development of the more applied technical side of the colloid domain shows quite a different pattern. Colloid science has always been central to many technological applications. We have alluded to some of them in previous chapters. Until recently, empirical knowledge outpaced scientific understanding so that technologies were achieved mainly by a trial-and-error process, using experience as a guideline. However, as our understanding of fundamental colloidal processes developed, theoretical concepts increasingly guided technological developments. Today, for example, in the emerging field of nanotechnology, current theories and modern instruments combine to tailor systems on the nanoscale for very specific purposes, as in the case of lipid nanoparticles used in the COVID vaccine.

This brief history of the development of the understanding of the colloid domain serves to define today's scientific challenges. By using our established knowledge of how individual molecules or atoms interact, we can relate two phenomena: how a collection of molecules behaves in the colloidal domain and how colloidal entities integrate to yield the property of a macroscopic system. While this challenge has its historic roots in physics and chemistry, we find the most intriguing opportunities in biology. The living cell is a true miracle of organization in the colloidal domain! From a technological point of view, mimicking biological processes will lead to the design of new products and processes. It is also clear that scientific advances open up new conceptual worlds in the colloidal domain.

12.2 Quantum Mechanics, Statistical Mechanics, and Thermodynamics Provide the Conceptual Basis for Describing the Equilibrium Properties of the Colloidal Domain

Let us analyze and synthesize the conceptual framework used in the first 11 chapters of this book from the perspective of a balance between the reductionist and the holistic views we have described. We start with the individual atoms and molecules that build the colloidal entities. These, in turn, interact to determine the macroscopic properties of a system at a finite temperature.

Quantum mechanics provides the basis for understanding molecular behavior as well as spectroscopy. In the text, we quoted an explicit quantum mechanical result only once, in Eq. (5.2.1), describing the dispersion interaction between two molecules. In our discussion of electrostatic intermolecular interactions, we took molecular charge distribution as a given from quantum mechanics, and the interaction potentials shown in Figure 3.5 were obtained from quantum mechanical calculations. Thus, although our holistic approach to the description of the colloidal domain is based on confidence in the quantum mechanical description of molecular behavior, in actuality, we have made extensive use of the results of the theory without going into a detailed examination of it.

Equilibrium statistical mechanics provides the conceptual framework for relating properties of individual molecules to those of a colloidal entity or of a macroscopic system. Elementary textbooks usually use quantum mechanics as the basis for introducing statistical mechanics. When we use statistical mechanics to describe colloidal systems, the quantum mechanical perspective introduces an unnecessary complication that easily leads to confusion, and in fact, the older formulation of the theory based on classical mechanics is actually more useful. In Eq. (1.1.3), we introduced one of the fundamental relations of statistical mechanics, due to Boltzmann, relating entropy to the number of accessible states. Combining this relation with the lattice model of a liquid, we arrived at an explicit expression for a key concept used throughout the text: the ideal entropy of mixing (Eq. 1.1.8). Using the lattice model to describe a liquid appears artificial because we know that a liquid is a very disordered system in which molecules do not occupy specific sites. Clearly, it serves as a shortcut to avoid a more complex description, but we pointed out a fundamental property of the lattice model that leads to Eq. (1.1.8); molecules A and B can exchange positions without affecting the energy.

In Section 1.2.5, we introduced another key statistical mechanical relation: the Boltzmann distribution. In its most general form, we can write that equilibrium as the probability $P(x_i)$ for the variable x having a value x_i.

$$P(x_i) = \frac{\exp\{-F(x_i)/k_B\mathbf{T}\}}{\int \exp\{-F(x)/k_B\mathbf{T}\}dx}$$

where $F(x)$ is generally the conditional free energy with x specified to x_i. In a simple external potential, the free energy $F(x)$ reduces to the potential energy as in Eq. (1.2.26). In the case of a gravitational field, the Boltzmann distribution is exact and straightforward to apply. In Chapter 3, we introduced some complications. To arrive at the free energy of the average dipole–dipole interaction of Eqs. (3.3.9) and (3.3.10), we had to average over the Boltzmann distributions of the orientations of the two molecules. Even more problematic is the ion distribution entering the Poisson–Boltzmann equation because the electric potential is set up partly by the ions whose positions we want to describe. In this case, due to this coupling between the electrostatic potential and the ion distribution, the Boltzmann distribution is only an approximation.

The Boltzmann distribution also entered into our description of dynamic processes involving the passing of free energy barriers. In Section 2.3.3, we saw that the main quantity determining the nucleation rate was the Boltzmann factor for the state at the free energy barrier. A similar dependence arose in the problem of slow particle coagulation in Section 8.3.2 and in the theory of emulsion droplet coalescence in Section 11.3.6. The rates for these processes can be understood qualitatively as a product of the Boltzmann factor at the barrier and the intrinsic rate at this barrier.

Thermodynamics is a theory that, in itself, has little to offer in terms of explanation and understanding. However, when we combine it with results from other sources, notably statistical mechanics, it provides a simple and fruitful route toward the macroscopic world. In Section 1.1, we cited the most useful thermodynamic equalities, which captures the conflict between energy and entropy so often observed in colloidal systems. From Eq. (1.1.5), we derived, within the regular solution model, expressions for chemical potentials and illustrated key thermodynamic concepts such as osmotic pressure and standard state.

Chapter 2 introduced another central concept that played a major role throughout the book: surface tension or surface free energy. Although strictly defined only in the macroscopic limit, surface tension is also used for describing objects in the colloidal domain. For example, our discussion of nucleation in Section 2.3 illustrated an ongoing dilemma of a continuum versus a molecular description.

Having introduced chemical potential, osmotic pressure, and surface tension in the first two chapters, we used these concepts repeatedly in subsequent chapters. In Chapter 5, we showed how osmotic pressure relates intimately to the force between planar surfaces. The concentration dependencies of osmotic pressure also appear as the fundamental thermodynamic quantity measured in a light scattering experiment.

Osmotic pressure is a measure of the chemical potential of a solvent. It is related to the chemical potential of a solute through the Gibbs–Duhem relation (Eq. 1.2.22), but in many applications, it is more straightforward to use the solute chemical potential directly. In Chapter 2, the surface excess was related to the change in surface tension with changes in the chemical potential in the Gibbs adsorption isotherm. By sticking to the basic equation, we were able to develop a simple explanation of two features of the SDS system. First, only a small change in the surface tension occurs with increasing concentration above the CMC because the chemical potential stays virtually constant. Second, the contributions to the chemical potential from both the surfactant anion and the sodium counterion must be considered in calculating the surface excess below the CMC.

In Chapter 4, we used the chemical potential again in our description of the equilibria leading to micelle formation. We also used it extensively to account for phase equilibria in Sections 1.1, 2.3, 5.3, and 7.2, and throughout Chapter 10. The conceptual notion that the chemical potential of all components must be equal for phases in equilibrium is a powerful one.

12.3 Intramolecular, Intermolecular, and Surface Forces Determine the Equilibrium Properties and Structure of Colloidal Systems

In this book, we have mainly considered three different types of colloidal systems: self-assembled aggregates of surfactant molecules, polymer systems, and solid colloidal particles. They differ by the forces that keep the colloidal entity together. The simplest case is the solid particles, which are fixed entities kept intact by the same forces that act in a solid. These may be crystals held together by a combination of covalent and ionic bonds (as in clay or metal oxides) or by van der Waals interactions (as in latex particles). A key feature is that the internal degrees of freedom largely remain fixed within the particle, which has a defined size and shape. In a linear polymer, the monomer units are covalently linked, but a large configurational degree of freedom still exists due to the rotations around the links between the monomers. Finally, self-assembled aggregates owe their cohesion to intermolecular forces, and changing shape and size by incorporating or expelling molecules is always an alternative.

Even though covalent bonds are ubiquitous in the colloidal domain, we took their existence and properties for granted and started our discussion from intermolecular interactions in Chapter 3. This strategy results in an enabling simplification, since only a few types of intermolecular interactions are relevant. Overlap repulsion prevents different atoms from coming too close to one another. This force typically determines the density of a liquid and is also responsible for the effects of molecular shape. Second, a short-range dispersion attraction due to charge correlations is always present, varying as the inverse sixth power of the separation between the molecules. These two interaction forces operate in all molecular systems. What really differentiates one molecule from another is the electrostatic interaction of the direct or the induced type. In Chapter 3, we divided electrostatic interactions into a number of types, depending on their distance dependence. This division led to concepts such as ion–ion, dipole–dipole, ion–induced dipole, and dipole–induced dipole interactions. The hydrogen bond also belongs to this family, but illustrates a conceptual problem often encountered in the description of intermolecular interactions. Because the hydrogen bond is basically electrostatic in origin, it creates confusion when stating that a certain effect is due to hydrogen bonding rather than to a dipole–dipole interaction. The dipole–dipole interaction is a part of the hydrogen bond.

Conceptual difficulties associated with writing the total force as a sum of individual contributions become much more pronounced when we consider the net interaction between two molecules in a medium. Such interactions always represent a more or less complex combination of different intermolecular interactions. A simple illustration is the form of the interaction parameter w in the regular solution theory (Eq. 1.1.16), where w emerges as a combination of three different pairwise interactions. The hydrophobic interaction provides a more important example. Intuitively, it might appear strange that the attractive interaction between two apolar molecules in water is due to electrostatic effects. However, the main contribution to the hydrophobic free energy clearly comes from the strong cohesion in water. This, in turn, is caused by electrostatic effects—remembering that the hydrogen bond is of electrostatic origin. When we discuss observed phenomena in terms of different effective interactions, ensuring that these are really separate entities becomes a true problem.

When we discuss forces between colloidal particles, it is even more obvious that these emerge through a complex interplay between individual intermolecular interactions and entropy effects from the thermal averaging over these interactions. There is no unique subdivision of the total force into different components, and one must be very careful when comparing two conceptually different approaches.

To stress the important role of intermolecular interactions for the behavior of colloidal systems, we choose to begin by discussing self-assembled systems. In this case, the colloidal entity is formed by the action of intermolecular forces, and these same forces are responsible for the interaction between aggregates. Aggregate formation is due to a delicate balance between the solvophobic interactions and the entropy decrease associated with bringing the free monomers into an aggregate. We can study this process experimentally, and it is also sensitive to variations in intensive parameters such as temperature or salt concentration. Thus, a study of the aggregation process provides detailed knowledge of intermolecular interactions under circumstances that are relevant for many colloidal systems.

The importance of forces between colloidal particles becomes most apparent when we discuss the stability of colloidal sols, as in Chapter 8. In Chapter 10, we also demonstrated that these forces govern the phase equilibria, so the swelling of a lamellar phase is determined by the force between the lamellae. The issue becomes more involved when the shape or the size of the interacting colloidal entities varies. For surfactant water systems, the same forces that can induce aggregate deformations also lead to the formation of new phases and a richness of phase diagrams for surfactant systems. In addition to emulsions and foams, these same forces lead to deformations of the drops or bubbles at high packing fractions.

Polymers owe their solution behavior to a combination of forces on the molecular and colloidal levels. Take a protein as an example. Its conformation(s) are determined by specific interactions among the amino acid monomers and the amino acid–water interaction. Once folded, the protein will interact with other similar proteins in the solution. For a stable system, this interaction should be repulsive, but aggregation can occur in some cases. A living cell contains a large number of colloidal components, and for proper function, aggregation processes must be controlled in detail. In fact, many pathological effects are caused by protein aggregation. Well-known examples include sickle cell anemia, eye cataracts, BSE (mad cow disease), and Alzheimer's disease.

12.4 Crucial Interplay Between the Organizing Energy and the Randomizing Entropy Governs the Colloidal World

We can understand that many phenomena in the colloidal domain result from a balance between an energy that strives toward a minimum of an ordered state and an entropy that has its maximum at the most disordered state of the system. This interplay is always at work in a molecular system at finite temperatures, but it is particularly apparent in colloidal systems, in which the energy has a greater chance of winning the battle due to the larger size of the colloidal particles relative to molecules.

The simplest example of this interplay that we considered was the concentration profile in a gravitational field. Gravitational energy is minimized when all particles collect at the bottom of the container (test tube). Due to the nonzero temperature, however, the equilibrium state is one with an exponential concentration profile in the gravitational potential, according to the

Boltzmann distribution. In accessible fields, the gravitational or centrifugal forces exerted on small molecules are so small that their effects are negligible. For colloidal particles and macromolecules in an ultracentrifuge, however, the accumulation at one end of the test tube can be more or less complete. Even Earth's gravity can be strong enough to cause colloidal particles to settle (or float).

One virtue of the regular solution model presented in Section 1.1 is that it is formulated so that the balance between energy and entropy emerges with clarity. Thus, for interaction parameters smaller than 2 $k_B\mathbf{T}$, the two liquids are miscible at all proportions, while for larger interactions—at least for some compositions—the energy term is strong enough to cause a separation into two phases. For polymer solvent systems, we encountered an analogous energy–entropy balance, but here the main contribution to the entropy comes from the intramolecular configurations of the polymer, which provides a somewhat smaller contribution to the entropy of the solution. According to the Flory–Huggins theory, therefore, a phase separation occurs already at a monomer–monomer interaction of $w = (1/2)\, k_B\mathbf{T}$ in the limit of a long polymer. However, if we count the interaction per pair of polymer molecules, it can be very large indeed, and the polymers are still soluble.

For colloidal particles, it is only the three translational degrees of freedom that contribute to the entropy of mixing, and such solutions are only stable if there is a repulsive interaction or an unusually weak attraction. Since we usually consider low concentrations, the interaction contact can be in the range $10 - 20\ k\mathbf{T}$ before precipitation occurs; but for typical particle sizes, the van der Waals interaction is substantially larger since we are adding contributions from many molecular units contained within each particle.

For electrostatic interactions, we have also seen how the balance between energy and entropy results in intricate phenomena. In Section 3.4, we analyzed how angle-dependent ion–dipole and dipole–dipole interactions are changed when we allow for a thermally excited rotation of the molecules. A perfectly isotropic rotation would lead to a zero average of the interaction, but the fact that the Boltzmann factor gives a higher weight to low-energy orientations creates a net attractive Keesom interaction. However, in the limit of high temperatures, the exponent in the distance dependence is doubled so that the dependence of the dipole–dipole interaction changes once the thermal average is made.

For solutions with charged species, the entropy of the ions provides a crucial contribution to the free energy of the system. The charging of a surface most clearly illustrates the energy–entropy balance. Take as an example a mica surface cleaved along the plane of the potassium ions. On a virgin surface, potassium ions are bound by electrostatic forces. As the surface is immersed in water, the potassium ions can be solvated by the water molecules, and they will tend to leave the surface to explore the freedom in the solution, i.e., to increase their entropy. However, they leave behind a net-charged surface, which creates a long-range electrostatic field that holds the ions back. Thus, not all potassium ions can leave the surface. A balance is reached when the increase in the TS term of the free energy due to the counterion entropy is balanced by the energy increase associated with augmenting the charge by one unit. When we see the source of the charged surface in this way, it becomes very natural to describe the force between a charged and a neutral surface as due to a constraint decreasing the entropy of the counterions expressed in terms of the ion concentration at the neutral wall (Eq. 5.1.12). It is then a small step to realize that the force between similarly charged walls also has an entropic cause, and that the quantitative expression contains the ion concentration at the midplane. In Sections 3.8 and 5.1, we described the electrostatic free energy explicitly as a sum of a direct interaction and a counterion entropy term. In our opinion, this helps considerably in the conceptual understanding of electrostatic interactions in colloidal systems. This conceptual picture also leads to the realization that there can be important correlations between the counterions of two similarly charged surfaces. The average ion distributions of the two halves have no net electrical interaction, but for every configuration of the ions in one half, 50% of the configurations of the other half are attractive, while the other 50% are repulsive. By the fact that the attractive configurations have a higher Boltzmann weight, the result is a net attractive contribution to the force. We see here a more complex manifestation of the basic mechanism behind the attractive Keesom interaction between rotating dipoles.

So far, we have treated energy and entropy as clear-cut concepts and assumed that we could divide the free energy simply into these two kinds of contributions. However, two complications arise. Thermodynamics knows two kinds of free energy: Helmholtz free energy **A** and Gibbs free energy **G**. In a condensed phase, we have been working within the approximation that we can neglect the compressibility of the systems. In that case, **G** and **A** differ by a trivial term $\mathbf{p}V$. However, if gases are present or we want to consider pressure effects explicitly, we must distinguish between the two types of free energy. For example, describing the nucleation of a bubble in a liquid is trickier than the opposite case of the nucleation of a liquid drop in a gas.

A second, more problematic, complication in separating energy and entropy stems from our use of effective interactions. The thermodynamic definition of internal energy, enthalpy, and entropy is very strict. However, in our pursuit of a simultaneous molecular and thermodynamic description, we repeatedly work with statistical mechanical models, which are invariably simplified and include only a part of the molecular degrees of freedom. The effect of the remaining degrees of freedom is subsumed in effective potentials, as we did when we introduced the relative permittivity in the ion–ion interaction in a medium in Section 3.5. In this case, the energies that go into the statistical mechanical problem assume the character of free energies. Thus, when we compare with the thermodynamic internal energies and entropies, we must bear in mind a conceptual discrepancy. For example, we saw in Eq. (3.5.16) that from a thermodynamic point of view, the entropy contribution totally dominates the electrostatic free energies in an aqueous medium because even the ion–ion interaction contains a large entropy part, which is due to the temperature dependence of the relative permittivity.

The interplay among internal energy, effective energy, and entropy causes particularly large conceptual problems in the description of the hydrophobic interaction. From a thermodynamic point of view, the entropy contributions dominate the interaction, but this perspective obscures the molecular cause. As we stated earlier, the hydrophobic interaction has its basic molecular source in the strong electrostatic interaction between water molecules. That the effective interaction becomes entropy-dominated at room temperature is due to a combination of the general free energy character of averaged orientation-dependent interactions and the fact that water molecules outside an apolar solute can interact more optimally with a reduced number of neighboring water molecules.

12.5 The Dynamic Properties of a Colloidal System Arise from a Combination of the Thermal Brownian Motion of the Individual Particles and the Collective Motion of the Media

Understanding the interactions between individual molecules and particles forms the basis of understanding both the equilibrium and dynamic properties of a colloidal system. We have discussed how the equilibrium properties follow from the interactions. With dynamic properties, the relation to the interactions is more intricate. We have focused exclusively on systems with sufficiently large masses at sufficiently high temperatures that they behave according to the laws of classical physics. For this case, there is a fundamental simplification that is sometimes neglected, which states that equilibrium properties are independent of dynamics. In quantum systems, kinetic and potential energies are intimately coupled. Decoupling them in the classical regime enables us to discuss equilibrium properties without explicitly considering the dynamics.

Basically, we have considered dynamics on three levels: molecular Brownian diffusion, chemical kinetics, and hydrodynamic motion. The simplest but most fundamental is the Brownian diffusional motion of individual molecules or colloidal particles. We introduced this concept in Section 1.7 and showed that the mean square displacement of a molecule diffusing in a homogeneous medium increases linearly with time. Brownian motion is the dynamic mechanism that produces disorder or mixing. External fields or molecular interactions can restrain the effect of Brownian motion, and in the presence of inhomogeneities, the equation for normal diffusion (Eq. 1.7.6) transforms into the equation for diffusion in a gradient of the chemical potential (Eq. 1.7.7).

We applied the diffusion picture in Section 8.3.1 in describing the rate of association in an unstable colloidal system. In the presence of a barrier, we could still apply the same point of view, but the concentration gradient was replaced by the more general concept of a gradient in chemical potential (as in Eq. 1.7.7) as the driving force, an approach we also used when we analyzed passive molecular transport across lipid membranes. For the Ostwald ripening, we also considered the gradient in the chemical potential as the driving force for the transport of individual molecules between emulsion droplets or foam bubbles.

We obtained a more abstract description of the dynamics of individual molecules in chemical kinetic schemes through the analysis of micelle formation in Eq. (4.2.5). Here rate constants appear as parameters, and the challenge is to describe the dynamics of an overall complex process in terms of the kinetic characterization of the individual steps. We also encountered the same approach in the description of a nucleation process in Sections 2.3.3 and 8.2.3, we used it to analyze the buildup of aggregates of colloidal particles.

The colloidal domain falls between the world of small molecules and the macroscopic world. To get a complete picture, we must shift between these two perspectives. Using the more macroscopic point of view, we described dynamics in terms of the concepts of fluid mechanics. This process led to Stokes' law for the diffusion of a sphere in a medium of small molecules (Eq. 1.7.17). It also introduced the Reynolds number used in Sections 5.8 and 9.5. In Section 5.8, we discussed how hydrodynamic flow could generate a transient force among particles. Many applications of colloids involve the modification of the flow properties of the medium, and in Section 7.2.3, we introduced the basic rheological concepts used to characterize the hydrodynamic properties of a system.

Index

d

e

q

r